A Textbook of
PRACTICAL PHYSICS
(for Degree Classes of Indian Universities)

Late INDU PRAKASH
Life member, Indian Association of Physics Teachers and Indian Physics Association, Formerly Head of the Department of Physics, Christ Church College, Kanpur

Late RAM KRISHNA
Head of the Department of Physics N.R.E.C. College, Khurja

Dr. A. K. JHA
M.Sc. (DU), Ph.D. (IIT, Delhi)
Department of Applied Physics
Delhi College of Engineering
(University of Delhi)
Delhi

KITAB MAHAL

First Edition	:	1950
Fully Revised and Enlarged Edition	:	2007
Reprinted	:	2008
Fully Revised and Enlarged Edition	:	2011
Reprint	:	2024

ISBN : 81-225-0416-7

Price : ₹650.00

Head Office:

Kitab Mahal Publishers
4655/21, Ansari Road (G. F.)
Daryaganj, New Delhi-110002
Phone : 011-23273230/43526875
(WhatsApp) 9335107914
Email : kitab_mahal@hotmail.com

Published By: Kitab Mahal, 4655/21, Ansari Road, (G. F.) Daryaganj, New Delhi-110002
Phone : +91 9350075776 Email : kitabmahalpublishers@gmail.com
Visit : www.kitabmahalpublishers.com
Printed by : Avatar Offset, 43/2, Site-IV, Sahibabad Industrial Area, Ghaziabad (U.P.)

Preface

It gives us great pleasure in bringing out the thoroughly revised and enlarged edition of this book. Many new Chapters with experiments on semiconductor, *p-n* junction diode, transistor, *e/m*, nuclear physics—GM counter, photoelectric effect, CRO, etc. have been included. It is hoped that along with Vol. I, the book covers all the topics of undergraduate Physics, and students of B.Sc. (H)/(G), B.E., B.Tech and Diploma students will find the book very useful. Complete and comprehensive theoretical treatment of each topic is the strength of this book and it will be of great help to the students for oral/*viva-voce* examinations.

Suggestions and constructive criticism for the improvement of the book are solicited. Despite our best efforts, some mistakes might have crept in. Kindly, bring them to our notice. It will be thankfully acknowledged.

Enjoy the process of experimentation and cultivate the habit of devotion to your practical work. If you do it successfully, the foundation of a good Scientist/Engineer will be laid inside you.

I thank Sh. I.K. Agarwal and Sh. Raghvendra Agarwal for their keen interest taken in revision and upgradation of the book.

I acknowledge the patience and perseverance of my family members which enabled me to spend time over the years in revision of the book.

I wish reader students a bright future.

Dr. A.K. Jha

PREFACE TO THE FIRST EDITION

The book is self-contained and the treatment followed in it is thoroughly modern. Every experiment is preceded by the just necessary and relevant theory which enables the student to grasp its underlying principles. Detailed instructions about each experiment are given in elaborate manner to enable the student to understand thoroughly the full experimental procedure. A brief description of the actual apparatus illustrated by diagrams has been given where necessary. The various precautions to be observed in an experiment have not only been emphasised in the method but have also been exhaustively discussed separately. The observation tables illustrate the sequence of the method and help the student to record his observations in an orderly and scientific manner. The criticism of the method given at the end of each experiment together with the exhaustive list of questions given at the close of each chapter and thoroughly dealt with in the text are the special features of the book. These are of considerable help to the student in his preparation for the *viva voce* test.

The book also includes practical details of description and working of apparatus of general use in the laboratory and a number of intelligent exercises based directly on experiments.

I.P.
R.K.

CONTENTS

Section 1

MECHANICS AND GENERAL PROPERTIES OF MATTER

Section 2

SOUND

Section 3

OPTICS

Section 4

HEAT

Section 5

MAGNETISM AND ELECTRICITY

Preface

It gives us great pleasure in bringing out the thoroughly revised and enlarged edition of this book. Many new Chapters with experiments on semiconductor, *p-n* junction diode, transistor, *e/m*, nuclear physics—GM counter, photoelectric effect, CRO, etc. have been included. It is hoped that along with Vol. I, the book covers all the topics of undergraduate Physics, and students of B.Sc. (H)/(G), B.E., B.Tech and Diploma students will find the book very useful. Complete and comprehensive theoretical treatment of each topic is the strength of this book and it will be of great help to the students for oral/*viva-voce* examinations.

Suggestions and constructive criticism for the improvement of the book are solicited. Despite our best efforts, some mistakes might have crept in. Kindly, bring them to our notice. It will be thankfully acknowledged.

Enjoy the process of experimentation and cultivate the habit of devotion to your practical work. If you do it successfully, the foundation of a good Scientist/Engineer will be laid inside you.

I thank Sh. I.K. Agarwal and Sh. Raghvendra Agarwal for their keen interest taken in revision and upgradation of the book.

I acknowledge the patience and perseverance of my family members which enabled me to spend time over the years in revision of the book.

I wish reader students a bright future.

Dr. A.K. Jha

PREFACE TO THE FIRST EDITION

The book is self-contained and the treatment followed in it is thoroughly modern. Every experiment is preceded by the just necessary and relevant theory which enables the student to grasp its underlying principles. Detailed instructions about each experiment are given in elaborate manner to enable the student to understand thoroughly the full experimental procedure. A brief description of the actual apparatus illustrated by diagrams has been given where necessary. The various precautions to be observed in an experiment have not only been emphasised in the method but have also been exhaustively discussed separately. The observation tables illustrate the sequence of the method and help the student to record his observations in an orderly and scientific manner. The criticism of the method given at the end of each experiment together with the exhaustive list of questions given at the close of each chapter and thoroughly dealt with in the text are the special features of the book. These are of considerable help to the student in his preparation for the *viva voce* test.

The book also includes practical details of description and working of apparatus of general use in the laboratory and a number of intelligent exercises based directly on experiments.

I.P.
R.K.

Introduction

Importance of Practical in Physics

If we go deep into the history of human progress we find that, in every phase of the progress, one thing, which is common, and therefore, which has made this progress possible, is the application of human intelligence, in overcoming his physical shortcomings. Man began by exploiting the superior abilities of lower animals and gradually succeeded in inventing machines infinitely more efficient than any animal in their capacity for work.

With the success in inventing machines capable of doing miracle in the field of physical work, man soon realized that these machines will be equally useful in the field of knowledge, particularly in the field of physical knowledge. History of progress of physical sciences and the ever-increasing applications of instruments in the fields of other branches of knowledge amply justify the present age being called the Age of Machines, as they have influenced not only our capacity of doing, but also of thinking and willing.

If we carefully study the progress of Physics we find that the greatest contribution in this progress has been made by the inventions of more and more refined instruments of measurement. This had to be done because it was early recognised that the information of a phenomenon given to us by our unaided senses was so defective that it could not be relied upon for true knowledge. Since consistency is the criterion of truth and since sensual impressions are seldom consistent, it follows that knowledge based on unaided senses cannot always be true.

Our knowledge of a phenomenon is said to be true not only when the same observer agrees in his own experience of the phenomenon at all times but also when all observers agree in their experiences of that phenomenon under like circumstances. There were two main reasons why there may not be an agreement between different experiences of the same phenomenon :—

(i) Our senses being parts of a living or ever-changing body, it is not possible to keep a sense in a steady state or to reproduce that state at will nor it is possible to bring the senses of different observers in identical states.

(ii) It is not possible with the help of unaided senses alone to ensure that a given phenomenon is reproduced under identical conditions.

For true knowledge, therefore, we require some device which is more reliably consistent than our senses in its indications. Such devices are the various measuring instruments invented by scientists to make their knowledge of the physical world as free from defects as is humanly possible.

The following general scheme of a scientific investigation shows how experiment is the foundation of the development of a science. In this development we start with observations and proceed step by step, almost in an endless spiral, in which every new discovery, predicted sometimes by a theory, brings us nearer the truth.

Observation	(Qualitative)	of the phenomenon as a whole generally with unaided senses.
Experimentation	(Quantitative)	with suitable instruments finding effects of various factors determining the phenomenon.
Generalisation or Law Formulation	Co-ordination of results of all cases of the same phenomena (Inductive)	to find the answer to the question, (How ?) the various factors that influence the phenomenon and express the answer in a formula or a graph.
Formulation of Theory	Co-ordination of results of allied phenomena	to find the answer to the question (Why ?) the phenomenon obeys this discovered law and no others.
Prediction of new phenomena	Deduction of particular consequences of the more general proposition of the theory.	provides verification of the theory as well as discovery of newer facts of greater utility.
New experimentation		with better instrument sometimes specially designed for the purpose.

Observation of new Facts
Formulation of a new Law
Formulation of a new Theory or modification of the old one.
The continuation of the process.

In fact the aim of science is the formulation of a law governing the phenomena the science aims to study; because mere observation of isolated facts is not science. We have science only when we have discovered a proposition which includes all the possible infinite cases of the same class. The discovery of a law is in fact the discovery of the hidden element of *oneness* in the apparent *manyness* of the phenomenon. A theory aims at introducing a further element of oneness in our knowledge by showing a sort of oneness in apparently different types of phenomenon. From this point of view the aim of science, in fact, of all knowledge is to show that the world of our common experience is not what it seems to be; it is not so complex as it appears. On the other hand, it is simple and one complete whole. It is in reality a Universe and not a Multiverse.

Foundation of a science, particularly of Physics is, therefore, experiment and the aim of a carefully conducted experiment in Physics is the discovery of a law governing some phenomenon or the verification of a law derived from a theory which is acceptable to the physicist only when the experiment approves it.

As a college student most of the times objective of your experiments is to verify the theories learnt. One must perform the experiment carefully and take the observations honestly even if the observations are not as expected. Sometimes students manipulate the readings to get the expected result. This is not only academic dishonesty but develops into a bad habit that should be avoided. It you are not getting observations as expected, record whatever you are getting and discuss it with your teacher. If there is no manual mistake, some instrument may be faulty and in this way you can detect the fault and help in rectifying the fault of the instrument.

"Experimentation and Errors"

With a view to make their investigations and lead us as close to the truth as possible, physicists have been not only trying to invent more and more perfect instruments but have also developed the theory of errors which suggests methods of eliminating from their investigations the possible errors.

The theory of errors starts with two assumptions :—

(*i*) That every observer is more or less careless and so liable to commit mistakes in reading an instrument. Therefore, every reading taken with the same measuring instrument is more or less erroneous; the finer the instrument, the greater the chance of divergence in the several readings. Such errors are called *chance errors*, and their magnitude depend on the quality of the instrument as well as that of the observer. If the instrument is free from defects and the observer is careful, the errors will be small.

(*ii*) That every instrument is more or less defective and so its indications are not free from errors. Such errors due to some inherent defects in the instrument are called *instrument or constant errors.*

Aim of this theory is, therefore, to devise means to get the right result from all these wrong observations. It is a sort of miracle the theory aims to achieve, because in brief the theory is :

Given—A set of erroneous observations.

Required—To find a result free from error.

With the help of this theory a physicist can either eliminate these errors or find out the magnitude of the probable errors which depend on the quality of the instrument at his disposal.

As the first kind of error depends on chance, its distribution follows the same law as any other chance event follows. Measurement of a quantity by an instrument is often compared to the hitting of a fixed target with a rifle, and then divergence of the shot from the exact point on the target is compared to the error of observation. Maxwell has theoretically derived a formula which gives the nature of the distribution of velocity amongst the molecules of a gas; and this formula applies, with necessary changes, to the distribution of chance errors amongst various observations. Results of Maxwell's theory are based on the mathematical theory of probability which is truly applicable only to a group of a large number of individual independent events. This will apply to a set of sufficiently large number of observations taken independently without any bias of the previously taken observations.

It can be easily shown that the arithmetic mean of such a large number of observations is the actual correct value of the quantity. Let the unknown true value of the quantity be x_1, x_2, x_3,.......x_n, so that e_1, e_2, e_3, ...e_n, are the errors of the corresponding observation. Hence we have,

$$x_1 = x + e_1$$
$$x_2 = x + e_2$$
$$x_3 = x + e_3$$
$$\dots\ \dots\ \dots$$
$$\dots\ \dots\ \dots$$
$$x_n = x + e_n$$

$$\therefore \quad (x_1 + x_2 + x_3 + \dots\ x_n) = nx + (e_1 + e_2 + e_3 + \dots + e_n)$$

since e_1, e_2, e_3.......e_n are all chance errors, they should be so distributed that for any error, say, $+e_1$ there should be an error $-e_1$, for there is no reason why mere chance should favour positive errors, and not negative errors. Hence,

$$e_1 + e_2 + e_3 + \dots e_n = 0$$

$$\therefore \quad (x_1 + x_2 + x_3 + \ldots + x_n) = nx$$

or
$$x = \frac{1}{n}(x_1 + x_2 + x_3 + \ldots x_n)$$
= arithmetic mean of observations.

Hence arithmetic mean of a large number of observations is the correct value of a reading. If we call this x_m, then error of any observation say r th one, is defined by,

$$e_r = x_m - x_r$$

A reading x_r is said to be precise or sharp if its error e_r is small or zero. Precision or freedom from these chance errors depends on the carefulness of the observer as well as the sensitivity of the instrument.

As the number of readings is usually not very large, $e_1 + e_2 + e_3 + \ldots e_n \neq 0$ and so the mean of errors $\frac{1}{n}(e_1 + e_2 + e_3 + \ldots e_n) = e_m$ is a small quantity and so the actual value $x \neq x_m$ but $= x \pm e_m$.

Three kinds of average values of errors are used :

(*a*) Arithmetic mean, $e_m = \frac{1}{n}(e_1 + e_2 + e_3 + \ldots + e_n)$

(*b*) Root mean square $e_r = \frac{1}{n}\sqrt{e_1^2 + e_2^2 + \ldots e_n^2}$

(*c*) Probable error $e_p = 0.6745\ e_m$.

If a given error, e, occurs in *f* observations, then *f* is called the frequency of its occurrence and f/n is called the relative frequency, in the total number of *n* observations. The most probable error is that for which f/n has the highest value and its relation with the mean value is shown to be $e_n = 0.674\ e_m$.

To eliminate these chance errors, therefore, the experimenter should take as large a number of observations as possible and should keep his mind perfectly free from any bias.

On the other hand, elimination of instrumental error is not quite so easy as that of the chance errors. For the elimination of errors due to some likely defects in the make of the instrument, the experimenter has to first carefully study the various factors which determine the two chief qualities of the instrument, *viz.*, the sensitivity and the accuracy, and then see which part of a given instrument affects these and in what manner. *Accuracy* of an observation is freedom from these instrumental errors.

Sensitivity of an instrument is measured either by the smallest amount of the difference in the quantity the scale of the instrument is capable of measuring or by the number of scale divisions which a given small difference in the quantity to be measured will produce. For example, sensitivity of a balance may be defined either as the number of m. gm. required to produce a shift of the pointer through one division or by the number of divisions the pointer will shift on the addition of an extra 1 m. gm. weight on one of the pans. Usually one of these gives insensitivity rather than sensitivity. In the example of the balance the first method gives insensitivity because a balance which requires a larger weight to shift the pointer through one division is less sensitive.

Sensitivity and accuracy of an instrument usually go against each other, for, if we try to increase one, the other is automatically decreased. In rare cases we may have both improved simultaneously.

Log and Proportional Error

The final result of the experiment is usually calculated from a set of observations taken with a number of measuring instruments and connected by means of a formula. It can be shown that each of these observed quantities does not influence the result equally, someone influences more than the other. If, for example, a quantity X, depends on other quantities A, B, C and D in the manner given by the equation

$$X = \frac{A\,B^2C^3}{D^4}$$

Then, taking logarithm on both sides and differentiating, we get

$$\frac{dX}{X} = \frac{dA}{A} + 2\frac{dB}{B} + 3\frac{dC}{C} - 4\frac{dD}{D}$$

The small quantities dX, dA, dB,....... are considered as the errors in the measurements of X, A, B,..... When calculating composite error, we assume the worst case of all errors being in the same direction and hence use positive sign on right hand side even though it should be negative from a purely mathematical consideration. Therefore, we have

$$\frac{dX}{X} = \frac{dA}{A} + 2\frac{dB}{B} + 3\frac{dC}{C} + 4\frac{dD}{D} \quad ..... \text{(i)}$$

The quantity dX gives the *maximum log error* in the measurement of X.

We observe from the above that for given possible errors in A, B, C and D, the effect on the final result is greater for the factor having highest power and that (the factor) having smaller magnitude. For example, if D is the smallest quantity having highest power, the measurement of D should be most carefully made.

A few examples are given below to illustrate the application of equation (i) for the purpose of calculating the extreme error.

I. Determination of Y by bending of a beam

When a beam of length *l*, breadth *b* and depth *d* is loaded in the middle by a weight *mg*, the depression is *e* and the value of the Youngs modulus of the material of the beam is given by

$$Y = \frac{mgl^3}{Abd^3 e}$$

Taking logarithm and differentiating and assuming that the errors have such signs that they all add up, we have, for the extreme proportionate error in *Y*,

$$\frac{dY}{Y} = \frac{dm}{m} + \frac{dg}{g} + \frac{3dl}{l} + \frac{db}{b} + \frac{3dd}{d} + \frac{de}{e}$$

Here we reproduce the values of the various quantities, involved in the above equation, from actual observations.

m = 1500 gm.	dm = 3 gm., assuming an error of 1 gm. per 1/2 kg.
g = 980 cm./sec.^{2}	dg = 1 cm./sec.^{2} (assumed)
l = 86.2 cm.	dl = 0.1 cm. (Least count of metre scale)
b = 2.54 cm.	db = 0.01 cm. (L. C. of callipers)
d = 0.602 cm.	dd = 0.001 cm. (L. C. of screw gauge)
e = 0.1674 cm.	de = 0.0005 cm. (L. C. of spherometer)

Hence, on substituting these values in the equation above,

$$\frac{dY}{Y} = \frac{3}{1500} + \frac{1}{980} + \frac{3 \times 0.1}{86.2} + \frac{0.01}{2.54} + \frac{3 \times 0.001}{0.602} + \frac{0.0005}{0.1674}$$

$$= 0.0020 + 0.0010 + 0.0034 + 0.0038 + 0.0050 + 0.0030$$

$$= 0.0182$$

If calculated $Y = 12.0 \times 10^{11} \frac{\text{dyne}}{\text{cm}^2}$, then maximum log error

$$dY = 12 \times 0.0182 \times 10^{11} \frac{\text{dynes}}{\text{cm}^2}$$

$$= 0.2184 \times 10^{11} \frac{\text{dynes}}{\text{cm}^2}$$

Also, the proportionate error in *Y* being 0.0182, the percentage error is

$$Y = 0.0182 \times 100$$

$$= 1.82\%$$

The major contribution to this error is the proportionate error of 0.0050 in the measurement of the depth of the beam, *i.e.*, a percentage error of 0.5%, nearly one-third of the total error is contributed by d*d*. Hence it should be measured very carefully and a number of times with a screw gauge.

II. Determination of modulus of rigidity by Maxwell's Needle

The expression for the modulus of rigidity of the material of a wire in Maxwell's needle experiment, is

$$n = \frac{2\pi l\,(m_2 - m_1)\,L^2}{r^4\,(T_1^2 - T_2^2)}$$

$$= \frac{32\pi l\,(m_2 - m_1)\,L^2}{d^4 (T_1 + T_2)\,(T_1 - T_2)}$$

where d is the diameter of the wire (=2r, r being the radius).

As before, we get for the extreme proportionate error in n.

$$\frac{dn}{n} = \frac{dl}{l} + \frac{d(m_2 - m_1)}{(m_2 - m_1)} + \frac{2dL}{L} + \frac{4.dd}{d} + \frac{d(T_1 + T_2)}{(T_1 + T_2)} + \frac{(T_1 - T_2)}{(T_1 - T_2)}$$

Following are the observations in an experiment :

		Least count of:
l = 58.0 cm.	dl = 0.1 cm.	metre scale
$(m_2\text{-}m_1)$ = 310.5 gm.	dm = 0.5 gm.	Kilo-balance
L = 29.9 cm.	dL = 0.1 cm.	Metre scale
d = 0.090 cm.	dd = 0.001 cm.	Screw gauge
T_1 = 8.46 sec.	$dT_1 = dT_2 = \frac{0.2}{33}$	If 30 vibrations have been timed with a stop watch of least
T_2 = 5.90 sec.	= 0.0066	count 0.2 sec.

Hence,

$$\frac{dn}{n} = \frac{0.1}{68} + \frac{0.5}{320.5} + 2 \times \frac{0.1}{29.9} + 4 \times \frac{0.001}{0.090} + \frac{132}{14.36} + \frac{0.0132}{2.56}$$

$$= 0.0015 + 0.0016 + 0.0033 + 0.0444 + 0.0009 + 0.0051$$

$$= 0.0567$$

The proportionate error in n being 0.0567, the percentage error in the determination of n by this method is 5.67%.

The major contribution to this error is 4.4% by the error in the measurerment of its diameter *d*. Hence the diameter should be carefully measured with a screw gauge and consistent with the working of the apparatus, a wire of somewhat greater diameter should be used.

III. Determination of 'g' by bar pendulum

The acceleration due to gravity is given by :

$$g = 4\pi^2 \frac{L}{T^2}$$

where L is the length of the equivalent simple pendulum. We have for the extreme proportionate error in g.

$$\frac{dg}{g} = \frac{dL}{L} + 2.\frac{dT}{T}$$

Now take that the graph between length and time has been plotted on a cm. graph paper and that the following are the scales chosen on it.

For length, 1 mm. represents 0.5 cm.

For time, 1 mm. represents 0.01 sec.

Hence $dL = 0.5$ cm. and $dT = 0.1$ sec.

$L = 92.0$ cm. and $T = 1.92$ sec.,

$$\frac{dg}{9} = \frac{0.5}{92.0} + 2 \times \frac{0.01}{1.92} = 0.0054 + 0.0104$$

$$= 0.0158$$

Hence the extreme proportionate error in g is 0.0158 or 1.58%. This large error partly arises because of the limitations of the graph paper in representing the length and the time to the accuracy to which they are measured. Length is measured correct to 0.1 cm. whereas on the graph the minimum length which can be represented is 0.5 cm. Similarly if the periodic time is measured with a stop watch of least count 0.1 sec. by timing 50 oscillations of the pendulum, $dT = 0.1/50 = 0.002$ sec., whereas on the graph the minimum time represented is 0.01 sec. Thus there is a five-fold loss in accuracy in the representation of both length and time on the graph paper :

If it were possible to retain the observed accuracies in the representation of length and time on large sheet of graph paper

$$\frac{dg}{g} = \frac{0.1}{92} + \frac{2 \times 0.002}{1.92} = 0.00308$$

i.e., the error in g is 0.308%.

IV. Determination of Coefficient of Viscosity of water by Poseuille's method

The co-efficient of viscosity is given by

$$\eta = \frac{\pi a^4 \rho g h t}{8 Vl}$$

$$= \frac{\pi d^4 \rho g h t}{128 Vl}$$

where d is the diameter of the tube = ($2a$, a being its radius).

Assuming ρ and g known to a high degree of accuracy, we have, for the extreme proportionate error in η.

$$\frac{d\eta}{\eta} = 4.\frac{dd}{d} + \frac{dh}{h} + \frac{dt}{t} + \frac{dV}{V} + \frac{dl}{l}$$

Below we reproduce the observations in an experiment :

		Lest count of;
d = 0.130 cm.	dd = 0.001 cm.	Vernier microscope
h = 25.4 cm.	dh = 0.1 cm.	Metre scale fitted to read pressure head.
t = 120 sec.	dt = .1 sec.	Stop-clock
V = 40.0 c.c.	dV = 1 c.c.	Graduated cylinder
l = 40.0 cm.	dl = 0.1 cm.	Metre Scale.

Substituting these values in the equation above,

$$\frac{d\eta}{\eta} = \frac{4 \times 0.001}{0.130} + \frac{0.1}{25.4} + \frac{1}{120} + \frac{1}{40.0} + \frac{0.1}{40.0}$$

$$= 0.0308 + 0.0039 + 0.0083 + 0.0250 + 0.0025$$

$$= 0.0705$$

Hence the extreme proportionate error in the determination of the co-efficient of visocosity η is 0.0705, *i.e.*, the percentage error is 7.05% and the major contribution to this is 3.08% by the error in the measurement of the diameter of the capillary tube and 2.5% by the error in the determination of the volume of water collected. A graduated cylinder having a least count of 0.5 c.c. would reduce this error by 1.25%.

V. Determination of the frequency of A.C. mains with a sonometer

The frequency of A.C. mains is given by

$$n = \frac{1}{2l}\sqrt{\frac{M}{m}} = \frac{1}{2l}\sqrt{\frac{M}{M'/L}}$$

where l is the length of a loop of the wire, vibrating with the frequency of A.C. mains, under a tension Mg and having a mass per unit length m as determined by finding the mass M′ of a length L of the wire.

Taking g as known to a high degree of accuracy, the extreme proportionate error in n is given by

$$\frac{dn}{n} = \frac{dl}{l} + \frac{1}{2}.\frac{dM}{M} + \frac{1}{2}.\frac{dM'}{M'} + \frac{1}{2}.\frac{dL}{L}$$

From a set of observations in this experiment, we have

		Least count ;
l = 60.8 cm.	dl = 0.1 cm.	Metre scale
M = 412.2 gm.	dM = 0.1 gm.	Balance
M' = 1.0874 gm.	dM' = 0.0002 gm.	Analytical balance
L = 100.0 cm.	dl = 0.1 cm.	Metre scale.

Hence,

$$\frac{dn}{n} = \frac{0.1}{60.8} + \frac{1}{2} \times \frac{1.0}{12.2} + \frac{1}{2} \times \frac{0.0002}{10874} + ½ \times \frac{0.1}{1.000}$$

$$= 0.0016 + 0.00012 + 0.00009 + 0.0050 = 0.0068$$

Thus the extreme error in n is 0.68%

VI. Determination of λ by Biprism

The wavelength of light is given by

$$\lambda = \beta \frac{2d}{D} = \beta \frac{\sqrt{d_1.d_2}}{D}$$

where β is the fringe width, d_1 and d_2 are the separations of the two images (of the virtual coherent sources) formed by the lens in its two positions and D, is the distance of the micrometer eyepiece from the slit.

Taking logarithm and differentiating, we get for the extreme proportionate error in λ.

$$\frac{d\lambda}{\lambda} = \frac{d\beta}{\beta} + \frac{dD}{D} + \frac{1}{2}\cdot\frac{dd_1}{d_1} + \frac{1}{2}\cdot\frac{dd_2}{d_2}$$

In an experiment the following observations were recordded :

		Least count of;
β=0.0177 cm.	dβ=0.001 cm.	Micrometer screw of eyepiece
D = 79.1 cm.	dD=0.1 cm.	Bench scale,
d_1=0.380 cm.	$dd_1 = dd_2$ = 0.001 cm.	Micrometer screw of the eyepiece.
d_2 = 0.134 cm.		

Substituting these values in the equation above

$$\frac{d\lambda}{\lambda} = \frac{0.001}{0.0177} + \frac{0.1}{79.1} + \frac{1}{2} \times \frac{0.001}{0.380} + \frac{1}{2} \times \frac{0.001}{0.184}$$

$$= 0.0056 + 0.0013 + 0.0013 + 0.0052 = 0.0134$$

Hence the percentage error in λ is 1.34%, *i.e.*,

$$d\lambda = \lambda \times 0.0134 = 5893 \times 0.0134 = 79 \text{ A}^{\circ}.$$

If all the errors add up, the maximum possible error in the measurement of λ in the above set of observations is 79 A°.

The error in the measurement of β is 0.56%, while evaluation of $2d$ contributes an error of (0.13, in d_1 and 0.52% in d_2) 0.65%. Hence β should be measured at as large a value of D as permissible so that the proportionate error in β is reduced (because β increases with the increase in D).

VII. Determination of λ with a diffraction grating

The wavelength of light is given by

$$\lambda = \frac{(a+b)\sin\theta}{n}$$

where $(a+b)$ is the grating element, θ the angle of diffraction of light of wavelength λ in the order n. The grating element is known to a high degree of accuracy and n is correctly observed, hence taking logarithm and differentiating, the proportionate error in λ is given by

$$\frac{d\lambda}{\lambda} = \frac{\cos\theta}{\sin\theta}\,.\,d\theta = \cot\theta\,.\,d\theta$$

Using the mercury green line (λ = 5461 Å), the angle of diffraction in the first order with a grating having 14941 lines per inch, is

$$\theta = 18^\circ\ 21'\ ;\quad d\theta = 1 \text{ min.} = 2.909\times 10^{-4} \text{ radian.}$$

$$\frac{d\lambda}{\lambda} = \cot 18^\circ..\times 2.909\times 10^{-4}$$

$$= 0.0119\times 2.909\times 10^{-4} = 0.00087$$

Thus the percentage error in the determination of λ is 0.00087 × 10 = 0.087%.

The error dλ in the determination of λ is

$$5461 \times \frac{0.087}{10} = 4.77\text{A}^\circ.$$

Hence the error of each 1' in the reading of θ causes an error of 4.77A° or 5A° approx. The obvious conclusion of this discussion is that the angle of diffraction θ should be carefully measured. It should also be borne in mind that setting of the grating normal to the incident pencil is equally important.

VIII. Mean Coeffcient of apparent expansion of water with a Weight Thermometer

The equation used for determining the mean coefficient of apparent expansion of a liquid, say water, is

$$\gamma_a = \frac{m_1 - m_2}{m_2(t_2 - t_1)}$$

The extreme proportionate error in γa is given by

$$\frac{d\gamma_a}{\gamma_a} = \frac{d(m_1 - m_2)}{(m_1 - m_2)} + \frac{dm_2}{m_2} + \frac{d(t_2 - t_1)}{(t_2 - t_1)}$$

The following are the observations in an experiment with water giving a mean value of γ_a as 5.1×10^{-4} per °C.

$m_1 = 6.6958$ gm. $\quad d(m_1 - m_2) = 0.0002$ gm.

$m_2 = 6.4224$ gm. $\quad dm_2 = 0.0001$ gm.

$t_1 = 16.5$° C $\quad d(t_2 - t_1) = 1$ °C

$\therefore \quad m_1 - m_2 = 0.2734$ gm. and $t_2 - t_1 = 83.0$ °C

Substituting the above values.

$$\frac{d\gamma_a}{\gamma_a} = \frac{0.0002}{0.2734} + \frac{0.0001}{6.4224} + \frac{1}{83}$$

$$= 0.0007 + 0.00001 + 0.01204 = 0.0128$$

Hence the percentage error in γ_a is 1.28% of which the major contribution is by the measuremnet of temperature difference with a thermometer reading upto $\frac{1}{2}$ °C. We have taken $d(t_2 - t_1)$ as 1°C assuming that the errors in the measurement of t_1 and t_2 (each being, of $\frac{1}{2}$ °C) are such as to add up.

From the above analysis it is at once apparent that the use of an analytical balance for the different weighings reduces the errors in $(m_1 - m_2)$ and m_2 considerably and that the use of a thermometer reading to 1/2°C would reduce the error in the measurement of $(t_2 - t_1)$.

IX. Thermal Conductivity of Copper by Searle's apparatus

The expression used for determining the thermal conductivity by Searle's apparatus is,

$$K = \frac{m(\theta_4 - \theta_3)\,d}{\pi r^2(\theta_1 - \theta_2)} = \frac{M(\theta_4 - \theta_3)\,d}{t\pi r^2(\theta_1 - \theta_2)}$$

where m is the rate of flow water as determined by collecting a mass M gm. in a time t secs; the other letters have their usual significance.

The extreme proportionate error in K is given by

$$\frac{dK}{K} = \frac{dM}{M} + \frac{d(\theta_4 - \theta_3)}{(\theta_4 - \theta_3)} + \frac{dd}{d} + \frac{dt}{t} + \frac{2dr}{r} + \frac{d(\theta_1 - \theta_2)}{(\theta_1 - \theta_2)}$$

The following are the observations for this experiment : Volume of water collected = 10.0 cc ; Least count of graduated cylinder = 0.5 cc.

$\therefore \quad M = 10.0$ gm. $\quad dM = 0.5$ gm.

$\theta_4 = 58.9$ °C $\quad d\theta_4 = d\theta_3 = 0.2$°C, the least count of the

$\theta_3 = 31.2^\circ C$		thermometers
d = 10.50 cm.	dd = 0.01 cm.	L.C. of callipers
r = 2.54 cm.	$dr = \frac{1}{2} \times 0.001$cm.,	L.C. or screw gauge being 0.001 cm.
t = 120 sec.	dt = 0.2 sec.	
$\theta_1 = 90.2^\circ C$	$d\theta_1 = d\theta_2 = 0.2^\circ C$	L.C. of the thermometers.
$\theta_2 = 84.8^\circ C$		

Substituting the above values

$$\frac{dK}{K} = \frac{0.5}{10.0} + \frac{0.04}{(58 - 31.2)} + \frac{0.01}{10.50} + 2 \times \frac{1}{2} \times \frac{0.001}{2.54} + \frac{0.2}{1.20} + \frac{0.04}{(90.2 - 84.8)}$$

$$= 0.05 + 0.0015 + 0.0009 + 0.0018 + 0.0016 + 0.0078 = 0.0636$$

Hence the extreme proportionate error in K is 0.0636, i.e., the percentage error in K is 6.36% of which 5% is contributed by the measurement of M and 0.78% by the measurement of $(\theta_1 - \theta_2)$

From the above it is evident that the student made the mistake of having too low a rate of flow of water and thus the error in its measurement amounted to 5% ; this low rate of flow also accounts for the small temperature difference between θ_1 and θ_2, which gives an error of 0.78%. Hence the rate of flow should have been more. This helps in reducing the percentage error in K. This is also rendered evident from another set of observations on the same rod.

Other factors being same, the altered observations are

M = 44 gm. ;	$\theta_1 = 83.6^\circ C_1$	$\theta_2 = 70.4^\circ C$
	$\theta_4 = 45.4^\circ C_2$	$\theta_3 = 31.0^\circ C$

Making the calculation of the error in K we have

$$\frac{dK}{K} = \frac{0.5}{44} + \frac{0.004}{(45.4 - 31.0)} \frac{0.01}{10.50} + 2 \times \frac{1}{2} \times \frac{0.001}{2.54} + \frac{0.2}{120} + \frac{0.04}{(83.6 - 70.4)}$$

$$= 0.0113 + 0.0027 + 0.0009 + 0.0018 + 0.0016 + 0.0030 = 0.0213$$

Hence the percentage error in K is 2.13%.

X. Determination of H with Vibration and Deflection magnetometers

The equation employed for calculating H from the observations is

$$H = \frac{2\pi}{T(d+l)(d-l)} \left(\frac{2 I d}{\tan \theta} \right)$$

The extreme proportionate error in H is given by

$$\frac{dH}{H} = \frac{dT}{T} + \frac{d(d+l)}{(d+l)} + \frac{d(d-l)}{(d-l)} + \left[\frac{dl}{l} + \frac{dd}{d} + \frac{\sec^2 \theta d\theta}{\tan \theta} \right]$$

From the following observations in an experiment, we can calculate the error.

$T = 3.14$ sec. $dT = \frac{0.2}{50}$ sec. (as 50 vibrations were timed)

$d = 18.5$ cm. $dd = 0.1$ cm.

$l = 2.505$ cm., $dl = 0.2$ cm., assuming this to be the uncertainty in locating the poles.

Hence, $d(d \pm l) = 0.3$ cm.

$\theta = 40^\circ$ $d\theta = 1^\circ = 0.0175$ radian

$m = 55.750$ gm. $dm = 0.001$ gm.

$a = 5.10$ cm. $da = db = 0.01$ cm., least count of vernier callipers.

$b = 1.60$ cm.

Since $I = m. \frac{a^2 + b^2}{12}$

$$\therefore \quad \frac{dI}{I} = \frac{dm}{m} + \frac{2da}{a} + \frac{2db}{b}$$

$$= \frac{0.001}{55.750} + \frac{2 \times 0.01}{5.10} + \frac{2 \times 0.01}{1.66} = 0.0269$$

Hence,

$$\frac{dH}{H} = \frac{0.2}{50 \times 3.14} + \frac{0.3}{21} + \frac{0.3}{16} + 1\left[0.0269 + \frac{0.1}{18.5} + \frac{0.0175}{(0.7760)^2 \times (0.8391)}\right]$$

$$= 0.0013 + 0.0143 + 0.0187 + 0.0134 + 0.0029 + 0.0173 = 0.0679$$

Thus the percentage error in the measurement of H is 6.79%. The major contributions to this error are due to the uncertainty in locating the poles, and 1.73% in the measurement of θ.

If θ were 25°

$$\frac{d(\tan\theta)}{\tan\theta} = \frac{\sec^2\theta d\theta}{\tan\theta} = \frac{0.0175}{(0.9063)^2 \times (0.4663)} = 0.04569$$

Hence the error contribution due to this factor would have been $1/2 \times 0.04569 = 0.0228$, *i.e.*, 2.28% as against 1.73% when θ is 40°. It is, therefore, necessary to have θ in the neighbourhood of 45° consistent with the fact that d is taken large. Taking d large reduces the errors in the factors involving $(d + l)$ and $(d - l)$.

XI. Determination of difference in two nearly equal resistances with a Carey Foster's Bridge.

In a Carey Foster's Bridge

$$X - Y = (l'_2 - l'_1)\,\rho$$

Where

$$\rho = \frac{R}{(l_2 - l_1)}$$

Hence,

$$X - Y = \frac{R}{(l_2 - l_1)} \times (l'_2 - l'_1)$$

The proportionate error in (X – Y) is given by

$$\frac{d(X-Y)}{X-Y} = \frac{dR}{R} + \frac{d(l_2 - l_1)}{(l_2 - l_1)} + \frac{d(l'_2 - l'_1)}{(l'_2 - l'_1)}$$

In an experiment the observations were

$$\left.\begin{array}{ll} R = 1.00\ \Omega & dR = 0.001\Omega \\ (l_2 - l_1) = 49.8\text{ cm.} & d(l_2 - l_1) = 0.2\text{ cm.} \\ (l'_2 - l'_1) = 4.5\text{ cm.} & d(l'_2 - l'_1) = 0.2\text{ cm.} \end{array}\right\} dl_1 - dl_2 = 0.1\text{ cm}$$

Substituting the values above

$$\frac{d(X-Y)}{X-Y} = \frac{0.001}{1.00} + \frac{0.2}{49.8} + \frac{0.2}{4.5}$$

$$= 0.001 + 0.004 + 0.0444$$

$$= 0.0494$$

Hence the proportionate error in (X – Y) in 0.0494, *i.e.*, the percentage error in the determination (X – Y) is 4.94%.

Three points are worth noting here.

(1) In order to determine ρ choose a large value of R to have $(l'_2 - l'_1)$ large; then alone the proportionate error in the determination of $(l'_2 - l'_1)$ would be less.

In the above experiment, with R = 0.1 ohm, $(l'_2 - l'_1)$ would be 3 cm., and the proportionate error in its measurement would be 0.2/5 = 0.04, *i.e.*, the percentage error in its determination would be 4%. It is ten times the value of 0.4% with R = 1 ohm.

(2) From the observations above, it is apparent that the *difference* between X and Y is of the order of 0.09 ohm. The total extreme error is 4.94% of which that contributed by the measurement of $(l'_2 - l'_1)$ is 4.4%. Hence the error in the difference of the resistances, X and Y is (0.09× 4.94)/100 = 0.0044 ohm. Thus this error's contribution of 4.4% in the measurement of $(l'_2 - l'_1)$ need not be alarming, if this error is 4.4% of the *difference*.

This point is further illustrated by taking a value of $(l'_2 - l'_1)$ as 1.5cm., when the difference between the two resistances is of the order of 0.03 ohm. In this case

$$\frac{d(X-Y)}{X-Y} = 0.001 + 0.004 + \frac{0.2}{1.5} = 0.1383$$

i.e, the percentage error in the measurement of (X - Y) is as high as 13.83%. But the difference is of the order 0.03 ohm, and hence the error in the difference (X – Y) is $0.03 \times \frac{13.83}{100} = 0.0041$ ohm.

(3) The method is suitable for finding the difference in the value of two very nearly equal resistances, for then the error's contribution in the determination of ρ becomes negligible.

XII. Comparison of two resistances with a potentiometer

We know that

$$\frac{R_1}{R_2} = \frac{l_1}{l_2}$$

In an experiment, l_1 = 918.1 cm., l_2 = 900.5 cm. $dl_1 = dl_2$ = 0.1 cm. Hence the proportionate error in the determination of the ratio R_1/R_2 is

$$\frac{d(R_1/R_2)}{R_1/R_2} = \frac{dl_1}{l_1} + \frac{dl_2}{l_2} = \frac{0.1}{918.1} + \frac{0.1}{900.5}$$

$$= 0.000108 + 0.000111$$

$$= 0.00019 = 0.00022$$

Thus the percentage error is 0.022% and it is almost equally distributed between the two factors. The method should be applied for the comparison of two nearly equal resistances.

XIII. Determination of E.C.E. of Copper using a Tangent Galvanometer

The electro-chemical equivalent of copper is given by

$$z = \frac{m}{Kt \,.\, \tan\theta}$$

The proportionate error in z is given by

$$\frac{dz}{z} = \frac{dm}{m} + \frac{dK}{K} + \frac{dt}{t} + \frac{d(\tan\theta)}{\tan\theta}$$

In an experiment the observations gave

$m = W_2 - W_1$ = 99.6752 – 99.6102 gm. = 0.3650 gm.   dm = 0.0002 gm., m being the dif. of two weighings,

K is known to a high degree of accuracy as H, *n* and *r* are known fairly accurately ; *r* is usually supplied by the manufacturer. Hence the factor dK/K can be left out.

t = 1200 sec.   dt = 1 sec.

$\theta = 42^\circ$ $d\theta = 1^\circ$ = 0.0175 radian

Substituting the values in the equation above

$$\frac{dz}{z} = \frac{0.0002}{0.3650} + \frac{1}{1200} + \frac{\text{Sec}^2\ 42 \times 0.0175}{\tan 42}$$

$$= 0.00054 + 0.00083 + 0.03416 = 0.0356$$

Hence the extreme percentage error in z is 3.56%, the major contribution being due to an error in the observation θ. As the weighings are done on a chemical balance, d*m*/*m* factor contributes 0.054% error only. Using a stop watch of least count 0.2 sec., would not in any way improve the result as the error contribution by stop clock of least count 1 sec. is 0.083 when the duration for which the current is passed is 20 minutes. If the duration be raised to 30 minutes, it would be less. We therefore conclude that great care must be exercised in the observation of θ, the angle of deflection. It is assumed, however, in the discussion that the setting of the galvanometer is correct.

GRAPHICAL REPRESENTATION OF DATA

I. Advantages of Graphs—The presenting of data by a graphical method is an adaptation of the principles of Descarte's analytical geometry. Whereby numerical values are represented in geometrical form by the length of a line, the area of a surface etc.

It may conveniently be said in favour of graphs that

(*i*) They appeal to the attention of the reader,

(*ii*) They permit an easy reference to data,

(*iii*) They facilitate comparisons of values,

(*iv*) They reveal some significant features in a set of data—*e.g.*—presence of maxima or minima or points of inflection—which may not be easily noticed in a survey of the data.

II. Choosing the Co-ordinate scales—For the under-graduate students the purpose of plotting a graph is limited. His job is not to investigate whether or not a relationship exists between the variables considered, and if such a relationship exists, to determine its mathematical form. His job is to show that the relationship as embodied. In the mathematical equation which he is using in his experimental work, does exist ; he may further use the graph as a tool for making some determinations.

Further the under-graduates have not to make any choice about the type of graph paper to be used in the graphical representation of the experimental data. He has usually to use the rectangular graph paper which has either mm. divisions or

1/10 divisions. The main problem before him is the *choice of the co-ordinate scales.*

"A poor choice of scales for the co-ordinates, more than any other single factor, will make an otherwise acceptable graph unsatisfactory as a tool." This being the case, the need of suitability rules is evident. Certain general rules may be stated as follows :

It is customary to choose the scale for the independent variable along with X–axis.

This is an established custom. Which of the two variables is independent is usually greatly influenced by the experimental procedure adopted. While taking observations in an experiment, the values of one quantity are arbitrarily fixed, and the corresponding values of the other quantity are observed; in such a case it is evident that the former quantity is to be regarded as an independent variable.

In the experiment for the determination of difference in two nearly equal resistances by C.F. Bridge, the formula used is

$$R_1 - R = \rho\,(l_1 - l_2)$$

where l_1 is the length of the bridge wire at the balance with R_1 in left gap and R in the right, l_2 the corresponding length when R_1 and R are interchanged, ρ is the resistance per unit length of the bridge wire.

In the experiment under reference R is varied arbitrarily by using suitable resistance in the decimal ohm box and the corresponding diference $(l_1 - l_2)$ is computed by observing l_1 and l_2. R is therefore an independent variable while the corresponding value of $(l_1 - l_2)$ is a dependent variable. The graph (G. A.) has been plotted with R along X-axis and $(l_1 - l_2)$ along Y-axis. The data for the graph have been taken from the table below.

TABLE—*For Measurement of* R_1

R	l_1	l_2	$(l_1 - l_2)$
2.4	56.0	47.9	8.1
2.5	54.4	42.5	4.9
2.6	52.8	51.2	1.7
2.7	51.2	52.8	– 1.6
2.8	49.2	54.4	4.5
2.9	48.2	55.9	7.7

2. *The choice of the scale should be such that the co-ordinates of any point on the graph may be ascertained easily and quickly.*

For rectangular graph paper with successive main lines divided into ten parts (each part equal to 1 mm. in a cm. graph sheet or 1/10 in an inch graph sheet), the scales convenient to use are those in which the distance between two consecutive main lines represents a difference in value of 1, 2, 4 or 5 units or these values multiplied by 1 on where *n* is an integer. It would certainly be very inconvenient to plot the points and also troublesome to read the co-ordinates of a point on the plot, if the distance between the consecutive main lines repressented a difference of 3, 6, 7, 9..............units.

CAREY FOSTER BRIDGE

Determination of an unknown Resistance R_1

$[R_1 (l_1 - l_2)]$ GRAPH

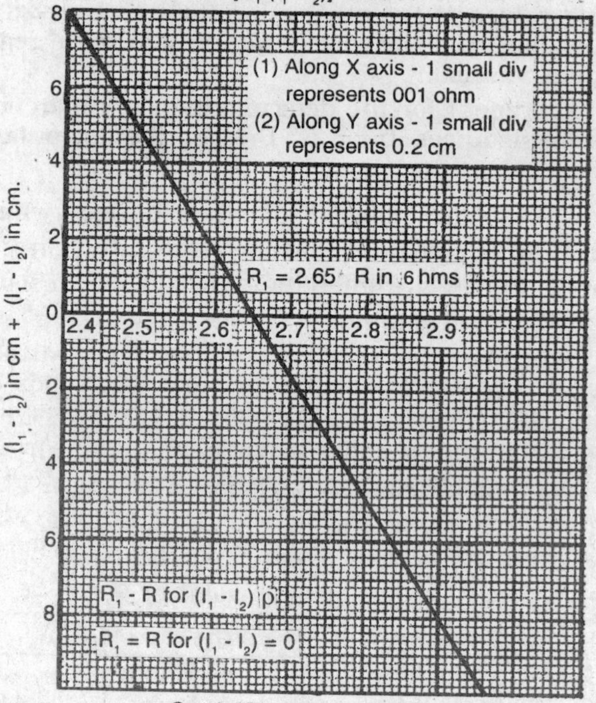

Graph (G.A.) between R_1 $(l_1 - l_2)$

Thus in the graph (G. A.) between R and $(l_1 - l_2)$ the choice of scales along the two axes is

(*i*) 0.1 ohm., difference between two consecutive *main* lines along the X-axis;

(ii) 1 cm., difference between two consecutive *main* line along the Y-axis.

3. *The choice of the scale should be such that the resultant curve is as extensive as the sheet permits but subject to the condition that "uncertainties of measurements are not made thereby to correspond to more than one or two of the smallest divisions."*

The co-ordinate units need not usually start from zero values. It is customary to begin the scale for each variable from the lowest rounded value in the data and end it at the highest value or just greater than the highest value.

The above graph has been drawn by taking the above rule into consideration with the exception that $(l_1 - l_2)$ has a starting point zero it is necessary in this case because $(l_1 - l_2)$ has both +ve and – ve values for different values of R.

Supposing there were observations only for +ve values of $(l_1 - l_2)$, taken then the starting point for $(l_1 - l_2)$ would have to be taken as zero, because the calculation of R_1, the unknown resistance, is based on finding the value R for $(l_1 - l_2) = 0$. For, from the equation,

$$R_1 - R = \rho\,(l_1 - l_2)$$

It is evident that as R is altered, $(l_1 - l_2)$ is correspondingly altered, R and ρ being constants. A plot of R and $(l_1 - l_2)$ is expected to be a straight line, as shown in graph GA.

When $(l_1 - l_2) = 0$.

$$R_1 = R$$

i.e., the intercept of the plot line on the R axis for a value $(l_1 - l_2) = 0$, gives the value of R_2. Hence the strting point for $(l_1 - l_2)$ has to be taken zero in this case of plotting the graph.

We conclude therefore that except where necessary we begin the scale of each variable from the lowest rounded value. For R we begin the scale with 2.4 ohm but for $(l_1 - l_2)$ we start with zero at the origin because it is necessary to do so.

Secondly, where the sheet is large enough we could have spread the $(l_1 - l_2)$ scale but it would have been incorrect to spread the scale such that each small division in the graph reads, say, 0.05 cm., for the least count of metre scale used to read l_1 or l_2 is 0.1 cm.

4. *Scales are to be marked properly by labelling the main co-ordinate lines with values which they represent.*

It should be noted that

(a) The numbers used in marking the scale on the main lines should contain as many significant figures as the data justify.

(b) For a satisfactory scale designation it is necessary to have along each axis the name of the quantity represented and the units in which it is measured.

As shown in graph GA, along X-axis it is labelled as, R in ohm along Y-axis, $(l_1 - l_2)$ in cm.

5. *Plotting of the points on the graph sheet should be done very carefully and each point should be indicated by a cross (X) or a circle (O).*

Where two plots on the same sheet interesect or are close together, points of the one may be marked with cross (X) and of the other with a circle (O).

6. *Fitting a curve to the plotted points.*

The curve should be smooth and should pass as close as reasonably possible to all the plotted points.

When the number of observations is fairly large, of the points which do not lie on the curve but close to it, about one half should fall on one side of the curve and the other half on the other side of the curve.

A point lying too remote from the best fit curve suggests obviously a mistake in the taking of the observation corresponding to this.

7. *Appropriate captions*

An appropriate and descriptive caption accompanying a graph should clearly indicate what the graph is intended to show.

As the primary aim of a scientific investigator is the discovery of truth, he should take all possible care to eliminate all those factors that tend to distort truth. One of the most important factors that is responsible for this distortion is the mental prejudice and habit of carelessness of the investigator, Defects of instruments are only of secondary importance. Most of the poineer workers in the fields of scientific investigations did not use very perfect instruments, for which like a bad carpenter our students are seen quarrelling. Every serious student and teacher of Physics knows well that a student learns most when he works with an imperfect instrument, for it develops resourcefulness in the worker.

For these and other considerations a student of practical physics will do well to begin his work after carefully going through these few pages and thus putting himself in the proper scientific attitude of mind of an original investigator. It is for want of this attitude of mind that a student sometimes does not feel interested in this work. A student will, therefore, do well if he carefully reads and tries to follow throughout his work in the laboratory, the rules laid down below :

1. The aim of experimental science being the discovering of truth, it should be carefully borne in mind by every student of Physics that a perfectly open and unbiased state of mind should be maintained during experimental work.

2. Before beginning any of the experimental work, the students should make a self-study of the work and thus have a clear idea of the aim of his work and of the way how he is going to realize that aim.

3. Fore-thought should be brought to bear upon every phase of work in hand. It counts more than anything else in the saving of time, in the prevention of trouble, and in the assurance it gives of satisfactory work.

4. If no plan or diagram of the arrangement of the apparatus is provided, one should be carefully made and verified before actually setting up the apparatus.

5. A sufficiently complete description of every piece of apparatus as also the order of taking observation, should be recorded in complete details in a carefully pre-arranged form.

6. In all cases as many observations as possible should be taken and entered in a suitable tabular form to be so planned by the student that it gives all the details of his observations and in certain cases also the results of his calculation.

7. These observations and results are to be directly entered in the book in which the final fair report of the work has to be written.

8. In most of his experiments, the student has to determine in the value of a physical quantity which is not directly observed but is calculated from the observations of other quantities that are directly measured. It often happens that several of these directly measurable quantities are capable of more accurate determination than others, so it becomes desirable to previously determine the degree of accuracy to which each quantity should be measured. For this the student should refer to Eq. (i) and the following examples which illusrate its use.

9. In the calculation of results also, no advantage will be found to be gained by using more than just the necessary number of figures. Thus the use of contracted and approximate methods

of calculations, and also the use of log tables upto a limited number of figures is desirable.

10. The student should make a free and frequent use of graph in the representation of his observations and also in the interpretation of their result. A graph gives a deeper insight into the phenomenon than even the tabular form.

The record of the experiment should be entered in the following form :

Date *Experiment No.* *Page*

1. A clear and concise statement of the object of the experiment.
2. A discussion of the theory involved in the experiment resulting in the formula from which the quantity is to be calculated from the direct observation of other quantities. In certain cases it may suffice to give the formula and the meaning of each symbol used in it.
3. A list of the various pieces of apparatus actually used.
4. A clear description of the assemblage of the apparatus with a linear, self-explanatory diagram neatly and carefully drawn.
5. All original observations in a neatly tabulated form.
6. Calculations.
7. Method of taking observations.
8. Precautions that were actually taken to be stated with reasons.
9. Result and percentage Error. Never omit unit.
10. Discussion of the result and criticism of the apparatus.

SECTION I

MECHANICS
AND
GENERAL PROPERTIES OF MATTER

1

Measurement of Fundamental Quantities

1.1 Choice of Units

The most important step in the progress of every science is the measurement of different quantities, for, accurate knowledge about physical phenomena can be acquired only through careful measurements. Measurement is essentially a process of comparison and consists in finding out the number of units contained in the quantity to be measured. In general all measurements are expressed in a phrase composed of two members :

(*a*) *The pure number or numeric which is the measure of the quantity in terms of the unit employed.*

(*b*) *The name of the unit according to the nature of the quantity.*

The measure of a physical quantity varies according to the size of the unit employed, but *the product of the numeric and the size of the unit remains constant.* Thus, if n_1 is the numeric of a given physical quantity in terms of a unit $[u_1]$, and if n_2 is the numeric of the same quantity in terms of another unit $[u_2]$, then

$$n_1 [u_1] = n_2 [u_2]$$

or

$$n_2 = n_1 \frac{[u_1]}{[u_2]}$$

For example, a length measured in feet is found to be, say 6 ft., and the same length when measured in yards is 2 yds. Here 2 [Yard] = 6 [Foot]. Thus *the measure of a physical quantity is inversely proportional to the size of the unit employed.* The expression $[u_1]/[u_2]$ is termed the *change ratio* for the conversion from the first unit to the second.

In order that the numeric expressing the measure may be intelligible, the magnitude of the unit must be known. This leads to the conventional choice of certain magnitudes as units of measurement, and any other magnitude is then simply expressed by a number which tells how many times the appropriate unit is contained in the magnitude considered.

Now each type of physical quantity requires its own unit, but these units are not necessarily independent, and it is desirable that the number of different kinds of units introduced into our measurements should be as small as possible. But a proper scientific unit should satisfy the following conditions :

(*a*) It must be well defined, *i.e.*, its value should be fixed.

(*b*) Its size must be chosen according to the quantities intended to be measured with it.

(*c*) It should not change with time.

(*d*) It should not readily change with changing conditions such as temperature and pressure, and if such changes do take place their law of variation should be accurately known.

(*e*) It should be easily comparable with other similar units.

(*f*) It should be easily reproducible.

1.2. Fundamental Quantities

There are certain relations which exist between the various physical quantities and by utilising these relations it is possible to select the units in a limited number. Actually speaking, physical quantities met with in Mechanics may all be expressed in terms of or defined with relation to three selected quantities, *length*, *mass* and *time*, called the three *fundamental quantities of Mechanics*, because they are simple to understand and convenient to work with, as they enable exact measurement and standards of length and mass can be easily copied and compared with the original standards.

1.3. Systems of Units

A complete system in which certain fundamantal units are used and in which all other units are derived from the fundamental units, is termed an *absolute system* of units. There are two absolute systems in common use : (*a*) the C. G. S. (*centimetre, gramme, second*) system, and (*b*) the F. P. S. (*foot, pound, second*) system, the former being more important as it is very widely used, especially in *scientific work*.

(a) C. G. S. system—The unit of length is the *centimetre* and is equal to $\frac{1}{100}$ of a metre which is *defined as the quite arbitrary distance between two marks on a bar made of an alloy of 90% platinum and 10% iridium kept at 0°C, and preserved in the International Bureau of Standards at St. Cloud, near Paris*. In order to preserve the length of metre in the case of loss or alteration, it has now been evaluated in terms of wavelength of light, especially because technical measurements of length are carried out to an increasing degree with the help of light waves. The metre is now defined as equal to 1,650, 763.73 wave lengths in vacuum of the orange-red radiation of the Krypton-86 atom.

The unit of mass is the *gramme*, and is equal to $\frac{1}{1000}$ of **a kilogramme** *which is the mass of a piece of platinum called the "Kilogramme des Archives" made by Borda in accordance with a decree of the French Republic, and preserved at the Bureau of Metric Standards.* A kilogramme is very nearly equal to mass of 1000 cubic centimetres of pure water at 4°C.

The unit of time is the *mean solar second* which is $\frac{1}{86400}$ of a mean solar day. *The time which elapses between two consecutive transits of the sun across the meridian at any place on the earth's surface is called the solar day of the place, and its average* value for one year is called the *mean* solar day. In practice the mean solar day is based on the sidereal day. The latter is defined as the time taken between two successive passages of the same fixed star across the meridian of the place of observation. A year contains 366.25 sidereal days, but only 365.25 solar days, and hence the *mean* solar second is 366.25/365.25 of the sidereal second.

Presently *atomic time* is used as standard of time. According to this system, *one second is equal to the time interval or duration of 9,192,631,770 periods of radiation corresponding to the transition between the two hyperfine lines (4,0 – 3, 0) of the ground state of the caesium-133 atom.*

(b) F. P. S. system—The unit of length is the *foot*, and is equal to $\frac{1}{3}$ of a yard which *is defined as the straight distance between the transverse lines in two gold plugs on a bronze bar kept at 62° F and preserved in the Exchequer Officer, London.* The unit of mass is the **Pound avoirdupois** and *is the mass of a piece of platinum preserved in the Office of the Exchequer and marked "P. S. 1844, 1 lb".* The unit of time is the same as that in the C. G. S. system.

(c) The SI system of units—The units in this system are :

(*i*) *The Metre.* The SI unit of length is metre. The present unit of the metre is defined in terms of the wavelength of radiation.

"The metre is defined as the length equal to 1650, 763.73 wavelengths in vacuum of the radiation corresponding to the spectral line omitted during the transition between the levels $2p_{10} \longrightarrow 5d_5$ of krypton - 86 atom.

(ii) *The Kilogramme*—The SI unit of mass is the kilogramme (kg). One kilogramme is equal to mass of a Pt-Ir cylinder (90 : 10) (of equal diameter and height) kept at Paris. The Indian National Prototype kilogramme is in the custody of National Physical Laboratory (NPL), New Delhi.

(iii) *The Second* —The SI unit of time is the second. The present value of a second is defined as above (in CGS system).

All these units were initially the units of the MKS system of unit. Later SI (SYSTEME INTERNATIONAL D' UNITE'S) was introduced with inclusion of three more basic units. These are :

(iv) *The Candela* —This is the SI unit for light intensity.

(v) *The Ampere* —This is the SI unit for electric current.

(vi) *The Kelvin* —This is the SI unit for temperature.

In India, National Physical laboratory (NPL), New Delhi, is entrusted with preservation and maintenance of National standards.

1.4 Units for Microscopic and Macroscopic Quantities

Various multiples and sub-multiples of the fundamental units have been chosen, the larger units to measure greater and the smaller units for lesser magnitudes, in order that the numerical value of the quantity may be a number neither too large nor too small *for convenience*. The following table gives a list of the various units of length and their equivalents.

(A) Units for our daily use on the earth—

(i) *Metric Units—*

Metre	...	...	1 m.	= 39.37 in.
				= 3.28 ft.
				= 1.09 yd.
Kilometre	...	...	1 km.	= 0.62 mile.

(ii) *British Units—*

Mile	...	...	1 mi.	= 1609.34 m.
Yard	...	...	1 yd.	= 0.91 m.
Foot	...	...	1 bl.	= 0.30 m.
Inch	..	..	1 in.	= 2.54 cm.
Mil.	...	...	1 mil.	= 0.001 in.

(B) Astronomical scale

The astronomical unit is equal to the semi-major axis of the earth's orbit.

Astronomical unit = 1.495×10^8 km.
= 9.289×10^7 mi.

Parsec. ... = distance at which the astronomical unit subtends 1 second (1″)
= 2.06×10^3 astr. units approx.
= 3.083×10^{13} km.
= 1.91×10^{13} mi.

Light year ... = the distance travelled by light in 1 year (velocity of light = 2.998×10^{10} cm./sec)
= 0.31 parsec.

(C) Molecular and Atomic scale

Fermi ... ... = 10^{-13} cm.

Angstrom unit ... 1 A° = 10^{-8} cm.

Micron ... ... 1 μ = 10^{-3} mm.

Millimicron ... 1 m μ = 10^{-6} mm.

1.5 Metre Scale

The simplest method of measuring a length is by means of a metre scale, but accuracy of measurement in this case is very much limited since the dividing lines have a finite thickness and naked eye cannot correctly estimate less than 0.5 mm. To measure a length, two observations have to be made, one at each end of the length to be measured, and hence with a metre scale the accuracy is not greater than about 1 mm. For greater accuracy an ingenious device called *vernier* is employed.

1.6 Vernier

It consists of an auxiliary scale called *vernier* scale which slides along an ordinary scale called the *main scale*, the divisions of *the vernier scale being either a little longer or a little shorter than the divisions of the main scale.*

Referring to Fig. 1.1, let AB be the *main scale* and CD the vernier which slides alongside AB. Let a rod PQ whose length is to be measured be placed along the main scale with its one end P *in line* with the *zero* of the *main scale* and with its other end Q coinciding with the zero of the *vernier scale.* Then the length

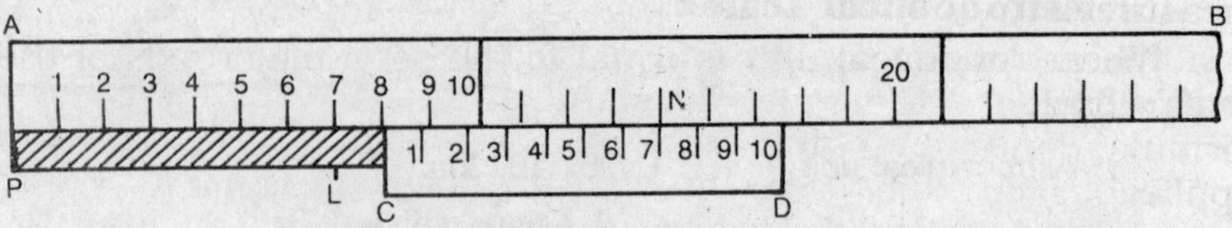

Fig. 1.1

of the rod is evidently equal to seven main scale divisions+the fraction LM of the eighth division. In this position of vernier, the seventh division of vernier is coincident with the fourteenth division of the main scale. Hence the fraction LM which is equal to LN–MN = value of 7 main scale divisions+value of 7 vernier scale divisions. Hence,

Length of the rod = Value of 7 main scale divisions + 7 (*difference* between the values of one main scale and one vernier scale division).

The difference between the value of one main scale division and that of one vernier scale division is constant for a particular vernier and is called its **vernier constant** *and since it is the least distance which can be accurately measured with the vernier, it is also referred to as the* **least count** *of the vernier.* Thus

Length of the rod = Value of 7 main scale divisions + 7 × (*vernier constant*).

Hence the rule for measuring a length with a vernier is as follows :

Find out the value of the divisions on the *main scale* upto that *next* before the zero of the vernier scale; then find the number of the vernier division which is *in line* with a scale division and multiply it by the *vernier constant.* Add this product to the reading on the main scale. The result gives the distance from the zero of the main scale to the zero of the vernier scale.

Method of Finding Vernier Constant—Let S and V be the values of one *main scale* and one vernier scale divisions respectively, and let *(n – 1) divisions of the main scale* be equivalent to *n* divisions of the *vernier.* Then evidently

$$(n - 1)\,S = nV$$

or $$nS - nV = S$$

whence $$(S - V) = S/n$$

$$\textbf{Vernier Constant} = \frac{\text{Value of smallest division on the main scale}}{\text{Total number of divisions on the vernier scale}}$$

The *same* principle applies to *circular* verniers used for the measurement of small angles, *e.g.*, in sextant, spectrometer, *etc.*

When the vernier division is slightly *longer* than the main scale division, the vernier constant is obtained from the *same* formula, and the same method of reading the vernier is applicable. In this case the vernier divisions are written in *reversed* order, *i.e.*, the zero of the vernier is *farthest* from the zero of the main scale. This type of vernier is known as *reverse-reading* or *backward-reading* vernier.

1.7 Vernier Callipers

This instrument employs the principle of vernier and is used for the measurement of small lengths and internal and external diameters of bodies. As illustrated in Fig. 1.2, the *main scale* S is engraved on a steel strip and the *vernier* V on a frame D which slides along the strip. A steel jaw A is fixed at one end of the strip and at right angles to it, and a *similar* jaw B is fixed *at right angles* to the frame carrying the vernier. A clamp screw is also provided to fix the vernier to the scale as and when desired.

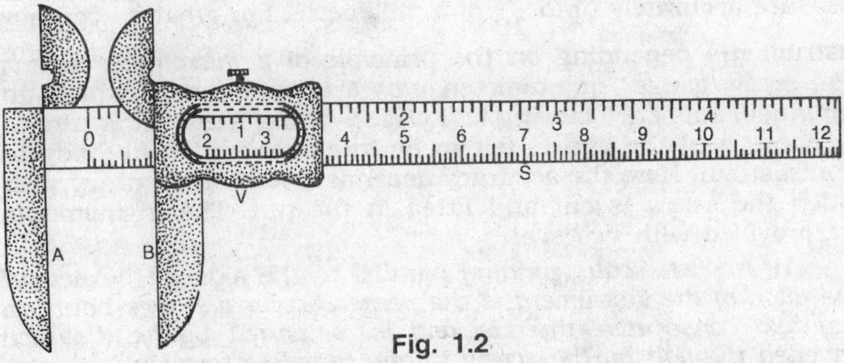

Fig. 1.2

The construction of the instrument is such that when the two jaws are in contact, the zero of the vernier scale coincides with the zero of the main scale so that, if a body whose length is to be measured is placed lengthwise between the jaws, the reading of the vernier gives the length of the body. Care should be taken not to press the jaws too much to compress the body.

Zero error and its correction—In explaining the working of vernier callipers it was assumed that when the two jaws are in contact, the zero of the vernier is *opposite* to the zero of the main scale. But on account of its faulty use or construction, it is seldom the case, *This error of the non-coincidence of the zeros of the main scale and the vernier scale, the two jaws coinciding, is called the zero error of the instrument and its value is equal to* *the reading on the instrument when the two jaws meet together without anything between them.*

If the zero of the vernier scale is to the right of the zero of the main scale, the two jaws coinciding, the *zero error* is said to be *positive* and the reading given by the instrument is *more* than the actual length of the body. If, however, the zero of the vernier lies to the left of the zero of the main scale *zero error* is *negative*, and the reading of the instrument is *less* than the actual length of the body. In general, the zero error is always subtracted *algebraically, i.e.*, with its proper sign from the reading of the instrument to get the correct length of a body.

Note that when the zero error is *negative*, the number of vernier division coincident with a scale division is *subtracted* from the *total* number of divisions on the vernier scale before multiplying it by the vernier constant.

1.8. Micrometer Screw

Vernier callipers usually employed in the laboratory can measure accurately upto $\frac{1}{10}$ of a millimetre, For greater accuracy instruments depending on the principle of a *micrometer screw, e.g.*, screw gauge, spherometer, *etc.*, are employed in which an *accurately* cut screw moving in a closely-fitting fixed nut is turned by hand until the end is felt to be in contact with the body to be measured. Here the accuracy depends upon the accuracy with which the screw is cut and fitted in the nut. The instruments are provided with two scales.

(1) A *pitch scale, running* parallel to the axis of the screw. *The pitch of the instrument is the perpendicular distance between any two consecutive threads and is measured by he distance travelled through by the screw in one complete rotation.*

(2) A *head or circular scale* round the edge of the screw head.

The motion of translation of the screw is directly proportional to the amount of rotation that is given to it. The *complete* rotations of the screw are read on the *pitch scale* and the *fraction* of a rotation on the *head scale* so that sum of the readings of these two scales is the distance advanced by the screw. *The* **least count,** *i.e., the smallest distance which can be accurately measured by the screw is the value of one division on the head scale and is equal to the pitch of the screw divided by the total number of divisions on the head scale.*

1.9. Screw Gauge

Referring to Fig. 1.3. a *screw gauge* consists of a measuring

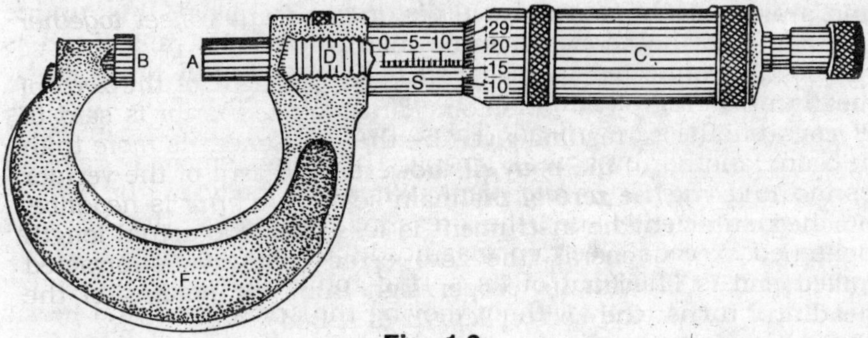

Fig. 1.3

screw D which works in a nut fixed inside a tubular hub S attached *at right angles* to one end of a U-shaped frame F. The screw head or thimble C often *milled* at its end to ensure a grip moves over the outside of the hub and is attached to the screw spindle at the right hand end. By turning the screw head the *planed* end A of the screw D moves and makes a wider or narrower gap between it and the planed face of a steel plug B attached to the frame F. On the outside of the hub there is the *pitch scale* divided into millimetres or half millimetres according as the pitch of the screw. Its base line running along the hub is called the *line of graduation*. The equidistant marks on the left hand *bevelled* edge E of the screw head forms the *head scale* which measures incomplete turns of screw.

To use the instrument, first the value of the *smallest* division of the *pitch scale* is found out. Then the screw is given a number of *complete rotations and the distance advanced by the screw in one complete rotation is determined which gives the value of pitch of the screw*. The pitch is divided by the total number of divisions on the head scale which gives the *least count* of the screw gauge. Next the body whose length is required is placed in the gap and the screw is turned slowly till the object is just held. No attempt should be made to force or hurry the movement of the screw. The pitch scale reading gives the value of complete rotations of the screw and the head scale reading the value of fraction of a rotation so that the sum of these two readings gives the length of the object.

Care must be taken not to screw up tightly otherwise the readings will be inconsistent. The turning of the screw is stopped when the milled end just shows a tendency to "slip". If the screw is tightened too much the threads of the screw will be damaged and frame of the instrument distorted. To avoid this undue tightening of the screw, in some instruments the milled end is not rigidly connected to the head of the screw but engages, by means of a ring of flat-shaped cogs, with a corresponding ring of cogs on the screw head. These two sets of cogs are pressed into one another by a weak spring. This arrangement is called a *spring and ratchet arrangement.* When the body in the gap is pinched sufficiently, *i.e.*, the pressure with which the screw is tightened exceeds a certain magnitude, the spring gives and the milled end is lifted out of its spring connection with the screw head and turns freely without moving the screw head and hence the measuring screw.

Zero Error—In a perfect instrument, when the gap is reduced to zero, the zero of the head scale lies along the line of graduation and the bevelled edge against the zero of the pitch scale. If it is not so the instrument possesses a *zero error* and *its value is equal to the reading of the instrument when the gap is closed.* The zero error is *positive* when the bevelled edge is to the *right* of the zero-line of the main scale and *negative* when to the *left*, the gap being closed.

When the readings are negative the head scale observations should be subtracted from the total number of divisions on the head scale, before multiplying them by the least count. To get the actual length of the object in the gap, the zero error is always subtracted *algebraically* from the reading of the instrument.

Back-Lash—Unless very perfectly constructed, there is an appreciable amount of *slackness* in the fit of the screw and the nut *owing to wear between them.* This makes the screw wobble to a certain extent in the grooves of the nut, and on account of this small amount of play the screw presses itself sometimes more tightly, sometimes less tightly in the grooves of the nut. Hence when the screw has been adjusted by turning it in one direction and then turned in the opposite direction, the head moves through an *appreciable* angle before the screw begins to move. Thus the reading on the head scale for any position of the screw when adjusted by turning to the right may be noticeably different from that obtained by turning to the left. This error in the readings is called *error due to back lash* of the screw, and can be avoided to a considerable extent by turning the screw always in the *same direction* when making the final adjustment to any particular position.

1.10 Spherometer

This instrument employes the principle of *micrometer screw* and as shown in Fig. 1.4, consists essentially of a fine and accurate, pointed micrometer screw which can turn in a nut fixed to a metal frame provided with three pointed legs A, B and C of *equal* length and arranged at the corners of an *equilateral* triangle. The axis of the screw is *perpendicular* to the plane ABC, called the *reference or ground plane* of the spherometer, and passes through the centre of the *circumscribed* circle of the triangle. The screw carries at its top, below the milled heed, a circular disc the circumference of which is divided into equal parts which form the *head scale.* To the side of the disc and nearly in contact with its rim is the *pitch scale* fixed to the frame with its edge *parallel* to the axis of the screw, and graduated in divisions equal to the pitch of the screw.

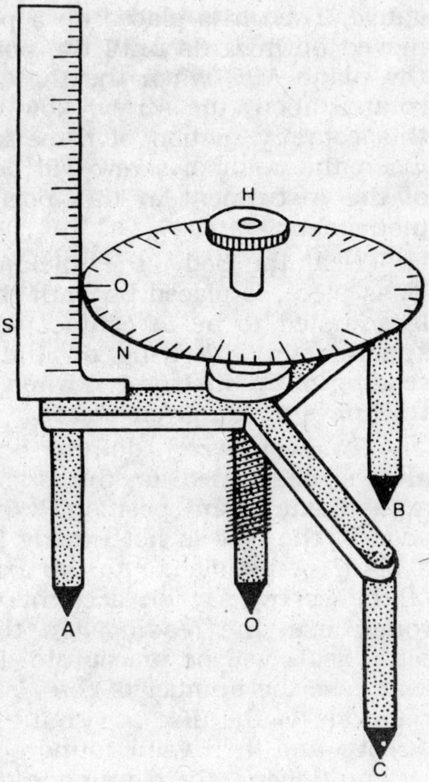

Fig. 1.4

When the instrument is placed with its three legs on an accurately plane surface and the screw is turned downwards until its bottom end touches the surface, the head scale as well as the pitch scale should register zero, if the instrument is perfect. If it is not the case, there is some *zero error* in the instrument and *its value is equal to the reading of the instrument when the tip of the screw lies in the plane* ABC. The *zero error* is *positive* when the disc is *above* the zero line of the pitch scale and *negative* when *below.* When the reading is negative the head scale observation is subtracted from the total number of divisions on the head scale before multiplying it by the least count. To correct a reading for zero error, the latter must always be subtracted *algebraically* from the former.

To measure a small thickness with the instrument first its pitch and least count are determined as in the case of a screw gauge. Then it is placed on a plane glass plate, and its screw is moved downwards until the point of the screw is slightly below the plane ABC when the three legs of the spherometer can be rotated about the screw. Now the screw is slightly raised until this rotatory motion of three legs about the screw *just* ceases when the point of screw will be in the plane ABC. The reading of the instrument in this position gives the zero error of the instrument.

Next the body, the thickness of which is required, *e.g.*, a glass piece, is placed beneath the screw and the tip of the screw is adjusted to be in contact with it with the three legs of the spherometer *still* lying on the former plane glass plate. The reading of the instrument when corrected for zero error gives the thickness of the given object.

To avoid error due to *back-lash* the screw head should always be turned in the *same* direction when making final adjustment for any position. If the edge of the disc is not exactly in a plane perpendicular to the axis of the screw, say on account of rough use, the readings on the pitch scale will be inaccurate. In that case the number of complete rotations of the disc are counted directly and their value found out by multiplying their number by the pitch of the screw.

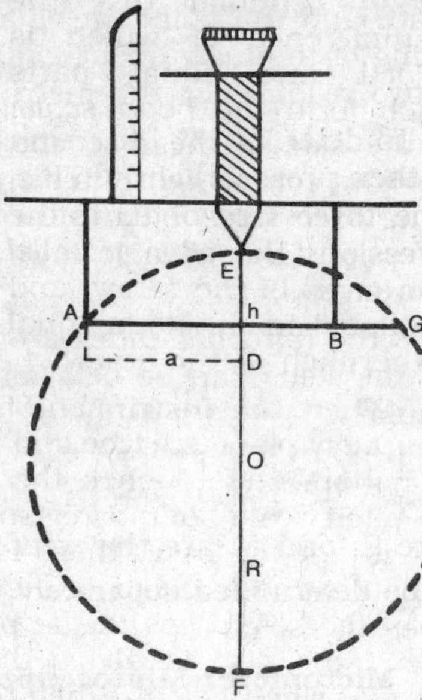

Fig. 1.5

Spherometer can also be used to determine the *radius of curvature* of a spherical surface and hence its name. Referring to Fig. 1.5 which represents a *vertical* section of a spherical surface by a plane passing through the centre of the sphere of which the surface forms a part, and through a diameter of the circle through ABC, when the points of the three legs of the spherometer and the screw lie on the surface, we have from the *geometry* of the figure

$$h\,(2R - h) = a^2$$

Whence $$R = \frac{a^2}{2h} + \frac{h}{2} \tag{1.1}$$

If l be the length of each side of the equilateral Δ ABC, $l/2 = a \cos 30°$. Thus $a = l/\sqrt{3}$, and substituting in Eq. (1.1), have

$$R = \frac{l^2}{6h} + \frac{h}{2}$$

from which R can be calculated if l and h are determined. The quantity $h/2$ on the right hand side of the above equation is often so small compared to $l^2/6h$ that it can be neglected. In that case

$$R = \frac{l^2}{6h} + \frac{a^2}{2h}$$

To determine h, the spherometer is placed on the spherical surface and the screw moved up (convex) or down (concave) until it just touches the spherical surface, and the reading of the instrument is taken. Next the spherometer is placed on a plane surface and the point of the screw is made to touch it, and the reading of the instrument is again taken. The difference between the two readings of the instrument gives the value of h.

To determine l, the instrument is placed on a sheet of paper and then pressed lightly. This produces impressions of the points of the three legs of the spherometer on the paper. From these impressions the triangle is drawn and its mean side found out which gives l.

If the *refractive index* μ of a lens be known, the *focal length* f of the lens can be determined with a spherometer using the formula

$$\frac{1}{f} = (\mu - 1)\left(\frac{1}{R_1} - \frac{1}{R_2}\right)$$

where R_1 and R_2 are the *radii of curvature* of its two faces which can be determined separately with the spherometer as described above.

1.11 Micrometer Microscope

It consists of a compound microscope whose eye-piece is fitted with a *micrometer screw* (Fig. 1.6) by means of which a fine thread or two parallel threads separated by a small distance can be moved in the field of view of the microscope, the rotation of the screw being indicated by a scale drum divided into 100 divisions and attached to a small wheel fixed to the screw.

The objective of the microscope forms a real image of the small object to be measured, coincident with the fine thread on which the eye-piece is previously focussed. The fine thread is then moved from one end of the object to the other by means of the screw. The distance travelled through by the screw as indicated by the divisions on the scale drum gives the size of the object. The absolute value of one division of the micrometer depends upon the magnification of the microscope and is obtained by observing the image of a fine scale near the focus of the eye-piece and then determining the number of rotations of the wheel which are equivalent to one millimetre. The pitch of the screw is generally $\frac{1}{2}$ mm., and the head is divided into 100 equal parts. If the magnification of the microscope is 10, the value of one divisions of the micrometer screw in this case is $\frac{1}{1000}$ mm.

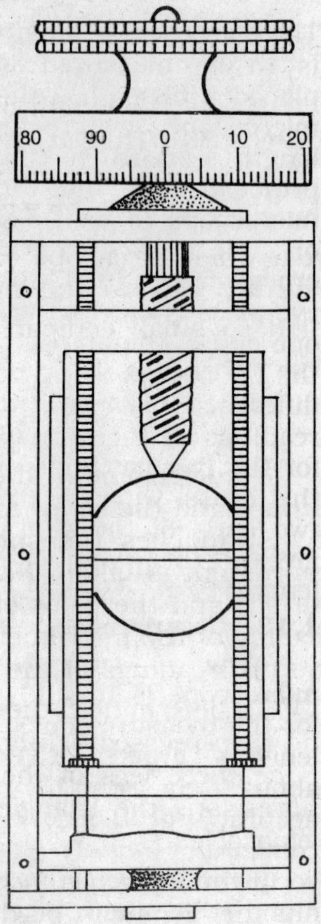

Fig. 1.6

It is chiefly used for the measurement of diameter of very fine capillary tubes and the minute motions of the gold leaf of an electroscope. While taking a reading care should be taken to avoid error due to *back-lash*.

1.12 Travelling Microscope

It is an instrument for measuring small lengths, *e.g.*, the diameter of Newton's rings (see interference in light), narrow tubes, etc. and consists of a compound microscope mounted vertically or horizontally on a rigid metal frame so that it can be moved in a direction at right angles to its axis by means of a screw or a *rack and pinion* arrangement.

In another type called the *vernier microscope* (Fig. 1.7), the displacement of the microscope is measured with a vernier which moves with the microscope along a scale fixed to the instrument.

The object whose length is to be measured is placed beneath the microscope with its length parallel to the principal scale, and the microscope is focussed on it. Next the cross-wire in the eye-piece is moved from one end of the object to the other. The difference between the readings of the vernier for the two positions of the cross wire at the two ends of the object gives the required length.

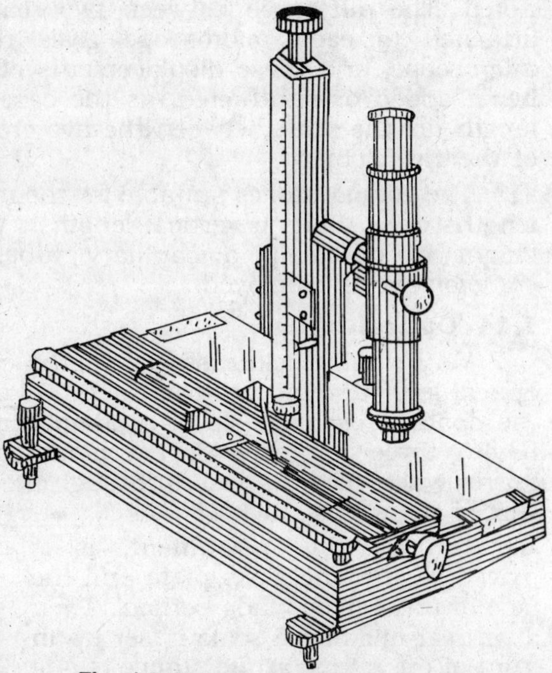

Fig. 1.7 The Travelling Microscope

1.13. Comparator

The travelling microscope is unsuited for the measurement of lengths greater than about 5 cm. Lengths uptc say, one metre are measured upto an accuracy of 0.001 mm. by means of an instrument called *comparator* which consists essentially of two micrometer or vernier microscopes mounted on separate carriages which move along the same slides so that the distance between the microscopes can be varied.

The object whose length is to be measured is placed horizontally below the microscopes with its length parallel to the slides and the microscopes are adjusted so that their cross-wires coincide with the end fo the length to be measured, and the readings of the micrometres attached to the two microscopes or verniers are taken. Now the object is removed and replaced by a standard scale, the level of the scale being so adjusted that its graduations are brought into focus without disturbing the microscopes. The cross-wire of each microscope will in general lie between two scale marks. Now each microscope is moved by means of he screw towards the nearer mark until its cross-wire coincides with the mark. The readings on the two micrometers or verniers and the length l of the scale between two marks which were in coincidence with the cross-wires are also noted. The difference between two readings of the micrometer attached to each microscope gives the displacement of the microscope, and these displacements of the two microscopes on being added or subtracted, as the case may be, to or from the length l of the scale between the two cross-wires gives the length of the given object.

The comparator is suitable for the measurement of *horizontal* lengths only. When a vertical length is to be measured, *e.g.*, the height of liquid in a capillary tube, an instrument called *cathetometer* is employed.

1.14 Optical lever

Very small thicknesses or depths can be measured with an optical lever upto an accuracy of 0.005 cm. As shown in Fig. 1.8, the *optical lever* consists of a plane mirror mounted on a stand having three pointed legs. The plane containing the face of the mirror when vertical, passes through the line joining the two hind legs of the stand.

To use the instrument, it is placed on a plane glass plate and the face of the mirror made *vertical*. Then a vertical millimetre scale is set up in front of the mirror at a distance of one or two metres from it and its image in the mirror is viewed through a telescope and the position of the *horizontal* cross-wire of the telescope on this image of the scale noted. *The distance of the telescope from the scale should be as small as the focussing of the telescope allows in order that the magnification may be as great as possible.* Then the object whose thickness is to be measured, say a piece of microscope cover glass, is placed below the front leg of the optical lever, the hind legs *still* resting on the plane glass plate. This will make the mirror tilt about the line joining the two hind legs and image of the scale in the mirror will be seen to be displaced past the horizontal cross-wire of the telescope. The position of the cross-wire on the image of the scale in the mirror is again noted. The difference between the two positions of the cross-wire on the image of the scale gives the displacement of the image past the cross-wire due to the introduction of the cover glass under the front leg of the optical lever.

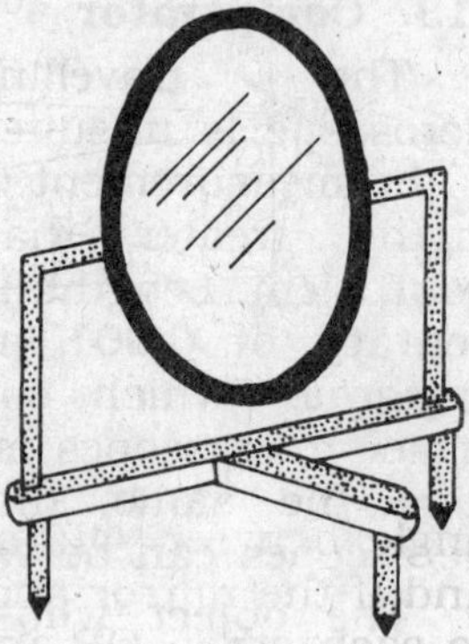

Fig. 1.8

Now, if a plane mirror is turned through an angle θ about an axis perpendicular to the plane of incidence and if the direction of the reflected ray remains *unchanged*, the incident ray must have changed through an angle 2θ. Thus referring to Fig. (1.9), if PO and QO′ be the rays for the two positions AB and AB′ of the mirror which proceed after reflection along OR, the

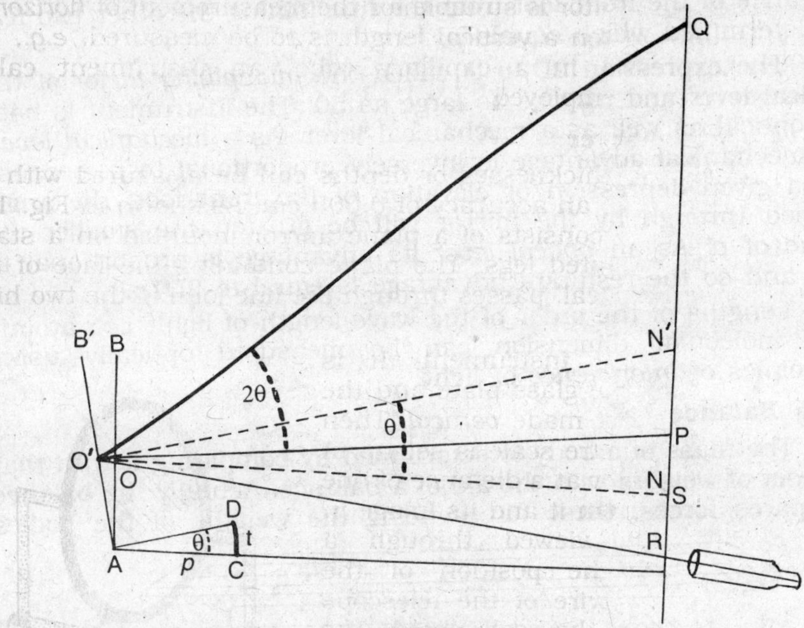

Fig. 1.9

angle between OP and O′Q is 2θ. If the angle θ be very *small* and if the mirror and the scale were initially very nearly *parallel* to each other, OP and O′Q are very nearly *equal* to OS. Then, if PQ = x, we have

$$2\theta = \frac{PQ}{O'Q} = \frac{PQ}{OS} = \frac{x}{D}$$

or $$\theta = \frac{x}{2D} \tag{1.3}$$

where D = distance between the mirror and the scale.

The thickness of the cover glass CD is given by

$$t = \theta \times p \tag{1.4}$$

where p = AC (lever arm) = the distance of the *front* foot of the optical lever from the line joining the two hind feet.

Thus, from Eqs. (1.3) and (1.4), we have

$$t = \frac{xp}{2D} \tag{1.5}$$

To measure *p*, the *lever arm*, the optical lever is placed on a sheet of paper and then pressed lightly thus producing impressions of its feet on the paper. From these impressions the distance of the front foot from the line joining the other two feet is determined which equals *p*.

The expression 2D/*p* is called the *multiplying factor* of the optical lever and may be as large as 50. The instrument is both an optical as well as a mechanical lever. As a *mechanical lever*, its mechanical advantage is inversely proportional to *p*, because for a given depression or elevation of the front foot, the angle turned through by the mirror will be greater, the smaller the value of *p*. As an *optical lever* its advantage is proportional to 2D, and so the resultant advantage is equal to 2D/*p*.

Lengths of the order of the wave-length of light, *i.e.*, atomic and molecular dimension can be measured optically using principles of *interference* of light.

1.15 Balance

The mass of a body is determined by comparing it with that of a set of weights with the aid of a balance. Actually the balance compares forces, these forces being the weigths of the bodies

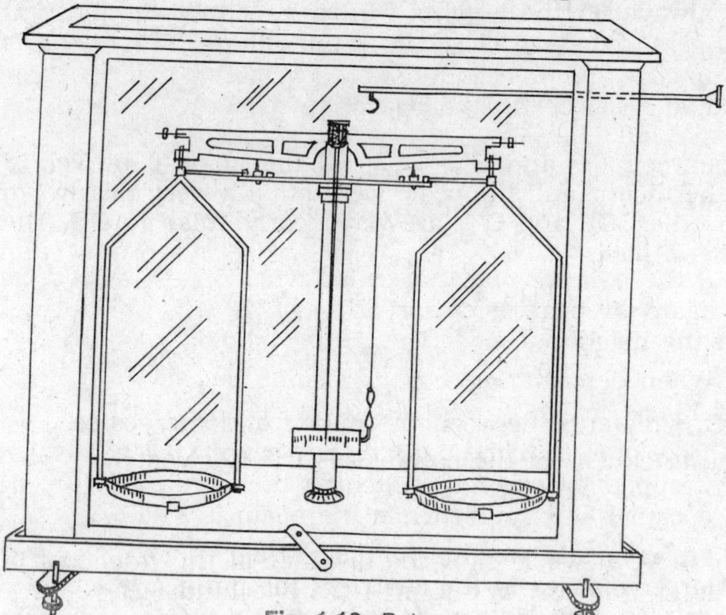

Fig. 1.10 Balance

placed in the pans, but as the weights are proportional to the masses if weighed at the same place, the balance serves for the comparison of masses.

1.16 General Precautions in the Use of a Balance

The following precaution should be observed during the operation of weighing if the balance is to maintain its efficiency while in continual use.

1. Before starting to weigh the object, see that the scale-pans are *clean* and *dry*. If necessary they may be dusted with a camel-hair brush. Make sure that the rider is off the beam and the stirrups are in their proper positions over the knife-edges. Also see with the help of the plumb line or the spirit level that the balance is *correctly* levelled.

2. Always determine the *resting point* or zero of the *unloaded* balance.

3. Always place the body to be weighed in the left pan and the standard masses in the right.

4. Always keep the weight box near the *right* hand end of the balance base and make sure that all the weights are present in it.

5. Always lift the weights with a forcep. Never touch them with fingers. Always pick up milligramme weights by the turned-up edge. On no account should weights be left lying about, they should be either in the pan or in the weight box.

6. Always arrest the balance before adding or removing a mass from the pan and do so slowly and without jerking when the pointer is passing through its mean position. This avoids deforming of the knife-edges.

7. If the balance does not swing when released, arrest and release again or beat some air down on a pan by moving the hand in the neighbourhood. The pointer should not be touched.

8. When determining a resting point always lower the case of the balance box to prevent disturbance due to air draughts.

9. Take care that the pans do not swing. Swinging of pans may be stopped by carefully touching the pans with the finger or with a camel-hair brush while the beam is arrested.

10. Always add weights in *descending* order of magnitude placing larger masses in the centre of the pan. To test the want of balance after the addition or removal of a weight, the balance

need not be raised to the full extent until weights less than a gramme are added or removed.

11. Never load the balance with weights greater than its *maximum* capacity which is generally about 250 gms.

12. Never weigh a body when hot otherwise the air currents will be set up which will affect the accuracy of weighing. All substances liable to damage the pans, *e.g.*, acids, must be put in appropriate vessels. All bodies to be weighed should be dry, and drops of water sticking to them should be removed before putting them in the pan.

13. When the body has been weighed, always count the masses whilst on the scale-pan and their magnitude added up and checked by noting empty spaces in the weight box. To this add the amount indicated by the rider. Finally replace the weights in the box.

14. Never leave the case of the balance box or that of the weight box open when weighing has been finished.

1.17. Spring Balance

The spring balance measures weights by balancing them against the extension of a spring. *The extension of a spring within elastic limit is proportional to the force applied to it* and hence knowing the extension produced by unit force, the weight of a body can be deduced from the extension produced by it when suspended at the end of the spring. But as the weight of a body is different at different places on the surface of the earth on account of variation of *g*, the spring balance will give different results for the same mass at different places. The spring balance, however, gives a direct reading of the weight of the body, is quick in action and is compact and portable.

1.18. Instruments used in the Measurement of Time

(a) Clocks and Watches—Time is generally measured by means of clocks or watches in which a wound-up spring or lifted weight is compelled to move very slowly and regularly by some controlling device, *e.g., a vibrating pendulum* or *a balance wheel*. Short intervals of time can be measured with stop-clocks and stop-watches which can be started and stopped at will by pressing a lever. Ordinarily a stopwatch can measure upto 1/10 of a second but we can get pocket stopwatches which can read upto 1/50 or 1/100 of a second.

(b) Metronome—This instrument consists of a pendulum with a loud percussion device. The time of swing of the pendulum can be increased or decreased by sliding a small weight up or down on the upper part of the pendulum rod so that the instrument can be adjusted to beat seconds, half-seconds, etc. Behind the sliding weight is fixed a scale on which can be read the number of strokes made by the metronome per minute, when the weight is opposite the respective mark on the scale. In some metronomes a bell is struck, say every second or half-a-second as arranged.

(c) Tuning fork—Small intervals of time, say $\frac{1}{100}$ second can be measured with a tuning fork of *known* frequency. A light metallic style is attached to one prong of the tuning fork and is arranged to touch a smoked surface. When the fork is allowed to vibrate and the smoked surface moves beneath the style, the latter traces out a *wavy curve* on the smoked surface. If λ is the *wave-length* of the waves thus traced out and n the *frequency* of the fork, $v = n\lambda$, where v is the *velocity* with which the smoked surface moves. The fork is kept in vibration generally by an electromagnet.

(d) Chronometer—It is a large watch which measures time with great accuracy. The mechanism is worked by a spring controlled by a compensated balance wheel.

ORAL QUESTIONS

VERNIER

1. What is a vernier and why is it so called ?
2. What is the principle of a vernier ?
3. Define vernier constant and explain the method of determining it. What is the value of vernier constant of your instrument ?
4. If 25 div. of the vernier scale coincide with 49 div. of the main scale, each main scale div. being 1/2 mm., what is the vernier constant ?
5. What is a vernier callipers ?
6. What precautions will you observe while taking a reading with a vernier callipers ?
7. What is meant by "zero error"?
8. How do you find its value and determine its sign?
9. How do you correct a reading of the vernier for zero error ?

SCREW GAUGE

1. Define the "least count" and the "pitch" of a screw gauge. How are they related to each other ?
2. What is meant by zero error of a screw gauge ?
3. How will you determine its value and sign for your instrument ?
4. What is meant by "back-lash" of a screw ?
5. How can you avoid error due to back-lash in your observation ?
6. What is a spring and ratchet arrangement ?

SPHEROMETER

1. What is a spherometer and why is it so called ?
2. What is the principle of action of a spherometer ?
3. How do you determine the zero error in a spherometer ?
4. What is the most reliable test for finding that all the legs of the spherometer and the tip of the micrometer screw lie in the same plane ?
5. What are the uses of this instrument ? Why are the legs and the screw of a spherometer made of steel of gun-metal and not of brass ?
6. Why is the brass disc bevelled ?
7. Differentiate between curvature and radius of curvature.
8. How will you test whether a given surface is spherical ?
9. Is it necessary that the tip of the micrometer screw should pass through the circumcentre of the triangle formed by joining the legs of the spherometer ?
10. Which of the two quantities l and h should you measure more accurately and why ?
11. How can you determine the focal length of a mirror and a lens with a spherometer ?
12. Which will you prefer : a disc of small diameter or one of a large diamcter and why ?
13. What qualities should you look for in a good spherometer ?

MICROSCOPE

1. Enumerate the various types of microscopes used in the measurement of length. What is a micrometer eye-microscope and how does it work ?

2. How can you find out its least count ?
3. What is a vernier microscope and what is it used for ?
4. How do you avoid error due to back-lash in this type of microscope while taking observations ? What it a cathetometer and what it it used for ?

OPTICAL LEVER

1. What is an optical lever ?
2. What is it used for ?
3. Explain its action. How far does it act as an optical lever and how far as a mechanical lever ?
4. Why should the distance between the scale and the telescope be small ?
5. How will you use an optical lever for the measurement of radius of curvature of a spherical surface ?

BALANCE

1. Do you measure mass or weight with a beam balance ?
2. How do you distinguish between mass and weight ?
3. What is the principle of a beam balance ?
4. State the requisites of a good balance.
5. What do you understand by sensitivity of a balance ?
6. What are the factors on which the sensitivity of a beam balance depends?
7. Why are knife edges made of agate and why do they rest on agate plates?
8. Why do the knife edges lie in one horizontal plane and why are they parallel to one another ?
9. What do you understand by zero or resting point of a balance ?
10. If a balance is not true how can you find out the correct mass of a body with it ?
11. What is a spring balance ?
12. Do a beam balance and a spring balance work on the same principle ?
13. Do you measure mass or weight with a spring balance ?
14. Will the readings of a spring balance with a given mass be the same at all places ?

4

2

Moment of Inertia and Acceleration Due to Gravity

2.1. Moment of Inertia. From *Newton's laws of motion* we know that force and mass have *opposite* effects on translatory motion, *the force causing and the mass opposing acceleration.* Laws *similar* to these laws have been found to be obeyed in the case of *rotatory* motion, but with the following modifications :—

(*a*) The place of a single force is taken by the couple whose effectiveness or *moment* is found to depend not only on the magnitude of the forces constituting it but also on their distance apart.

(*b*) The place of ordinary mass is taken by a quantity called the *effectiveness* of the mass or more technically the *moment of inertia* of the body about the axis of rotation. Just as the effectiveness of the driving force or rather the driving couple is found to increase directly as the distance of the force from the axis of rotation, so also the effectiveness of the opposing mass is found to increase even more rapidly (varying as the *square* of the distance) with the distance of the mass from the axis of rotation. Hence we say that :—

The moment of inertia of a particle of mass m distant r from the axis of rotation is mr^2 *or the moment of inertia of a body consisting of particles of masses* m_1, m_2, m_3.......... *at distances* r_1 r_2, r_3.........*respectively from the axis of rotation is* $m_1 r_1^2 + m_2 r_2^2 + m_3 r_3^2 +$ ...., *or more simply* Σmr^2.

Thus the moment of inertia of a body about an axis depends upon

(*i*) *The mass of the body,* and

(*ii*) *The distribution of the mass about, and its distance from the axis of rotation.*

The G.G.S. unit of moment of inertia is *gm.* × *cm*^{2} and the F.P.S. unit is *lb.* × *ft*^{2}. and its M.K.S. unit is kg. × m^2.

The moment of inertia plays the same part in rotatory motion as does ordinary mass or inertia in translatory motion. In fact it will be more helpful if we distinguish between these two by calling them rotatory and translatory inertias. The dynamical definition of mass or translatory inertia is derived from *Newton's second law of motion*,

$$m\frac{d^2x}{dt^2} = \text{Force of } \textit{inertial reaction.}$$

From this we see that, if $\frac{d^2x}{dt^2} = 1$

$$m = \text{Force of } \textit{inertial reaction}$$

or *the mass is the measure of the force of inertial reaction the body offers to unit acceleration.*

Extending the application of this law to rotatory motion with the modifications given, we have

$$I\frac{d^2\theta}{dt^2} = \text{Moment of couple due to } \textit{inertial reaction}$$

If, however, $\frac{d^2\theta}{dt^2} = 1$, we get

I = Moment of couple due to *inertial reaction*

Thus *moment of inertia of a body rotating about a given axis is the measure of the moment of couple of inertial reaction the body offers when it is made to undergo unit angular acceleration.*

2.2 Radius of Gyration. From the definition of moment of inertia we know that

$$I = \Sigma\, mr^2$$

Now, if the *whole* mass M of the body were concentrated at a point distant k from the axis of rotation,

$$I = Mk^2$$

where k is called *radius of gyration* of the body about the axis of rotation.

Thus $$k^2 = \frac{I}{M} = \frac{\Sigma\, mr^2}{\Sigma\, m}$$

The **radius of gyration** *of a body about an axis of rotation is the radial distance at which the mass of the body could be concentrated without altering the moment of inertia of the body about that axis.*

The unit of radius of gyration on the C.G.S. system is *cm*. The magnitude of radius of gyration of a body about an axis gives us an idea of the distribution of moment of inertia in space and also the effect of the mass distribution in space, on moment of inertia.

2.3. Calculation of Moment of Inertia. The magnitude of moment of inertia of a body about an axis can be found by simple integration if the body is of a regular geometrical shape, and in many cases its evaluation is simplified by the application of the following two general theorems :—

(1) Theorem of Parallel Axes. *The moment of inertia of a body about an axis is equal to its moment of inertia about a parallel axis through its centre of gravity, together with the product of the mass of the body and the square of the distance between the two axes.*

(2) Theorem of Perpendicular Axes. *The sum of the moments of inertia of a plane lamina about any two mutually perpendicular axes in its own plane is equal to its moment of inertia about a third axis perpendicular to its plane and passing through the point of intersection of the first two axes.*

Below are given the values of moments of inertia in some simple and important cases of bodies of regular geometrical shape.

Values of Moments of Inertia

Thin Circular Disc : *radius r, mass M* :

(*i*) About an axis through the centre, and perpendicular to its plane ... ... $\frac{1}{2}Mr^2$

(*ii*) About a diameter as axis ... $\frac{1}{4}Mr^2$

Circular Ring : radii r_1, r_2, mass M ; ...

About an axis through its centre and perpendicular to its surface ... ... $\frac{1}{2}M(r_1^2 + r_2^2)$

Right Cylinder : length l, radius r, mass M ;

(*i*) about its own axis ... ... $\frac{1}{2}Mr^2$

(*ii*) About an axis through its middle point and perpendicular to the axis of the cylinder ... $M\left(\frac{l^2}{12} + \frac{r^2}{4}\right)$

Rectangular Lamina : *sides a, b, mass M* ;

About an axis through its centre and perpendicular to its surface ... $M\left(\frac{a^2 + b^2}{12}\right)$

Thin Uniform Rod : *length l, mass M;*

About an axis through its centre and perpendicular to its length ... ... $\frac{Ml^2}{12}$

Sphere : *radius r, mass M* :

About a diameter ... ... $\frac{2}{5}Mr^2$

2.4. Torsional Vibrations. If a body suspended by means of a *fairly thin* wire be given a slight rotation in a horizontal plane and then released, the *restoring couple due to torsional reaction* of the suspension wire is proportional to the value of the twist θ for *quite large* angles of twist. If I is the *moment of inertia* of the body about the wire as axis and $\frac{d^2\theta}{dt^2}$ is the angular acceleration, the torque due to inertial reaction is $I\frac{d^2\theta}{dt^2}$, and the torque due to *elastic reaction* being proportional to the angle of twist, is $c\theta$, where c is the restoring couple due to *torsional reaction* per *unit* radian twist. Since no external force is acting on the body while it is oscillating, we have from *Newton's third law of motion,*

$$I\frac{d^2\theta}{dt^2} + c\theta = 0$$

or

$$\frac{d^2\theta}{dt^2} + \frac{c}{I}\theta = 0$$

The motion of the body is, therefore, *simple harmonic* and the period is given by

$$T = 2\pi\sqrt{I/c} \tag{2.1}$$

The torsional vibrations are isochronous whatever the angle of vibrations, within wide limits provided the elastic limit is not exceeded, and in this respect differ from pendular oscillations which are isochronous for *small* amplitudes only.

2.5. Pendular Oscillations. When a rigid body is allowed to oscillate in a *vertical* plane about a *horizontal* axis passing through the body, it constitutes what is called a *compound or rigid or physical pendulum* and its oscillations are known as *pendular oscillations.*

The point in which the axis of rotation meets the vertical plane through the C. G. of the pendulum is called its centre of suspension.

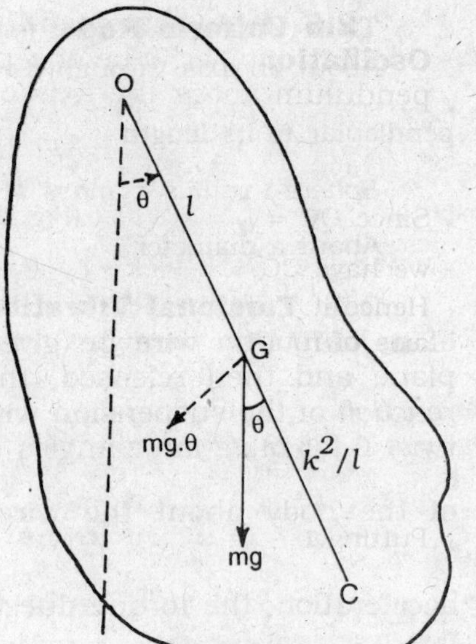

Fig. 2.1

Let the Fig. 2.1 represent the *vertical* section of a rigid body taken through its C. G. Let O be the *centre of suspension* and θ the angular displacement at time *t*, from the equilibrium position. Then, if OG = *l*, the *restoring couple* due to gravity = $mgl \sin \theta = mgl. \theta$, if θ is small.

The couple of *inertial reaction* is $I \dfrac{d^2\theta}{dt^2}$, where I is the *moment of inertia* of the pendulum not about its natural axis through its C. G., but about the *actual* axis through O, and so $I = I_0 + ml^2 = mk^2 + ml^2 = m(k^2 + l^2)$ where *k* is the *radius of gyration* about a *parallel* axis through G. Equating the sum of these two equations to zero, we get from *Newton's third law of motion*

$$I \frac{d^2\theta}{dt^2} + mgl\,\theta = 0$$

or

$$\frac{d^2\theta}{dt^2} + \frac{mgl}{I}\,\theta = 0$$

This represents a *S. H. M.* of *period*

$$T = 2\pi\sqrt{\frac{I}{mgl}} = 2\pi\sqrt{\frac{k^2 + l^2}{lg}}$$

or

$$T = \sqrt{\frac{(k^2/l) + l}{g}} \qquad (2.2)$$

The above formula is similar to that for a *simple pendulum* in which the simple length *l* is replaced by a more complex one $(k^2/l + l)$. *A simple pendulum with this length will have the same period as the compound pendulum and so it is called the* **equivalent simple pendulm.** *The point C distant k^2/l from G is called the* **centre of oscillation.**

2.6. Interchangeability of Centres of Suspension and Oscillation. Let T be the period of oscillation of the compound pendulum about the axis of suspension at O, then

$$T = 2\pi \sqrt{\frac{(k^2/l) + l}{g}}$$

Since OC = L = $k^2/l + l$ (Fig. 2.1), and OG = l

we have CG = $L - l = k^2/l$ (2.3)

Hence if T′ be the period of the oscillation of pendulum about the axis of rotation through C,

$$T' = 2\pi \sqrt{\frac{\frac{k^2}{L-l} + (L-l)}{g}}$$

Putting $L - l = k^2/l$ from Eq. (2.3), we have

$$T' = 2\pi \sqrt{\frac{l + (k^2/l)}{g}} = T$$

Hence the period of the pendulum will be the *same* whether the pendulum is suspended from O or C, *i.e., the centres of suspension and oscillation are interchangeable.*

2.7. Variation of T with l — Minimum and Maximum periods. Differentiating T with respect to l in Eq. (2.2), we have

$$\frac{dT}{dl} = \frac{2\pi}{\sqrt{g}} \times \frac{1}{2}\left(\frac{k^2 + l^2}{l}\right)^{-\frac{1}{2}} \left(-\frac{k^2}{l^2} + 1\right)$$

For *minimum* and *maximum* values of T, $dT/dl = 0$

whence $l = \pm k$

To show that T is *minimum* it is enough to show that for this value of l, $\frac{d^2T}{dl^2}$ is *positive*. The length of the *equivalent simple pendulum* in this particular case = $2k$.

T is *infinite* when $l = 0$, *i.e., when the axis passes through the C. G.* In this case the body is in *neutral equilibrium* and if it is displaced from any position and then released, gravity has no tendency to make it return to that position. T is also *infinite* when $l = \infty$. For any particular value of T greater than the minimum, there will be two values of l on one side of G and collinear with it, one l_1 *greater* than k and the other l_2 *less* than k.

Squaring Eq. (2.2) and rearranging in *descending* powers of l, we get

$$4\pi^2 l^2 - T^2 gl + 4\pi^2 k^2 = 0$$

or $$l^2 - \frac{T^2 g}{4\pi^2} l + k^2 = 0$$

This is a *quadratic* equation and gives *two* values for l,

$$l_1 = \frac{\frac{gT^2}{4\pi^2} + \sqrt{\frac{g^2T^4}{16\pi^4} - 4k^2}}{2} > k$$

and $$l_2 = \frac{\frac{gT^2}{4\pi^2} - \sqrt{\frac{g^2T^4}{16\pi^4} - 4k^2}}{2} < k$$

Similarly there will be two more points on the *other* side of G for which the value of T will be the *same*.

Now $$l_1 + l_2 = gT^2/4\pi^2 \text{ and } l_1 l_2 = k^2$$

$$\therefore \quad T = 2\pi\sqrt{\frac{l_1 + l_2}{g}}$$

i.e. the length of the *equivalent simple pendulum* = $(l_1 + l_2)$ and $k = \sqrt{l_1 l_2}$

Now, if two circles with centre G (Fig. 2.2) and radii equal to l_1 and l_2 are drawn, any axis *parallel* to the suspension axis at O taken on either of these circles would give the *same* value for T. Hence *there will be an infinite number of points in a compound pendulum for which T will be the same.*

Any *straight* line passing through G will cut these two circles *in four* points. Hence *in a compound pendulum there are four points collinear with the C. G., the periods of oscillation about which are equal.*

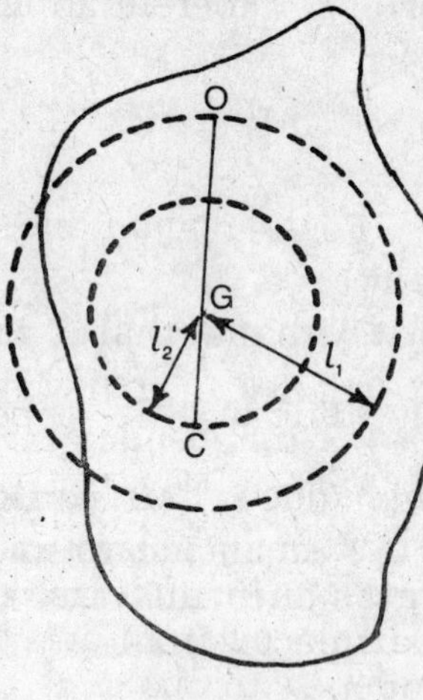

Fig. 2.2

Experiment 2.1

Object. To determine the moment of *inertia* of a given body by means of *Inertia Table* with the help of an auxiliary body whose moment of inertia can be calculated from its dimensions.

Apparatus. Inertia Table, an auxiliary body of regular geometrical shape (say, a right cylinder), stop-watch, vernier callipers, weight box and a spirit level.

Description of Inertia Table. Referring to Fig. 2.3, an *inertia table* consists of a circular aluminium table about 6" in diameter suspended by means of a wire fixed at the top to a chuck fitted to a crossbar between two pillars standing vertically on a heavy iron base provided with three levelling screws. The table is made of aluminium in order that it may be *light* so that when a body is placed upon it there may be *considerable* difference between the periods of the combination and that of the table alone. On the upper face of the table *concentric circles* are described which facilitate the *symmetrical* loading of the table with a body. Near the circumference on the table, a *concentric groove* is cut in which three *balancing weights* may be placed which help in setting the table *horizontal*. The whole apparatus is enclosed in a box with glass sides which serve to protect it from air draughts.

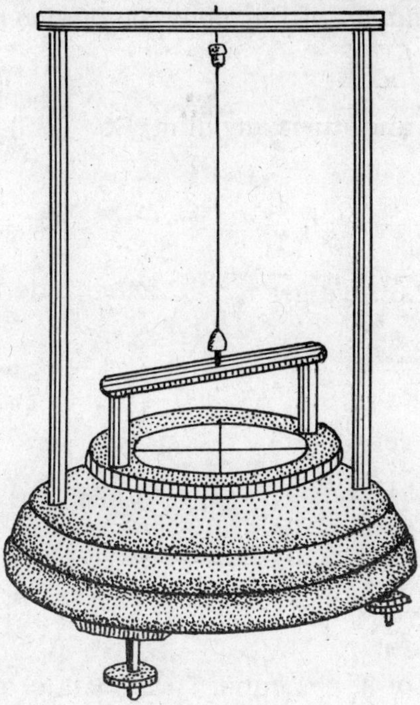

Fig. 2.3

Theory. Let the aluminium table be given a slight rotation in a *horizontal* plane and then released and let T_0 be its *period of oscillation*. Then, if I_0 is the *moment of Inertia* of the table with its two small pillars and the cross-bar between them about the wire as the axis. and c is the *restoring couple* per *unit* radian twist due to *torsional reaction*, we have from Eq. (2.1)

$$T_0 = 2\pi \sqrt{I_0/c} \tag{2.4}$$

Now let a body of known moment of inertia I_1 be placed *centrally* on the inertia table with its C.G. on, and axis of rotation coinciding with the axis of the wire, let T_1 be the *period of oscillation of the combination*. Then

$$T_1 = 2\pi\sqrt{\frac{I_0 + I_1}{c}} \qquad (2.5)$$

Now let the body of known moment of inertia be removed and the *experimental* body of *unknown* moment of inertia I_2 be placed on the table with its *specified axis* in the axis of rotation. Then, if T_2 is the *period of oscillation* of the *new combination*

$$= 2\pi\sqrt{\frac{I_0 + I_2}{c}} \qquad (2.6)$$

Squaring Eqs. (2.4) and (2.5) and then dividing Eq. (2.3) by Eq. (2.4) we get

$$\frac{T_1^2}{T_0} = \frac{I_0 + I_1}{I_0} = 1 + \frac{I_1}{I_0} \qquad (2.7)$$

Similarly from Eqs. (2.4) and (2.6), we get

$$\frac{I_2}{I_0} = \frac{T_2^2 - T_6^2}{T_0^2} \qquad (2.8)$$

Dividing Eqs. (2.8) by (2.7), we get

$$\frac{I_2}{I_1} = \frac{T_2^2 - T_0^2}{T_1^2 - T_0^2}$$

or

$$I_2 = \frac{t_2^2 - T_0^2}{T_1^2 - T_0^2} \times I_1 \qquad (2.9)$$

Thus observing the values of T_0, T_1 and T_2 the value of I_2 can be calculated while that of I_1 being already known from its dimensions and mass.

Method. Level the heavy iron base of the inertia table by the levelling screws and test the levelling with a spirit level. With the help of a plumb-line or the spirit levels adjust the positions of the *balincing weights* in the concentric groove so that the aluminium table is *horizontal*. Set the aluminium table oscillating by slightly rotating it in horizontal plane and then releasing it. Take care not to produce any up and down or pendular oscillations in this process. Determine *twice* the time taken by a large number of complete oscillations, say 40 by means of an accurate stop-watch, and from it find out the value of T_0. Repeat the observations with the number of oscillations *altered*, say to 50, and find out the *mean* value of T_0.

To facilitate counting of oscillations a *reference* mark may be made on the inertia table and a *vertical* pointer set up just *opposite* to it on the heavy iron base.

Next taking the help of concentric circles, place *centrally* on the aluminium table the *auxiliary* body of *known* moment of inertia I_1, viz., the right cylinder, and move it to and from until the horizontality of the table is secured as tested with the spirit levels or the plumbline, *without disturbing the balancing weights*. Then the C. G. of the cylinder will lie on, and its axis will coincide with, the axis of the wire. Set the combination in oscillation and observing the time for at least *two sets* of oscillations as above, determine the mean value of T_1. Note that T_1 will be *greater* than T_0.

Replace the auxiliary body by the *experimental* body of *unknown* moment of inertia I_2 with the *specified axis* coinciding with the axis of the wire. If this tilts the table to one side as would be the case when the specified *axis does not* pass through the C. G. of the experimental body *restore* the horizontality of the table by displacing the balancing weights in the concentric groove. *So long as the horizontality of the table is maintained*, displacement of the balancing weights in the concentric groove will not alter the moment of inertia I_0 of the table. This only shifts the position of C.G. of the table. It may, however, be clearly noted that if the specified axis *does pass* through the C. G. of the experimental body, its position on the table should be adjusted *without disturbing the balancing weights*. Set the combination in oscillation and determine as above the value of T_2. The value of T_2 will also be *greater* than T_0.

Weigh the right cylinder and determine its diameter by means of vernier callipers, and calculate its moment of inertia about the axis of rotation. Using this value for T_1 and substituting the mean values of T_0, T_1 and T_2 in Eq. (2.9), calculate the value of T_2.

Sources of error and Precautions. (1) The aluminium table should always remain *horizontal* so that its moment of Inertia I_0 about the axis of rotation remains *unaltered* throughout the *whole* experiment.

In case the horizontality of the table is tested with spirit levels, the two spirit levels used for setting the table horizontal should be *permanently* fixed on it at right angles to each other, otherwise their removal from the table will no longer leave it horizontal.

(2) The bodies should be so placed on the aluminium table that the axis about which the moment of inertia is known or to be determined coincides with the axis of the suspension wire

without affecting the horizontality of the table, and *unless the axis does not pass through the C. G. of the body, this adjustment must be done without disturbing the balancing weights.*

(3) In case the specified axis about which the moment of inertia of the experimental body is to be determined *does not* pass through the C. G. of the body, the determination of period of the combination of the table and the *auxiliary* body should be done *first* and the determination of period of the combination of the table and the experimental body afterwards, for the adjustment of the experiment body on the table would then require a shifting of the balancing weights in the concentric groove whereas for the adjustement of the auxiliary body on the table, the balancing weights must remain *undistrubed.*

(4) The motion of the table should be purely *rotational* in a horizontal plane. Up and down, and pendular oscillations must be completely checked.

(5) As in the expression for I_2 the periods occur raised to the *second* power, they must be accurately determined by timing a *large* number of oscillations with a stop-watch reading upto 1/10 sec.

(6) The suspension wire should not be twisted *beyond* elastic limit otherwise the restoring couple due to *torsional reaction* will not be proportional to the angle of twist.

(7) The auxiliary body whose moment of inertia is to be calculated from its dimensions must be of *uniform* density throughout, *e.g.*, brass.

(8) There should be no kinks in the suspension wire and it should be fairly thin and long.

Observations

(A) (i) Mass of the cylinder (auxiliary body) = gm.

(ii) Measurement of diameter of the cylinder

Vernier constant = cm.

Zero error = cm.

S. No.	Reading along any diameter a cm.	Reading along perp. diameter b cm.	$\frac{a+b}{2}$ cm.
1 2 3			
Mean uncorrected diameter			

(B) Determination of T_0 , T_1 and T_2

Least count of stop-watch = sec.

Set No.	No. of oscillations	Time Taken						T_0	T_1	T_2
		Inertia Table alone		Inertia Table + auxiliary body		Inertia Table + body of unknown moment of inertia				
		Min.	Sec.	Min.	Sec.	Min.	Sec.	Sec.	Sec.	Sec.
1										
2										
							Mean			

Calculation

Mean *corrected* diameter = cm.

Mean radius of the cylinder = cm.

Moment of inertia I of the cylinder about its *own* axis

$= Mr^2/2$

$=$ gm.$\times$ cm^2.

$$I_2 = \frac{T_2^2 - T_0^2}{T_1^2 - T_0^2} \times I_1$$

$=$ gm. $\times$ cm^2

Result. The moment of inertia of the experimental body = gm. $\times$ cm^2 = kg$\times$ m^2.

Note. If the experimental body is of *regular geometrical* shape, calculate its moment of inertia from a knowledge of its mass and dimensions and compare it with the value obtained *experimentally* from inertia table.

Criticism of the Apparatus. The experiment when performed carefully yields good result. The accuracy of the result depends *chiefly* upon

(*i*) *the largeness* of difference between the periods of the combination and that of the inertia table *alone*, and

(*ii*) the accuracy in the determination of these periods.

The first condition is fulfilled by making the inertia table *light* so that its moment of inertia may be small when compared to that of the combination.

The second condition is fulfilled by timing a *large* number of oscillations by a stop-watch of *small* least count and by making the periods *large*. The periods can be increased by increasing the length of the wire and decreasing its radius so that the torsional couple per unit radian twist may be small. But the larger and thinner the wire, more, are the chances for the formation of kinks and the shorter and thicker the wire greater is the value of the couple per unit twist due to torsional reaction and hence lesser is the period. In practice a wire about 50 cm. long and about $\frac{1}{10}$ cm. in diameter should be used.

For still greater accuracy, the times of oscillation of bodies should be recorded with the help of a chronometer and the oscillations should be observed through a telescope. The accuracy is, however, limited as the restoring couple per unit twist due to torsional reaction, assumed to be constant, *varies* with the amount of load suspended by means of the wire.

The apparatus is particularly useful in determining the moment of inertia of an *irregular* body. But, if the specified axis does not pass through the C. G. of the body, the success of the experiment would depend upon whether

(*i*) the experimental body can be placed in *stable equilibrium* on the table with the specified axis coinciding with the axis of the suspension wire, and

(*ii*) the balancing weights are *heavy enough* to restore the horizontality of the table when the position of the experimental body is adjusted on it.

Experiment 2.2

Object. To determine the *moment of inertia of a flywheel* about its own axis of rotation.

Apparatus. The flywheel, weight box, thread, stop-watch, metre scale and vernier callipers.

Description of Apparatus. As illustrated in Fig. 2.4., flywheel is simply *a heavy wheel with a long axle* which when properly supported in bearings may remain at rest in *any* position, *i.e., its centre of gravity lies on the axis of rotation.* Its moment of inertia can be determined experimentally by setting it in motion with a known amount of energy.

Theory. The flywheel is mounted in its bearings with its axle *horizontal* and at a suitable height from the ground; and a string carrying a suitable mass *m* at its one end and having a length *less* than the height of the axle from the ground, is wrapped *completely* and *evenly* round the axle. When the mass

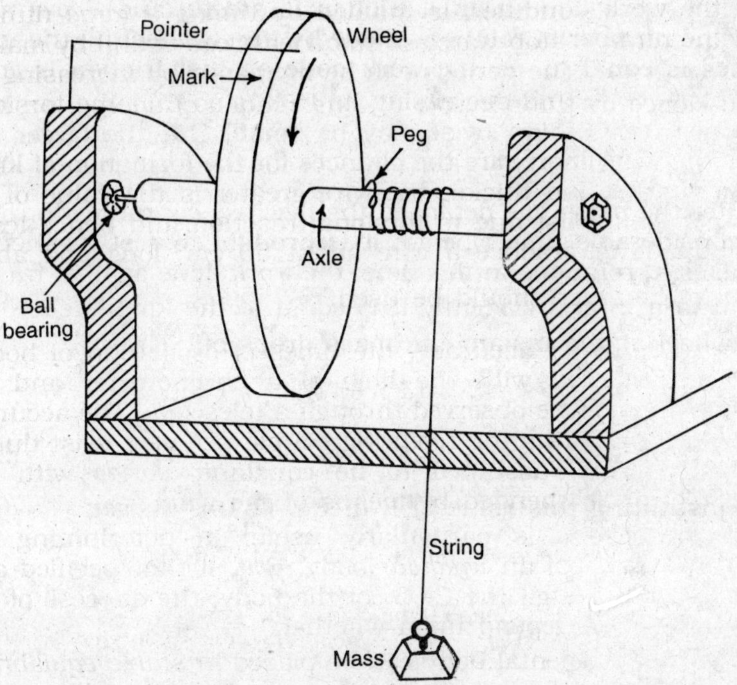

Fig. 2.4

m is released, the string *unwinds* itself, thus setting the flywheel in rotation. As the mass m descends further and further the rotation of the flywheel goes on *increasing* till it becomes maximum when the string leaves the axle and the mass drops off.

Let *h* be the distance fallen through by the mass *before* the string leaves the axle and the mass drops off, and let v and ω be the linear velocity of the mass and angular velocity of the flywheel respectively *at the instant the mass drops off*. Then, as the mass descends a distance *h*, it loses *potential energy mgh* which is used up

(*i*) partly in providing *kinetic energy of translation* $\frac{1}{2}mv^2$

(*ii*) partly in giving *kinetic energy of rotation* $\frac{1}{2}I\omega^2$ *to the flywheel* (where I is the moment of inertia of the flywheel about the axis of rotation) and

(*iii*) partly in doing *work against friction*.

If the work done against friction is *steady* and F per turn, and, if the number of rotations made by the flywheel till the mass detaches is equal to n_1, the work done against friction is equal to n_1 F. Hence by the *principle of conservation of energy*, we have

$$mgh = \frac{1}{2}mv^2 + \frac{1}{2}I\omega^2 + n_1 F \qquad (2.10)$$

After the mass has detached the flywheel continues to rotate for a considerable time *t* before it is brought to rest by friction. If it makes n_2 rotations in *this* time, the work done against friction is equal to n_2 F and evidently it is equal to the kinetic energy of the flywheel at the instant the mass drops off. Thus

$$n_2 F = \frac{1}{2}I\omega^2$$

or

$$F = \frac{1}{2}I\omega^2/n_2 \qquad (2.11)$$

Substituting this value of F in Eq. (2.10), we get

$$mgh = \frac{1}{2}mv^2 + \frac{1}{2}I\omega^2 + \frac{1}{2}I\omega^2\, n_1/n_2$$

whence

$$I = \frac{2mgh - mv^2}{\omega^2(1 + n_1/n_2)}$$

If *r* be the radius of the flywheel,

$$v = r\omega$$

∴

$$I = \frac{2mgh - mr^2\omega^2}{\omega^2(1 + n_1/n_2)}$$

or

$$I = \frac{m(2gh/\omega^2 - r^2)}{1 + n_1/n_2} \qquad (2.12)$$

After the mass has detached, its angular velocity decreases on account of friction and after some time *t*, the flywheel finally comes to rest, At the time of detachment of the mass the angular velocity of the wheel is ω and when it comes to rest its angular velocity is zero. Hence if the force of friction is *steady*, the motion of the flywheel is *uniformly retarded* and the *average* angular velocity during *this* interval is equal to ω/2. Thus

$$\omega/2 = 2\pi n_2/t$$

or

$$\omega = 4\pi n_2/t \qquad (2.13)$$

Thus observing the time *t* and counting the rotations n_1 and n_2 made by the flywheel, its moment of inertia can be calculated from Eqs. (2.12) and (2.13).

Method. Take a string of length *less* than the height of the axle from the floor. Make a loop at its one end and tie a suitable mass at its other end. Slip on the loop to the small peg projecting on the axle of the flywheel and wrap the string *evenly* round the axle counting the number of turns wound. When almost the whole string has been wound, stop at a stage when the projecting peg is *horizontal* and make a mark on the string where its contact with the axle just ceases. This ensures that a whole number of turns of string has been wound on the axle.

Sometimes instead of a peg there is a hole on the axle in which a brass pin can be fitted which serves the purpose of a peg. To facilitate counting of rotations on the flywheel, a *reference* mark may be made on its rim *opposite* to the *horizontal* pointer fixed to the structure on which the flywheel is mounted. Sometimes there is inbuilt counter in which the number of rotations can be read.

Let the mass be released. Count the number of rotations n_1 the flywheel makes *before* the loop comes off the peg and the mass drops off. *The number* n_1 *must be equal to the number of turns of the thread wound on the axle.* Carefully start the stop watch at *the moment the mass detaches* and also continue to count the number of rotations n_2 the flywheel makes before it comes to rest. Stop the stop-watch when the flywheel finally comes to rest and thus determine the value of t, *the interval for which the flywheel continues to rotate after the detachment of the mass.*

Measure by a metre scale the length of the string between the loop and the *mark* at the other end which gives h, the distance *descended* by the mass. With the help of a vernier callipers measure the diameter of the axle of the flywheel at a number of point along its length and calculate the *mean* radius of the axle.

Repeat the experiment with different masses and strings of different lengths and take at least three sets of observations, and in each set for the same value of m and h take at least three observations, for n_1, n_2 and t. If the values of n_1, n_2 and t differ slightly for the *same* values of m and h, calculate their *mean* values. Calculate the moment of inertia of the flywheel for *each set* of observations *separately* and then find out the *mean* value of moment of inertia of the flywheel about its axis of rotation.

Sources of error and precautions. (1) The length of the string should be always *less* than the height of the axle of the flywheel from the floor so that it may leave the axle *before* the mass strikes the floor.

(2) The loop slipped over the peg should be *quite loose* so that when the string has unwound itself, it must leave the axle and there may be no tendency for it to rewind in the opposite direction.

(3) The string should be *evenly* wound on the axle, *i.e.*, there should be no overlapping of, or a gap left between, the various coils of the string.

(4) To ensure winding to *whole* number of turns of string on the axle the winding should be stopped, when almost complete the projecting peg is *horizontal.*

(5) To determine h measure only the length of the string between the loop and the mark at the *other* end where the string left the axle before the start of the flywheel.

(6) The string used should be of *very small* diameter compared with the diameter of the axle. If the string is of appreciable thickness *half* of its thickness should be added to the radius of the axle to get the *effective* value of r.

(7) The friction at the bearings should not be great and the mass tied to the end of the string should be sufficient to be able to overcome the bearing-friction and so to start falling of *its own accord.*

(8) Take extra care to start the stop-watch *immediately* the string leaves the axle.

(9) The diameter of the axle should be measured at a number of points along its length and at each point two readings of diameters *at right angles* to one another should be taken.

Observations. (*A*) *Measurement of the diameter of the axle*

Vernier Constant = cm.

Zero Error = cm.

Set No.	Reading along any diameter a cm.	Reading along perpendicular diameter b cm.	$\frac{a+b}{2}$ cm.
1 2 3			
Mean uncorrected diameter			

(B) Measurement of h, n_1, n_2 *and t*

Least count of stop-watch = sec.

Set No.	m gm.	b cm.	n_1	n_2	t Sec.	Mean		
						n_1	n_2	t Sec.
1								
2								
3								

Calculations

Mean *corrected* readius of the axle = cm.

$\omega = 4\pi n_2/t$ = Sec^{-1}

$$I = \frac{m(2gh/\omega^2 - r^2)}{1 + n_1/n_2}$$

$$= \quad \text{gm.} \times \text{cm.}^2 = \quad \text{kg } m^2$$

(**Note** :—Make similar calculations for I from *other* sets of observations)

Mean value of moment of inertia of flywheel = kg m.^{2}

Result. The moment of inertia of the flywheel about its axis of rotation = kg m^2

Criticism of the Method. In this method the *exact instant* at which the mass drops off cannot be correctly found out and hence the values of n_1, n_2 and t cannot be determined very accurately. The angular velocity ω of the flywheel at the instant the mass drops off has been calculated from the formula $\omega = 4\pi n_2/t$ on the assumption that the force of friction remains constant while the angular velocity of the flywheel decreases from ω to zero. But as the friction is less at greater velocities, we have no justification for this assumption. Hence for more accurate result, ω should be measured by a method in which no such assumption is made, *e.g.*, with a tuning fork.

Experiment 2.3

Object. To determine, by the *method of oscillations, the moment of inertia* of a rectangular lamina about an axis perpendicular to its plane passing through its centre.

Apparatus. A metallic rectangular lamina with several holes, a horizontal knife-edge attached to a rigid support, a plumb-line, a stop-watch and a metre scale.

Theory. Let a rectangular lamina be allowed to oscillate in a *vertical* plane about a *horizontal* axis passing through it, then, as the lamina constitutes a *rigid pendulum, its period of oscillation for small* amplitude is, from Eq. (2.2), given by

$$T = 2\pi\sqrt{\frac{(k^2/l) + l}{g}} \qquad (2.14)$$

where l is the distance of the point of suspension from the C.G. and k is its *radius of gyration about a parallel* axis through the C.G.

Squaring Eq. (2.14) and then rearranging, we get

$$T^2 l = \left(\frac{4\pi^2}{g}\right) l^2 + \left(\frac{4\pi^2}{g}\right) k^2 \qquad (2.15)$$

When

$$k^2 = \left(\frac{T^2 g}{4\pi^2}\right) l - l^2 \qquad (2.16)$$

This equation can be used to calculate k, if the period of oscillation T for a known value of l is detrmined. Then knowing the mass m of the lamina, its moment of inertia I_0 about an axis perpendicular to its plane and passing through its centre can be evaluated from the equation

$$I_0 = mk^2 \qquad (2.17)$$

A more *accurate* value of k can be obtained *graphically*. Referring to Eq. (2.15), if $T^2 l$ is plotted against l^2 the graph will be *straight line* as shown in Fig. 2.5; it is also evident from the same equation that when $T^2 l = 0$, $k^2 = -l^2$. Hence the *negative intercept* OP on the X-axis directly gives the value of k^2.

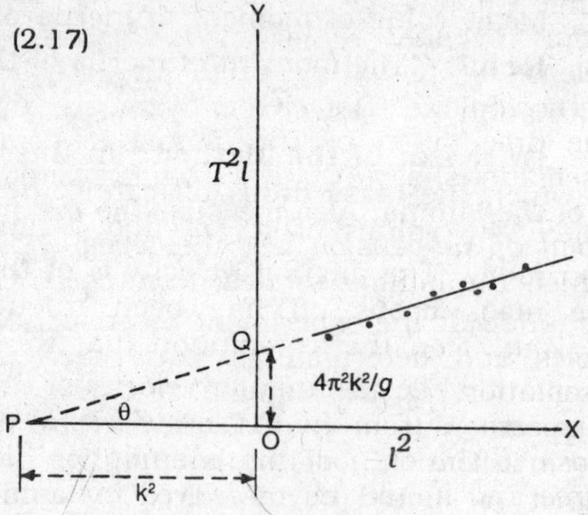

Fig. 2.5

The moment of inertia of the lamina can also be calculated from the formula

$$I_0 = m\left(\frac{l^2 + b^2}{12}\right) \qquad (2.18)$$

where l is its length and b the breadth, and its value compared with that obtained *experimentally*.

Method. Set up a knife-edge K horizontally on a rigid support, as shown in Fig. 2.6. Take a metallic rectangular lamina L with several holes in it and suspend it carefully in a *vertical* plane from the knife-edge by allowing the knife-edge to pass through one of the holes in the lamina. Then, with the help of a plumb-line, draw a *vertical line* on the surface of the lamina passing through the *point of suspension*. Repeat this operation by suspending the lamina from other holes. The point of interesction of the lines drawn on te surface of the lamina gives the position of its *centre of gravity*.

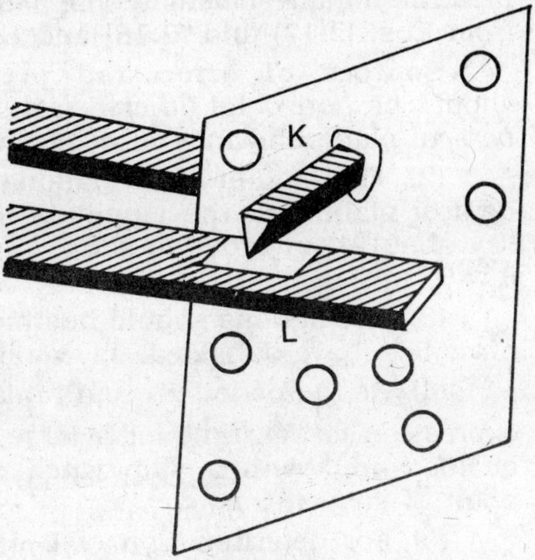

Fig. 2.6

Next, suspending the lamina about the knife-edge from any one of the holes, displace it *slightly* aside and then release it. It will begin to oscillate about the knife-edge in a *vertical* plane. Determine by means of an accurate stop-watch the time taken by the lamina to complete a *large* number of oscillations, say 100 and from it calculate the *period of oscillation* T of the lamina. Also measure the distance l of the corresponding *point of suspension i.e.*, the *sharp-edge* of the knife-edge about which the lamina rotates, from the C.G. of the lamina.

Repeat the experiment by suspending the lamina from *other* holes and determine in each case, as before, the period of oscillation of the lamina and the distance of the point of suspension from its C.G. It may be noted that for holes *very close* to the C.G. of the lamina, the period of oscillation will be *large* and hence only a *few* oscillations can be timed. Using Eq. (2.16), calculate the value of k^2 from *each* set of values for T and

l and then find the *mean* value of k^2.

Next plot a graph taking l^2 along the X-axis and the corresponding values of T^2l along the Y-axis. The graph will come out to be a *straight line as illustrated* in Fig. 2.5. Produce this straight line *backwards* cutting the—*ve side* of X-axis at P. Then measure the—*ve intercept* OP on the X-axis. This gives the value of k^2.

Now weigh the lamina and also determine its length and breadth. Finally calculate the *moment of inertia* of the lamina from Eqs. (2.17) and (2.18) and compare the two values.

Sources of error and precautions. (1) The knife-edge should be *horizontal* otherwise the lamina will not oscillate in a *vertical* plane and may also tend to slip off the knife-edge.

(2) The motion of the lamina should be confined only to a *vertical* plane and the amplitude of oscillation should be *small*, say about 4° in order to satisfy the condition of theory that $\sin\theta \simeq \theta$.

(3) Oscillations should be timed only when any irregularities of motion have subsided, say after 3 or 4 oscillations.

(4) As the period T occurs raised to the *second* power in the expression for k^2, time for a *large* number of oscillations should be measured with a stop-watch of small least count for each point of suspension.

(5) For determination of l, distance of *point of suspension* and *not* of the corresponding hole, should be measured from C. G. of the lamina.

(6) The graph between l^2 and T^2l should be drawn on a *large* graph paper and using the *entire* space available, the scales for the two axes should be so chosen as to represent *fully* the observed accuracies in the quantities to be plotted.

(7) The graph between l^2 and T^2l should be *smoothly* drawn and the points should be *evenly* distributed about the line and should lie as *close* to it as possible.

(8) The straight lines drawn on the surface of the lamina with the help of the plumb-line should intersect at *one* point only, *viz*., the C.G. of the lamina.

Observations and calculations

(*A*) (i) *Mass of the lamina* = gm.

(*ii*) *Length of the lamina* = cm.

(*iii*) *Breadth of the lamina* = cm.

(B) *Measurement of T and l*

Least count of the stop-watch = sec.

No.	Distance of point of suspension from C.G., cm.	No. of oscillations.	Time taken		Period T sec.	k^2 cm^2	l^2 cm^2	$T^2 l$ cm Sec^2.
			Min.	Sec.				
1								
					Mean			

From 1st set

$$k^2 = \frac{T^2 g}{4\pi^2} l - l^2 \tag{2.19}$$

$= \quad cm.^2$

(N. B.—Make similar calculations for k^2 from *other* sets of observations).

$\therefore$ *Mean* value of k^2 = cm.^{2}

Hence $I_0 = mk^2$

$=$

$= \quad gm. \times cm.^2$

From Graph (2.5) $k^2 = -$ *ve intercept* OP on the X-axis

$= \quad cm.^2$

$\therefore$ $I_0 = mk^2$

$=$

$= \quad gm. \times cm.^2$

Also $I_0 = m\left(\frac{a^2 + b^2}{12}\right)$

$=$

$= \quad gm. \times cm.^2$

Results. The moment of inertia of the given rectangular lamina about an axis perpendicular to its plane and passing through its centre obntained *experimentally.*

(i) *by calculations* = gm. × cm.^{2} = kg m^2

(ii) *by graph* = gm. × cm.^{2} = kg m^2

and as calculated from the knowledge of its mass and dimensions

$$= \quad \text{gm.} \times \text{cm.}^2 = \quad \text{kg m}^2$$

Criticism of the Method. On account of holes bored in the lamina the value of moment of inertial obtained by this method is always *less* than its value calculated from the -formula $I_0 = m(a^2 + b^2)^{12}$. The knife-edge attachment described in the experiment does not guarantee perfect horizontality of the knife-edge and hence the oscillations of the lamina may not be confined strictly to a vertical plane. This difficulty can be removed by the use of a knife-edge provided with levelling screws as illustrated in Fig. 2.7.

The method can also be used to determine the value of g in the laboratory.

Experiment 2.4

Object. To study the *variation* of T with l for a compound pendulum and to determine

(1) the *radius of gyration* of the bar about an axis passing through its C. G. and perpendicular to its length and to verify the result by calculation of its radius of gyration from its dimensions.

(2) the *position of the C.G.* of the bar, and

(3) the value of *acceleration due to gravity* in the laboratory.

Apparatus. Bar form of compound pendulum, knife-edge attached to a rigid support, metre scale, stop-watch and a spirit level.

Description of Apparatus. Referring to Fig. 2.7, the bar form of compound pendulum consists of a brass bar about a metre long and about 2 cm. wide. Along the length of the bar from one end to the other, a series of circular holes, 5 or 6 mm. in diameter, are drilled with their centres at *equal* distances about 2 cm. apart. These holes allow the bar to be suspended from a knife-edge which passes through the hole. The knife-edge is simply a piece of hard steel ground to a sharp edge, and is driven into the edge of a piece of wood which rests

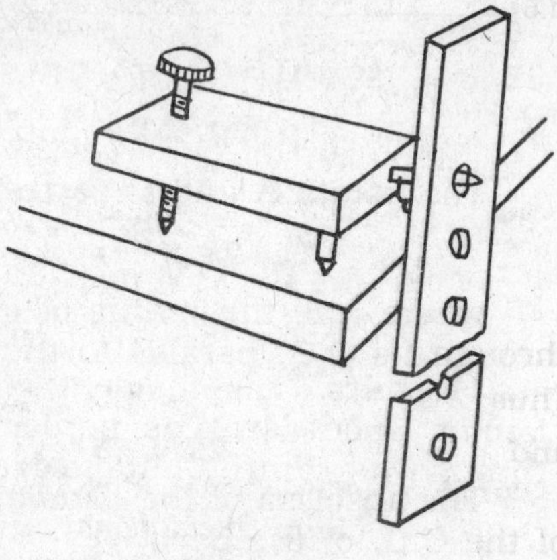

Fig. 2.7

on three screws. The two front screws are fixed while the third screw which is behind the first two, is an *adjustable* levelling screw with the help of which the knife-edge is made *horizontal*.

Sometimes, instead of a fixed knife-edge two *movable* knife-edges are provided which can be fitted successively in various holes, one on each side of C.G. at *equal* distances from it. In such a case the pendulum is allowed to oscillate about any one knife-edge by resting the latter on a horizontal rigid support.

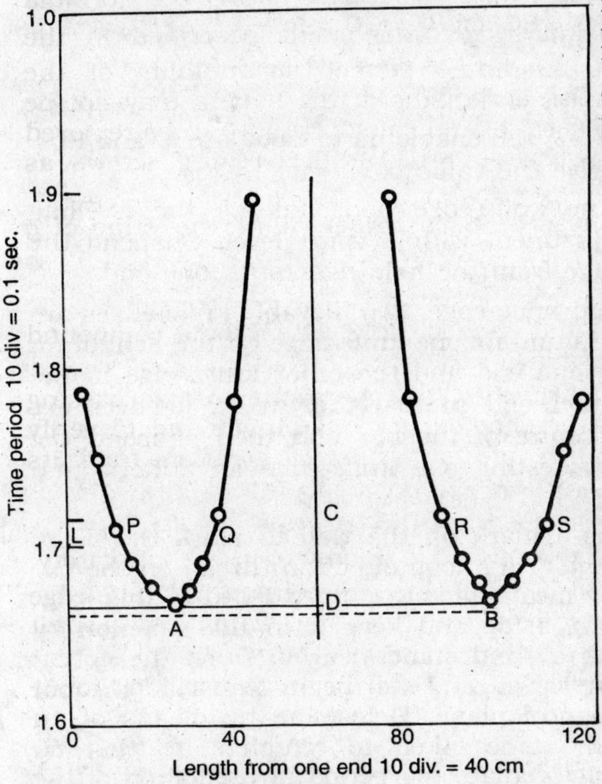

Fig. 2.8

Theory. If the bar be allowed to oscillate about horizontal knife-edge passing in each of the holes in turn, beginning with that nearest to one end, and a graph plotted taking the periods of oscillation as ordinates and corresponding distances of the axis of suspension from the end of the bar as abscissae, the graph will consist of two *symmetrical* branches as shown in Fig 2.8.

The points A and B correspond to the *minimum* period

$$T_0 = 2\pi \sqrt{\frac{2k}{g}} \tag{2.20}$$

where k is the radius of gyration of the bar about an axis through its C.G. parallel to the axis of suspension.

Thus $T_0 = MO$

and $2k = AB$ (2.21)

The abscissa of the *middle* point of AB will give the position of the C.G. of the bar.

Any line above A or B *parallel* to the *distance-axis* will cut the curve in *four* points such as P, Q, R and S which are placed by pairs P, S and Q, R *symmetrically* about the *central* line. If T be the period of oscillation corresponding to these ponits, we have

$$T = 2\pi \sqrt{\frac{l_1 + l_2}{g}} \qquad (2.22)$$

where T = LO, l_1 = PC = SC and l_2 = QC = RC

or $l_1 + l_2$ = PR = SQ

Also $k = \sqrt{l_1 l_2} = \sqrt{PC.\ QC} = \sqrt{RC.SC}$ (2.23)

Eqs. (2.20) and (2.22) will enable us to calculate g, and Eqs. (2.21) and (2.23) will give the value of k.

Method. Make the knife-edge *horizontal* by the levelling screws testing the adjustment with a spirit level. Suspend the bar about the knife-edge from the hole *nearest* to one end.

If instead of a fixed knife-edge, two movable knife-edges are provided with the pendulum, fix one knife-edge on the pendulum in the hole nearest to one end and the other knife-edge in the hole nearest to the other end so that the two knife-edges are *equi-distant* from the centre of the bar and then suspend the pendulum vertically by resting one knife-edge on a horizontal rigid support.

Place a pointer or a mark on the wall to mark the *mean* position of the pendulum. Tie a loop of cotton thread on the bar at its lower end and by means of cotton thread tied to this loop, displace the bar *slightly* aside and keep it in this position by attaching the cotton to a *fixed* stand nearby. Burn the cotton when the bar will be released and will begin to oscillate about the knife-edge in a *vertical* plane. Determine by means of an accurate stop-watch the time taken to complete at least 50 oscillations and from it calculate the period of oscillation of the bar. Also measure from one *end*, the distance of the *point of suspension, i.e.*, the *sharp* edge of the knife-edge about which the pendulum rotates.

Repeat the experiment and determine the period of the bar when suspended by passing the knife-edge in each of the holes in turn, by observing the time taken to complete at least 50 oscillations. However, when the centre of the bar is approached, only a *few* oscillations can be timed as the period gets very long. *At the central hole the period will be too long to be observed and the bar may even refuse to vibrate steadily, as the bar is balanced here and its weight has no component about the fulcrum which now passes through C. G.*

Next plot a graph between the periods T and the corresponding distances cf the points of suspension from the end of the bar, taking the periods as ordinates and the distances as abscissae. The graph will be as shown in Fig. 2.8 and will consist of two *symmetrical* branches. Join A and B and bisect AB at D. Then the *abscissae* of D will give the position of C. G. of the bar. Draw a straight line above A or B perpendicular to the time-axis cutting the curve in four points P, Q, R and S. Measure PC, SC, QC, RC, and LO : and also AB and MO. Calculate the value of k and g from Eqs. (2.20), (2.21), (2.22) and (2.23).

Sources of error and precautions. (1) The knife-edge should be *horizontal* otherwise the bar may tend to slip off the knife-edge and also may not oscillate in a vertical plane.

If two *movable* knife-edges are provided with the pendulum, one knife must be fixed on *each* side of C. G. and in all observations for period, the knife-edges must be *equidistant* from the centre of the bar so that the position of C.G. remains *unaltered.*

(2) Begin noting time of oscillations when any irregularities of motion have subsided, say, after 5 or 6 oscillations.

(3) The amplitude fo oscillation should be *small*, not greater than 4°, so that the motion of bar may satisfy the condition of theory that $\sin\theta \simeq \theta$.

(4) Take the time accurately, correct to 1/10 of a second, for at least 50 oscillations for each of the holes except for the holes near to the middle point of the bar where only a few oscillations can be timed. However, for holes giving *minimum* period and for two holes on either side of them at least 100 oscillations should be taken so as to get very exact points on the curve in this part.

(5) For determination of position of point of suspension, the distance of *sharp* edge of the knife-edge and not of middle point of hole, should be measured from one end.

(6) The graph between l and T should be drawn on a *large* graph paper and using the *entire* space available, the scales of the two axes should be so chosen as to represent fully the observed accuracies in the quantities to be plotted.

(7) The graph drawn should be smooth and symmetrical.

Observations. (*A*) *Measurement of* T and l.

Least count of stop-watch = sec.

No. of hole	Distance of *point of suspension* from one end cm.	No. of oscillations	Time		Period T sec.
			Min.	Sec.	

[*B*] *Length of the bar* = cm.

Calculations. From Graph (2.8)

PC = cm.

SC = cm. $\therefore$ Mean l_1 = cm.

QC = cm. $\therefore$ Mean l_2 = cm.

RC = cm.

and T = LC = sec.

Also AB = cm and MO = sec.

(*i*) The *position of* C. G. of the bar from one *end* = abscissae of D = cm.

(*ii*) 2k = AB = cm.

$\therefore$ k = cm.

Also $k = \sqrt{l_1 l_2}$ = = cm.

Value of k caculated from the dimensions of the bar $\sqrt{l_1 l_2}$ = cm.

(*iii*) T_0 = MO = sec. and $2k$ = AB = cm.

$$\therefore \quad g = \frac{2\pi^2}{T_0^2} \times k^2 = \quad \text{cm./sec.}^2$$

Also T = LC = sec. and $l_1 + l_2$ = cm.

$$\therefore \quad g = \frac{4\pi^2}{T^2}(l_1 + l_2) = \quad = \quad \text{cm/sec}^2$$

$\therefore$ Mean value of g = cm. / sec.^{2} = m/sec^2.

Result. The graph between T and l is as shown in Fig. 2.8 and consists of two *symmetrical* branches. The points A and B correspond to *minimum* period. T = ∞ when l = 0 or ∞ and for any *particular* period there are two values of l, one greater than k and the other less than k.

(1) The *position of the C.G.* from one end of the bar = cm.

(2) The *radius of gyration* about an axis through the C. G. of the bar and perpendicular to its length as *experimentally* determined

From Graph (2.8) = cm.

and as calculated from its dimensions = cm.

(3) The *value of acceleration due to gravity* = cm./sec.^{2} = m/sec^2.

Criticism of the method. This method gives a fairly satisfactory value of g. Since at the points corresponding to minimum period on T and l curve, a slight error in the measurement of l produces the *least* effect on the value of T, the value of g obtained from these points will be *more accurate* than that calculated from any *other* point of the curve.

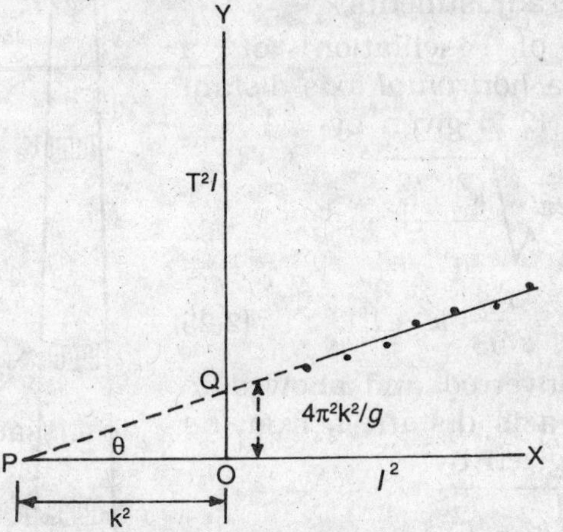

Fig. 2.9

An alternative method for determination of k and g is to plot T^2l against l^2. Then as the expression

$$T = 2\pi\sqrt{\frac{\frac{k^2}{l}+l}{g}}$$

the *period* of the pendulum can be written as—

$$T^2 l = \left(\frac{4\pi^2}{g}\right) l^2 + \left(\frac{4\pi^2}{g}\right) k^2$$

It is obvious that the graph will come out to be *a stright line* as illustrated in Fig. 2.9. Since when $T^2l = 0$, $k^2 = -l^2$, the –*ve intercept* OP on the X-axis directly gives the value of k^2.

The slope of the straight line graph is given by

$$\tan\theta = \frac{OQ}{OP} = \frac{4\pi^2}{g}$$

when
$$g = \frac{4\pi^2}{\tan\theta} \qquad (2.24)$$

Experiment 2.5

Object. To determine the value of *acceleration due to gravity in the laboratory with a Kater's pendulum.*

Apparatus. Kater's pendulum, a simple pendulum, a reading microscope, a telescope, and an accurate stop-watch.

Kater's reversible pendulum. As depicted in Fig. 2.10, *Kater's pendulum* essentially consists of a metal rod weighted at one end so that the C. G. is *much nearer* at one end than the other, and having two fixed or movable knife-edges K_1 and K_2. On the rod are two *adjustable* weights W and S to make the period of the pendulum about the two knife-edges the *same*, the smaller weight S being used for fine adjustments.

Theory. The *period* of oscillation of a compound pendulum about a *horizontal* axis distant l_1 from its C. G. is from Eq. (2.2) given by

$$T_1 = 2\pi \sqrt{\frac{k^2 + l_1^2}{l_1 g}}$$

whence
$$T_1^2 \, l_1 = \frac{4\pi^2}{g} (k^2 + l_1^2) \qquad (2.25)$$

If the pendulum were inverted and allowed to oscillate about a horizontal axis distant l_2 from its C. G., the new period T_2 is given by

$$T_2 = 2\pi \sqrt{\frac{k^2 + l_2^2}{l_2 \, g}}$$

whence
$$T_2^2 = \frac{4\pi^2}{g} (k^2 + l_2^2) \qquad (2.26)$$

Subtracting Eq. (2.26) from Eq. (2.25), we get

$$T_1^2 l_1 - T_2^2 l_2 = \frac{4\pi^2}{g} (l_1^2 - l_2^2) \qquad (2.27)$$

If $T_1 = T_2 = T$, this reduces to

$$T^2 = \frac{4\pi^2}{g} (l_1 + l_2)$$

$$T = 2\pi \sqrt{\frac{l_1 + l_2}{g}}$$

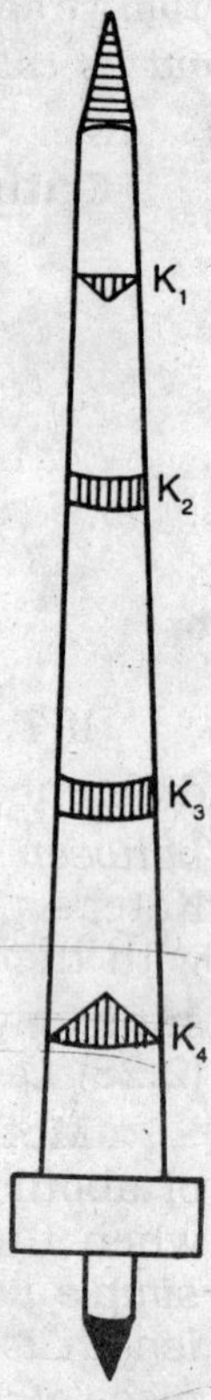

Fig. 2.10

Thus, *if two points collinear with the C. G. and on opposite sides of it are found about which the periods of the pendulum are equal, the distance between them will be equal to the length of the equivalent simple pendulum*, and can be used for the accurate determination of *g*.

This process of adjusting the periods to equality is very tedious and difficult. As pointed out by Bessel it is not absolutely necessary, for an accurate calculation can be made if they were *almost equal*. Now Eq. (2.27) can also be written as

$$\frac{4\pi^2}{g} = \frac{l_1 T_1^2 - l_2 T_2^2}{l_1^2 - l_2^2}$$

Putting $T_1^2 = T^2 + \alpha^2$, and $T_2^2 = T^2 - \alpha^2$ we have

$$\frac{4\pi^2}{g} = \frac{l_1 (T^2 + \alpha^2) - l_2 (T^2 - \alpha^2)}{l_1^2 - l_2^2}$$

$$= \frac{T^2}{l_1 + l_2} + \frac{\alpha^2}{l_1 - l_2}$$

or

$$\frac{4\pi^2}{g} = \frac{T_1^2 + T_2^2}{2(l_1 + l_2)} + \frac{T_1^2 - T_2^2}{2(l_1 - l_2)} \qquad (2.28)$$

If T_1 and T_2 are *very nearly equal and* there is *a considerable* difference between l_1 and l_2 *i.e., the C. G. is not nearly midway between the knife-edges,* (which is an *essential* feature of the Kater's pendulum) the second term will be very small compared with the first. Hence an *approximate* knowledge of $(l_1 - l_2)$ will be sufficient and so it need not be measured very accurately. Eq. (2.28) can then be used to calculate g.

Method. Fix the knife-edges, if movable, firmly at a distance of about 5 cm. from each end of the pendulum taking care that when they are fixed, the two are *parallel* to each other. Take a *simple pendulum* of an *adjustable* length, and arrange so that *its length is equal to the distance between the knife-edges.* Take a piece of a thin strip of paper and paste it on the Kater's pendulum along its length. Set up the Kater's pendulum behind the simple pendulum with one of its knife-edges resting on glass or agate plate on a firm bracket. When the two pendulums are at *rest,* draw a fine vertical line on the paper just behind the thread of the simple pendulum. till their images coincide with the vertical cross-wire of the telescope.

Use the plam of your hand to push both pendulum a very short distance aside and then withdraw the hand so that both pendulum start oscillating *simultaneously.* Note whether one pendulum gains on the other ; and if after a few oscillations the two pendulums get out of step, alter the positions of the sliding weights on the bar of the Kater's pendulum. Repeat the experiment and adjust the positions of the weights on the bar by trial so that the two pendulums swing together, the final adjustment being made with smaller weight. This will be the case when the thread of the simple pendulum and the vertical line on the paper attached to the Kater's pendulum appear to move as one, *i.e., the images of the two cross the cross-wire of the telescope together at the same instant.* When satisfied that they do swing together, invert the Kate's pendulum and allow it to oscillate about the other knife-edge. The two pendulums should still swing together.

Now remove the simple pendulum. Tie a loop of a cotton thread round the Kater's pendulum at its *lower* end, and by other cotton thread tied to this loop, draw *slightly* aside the Kater's pendulum and keep it in this position by attaching the thread to a *fixed* stand nearby. Burn the cotton thread. The pendulum will be released and will begin to oscillate. When the vertical line on the paper attached to the pendulum *just crosses* the *vertical cross-wire* of the telescope, start your stop-watch, and go on counting the number of oscillations till 200 oscillations have just been completed when the stop-watch is stopped. The stop-watch gives the time for 200 complete oscillations from which calculate the period of oscillation of the pendulum. Repeat the experiment and taking at least three sets of observations about one knife-edge, calculate the *mean* period about *each* knife-edge in turn. The two periods will be *very nearly equal.*

Next balance the Kater's pendulum on a *horizontal* blunt knife-edge and mark the position of the C.G. of the pendulum. Now place the pendulum on a *horizontal* table and support a steel scale alongside so that the *divided surface of the scale is on a level with the edges of the knife-edges.* Focus a reading microscope on the scale and determine the position of the two knife-edges and the C.G. of the pendulum. Calculate the values of l_1 and l_2 and substituting them and the mean values of T_1 and T_2 in Eq. (2.28), find out the value of *g*.

Sources of error and precautions. (1) The two knife-edges should be *parallel* to each other.

(2) The motion of the pendulums should be in a vertical plane. While timing the oscillations of the Kater's pendulum, first two or three oscillations may be left untimed so that any possible irregularity of motion may be eliminated.

(3) The amplitude of vibration should not exceed 3° to 4° so that the motion of the pendulum may satisfy the condition of theory that $\sin\theta \simeq \theta$.

(4) A *very accurate* stop-watch should be used for timing the oscillations and it must give results correct to one-tenth of a second.

(5) For the determination fo the periods about the two

knife-edges at least three sets of 200 *small* swings should be taken about each knife-edge.

(6) The distance between the knife-edges should be measured accurately, correct to a millimetre, and while making measurement the divided surface of the scale must be *in level* with the edges of the knife-edges.

Observations. (A) *Measurement of* T_1 and T_2.

Least count of the stop watch = sec.

S. No.	No. of oscillations	Time about *one* knife-edge K_1		Time about second knife-edge K_2		T_1 Sec.	T_2 Sec.
		Min	Sec.	Min.	Sec.		
1 2 3							
				Mean			

(B) *Measurement of* l_1 *and* l_2

(*i*) Position of knife-edge K_1 on the bar = cm.

(*ii*) Position of C. G. on the bar = cm.

(*iii*) Position of knife-edge K_2 on the bar = cm.

Calculations, $(l_1 + l_2)$ = cm.

l_1 = cm. and l_2 = cm.

$\therefore$ $(l_1 - l_2)$ = cm.

$$\therefore \quad \frac{4\pi^2}{g} = \frac{1}{2}\left[\frac{T_1^2 + T_2^2}{l_1 + l_2} + \frac{T_1^2 - T_2^2}{l_1 - l_2}\right]$$

$$\therefore \quad g = \left[\frac{8\pi^2}{\dfrac{T_1^2 + T_2^2}{l_1 + l_2} + \dfrac{T_1^2 - T_2^2}{l_1 - l_2}}\right]$$

= cm. / sec.^{2} = m/Sec^2.

Result. The value of *acceleration due to gravity* at .... = cm/sec^2 = m/sec^2

Standard value = m/sec^2

Error = %

Criticism of the Apparatus. The experiment when performed carefully gives very accurate value of *g*. If, however, more accurate value of g is required, as is needed in geological survey, periods should be determined by the *method of coincidences* and corrections should be applied for (*a*) effects of the medium on the motion of the pendulum (*b*) curvature of the knife-edges, (*c*) yielding of the support and (*d*) finite arc of swing.

ORAL QUESTIONS

MOMENT OF INERTIA

1. Define moment of inertia and explain its physical significance. What is the unit of moment of inertia ?
2. What do you understand by radius of gyration ?
3. What is its physical significance ?
4. What is its unit ?
5. How does the value of moment of inertia of a body depend upon its mass, its distribution about and its distance from, the axis of rotation ?
6. State the theorem of parallel and perpendicular axes and point out their use in calculations of moment of inertia. About which axis is the moment or inertia of a body the least and why ?

INERTIA TABLE

1. What is an inertia table and why is it so called ?
2. Why is it made of aluminium ?
3. Why should it be light ?
4. Explain how you can determine the moment of inertia of an irregular body with the help of inertia table. What are the forces acting on the inertia table when it is oscillating ?
5. What type of oscillations does the inertia table execute ?
6. Compare torsional oscillations with pendular oscillations. Is there any restriction on the amplitude of oscillation in this case ?
7. Is the motion of inertia table S. H. M. ?
8. Define S. H. M. and state its chief characteristics. What adjustments do you make before allowing the table to oscillate ?
9. Why are balancing weights provided with the inertia table ?
10. When the balancing weights are disturbed in the groove, what changes—the C.G. or the moment of inertia or both ?

11. Instead of a circular groove for placing balancing weights, can there be three straight groves meeting at the centre and equally inclined to each other?
12. About which axis do you find the moment of inertia of a body with inertia table ?
13. What is the functions of concentric circles on the aluminium table ?
14. Why should you measure the periods very accurately in this experiment ?
15. How can the periods be made large ?
16. What should be the dimensions of the wire so that you may get a large value of periods as well as the wire may be free from any kiks ?
17. Do you think that measuring periods by chronometer and observing the oscillations by telescope can in any way increase the accuracy of your observations ?

FLYWHEEL

1. What is a flywheel and why has it this particular shape ?
2. Where is its mass concentrated and why ?
3. What happens to the potential energy of the mass when the string unwinds itself ?
4. Why should the length of the string be less than the height of the axle of the flywheel from the floor and why should the loop slipped over the peg be quite loose ?
5. What is the harm if the various coils wound on the axle overlap each other ?
6. Why should the diameter of the string used be small ?
7. If it is not so what correction is necessary to get the effective value of r ?
8. What should be the magnitude of the mass to be attached to the string?
9. If the friction at the bearings is too great, how can you deerease it ?
10. Do you measure ω of the flywheel very accurately by this method ?

DETERMINATION OF g

1. What is meant by g ?

2. What is the unit ?
3. Distinguish between gravity and gravitation ?
4. How many methods of measuring *g* do you know and which of them is the most accurate and why ?
5. How does *g* vary from place to place with latitude, depth below or height from the surface of earth ?
6. What is the value of g at the centre of the earth ?

MOMENT OF INERTIA OF RECTANGULAR LAMINA

1. How can you find moment of inertia of a rectangular lamina by the method of oscillations ?
2. What precautions do you take in this experiment. ?
3. Why should the knife-edge be horizontal ?
4. Why should the amplitude of oscillation be small ?
5. How do you locate the C. G. of the lamina ?
6. discuss the accuracy of result obtainable by this method. Do you know a more accurate method for determination of moment of inertia of the lamina ?
7. Can you also determine *g* by this method if so, how ?

PENDULUMS

1. What is a simple pendulum and why is it so called ?
2. State the laws of oscillation of a simple pendulum.
3. What are the various defects in a simple pendulum and how can they be overcome ?
4. What is a second's pendulum ? Does the period of a simple pendulum depend upon the size and material of the bob and its temperature ?
5. What is a compound pendulum and what advantages does it possess over a simple pendulum ?
6. How does its period of oscillation depend upon the distance of the point of suspension from its C. G. ?
7. When will it be maximum or minimum ?
8. What are centres of suspension and oscillation ?
9. Are they interchangeable ?
10. About how many points in a compound pendulum are the periods the same ?
11. What is an equivalent simple pendulum and what is its length equal to ?

12. How will you use a compound pendulum to determine g ?
13. How can you determine the value of radius of gyration and the position of C. G. of a bar form of compound pendulum by plotting a graph between T and l ?
14. Why do you make the knife-edge horizontal ?
15. What is Kater's pendulum ?
16. How do you use it to determine g ?
17. Why should the knife-edges be parallel to each other ?
18. Why should the amplitude of oscillation be kept small ?
19. Is it necessary that the periods about the two knife-edges be exactly equal ? If not, how will you modify your method of determining g from Kater's pendulum ?
20. Why should the position of C. G. in this method be not very accurately determined ?
21. What are the sources of error in experiments with a compound pendulum and what corrections are necessary to eliminate them ?

3

Elasticity

3.1. Elasticity. When a system of forces or couples *in equilibrium* act on a body, there is in general either a change of shape or size of the body. *When the body regains completely its original shape and size after the deforming forces have been removed, it is said to be* **perfectly elastic**, while if it *completely retains* its modified form, it is said to be *perfectly plastic and the property by virtue of which it resists and recovers from change of shape or size is called* **elasticity** *of the body*. There is no perfectly elastic or plastic body. Actual bodies lie between these two extremes but, all bodies are nearly perfectly elastic if the deformation is small.

3.2. Stress. When a body is deformed by the application of external forces, it offers a resistance to the change of its shape or size, because attractive or repulsive forces between the molecules of the body come into play which will tend to move them back to their original positions. These forces are *equal* in magnitude to the external forces but act on the *opposite* side of the plane over which the external forces have been applied. *The magnitude of these attractive or repulsive forces between the molecules of the body per unit area is called* **stress.** When the stress is *normal* to the area it is called a *normal stress*, and may be classified as a *pressure* or a *tension* according to its direction; and when it is *tangential* to the area, it is called a *tangential or shearing stress*. M.K.S. unit of stress is N/m^2 while that in C.G.S. it is $dyne/cm^2$.

3.3. Strain. *The change of shape or the fractional change of size of a body by a given set of forces or couples is called* **strain**, and its character evidently depends upon the nature of the force producing it. For a change of length the strain is called *linear strain* and for a change of volume the strain is called *volume strain*. When there is a change of shape, the strain is called *shearing strain* or simply *shear*. Strain has no unit.

3.4. Shear. Referring to Fig. 3.1. consider two *parallel* planes AB and CD drawn in a body at a certain distance apart and let all the particles in the plane AB remain *fixed* in position. If now every particle in the plane CD be moved in that plane through the same small distance and in the same direction the

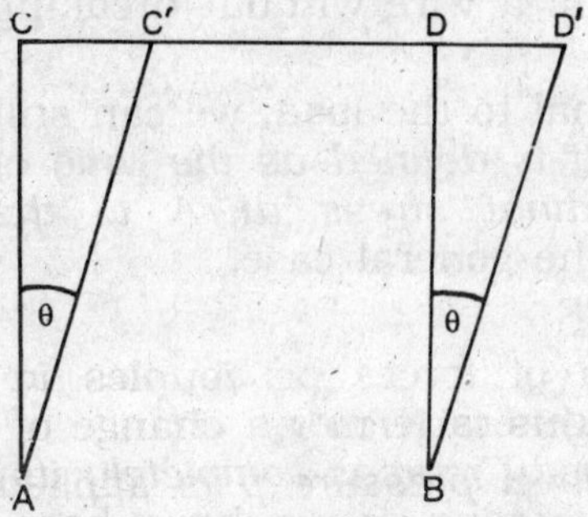

Fig 3.1

part of the body between the planes is said to be *sheared.* The shear is measured by the angle θ between CA and C′ A. If the displacement of any particle between the planes AB and CD be *parallel* to that of the particles in CD and *proportional* to the distance of the particle from the fixed plane AB there is no change of the volume of the body between the two planes and the strain is called a *uniform* shear. A plane through any point P between the planes AB and CD and parallel to them is called the *plane of the shear* at P.

3.5. Hooke's Law and Modulus of Elasticity. *Provided that the strain is small, the stress opposing it is proportional to the strain.* This statement of proportionality of stress to the corresponding strain is called *Hooke's Law,* and is the basis of the theory of elasticity. *Within the limits to which Hooke's law is applicable, the ratio of any stress to the corresponding strain is called* **modulus of elasticity**. Its value is independent of the value of the stress or the strain and depends only on the nature of the material and its other conditions. However, it is different for different types of strains. The M. K. S. unit of modulus of elasticity is N/m^2 while that in CG.S. it is dyne/cm^2.

There are three moduli in common use.

(a) Young's Modulus. *The ratio of longitudinal stress to longitudinal strain, within elastic limits to which Hooke's law is applicable, is called the modulus of longitudinal elasticity or* **Young's Modulus** *of the material of the body.* Suppose a wire of length *L* and cross-section *A* is stretched an amount *l* by a force *F* acting along its lengh, then in the *equilibrium* state when stress has become *equal* to *F/A,* the applied force per unit area, we have

$$Y = \frac{F/A}{l/L} = \frac{FL}{Al} \tag{3.1}$$

where *Y* is *Young's modulus* for the material of the wire. If *A* = 1, *l* = *L*, Y will be equal to F. Therefore, *if we could imagine a wire of unit cross-sectional area pulled out by a force to twice its length, this force would measure Young's modulus for the material if Hooke's law held for such a great strain.* Hooke's law fails if the elongation exceeds 1/1000 cm. per cm. and hence in practice it is impossible to double the length of a wire without breaking it.

If the elongation is not proportional to the load, we can still speak of Young's modulus but then *it is defined as the ratio of an infinitesimal increase of longitudinal stress dF/A to the corresponding strain dL/L.* Hence in the general case,

$$Y = \frac{L}{A} \cdot \frac{dF}{dL}$$

The M.K.S. unit of Young's modulus is N/m^2.

(b) Bulk Modulus. When a *uniform pressure p* is applied over the whole surface of a body, it produces a *uniform compression* such that *each* elementary cube of the body is compressed in the *same* proportion. *If the pressure is not too great, the compression is proportional to the pressure, and the ratio of the pressure to the volume strain is called* **bulk modulus** *or volume elasticity of the material of the body.* If a pressure *p* causes the volume of the body to diminish from *V* to $(V - v)$, the compression or *volume strain* is v/V and hence the *bulk modulus* is given by

$$K = \frac{p}{v/V} = \frac{pV}{v}$$

If the compression be not proportional to the pressure, we can still speak of bulk modulus, but then *it is defined as the ratio of an infinitesimal increment of pressure from p to (p + dp) to the corresponding volume strain – dV/V.* Hence in the general case

$$K = \frac{dp}{-dV/V} = -V\frac{dp}{dV}$$

The M.K.S. unit of bulk modulus is N/m^2.

The only stress which a liquid or a gas can permanently sustain is a *hydrostatic pressure.* Hence bulk modulus is the *only* elastic constant for fluids, and it is often spoken of as the *elasticity* of the liquid or the gas.

(c) Shear Modulus or Rigidity. Let PQRS, LMNO be a cube fixed at the bottom (Fig. 3.2) and let a force F be applied in the plane of the upper face of the cube, the force being *parallel* to PQ. The force constitutes a *tangential stress* on the upper face of the cube, hence the cube would be distorted into the position P′ Q′ R′ S′, LMNO.

When the strain is small, the stress is proportional to it and the ratio of tangential stress to the corresponding shear is called **modulus of rigidity** *of the material of the cube.* Thus, if the *tangential stress* be F/A and the corresponding *shear* θ radians, the *modulus of rigidity* is given by

$$n = \frac{F/A}{\theta} = \frac{F}{A\theta}$$

the modulus of rigidity is, therefore, measured by the tangential stress which would be developed for a unit shear, i.e., a shear of one radian or 57° 18′, if Hooke's law held for so great a strain. Hooke's law fails if the shear exceeds 1/200 radian or 1/3 degree.

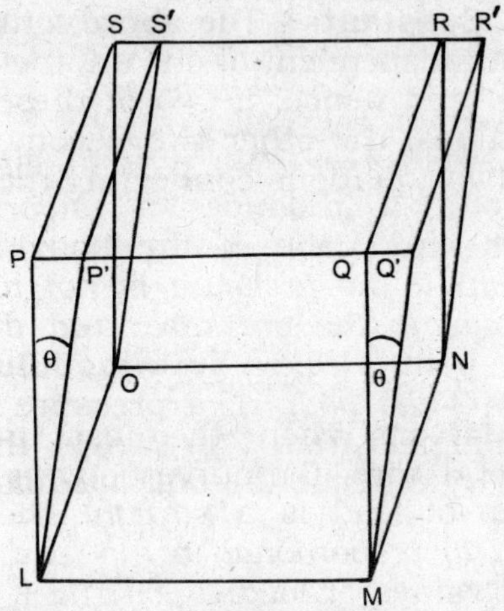

Fig 3.2

If the shear be not proportional to the shearing stress, we can still speak of the rigidity of the material, but then it is *defined as the ratio of an infinitesimal increase of shearing stress to the corresponding increase of the angle of shear.* Hence in the general case,

$$n = \frac{dF/A}{d\theta} = \frac{1}{A} \cdot \frac{dF}{d\theta}$$

The unit of modulus of rigidity is N/m^2.

(d) Poisson's Ratio. In the case of a wire elongated by the application of two equal and opposite forces acting along its length we have assumed that the strain is only a change in length and no change in the diameter of the cross section occurs. In practice, however, the strain of the wire is not so simple. Possion found that in every case of such an elongation there is a contraction of the lateral dimension. *Within elastic limits there is a complete proportionality between the lateral strain and the longitudinal strain and the ratio between the lateral strain and longitudinal strain is called* **Poisson's ratio**. Thus, if the original diameter and length of a wire or rod be D and L respectively and increase in length due to a given longitudinal force is l and diminution of diameter d, then Possion's ratio is given by

$$\sigma = \frac{d/D}{l/L} = \frac{Ld}{lD}$$

or expressed as a differential co-efficient

$$\sigma = \frac{L}{D} \cdot \frac{dD}{dL}$$

Since both the lateral and longitudinal strains are pure numbers, Poisson's ratio has no unit. The value of Posson's ratio depends only on the nature of the material, and for homogeneous isotropic materials its value lies between -1 and $\frac{1}{2}$.

3.6. Relation between Elastic Constants. The above four elastic constants of a material are not independent of one another but are connected by two equations, and hence, if two of these constants are given any arbitrary values, the other two become determined by these equations. The two equations connecting the four elastic constants are

$$Y = \frac{9\,nk}{3k + n}$$

and

$$\sigma = \frac{3k - 2n}{6k + 2n}$$

3.7. Elastic Properties of Matter. When the load is continuously increased in the case of a wire, the curve between load and extension for *small* values of load is a *straight line* indicating that *the extension is directly proportional to the load, i.e.,* Hooke's law is obeyed for small values of load. Referring to Fig. 3.3., *the point P upto which Hooke's law is obeyed is termed the* **limit of proportionality**, and is determined experimentally by plotting values of the applied load and the resulting elongation. *The point upto which the curve remains a straight line gives the limit of proportionality of the specimen.*

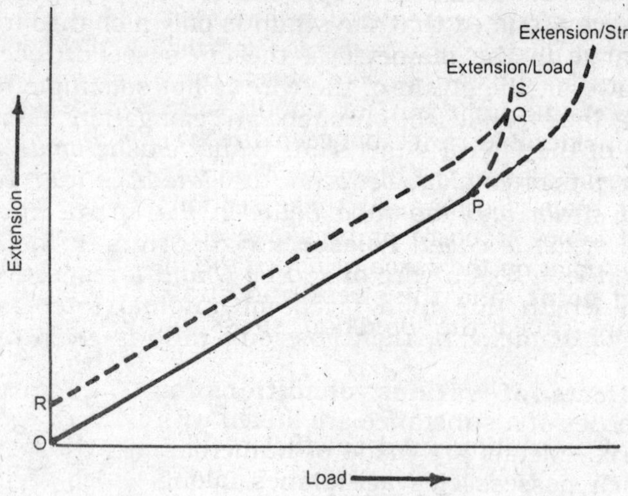

Fig 3.3

When the load is further increased, a point known as the elastic **limit** of the substance is reached. *Upto this point the wire regains its original length when the load is removed, i.e.*, it is perfectly elastic within this limit. The limit of proportionality and the elastic limit is in general *not* the same. When the wire is loaded beyond elastic limit, say upto Q, it does not regain its original length on the removal of the load and a large part of the extension represented by OR in the figure is *permanent*. The wire is then said to have received a **permanent set**. Lead wire is very inelastic because comparatively much smaller loads produce a permanent set. Sometimes the *elastic limit is defined as the force which produces the maximum recoverable deformation.* In this case to find its value, the specimen is loaded and unloaded with different loads and its dimensions measured after the removal of load till the recovery on unloading is *just not complete, i.e.*, a definite permanent set is formed. Then the amount of permanent set is plotted against the load and from the curve an estimate of the load is made at which the permanent set begins.

On further loading the wire a point S is reached on the curve beyond which the extension increases very rapidly for only a small increase in load. Above S the curve becomes practically *vertical* and the wire continues to elongate without any increase of load, *i.e.*, the material the wire *flows* beyond S. The wire can no longer support so large a load as before, as on account of considerable thinning of the wire, the stress is greater than before and if it is to be prevented from being broken, the load must be diminished; thus the curve turns towards the extension axis. The relation between stress and extension is represented by the dotted curve. *The point where the wire begins to 'run' is called the* yield point. For brittle substances like glass, the yield point is the same as the elastic limit but for ductile substances, *e.g.*, lead, there is a considerable range between the two.

After the yield point the diameter of certain sections decrease more rapidly than of others. At these sections the stress is the greatest and hence fracture or breakage occurs at one of these sections. *The point on the curve at which the wire breaks is called the* **breaking point**, and the stress and strain corresponding to this point are called the *breaking stress and breaking strain* respectively.

3.8. Effects of various conditions on Elasticity. The elastic properties of a substance are linked with the fine structure of the material. Metals are made of numerous aggregates of fine crystals which possess cleavage planes, along which the atoms forming the crystals may easily slide over each other. When the metal is overstrained, rolled or hammered, the cleavage planes are distributed indifferently in all directions, and hence the elastic limit of the material is extended. But when the metal is annealed, *i.e.*, first heated and then allowed to cool slowly, the cleavage planes are distributed in definite directions, and hence the elastic limit is lowered and the metal becomes soft or plastic. When the annealed metal is stretched, it again becomes highly elastic. The 'hard drawn' or 'cold worked' wire, *i.e.*, the wire which has been stretched by drawing it through a die (a small hole in a steel plate called a Wurtel plate) possesses still higher elastic limit.

If an impurity is present in a metal, it forms a sort of cement and unites the crystal aggregates, thus lowering or raising the elastic limit according as the cement is more plastic or more elastic than the metal. For example, an addition of 2% of potassium to gold increases the elasticity twenty times, and the addition of a little carbon to molten iron forms a carbide Fe_3C which unite the crystal aggregates of iron and produces steel which is much more elastic than pure iron.

Further, if an elastic substance is heated to a high temperature, it becomes plastic. For example, carbon filament of an electric glow lamp is highly elastic at ordinary temperature and becomes almost plastic when it is heated by an electric current and can be distroted by a magnet. If lead is cooled in liquid air it becomes elastic and rings like steel when struck. The elasticity of a body changes with temperature and, in general, the values of elastic moduli decrease with the rise of temperature. But for nickel steel alloys, *i.e.*, el-invar, etc., which are used in the construction of hair spring of watches and chronometers, the value of Young's modulus is independent of temperature, and in some cases it even rises with the rise of temperature.

(*a*) elastic after-effect. If a fine fibre of quartz is twisted within elastic limit and then released, it regains its original condition *immediately* the twisting force is removed. Other substances particularly glass, take some time to recover their shape when deforming forces are removed though they will do so eventually. *This delay in recovering the original condition is called* **elastic after-effect**.

On account of promptness in recovering, quartz fibres are extremely valuable in the measurement of small torsions, *e.g.*, in Boys' apparatus for the determination of G, quadrant electrometers, and sensitive moving coil galvanometers. This elastic after-effect is also absent in phosphor bronze and silver, and hence they are good substitutes for quartz.

(*b*) Elastic Fatigue. When a wire is vibrating continuously for some days, *the rate at which the vibrations die away is much greater than when the wire is fresh.* This is due to **elastic fatigue**. The wire gets tired and can continue vibrating with difficulty. If it is allowed to rest for some time, its original condition is recovered.

(c) Elastic Hysterestis. Elastic hysteresis is a phenomenon similar to that of magnetic hysteresis. These phenomena are presented when a sample is subjected to rapid *cyclic* variations of mechanical or magnetic strain so that the strain in the sample cannot keep pace with the varying external influences.

Experiment 3.1

Object. To determine the *Poisson's ratio* for rubber.

Apparatus. Rubber tube with metal sleeve and rubber stopper, burette, 250 gm. slotted weights, metre scale and a pointer.

Theory. Let a rubber tube suspended *vertically* be loaded at its lower end with a *small* weight. This stretches the rubber tube *slightly* with a consequent *increase* in its length and internal volume. Let V be the original volume of the tube. A its area of cross-section, L its length and D its diameter. Then, if for a small increase of volume dV, the corresponding increase in length is dL and the decrease in area is dA, we have

$$V + dV = (A - dA)\ (L + dL)$$

$$= AL + A.dL - L.dA - dA.dL$$

Putting $AL = V$ and neglecting $dA\ dL$, a product of two *small* quantities, we have

$$dV = A.dL - L.dA$$

Now $$A = \pi\ (D/2)^2 = \pi D^2/4$$

Differentiating $$dA = \frac{\pi D}{2}\ dD = \frac{2A}{D}\ dD \qquad (3.2)$$

Substituting this value of dA in Eq. (3.2), we get

$$dV = A.dL - \frac{2AL}{D}.dD$$

or $$\frac{dV}{dL} = A - \frac{2AL}{D} \cdot \frac{dD}{dL}$$

whence $$\frac{LdD}{DdL} = \frac{1}{2}\left(1 - \frac{1}{A}\frac{dV}{dL}\right)$$

or from Sec. 3.5d, $$\sigma = \frac{1}{2}\left(1 - \frac{1}{A}\frac{dV}{dL}\right) \qquad (3.3)$$

This equation can be employed to calculate σ, if other quantities are determined.

Method. Suspend *vertically* a rubber cycle-tyre tube BA (Fig. 3.4) of about one metre in length. Close it securely at its upper end with a rubber stopper and seccotine, and at its lower end with a metal sleeve carrying a hook H. Bore the stopper at the upper end of the tube to carry a burette C held vertically. Set up a metre scale S vertically in front of the tube and attach a small horizontal pointer P to the metal sleeve B so as to be just *not* in contact with the scale.

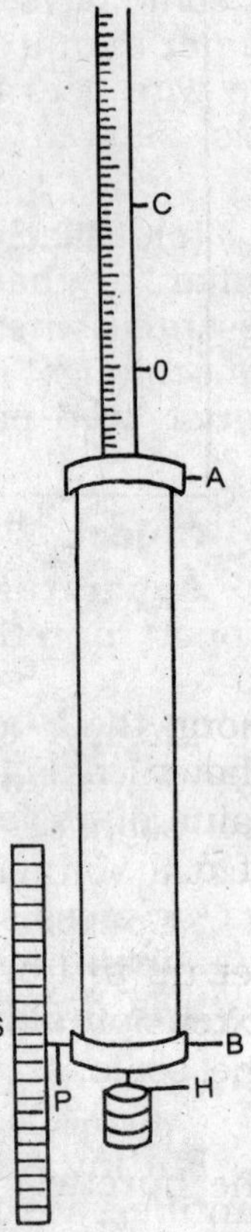

Fig 3.4

Pour water into the burette until the whole of the rubber tube and about 4/5 of the burette are filled with it. Allow the apparatus to settle down for about 5 minutes. Measure the diameter of the rubber tube at a number of points with a vernier callipers and find its mean value, and thus calculate the area of cross-section of the tube.

Next note down the position of the water meniscus inside the burette and the position of the pointer P on the scale S. Place *gently* a load, of say 250 gm., on the hanger at the lower end of the tube and wait for about 5 minutes. Again note down the readings of the burette and the pointer P on the scale S. The difference between the two readings of the burette gives the change in volume *dV* for the load on the hanger and the difference between the two readings of the pointer P on the scale gives the corresponding change in length.

Now increase the load on the hanger in equal steps of, say 250 gm., till maximum permissible load within elastic limit is reached, taking down the readings of the burette and the pointer P after addition of each load when the apparatus has settled down. Then decrease the load on the hanger in the *same* stages and repeat the above observations till the load on the hanger is reduced to zero. Take the mean of the two readings of the burette for the same load on the hanger obtained with increasing and decreasing load, and then subtracting the mean reading for zero

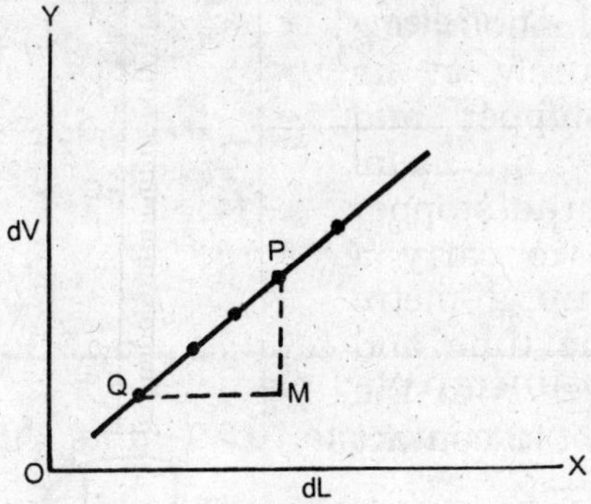

Fig 3.5

load on the hanger from the mean reading for any load, calculate the change in volume dV of the rubber tube for various loads on the hanger. Similarly calculate the value of corresponding change in length dL for the various loads. Then calculate the value of dV/dL for *each* set of observations *separately* and find its mean value of σ . Finally calculate σ from Eq. (3.3).

Next plot a graph taking dL along the X-axis and the corresponding value of dV along the Y-axis. This will come out to be a straight line PQ as shown in Fig. 3.5, and its *slope* PM/QM will give the *average* value of dV/dL. Use this in Eq. (3.3) to calculate again the value of σ.

Sources of error and precautions. (1) The vertical scale set up in front of the tube should be arranged to give maximum extension allowable and the pointer should be just **not** touching the scale.

(2) There should be no air bubble inside the rubber tube or the burette.

(3) Diameter of the rubber tube should be measured at a number of points along its length and at each point two diameters *at right angles* to each other should be measured.

(4) Weights should be placed or removed *gently* and in *equal* steps of, say 250 gm.

(5) After each addition or removal of load wait for about 5 minutes before taking observations in order to allow the apparatus to settle down to new conditions of stress and strain.

(6) The load suspended at the lower end of the rubber tube should not exceed the *maximum load permissible within elastic limit.*

Observations. (A) *Measurement of diameter of the rubber tube.* Vernier constant = cm.

Zero error = cm.

Serial No.	Reading along any diameter α cm.	Reading along perp. diameter β cm.	$\frac{\alpha+\beta}{2}$ cm.
1			
2			
3			
4			
5			

Mean uncorrected diameter =

[B] *Determination of change in volume dV and the corresponding change in length dL.*

Sl. No.	Load on the hanger gm.	Reading of burette c.c. Load increasing x	Load decreasing y	Mean $\frac{x+y}{2}$	Chane in volume c.c	Reading of pointer P on scale S cm. Load-inreasing x	Load-inreasing y	Mean $\frac{x'+y'}{2}$	Chane in length dL cm.	$\frac{dV}{dL}$
1.	0			a				a'		
2.	250			b	$b—a$			b'	$b' - a'$	
3.	500			c	$c—a$			c'	$c' - a'$	
4.	750			d	$d—a$			d'	$d' - a'$	
5.	1000			e	$e—a$			e'	$e' - a'$	
6.	1250			f	$f—a$			f'	$f' - a'$	
									Mean=	

Calculations. Mean *corrected* diameter of the rubber tube = cm.

$\therefore$ Area of cross-section of the tube $A = \pi D^2/4 =$ sq. cm.

$$\sigma = \frac{1}{2}\left(1 - \frac{1}{A} \cdot \frac{dV}{dL}\right)$$

From *graph* (3.6)

PM = c.c. QM = cm.

$$\frac{dV}{dL} = \frac{PM}{QM} = \quad \text{sq. cm.}$$

and

$$\sigma = \frac{1}{2}\left(1 - \frac{1}{A}\frac{dV}{dL}\right)$$

$$=$$

$$=$$

Result. *The value of Poisson's ratio* for rubber as obtained *experimentally.*

(*i*) by *calculations* =

(*ii*) by *graph* =

Standard of σ =

∴ Error = %

3.9. Bending of beams. When a beam is bent by an applied couple, *its longitudinal filaments are lengthened on the convex side and shortened on the concave*, and thus the *resistance* which the beam offers to bending, will depend upon *Young's modulus* for the material. If the beam be *straight* and of *uniform* cross-section, it will have a *plane of symmetry* parallel to its length which interesects every *transverse* section in a *straight* line, which is an *axis of symmetry* for that section. *When the bending is uniform all the longitudinal filaments are bent into circular arcs in planes parallel to the plane of symmetry which may then be called the* **plane of bending**, *and the centres of curvature of all the longitudinal filaments lie on a straight line perpendicular to the plane of bending which may be called the* **axis of bending**. In an uniform bending all particles which lay in transverse planes before the bending of the beam, lie after bending in corresponding planes, passing through the axis of bending. These planes, therefore, cut the filaments at right angle.

In the *middle* of the beam is a surface such that the particles in it are at their *normal* distance apart, *i.e.*, the filaments lying in this surface are neither shortened nor lengthened but remain *constant* in length. This surface is called the **neutral surface**. In the unstrained position of the beam it becomes a plane surface. *The filament of this unstretched layer which is in the plane of symmetry of the bent beam is called the neutral axis or the* **neutral filament.** Neutral surface is perpendicular to the plane of bending and the neutral filament passes through the centre of gravity or centroid of each transverse section of the beam.

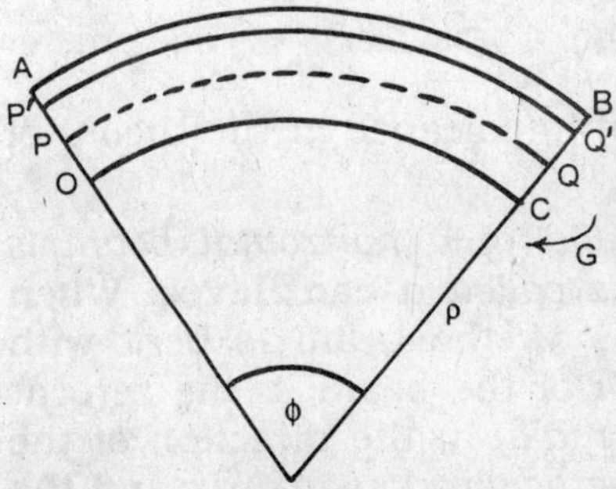

Fig 3.6

Referring to Fig. 3.6, let ρ be the radius of curvature of the neutral axis PQ. Considering a filament P′Q′ of the beam at a distance y (=PP′) from the neutral axis and hence (ρ + y) from the axis of bending, we have

$$P'Q' = (\rho + y)\ \phi$$

The length of this filament was equal to $PQ = \rho\phi$ *before* the beam was bent so that the extension of this filament is

$P'Q' - PQ = (\rho + y)\phi - \rho\phi = y\phi$ and its elongation, or extension per *unit* length is, therefore, given by

$$e = y\phi/\rho\phi = y/\rho \tag{3.4}$$

If the area of cross section of the filamemt is α and if Y is *Young's modulus*, the *stress* across the transverse section of the filament is Yy/ρ and hence the force on this section of the filament $Y\alpha y/\rho$ and its *moment* about an axis through its centroid at right angles to the plane of bending is $Y\alpha y^2/\rho$ and hence the sum of all these moments over the entire cross-section is $= \Sigma Y\alpha y^2/\rho$.

This moment is called the **bending moment** *at the section,* and is equal to the moment of the *external* bending couple applied in the *equilibrium state.*

Thus *bending moment* $G = \Sigma Y\alpha y^2/\rho = (Y/\rho)\ \Sigma\alpha y^2$ (3.5)

The expression $\Sigma\alpha y^2$ is analogous to the moment of inertia about the axis $y = 0$, the difference being that instead of mass of the element its area has been taken. It is, therefore, called *geometrical moment of inertia* of the strained transverse section of the beam about an axis through its centroid and perpendicular to the plane of bending. If $\Sigma\alpha y^2$ be denoted by I,

$$G = YI/\rho \tag{3.6}$$

If A is the area of cross-section and K its *radius of gyration* $I = AK^2$ and

$$G = Y\ AK^2/\rho$$

This equation is of fundamental importance in the theory of the bending of beams.

3.10. Cantilever. *When one end of a horizontal beam is fixed and the other end is free, it is called a* **cantilever**. When the free end is loaded with a mass M, the beam is bent with curvature changing along the length of the beam, being zero at the fixed end. Taking X-axis *horizontally* in the direction of the length of the *unbent* beam, Y-axis *vertically downwards* and the origin of co-ordinates at the *fixed end* O of the beam (Fig. 3.7). The *radius of curvature of the natural axis* at a distance x from O where the depression is y, is given by

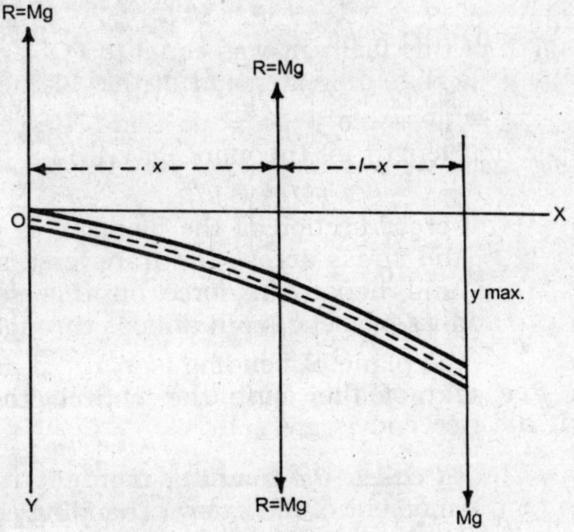

Fig 3.7

$$\frac{1}{\rho} = \frac{d^2y/dx^2}{\left[1 + \left(\frac{dy}{dx}\right)^2\right]^{3/2}}$$

It the bending is small $\left(\frac{dy}{dx}\right)^2 < <1$ and hence $\frac{1}{\rho} = \frac{d^2y}{dx^2}$.

Considering the equilibrium of transverse section of the beam at (x, y), the *moment of the external couple is Mg* $(l - x)$, where l is the length of the beam, the weight of the beam being neglected and the bending moment is

$$\frac{YI}{\rho} = YI \frac{d^2y}{dx^2}$$

or $$YI \frac{d^2y}{dx^2} = Mg\ (l - x)$$

or $$\frac{d^2y}{dx^2} = \frac{Mg}{YI}\ (l - x)$$

Integrating once, we get

$$\frac{dy}{dx} = \frac{Mg}{YI}\left[lx - \frac{x^2}{2}\right] + C_1$$

When $x = o$, $\frac{dy}{dx} = 0$, $\therefore$ $C_1 = 0$

Hence $$\frac{dy}{dx} = \frac{Mg}{YI}\left(lx - \frac{x^2}{2}\right)$$

Integrating again, we get

$$y = \frac{Mg}{YI}\left(\frac{lx^2}{2} - \frac{x^3}{6}\right) + C_2$$

When $x = 0$, $y = 0$, $\therefore$ $C_2 = 0$

Hence $$y = \frac{Mg}{YI}\left(\frac{lx^2}{2} - \frac{x^3}{6}\right)$$

At the *free end* of the cantilever $x = l$, therefore the depression at the free end is given by

$$\delta = \frac{Mgl^3}{3YI} \tag{3.7}$$

For a rectangular beam of breadth b and depth d (vertical), the area of cross-section of the beam is equal to bd and the *geometrical moment of inertial* I about the line through the centroid of the section and perpendicular to the plane of bending is equal to $bd \times d^2/12 = bd^3/12$. For a circular beam of radius r, $I = \pi r^4/4$.

$\therefore$ $$\delta = \frac{4Mgl^3}{Ybd^3} \quad \text{(rectangular)} \tag{3.8}$$

and $$\delta = \frac{4Mgl^3}{3\pi Yr^4} \quad \text{(circular)} \tag{3.9}$$

3.11. Beam supported at its end and loaded at the middle. Let a beam be supported on two knife-edges at a distance l apart in a *horizontal* plane with *equal* length of the

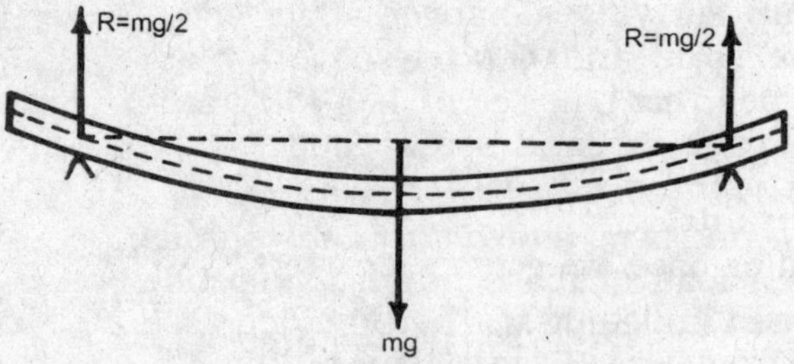

Fig 3.8

beam projecting beyond the knife-edges. If the beam be loaded at the middle with a mass *m*, each knife-edge will exert a *vertical upward* force of *reaction mg*/2 on the beam. The depression will be *greatest* at the middle point O of the beam (Fig. 3.8) and in the neighbourhood of O, the beam will be *horizontal*. Hence, if the beam be divided into two equal portions of length *l*/2 by a transverse section through O, each portion will be equivalent to a *cantilever* of length *l*/2 with a force *mg*/2 acting upwards at its free end. Hence the depression at the middle of the beam is found out by putting *mg*/2 in piace of *Mg* and *l*/2, in place of *l* in Eq. (3.7) for the depression of a cantilever.

Thus
$$\delta = \frac{(Mg/2)(l/2)^3}{3YI}$$

$$= \frac{mgl^3}{48YI} \qquad (3.10)$$

and hence for a rectangular beam of breadth b and depth d (vertical)

$$\delta = \frac{Mgl^3}{4Ybd^3} \qquad (3.11)$$

and for circular beam of radius *r*,

$$\delta = \frac{mgl^3}{12\pi Yr^4} \qquad (3.12)$$

Equations (3.11) and (3.12) can be used to determine the value of *Young's modulus* for the material of a beam by observing the value of the depression produced at the centre by a known load.

Experiment 3.2

Object. To determine the value of *Young's modulus* of............ from the flexure of a beam supported symmetrically on near to its ends and loaded at its middle point.

Apparatus. A bar of the material whose *Young's modulus* is required, two knife-edges fixed in rigid supports in a horizontal plane a stirrup carrying a hanger at its end, optical lever with the millimetre scale and the telescope (or a micrometer screw with electric bell, etc.), a set of half-kilogram weights, a metre scale, vernier callipers, and screw gauge.

Theory. Let a *uniform* rectangular bar supported *symmetrically* on two *horizontal and parallel* knife-edges be depressed by a load of mass *m* gm., suspended from the beam *midway* between knife-edges. Then the depression δ of the *middle* of the bar is from Eq. (3.11) given by

$$\delta = \frac{mgl^3}{4bd^3Y} \tag{3.11}$$

where Y is *Young's modulus* for the material, l the length, b the breadth, and d the thickness (vertical) of the bar.

The depression δ being *very small* is measured with

(i) *an optical lever* or

(ii) *a micrometer screw.*

Method. Measure with a *screw guage* the thickness of the experimental bar at several places along its length and find out the *mean thickness* d in cm., corrected for zero error of the instrument. Similarly take a number of readings for the breadth of the bar with a *vernier callipers* and calculate the *mean breadth* in cm. Next place a meter scale on the two knife-edges with its graduated surface vertical, and with it measure the distance between the knife-edges. Then place horizontally on the knife-edges the experimental bar with its *smallest* edge vertical and with its *equal* lengths projecting beyond the knife-edges. The distance between the knife-edges already measured gives the length l of the bar between the knife-edges. Next suspend the hanger for placing weights from a stirrup resting on the bar *midway* between the knife-edges and also *parallel* to them.

(a) Optical lever arrangement. As illustrated in Fig. 3.9. place the optical lever on a fixed support behind the bar with its two legs on the support and with its *front leg* resting on the top of the stirrup at its *middle* point and adjust the force of the mirror to be *vertical* and also *parallel* to the length of the bar. Set up a vertical millimetre scale in front of the mirror at a distance of about one metre : and close to the scale at the same

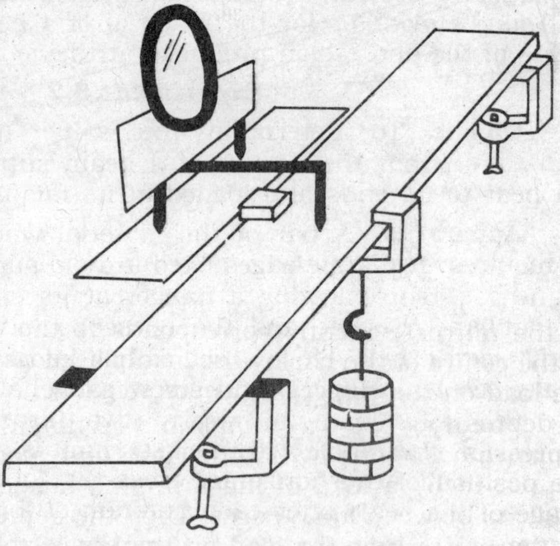

Fig 3.9

height as the mirror, place a telescope firmly clamped in a rigid stand. Focus the eyepiece on the horizontal cross-wire of the telescope till it is distinctly seen. Then focus the telescope on the image of the scale in the mirror by turning round the mirror *slightly*, if necessary, about a horizontal axis. *In focussing the telescope it is sometimes helpful to first focus it on the mirror itself and then pushing the tube of the telescope in until the image of the scale appears in view*. Note down the position of the horizontal cross-wire of the telescope on the image of the scale.

Next *gently* place a load of, say, half a-kilogramme on the hanger, which depresses the bar slightly and hence the mirror will be *slightly* titled about the line joining its two hind legs and the image of the scale in the mirror will be displaced past the cross-wire of the telescope. Again note down the position of the horizontal cross-wire of the telescope on the image of the scale. The difference between the two readings of the position of the horizontal cross-wire of the telescope on the image of the scale

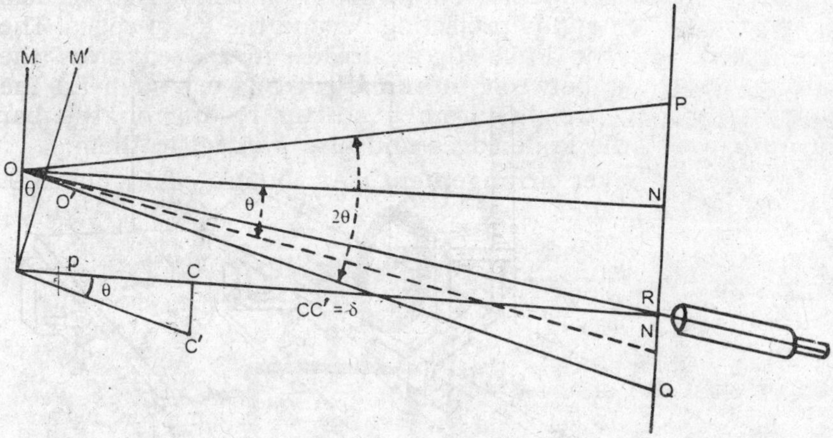

Fig.3.10

in the mirror evidently corresponds to the depression produced at the centre of the bar by the load placed on the hanger. Increase the load on the hanger by *equal* steps of half a-kilogramme until a depression equal to nearly two-thirds of the maximum depression obtainable within elastic limit is reached, noting down the position of the horizontal cross-wire of the telescope on the image of the scale after each addition of load.

Now decrease the load on the hanger by the *same* stages to zero noting down the position of the cross-wire on the image of the scale in the mirror after the removal of each load. Find out the mean of the two readings of the position of the cross-wire on the image of the scale in the mirror for various loads, taken with increasing and decreasing loads. From these means, calculate the mean shift x of the image of the scale past the cross-wire of the telescope for a load.

Measure the distance D between the mirror and the scale. To measure p, *i.e.*, the *lever arm*, place the optical lever on a sheet of paper and by pressing it *lightly* produce impressions of its feet on it. From these impressions determine the *perpendicular* distance of the front foot of the optical lever from the line joining the two hind legs which gives p. As indicated in Sec. 1.15, now calculate the value of δ from the formula

$$\delta = xp/2D \tag{3.12}$$

Finally calculate the value of Y from Eq. (3.11).

(b) Micrometer screw arrangement. Determine the *pitch* and *least count* of the *micrometer screw* and then set it up as illustrated in Fig. 3.11 making the connections for the electric

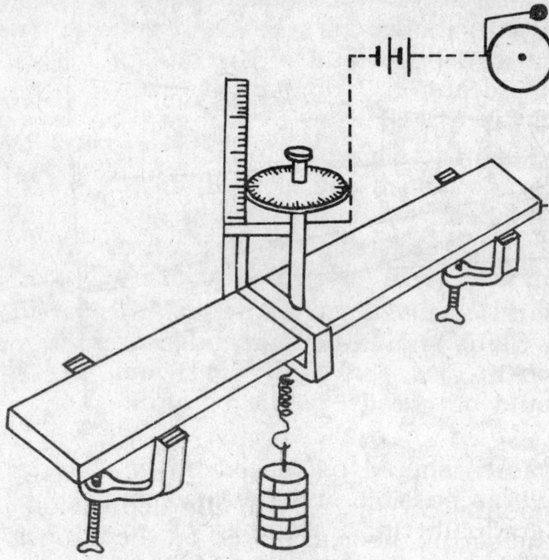

Fig 3.11

bell as shown by the dotted lines. Lower the micrometer screw and note down its reading when its tip *just* touches the bar as revealed by ringing of the bell. Place gently a load of, say half-a-kilogramme on the hanger. This depresses the base slightly. Lower the micrometer screw till it is again *just* in contact with the bar and note down its reading. The difference between the two readings of the micrometer evidently gives the depression produced at the centre of the bar by the load placed on the hanger. Increase the load on the hanger by *equal* steps of half-a-kilogramme until maximum permissible load is reached, taking the reading of the micrometer screw after each addition of load.

Now keeping the load on the hanger unchanged, lower the screw a little further and then raise it by turning it back till the tip of the screw is *just not* in contact with the bar. This will be indicated by just stopping of the ringing of the bell. The reading of the micrometer in this position will be the same as, or nearly equal to, its previous reading depending upon the amount of *back-lash* present in the screw.

Next unload the hanger in the *same* stages till the load on the hanger is reduced to zero, taking down the reading of the micrometer screw after removal of each hoad. Find the mean micrometer reading for various loads from the corresponding readings taken with increasing and decreasing loads and then calculate the mean depression δ of the bar for a particular load as indicated in precaution 7 of Expt. 3.1. Finally, calculate the value of Y from Eq. (3.11).

Sources of error and precautions. (1) Since in the case of metals, Hooke's law fails if the elongation exceeds 1/1000 cm. per cm., the highest strained filament of the bar. *i.e.*, the *outer-most* one should not be strained beyond this limit. For l = 80 cm., d = 0.5 cm., *maximum* value of δ = 1.06 cm.

(2) To secure *symmetrical* loading of the bar, the hanger used to carry weights should be suspended from a stirrup of the shape as shown in Figs. (3.9 and 3.11) and the knife-edge of the stirrup should be *parallel* to the two knife-edges supporting the bar.

(3) The masses should be placed on or removed from the hanger as *gently* as possible and the load on the bar increased or decreased gradually in a number of *equal* steps of, say, half-a-kilogramme.

(4) While calculating the mean depression from the mean micrometer (or optical lever) readings for increasing and decreasing loads, the student should take care *not to take successive differences between them.*

(5) As the thickness (vertical) of the bar occurs raised to the *third* power in the expression for Y and is also a *small* quantity,

many observations for it should be taken at a number of points along the length of the bar with a *screw gauge.*

For optical lever only

(6) Initially the face of the mirror should be very nearly *parallel* to the scale and the distance of the telescope from the scale should be as small as the focussing of the telescope allows in order that the magnification may be as *large* as possible.

(7) As the *lever arm, i.e.,* the distance of the front foot of the optical lever from the line joining the other two is small, it should be accurately determined as described above.

For micrometer screw only

(8) While taking readings of the micrometer screw the eye should be in the *same* level as the upper face of the divided head of the screw, lest a mistake should creep in due to parallax between the vertical scale and the divided head of the screw.

(9) To avoid *back-lash error* in the micrometer screw, it should always be turned in the *same* direction before taking a reading.

Observations

(*A*) (*i*) *Length of the bar between the knife-edges* = cm.

(*ii*) *Measurement of thickness (vertical) and breadth of the bar.*

Least count of the *screw gauge* = cm.
Its zero error = cm.
Vernier constant = cm.
Its zero error = cm.

S. No.	Thickness cm.	Breadth cm.
Mean		

(*B*) *Measurement of depression of the bar*

For optical lever only
(*i*) *Lever arm, i.e.,* the distance *p* of the front foot of the optical lever from the *line joining the other two* = cm.
(*ii*) Distance *D* of scale from the mirror = cm.

for micro-
Pitch of the screw = cm.

meter screw only

No. of divisions on circular head =

∴ *Least count* of the micrometer = cm.

S. No.	Load on the pan gm.	Micrometer screw reading (or position of cross-wire on image-scale)			Depression for 1,500 gm. cm.
		Load increasing x cm.	Load decreasing y cm.	Mean $\frac{x+y}{2}$ cm.	
1	0				
2	500				
3	1000				
4	1500				
5	2000				
6	2500				
				Mean	

Calculations *for optical lever only*

Multiplying factor of the optical lever $p/2D$ = =

$\delta = \frac{xp}{2D}$ = cm.

$$Y = \frac{Mgl^3}{4\,bd^3\,\delta} = \quad = \text{dynes/cm.}^2 = \text{N/m}^2.$$

Result. The value of Young's modulus for the material of the bar ...... = N/m^2.

Standard value = N/m^2.

∴ Error = %

Criticism of the method. By means of *optical lever* arrangement the depression of the bar can be magnified *several* times and hence can be measured sufficiently accurately. With the distance between the scale and the mirror equal to about 2 metres, depressions can be measured with accuracy of 0.002 cm. For more accurate results the depression should be measured with a *micrometer screw*. If the pitch of the screw is $\frac{1}{2}$ mm. and the head is divided into 100 divisions, depressions can be measured with an accuracy of 0.0005 cm.

The result is affected by error due to force of friction between the bar and the knife-edges. When the load on the hanger is changed, the bar slides *slightly* over the knife-edges and its motion is opposed by a horizontal force of friction. But since the depression of the centre of the bar for a given load is practically the same whether the bar is being gradually loaded or unloaded the effects of this horizontal force of friction are small and can be neglected.

3.12. Torsion of a cylinder. Let a cylinder of length l and radius r be fixed at one end and twisted at the other by means of a couple of *moment* τ, its axis coinciding with the axis of the cylinder. Each cross-section of the cylinder is rotated about the axis of the cylinder by an amount proportional to its distance from the fixed end and to the value of the applied couple, the angle of twist being greatest at the free end and decreasing as the fixed end is approached. The twisting produces no change in either the length or the radius of the rod, and the value of twist for any cross-section of the rod is independent of the direction of the couple applied since reversing the direction of the couple will only twist the rod in the opposite direction. This, therefore, is an example of pure shear. But the shear is *not uniform* for, in a cross-section of the rod, the displacement of any radius is greatest at the rim and decreases to zero at the centre.

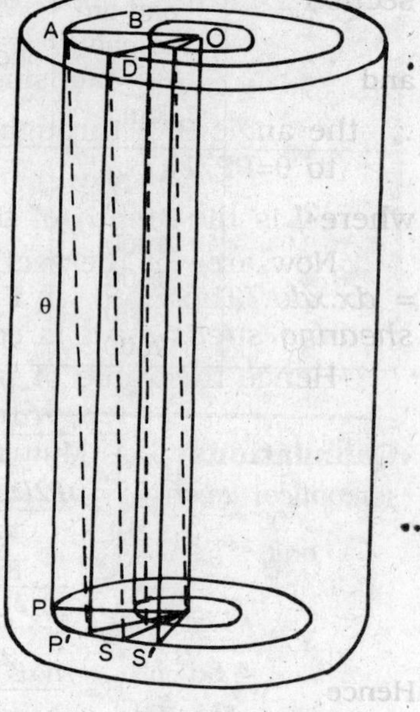

Fig 3.12

Let the cylinder be divided into a *very large* number of *coaxial* hollow cylinders, each of a very small thickness and consider a hollow cylinder of outer and inner radii x and $(x - dx)$ respectively.

This cylinder of thickness dx may be divided into a large number of parallelopipes, a typical one of which contained between the radii AO and DO of the cylinder containing an angle AOD = $d\alpha$ is shown in Fig. 3.13.

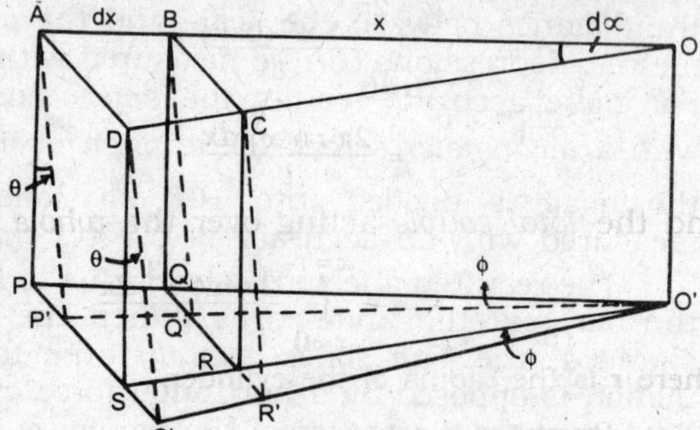

Fig 3.13

The top of this upright is ABCD and the bottom is PQRS. When the cylinder is twisted by a couple applied at the lower end, the upper section fo the upright remains fixed whilst lower section PQRS moves round the axis through an angle ϕ . Let section PQRS be displaced to P′Q′R′S′. Then

$$PO'P' = SO'S' = \phi$$

and $$PP' = x\phi, \text{ since } PP'/x = \phi$$

∴ the angle PAP through which the upright has moved is equal to $\theta = PP'/l = x\phi/l$,

where l is the length of the cylinder.

Now area of the section PQRS or ABCD is equal to PS× PQ = $dx.xd\alpha$. Therefore, if F be the *tangential force* on PQRS, the *shearing stress* on it is equal to $F\ (dx.xd\alpha)$.

Hence the value of *modulus of rigidity* n is given by

$$n = \frac{F/(dx.x.d\alpha)}{\theta}$$

$$= \frac{F/(dx.x.d\alpha)}{x\phi/l}$$

$$= \frac{Fl}{\phi\, x^2 dxd\alpha}$$

Hence $$F = \frac{n\phi x^2 dxd\alpha}{l}$$

The *moment* of this force about the axis of the cylinder is equal to $F.x = n\,\phi x^3 dxd\alpha/l$. Hence the *torsional couple* acting across the small area PQRS is $n\phi x^3 dxd\alpha/l$. Therefore the couple acting over the *annular* portion of the cross-section fo the rod is given by

$$\int_{\alpha=0}^{\alpha=2\pi} \frac{n\phi\, x^3\ dxd\alpha}{l}$$

$$= \frac{2\pi\ n\phi\ x^3\ dx}{l}$$

and the *total couple* acting over the *whole* cylinder is

$$\tau = \int_{x=0}^{x=r} \frac{2\pi\ n\,\phi\ x^3\ dx}{l}$$

where r is the radius of the cylinder.

$$\therefore\ \tau = \frac{2\pi\ n\phi.r^4}{l} = \frac{\pi\ nr^4\phi}{2l} \qquad (3.14)$$

If the cylinder is hollow and the inner and outer radii are r_1 and r_2,

$$\tau = \int_{r_1}^{r_2} \frac{2\pi n\phi\, x^3\, dx}{l} = \frac{\pi n\phi}{2l}\,[r_2^4 - r_1^4]$$

The quantity τ/ϕ is called the *torsional rigidity* or the restoring couple per unit twist due to torsional reaction, and its value in the case of a solid cylinder is equal to $\pi\, nr^4/2l$. By noting down the value of twist for a known couple, Eq. (3.14) can be used to determine the value of rigidity n.

Experiment 3.3

Object. (*a*) To verify that (*i*) the angle of twist for a *given length* of a cylindrical rod or wire is proportional to the couple applied and (*ii*) the angle of twist produced by a *given couple* is proportional to the length of the rod or wire twisted, and

(*b*) To calculate the value of *modulus of rigidity* for the material of the rod or wire ... ... — **statical method.**

Apparatus. Torsion apparatus, meter scale, $\frac{1}{2}$ kilogramme weights, a vernier callipers and a screw gauge.

Horizontal Torsion Apparatus

Description of Apparatus. As illustrated in Fig. 3.14, one end of the rod to be twisted is fixed to a block A and the other end is attached to a steel axle which is fixed to the pulley B. Both the bearing and the block A are firmly fixed to a stout base-board. The couple is applied to the rod by means of load placed in a pan suspended by a cord wound round the pulley B and the twists produced at two points at a known distance apart are measured by pointers P_1 and P_2 which can be clamped to the rod at any point and move over circular scales graduated in degrees.

Theory. Let a load of mass M when placed in the pan *change* the angle of twist by ϕ radians at a distance l from the *fixed end* of the rod. Then, if R is the radius of the pulley over which the cord is wound, the couple acting on the rod is MgR. This couple is balanced by the couple of *reaction due to elasticity* of the rod whose value is $\pi nr^4\phi/2l$. where n is the *modulus of rigidity* of the material of the rod and r its radius. Thus

$$\pi nr^4\phi/2l = MgR$$

whence

$$n = \frac{2MgRl}{\pi r^4\phi}$$

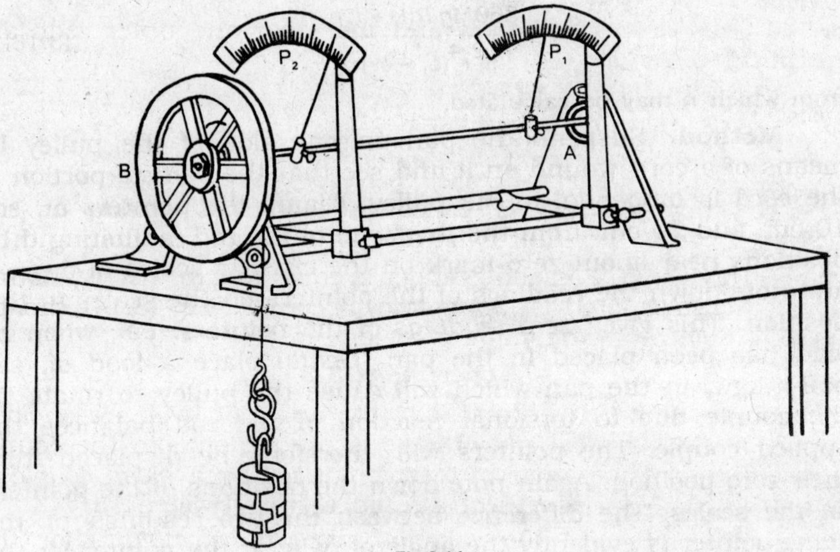

Fig 3.14

If the twist ϕ be measured in degrees, the expression for n is modified to

$$n = \frac{360MgRl}{\pi^2 r^4 \phi} \tag{3.15}$$

The value of n obtained from the above expression is subject to slight error on account of uncertainty of the *exact* position of the end of the rod inside the block A in which it is fixed. This error is, however, eliminated by determining the twists at two points of the rod at a *known* distance apart. For, if $(l_1 + x)$ and $(l_2 + x)$ are the actual lengths of the rod from its fixed end twisted through angles ϕ_1 and ϕ_2 respectively, we have from Eq. (3.15).

$$n = k\frac{(l_1 + x)}{\phi_1}$$

where $k = 360\ Mg\ R/\pi^2 r^4$

or $$n\phi_1 = k(l_1 + x)$$

similarly $$n\phi_2 = k(l_2 + x)$$

Subtracting, we get

$$n(\phi_1 - \phi_2) = (l_1 - l_2)$$

$$\therefore \quad n = k\frac{l_1 - l_2}{\phi_1 - \phi_2}$$

or
$$n = k\frac{360Mg\,R(l_1 - l_2)}{\pi^2 r^4(\phi_1 - \phi_2)} \tag{3.16}$$

from which n may be calculated.

Method. Suspend the pan on one side of the pulley by means of a cord wound on it and see that the *vertical* portion of the cord is *tangential* to the pulley. Clamp the pointers at, say 10 cm. and 20 cm. from the *fixed end* of the rod, adjusting their positions near about zero mark on the circular scales of degrees and note down the readings of the pointers on the scales in this position. This gives *zero readings* of the pointers, *i.e.*, when no load has been placed in the pan. *Gently* place a load of, say half-a-kgw, in the pan which will cause the pulley to rotate till the couple due to torsional reaction of the rod balances the applied couple. The pointers will, therefore, be displaced from their zero position. Again note down the positions of the pointers on the scales. The difference between the two readings of the *same* pointer is evidently the angle of twist at the pointer for the couple due to the load in the pan.

Now gradually increase the load in the pan in *equal* steps of half-a-kgw, till the maximum permissible load is reached taking down the readings of the pointers on the scales after each addition of load. Next decrease the load in the pan in the *same* stages to zero noting down the readings of the pointers on the scales after the removal of each load. The readings of the *same* pointer on the scale for a certain load in the pan taken with increasing and decreasing load should agree very closely. If they do not, *the rod was not clamped sufficiently firmly between the clamps so that when the load was placed in the pan it turned a little in the clamp.* In this case the readings will have to be rejected and a new set taken after having clamped the rod firmly. From a satisfactory set of observations thus obtained deduce the values of the angle of twist for various loads in the pan by subtracting the mean of the two readings for zero load from the mean of the two readings for any load.

Next suspend the pan on the *other* side of the pulley and repeat the above observations. Find out the mean values of twist for various loads from these two sets of observations for the same value of length of the rod. *These values of twist will be free from any error due to slight wrong centring of the pulley on the axis of the rod.*

Now measure the length of the rod from the fixed end upto

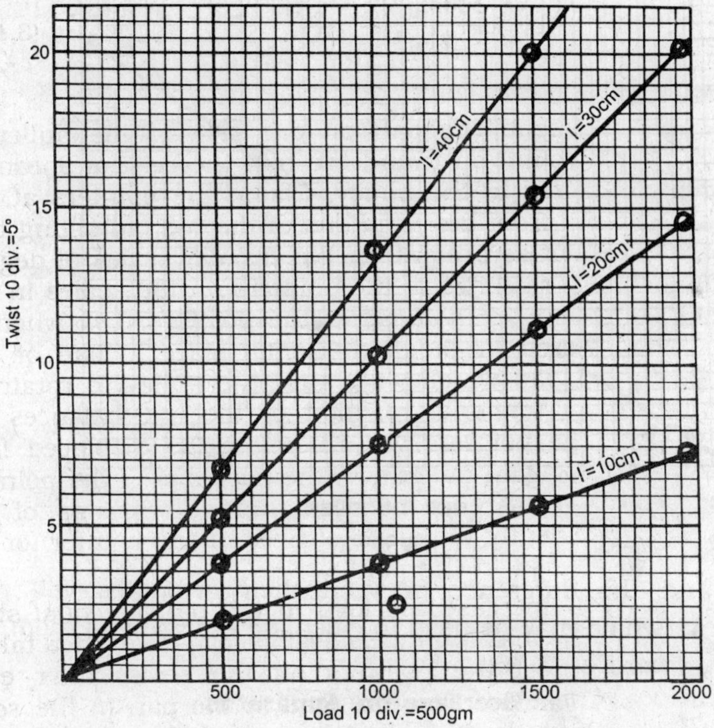

Fig. 3.15

the pointers and alter the positions of the pointers on it and take similar sets of observations for at least *three* other positions of the pointers, say, at 30 or 40 cm. from the fixed end.

Next find out the *pitch*, the *least count* and the *zero error* of a screw gauge and with it measure the diameter of the rod. Now measure the diameter of the pulley with callipers and find out its mean radius R.

Next plot a graph (Fig. 3.15) between the corresponding values of twist ϕ and load M for *each* value of the length of the rod, taking loads as abscissae and the angles of twist as ordinates. *The graph between* ϕ *and* M *for each value of* l *will be a straight line passing through the origin.* From the graph, find out the *best* value for ϕ/M for *each* value of l.

Next plot another graph (Fig. 3.16) showing the corresponding values of ϕ/M and l. Here again *the points should lie on a straight line passing through the origin.* Select two points on this graph. The difference between their abscissae will give

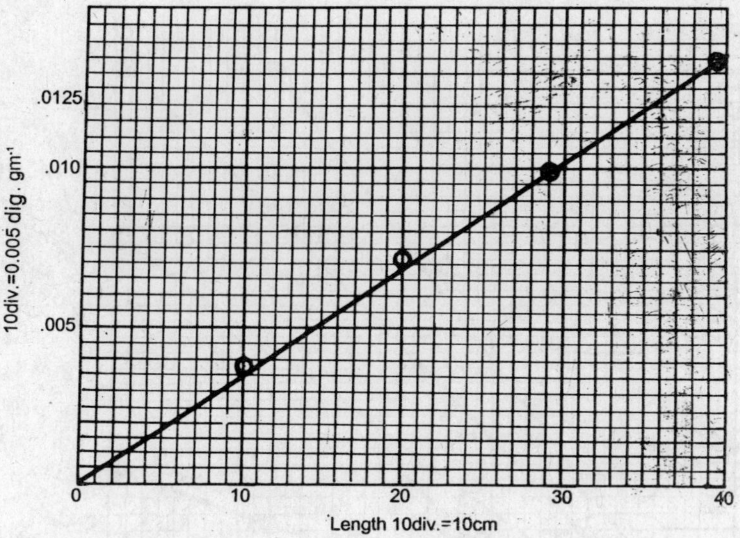

Fig 3.16

$(l_1 - l_2)$ and the difference between their ordinates will give $(\phi_1 - \phi_2)/M$ which may be used to calculate the value of modulus of *rigidity from* Eq. (3.16).

Vertical Torsion Apparatus

Description of Apparatus. This apparatus is due to Barton and is illustrated in Fig. 3.17. A wire of the material to be tested is fixed at the top to a rigid support and at the bottom to a *heavy* cylinder C of brass. *The cylinder* being heavy keeps the wire vertical. Two parallel flexible threads leave opposite sides of the cylinder *tangentially*, and passing over two small *frictionless* pulleys A and B, carry at the other ends pans of *equal* weights. *Equal* loads are placed in the two pans and the twists produced at three known distances from the fixed end of the wire are measured by means of *double-end* horizontal pointers P_1, P_2 and P_3 which move over scales graduated in degrees.

Theory. If the load placed in *each* of the pans is *M* and the *change* in twist produced at a distance *l* from the *fixed* end of the wire of radius *r* is $\phi°$, then the couple due to equal loads *M* in the two pans is *Mgd*, where *d* is the diameter of the heavy brass cylinder and the couple due to *torsional reaction* is $\dfrac{\pi n r^2 \phi \pi}{2l \times 180}$. Equating these two couples

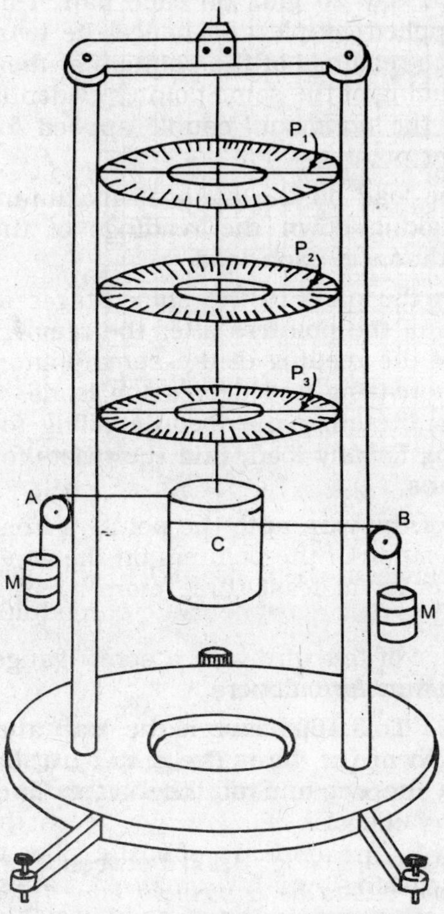

Fig 3.17

$$\frac{\pi^2 n r^4 \phi}{360 l} = Mgd$$

whence $$n = \frac{360\ Mgdl}{\pi^2 r^4 \phi}$$

The uncertainty in the exact *position* of the wire where it is clamped can be eliminated by noting down the twists ϕ_1 and ϕ_2 for two lengths l_1 and l_2. Then evidently the twist for the length $(l_1 - l_2)$ of the wire will be $(\phi_1 - \phi_2)$ and hence the above expression for n becomes

$$n = \frac{360\ Mgd\,(l_1 - l_2)}{\pi^2 r^4 (\phi_1 - \phi_2)}$$

This formula can be used to calculate n.

Method. Level the base of the Barton's apparatus by means of the levelling screws, testing the levelling with a spirit level or a plumb line arrangement if the latter is incorporated in the apparatus. Adjust the pulleys A and B to be at the *same level* and see that the strings leaving the cylinder in opposite directions are *parallel* and also *tangential* to the cylinder and the pulleys. Clamp the pointers at say 40 cm., 60 cm., and 80 cm., from the *fixed* end of the wire, adjusting their positions near about zero mark on the circular scales. Note down the *zero* readings of the pointers at their two ends, *i.e.*, with no load in the pans, and take their mean. *The mean of the two readings at the ends of the same pointer will be free from any error due to the eccentricity of the axis of the wire with respect to the circular scale.*

Next place *gently* a load of, say 20 gm., on each pan. This will change the value of the applied couple and hence the twist in the wire. Again note down the readings of the pointers at their two ends. The change in the reading of the same pointer evidently gives the twist in the wire by the *additional* couple applied for the length of the wire upto *that* pointer.

Now gradually increase the load on the pans till maximum permissible load is reached taking down the readings of the pointers at both ends after addition of each load.

Next decrease the load on the pans in the *same* stages to zero, noting down the readings of the pointers after the removal of each load. Take the mean of the reading of the *same* pointer for various loads taken with increasing and decreasing loads, if they agree closely, and then subtracting the mean reading for zero load from the mean reading for any load, find the twists for various loads placed on the pans.

Now measure the lengths of the wire upto the pointers from its fixed *end* and altering the position of the pointers on the wire, take similar sets of observations for at least three more lengths of the wire.

Next determine the diameter of the wire with a screw gauge and that of the cylinder with a vernier callipers.

Plot a graph as shown in Fig. 3.16, between the load and the twist for *each* value of length of the wire. *The graph will be a straight line passing through the origin.* From the graph find the best value of ϕ/M for each value of l.

Next plot a graph as illustrated in Fig. 3.17, between ϕ/M and l. *The graph will again come out to be a straight line.* Take two convenient points on this line and find out the difference between their abscissae which gives $(l_1 - l_2)$ and also the difference between their ordinates which gives $(\phi_1 - \phi_2)/M$. Then calculate the value of modulus of rigidity from Eq. (3.17).

Sources of error and precautions common to both types of apparatus. (1) The rod or wire should be clamped *firmly* otherwise when loads are placed in the pan, it will turn a little in the clamps and the observations for angle of twist will be wrong.

(2) The *vertical portion* of the string should always be *tangential* to the pulley so that the force due to the load in the pan may always act along a tangent to the pulley.

(3) For all metals Hooke's law holds only for *small* shear and ceases to apply when the shear exceeds 1/200 radians or 1/3°, either because the metal breaks before this shear is reached (*e.g.*, hard steel) or because it blow (*e.g.*) lead. In practire the shear should not exceed 1/600 radians or 1/9°. Hench angle of twist $\phi = l\theta/r < l/9r$ degrees. When l = 50 cm., r = 0.2 cm., ϕ (max.) = 28°.

(4) The masses should be placed or removed from the pan *gently* and the load in the pan should be increased *gradually* to a maximum in a number of *equal* steps and should be decreased in the *same* stages to zero.

(5) As the radius of the rod or wire occurs raised to the fourth power in the expression for n and is a *very small* quantity, the diameter should be measured at a number of points along the length of the rod or wire and at each point two measurements should be taken along the two *mutually perpendicular* diameters.

(6) The graphs whould be smoothly drawn. The representative point should be distributed about the straight lines as evenly as possible and should lie as close to them as possible.

For horizontal torsion apparatus only

(7) The zero position of the pointers on the scales should be adjusted *by placing the pan on the table with no load in it.* In this case the zero positions of the pointers will be the *same* when the cord is on either side of the pulley.

(8) To eliminate any small *error of wrong centring* of the pulley, observations for angles of twist should be taken by twisting the rod in *both* directions and taking the mean of the results.

(9) If the string wound over the pulley is of *appreciable* thickness, half of its thickness whould be added to the radius of the pulley to get the *effective* value of R.

For vertical torsion apparatus only

(10) The base should be levelled so that pillars and the pulleys are vertical.

(11) The strings leaving the cylinder in opposite directions should the *parallel* and *tangential* to the cylinder.

(12) Both ends of the pointer should be read. The mean reading will be free from *error due to eccentricity* of the axis of the wire with respect to the circular scale.

Observations

For horizontal torsion apparatus only

(*A*) *Measurement of angle of twist* ϕ

(*i*) Length of the rod = cm.

S. No.	Load M on the pan gm.	*Cord to the left of pulley* Reading of pointer in degrees				*Cord to the right of pulley* Reading of pointer in degrees				Mean twist ϕ in degrees
		with load increasing x	with load decreasing y	$\frac{x+y}{2}$	Twist	with load increasing x'	with load decreasing y'	$\frac{x'+y'}{2}$	Twist	
1	0									
2	500									
3	1000									
4	1500									
5	2000									
6	2500									

(N. B.—Prepare such tables for various lengths of the rod).

(*B*) Diameter of the pulley (*i*) cm. (*ii*) cm.

(*iii*) cm. (*iv*) cm.

Mean = cm.

For vertical torsion apparatus only

(*A*) *Measurement of angle of twist* ϕ

(*i*) Length of the wire = cm.

S. No.	Load M on the pan gm.	Reading of the pointer in degrees							Mean twist ϕ in degrees
		with increasing load			Twist	with decreasing load			
		Reading at left end x	Reading at right end y	$\frac{x+y}{2}$		Reading at left end x'	Reading at right end y'	$\frac{x'+y'}{2}$	
1	0								
2	20								
3	40								
4	60								
5	80								
6	100								

(*B*) *Diameter of the cylinder* (*i*) cm. (*ii*) cm.

(*iii*) cm. (*iv*) cm.

Mean = cm.

(C) *Measurement of diameter of the rod or wire*

Least count of screw gauge = cm.

Zero error = cm.

S. No.	Reading along any diameter a cm.	Reading along perpendicular diameter b cm.	$\frac{a+b}{2}$ cm.
1. 2. 3. 4. 5. 6.			

Mean uncorrected diameter = cm.

Calculations

From *Graph* (3.16)

Piont on the graph	Length l of the rod or wire in cm.	ϕ/M degrees. gm^{-1}
A B C D		

From the points P and Q on Graph (3.17)

$(l_1 - l_2) =$ cm.

$(\phi_1 - \phi_2)/M =$ degree $gm.^{-1}$

For horizontal torsion apparatus only
- Mean *corrected* radius of the rod = cm.
- Mean radius of the pulley = cm.
- $n = \dfrac{360\ MgR\ (l_1 - l_2)}{\pi^2\ r^4\ (\phi_1 - \phi_2)}$
- = dynes/cm^2 = N/m^2.

For vertical torsion apparatus only
- Mean *corrected* radius of the wire = cm.
- Mean radius of the cylinder = cm.
- $n = \dfrac{360\ Mgd\ (l_1 - l_2)}{\pi^2\ r^4\ (\phi_1 - \phi_2)}$
- = dynes/cm^2 = N/m^2.

Result. The graph between ϕ and M for a given value of l

and between ϕ/M and l are *straight lines* passing tyhrough the origin showing that the angle of twist is proportional to the value of the couple applied, and that if l is varied, the value of twist for a definite couple is proportional to the length of the rod or wire.

The value of modulus of rigidity for the material of the rod or wire (.........) = N/m^2.

Standard value = N/m^2

Error = %

Criticism of the apparatus. Both types of apparatus yield satisfactory results. The angle of twist is practically proportional to the length of the rod or wire as well as to the applied couple. The slight irregularities in the graphs are due to the *non-uniformity* of the material.

The horizontal torsion apparatus suffers from two defects :

(*i*) On account of a single-end pointer error due to eccentricity of the axis of the rod with respect to the circular scale cannot be eliminated.

(*ii*) The apparatus carries only one pulley by means of which only a single force can be applied to the end of the rod. The force produces a side pull on the rod, which hinders the free twisting of the rod.

The apparatus is very suitable for comparatively thick specimens. In that case the diameter being comparatively large is susceptible to accurate measurements.

Both the errors due to eccentricity of the axis of the wire with respect to the circular scale and the side pull on it are eliminated in Barton's vertical torsion apparatus. With this apparatus comparatively thin specimens can be tested. The value of twist for a given couple with it is also much larger than with the horizontal torsion apparatus.

Experiment 3.4

Object. To determine the value of modulus of *rigidity* of the material of a given wire (... ... ...) with *Torsion Pendulum*—**Dynamical method**.

Apparatus. Torsion pendulum, a fairly thin and long wire of the material to be tested, clamps and chucks, a stop-watch, an auxiliary body, say a cylinder, screw gauge, vernier callipers, metre scale, spirit level, a balance and a weight box.

Description of Apparatus. The torsion pendulum is illustrated in Fig. 3.18. One end of a fairly thin and long wire is

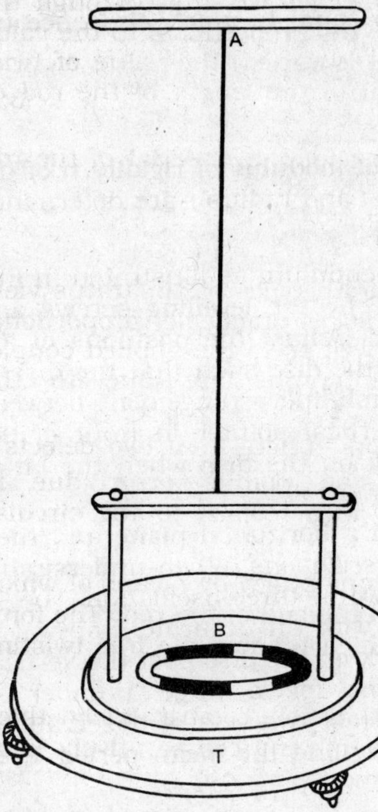

Fig 3.18

clamped at A to a rigid support and the other end is fixed in the centre of a projection coming out of the central portion of a circular dise B, of aluminium or brass. On the upper face of the disc are described concentric circles and a concentric groove is cut in which three balancing weights can be placed. Beneath the disc is a heavy iron table T provided with three levelling screws. The plumb-line arrangement between the disc and the table serves to test the horizontality of the disc.

Theory. When the disc is rotated in a *horizontal* plane and then released, it executes *torsional vibrations* about the wire as the axis. If I_0 be the moment of inertia of the disc with its projection about the wire as the axis, its *period of oscillation* is, from Eq. (2.1). given by

$$T_0 = 2\pi \sqrt{I_0 / c}$$

where c is the *torsional couple* per *unit* radian twist.

If an auxiliary body of *known* moment of inertia I is placed *centrally* upon the disc, then the *period of osciallation* T of the *combination* is given by

$$T = 2\pi \sqrt{\frac{I_0 + I}{c}} \qquad (3.19)$$

Squaring Eqs. (3.18) and (3.19) and then subtracting Eq. (3.18) from Eq. (3.19), we get

$$T^2 - T_0^2 = \frac{4\pi^2 I}{c}$$

But from Eq. (3.14), c is also equal to $\pi nr^4/2l$. Hence the above equ, yields

$$n = \frac{8\pi Il}{(T^2 - T_0^2)\, r^4} \tag{3.20}$$

From this equation the value of modulus of rigidity n of the wire can be calculated, if its length l and radius r are determined and the periods T_0 and T observed.

Method. Set up the torsion pendulum as illustrated in Fig. 3.18. Level the heavy iron table T by the levelling screws and test the levelling with a spirit level. Adjust the positions of the balancing weights in the groove in the disc such that the disc is horizontal as indicated by the plumb-line arrangement between the disc and the table. Place a vertical pointer in front of the disc and just behind it put a mark on the disc when the latter is at rest.

Next rotate the disc slightly in a horizontal plane and then release it. It will perform *torsional oscillations*. When undersirable motions have subsided, say after two or three oscillations, begin timing the oscillations of the disc with an accurate stop-watch by observing the transits of the mark on the disc past the vertical pointer. Determine *twice* the time for a *large* number of oscillations say, 30 and taking at least two sets of observations with different number of oscillations, find the mean period T_0 of the disc.

Now place a right cylinder *centrally* upon the disc and determine, as before, the period T of the combination. Note T will be *greater* than T_0.

Next measure the length of the wire between the clamps with a metre scale and the diameter of the wire at *several* places with a screw gauge. Also weigh the cylinder in a physical balance and find its diameter with vernier callipers.

Calculate the moment of inertia of the cylinder from the formula $I = MR^2/2$ and then the value of modulus of rigidity of the wire from Eq. (3.20).

Sources of error and precautions. (1) The disc should always remain *horizontal* so that its moment of inertia I_0 remains *unaltered* throughout the experiment. Consequently *the balancing weights, when once adjusted, should not be disturbed in subsequent observations for T.*

(2) The motion of the torsion pendulum should be purely

rotational in a horizontal plane. Pendular or up and down motions should be completely checked.

(3) The suspension wire should be free from kinks and should be fairly thin and long, say about 70 cm., and 0.1 cm. thick, so that torsional rigidity may be small and hence the periods of the pendulum *large*.

(4) The wire should not be twisted beyond elastic limit otherwise the torsional couple will not be prooprtional to the value of the twist.

(5) The auxiliary body (cylinder) should be of uniform density throughout *e.g.*, of brass, and should be placed *centrally* upon disc so that its axis is coincident with the axis of the suspension wire.

(6) As the periods occur raised to the *second* power in the expression for *n*, they must be measured very accurately by timing a large number of oscillations with a stop-watch reading upto 0.1 sec.

(7) As the radius of the wire occurs raised to the *fou th* power in the expression for *n* and is a *small* quantity, the diameter of the wire must be measured very accurately with a screw gauge. Readings should be taken at several points along the length of the wire and at each point two *mutually perpendicular* diameters should be measured. The diameter of the cylinder should be similary measured with a vernier callipers.

Observations

(*A*) *Determination of periods T and T_0*

Least count of stop-watch = sec.

S. No.	No. of oscillations	Determination of T_0			Determination of T		
		Time taken		T_0 sec.	Time taken		T sec.
		min.	sec.		min.	sec.	
Mean					Mean		

(*B*) (*i*) Length of the wire = cm.

(*ii*) Measurement of diameter of the wire

L. C. of the screw gauge = cm.

Zero error = cm.

S. No.	Reading along any diameter *a* cm.	Reading along perp. diameter *b* cm.	$\frac{a+b}{2}$ cm.
	Mean *uncorrected* diameter		

(*C*) (*i*) *Mass of the cylinder* = gm.

(*ii*) *Measurement of diameter of the cylinder*

Vernier constant of the callipers = cm.

Zero error = cm.

Diameter of the cylinder (*i*) cm. (*ii*) cm. (*iii*) cm. (*iv*) cm.

Mean diameter = cm.

Calculations

Mean *corrected* diameter of the wire = cm.

∴ Radius of the wire = cm.

Mean *corrected* diameter of the cylinder = cm.

∴ Radius of the cylinder = cm.

$$I = \frac{MR^2}{2}$$

$$=$$

$$= \quad \text{gm.} \times \text{cm.}^2$$

$$n = \frac{8\pi I l}{(T^2 - T_0^2)\, r^4}$$

$$= \text{dynes/cm}^2 = \quad \text{N/m}^2$$

Result. The *modulus of rigidity* of the material of the given wire (... ....)

$n =$ N/m^2.

Standard value = N/m^2.

∴ error = %

Criticism of the Method. The method yields quite satisfactory results. For still greater accuracy the periods should be determined with a chronometer and the oscillations observed through a telescope. There is no restriction on the amplitude of oscillation, particularly when the wire is fairly thing and long, provided the elastic limit is not exceeded.

The method suffers from two defects :

(*i*) The torsional rigidity c does not remain constant as the radius of the wire changes with the load carried at the end of the wire.

(*ii*) The moment of inertia I of the auxiliary body is calculated from its geometerical shape on the assumption of *uniform* density of the material throughout, but this condition is not complied with absolutely in practice since even the cylinders of rolled brass are not perfectly homogeneous.

Both these sources of error are, however, eliminated by the use of a body of special construction called Maxwell's *needle.*

Experiment 3.5

Object. To determine the value of *modulus of rigidity* of the material of a given wire (..........) with *Maxwell's needle*—**dynamical method.**

Apparatus. Maxwell's needle, a fairly thin and long wire of the material to be tested, clamps and chucks, screw gauge, a metre scale, a stop-watch, a balance and a weight box.

Description of Apparatus. As illustrated in Fig. 3.19, Maxwell's needle consists of a hollow brass tube of length L

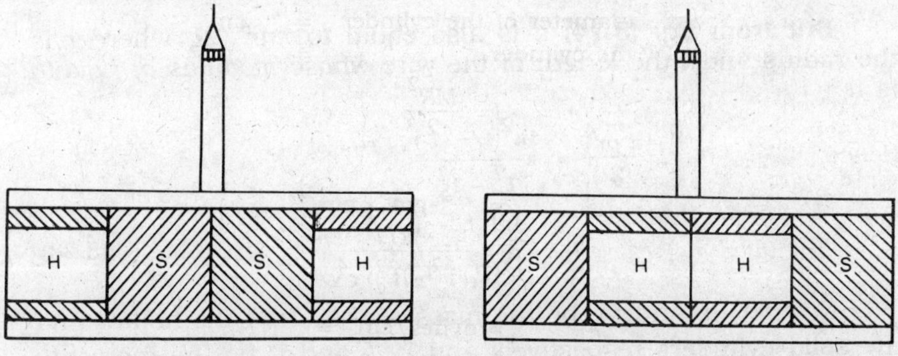

Fig 3.19

suspended horizontally by a vertical wire attached to its middle point and carrying within it four cylinders made of *rolled brass* and *each* of length $L/4$. The four cylinders are *identical* in dimensions. The two of them are *solid* and equal in mass, and the remaining two are *hollow* and also equal in mass.

Theory. Let the two hollow cylinders be placed in the middle and the solid ones at the two ends of the tube (Fig. 3.19*a*) and let the combination be slightly rotated in a horizontal plane and then released. The body will then execute S. H. M., about the wire as the axis and the *period of oscillation* is given by

$$T_1 = 2\pi\sqrt{I_1/c} \tag{3.21}$$

where I_1 is the *moment of inertia of combination* about the wire as the axis and c the *restoring* couple per *unit* twist due to *torsional reaction.*

Now let the positions of the hollow and solid cylinder be interchanged so that the solid cylinders are now in the middle (Fig. 3.19 *b*). Then, if I_2 is the moment of inertia of the *new* combination about the axis of rotation, the *new period of oscillation* is given by

$$T_2 = 2\pi\sqrt{I_2/c} \tag{3.22}$$

Squaring Eqs. (3.21) and (3.22) and subtracting Eq. (3.22) from Eq. (3.21), we get

$$T_1^2 - T_2^2 = \frac{4\pi^2}{c}(I_1 - I_2)$$

whence

$$c = \frac{4\pi^2(I_1 - I_2)}{T_1^2 - T_2^2}$$

But from Eq. (3.14) c is also equal to $\pi nr^4/2l$, where r is the radius and l the length of the wire whose *modulus of rigidity* is n

Hence

$$\frac{\pi nr^4}{2l} = \frac{4\pi^2(I_1 - I_2)}{T_1^2 - T_2^2}$$

or

$$n = \frac{8\pi l(I_1 - I_2)}{r^4(T_1^2 - T_2^2)} \tag{3.23}$$

Now let m_1 and m_2 be the masses of each of the hollow and the solid cylinders respectively and I_0, I' and I'' be the moments of inertia of the hollow tube, the hollow cylinder and the solid cylinder respectively about a vertical axis passing through their *middle* points. Then, if L is the length of the *hollow* tube,

$$I_1 = I_0 + 2I' + 2m_1\left(\frac{L}{8}\right)^2 + 2I'' + 2m_2\left(\frac{3L}{8}\right)^2$$

and $$I_2 = I_0 + 2I' + 2m_2\left(\frac{L}{8}\right)^2 + 2I' + 2m_1\left(\frac{3L}{8}\right)^2$$

Hence $$I_1 - I_2 = 2m_2\left[\left(\frac{3L}{8}\right)^2 - \left(\frac{L}{8}\right)^2\right] - 2m_1\left[\left(\frac{3L}{8}\right)^2 - \left(\frac{L}{8}\right)^2\right]$$

$$= (m_2 - m_1)L^2/4$$

Putting this value of $(I_1 - I_2)$ in Eq. (3.23)

$$= \frac{2\pi l\,(m_2 - m_1)L^2}{r^4\,(T_1^2 - T_2^2)} \tag{3.24}$$

This expression can be used to find the value of modulus of rigidity n of the material of the suspension wire.

Method. Suspend a wire of about 50 cm, length and 0.1 cm. in diameter of the material whose modulus of rigidity is required from a rigid support, and to its lower end fasten with a suitable chuck and a clamp the hollow tube of the Maxwell's needle at its *middle* point so that it may remain *horizontal.* Place all the four cylinders of the Maxwe'l's needle inside the hollow tube, the two hollow cylinders being in the middle and *no portion of the solid cylinder should be projecting outside the tube.* Then fix a light pointer in the middle of the needle and place a strip of mirror with a straight line mark below the pointer in the equilibrium position.

Rotate the Maxwell's needle slightly in a horizontal plane and then release it, it will perform *torsional oscillations.* Begin timing the oscillations of the Maxwell's needle with an accurate stop-watch by observing the transits of the pointer past the mark on the mirror when viewed vertically. Determine *twice* the time for a large number of oscillations, say 30 and take at least two sets of observations with different number of oscillations and thus find out the mean value of T_1 .

Now interchange the positions of the hollow and solid cylinders fitting them properly inside the tube. Again set the Maxwell's needle oscillating in a horizontal plane and determine the period T_2 as before. T_2 will be *less* than T_1 .

Next measure the length of the hollow tube and that of the wire between the two clamps by a metre scale. Then measure with a screw gauge the diameter of the wire. Now to find out $(m_2 - m_1)$ put one solid cylinder in the left pan of a physical balance and one hollow cylinder in its right pan, and by putting *extra* weights in the right pan, find the difference between the masses of the two cylinders. Similarly determine the difference between the masses of the other two cylinders. From these two weighings, find out the mean value of $(m_2 - m_1)$. If a balance of high capacity and suitable sensitivity is available, the value of $(m_2 - m_1)$ may be found by a single weighing. Finally calculate the value of modulus of rigidity of the material of the wire from Eq. (3.24).

Sources of error and precautions. (*i*) The two sets of cylinders should be exactly *identical* and the hollow tube should be clamped exactly in the middle.

(2) The Maxwell's needle should always remain *horizontal* so that the moment of inertia of the hollow tube about the axis of rotation remains unaltered throughout the whole experiment. Hence while placing the cylinders inside the tube, no portion of them should be left projecting outside the hollow tube.

(3) The motion of the Maxwell's needle should be purely rotational in a horizontal plane. All undesirable motions (up and down, or pendular) should be completely checked.

(4) As in the expression for *n* the periods occur raised to the *second* power, they must be carefully measured by timing a *large* number of oscillations with an accurate stop-watch upto an accuracy of say, 0.1 of a second.

(5) The wire should not be twisted beyond elastic limit otherwise the restoring couple due to *torsional reaction* will not be proportional to value of the twist.

(6) There should be no kinks in the wire. The wire should be fairly *long* and *thin* particularly when the rigidity is high so that the restoring couple per unit twist due to torsional reaction may be small and hence the period of oscillation of the Maxwell's needle is large.

(7) In the expressional for *n* the radius occurs raised to the *fourth* power and is a *very small* quantity usually of the order of 0.1 cm. Hence the diameter must be measured very accurately. Readings should be taken at *several* points equally spaced along the wire and two diameters at *right angles* to each other should be measured at each point, care being taken not to compress the wire in taking the readings.

Observations

(A) *Determination of* T_1 and T_2

Least count of stop-watch = sec.

S. No.	No. of oscillations	Determination of T_1			Determination of T_2		
		Time taken		T_1 Sec.	Time taken		T_2 Sec.
		Min.	Sec.		Min.	Sec.	
1 2 3 4 5							
			Mean			Mean	

(B) (i) *Length of the wire* = cm.

(ii) *Measurement of diameter of the wire*

Least count of screw gauge = cm.

Zero error = cm.

S. No.	Reading along any diameter a cm.	Reading along *Perpendicular diameter* b cm.	$\frac{a+b}{2}$ cm.
1 2 3 4			
	Mean *uncorrected* diameter		

Mean *corrected* diameter = cm.

(C) (i) *Length of the hollow tube* = cm.

(ii) *Determination of* $(m_1 - m_2)$

Difference between the masses of one solid and one hollow cylinder = gm.

Difference between the masses of *other* pair of solid and hollow cylinder = gm.

Mean value of $(m_2 - m_1)$ = gm.

Calculations

Mean radius of the wire = cm.

$$n = \frac{2\pi l\,(m_2 - m_1)\,L^2}{r^4(T_1^2 - T_2^2)}$$

$=$

$=$ dynes/cm^2 $=$ N/m^2.

Result. The value of modulus of rigidity for the material of the wire (...........)

$=$ N/m^2

Standard value $=$ N/m^2

Error $=$ %

Criticism of the apparatus. The apparantus yields results better than those obtainable with torsion pendulum. The load suspended at the end of the wire does not change during the performance of the experiment, and hence the torque per unit radian twist due to torsional reaction fo the wire remains constant throughout the experiment. The evaluation of $(I_1 - I_2)$ reduces to the determination of the difference between the masses of a solid and a hollow cylinder and the length of the hollow tube of the Maxwell's needle, which can be done easily and accurately. The accuracy of result *chiefly* depends upon

(*i*) The accuracy in the measurement of r,

(*ii*) The largeness of difference between T_1 and T_2 and the accuracy in their measurement.

To make $(T_1 - T_2)$ large, the hollow tube of the Maxwell's needle is made of small moment of inertia and the difference in the masses of solid and hollow cylinders is kept large.

It may be noted that as the wires are made by drawing the metal through a small circular aperture, the outer layers of the wire are tough and close-grained and therefore possess a higher rigidity than that of the inner core. Hence *the value of rigidity determined from experiments on a fine wire is greater than that determined on a thick wire of the same material.*

Further using a given wire, the *value of rigidity determined dynamically is generally somewhat larger than the value determined by a statical method.* This is due partly to the fact that a wire, under the action of a torque, does not at once acquire its final twist but the twist increases with time. With the dynamical method the torque is applied for a brief period as compared with the statical method and consequently the torque per unit twist and hence the value of the rigidity in the dynamical method has a larger value than that in the statical method in which the torque is applied for a longer time.

Experiment 3.6

Object To determine the value of *Y*, *n* and σ for the material of a wire (..........) by *Searle's method.*

Apparatus. Two equal inertia bars made of brass about 30 cm. long and of square cross section of about 2 cm. wide with the necessary screws and clamps, a wire of about 30 cm. length and about 1 mm. diameter of the material to be tested, an accurate stop watch, a screw gauge, a vernier callipers, thread, metre scale, a weight box and a balance.

Theory. Referring to Fig. 3.20 two equal *inertia bars* AB and CD of square section are joined at their centres by a fairly

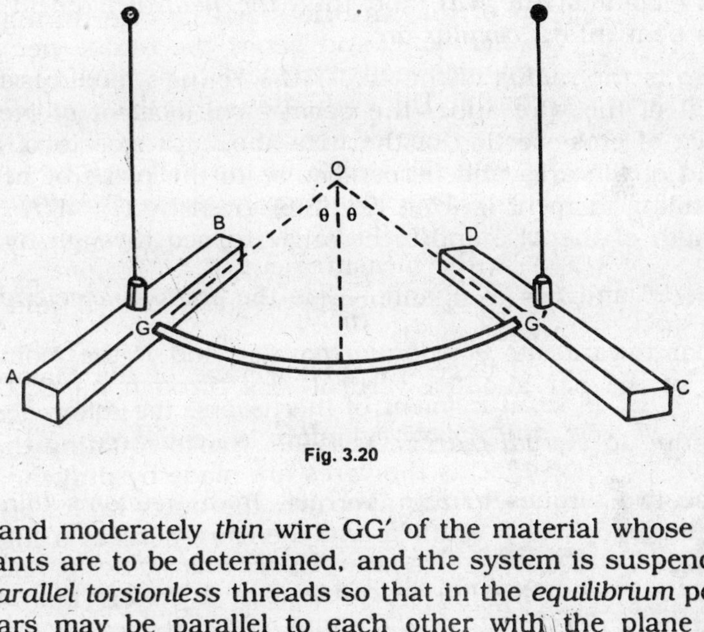

Fig. 3.20

short and moderately *thin* wire GG′ of the material whose elastic constants are to be determined, and the system is suspended by two *parallel torsionless* threads so that in the *equilibrium* position the bars may be parallel to each other with the plane ABDC *horizontal.* If the two bars be turned through *equal* angles in opposite directions and be then set free, the bars will execute *flexural vibrations* in a *horizontal* plane with the *same* period about their supporting threads.

When the amplitude of vibration is *small,* the wire is only slightly bent and the distance GG′ between the ends of the wire measured along the straight line will never differ perceptibly from the length of the wire so that the distance between the lower ends of the supporting threads remains practically constant and hence the threads remain vertical during the oscillations of the bars, and there is thus no horizontal component of tensions in the threads acting on the wire.

The mass of the wire is negligible compared with that of the bars so that the motion of G and G′ at right angles to GG′ may be neglected. Further, since the horizontal displacements of G and G′ are very small compared with the lengths of the supporting threads, the vertical motion of G and G′ is also negligible. The centres of gravity of the bars, therefore, remain at rest, and hence the action of the wire on either bar and vice versa is simply a couple which, by symmetry, must have a vertical axis. The moment of this couple called the "*bending moment*" is the same at every point of the wire and thus the *neutral filament* of the wire is bent into a *circular arc*.

If ρ is the radius of the arc, Y the Young's modulus for the material of the wire and I the *geometrical moment of inertia* of the area of cross-section of the wire about an axis through the centroid of the area and perpendicular to the plane of bending, the bending moment is from Eq. (3.6) given by $G = YI/\rho$. If l is the length of the wire and θ the angle turned through by *either* bar, $\rho = l/2\theta$ and $G = \frac{2YI\theta}{l}$ and $\frac{d^2\theta}{dt^2}$ is the angular acceleration of each bar towards its *equilibrium position* and K the moment of inertia of the bar about a vertical axis through its *C. G.*, the torque due to *inertial reaction is* $K\frac{d^2\theta}{dt^2}$. Hence equating the sum of these two torques to zero, we get, from *Newton's third law*, the equation

$$K.\frac{d^2\theta}{dt} + \frac{2YI\theta}{l} = 0$$

or

$$\frac{d^2\theta}{dt^2} + \frac{2YI\theta}{Kl} = 0$$

The motion of the bars is, therefore, *simple harmonic* and hence the *period of flexural vibrations* is given by

$$T_1 = 2\pi\sqrt{\frac{Kl}{2YI}}$$

whence

$$Y = \frac{2\pi^2 Kl}{T_1^2 I}$$

If the radius of the wire be r, $I = \frac{1}{4}\pi r^4$ and hence

$$Y = \frac{8\pi K l}{T_1^2 r^4} \tag{3.25}$$

If the length and breadth (horizontal) of the bar be a and b respectively.

$$K = M\left(\frac{a^2 + b^2}{12}\right) \tag{3.26}$$

where M = mass of the bar.

Now the suspensions of the bars are removed and one of the bars is fixed horizontally on a suitable support (Fig. 3.21), while the other is suspended from the now vertical wire. If the wire is twisted through an angle and the bar allowed to execute torsional oscillations, the period of oscillations is given by

$$T_2 = 2\pi\sqrt{K/c}$$

where c is the restoring couple per *unit* radian twist due to *torsional reaction* of the wire and is equal to $\pi n r^4/2l$, where n is the modulus of rigidity for the material of the wire. Thus

$$c = \frac{\pi n r^4}{2l} = \frac{4\pi^2 K}{T_2^2}$$

whence

$$n = \frac{8\pi K l}{T_2^2 r^4} \tag{3.27}$$

Dividing Eq. (3.25) by Eq. (3.27), we get

$$\frac{Y}{n} = \frac{T_2^2}{T_1^2} \tag{3.28}$$

Now $Y = 2n(1 + \sigma)$, where σ is the Poisson's ratio. Hence

$$\sigma = \frac{Y}{2n} - 1$$

$$\sigma = \frac{T_2^2}{2T_1^2} - 1 \tag{3.29}$$

Method. Take two cotton threads of about 60 cm. length and by means of them suspend the two inertia bars from a rigid support so that the axis of the bars and the wire, when the system is at rest, lie in the same horizontal plane as illustrated in Fig. 3.21. Then set up a pointer close to the end of one of the bars and make a mark on the bar in line with the pointer when the bar is at rest.

Now draw the two ends B and D of the bars *slightly* together by a loop of thin cotton thread slipped over them and bring the system thus constrained carefully to rest. Then burn the thread, the bars will be released and commence to oscillate regularly. The amplitude of vibration should be small, about 3°. Note down *twice* with an accurate stop-watch the time for a large number of oscillations, say 20 and find the value of the period T_1. Take at least two sets of observations for T_1 with different number of oscillations and from them calculate the mean value of T_1.

Next remove the cotton threads supporting the bars and clamp one of the bars to a suitable support as shown in Fig. 3.21 so that the wire is vertical. Twist the wire by rotating the bar at its end in a horizontal plane and then release it. The bar will execute *torsional oscillations*.

Twice note down the time taken by the bar to complete again a *large* number of oscillations, say 20, and from it calculate the value of T_2. Take at least two sets of observations for T_2 with different number of oscillations and from them find out its mean value.

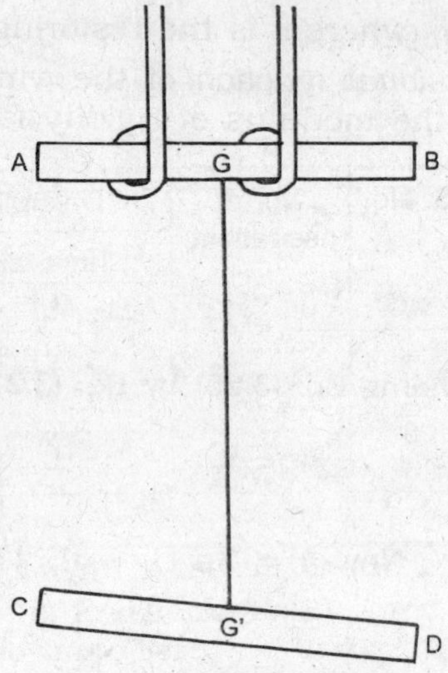

Fig 3.21

Now measure with a screw gauge the diameter of the wire and thus find out the mean radius *r*. Next measure the length of the wire and the lengths of the inertia bars by a scale. Now measure the breadth of each bar with vernier callipers. Finally weigh the bars separately in a balance and determine the masses. *For very accurate work mass of each bar is found before the hole is bored in it and before the hook is fixed to it, and the value of this mass is stamped on the bar.*

Calculate the moments of inertia of each bar from Eq. (3.26) and thus find the *mean* value of *K*. Having determined *K*, *l*, *r* and the periods T_1 and T_2, calculate the values of *Y*, *n* and σ from Eqs. (3.25), (3.27) and (3.29).

Sources of error and precautions. (1) While taking observations for period T_1 of *flexural vibrations*, the amplitude of oscillation should be *small*. In that case threads will remain vertical and there will then be no horizontal component of the tension in the threads acting on the wire. Further the wire will also be not strained beyond elastic limit.

(2) All undesirable motions (up and down or pendular) should be completely checked.

(3) The periods occur raised to the *second* power in the expression for *Y*, *n* and σ. Hence they must be determined very accurately by timing a *large* number of oscillations with an accurate stop-watch upto the smallest value measurable with it.

(4) While determining period T_2 for *torsional oscillations*, the amplitude of oscillation may be kept *quite large*, but the wire *should not be twisted beyond elastic limit.*

(5) Since the radius of wire occurs in the *fourth* power in the formulae for calculation of elastic constants and is a *small* quantity, readings for diameter should be taken with a screw gauge at a number of points equally spaced along the wire, two *perpendicular* diameters being measured at each place taking care not to compress the wire while making measurements.

Observations

(A) *Determination of* T_1 and T_2

Least count of stop-watch = sec.

S. No.	No. of oscillations	Determination of T_1			Determination of T_2		
		Time taken		T_1 Sec.	Time taken		T_2 Sec.
		Min.	Sec.		Min.	Sec.	
1 2 3 4							
		Mean			Mean		

(B) (*i*) *Length of the wire* = cm.

(*ii*) *Measurement of diameter of the wire*

Least count of screw gauge = cm.

Zero error = cm.

S. No.	Reading along any diameter *a* cm.	Reading along any diameter *a* cm.	$\frac{a+b}{2}$ cm.
Mean *uncorrected* diameter			

Mean *corrected* diameter = cm.

(C) (i) *Mass of the inertia bar* = cm.

(ii) *Length of the inertia bar* = cm.

(iii) *Breadth of the inertia bar* = cm.

(N. B.—Record the mass and dimensions of the second inertia bar also, if the two bars are not exactly identical).

Calculations. *Mean* radius of the wire = cm.

$$K = M\left(\frac{a^2 + b^2}{2}\right)$$

=

= gm. × cm.^{2}

$$Y = \frac{8\pi Kl}{T_1^2 r^4}$$

=

= dynes/cm^2 = N/m^2

$$n = \frac{8\pi Kl}{T_1^2 r^4}$$

=

= dynes/cm^2 = N/m^2

and $$\sigma = \frac{T_2^2}{2T_1^2} - 1$$

=

=

Results. The values of *elastic constants* for the material of the wire (................) are

Y = N/m^2

n = N/m^2

and σ =

Criticism of the apparatus. The second method of determining Young's modulus with Searle's apparatus using micrometer screw for measurement of extension requires the use of a wire several metres long while in determining modulus of rigidity by, Maxwell's needle only a short length of the wire is used. In Searle's method for the determination of the elastic constants using inertia bars, the values of both Y and n can be determined for the *same short* length of a wire so that the value of bulk modulus obtained from these values of Y and n will be more accurate.

Further, the value of σ can be independently determined without any knowledge of the values of Y and n. The ratio of Y and n and the value of σ have been expressed in terms of the two very accurately determinable quantities T_1 and T_2 and a value of the radius of the wire which is a very serious source of error is not required.

The value of T_1 is, however, subject to some error as the undesirable motions of the bars cannot be completely checked.

Experiment 3.7

Object. To determine the restoring force per unit extension of a spiral spring by *statical* and *dynamical* methods and also to determine the mass of the spring.

Apparatus. A spiral spring, a pointer and a scale-pan, half-a-kilogramme slotted weights and a stop-watch.

Theory. Let a spiral spring, suspended vertically from a rigid support, be stretched vertically downward by a load of mass M applied at its lower end. If the extension of the spring is *small*, the force of *elastic reaction* F is, from *Hooke's law*, proportional to the extension l, or

$$F = Kl$$

where K is a constant giving a measure of the *stiffness* of the spring and is called the *restoring force per unit extension* of the spring.

But in the *equilibrium state*.

$$F = Mg$$

$\therefore$ $$Kl = Mg$$

or $$K = \frac{Mg}{l} \quad (3.30)$$

from which K may be calculated, if the extension l for a known load M at the end of the spring is determined.

If the mass at the end of the spring is now displaced vertically downward and then released, then for *small* oscillations, the *restoring force* at any instant is proportional to the displacement x, *i.e.*, $F = Kx$. If $\frac{d^2x}{dt^2}$ is the acceleration of the mass at that instant, the *inertial reaction* of the system is $(M + m)\frac{d^2x}{dt^2}$; where m is the *effective* mass of the spring. Equating the sum of these forces to zero we get from *Newton's third law of motion*, the equation

$$(M + m)\frac{d^2x}{dt^2} + Kx = 0$$

or
$$\frac{d^2x}{dt^2} + \frac{K}{M+m}x = 0$$

This equation represents a *S. H. M.* whose *period* is given by

$$T = 2\pi\sqrt{\frac{M+m}{K}} \qquad (3.31)$$

If the experiment is performed, with two masses M_1 and M_2 suspended successively at the end of the spring and the respective periods are T_1 and T_2 we have

$$T_1 = 2\pi\sqrt{\frac{M_1 + m}{K}} \qquad (3.32)$$

and
$$T_2 = 2\pi\sqrt{\frac{M_2 + m}{K}} \qquad (3.33)$$

Squaring Eqs. (3.32) and (3.33) and subtracting, we get

$$T_1^2 - T_2^2 = \frac{4\pi^2}{K}(M_1 - M_2)$$

whence
$$K = \frac{4\pi^2(M_1 - M_2)}{T_1^2 - T_2^2} \qquad (3.34)$$

from which K may be calculated.

Squaring Eqs. (3.32) nd (3.33) and then dividing Eq. (3.32) by Eq. (3.33) we have

$$\frac{T_1^2}{T_2^2} = \frac{M_1 + m}{M_2 + m}$$

whence on simplification,

$$m = \frac{M_1 T_2^2 - M_2 T_1^2}{T_1^2 - T_2^2}$$

It can be shown that the mass of the spring m' is *three* times the *effective* mass m of the spring.

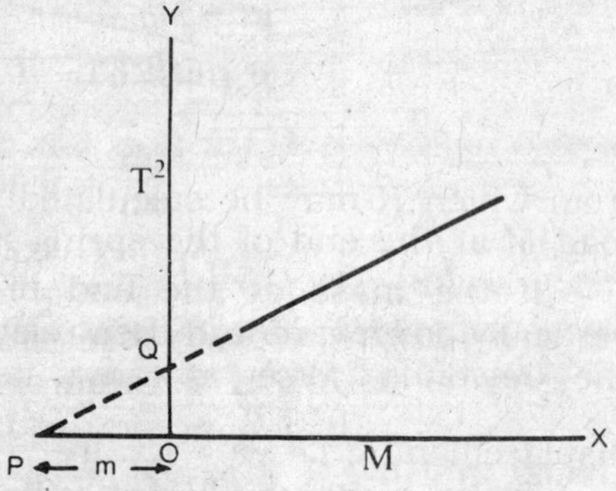

Fig 3.22

$$\therefore \text{ Mass of the spring } m' = 3\left[\frac{M_1 T_2^2 - M_2 T_1^2}{T_1^2 - T_2^2}\right] \qquad (3.35)$$

Squaring and rearranging Eq. (3.31) gives

$$T^2 = \left(\frac{4\pi^2}{K}\right) M + \left(\frac{4\pi^2}{K}\right) m \qquad (3.36)$$

A plot of T^2 against M gives a *st. line* as shown in Fig. 3.22. The *–ve* intercept OP on the X-axis equals m and may be used to find m graphically.

Method. Suspend *vertically* a spiral spring A from a rigid support as shown in Fig. 3.23 and fasten at its lower end a small scalepan B. Attach a small *horizontal* pointer P to the scale-pan and set up a millimeter scale S in front of it with the pointer *just not* touching the scale.

Statical experiment. Note down the *zero* reading of the pointer, *i.e.*, with no load in the scale-pan. Place *gently* in the pan a load of, say, half-a-kilogramme. This stretches the spring *slightly* and the pointer will move down on the scale. Again note down the reading of the pointer on the scale. The difference between the two readings of the pointer on the scale evidently gives the extension fo the spring for the load in the pan.

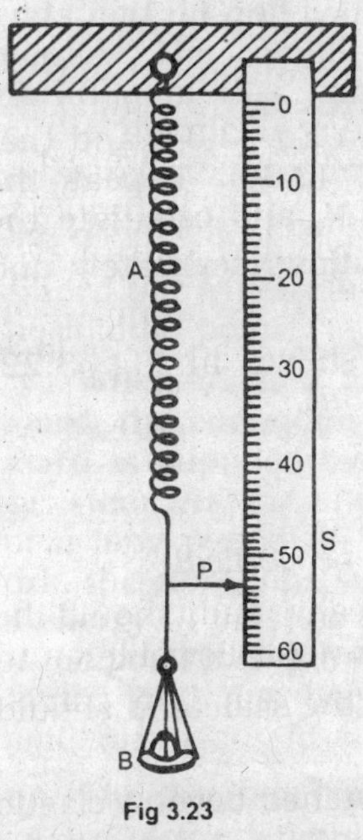

Fig 3.23

Gradually increase the load in the pan in *equal* steps of, say, half-a-kilogramme until the maximum permissible load is reached noting down the reading of the pointer on the scale after the addition of each load. Find the *mean* extension l for a certain known load M in the pan as indicated in the observation table and then calculate the value of restoring force per unit extension of the spring K from Eq. (3.31).

Plot a graph as illustrated in Fig. 3.24 between the load and the scale-readings, taking the load as abscissae and the corresponding scale readings as ordinates. The graph will be a straight line. Measure PR and QR and calculate K from the formula $K = \frac{QR}{PR} \cdot g$

Dynamical experiment. Place *gently* a load M_1 of, say half-a-kilogramme in the pan. Displace the pan *vertically* downward through a *small* distance and then release it. The load at the end of the spring will perform *simple harmonic* vertical oscillations. Taking the rest position of the pointer on the scale as the *reference point*, determine the time taken by the load to complete a *large* number of oscillations and thus find it period T_1. Increase the load in the pan to M_2, say 1 kgm, and determine as before the period T_2 of small oscillations of the load. Calculate the value of restoring force per unit extension of the spring K from Eq. (3.34) and that of mass of the the spring m' from Eq. (3.35). Repeat the experiment with *different* values of M_1 and M_2 and calculate the values of K and m' for each set of observations separately and then find the mean values of K and m'.

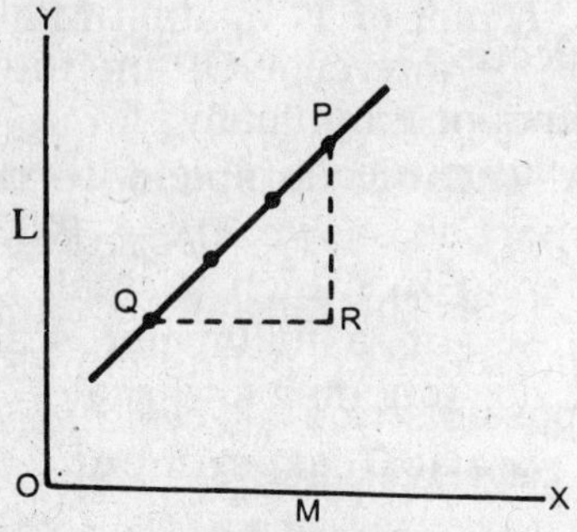

Fig 3.24

Plot a graph between T^2 and M as shown in Fig. 3.22. Measure *OP*. This gives *m* and then $m' = 3m$.

Sources of errors and precautions

Statical experiment

(1) The axis of the spiral spring must be *vertical*.

(2) The scale should be set up vertically and should be arranged to give almost the maximum extension allowable.

(3) The pointer should move *freely* over the scale and should be *just not* in contact with it.

(4) The spiral spring should not be stretched beyond elastic limit.

(5) The load in the scale-pan should be placed *gently* and should be increased in *equal* steps, say of half-a-kilogramme.

(6) While calculating the mean extension of the spring for a certain load, successive differences between consecutive readings of the pointer on the scale should not be taken.

Dynamical experiment

(7) The spiral spring should oscillate *vertically*.

(8) The amplitude of oscillation should be *small.*

(9) As in the expression for K and m' the periods T_1 and T_2 occur raised to the *sccond* power, they should be accurately determined by timing a *large* number of oscillations correct upto the value measurable with the stop-watch.

Common to both expts.

(10) The graph should be smoothly drawn and the respresentative points should be distributed as *evenly* about it as possible and as *close to it as possible.*

Observations. **Statical experiment**

(*A*) *Measurement of extension of the spring.*

S. No.	Lead in the pan gm.	Reading of pointer on the scale in cm.			Extension for 1500 gm cm.
		Load increasing	Load decreasing	Mean	
1.	0				
2.	500				
3.	1000				
4.	1500				
5.	2000				
6.	2500				
				Mean	

Dynamical experiment.

(*B*) Measurement of periods T_1 and T_2 for the loads M_1 and M_2

Least count of stop-watch = sec.

Set. No.	Load in the pan		No. of oscillations	Time taken with load		Periods		K dynes/ cm.	m' gm.
	M_1	M_2		M_1	M_2	T_1	T_2		
	gm.	gm.		min. sec.	min. sec.				
1	500	1000							
2	1500	2000							
3	2500	3000							
4.	3500	4000							
								Mean	

Calculations. Statical experiment

The restoring force per unit extension of the spring

$$K = \frac{Mg}{l}$$

$$= \quad \text{dynes/cm.} = \text{N/m}$$

From Graph (3.25),

PR = ; QR =

$$\therefore \quad K = \frac{QR}{PR} \times g = \quad \text{dynes/cm.} = \quad \text{N/m}$$

Dynamical experiment. I set

Restoring force per unit extension of the spring

$$K = \frac{4\pi^2 (M_1 - M_2)}{T_1^2 - T_2^2}$$

$$= \text{dynes/cm.} = \quad \text{N/m}$$

$$\text{Mass of the spiral spring } m' = 3 \left[\frac{M_1 T_2^2 - M_2 T_1^2}{T_1^2 - T_2^2} \right]$$

$$=$$

$$= \quad \text{gm} = \text{kg}$$

(*N. B.*—Similarly calculate the values of K and m' from *other* sets of observations).

∴ *Mean* value of K = dynes/cm = N/m

Mean value of m' = gm. = kg

From Graph (3.23)

m = OP = gm = kg

$m' = 3m$ = gm = kg

Results. The restoring force per unit extension of the spring as determined

by *statical* experiment = N/m

and as determined by *dynamical* experiment = N/m

The mass of spring = kg

ORAL QUESTIONS

ELASTICITY

1. Define stress and strain and state their units. How many types of strains do you know and how are they produced ?
2. What is shear and how is it produced ?
3. What is Hooke's law ?
4. What is modulus of elasticity ?

5. How many types of moduli of elasticity do you know ? Give their definitions. What happens when a wire is gradually loaded ?
6. What is limit of proportionality ?
7. Explain the terms : elastic limit, permanent set, breaking stress, yield point. What is the effect of temperature on elastic properties of a metal ?
8. How does the elastic limit of the metal change when it is overstrained, rolled, or hammered ?
9. What is the effect of annealing a metal ?
10. How are wires drawn ?
11. What is the effect of an impurity on elasticity ?
12. Do you know a substance for which the elasticity increases with temperature or is independent of it ?
13. What is elastic after-effect ?
14. Why are quartz fibres used in the measurement of small torsions ? What are their substitutes and why ?
15. What is elastic fatigue ? How does it affect your observations in experiments on elastic constants ?
16. What is elastic hysteresis ?

POISSON'S RATIO FOR RUBBER

1. What is Poisson's ratio ?
2. What are its units and dimensions ?
3. What is the value of σ for homogeneous and isotropic materials ?
4. How can you find out the Poisson's ratio for rubber ?
5. Why should there be no air bubble inside the tube ?
6. Why should you wait for about 5 min., after each addition or removal of a load before taking observations ?
7. Is there any limit to the load placed on the hanger ?
8. Can you use this method for finding Poisson's ratio for glass, if so how ?

BENDING OF BEAMS

1. How is bending of a beam related to Young's modulus ?
2. What is the nature of stress and strain in this case ?
3. How do applied forces bring about the stretching ?
4. What is neutral axis ?
5. What do you understand by plane of bending and axis of bending ?

6. Why is it not necessary to take into consideration the weight of the bar ?
7. Y refers to elasticity of size and not of shape, but bending is change of shape and not of size. How do you explain this anomaly ?
8. How does the shape change when the beam is bent uniformly ?
9. What is antielastic surface ?
10. How will you calculate σ from the radii of curvature of antielastic surface ?
11. What is a cantilever ?
12. How is your bar equivalent to two cantilevers ?
13. In what different ways you may measure the depression and which may be used for the best result ?
14. How will you use a micrometer to measure depression ?
15. How will you know that the tip of the screw is just in contact with the bar ?
16. How will you use an optical lever for measuring the depression ?
17. Why should the face of the mirror be initially very nearly parallel to the scale ?
18. Why should the distance between telescope and scale be small ?
19. How do you calculate the maximum permissible depression ?
20. Why do you take a bar of large length and small thickness ?
21. Why should the stirrup be made of this particular shape ?
22. Why do you load and unload the bar in small steps ?
23. How do you calculate the mean depression in this particular applications of your knowledge of Y.
24. Is the method statical or dynamical ?
25. Can you use a dynamical method for determining Y ?
26. State some practical applications of your knowlege of Y.

TORSION OF ROD OR WIRE

1. What do you understand by rigidity ?
2. How is the wire sheared in your experiment ?
3. Is it uniformly sheared ?

4. What is the nature of stress in this case ?
5. How and where is the tangential stress applied here ?
6. What do you understand by torsional rigidity of a wire ?
7. What is error due to the eccentricity of the axis of the rod with respect to circular scale ?
8. How does the mirror scale and telescope method eliminate this error ?
9. In the borizontal torsion apparatus,
   (*i*) What will be the effect of wrapping the string a number of times round the pulley ?
   (*ii*) Why should vertical portion of the string be tangential to the pulley ?
   (*iii*) Why should the zero position of the pointer be adjusted by placing the pan on the table ? and
   (*iv*) What is the maximum permissible twist in your case ?
10. How do you calculate it ?
11. Why should the masses be placed gently and in small steps in the pan ?
12. Why do you take the observations with string on both the sides of the pulley ?
13. If you cannot correctly locate the position of the fixed end of the rod, how will you modity your method of calculation of rigidity ?
14. Which will be better : a long thin rod or a short thick rod, and why ?
15. What is the nature of graph between ϕ and M and between ϕ/M and l ?
16. How do you use these graphs to calculate n ?
17. What are defects in your apparatus and how can they be removed ?
18. Which apparatus do you think gives better results : apparatus employing horizontal thick specimen or the apparatus using vertical thinner specimen and why ?

TORSION PENDULUM AND MAXWELL'S NEEDLE

1. What is a torsion pendulum ?
2. How can it be used to determine n ?
3. What are the assumptions made in the derivation of the theory of the method ?

4. Do you get good results from torsion pendulum ?
5. What is Maxwell's needle ?
6. How do you measure *n* with Maxwell's needle ?
7. Which is better : an ordinary pointer or a telescope and scale method of observing oscillations ?
8. Can you improve the pointer method to be nearly as good ?
9. What is parallax and how can you best remove it ?
10. Is it necessary that the oscillations should have small amplitude, if not, why ?
11. What would be the change in T if (*i*) *l* were doubled (*ii*) *r* were doubled (iii) I were doubled ?
12. Why do you measure the diameter so accurately ?
13. Which do you prefer, a long and thin wire or a thick short wire ?
14. Why should the Maxwell's needle be kept horizontal throughout the experiment ?
15. Why does the needle oscillate when released after twisting the wire ?
16. Do you get the same value of rigidity for a thin and a thick wire, if not, why ?
17. Which is better, statical or dynamical method ?
18. Do you get the same value for *n* from the statical and dynamical methods ? If not, why ?

SEARLE'S METHOD FOR ELASTIC CONSTANTS

1. What are the four elastic constants of a material ?
2. Define them and state their units.
3. What is the relation between them ?
4. How can you determine them by Searle's method ?
5. Do you know any other method of determining them ?
6. Which is better and why ?
7. What are the assumptions made in the theory of this method and how far are they fulfilled in practice ?
8. Distinguish between flexural and torsional vibrations. How do you start the bars oscillating and why ?
9. What should be the amplitude of vibration and why ?
10. What are the forces acting on the rod when they are oscillating ?

11. What precautions do you take in the measurement of periods ?
12. What precaution will you take in the measurement of diameter of the wire ? Are the moments of inertia of the two bars exactly equal ?
13. If not, how will you modify the method of taking observations ?

RESTORING FORCE AND MASS OF A SPRING

1. What is a spiral spring ?
2. How can you determine its restoring force per unit extension by statical and dynamical experiments ?
3. How can you determine the mass of the spring ?
4. Why should the spiral spring be susended exactly vertically ?
5. Why should the extension of the string be small ?
6. How will you calculate the mean extension in the statical experiment and why ?
7. Why should the amplitude of oscillation of the spring be small ?
8. Why do you measure period of spring so accurately ?
9. Do you get the same value of restoring force per unit extension of the spring from the statical and dynamical experiments ? If not, why ?

4

Surface Tension

4.1. Surface Tension—Referring to Fig. 4.1 let a liquid film be formed between two limbs of a bent wire PBQ and a *horizontal* wire PQ of length l placed across it. Since there is a certain tension in the film which tries to reduce its surface, the wire PQ will tend to move towards the film and hence a force F will be required to keep it *in equilibrium* position. If the experiment is repeated, with different lengths of PQ, the force required to keep PQ in equilibrium is found to depend, for a given liquid at a constant temperature, only on the *length* PQ. This shows that the *total force of tension in the surface of the film is proportional to the length.* Thus, since the film under consideration has two free surfaces,

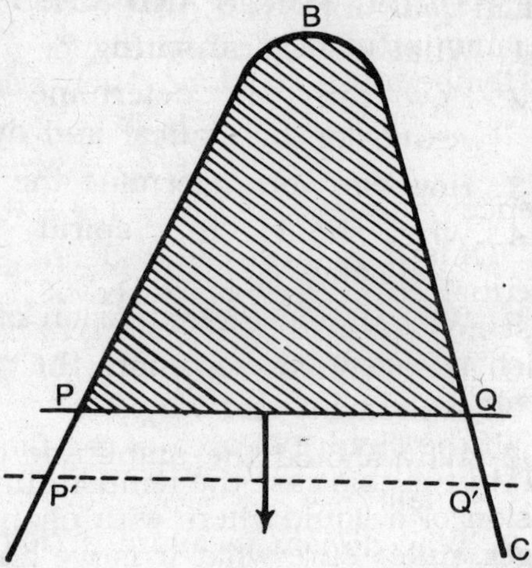

Fig. 4.1

$$F \propto 2l$$

or $$F \equiv 2lT \qquad (4.1)$$

where T is constant called the *surface tension* of the liquid. If $2l=1$, $F=T$. Hence **surface tension** *of a liquid is defined as the force per unit length acting on either side of a line drawn in the liquid surface in equilibrium, the direction of the force being tangential to the surface and perpendicular to the line.* The M.K.S. unit of surface tension is N/m.

Now while the temperature remains *constant*, let the wire be pulled out through a *small* distance x *to a parallel* position P′Q′. The film will be stretched. The work done in stretching against the force F is equal to $W=F.\ x=2TA$, where $2A=2lx$ is the *total* increase in the area of film. If $2A=1$, $W=T$. Thus, *the* **surface tension** *of a liquid is* **numerically** *equal to the work done in enlarging the* *surface of the liquid by unit amount under isothermal conditions.* This work done is stored in the surface molecules in the form of potential energy, and since the number of molecules in the surface is proportional to the area of the surface, the excess of potential energy in the surface molecules must be proportional to the area of the surface. The *excess of potential energy per unit area of the molecules in the surface is called* **'surface energy'.**

We have assumed above that the temperature of the film remains constant. But when the film is streched it is cooled and so heat must be supplied from outside to keep the temperature of film constant. Let E be the *surface energy* of the film and H the amount of *heat* supplied per *unit* area from outside to keep the temperature *constant.* Then evidently

$$E\ 2lx = W + H.\ 2lx$$

or $$E\ 2lx = T2lx + H.\ 2lx$$

whence $$E = T + H$$

Thus it is incorrect to define surface tension as the excess potential *energy per* unit area of the surface of a liquid. *Surface tension is equal to the mechanical part (E–H) of the surface energy* which is called the *free* energy of the surface or the *surface free energy.*

If the change of area of the film is supposed to be adiabatic, $E=T$: but in this case the temperature will vary, and as the surface tension of a liquid alters with change of temperature, we are no longer quite sure what it is we are defining.

Surface tension of a liquid is very sensitive to the presence of even small quantities of impurities on the surface of the liquid. The surface tension of pure liquid is greater than that of solution, but there is no simple law of variation of surface tension with concentration. The *surface tension of a liquid* **falls** *with rise of temperature.* For *small* ranges of temperature, the surface tension of a liquid *decreases linearly* with rise in temperature.

4.2 Angle of Contact—When a liquid is placed in contact with a solid, the liquid surface is in general curved, even when gravity is the only external force acting throughout the mass of the liquid. *The angle which the tangent plane to the liquid surface makes with the tangent plane to the solid surface drawn into the substance of the liquid at any point on the line of contact is called the* **angle of contact** *between the liquid and the solid.* Its value depends upon the nature of the liquid and the solid. For liquids having concave meniscus the angle of contact is acute *e.g.*, in the case of water and glass its value is about 8°; while for liquids having convex meniscus the angle of contact is obtuse, *e.g.*, in the case of mercury and glass its value is about 135°.

4.3. Excess of pressure inside a curved membrane—Refer to Fig. 4.2. When the free surface of a liquid is *plane* the resultant force due to surface tension on a molecule in the surface is zero since it is pulled *equally* in all directions, while if the surface is *curved* there is a resultant force of surface tension

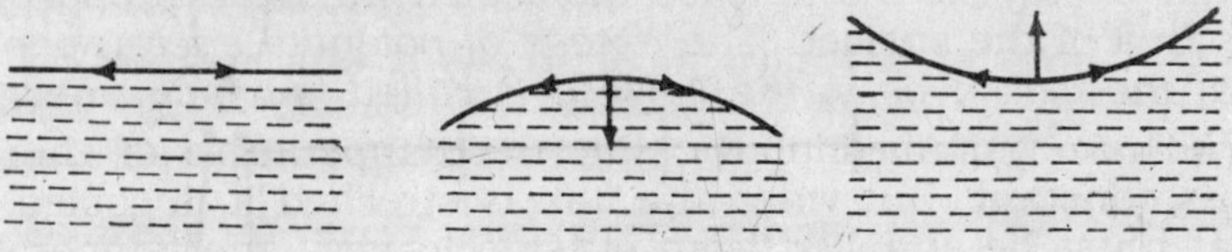

(*a*) (*b*) (*c*)

Fig. 4.2

which acts normally to the surface. If the surface is *convex* as shown in Fig. 4.2 (*b*), the resultant is directed *inwards* towards the interior of the liquid and increases the cohesion pressure (the attractive force between molecules of the liquid per unit area), while in the case of a *concave* surface as shown in Fig. 4.2 (*c*), the resultant is directed outwards away from the liquid and diminishes the cohesion pressure. On account of this resultant force of surface tension, *the curved surface of a liquid tends to contract.* Consequently to keep the surface in equilibrium, its tendency to contract must be neutralised by an *excess* of pressure on the concave side which of itself would produce an expansion.

In the case of a single surface of principal radii of curvature R_1 and R_2 the magnitude of this excess pressure on the *concave* side is given by

$$p = T\left(\frac{1}{R_1} + \frac{1}{R_2}\right) \tag{4.2}$$

If the surface is *spherical*, the two radii of curvature R_1 and R_2 are *equal* and we get $p=2T/R$ for a single surface, *e.g.*, a spherical liquid drop, an air bubble in a liquid, etc., and $p=4T/R$ for two surfaces, *e.g.*, a spherical soap bubble.

4.4. Rise of Liquid in a Capillary Tube—When a capillary tube open at both ends, is dipped vertically in a liquid, the surface of the liquid inside the tube is, in general, curved. If the meniscus in concave upwards as in the case of water, the pressure in the liquid just below the meniscus is *less* than the atmospheric pressure by an amount $2T/R$, where R is the radius of curvature of the meniscus and T is the surface tension of the liquid; and hence the liquid *rises* in the tube until the pressure of the liquid column in the tube above the outside liquid equals the deficit $2T/R$ of the pressure just below the meniscus in the final position.

When the menisucs is convex upward as in the case of mercury, the pressure in the liquid just below the meniscus is *greater* than the atmospheric pressure by an amount $2T/R$, and hence the liquid inside the tube is *depressed* until this excess pressure is equalised by the pressure exerted by the column of liquid outside the tube above the liquid level inside the tube.

Experiment 4.1

Object—To determine the *surface tension* of a liquid (....) by *Jaeger's method.*

Apparatus—Jaeger's apparatus, a glass tube of about 5 mm diameter, a microscope, a scale and a thermometer.

Description of apparatus—As illustrated in Fig. 4.3, the apparatus consists of a thin glass tube drawn out at its *lower end* to a *fine capillary* and fixed *vertically* in the experimental liquid. The glass tube is connected to a manometer M and also to a Woulf's bottle W fitted with a dropping funnel (or burette) containing water.

Theory—Let the capillary tube be dipped in the liquid, the latter will *rise* in it and the shape of the meniscus will be approximately *spherical.* If air is forced into the glass tube by

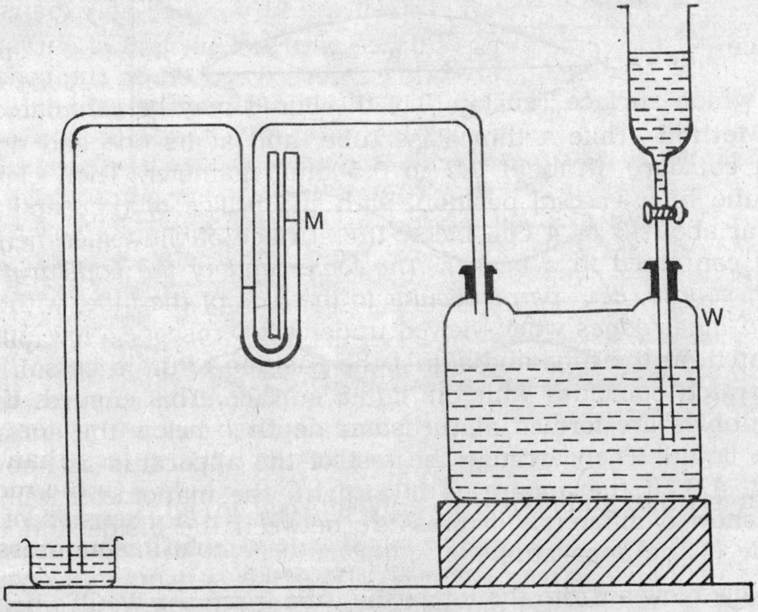

Fig. 4.3

dropping water from the funnel (or burette) into the Woulf's bottle, the surface of the liquid in the capillary tube is pressed downwards and as the pressure of air inside the tube increases the liquid surface goes on sinking lower and lower until finally a *hemispherical* bubble of radius r equal to that of the orifice of the capillary tube protrudes into the liquid below, the pressure inside the bubble being *greater* than that outside by $2T/r$, where T is the *surface tension* of the liquid.

At this stage the pressure in the manometer reaches a *maximum* and the bubble becomes unstable, for a slight increase in the radius decrease the internal pressure necessary to produce equilibrium, and as the external pressure is constant, the bubble breaks away. Just before the bubble breaks away, the pressure *inside it is* $Hgd+P$, where H is the *maximum* reading of the manometer, d the denisity of the liquid in the manometer and P the atmospheric *pressure*. The pressure *outside* the bubble is $hg\rho+P$, where h is the depth of the tip of the capillary tube below the surface of the liquid and ρ the density of the liquid. Thus, the *excess* pressure inside the bubble is

$$(Hgd + P) - (hg\rho + P) = \frac{2T}{r}$$

whence
$$T = \frac{rg}{2}(Hd - h\rho) \qquad (4.3)$$

from which surface Tension T of the liquid may be calculated.

Method—Take a thin glass tube, and at its one end draw out a *capillary* of about 0.2 to 0.5 mm. diameter. Then clamp the tube in a *vertical* position, with the orifice of the capillary tube at about 3 or 4 cm. below the surface of the experimental liquid contained in a beaker. *The lower end of the capillary is cut off square, i.e., perpendicular to the axis of the tube*, without *ragged inner edges* when viewed under a microscope. Now make a scratch on the tube and adjust the position of the tube so that the scratch coincides with the liquid surface. This ensures that the bubbles are formed at the *same* depth h below the surface of the liquid. Then arrange the rest of the apparatus as shown in Fig. 4.3. If the apparatus is airtight, the manometer will at once show a difference of pressure between the inside and the outside of the apparatus.

Now blow air into the top of the tube from the Woulf's bottle by allowing water to drop into it from the droppong funnel and adjust the rate of flow of water into the bottle so that a bubble is formed about every ten sceonds at the orifice of the capillary tube. The pressure indicated by the manometer rises until the bubble becomes unstable and then suddenly falls when the bubble breaks away. Note down the *maximum* difference H in heights of liquid in the two limbs of the manometer, Repeat the experiment several times and find out the *mean* value of H. Then measure the depth h of the orifice of the capillary tube from the surface of the liquid. Now alter the value of h and take at least three sets of observations for H with different values of h. Repeat the experiment with two more capillary tubes of slightly different bores.

Next, clamp the capillary tube in a *horizontal* position and focus a vernier or micrometer mocroscope on its orifice. Then coinciding the cross-wire with the *inner edge* of the tube at one side of the centre and then with its inner edge on the other side of its center, determine the diameter of the orifice. Now rotate the tube through *a right angle* and again determine its diameter. In this way take serveral observations for two *mutually perpendicular* diameters which will be practically the same if the bore of the tube is circular, and their mean value will denote the value of $2r$. Then calculate the value of surface tension of the experimental liquid for each set of observations separately and find out the *mean* value of surface tension. Finally, note down the temperature of the experimental liquid.

Sources of error and precautions—(1) There should be no leakage in the apparatus. It is distinctly advantageous to have the apparatus in *one piece* avoiding the use of rubber joints.

(2) The manometer should contain xylol so that the height H may be large; the density of xylol is *less* than that of water.

(3) To damp the oscillations of the liquid in the manometer, its open end may be drawn out to a capillary tube.

(4) Diameter of the capillary tube must be measured *only at the orifice dipping in the liquid.*

(5) The orifice of the capillary tube should be *circular* and *very small*, about 0.3 mm, in diameter, so that the maximum pressure in a bubble may occur when it is hemispherical.

(6) While measuring the diameter of the orfice of the capillary tube, several observations of *mutually perpendicular* diameters should be taken. This reduces the error due to ellipticity of the cross-section to *minimum.*

(7) In the case of a *micrometer* microscope error due to backlash should be avoided while taking observations for diameter.

(8) The capillary tube should be *clean*. All traces of grease must be carefully removed as they are fatal to surface tension experiments.

(9) The bubbles should be formed *singly* and *slowly*, say, at the rate of one in about ten seconds, otherwise the maximum pressure registered by the manometer will not be independent of the rate of bubble formation. The first condition is satisfied by keeping the air space in the Woulf's bottle *small* so that when one bubble breaks away from the end of the capillary tube, the pressure inside the apparatus is reduced *appreciably* below the maximum required to blow the bubble. The second condition is fulfilled by adjusting the flow of liquid from the dropping funnel into the Woulf's bottle.

(10) As the surface tension of a liquid depends upon its temperature, the temperature of the experimental liquid should be recorded.

Observations—[*A*] *Measurement of h and H.*

Set	Depth of Orifice *b* cm.	Manometer Readings			Mean H cm.
		Upper level of liquid cm.	Lower livel of liquid cm.	H cm.	
1.					
2.					
3.					

[*B*] *Temperature of experimental liquid* = °C

[*C*] (*i*) *Density of experimental liquid* = *gm/c.c.*

(*ii*) *Denisity of manometric liquid* = *gm/c.c.*

[*D*] *Measurement of diameter of orifice*
Least count of the microscope= *cm.*

S. No.	Observation for position of cross-wire at the section of the orifice of capillary tube						
	Reading along *any* diameter			Reading along *prep.* diameter			$\frac{a+b}{2}$ cm.
	At one end of Section cm.	At the other end of section cm.	Diameter *a* cm.	At one end of section cm.	At the other end of section cm.	Diameter *b* cm.	
1.							
2.							
3.							
					Mean diameter		

Calculations

Mean radius of the orifice of the capillary tube = cm.

Set 1. $T = \frac{rg}{2}[Hd - h\rho]$

=

= dynes/cm. = N/m

(*N.B.*—Make similar calculations for T from *other* sets of observations).

∴ *Mean* value of surface tension = dynes/cm = N/m

Result—The *surface tension* of the experimental liquid (...) at...°C = N/m

Standard value = N/m

Error = %

Criticism of the Method—This method has a number of advantages:

(1) The internal radius of the capillary tube has to be determined at the orifice where it is easily accessible and errors due to irregularity of the bore of tube are avoided.

(2) The bubbles are formed inside the liquid where temperature can be easily controlled and measured. Hence the method can be readily used to study the variation of surface tension with temperature, and the slope of the surface tension-temperature curve, if drawn, can be used to determine molecular aggregation, *i.e.*, the number of atoms in the liquid molecule. A capillary tube of platinum is used at very high temperature.

(3) The liquid-air interface is being continually renewed so that the effects of contamination of the surface, a factor which is fatal to the accuracy of measurements of surface tension, are reduced to a minimum. The capillary portion of the tube can be quite short and so it is easy to clean and keep clean.

(4) It can be used to determine the surface tension of molten metals also.

(5) It is particularly suited for the investigation of variation of surface tension of a solution with the concentration of the dissolved substance.

The method is, however, not very suitable for absolute determination of surface tension of a liquid. The assumption that the maximum pressure in a bubble occurs when it is hemispherical is true only for tubes of infinitely small diameter. When the pressure inside the bubble is maximum, the radius of the bubble is not exactly equal to that of the orifice of the capillary tube.

The theory assumes static conditons in the experiment and that the orifice of the tube is circular and horizontal and we do not know how far these two conditons are complied with in practice. Lastly, the measurement of pressure difference H is not susceptible of very great accuracy as it is a small quantity.

4.5 Surface Tension by the Method of Ripples—The velocity of propagation of waves passing over the surface of a liquid depends on *gravity* as well as *surface tension* of the liquid. If the amplitude of the waves is small, each particle of the liquid in the waves describes a circular path in a *vertical plane*, the waves advancing through a distance equal to the wavelength λ during one complete revolution of the particle. For waves proceeding from left to right, the circles are described in the *anti-clockwise* direction.

Let Fig. 4.4 represent a vertical section of the liquid taken parallel to the direction of propagation of the waves. Let C be

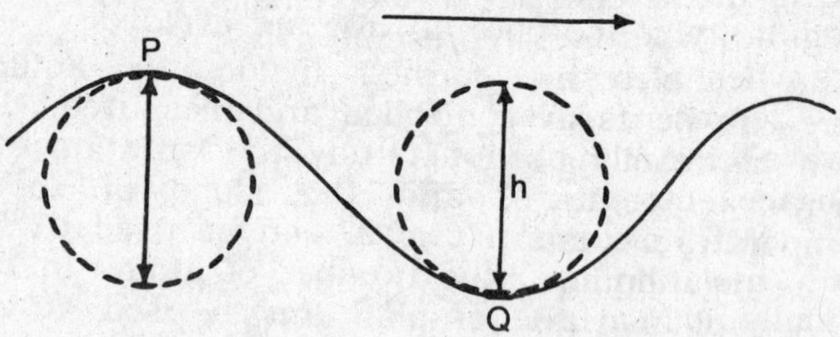

Fig. 4.4

the velocity of the waves in the horizontal direction conditioned by *gravity* alone. Also let T be the period of revolution of a particle in the waves and r the radius of the circular patch described by it. Then the instantaneous horizontal velocity at a crest P is given by

$$v_1 = C - \frac{2\pi r}{T} \tag{4.5}$$

While at the trough Q it is given by

$$v_2 = C + \frac{2\pi r}{T} \tag{4.6}$$

Assuming that the increase in velocity of the particle is due to its fall under the action of gravity through a height $h=2r$, we have

$$v_2^2 = v_1^2 + 2g.\ 2r$$

$$v_2^2 - v_1^2 = 4\ gr$$

Squaring Eqs. (4.4) and (4.5) and then subtracting, we have

$$v_2^2 - v_1^2 = \frac{8\pi rC}{T} \tag{4.7}$$

Comparing Eqs. (4.6) and (4.7) we have

$$4gr = \frac{8\pi rC}{T}$$

whence $\quad C = \frac{gT}{2\pi} = \frac{g\lambda}{2\pi C}$

since $\quad T = \lambda/C.$

hence $\quad C = \sqrt{\frac{g\lambda}{2\pi}} \qquad (4.8)$

This represents the velocity of propagation of the waves as determined by gravity alone.

To account for the effect of surface tension of the liquid, let us refer to Fig. 4.5 where AB represents a *vertical* section of the liquid surface when at rest, and ACD represents/a harmonic wave travelling over the surface. Let the displacement MN of a particle on the liquid surface, originally at N at a distance x from some arbitrary origin, at any instant be given by

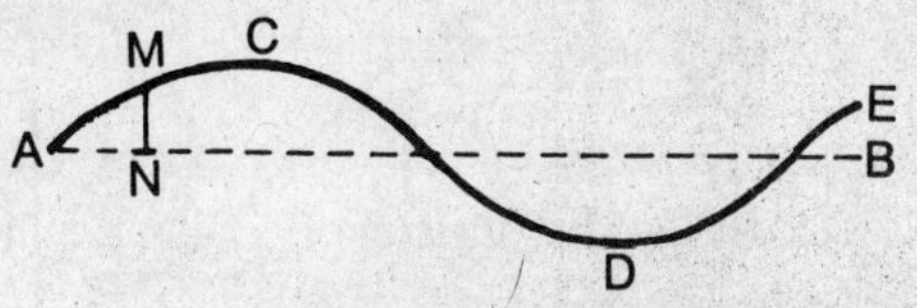

Fig. 4.5

$$y = a \sin\left(\frac{2\pi x}{\lambda} + \alpha\right) \tag{4.9}$$

where *a is the amplitude* and α the *initial phase.*

The vertical pressure at N due to the action of gravity alone is increased by *yρg* where ρ is the denisty of the liquid, while the excess pressure at M on the concave side due to the surface tension directed normally outwards to the surface is *T/R where T* is the surface tension of the liquid and *R* the radius of curvature of the liquid surface at M in the plane of the paper. The other radius of curvature of the surface at M perpendicular to the plane of the paper is infinite since the system of waves is *cylindrical.* If the amplitude of waves is very small in comparison to the wavelength as is practically the case in experiments on ripples, the normal pressure at M is *sensibly vertical,* so that the total increase in the vertical pressure at N when the waves pass over the liquid surface is given by

$$p = \gamma\,\rho g - \frac{T}{R} \tag{4.10}$$

Now $$R = \frac{\left[1 + \frac{dy^2}{dx}\right]^{3/2}}{\frac{d^2y}{dx^2}}$$

But as $\alpha << \lambda, \frac{dy}{dx} >> 1$

or $$\frac{1}{R} = \frac{dy^2}{dx^2}$$

Differentiating Eq. (4.9) twice, we have

$$\frac{d^2y}{dx^2} = -\frac{4\pi^2 y}{\lambda^2}$$

$$\frac{1}{R} = -\frac{4\pi^2 y}{\lambda^2}$$

Putting the value of $\frac{1}{R}$ in Eq. (4.10), we get

$$p = y\rho g + \frac{4\pi^2 yT}{\lambda^2}$$

or $$p = y\rho\left(g + \frac{4\pi^2 T}{\lambda^2 \rho}\right)$$

It is evident from this equation that the effect of surface tension is, as it were, to change the *effective value* of acceleration due to gravity from g to $\left(g + \dfrac{4\pi^2 T}{\lambda^2 \rho}\right)$. Hence, if u is the velocity of propagation of the waves under the combined influence of gravity and surface tension, we have by substituting the augmented value of g in Eq. (4.8).

$$u = \sqrt{\frac{\lambda}{2\pi}\left(g + \frac{4\pi^2 T}{\lambda^2 \rho}\right)} \tag{4.11}$$

It is easily seen from this equation that $u = \infty$ both when λ=0 and when $\lambda = \infty$. Between these values of λ there is a minimum value of u. Since u is equal to the square root of sum of two terms whose product $\dfrac{gT}{\rho}$ is constant, it follows that u will be minimum when the two terms are equal, *i.e.*,

$$\frac{\lambda g}{2\pi} = \frac{2\pi T}{\lambda \rho}$$

whence the critical wavelength corresponding to the minimum velocity u_m is given by $\lambda_c = 2\pi\sqrt{\dfrac{T}{g\rho}}$

Substituting this value of λ in Eq. (4.11) we have the minimum velocity

$$u_m = \sqrt{\frac{1}{2\pi}.\, 2\pi\left(\frac{T}{g\rho}\right)^{1/2}\left(g + \frac{4\pi^2 T}{4\pi^2 T\rho/g\rho}\right)}$$

$$= \sqrt{\left(\frac{T}{g\rho}\right)^{1/2}(g+g)}$$

$$= \sqrt{2}\,.\sqrt{\frac{Tg}{\rho}}$$

For water taking T =75 dyne/cm. and ρ = 1 gm. per cc. λ_c=1.7 cm. and u_m= 23 cm/sec.

Thus, no waves which have a velocity smaller than 23 cm. per sec. can travel over the surface of water.

When $\lambda \le \lambda_c$, the first term in Eq. (4.11) becomes less and less important as the wavelength decreases. Consequently, in such a case

$$u = \sqrt{\frac{2\pi T}{\lambda p}}$$

Such *disturbances having wavelength smaller than the critical wave length are usually called ripples.* The velocity is mainly determined by *surface tension* and *diminishes with* increasing wavelength.

When $\lambda > \lambda_c$, the first term in Eq. (4.11) becomes more and more predominant as λ increases and hence in such a case

$$u = \sqrt{\frac{\lambda g}{2\pi}}$$

Such *disturbances whose wavelength exceeds the critical wavelength are known as waves.* They are propagated chiefly by gravity and their velocity increases with increasing wavelength.

Use of Eq. (4.11) is made to determine the surface tension of a liquid by measuring the wavelength λ of ripples of known frequency n excited on the surface of the liquid. For, then

$$u = n\lambda = \sqrt{\frac{\lambda}{2\pi}\left(g + \frac{4\pi^2 T}{\lambda^2 \rho}\right)}$$

whence $$T = \frac{n^2\lambda^3\rho}{2\pi} - \frac{\lambda^2\rho g}{4\pi^2} \qquad (4.12)$$

***Experiment* 4.2**

Object—To determine the surface tension of water by the method of ripples.

Apparatus—Electric vibrator, neon lamp, water trough and low power travelling microscope.

Theory— We have from Eq. (4.12)

$$T = \frac{n^2\lambda^3\rho}{2\pi} - \frac{\lambda^2\rho g}{4\pi^2}$$

where T is the surface tension of water.

λ is the wavelength of the ripples generated on the surface of water by a source of frequency n,

ρ is the density of water.

Method—Attach a pin at the free end of the rod of the electrical vibrator. Adjust the length of the vibrating rod such that its free end vibrates with a maximum amplitude when A.C. is fed into the coil of the vibrator. Under this condition the frequency of vibration of the rod is equal to the frequency of the alternating current. (For working of the electrical vibrator refer to Expt. 3.2)

Take water in a rectangular enamelled trough about 25 cm. long. 15 cm. wide and 5 cm. deep. Set the electrical vibrator in such a manner that the pin, attached to the free end of the rod, is vertical and that it touches the surface of water in the trough.

Switch on the current when the pin or the dipper alternately moves in and out of water, thus exciting trains of ripples on its surface. These, on reflection from the walls of the trough give rise to stationary ripples.

To observe these ripples for the purpose of measurement of their wavelength, illuminate the surface of water with a neon lamp operated on A.C., mains. The ripples then appear stationary due to stroboscopic effect.

Focus the low power microscope on the ripples. Adjust the crosswire on one of the ripples. Determine the least count of the microscope and take the reading corresponding to this ripple. Take the readings by moving the microscope in succession over the other ripples. From a measurement so made over 8 or ten ripples, compute the mean wavelength λ of *the ripples.*

Taking the value of n =50 Hz., ρ =1 gm. per c.c. and g = 980 cm/sec. and having determined λ, calculate T from Eq. (4.12].

Sources of error and precautions—1. The dish containing water and the tip of the pin must be clean. These should be washed with nitric acid and caustic soda in sucession and then with running water. Then take some water in the dish for experimental work.

2. The frequency of the electrical vibrator must be the same as that of the supply mains (50 Hz) ; to achieve this the length of the vibrating rod is so adjusted that its free end vibrates with a maximum amplitude, as judged visually.

3. It is necessary to use a low power microscope for measurement with the ripples. A micrometer eyepiece mocroscope would serve better.

4. Record the temperature of water with a thermometer.

Observations—*Known Data.*

n = 50 Hz. : ρ = 1 g. per c.c.

g = 980 cm/sec.^{2}

Temperature of water = ... ^0C

Determination of λ

Least Count of microscope = cm.

No. of Ripples	Reading of microscope	Difference for 5 ripples
1	a	
2	b	
3	c	f-a
4	d	I-b
5	e	h-c
6	f	i-d
7	g	j-e
8	h	
9	i	
10		
	Mean	

Mean value of 5 wavelengths = cm.

∴ Mean wavelength λ = cm.

Calculations. $T = \frac{\lambda^2\rho}{4\pi^2}\left(2\pi n^2\lambda - g\right)$

= dyne/cm = N/m

Result—The surface tension of water at ^0C = N/m

ORAL QUESTIONS

SURFACE TENSION

1. What do you understand by phenomenon of surface tension?
2. Why does liquid surface behave like a stretched rubber sheet?
3. What is the difference between the behaviour of the two liquids?
4. Do you know of any experiment which can illustrate the property of a liquid?
5. What is surface tension and what is its unit?
6. Compare surface tension and stress.
7. What do you understand by surface energy?
8. Distinguish between surface tension and surface energy.
9. What is the shape of a liquid surface when it is in contact with a solid?
10. When will the meniscus be convex or concave and why?

11. What do you understand by angle of contact?
12. What is its unit?
13. What is the value of the angle of contact in the case of (*a*) water and glass and (*b*) mercury and glass?
14. What are the different methods of determining surface tension? Compare them.

JAEGER'S METHOD

1. What is the theory underlying this method of measuring surface tension of a liquid?
2. Why is the pressure inside an air-bubble greater than that outside it?
3. Why should the apparatus be in one piece ?
4. Why does the liquid in the manometer rise and fall during the formation and breaking away of the bubble ?
5. Why do you record the maximum reading of the manometer?
6. Why should xylol be used in the manometer?
7. How can you damp the oscillations of liquid in the manometer?
8. How will you make sure that the bubbles are being formed at the same depth?
9. What should be the interval between the breaking away of one bubble and the formation of another and why?
10. Why should the bubbles be formed singly and slowly?
11. How will you accomplish the above adjustment? In measuring the diameter of the tube, why do you take readings along two mutually prependicular diameters?
12. What are the advantages of this method?
13. How can variation of surface tension with temperature be studied with the help of this apparatus?

CAPILLARY RISE

1. Why does water rise in a capillary tube and not in a wider tube?
2. Does every liquid rise in a glass tube? If not, why?
3. How will the rise be affected by (*a*) change in the diameter of the bore, (*b*) change in the shape of the bore, say, elliptical, (*c*) to bore becoming non-uniform and tube

conical, (*d*) including the tube at a small angle to the vertical and (*e*) if the length of the tube is less than the height required by the formula? Wherefrom does the energy come when the liquid rises in a capillary tube against gravity?

4. How can you determine surface tension of a liquid by measuring the rise of the liquid in a capillary tube?
5. Why do you measure the radius at the point up to which the liquid rises in the tube?
6. Why should you take reading along two mutually perpendicular diameters?
7. Why should the bore be uniform in the neighbourhood of the meniscus of the liquid in the tube?
8. Should the top of the capillary tube be open or closed?
9. Why should the tube be vertical and how will you adjust it to be so?
10. Why and how do you clean the capillary tube?
11. Why don't you use distilled water instead of tap water in your experiment?
12. What are the advantages of this method over other methods of measuring surface tension? What are its limitations ?

5

Viscosity

5.1. Definition of *Viscosity*—*When a liquid is allowed to flow through a narrow tube such that each particle of the liquid moves along the* **same** *path as its preceding particle, the motion is said to be orderly or* ***Stremline motion***. This occurs for *small* rates of flow only, *i.e.*, when the velocity of flow is below a certain limiting value called the *critical velocity*. When the velocity of the liquid reaches beyond the critical value, the motion becomes unsteady; the liquid no longer flows in a regular manner but flows in a sinuous course from side to side of the tube. This type of motion is called *turbulent motion.*

Now, referring to Fig. 5.1, consider a liquid flowing over a *fixed horizontal* surface *AB*. If the motion is *streamline* the liquid in contact with the fixed surface will be *at rest* and every other layer will move *parallel* to the fixed surface, the velocity of flow increasing up to a maximum on the liquid. Now, consider any *horizontal* plane *PQ* in the liquid. The molecules of the liquid immediately *above* this plane tend to *accelerate* the motion of the molecules lying in the plane, while the molecules of the liquid lying immediately *below* the place tend to *retard* the motion of the moelcules in it. Each layer of the liquid thus tends to exert on the one next to it a tangential backward force which tends to destroy the *relative* motion between the two layers. Consequently, if a relative velocity is to be maintained between parallel layers of the liquid an external force is required to overcome this backward drag. In the absence of this external force the relative

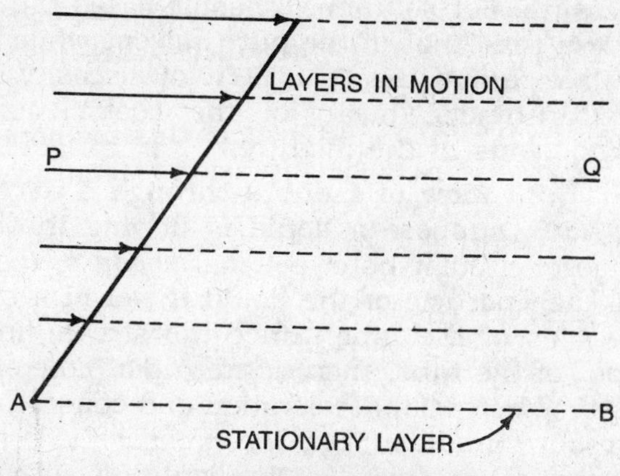

Fig. 5.1

motion between different layers ceases after some time. *The property by virtue of which a liquid opposes relative motion between its different layers is called internal friction or viscosity.*

From *Newton's* law *of viscous flow for* stream-line motion the *tangential viscous force* F between two *parallel layers* each of area *A*, a distance *dx* apart and moving with relative velocity *dv*, is *proportional* to the *area A* and to the *velocity gradient dv/dx*. Thus,

$$F = -\eta A\, dv/dx$$

where η is a constant called the *coefficient of viscosity* of the liquid and depends on the nature and physical conditions of the liquid. If $A=1$ and $dv/dx=1$, then $\eta=F$. Therefore, *the* **coefficient of viscosity** *of a liquid may be defined as the tangential backward force acting on unit area of a liquid layer moving in a region of unit velocity gradient.* The C.G.S. unit of coefficient of viscosity is dynes per sq. cm. per unit velocity gradient *i.e.*, gm., cm^{-1}, Sec^{-1}. The unit is called "poise" in honour of Poiseuille. The S.I. unit is ***deca-poise*** (1 = 10 poise = 1 NSm^{-2}).

The coefficient of viscosity of liquids decreases very rapidly with the rise of temperature and increases with the rise of pressure. For solutions in some cases the coefficient of viscosity is less than that of the pure solvent while in others it is greater. With mixtures the coefficient of viscosity is generally less than the arithmetic mean of the coefficients of viscosity of the components of the mixture.

5.2. Flow of a liquid through a narrow tube—Referring to Fig. 5.2, suppose a liquid is flowing through a *narrow* tube of *uniform* circular bore. Let *the motion of the liquid be orderly* and let each particle of the liquid travel in a straight line parallel to the axis of the tube. Since the stream-lines are *parallel* to the axis of the tube, there is no *radial flow* of the liquid and hence there is no radial acceleration and consequently there is no radial pressure gradient, *i.e., the pressure is constant over any cross-section of the tube.* As the liquid is *incompressible*, its amount passing across any cross-section of the tube in a given time is the *same*. Further, since the motion of the liquid is slow and steady, there is no slip between the liquid and the walls of the tube, *i.e., the layer of the liquid immediatley in contact with the walls of the tube is practically at rest.* The velocity of flow of the liquid increases as the axis of the tube is approached and is the same at points at the same distance from the axis.

Let *v* and *dv/dr* be the velocity and velocity gradient respectively at a distance *r* from axis of the tube. Then from

Newton's law the viscous drag on the layer of the liquid at a distance r from the axis of the tube given by

$$F = 2\pi rl\eta \frac{dv}{dr} \tag{5.1}$$

where l is the length of the tube and η the *coefficient of viscosity* of the liquid; and since dF/dr represents the *rate* of change of this force with r, the *viscous drag* on a parallel layer at a distance $(r+dr)$ from axis is

$$\left(F + \frac{dF}{dr}dr\right)$$

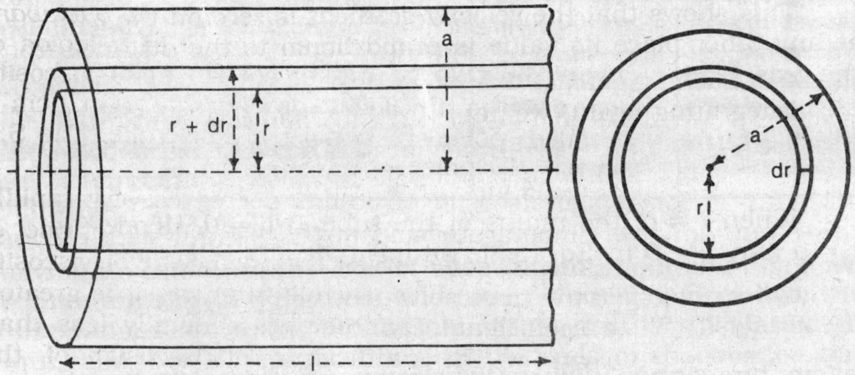

Fig. 5.2

Now, consider the flow of liquid between two *adjacent* cylindrical layers *coaxial* with the tube and of radii r and $(r+dr)$. The *inner* surface of the *inner* layer is pulled forward by the layer with it by a force F and the *outer* surface of the *outer* layer is held *back* by a force $\left(F + \frac{dF}{dr}dr\right)$ and hence the *resultant* force on this layer is $F + \frac{dF}{dr}dr - F = \frac{dF}{dr}dr$. *This resultant force is overcome by the difference in pressure between the ends of the tube.* If the pressure difference is p, then due to this pressure p the liquid flowing between the two layers exerts a forward force $p.\ 2\pi r.\ dr$ on the outer layer and hence

$$\frac{dF}{dr}dr = p.\ 2\pi r.\ dr$$

or

$$\frac{dF}{dr} = p.\ 2\pi r$$

Putting the value of F from Eq. (5.1), we have

$$\frac{d}{dr}\left(-2\pi r l \eta \frac{dv}{dr}\right) = p.2\pi r$$

Integrating, we get

$$-2\pi r l \eta \frac{dv}{dr} = p\pi r^2 + C_1$$

where C is a constant which equals zero, because when $r=0$, $dv/dr=0$ since at the axis v is maximum.

Thus, $$\frac{dv}{dr} = \frac{pr}{2l\eta}$$

This shows that the velocity gradient is zero on the axis and at any other place its value is proportional to the distance from the axis.

Integrating again, we get

$$v = -\frac{pr^2}{4l\eta} + C_2 \qquad (5.2)$$

When $r = a$, the radius of the tube, $v = 0$. Hence putting $r = a$ and $v = 0$ in Eq. (5.2), we have

$$0 = \frac{-pa^2}{4ln} + C_2$$

or $$C_2 = \frac{pa^2}{4l\eta}$$

and $$v = \frac{p}{4l\eta}(a^2 - r^2) \qquad (5.3)$$

Thus *the profile of the advancing liquid is a parabola.*

The volume of the liquid dQ flowing through the tube per second between the coaxial layers of radii r and $(r+dr)$ is given by

$$dQ = 2\pi\, rdr \times v$$

$$= 2\pi r dr \times \frac{p}{4l\eta}(a^2 - r^2)$$

or $$dQ = \frac{p\pi}{2\eta l}(a^2 - r^2)\, r\, dr$$

The *total* volume of the liquid flowing through the tube per second is, therefore, given by

$$Q = \int_0^a \frac{p\pi}{2\eta l}[a^2 - r^2]\, r\, dr$$

$$= \frac{\pi p a^4}{8 l \eta} \qquad (5.4)$$

This is *Poiseuille's formula* for the flow of a liquid through a narrow tube.

As we have seen above, *Poiseuille's formula* is valid only so long as the flow of liquid in the tube is *streamline*. The critical velocity is given by $\upsilon = K\eta/\rho a$, where ρ is the density of the liquid, η its viscosity, a the radius of the tube and K a constant called *Reynold's number* whose value for narrow tubes is in general equal to 1000 for all liquids. Thus motion of the liquid will be streamline at *low* velocities and for tubes of *small* diameter. Hence *Poiseuille's formula fails in the case of tubes of wide bore.*

Experiment 5.1

Object—To determine the *coefficient of viscosity* of water by measuring the rate of its flow through a *capillary* tube of *uniform* bore under *constant* pressure difference.

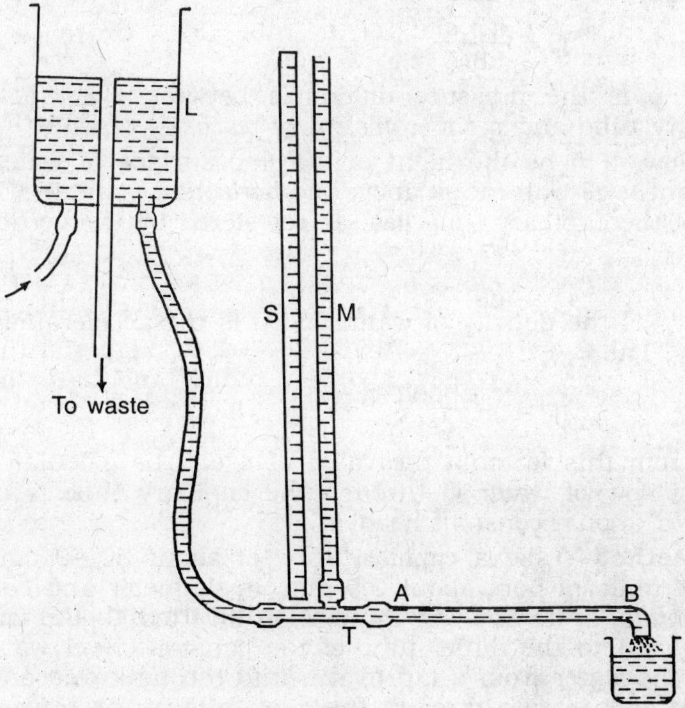

Fig. 5.3

Apparatus—Viscoity apparatus (Fig. 5.3), a capillary tube, a beaker, a graduated cylinder, a constant level water tank, a stop-watch, a microscope and a thermometer.

Description of apparatus—As illustrated in Fig. 5.3, the apparatus consists of a *capillary* tube through which water is allowed to flow from a *constant* level water tank. The tank consists of a brass cylinder of about 7.5 cm, length and of about 7.5 cm. internal diameter and is provided with inlet and outlet tubes near the sides and with a constant lever overflow tube in the centre. The ends of the tubes are *corrugated* for rubber tubing connections. The height of tank from the capillary tube can be varied by sliding it along the supporting rod and can be clamped in any desired position with screw clamps. The end of the outlet tube of the tank is connected through an inverted T piece to the capillary tube, the tail of T being connected to a vertical glass tube used to measure the head of water.

Theory—When water is allowed to flow *slowly* through a *uniform* capillary tube of length l and of radius a, its volume flowing per second is given by

$$Q = \frac{\pi \rho a^4}{8\eta l}$$

where p is the pressure difference between the ends of the capillary tube and η the *cofficient of viscosity* of water.

Now, if h be the hight of the *free* surface of water in the constant level water tank above the *horizontal* plane in which the axis of the capillary tube lies, as registered by the vertical glass tube

$$p = \rho g h$$

where p is the density of water and g is the acceleration due to gravity. Thus,

$$\eta = \frac{\pi a^4 \rho g h}{8 Q l} \tag{5.5}$$

From this formula the value of η can be calculated if the rate of flow of water Q through the capillary tube is observed under a *known constant* head h.

Method—Take a capillary tube of about 30-40 cm. length and of uniform bore about 0.5 mm. in diameter and connect it *horizontally* at its one end through an inverted T-piece as shown in Fig. 5.3 to the outlet tube of the constant level water tank. Allow the water from a tap to run into the tank and adjust its flow at such a rate through the inlet tube of the tank that the level of the water in the tank is maintained *constant*. Then watch the flow of water through the capillary tube and adjust it by altering the height of the tank from the capillary tube till the liquid leaves the tube in a *trickle*. Collect *sufficient* quantity of water trickling from the end of the capillary tube in a graduated jar during a carefully observed interval of time and find out its volume. From this calculate Q, *the volume of water flowing per second through the capillary tube. Note down the head of water* h, *i.e.*, the height of the level of water in the constant level water tank from the capillary tube as registered by the vertical glass tube.

Now alter the pressure difference by raising or lowering the water tank. Repeat the experiment with various values for h and calculate a mean value for h/Q. Note down the temperature of water.

Next, measure the length l of the capillary tube. Then mount the capillary tube in horizontal position and focus a vernier or micrometer microscope on one of its ends. Set the cross-wire *tangential* to the circular section on one side of its centre and note down the reading of the microscope. Then set the cross-wire tangential to the other side of the centre of the section and again note down the reading of the microscope. The difference between the readings of the microscope for the two positions of the cross-wire gives the diameter of the tube. Then rotate the tube through a right angle and again find out the diameter of the tube. Next focus the microscope on the other end of the tube and find out the two perpendicular diameters as above. The mean of these four readings for diameter gives the *average* diameter $2a$ of the capillary tube.

Having thus obtained a mean value for h/Q and determined l and a, calculate η from the Eq. (5.5).

Sources of error and precautions—(1) The bore of the capillary tube should be about 0.5 mm. for the *Poiseuille's formula* holds good in the case of *narrow* tubes only. With wider tubes the velocity of the liquid exceeds the critical velocity and the flow becomes turbulent.

(2) The capillary tube should be as *uniform* as possible in section; its uniformity can be tested, if desired, by measuring the length of a short thread of mercury in various positions in the tube.

(3) The capillary tube should be placed *horizontally*.

(4) Since the radius occurs raised to the *fourth* power in the *Poisuille's formula*, the diameter of the tube should be very accurately measured. Observations for diameter should be taken at both ends of the tube and at each end readings should be taken along two mutually *perpendicular* diameters.

In the case of a *micrometer* microscope, care should be taken to avoid error due to *back-lash* while taking observations for the measurement of diameter.

(5) Since *Poiseuille's foumula* is true for streamline motion only, the pressure difference between the ends of the capillary tube should be small so that the liquid may leave the tube in a *trickle*.

(6) For measurement of *Q*, a graduated cylinder of *small* least count should be used and *sufficient* quantity of water should be collected in it.

(7) While measuring the rate of flow of the liquid through the tube, the pressure difference between its ends must be *constant*. Hence the level of water in the constant level water tank should be maintained constant.

If the level of water in the vertical glass tube, used to measure *h*, changes as the water trickles down from the end of the tube, a small clean glass rod may be placed in contact with the water at the end of the tube.

(8) If the water runs back along the side of the capillary tube a little grease or vaseline should be smeared at the end of the tube, well clear of the bore.

(9) Since the viscosity of a liquid varies rapidly with temperature, the temperature of water should be observed with great care with a sensitive thermometer. In the neighbourhood of 15^0C, η for water decreases by about 3% for each degree rise of temperature.

Observations —[*A*]—*Measurement of h and Q.*

Least count of stop-watch = sec.

Least count of graduated cylinder = c.c.

Set No.	Determination of Q					h/Q
	h cm.	Volume of water collected c.c.	time taken sec.	Volume flowing per sec. Q. in c.c.	Mean value of Q. in c.c.	
1						
					Mean	

[*B*] (*i*) *Length of the capillary tube* = cm.

(*ii*) *Measurement of the diameter of the capillary tube Least* count of the microscope = cm.

Serial Number	Observations for position of cross-wire on the section of the tube											
	At one end of the tube						At the other end of the tube					
	Reading along any diameter			Reading along perp. diameter			Reading long any diameter			Reading along perp. diameter		
	One end of Section	Other end of Section	Diameter	One end of Section	Other end of Section	Diameter	One end of Section	Other end of Section	Diameter	One end of Section	Other end of Section	Diameter
	cm.	cm.	cm.	cm.	cm.	cm.	cm.	cm.	cm.	cm.	cm.	cm.
	Mean			Mean			Mean			Mean		

Average diameter of the tube = cm.

(*C*) Temperature of water = ^{0}C

Calculations —Mean radius of the tube = cm.

$$= \frac{\pi a^4 \rho g h}{8 Q l}$$

= dynes per cm.^{2} per unit velocity gradient

Result—The coeficient of viscosity of water at ^{0}C.

= dynes per cm.^{2} per unit velocity gradient

∴ *Standard* value = poise

Error = %

Criticism of the apparatus—The apparatus yields a fairly satisfactory result if the rate of flow of water through the capillary tube is kept *small*. For greater accuracy the diameter of the tube should be determined by introducing into it a column of mercury, measuring its length and then running out the mercury in a weighed watch glass, the second weighing of which gives the mass of mercury introduced into the tube and l' its length, $m = \pi a^4 l'/d$, whence $a = \sqrt{(m/\pi l' d)}$ where d is the density of mercury.

The difference of pressure between the ends of the capillary tube is not exactly equal of ρgh, for the velocity of the liquid increases as it approaches the entrance of the tube and its distribution becomes uniform only after a short length of the tube has been traversed. Correction can, however, be applied for it. This source of error is completely eliminated if the apparatus is modified as shown in Fig. 5.4 in which the pressure difference is measured directly by a manometer. The manometer is adjusted over two *fine* holes in the *capillary* tube, the distance of each hole from the nearest end of the tube being about cm. The result is that near the holes the velocity distribution becomes uniform.

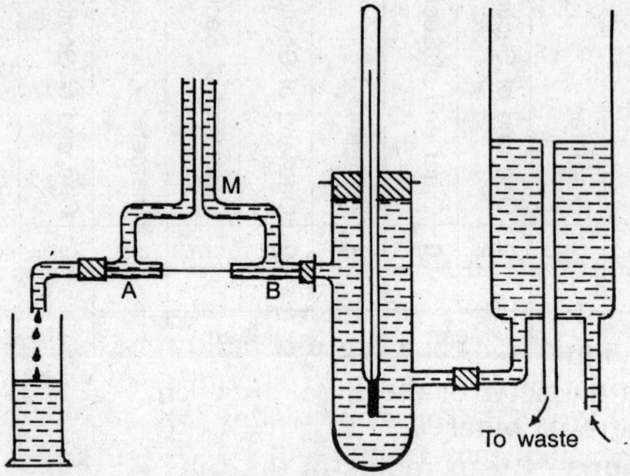

Fig. 5.4

The capillary flow method is not suitable for determining the viscosity of very viscous liquids, such as oils.

Experiment 5.2

Objects—To determine the *coefficient of viscosity* of a liquid, say, castor oil, by *Setoks' method.*

Apparatus—Falling-sphere viscometer (Fig. 5.6) steel ball-bearings, a thermometer, a stop-watch, a watch glass, a pen-knife and a scale.

Theroy—Let a very small sphere of radius a and denisty ρ, be allowed to fall *freely* from rest through a fluid of density σ and coefficient of viscosity η. At any instant the sphere is acted upon by two forces, *viz.*,

(1) A *viscous drag* $6\pi\eta a v$ directed *vertically upwards*, where v is the velocity of the sphere, and

(2) The *effective* weight $\frac{4}{3}\pi a^3 (\rho - \sigma)$ of the sphere directed *vertically downwards.*

The viscous drag increases with velocity v of the sphere and so long as it is less than the other force the velocity of the sphere increases. Finally, when the viscous drag equals the effective weight of the sphere, there is no resultant force acting on the sphere and consequently it acquires a *constant* velocity, called the *terminal velocity*. Thus, if V is the terminal velocity of the sphere, we have

$$6\pi \eta a V = \frac{4}{3}\pi^3 (\rho - \sigma)g$$

whence
$$\eta = \frac{2}{9}\frac{a^2 (\rho - \sigma)}{V} g \qquad (5.6)$$

If t is the time taken by a shpere to fall through a distance s when its velocity is constant $V = s/t$

and hence
$$\eta = \frac{2}{9}\frac{a^2 t (\rho - \sigma) g}{s} \qquad (5.7)$$

This equation can be used to calculate η.

If servaral spheres of different radii are allowed to fall through the liquid and their times of fall through the *same* height observed, a plot of a^2 aganist $1/t$ should give a *straight line as* illustrated in Fig. 5.5. The slope of this line will give an average value for $a^2 t$ which may be used in Eq. (5.7) to calculate η.

Description of Apparatus—The *falling-sphere viscometer* is shown in Fig. 5.6. A is a glass tube containing the experimental liquid. It is about 50 cm. long and 8 cm. wide, and is closed at the top by a cork bored to carry *centrally* a norrow glass tube B of about 0.5 cm. in diameter. It is through *this* tube B that the spheres are let fall into the liquid. This arrangement ensures that the motion of the spheres through the liquid is along the axis of the wider tube. The tube A is suspended

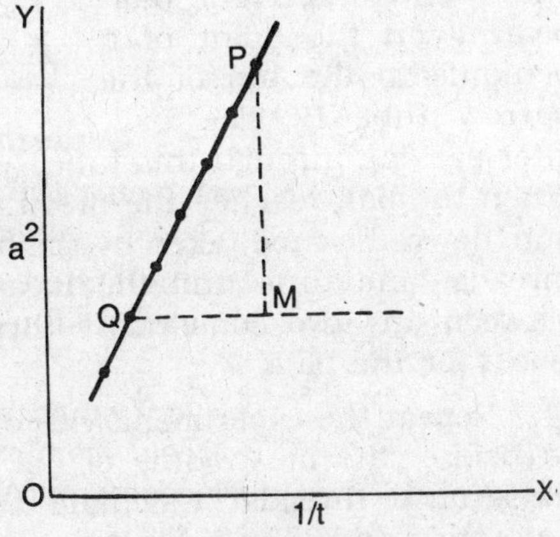

Fig. 5.5

in a large glass jar C containing water into which dip a stirrer S and a thermometer T. F is a funnel through which water can be poured into the jar.

Method—Take several steel ball-bearings of not more than 1/8 in diameter and measure the diameter of each ball accurately with a screw gauge. Then place the balls in the experimental oil taken in a watch-glass and roll them one by one until their surfaces are thoroughly wetted by the oil.

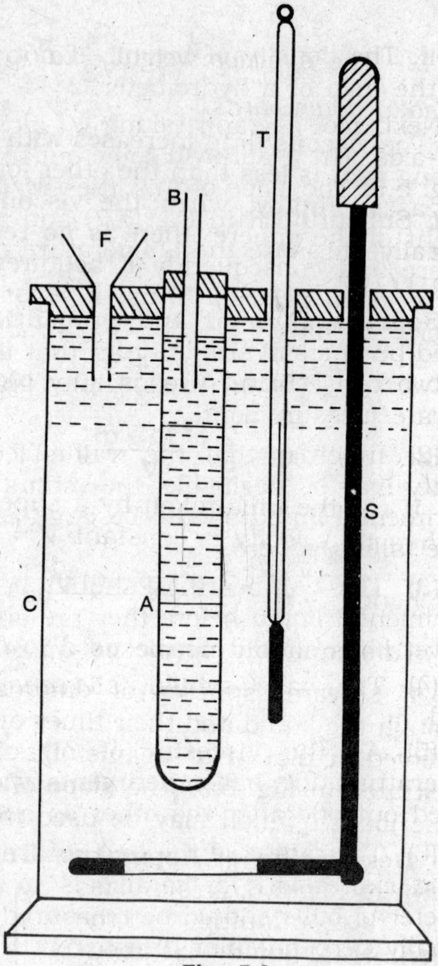

Fig. 5.6

Next set up the viscometer as shown in Fig. 5.6 and tie four cotton threads, preferably colured, around the wide tube A dividing it into five portions, the three central ones being adjusted to be *exactly* 10 cm. Then transfer one ball-bearing on the point of a penknife to the top of the narrow tube B. The ball, after passing through this tube, will fall *centrally* through the oil. When the ball reaches the first thread, start the stop-watch and note down the time taken by the ball to travel between successive threads, and thus find the mean value of the time of transit between any two consecutive threads. Then calculate the value of a^2t for the ball.

Repeat the experiment with other ball-bearings and find out, as before, the mean time of fall of each ball between any two *consecutive* threads. Calculate the value of a^2t for each ball separately and then take the mean. Use this mean value of a^2t in Eq. (5.7) to calculate the value of coefficient of viscosity of the oil. The density of the oil, if not known, can be determined with the help of a hydrometer.

Next, plot a graph taking $1/t$ along the X axis and a^2 along the Y-axis. The graph will come out to be a *straight line* as shown in Fig. 5.5. The slope of this line *PM/QM* gives the average value of a^2t. Substitute it in Eq. (5.7) and again calculate the value of η. Finally, measure the temperature of water in the jar correctly up to $1/10^0$C.

Sources of error and precautions—(1) The ball-bearings should be of small size in order that their time of transit between any two consecutive threads may be sufficiently large to admit accurate measurement.

(2) In order that the wall-effect of the value of terminal velocity may be negligible, the radius of the tube containing the experimental liquid should be large as compared with the radius of the falling sphere.

(3) The ball-bearings should be thoroughly wetted in the experimental liquid before they are allowed to fall through it and the wet balls should be picked up only by a pen-knife.

(4) The balls should fall *centrally* into the experimental liquid.

(5) As the viscosity of oil changes very rapidly with temperature, the ball-experiments with various balls should be carried out one after the other as quickly as possible.

(6) The radius of the ball occurs raised to the *second* power in the expression for η and is a small quantity. Hence the diameter of balls should be measured very accurately and in two mutually perpendicular directions.

(7) The temperature of the liquid should be measured with a sensitive thermometer reading up to $1/10\ ^0$C.

(8) The graph between a^2 and $1/t$ should be plotted on a large graph paper and the scales be so chosen as to represent fully the observed accuracies in the quantities to be plotted.

(9) The graph between a^2 and $1/t$ should be *straight line* and the various points should be evenly distributed about it and should lie as close to it as possible.

Observations and Calculations

[*A*] (*i*) *Density of the oil* = gm./c.c

(*ii*) *Density of the ball bearing* = gm./c.c.

(*iii*) *Temperature of the water-bath* = ^{0}C

[*B*] *Measurement of diameter of ball-bearings*

Least count of the screw = cm.

zero error = cm.

Ball	Diameter in any direction x cm.	Diameter in perp. direction y cm.	Mean Diameter $\frac{x+y}{2}$ cm.	Radius a cm.	a^2 $cm.^2$
A					
B					
C					
...					
...					
G					
H					

[*C*] *Measurement of time of fall*

Least count of stop-watch = sec.

Ball	Distance between consecutive threads cm.	Time of transit sec.	Mean distance s cm.	Mean time of fall t Sec.	l/t $sec.^{-1}$	a^2t $cm.^2$ sec.
A						
B						
...						
...						
G						
H						

Mean =

$$\eta = \frac{2}{9} \cdot \frac{a^2 t (\rho - \sigma) g}{s}$$

=

= poise

Average value of a^2t obtained from graph (Fig. 5.5)

$$\frac{PM}{QM} = \quad = \quad \text{cm.}^2 \text{ sec.}$$

$$\therefore \quad \eta = \frac{2}{9} \cdot \frac{a^2 t (\rho - \sigma) g}{s}$$

=

= poise

Result—The value of *coefficient of viscosity* of the liquid (......) at °C as obtained *experimentally*

(*i*) by calculation } = { dynes per cm. per unit velocity
(*ii*) by the use of graph } = { gradient.

Criticism of the method—It is a simple and convenient method for determination of viscosity of highly viscous liquids such as oils, and gives quite satisfactory results. The method can also be used to study the variation of viscosity of oils with temperature for which the water-bath has to be maintained at different temperatures. More accurate results can be obtained by use of very small spheres made of Rose's* metal and having radii between 0.002 cm. and 0.1 cm. The diameter of the spheres can then be determined with a micrometer eye-piece mocroscope.

Since Stokes' law holds for a infinite extent of a viscous fluid, two corrections due to the boundary conditions of the container have to be applied, if a very accurate value of η is required. Ladenburg had shown that the terminal velocity corrected for *wall-effect* is given by

$$V_\infty = V\,(1+2.4\ a/R)$$

where V is the observed value and R is the radius of the container.

Similarly, to correct for *end-effect*

$$V_\infty = V\,(1+3.3\ a/b)$$

where b is the total height of the liquid.

Experiment 5.3

Object—To determine the viscosity of water by variable pressure-head method.

Apparatus—A burette with pinch cock, capillary tube, stop watch, metre scale and microscope.

Theory—Consider a burette connected to a capillary tube with rubber tubing so that the tube is flush against the end of the burette. Let the burette be filled with water. If water be allowed to drain under its own hydrostatic pressure from the burette through the capillary as an outlet, then the pressure at any time is

$$p = h\,\rho g$$

where h is the vertical height of the liquid column, ρ the density and g is the acceleration due to gravity. The volume rate of flow can be written as

* Rose's metal Sb 25%, Pb 25% and Bi 50%, M.P. = 94°C

$$Q = -A\frac{dh}{dt} \tag{5.8}$$

where A is the cross-secional area of the burette.

From Poiseuille's equation we know that

$$Q = \frac{\pi R^4 p}{8 l \eta} = \frac{\pi R^4 h \rho g}{8 l \eta} \tag{5.9}$$

where p = $(h\rho g)$ is the pressure-head, l the length of the capillary tube of radius r and η the coefficient of viscosity.

Combining the above two equations, we have

$$-A\frac{dh}{dt} = \frac{\pi R^4 h\rho g}{8 l \eta}$$

$$\frac{dh}{h} = -\frac{\pi R^4 \rho g}{8 l \eta A}.dt$$

Upon integration and conversion to base 10 logs.

$$\log_{10} h = -\left(\frac{\pi R^4 \rho g}{2.303 \times 8 l \eta A}\right).t + C$$

$$= -\frac{Bt}{\eta} + C \tag{5.10}$$

where C is the constant of integration and where $B = \dfrac{\pi R^4 \rho g}{2.303 \times 8 l A}$ is a constant of the apparatus.

From the above Eq. (5.10) it is evident that a plot of $\log_{10} h$ versus t should be a straight line whose slope depends upon the viscosity of the liquid and the constant of the instrument.

From a graph of $\log_{10} h$ versus t, the slope (m) of the best line through the points is taken. Our equation relates

$$m = -\frac{B}{\eta}$$

where $$B = \frac{\pi R^4 \rho g}{2.303 \times 8 l A}$$

Thus, knowing m from the graph and the value of B from a knowledge of ρ, g and the constants of the apparatus, we can calculate η from the relation

$$\eta = -\frac{B}{m} \tag{5.11}$$

Method—Arrange the apparatus as depicted in Fig. 5.7. Fill the burette with distilled water and record its temperature.

Allow it to drain through the capillary. As the meniscus passes Zero ml. mark, start the stop-watch and record the time t as it passes each 5 ml. graduation h_x ($x = 0, 5, 10, 15.....h_x$) being the height of the x ml. mark from the level of the capillary tube.

Repeat the above set to get a series of measurements.

Measure the length of the capillary tube with a metre scale and its diameter with a microscope in the usual manner. Compute its mean radius r.

The cross-sectional area A of the burette is

$$A = \frac{50}{(h_0 - h_{50})}$$

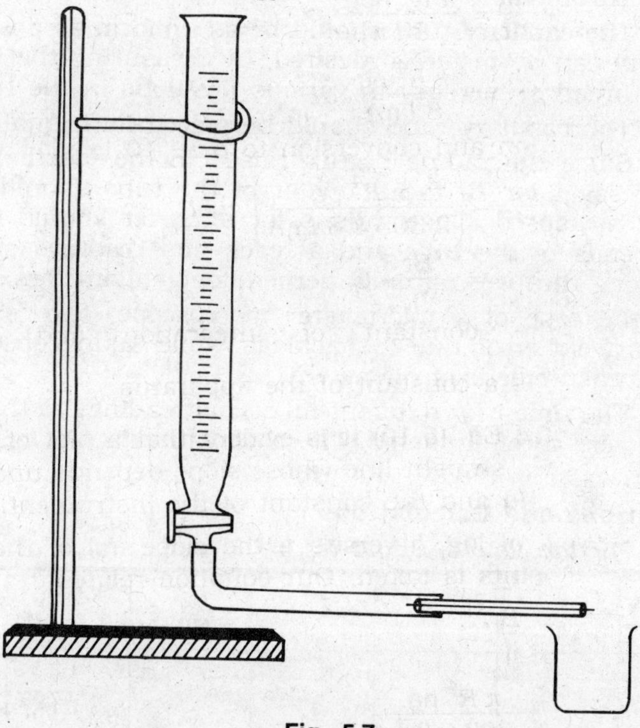

Fig. 5.7

Calculate this. This gives

$$h_x = h_0 - \frac{x}{50}(h_0 - h_{50})$$

From the above calculate *hoc* for the meniscus at every 5 ml. graduation and knowing t for these values of h_x, plot a graph between h_x and t.

Get the slope (m) of the best line through the points.

Calculate B from the equation :

$$B = \frac{\pi R^4 h\rho g}{2.303 \times 8\, l A}$$

and knowing m and B calculate η from the relation

$$\eta = -\frac{B}{m}$$

Sources of error and precautions—(1) The bore of the capillary tube should be about 0.5 mm. for Poiseuille's equation holds good in the case of narrow tubes only. With wider tubes the velocity of the liquid exceeds the critical velocity and the flow becomes turbulent.

(2) The capillary tube should be as uniform as possible. Its uniformity can be tested, if desired, by measuring the length of a short thread of mercury in various positions in the tube.

(3) The capillary tube should be placed horizontally.

(4) Since the radius occurs raised to the fourth power in the expression for B, the diameter of the tube should be very carefully measured. Observations for diameter should be taken at both ends of the tube and at each end readings should be taken along the two mutually perpendicular diameters.

In the case of a micrometer mocroscope, care should be taken to avoid error due to backlash while taking observations for the measurement of diameter.

(5) The time t for different meniscus readings to the burette must be very carefully noted. The observations must be repeated twice.

Observations and Calculations

1. *For the flow of liquid* h_{50} = cm.

ml. mark on burette	Time in Secs.			h_x cm.	log h_x	Point on Graph.
	I	II	III			
0 ml.						A
5 ml.						B
10 ml.						C
15 ml.						D
20 ml.						E
25 ml.						F
30 ml.						G
35 ml.						H
40 ml.						I
45 ml.						J

II. (*a*) $h_0 - h_{45} =$ cm.

$A =$ *sq. cm.*

(*b*) Length of the capillary tube = cm.

(*c*) *For diameter of Capillary tube*

S. No.	Reading along any diameter			Reading along a perpendicular diameter		
	one end	other end	diam.	one end	other end	diam.
1						
2						
3						
4						
5						
					Mean	

∴ Mean diameter of the capillary tube = cm.

∴ Mean radius of the capillary tube = cm.

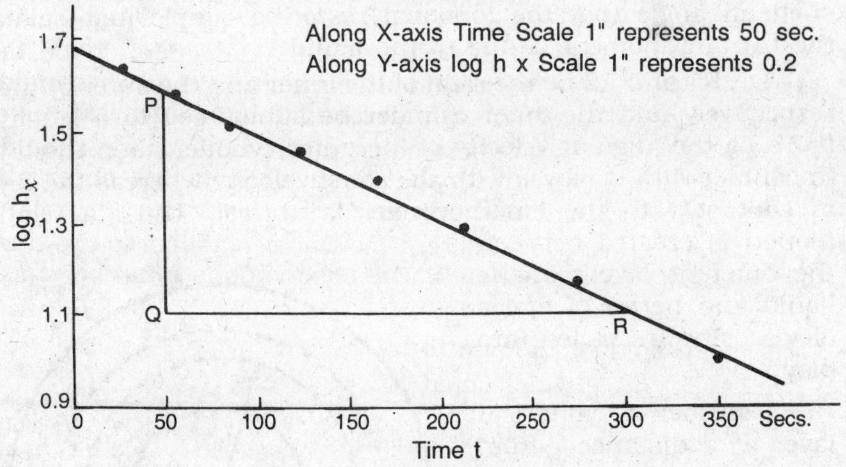

Fig. 5.8. Graph between log h_x and t

From the graph between log h_x and t

$$m = \frac{PQ}{QR}$$

∴ log m =

Also B $= \dfrac{\pi R^4 \rho g}{2.303 \times 8\, l A}$

∴ log B =

Hence $\log \eta = \log B - \log m$

$\therefore \quad \eta = $ Poise

Result—The coefficient of viscosity of waer at... °C = ... Poise

5.3. Rotating Cylinder Method—The value of the coefficient of viscosity of liquids like glycerine, water and of air is conveniently determined by means of a rotating cylinder. The apparatus consists of two coaxial metal cylinders. The outer cylinder is clamped on a turn-table which may be rotated with constant angular velocity by means of an electric motor. The inner cylinder is suspended coaxially with the outer cylinder by a phosphor-bronze suspension which carries a mirror to measure the angle of rotation of the inner cylinder by means of telescope and scale arrangement. The space between the inner and outer cylinders is filled with the experimental liquid. When the outer cylinder is set into rotation with constant angular velocity, a couple G is transmitted to the inner cylinder due to the intervening liquid. The couple rotates the inner cylinder through such an angle that the torsional restoring couple just balance the turning moment *G* due to the liquid

Let R1 and R2 be the radii of the inner and the outer cylinder respectively and the inner cylinder be submerged to a lenght *l*. If *Wa* be the angular velocity of the outer cylinder, then the layer in contact with it moves with the same velocity whereas the layer in contact with the inner cylinder is at rest. Thus, a relative motion is created between the different layers of the liquid and hence of forces of viscosity are called into play.

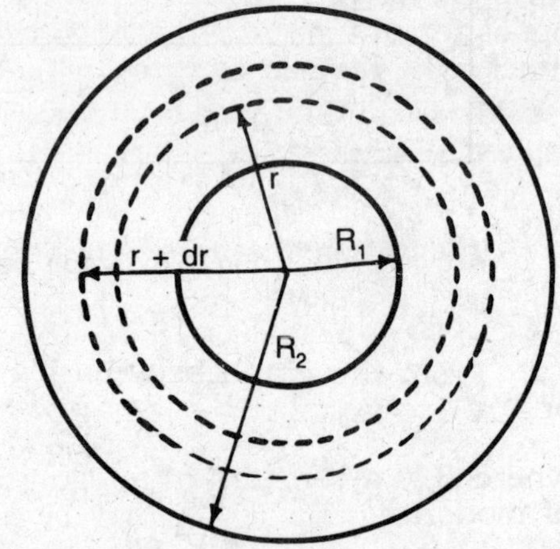

Fig. 5.9

Consider a cylindrical layer at a distance *r* from the axis of rotation, moving with an angular velocity *wn* then its linear velocity is

$$v = rw$$

Let the angular velocity of rotation increase by dw across the cylindrical shell between radii r and $r+dr$ as shown in Fig. 5.9. Then across the cylindrical shell

$$\text{Velocity gradient} = \frac{dv}{dr}$$

$$= \frac{d}{dr}(rw)$$

$$= w + r.\ \frac{dw}{dr}$$

Since the first term w on the right-hand side of this equation represents the angular velocity which the layer would have if the fluid were rotating like a rigid body, it is evident that the velocity gradient responsible for producing viscous drag on the layer under reference is $r.\ \frac{dw}{dr}$.

According to Newton's law the viscous forces acting tangentially at the surface of the cylindrical shell of liquid is

$$F = -\eta 2\pi rl.\ r.\frac{dw}{dr} = -2\pi\,\eta l.\ r^2.\ \frac{dw}{dr}$$

Taking the moment of this force about the central axis we get

$$G = F.r = -2\pi\,\eta\,l\,.r^2.\ \frac{dw}{dr}\ r$$

$$= -2\pi\,\eta\,l\,r^3.\ \frac{dw}{dr}$$

When the steady state is reached this couple G is constant throughout the liquid and must be equal in magnitude and opposite in sense to the steady external couple acting on the inner cylinder due to twist in the suspension wire.

The equation can be written as

$$G.\frac{dr}{r^3} = 2\pi\,\eta\,l\,dw$$

Integrating the above equation we have

$$\int \mathrm{G}\ \frac{dr}{r^3} = \int\ 2\pi\,\eta\,l\,.\,d\omega$$

or $$-\frac{G}{2r^2} + B = 2\pi\,\eta\,l\,\omega$$

where B is a constant of integration. To evaluate B the condition of motion $r = R_1$, $\omega = 0$ may be applied. This gives

$$B = \frac{G}{2R_1}$$

whence $-\frac{G}{2r}+\frac{G}{2R_1^2}=2\pi\, l\, \eta .\, \omega$

We know also that $r = R_2$, $\omega = \omega_0$, Then,

$$\frac{G}{2R_1^2}-\frac{G}{2R_2^2}=2\pi\,\eta\, l .\omega_0$$

$$\therefore \quad G = 4\pi\,\eta\,\omega_0 \,.\, \frac{R_1^2R_2^2}{(R_2^2-R_1^2)}\,.l$$

The restoring couple offered by the suspension fibre = θ C. where C is the couple per unit radian twist.

In the equilibrium position the two couples balance each other, hence,

$$G = C.\ \theta$$

or

$$\frac{4\pi\,\eta\, l\,\omega_0\ R_1^2\,R_2^2}{(R_2^2-R_1^2)} = C.\theta$$

In the above discussion we have not taken into account the torque on the base of the inner cylinder. To correct for this couple, the cylinder is submerged in the liquid for two different lengths. If $f(\beta)$ be the unknown couple over the base of the inner cylinder, then

$$G = \frac{4\pi\,\eta\, lw_0\ R_1^2\,R_2^2}{(R_2^2-R_1^2)} + f(\beta)$$

Let l_1 *and* l_2 be the two lengths of the inner cylinder in the liquid and θ_1 and θ_2 the steady deflections produced. Then

$$\frac{4\pi\,\eta\, l_1\,\omega_0\, R_1^2\,R_2^2}{(R_2^2-R_1^2)} + f(\beta) = C.\ \theta_1 \qquad (5.12)$$

and

$$\frac{4\pi\,\eta\, l_2\,\omega_0\ R_1^2\,R_2^2}{(R_2^2-R_1^2)} + f(\beta) = C.\ \theta_2 \qquad (5.13)$$

Subtracting Eq. (5.13) from Eq. (5.12) we have

$$\frac{4\pi\,\eta\, l\,\omega_0\ \ R_1^2\,R_2^2}{(R_2^2-R_1^2)}(l_1-l_2) = C.\ (\theta_1=\theta_2)$$

whence

$$\eta = \frac{C(R_2^2-R_1^2)\,(\theta_1-\theta_2)}{4\pi\omega_0\,R_1^2R_2^2\ (l_1-l_2)} \qquad (5.14)$$

To determine C, the torque per unit radian twist due to torsional reaction in the suspension wire, a hollow metal disc D

is provided whose moment of inertia I_1 about the axis of rotation can be calculated from a knowledge of its mass and dimensions. This inner cylinder is allowed to oscillate torisionally in air first alone and then with the disc placed centrally upon it. If T_0 and T_1 be the periods of oscillation in the two cases respectively and I_0 the moment of inertia of the cylinder about the axis of rotation, we have

$$T_0 = 2\pi \sqrt{\frac{I_0}{C}}$$

and $$T_1 = 2\pi \sqrt{\frac{I_0 + I_1}{C}}$$

From the above two equations we get

$$C = \frac{4\pi^2 I_1}{T_1^2 - T_0^2} \tag{5.15}$$

Substituting this value of C in Eq. (5.14) we have

$$\eta = \frac{4\pi^2 I_1}{T_1^2 - T_0^2} \times \frac{(R_2^2 - R_1^2)(\theta_1 - \theta_2)}{4\pi \omega_0 R_1^2 R_2^2 (l_1 - l_2)}$$

$$= \frac{\pi^2 I_1}{T_1^2 - T_0^2} \times \frac{(R_2^2 - R_1^2)(\theta_1 - \theta_2)}{\omega_0 R_1^2 R_2^2 (l_1 - l_2)} \tag{5.16}$$

If the mass of the disc placed over the cylinder be m and a and a its internal and external radii,

$$I_1 = m(a_2 + b_2)/2$$

Substituting this value of I_1 in Eq. above

$$\eta = \frac{m(a^2 + b^2)\pi}{2(T_1^2 - T_0^2)} \times \frac{(R_2^2 - R_1^2)(\theta_1 - \theta_2)}{\omega_0 R_1^2 R_2^2 (l_1 - l_2)} \tag{5.17}$$

Knowing all the factors on the right hand side of this equation, η can be calculated.

Experiment 5.4

Object—To determine the coefficient of viscosity of water by rotating cylinder method.

Apparatus—Rotating cylinder apparatus for viscosity determination, stop-watch, physical balance, telescope and scale arrangement for measuring angular deflections, vernier callipers and metre scale.

Description of apparatus—The apparatus consists mainly of a revolving table T, upon which is carried the outer cylinder B which can be easily clamped in a position coaxial with the spindle. The spindle can be rotated from 20 to 60 r.p.m by a

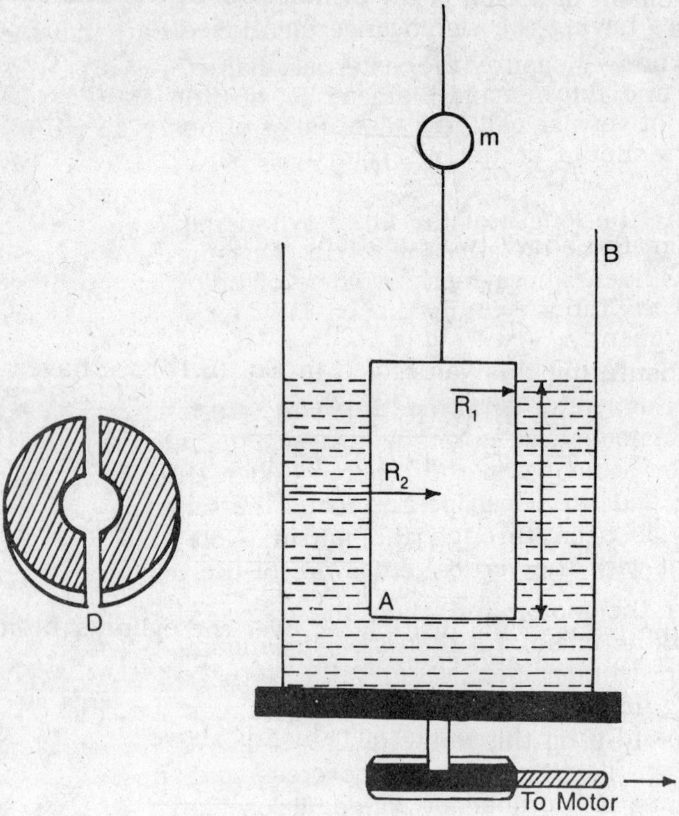

Fig. 5.10

small motor. The inner cylinder A is suspended in the fluid by means of a long and thin wire and carries a small plane mirror *m*. The steady twist produced in the wire when the outer cylinder is rotated at a constant speed is read by means of a telescope and scale arrangement. A hollow metal disc *D* usually of the same external diameter as that of the inner cylinder is also provided with the apparatus. An additional inner cylinder of the same diameter but of a different length is also supplied with the apparatus.

Theory—From Eq. (5.17) we have

$$\pi = \frac{m\,(a^2 + b^2)\,\pi}{2\,(T_1^2 - T_0^2)} \times \frac{(R_2^2 - R_1^2)\;(\theta_1 - \theta_2)}{\omega_0\,R_1^2\,R_2^2\,(l_1 - l_2)}$$

the letters having the significance mentioned earlier in Art. 5.3.

Method—Measure the external diameter R_1 of the inner cylinder and the internal diameter R_2 of the outer cylinder with the help of vernier callipers. A number of observations for these diameters should be taken. From these find the mean values of R_1 and R_2.

Along the length of the inner cylinder make scratches with a sharp point at distances of 2 cm. measured from the bottom of the cylinder. These marks will be of help in determining the height of the inner cylinder dipped in water during the course of the experiment.

Suspend the inner cylinder coaxially with the outer cylinder with the help of a long and thin wire. Place the telescope and scale arrangement at a suitable distance (usually one metre) from the mirror attached to the inner cylinder. Adjust the height of the scale and the telescope and focus the telescope on the scale divisions as seen through the mirror. Note the scale division coincident with the vertical crosswire of the telescope eyepiece.

Start the motor and allow the outer cylinder to rotate for half a minute or so and then stop the motor. As a result of this the inner cylinder being under the action of a torque starts oscillating about the suspension wire as axis. Look through the telescope and start the stopwatch when the scale division, which in the rest position was on the cross-wire, passes the cross-wire. Count a number of torsional oscillations of the inner cylinder and time these with the stop-watch. Compute the time period. Repeat the observations a number of times and calculate the mean value of the time period T_0.

Weigh the two pieces of the light brass annular disc, usually supplied with the apparatus, and find its mass *m*. Determine the internal and the external diameter of this disc. with a vernier callipers a number of times and find their mean values *a* and *b* respectively.

Place the two pieces of the brass disc symmetrically over the inner cylinder such that the suspension wire passes through the centre of the disc. As before allow the inner cylinder with the disc over it to oscillate torsionally about the suspension wire as axis. Looking through the telescope time a number of oscillations of this system with a stop-watch and compute its periodic time. Repeat the observations a number of times and calculate the mean value of the time period T_1.

Pour water in between the outer and the inner cylinder and with the help of a pipette carefully adjust the level of water to be on one of the scratches made on the inner cylinder. From this position of the mark determine the depth l_1 of the inner cylinder dipped in water.

With the outer cylinder stationary, oscillate the inner cylinder torsionally and when it comes to a position of rest, note the scale division x_1 which coincides with the vertical crosswire of the eyepiece.

Now start the motor and allow the outer cylinder to rotate with a constant angular velocity. As a result of this a orque G acts on the inner cylinder which therefore tends to rotate about the suspension wire as axis; the twist in the suspension wire provides the restoring torsional couple. As such the inner cylinder is rotated through an angle θ, the equilibrium position being attained when the torque G equals the restoring torsional couple. On looking through the telescope observe the deflected position of the inner cylinder by noting the scale division x_3 which coinclides with the vertical cross-wire of the eyepiece.

Evidently $(x_2 - x_1)$ gives the displacement on the scale whose distance measured from the morror is d cm. It is evident that θ_1, the angle of rotation of the mirror or the inner cylinder, can be calculated from the relation

$$\theta_1 = \frac{1}{2} \cdot \frac{x_2 - x_1}{d} \times \frac{180}{\pi} \text{ radian}$$

This is so because the reflected ray turns through twice the angle through which the mirror rotates. When this measurement has been made, stop the motor.

Add more water in between the inner and the outer cylinder and as before determine the depth l_2 to which the inner cylinder is immersed in water.

With the outer cylinder stationary, determine the rest position of the inner cylinder by noting the scale division x'' which coincides with the vertical cross-wire of the eyepiec.

Now, as before, start the motor and allow the outer cylinder to rotate at the same angular velocity. In looking through the telescope observe the deflected position of the inner cylinder by noting the scale division x'' which coincides with the vertical cross-wire of the eyepiece of the telescope. Compute θ_2 from the relation

$$\theta_2 = \frac{1}{2} \cdot \frac{x'' - x'}{d} \times \frac{180}{\pi} \text{ radian.}$$

This is for the length l_2 of the inner cylinder immersed in water.

As the outer cylinder is rotating, note the reading n_1 of the revolution counter and start the stop-watch. After the cylinder has made a number of revolutions so that its reading becomes n_2, stop the stop-watch. From the recorded time t seconds and the number of revolutions, $n_2 - n_1$, performed by the outer cylinder during this period, calculate its angular velocity ω_0 from the relation.

$$\omega_0 = \frac{2\pi(n_2 - n_1)}{t} \text{ radian/sec.}$$

Repeat these observations and find the mean value of ω.

Knowing now all the factors on the right hand side of Eq. (5.17), calculate η, the coefficient of viscosity of water.

Sources of Error and precautions—At one cross-section measure the diameter of the two cylinders along two perpendicular directions. Repeat the measurement in the same cross-section and at different cross-sections. Then find the mean values of R_1 and R_2. The inner and the outer diameter of the annular ring should be similarly measured.

2. The axis of the outer cylinder should be made vertical with the help of the plump line and levelling screens.

3. If the axis of the outer cylinder does not coincide with the axis of the inner cylinder, the distance between the two cylinders will not remain the same and as such an error is liable to be introduced. Whether the two axes are coincident or not, can be tested in the following manner. A thin rod smeared with powder is laid horizontally attached to the suspension of the inner cylinder such that it protrudes beyond the rim of the outer cylinder and is in contact with the rim. The outer cylinder is then rotated by hand. If the thin rod is touched by the outer cylinder only at one mark, the axes of the two cylinders are coincident. If the thin rod is touched by the outer cylinder at a

number of points as examined with the help of marks made by it on the powdered thin rod, the axes of the two cylinders are not coincident. For making the two axes coincident, move the whole arrangement which carries the outer cylinder suitably and test again with the help of the thin rod.

4. The inner cylinder should be suspended by a long and thin wire so that the value of θ is appreciable.

5. It must be very carefully noted that when water is poured in between the two cylinders, the inner cylinder does not float in it; if it does, it may be necessary to use a heavy annular ring placed over the inner cylinder to submerge it into water.

T_0 should be determined with this composite arrangement, of course with air between the cylinders. T_1 is then determined by placing another annular ring of known mass and dimensions.

6. T_0 and T_1 should be carefully determined by timing a large number of oscillations with a stop-watch reading upto 1/5th of a second.

7. The speed of rotation should be constant to obtain steady deflection of the inner cylinder and should be such that the motion of the liquid (water) between the two cylinders is stream-line

Observations—1. *Measurement of diameters.*

(1) ***Cylinders***

S.No.	Internal diam. of outer cylinder			External diam. of inner cylinder		
	Reading along		Mean	Reading along		Mean
	any diam	a perp. diam		any diam.	a perp. diam.	
1						
2						
3						
	Mean $2R_2$			Mean $2R_1$		

(2) ***Annular ring***

Similar table for $2a$ and $2b$, the outer and the inner diameters of the annular ring.

Mass of the ring (m) = gm.

II—*Measurement of Time Periods*

S. No.	T_0 (without Ring)			T_1 (with Ring)		
	No. of oscillations	Time taken secs.	T_0 sec	No. of oscillations	Time taken secs.	T_1 sec.
1						
2						
3						
4						
		Mean			Mean	

III—*For Measurement of* θ_1

Distance between scale and the mirror (d) = cm.

Length of inner cylinder dipped in water (l_1) = cm.

S.No.	Initial Position of scale reading (outer cylinder stationary) cm.	Final Position of scale reading (outer cylinder rotationg) cm.	Displacement cm.
1	x_1	x_2	$x_2 - x_1$
2			
3			
4			
		Mean	

IV—*For Measurement of* θ_2

Length of the inner cylinder dipped in water (l_2)= cm.

Distance between the scale and the mirror as before is d= cm.

Similar table as for θ_1

Mean displacement from this table... $x'' - x'$ = cm.

V—Determination of ω_0

S.No.	Initial Reading of counter n_1	Final Reading of counter n_2	No. of revolutions $n_2 - n_1$	Time taken t	Time period $T = \frac{t}{n_2 - n_1}$
1					
2					
3					
4					
				Mean	

Temperature of water during experiment = ^{0}C.

Calculations

Mean Radius of the outer cylinder (R_2) = cm.

Mean Radius of the inner cylinder (R_1) = cm.
Mean outer radius of the annular ring (a) = cm.
Mean inner radius of the annular ring (b) = cm.

$$\theta_1 = \frac{1}{2} \cdot \frac{x_2 - x_1}{a} \times \frac{180}{\pi}$$

= radian

Similarly,

$$\theta_2 = \frac{1}{2} \cdot \frac{x'' - x'}{d} \times \frac{180}{\pi}$$

= radian

$$\omega_o = \frac{2\pi}{T}$$

= radian per sec.

Substituting these calculated values and other observed values, we have

$$\eta = \frac{m(a^2 + b^2)\pi}{2(T_1^2 - T_0^2)} \times \frac{(R_2^2 - R_1^2)(\theta_1 - \theta_2)}{\omega_0 R_1^2 R_2^2 (l_1 - l_2)}$$

= Poise

Result—The coefficient of viscosity of water as determined by rotating cylinder apparatus at ... °C = Poise

Standard value at ... °C = Poise

Percentage Error =

ORAL QUESTIONS

VISCOSITY

1. Distinguish between streamline and turbulent motions.
2. What is critical velocity and on what factors does it depend?
3. Distinguish between viscosity and coefficient of viscosity. State the unit of coefficient of viscosity.
4. Compare viscosity with rigidity.
5. How does coefficient of viscosity change with temperature and pressure?

POISEUILLE'S METHOD

1. On what factors does the rate of flow of a liquid through a capillary tube depend?
2. Why does Poiseuille's formula not hold in the case of tubes of wide bore?
3. What are the conditions on which the validity of Poiseuille formula rests and how can they be fulfilled in practice?

4. Why should the capillary tube be of uniform bore and why should the bore be about 0.5 mm.?
5. How can the uniformity of the bore be tested?
6. Should the liquid leave the capillary tube in a trickle? If so, why?
7. How will you keep the pressure difference constant?
8. If the liquid runs back along the outside of the tube, what precautions will you take to stop it?
9. Which quantity in your formula for η will you measure most accurately?
10. Why do you measure two perpendicular diameters?
11. How can the diameter be determined more accurately by introducing a thread of mercury into the tube?
12. What limits the accuracy of the result in this experiment?
13. Why does the head of liquid at the inlet end of the tube does not represent the correct value of pressure difference? How can the apparatus be modified to get rid of this defect? Is this method suited to determine the viscosity of all liquids? If not, why?
14. Why do you record the temperature of the experimental liquid?

STOKES' METHOD

1. What is Stokes' law and what are its limitations?
2. How do you find out viscosity of oils by falling-sphere viscometer?
3. What is terminal velocity?
4. When does the falling sphere acquire a terminal velocity?
5. What precautions do you take in this expt.?
6. How is it ensured that the sphere falls centrally through the liquid?
7. Why are the spheres wetted in oil before allowing them to fall through the liquid?
8. How do you time the fall of spheres?
9. Why should the fall expts, with various spheres be carried out quickly one after another?
10. How can you study the variation of viscosity with temperature by means of this apparatus? Discuss the accuracy of result obtainable by the method.

6

Miscellaneous Experiments in Mechanics

Experiment 6.1

Object. To study simple harmonic under-damped oscillations.

Apparatus. A uniform rectangular wooden bar (a metre scale), suspension wire, weights, dampers and stop watch.

Theory. Any physical system exhibiting pure simple harmonic motion may be described by the following equation :

$$I \frac{d^2\theta}{dt^2} = -C\theta \tag{6.1}$$

However, this is only an ideal condition, for damping is invariably present in all practically realisable systems. Consequently another term accounting for the damping has to be included in Eq. (6.1) in order to make its application to a real system possible. As the damping is caused mainly by the resistive property of the fluid in which the entire system is immersed, the resistive force and hence the damping couple must be some function of velocity. This damping couple can within reasonable approximation be described as a linear function of velocity. Accordingly, Eq. (6.1) is modified to take the form

$$I. \frac{d^2\theta}{dt^2} = -K.\frac{d\theta}{dt} - C\theta \tag{6.2}$$

where $\left(-K.\frac{d\theta}{dt}\right)$ denotes the damping couple, the -ve sign signifying that it is always directed against the direction of velocity, and $-C\theta$ is the restoring couple as in the case of simple harmonic motion. The above equation could be rewritten as

$$\frac{d^2\theta}{dt^2} = -\frac{K}{I}.\frac{d\theta}{dt} - \frac{C}{I}\,\theta$$

or

$$\frac{d^2\theta}{dt^2} + \gamma.\frac{d\theta}{dt} + \omega_0^2\,\theta = \theta \tag{6.3}$$

where $\gamma = \frac{K}{I}$ and $\omega_0 = \sqrt{C/I}$

Thus it is observed that γ characterises the damping and ω_0 denotes the frequency of the system if there were no damping present. It may be noted that both γ and ω_0 have the same dimension—that of frequency.

One of the solutions of Eq. (6.3) is of particular relevance to the present discussion and hence only that one shall be considered. This is under the conditions of small damping,

$$\theta = \theta_0 \exp.(-\gamma t/2) \sin(\omega t+\phi) \tag{6.4}$$

where $$\omega^2 = \omega_0^2 - \frac{\gamma^2}{4} \tag{6.4a}$$

Figure (6.1) shows a plot of Eq. (6.4) for $\phi = \pi/2$. It is observed that the zeros of the curve are equally spaced with a

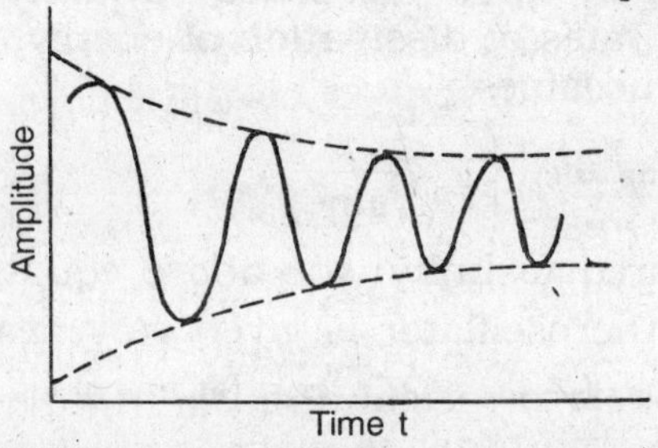

Fig. 6.1

separation $\omega \Delta t = \pi$ and hence the curve represents an oscillatory motion of gradually decreasing amplitude. The envelope of the damped oscillatory curve has also been depicted in the same figure. The curve is drawn for a case in which the decay of vibrations is rapid. If, however, the damping is small, the motion approximates to simple harmonic at constant amplitude over a number of cycles. Under such conditions, one could express the effect of damping in terms of an exponential decay of the total energy E. For, if $\gamma << \omega$ we can say that around time t the oscillations are well described over several cycles by S.H.M. of constant amplitude θ such that

$$\theta = \theta_0 \exp.(-\gamma t/2) \tag{6.5}$$

where $T = (1/\gamma)$ is known as the relaxation time for energy.

The total mechanical energy of a simple harmonic oscillator is given by

$$E(t) = \frac{1}{2} K \theta^2$$

Substituting for θ from Eq. (6.5) we obtain

$$E(t) = E_0 \exp.(-\gamma t) = E \exp.(-t/T) \tag{6.6}$$

Thus T, the relaxation time, is the time required for the energy to reduce to 1/e of its initial value.

Another parameter of interest is the *quality factor*, Q, of the oscillatory system, which may be *defined as 2π times the ratio of energy stored to the average energy loss per period.* The Q value of the oscillatory system is given by the ratio of ω_0 and γ Thus,

$$Q = \frac{\omega_0}{\gamma} = \omega_0 T \tag{6.7}$$

Q is a pure number, large compared to unity for oscillating systems with small rates of dissipation of energy. In terms of the Q value Eq. (6.4a) becomes

$$\omega^2 = \omega_0^2 \left(1 - \frac{1}{4Q^2}\right) \tag{6.8}$$

If Q is large compared to unity, the above equation yields $\omega = \omega_0$ and the motion of the oscillator is given very nearly by

$$\theta = \theta_0 \exp.(-\omega_0 t/2Q) \cos(\omega_0 t + \phi) \tag{6.9}$$

Q is closely related to the number of cycles of oscillation over which the oscillation amplitude falls by a factor *e*. Evidently it takes about Q/π cycles for the amplitude to fall by a factor *e*.

In terms of ω_0 and Q, the basic differential equation, equivalent to Eq. (6.2), becomes

$$\frac{d^2\theta}{dt^2} + \frac{\omega_0}{Q} \cdot \frac{d\theta}{dt} + \omega_0^2 \theta = 0 \tag{6.10}$$

In its present form it can be conveniently applied to a great variety of physical systems, both mechanical and non-mechanical.

Method—A uniform bar of rectangular cross-section about 1.5 metres long, resembling a common metre scale (which can also be employed for the purpose) is suspended horizontally with the help of a vertical suspension wire and weights as shown in Fig. 6.2. The suspension wire, about 2 metres in length, should be taut and free from kinks. The use of two weights of 500 gm. each is suggested for achieving the same. Another pair of weights of about 100 gm. each adjusts the time period nearly to a suitable value by varying the moment of inertia of the oscillating system. The process of adjusting the time period is supplemented by the use of a pair of dampers. The dampers are thin semi-circular aluminium plates whose size and position on the wooden bar determines the extent of damping.

By suitable adjustment of the above factors, the time period of the system is set at a reasonably large value, say, in the vicinity of 100 seconds.

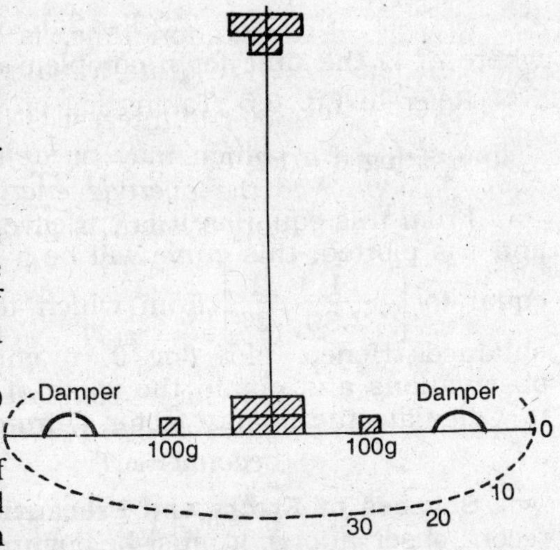

Fig. 6.2

The rest position of the wooden bar is recorded on a circular scale with markings at equal angle intervals 4° to 5° and having its centre on the axis of suspension. The system is given a large angular displacement and after an oscillation or two, the cbservations are taken for the change in angular displacement with time. For this purpose one of the ends of the wooden bar itself is used to record the position of angular displacement on the circular scale with time at regular intervals of say 10 seconds. This process is continued until the amplitude of oscillations falls to a reasonably low value, nearly one-sixth or so of that at $t=0$.

Use these readings to plot the time-displacement curve. The resultant curve would then resemble the one shown in Fig. 6.1. The exponential decay of the amplitude can be clearly demonstrated by smooth curves joining peaks of oscillations on either side of the mean position.

Obtain more sets of observations and the corresponding curves by varying the position of the adjusting weights or dampers or both. Variation of the length of the suspension wire is not suggested as it requires a corresponding change in the level of the chucknut from which the bar is suspended or of the table on which the circular scale is drawn. None of these changes is convenient and moreover the effect of this variation on the time period is far from marked.

From the above, compute the time period T as well as the angular frequency ω. This can be conveniently obtained as follows :

and $$T = \Delta t/n \qquad (6.11)$$

$$\omega = 2\pi/T = 2\pi n/\Delta T$$

where Δt is the time for n complete oscillations.

Refer to Eq. 6.5. Taking log on both sides of this, we get

$$\log \theta = \log \theta_0 - \frac{1}{2.30} \cdot \frac{1}{2T} \cdot t \qquad (6.12)$$

From this equation it is evident that if a curve between log θ and t is plotted, this curve will be a straight line having a slope equal to $\left[-\frac{1}{2.30} \cdot \frac{1}{2T}\right]$ from which the relaxation time may be obtained. Hence, plot (log θ, t) curve from the experimental observations and obtain the value of the relaxation time T from it. Calculate the quality factor Q from the equation.

$$Q = \omega_0 T$$

Sources of Errors and Precautions—(1) Before starting to record observations, it must be ensured that the motion is steady and in a horizontal plane.

(2) A reasonably large time period of the oscillating system is desirable so that at least 8 to 10 points are recorded on each cycle so as to enable one to draw a fairly accurate displacement-time graph.

(3) It is extremely important to ensure that there are no disturbing air currents present. These may alter the time period and can affect the steadiness of the system. These effects become more pronounced at small amplitudes of oscillation.

Observations and Calculations

Set 1.—Position of 100 gm. weights = cm.

Position of dampers = cm.

Table : Angular displacement vs. time

Time (sec.)	Displacement (div.)	Time (sec.)	Displacement (div.)	Time (sec.)	Displacement (div.)
0		210		410	
200		400		600	

From the data of above table, draw a time displacement graph and then the curves joining the peaks of the amplitude, on either side of the mean position. [This corresponds to the dotted line curve of Fig. 6.1].

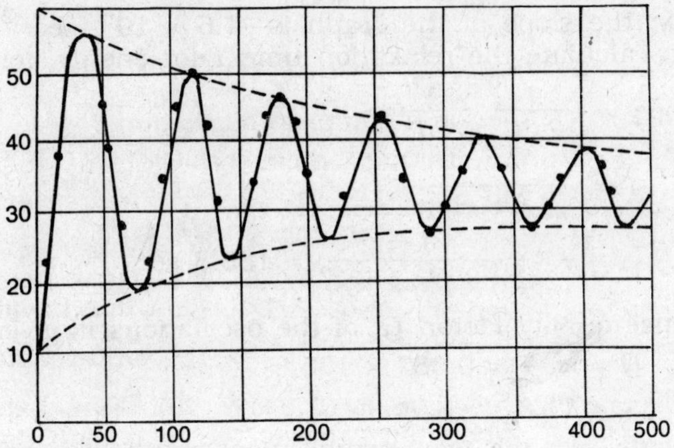

Fig. 6.3

With the help of Fig. 6.3,

(i) compute, n=say 50 oscillations, t=360 sec., say,

$$\therefore \quad T = \frac{\Delta t}{n} = \frac{360}{50} = 7.20 \text{ sec.}$$

$$\omega = \frac{2\pi}{T} = 0.087 \text{ sec.}^{-1}$$

(ii) Prepare the following table.

Time t sec.	Amplitude θ (t) (div)	$\log_{10} \theta(i)$	Slope of log θ vs. t curve
40	20	1.3010	Δ [log θ (t)] = 0.40
75	17	1.2304	Δt = 45 sec.
1.10	15	1.761	$\therefore$ slope = 1.6×10^{-3} sec.$^{-1}$
320	7		

and plot log θ (t) Vs. t graph. Compute the slope of this graph.

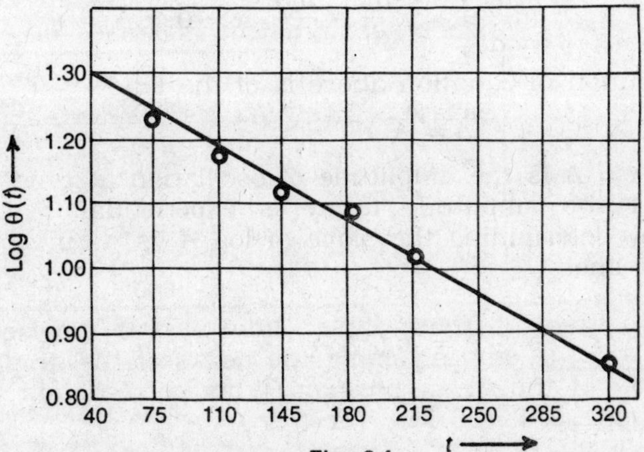

Fig. 6.4

Say, the slope of the graph is—1.6 × 10^{-2} sec.$^{-2}$. Use this value to calculate the relaxation time *T* for energy, for

$$\text{Slope} = -\frac{1}{2.30} \cdot \frac{1}{27}$$

whence $$T = -\frac{1}{4.6 \times \text{slope}}$$

$$= -\frac{1}{4.6 \times 1.6 \times 10^{-2}} = 135.8 \text{ sec.}$$

Hence quality factor, *Q*. of the oscillations is given by

$$Q = \omega_0 T = 0.087 \times 135.8$$

$$= 11.8$$

Result—For the given arrangement of weights and dampers, we have, for the damped oscillations.

(*i*) Angular frequency w=0.087 sec.$^{-1}$.

(*ii*) Relaxation time for energy, T = 135.8 sec.

(*iii*) Quality-factor of the oscillations, Q = 11.8

Exercise 1—To study the relaxation of a simple pendulum.

The oscillations of a simple pendulum are generally described as being simple harmonic—a statement which is valid with certain approximations only. No precise measurements are required to show that the amplitude of these oscillations falls off gradually and ultimately the oscillations die out. Of the various factors responsible for this energy dissipation, the most dominant one is the viscosity of air or the medium in which the pendulum is immersed. The assumption that the viscous force acting on the spherical bob of the pendulum is proportional to the instantaneous velocity yields the following equation of motion.

$$m\frac{d^2 x}{dt^2} + \gamma \frac{dx}{dt} + \frac{mg}{l} \cdot x = 0$$

The solution of equation above is of the type

$$A = A_0 e^{-\gamma t}$$

where $x = A$ is the amplitude of oscillation at time t. It is evident that the amplitude decay is exponential. γ can be computed by determining the slope of log A vs. t curve, which is a straight line.

For

$$\log_e A = \log_e A_n - \gamma t$$

whence $$\gamma = -2.30 \frac{\Delta (\log_{10} A)}{\Delta (t)}$$

And thus the relaxation time $\tau = \frac{1}{\gamma}$ can also be determined, and so also the quality factor Q which is given by

$$Q = \frac{2\pi}{T} \tau$$

where T is the period of oscillation.

For recording the observations, a simple pendulum with exceptionally long suspension cord (= 3 m.) is chosen. This has two main advantages—(*i*) the linear amplitude for the same angular displacement is greatly enhanced and (*ii*) the corresponding time period is large and hence it becomes more convenient to record the amplitude at equal gaps of, say, 5 vibrations.

The bob of the pendulum is displaced about 50 cm. from the mean position and released. After a few oscillations, start recording the amplitude of oscillation at a definite fixed gap of, say 5 oscillations. The process is continued till the amplitude falls to about 10 cm.

The $\log_{10} A$ vs. n curve drawn with the observations, n being the no. of oscillations, being linear confirms the assumption that viscous forces are linear function of the instantaneous velocity. The slope of the above curve is $\Delta (\log_{10} A)/ \Delta n$ which when divided by the time period T yields the value of $\Delta (\log_{10} A)/\Delta(t)$ necessary for evaluating γ, and hence Q.

Experiment 6.2

Object—To study the oscillations of a rubber band and a spring.

Apparatus—Rubber bands, a pan, weights of 50 gm. mounting arrangement, stop watch.

Theory—If to an ideal vertical spring of length l, we attach a mass m, there is an extension in the length of the spring by an amount Δl. Then, at equilibrium, the upward force p due to stretching of the spring is

$$p = - mg = - K.\Delta l \qquad (6.13)$$

where K is called the force constant (*i.e.* force required for unit extension) and is independent of extension Δl. Now suppose that the body of mass m is displaced slightly upwards a distance x from its equilibrium position. The corresponding extension is now $(\Delta l - x)$ and the resultant force acting on the body is

$$F = K(\Delta l - x)\ mg = Kx \qquad (6.14)$$

The resultant force F is thus proportional to the displacement x of the body from its equilibrium position. Therefore if such a system is made to oscillate in a vertical plane, it will exhibit simple harmonic motion with an angular frequency ω given by.

$$\omega = \sqrt{K/m}$$

and time period $$T = \frac{2\pi}{\omega} = 2\pi\sqrt{\frac{m}{K}} \qquad (6.15)$$

Correspondingly the expression for the potential energy of the spring will be

$$U = \frac{1}{2}Kx^2 \qquad (6.16)$$

From Eqs. (6.13) and (6.16), it is evident that (m, x) graph will be linear whereas (U, x) curve will be parabolic.

Also for two different masses m_1 and m_2 oscillating with same spring, the corresponding time periods T_1 and T_2 are in the ratio.

$$\frac{T_1}{T_2} = \sqrt{\frac{m_1}{m_2}} \qquad (6.17)$$

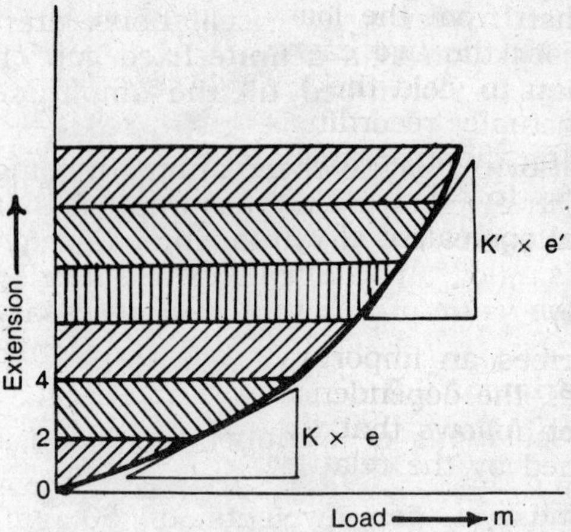

Fig. 6.5

However, the above discussion and the inferences drawn therefrom remain valid only as long as the load-extension characteristic is linear. Consequently these tend to lose their relevance and utility when applied to practical systems as rubber bands. In this case the load-extension curve is not linear but takes a shape similar to that shown in Fig. 6.5.

Quite evidently it can be stated that the force constant K does actually depend on the extension and is determined by the slope of the curve at any particular position.

Mathematically for any given extension x_o, the force constant is

$$Kx_0 = g\left(\frac{\Delta m}{\Delta x}\right)_{x_0} \qquad (6.18)$$

Therefore, in the case of two different masses oscillating on the same rubber band, the ratio of their time periods will no more be given by the simple relation (6.17). Instead of this, a different relation can now be found knowing that Kx_o is now a function of the extension x_o. It follows that

$$T_1 = 2\pi \sqrt{\frac{m_1}{Kx_0}} \tag{6.19}$$

$$T_2 = 2\pi \sqrt{\frac{m_2}{K' x'_0}}$$

Whence

$$\frac{T_1}{T_2} = \sqrt{\frac{m_1}{m_2}} \sqrt{\frac{K' x'_0}{Kx_0}} \tag{6.20}$$

Thus to experimentally verify Eq. (6.20), first we determine the equilibrium extensions x_0 and x'_0 corresponding to loads equal to m_1g and m_2g and then from the load-extension graph calculate the force constants for the two cases. These, in turn are substituted in this equation to yield the theoretical value of T_1/T_2. This is confirmed by actually recording the time periods in the two cases and comparing the experimental value of T_1/T_2 with the one obtained from Eq. (6.20).

Further, the mathematical equivalent of the curve in Fig. 6.4 can be written as

$$mg = f\,(x) \tag{6.21}$$

This equation also describes an important property of the curve under consideration, *i.e.*, the dependence of Kx_0, the force constant, on extension. It then follows that the potential energy of band, U_b, will be determined by the relation

$$U_b = \int_0^x f\,(x)\, dx = g \int_0^x m.\, dx$$

The integral in the above equation can be conveniently evaluated by the graphical method by determining the area enclosed between the curve and the extension axis. This is done by dividing the entire area into smaller zones by drawing lines parallel to the load axis. The areas corresponding to various zones are then evaluated separately and summed up. This multiplied by g gives the potential energy of the band. Ub for extension denoted by the uppermost line drawn parallel to the load axis.

From this the potential energy of the entire system comprising the rubber band and any given mass can also be computed. Thus, if U, U_b and U_m describe the potential energies of the entire system, the rubber band and the mass m respectively, we have.

$$U = U_b + U_m$$

But it is a commonly known fact that the gravitational energy of mass m is $-mg.x$. Thus,

$$U = U_b - mg.\ x$$

where x is the displacement from the equilibrium position due to a load mg. Since, for the rubber band. U_b also changes with the extension x, the graphical method of evaluating U is both preferable and practicable. The process has been shown in Fig. 6.5 where U_b and U_m curves are drawn independently from the observations; the resultant curve then depicting U.

Description and Method

A simple set up for the present experiment requires only a few similar rubber bands and a metre scale. One of the rubber bands is suspended from some rigid support and has a pan attached to it on the other end. The rubber bands commonly used in water balls may be employed although there is no restriction on the type of rubber band to be used. (We have used the cycle valve tube 50 cm. in length). The position of the pan is recorded by a pointer (a simple copper wire is used) which slides along a vertically mounted metre scale.

To begin with, suspend one of the rubber bands along with the pan and ensure that the movement of the pointer is smooth, *i.e.*, it is not obstructed by the scale coming in contact with either the pointer or the pan. Also record the reading of the pointer on the metre scale. Next, place a weight of 50 g. on the pan as a result of which the rubber band is stretched. Record the altered position of the pointer to get the increase in length of the band. Continue this process of loading in steps of 50 g and noting the extension until the stretched length becomes approximately $1\frac{1}{2}$ times the no load value. Make sure that the rubber band has not been stretched beyond the elastic limit. This is done by removing the weights one by one and noting the corresponding pointer readings every time. If these readings almost coincide with those observed during the loading process, the classic limit has not been exceeded.

Next continue loading the pan, in steps of 50 g. till the stretching becomes equal to the no-load length *i.e.*, length of the rubber band is almost doubled and check reproducibility of readings by deloading as before. Continue the same process gradually increasing the load in the pan in steps of 50 g. so that the rubber band is extended to 2.5, 3, 3.5 times its no-load length. This is done until the reproducibility is lost, *i.e.*, the band does not regain its original length on de-loading.

Now discard this band and suspend another piece of the band and attach the pan to its free end. Load it in steps of 50g. upto about 75% of the weight at which the elastic limit is exceeded (as obtained from the observations with the earlier rubber band). Again check reproducibility by de-loading and calculate extension for every 50 g. difference of weight.

Keeping stretching within elastic limit, load the pan, say, with 100 g. and record the equilibrium stretching. Stretch the band further by pulling down the pin slightly and then release it. The pan then oscillates vertically with the amplitude dying out pretty quickly. Record with the help of a sensitive stop-watch the time for a few oscillations and calculate the time period. Repeat the experiment with other values of load to get corresponding time periods.

Plot the load-extension graph which is shown in Fig. 6.6. Calculate U_b, the potential energy of the rubber band determining the area enclosed between the curve and the extension axis for different load increasing in regular steps. Draw a curve U_m on the same sheet and on the same scale. This is a straight line. The resultant of these two U_b and U_m gives the curve for U, the total potential energy of the spring-mass system.

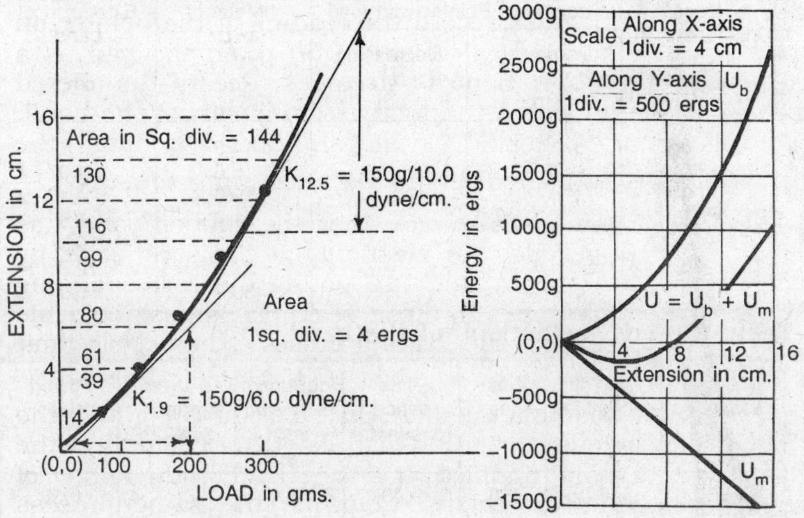

Fig. 6.6

Compare the time periods of the spring with masses m_1 and m_2 as loads with the calculated values of the same with the help of Eq. (6.20), for which the force constant has to be computed from the graph at extensions corresponding to m_1 and m_2.

Repeat the whole experiment with a spring.

Sources of error and Precautions—1. The rubber band should not be loaded beyond 75% of the load required for exceeding the elastic limit.

2. The time of oscillation of the loaded rubber band should be calculated from observations of oscillations with a sensitive stop watch reading to 0.1 sec. because the oscillations are few in number and die out pretty rapidly.

3. A smooth curve should be drawn for load-extension graph and the other graphs for the energy of the band, the mass and the composite system.

Observations and Calculations 1—With a rubber band 1. Original length of the band =

Table 1—For load-extension graph. [A new band to be taken].

S.No.	Mass suspended in gm.	Reading of Pointer with load		Mean. $\frac{a+b}{2}$ cm.	Extension of band cm.
		Increasing (*a*) cm.	Decreasing (*b*) cm.		
1.	0			14.9	0
2.	50			16.8	1.9
3.	100			19.0	4.1
4.	150			21.4	6.5
5.	200			24.3	9.3
6.	250			27.4	12.3
7.	300			30.9	16.0

Table 2—For oscillations of the band

S.No	Mass suspended gm.	No. of oscillations	Time sec.	Time period (observed)	Equilibrium extension from graph	K (from graph) dyne/cm.	Period (cal.)
1.	50	6	2.0	0.33	1.9	2.45×10^4	0.28
2.	150	7	3.6	0.5	6.5	1.91×10^4	0.55
3.	250	5	4.2	0.84	12.5	1.51×10^4	0.80

Table 3—Computation of U_b, U_m vs. extension a. m = 100 gm.

a cm.	U_b m ergs.	U_m for m = 100 g.m. in ergs.	U = U_b+U_m in ergs.
0	0	0	0
2	14 × 4 g = 56 g	– 200 g	– 144 g
4	(56 g + 39 × 4g) = 212 g	– 400 g	– 188 g
6	(212 g + 61 × 4g) = 456g	– 602 g	– 144 g
8	(456 g + 80 × 4 g) = 776 g	– 800 g	– 24 g
10	(776 g + 99 × 4 g) = 1172 g	– 1000 g	+ 172 g
12	(1172g + 116× 4 g) = 1636 g	– 1200 g	+ 436 g
14	(1636 g + 130 × 4 g) = 2156 g	– 1400 g	+ 756 g
16	(2156+ 144 × 4 g) = 2732 g	– 1600 g	+ 1152 g

For m = 50 g., $K_{1.9}$ =

m = 150 gm., $K_{6.5}$ =

$$T_1 = 2\pi \sqrt{\frac{m_1}{K_{1.9}}}$$

Repeat the table for experiment with a spring and make similar calculations.

Result—(*i*) The force constant for a rubber band is a function of the extension a, while for a spring it is independent of the extension a, so long as the elongation is within elastic limit.

(*ii*) Equation (6.20) has been verified.

(*iii*) U_{b1} U_m and U vs. extension, a, curves have been drawn.

Experiment 6.3

Object—To study the variation of moment of inertia, with variation in the distribution of mass with respect to the axis of rotation.

Apparatus—Inertia table with attached horizontal bars, two identical weights and a stop watch.

Theory—It is a well known fact that the concept of moment of inertia in rotational motion is as vital and important as that of mass in translational motion. However, a remarkable difference between the two is that the moment of inertia of a body, unlike its mass, is not its unique property but depends upon the choice of the axis of rotation, *i.e.*, on the distribution of mass about this axis, Mathematically

$$I = \Sigma mx^2$$

In the present experiment the validity of the above is confirmed by recording the variation of time period of torsional vibrations of an oscillating system as the mass distribution is altered, keeping the total mass unaltered.

An ordinary suspension system such as a suspension wire and a rectangular bar, which remains horizontal when at rest, can be used for the purpose. When it is made to oscillate about an axis coincident with the suspension wire, it has some fixed

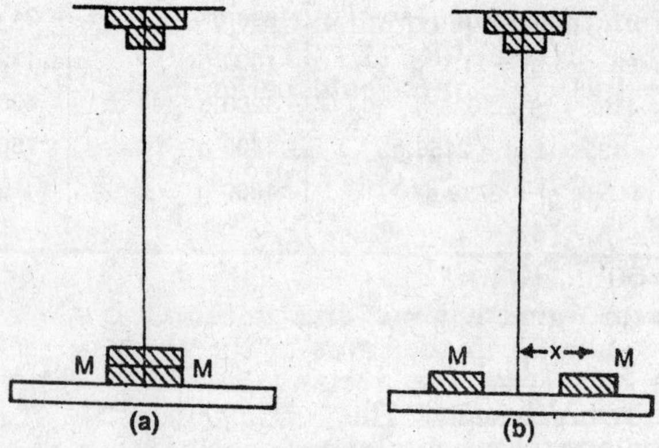

Fig. 6.7

time period depending on the length and radius of the suspension wire and the moment of inertia of the oscillating bar. If then another body of mass M is added to the system, the total moment of inertia is increased by an amount depending upon the position of this additional body with respect to the axis of rotation. In practice, however, two bodies of equal masses are required to be placed symmetrically about the axis of rotation so that the suspending bar may remain horizontal. Two different positions of the weights added to the system are shown in Fig. 6.7 (*a*) and (*b*). Let I_0 be the moment of inertia of the suspension system alone. Also let I_0' and I be the moment of inertia of oscillating system corresponding to Fig. 6.7 (*a*) and (*b*) respectively about the wire as the axis of rotation. Evidently.

$$I'_{s'} = I_s + 2I_0$$

where I_s is the moment of inertia of mass M placed symmetrically about the axis of rotation as shown in Fig. 6.7 (*a*). If now the two masses are shifted and their altered positions are as indicated in Fig. 6.7 (*b*), then we have, by the theorem of parallel axes,

$$I = I'_{s'} + 2\,Mx^2 = I_s + 2\,(I_0 + Mx^2)$$

The corresponding value of the period of oscillation is then

$$T + 2\pi \sqrt{\frac{I_s + 2\,(I_0 + Mx^2)}{C}}$$

where C is the couple per unit radian twist of the suspension wire $\left(= \frac{n\pi r^4}{2l}\right)$

From the above equation we have

$$T^2 = \frac{8\pi^2 M}{C} x^2 + \frac{4\pi^2}{C} (I_s + 2I_0)$$

It follows from the above equation that T^2 vs x^2 graph should be linear and has a slope ($8\pi^2 M/C$). Thus, by measuring the slope of the line the value of C could be computed. Thus,

$$\frac{\Delta\,(T^2)}{\Delta\,(x^2)} + \tan\theta = \frac{8\pi^2 M}{C}$$

whence $$C = \frac{8\pi^2 M}{\tan\theta}$$

Experimental arrangement and method—For obtaining more significant changes in the value of the time period, use is made of inertia table with two rectangular bars of aluminium attached to it symmetrically. This arrangement also permits placing of two balancing weights at x=0 position.

Before starting the actual observations ascertain that the inertia table is horizontal; for achieving this use could be made of the weights of the inertia table.

Allow the inertia table above to perform torsional vibrations and by timing 25 oscillations compute the corresponding time period. Next place the two identical weights at x=0 and again find the time period of the torsional vibrations. Repeat the experiment by changing the two identical weights symmetrically in steps of 5 cm.

From the observations plot a graph between T^2 and x^2, which comes out to be a straight line, thus verifying the basic theorem. $I = \Sigma\, mx^2$. From the graph determine the slope of the curve.

$$\tan\theta = \frac{\Delta\,(T^2)}{\Delta\,(x^2)}$$

and calculate C from the equation

$$C = \frac{8\pi^2 M}{\Delta\,(T^2)/\Delta\,(x^2)}$$ dyne cm. per unit twist.

In this experiment we assume the validity of T being proportional to $\sqrt{I}$ and then prove that $\Delta\, I \;\alpha\; x^2$

Sources of Error and Precautions—Same as in the experiment with Inertia Table.

Observations and Calculations—For the suspension system alone.

Value of M = gm

Time for 20 oscillations = ...sec. T_0 =....sec,

S. No.	Position of wts. from axis of rotation x (cm.)	Time for 20 oscillations	Time period T sec.	x^2 cm.^{2}	T^2 sec.^{2}
1	0			0	
2	5		16.9	25	286
3	10		21.0	100	441
4	15		26.4	225	697
5	20		32.0	400	1024
6	25		38.0	625	1444
7	30		44.0	900	1936
8	35		51.0	1225	2601

From the graph,

Scale Along X-axis 1div. = 200 cm^2

Along Y-axis 1div. = 500 sec^2

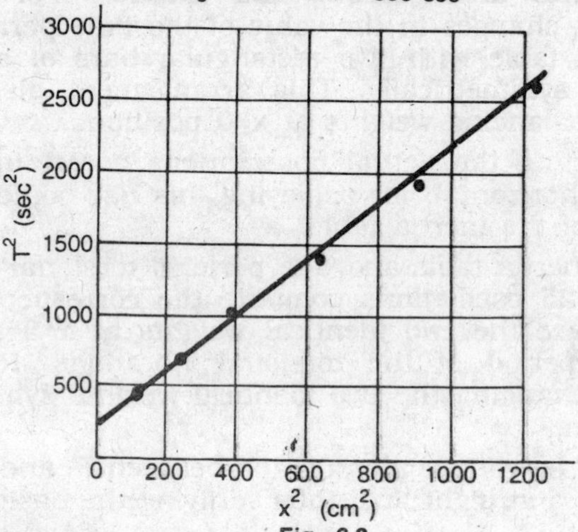

Fig. 6.8

$$\tan\theta = \frac{\Delta (T^2)}{\Delta (x^2)}$$

and hence $C = \dfrac{8\pi^2 M}{\tan\theta} =$

Result.—(*i*) The straight line graph between T^2 and x^2 proves $\Delta I \propto x^2$.

(*ii*) The value of couple per unit radian twist of the suspension wire = ...dyne, cm./unit twist.

SECTION II

SOUND

7

Some Fundamentals of Sound and Determination of Frequency

7.1. Definitions —When a system *in equilibrium is slightly* displaced, *reactive forces* are called into play to bring it back to its *original position.* But this condition is hardly reached by a single reversal of the displacement. The body, by virtue of its energy of motion, passes beyond the mean position till all its energy is converted to potential energy. It then returns towards the mean position and again overshoots this. In this way, the system vibrates or oscillates until it is finally brought to rest by external forces. *The periodic time or the* **period** *of vibration of the system is defined to be the time between two consecutive passages of the system in the same direction through the same position.* This is denoted by T.

The number of vibrations executed by the system per second is termed the **frequency** *of the system.* It is evident that if n be the frequency of vibration of a system having time period T.

$$n = 1/T \tag{7.1}$$

The unit of frequency is vibration per second or Hertz.

The **amplitude** *of vibration is said to be the maximum displacement the system undergoes.*

The **phase** *of a particle is determined by its state of displacement and motion of the instant under consideration.*

For the production of sound it is not sufficient to have a vibrating body. Certain other conditions have also to be satisfied. *e.g.*,

(1) To propagate the motion from the body in a state of vibration, a medium is necessary.

(2) As a result of the vibration of the body, disturbance must advance in the medium ; it should not simply produce a small local motion of the medium.

(3) The frequency of vibration must lie within the limits of audibility.

A body vibrating in air and producing sound causes a disturbance in air which on striking the sensitive membranes of our ears, gives us the sensation of sound. The disturbance in air consists of a series of *compressions* and *rarefactions* one after the other. The motion is a *longitudinal* wave-motion ; the particles oscillate in the same direction along which the wave advances.

The distance moved by the waves in the medium in one second is called the **velocity** *of sound in that medium.*

The distance advanced by the wave in one period or the distance between two consecutive particles in the same phase of vibration is called the **wavelength** and is related to frequency by the relation.

$$v = n\lambda \tag{7.2}$$

7.2. Progressive or Advancing Wave—*A* **progressive** *or an advancing wave is one that is formed when each particle of the medium in succession vibrates according to a law but with the condition that there is a constant phase difference between two consecutive particles.* Such a wave motion satisfies the following conditions :

(1) Each particle of the medium executes a *simple harmonic motion* about its mean position. For a *longitudinal wave*, the motion is *along* the line of advance of the wave; for a *transverse* wave the motion is *perpendicular* to the line of propagation of the wave.

(2) The phase of vibration varies continuously from particle to particle.

(3) For a longitudinal wave the particles are arranged along a straight line. However, if the displacements with proper signs are represented at right angles to the line of advance, the resulting displacement graph resembles with the positions of the particles of a transverse wave.

(4) Such an arrangement of particles, be it a longitudinal or transverse wave-motion, advances further with a uniform velocity without change of type.

(5) Particles in the same phase of vibration along the wave possess the same velocity and acceleration.

7.3. Formation of Beats—It is of special interest to study the superposition of two vibrations of *nearly the same* frequency. As is evident from Fig. 7.1, the resultant vibration indicated by the thick curve is of somewhat variable period. The amplitude

Fig. 7.1

also varies from point to point, although regularly. In the initial stages the amplitude is a maximum; it then gradually diminishes till the curve becomes almost flat and the amplitude then attains a minimum value. It then gradually increases to a maximum, again falls to a minimum ; such a variation is repeated further on. The resultant curve indicates alternations of amplitude and these are characterised by their regularity, being repeated at regular intervals of time. These variations arising from the superposition of two vibrations of nearly equal frequency are termed *beats*.

If the two curves are the displacement graphs of the sound waves issuing from two bodies vibrating with nearly the same frequency, the resultant amplitude in the initial stages then represents a maximum of sound. The intensity of sound goes on diminishing till it attains a minimum value. The maxima and minima of sound are repeated alternately. Hence what is heard is the alternate *waxing* and *waning* of sound. These variations are periodic and the frequency is the difference between the frequencies of the two vibrating bodies.

For, let the frequencies of two forks be n_1 and n_2 where $n_1 - n_2 = m$ a *small* number. When the waves arising out of the vibration of the forks are in phase, the resultant amplitude is a maximum, whereas when they are out of step, the resultant is a minimum. In one second, the fork with a higher frequency performs n_1 vibrations, the other with a lower frequency n_2 vibrations. The fork with a higher frequency thus performs $n_1 - n_2 = m$ vibrations per second more than the other. In one second one wave train is m wavelengths behind the other. In other words, m times the two wave-trains would have been in phase and the same number of times out of phase in one second. The frequency of beats is thus m per second.

The phenomenon of beats is of great importance in sound. It finds its application in setting two sources of sound in unison by the removal of beats between them.

7.4. Stationary Waves—Another case of special interest and often met within the branch of acoustics is the superposition of two wave trains traveling in opposite directions. We shall restrict the consideration to the simple case in which the two wave trains are of *equal* period and amplitude. Such a case is easily realisable in practice by getting an advancing wave reflected from a rigid wall; the reflection gives rise to a wave advancing in a direction opposite to that of the incident wave. Fig. 7.2 indicates the condition of the medium at various instants from 0 to $T/2$. From the time $T/2$ to T, the directions of resultant curves are

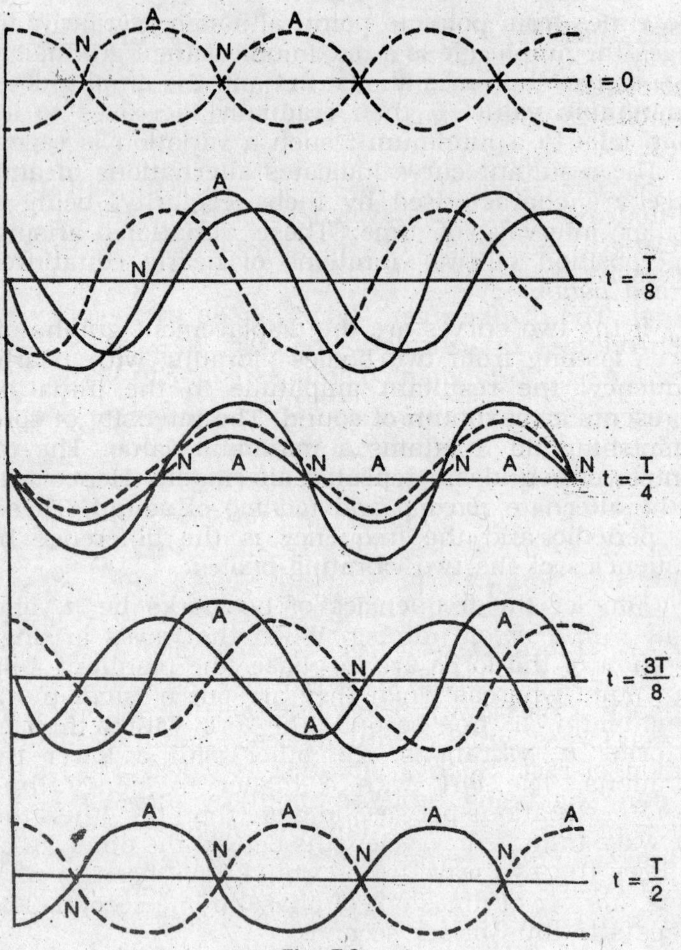

Fig. 7.2

simply reversed. On examination of the resultant curves we note that the points *N* and *A* are the places of absolute rest and maximum motion respectively. These are termed *nodes* and *anti-nodes* and their positions are fixed. The resultant wa'e motion is termed a *stationary wave*—a name which fitly describes its appearance. It has the following main characteristics:

(1) Each particle, with the exception of nodes, executes a simple harmonic motion.

(2) At the *nodes* the amplitude of vibration is *zero*. From point to point on the wave the amplitude varies continuously. Between two consecutive nodes the amplitude is of the same sign. Hence we can state that the amplitude of vibration of the particles

passes through positive, zero and negative values. Situated midway between consecutive nodes are points of maximum displacement ; these are called *anti-nodes*. The amplitude at the anti-node is twice that of each of the component progressive waves.

(3) The distance between two consecutive nodes or anti-nodes is half the wavelength of the stationary waves. This wavelength is equal to the wavelength of the component progressive waves.

(4) The displacement graph of the particles is a sine-graph which reduces to a straight line twice in each period. This form repeats itself at constant distances equal to a wavelength.

In addition to the above characteristics, common to both transverse and longitudinal stationary waves, the following characteristic is also established by *longitudinal stationary* waves:

(5) *The nodes of motion are the anti-nodes of pressure and vice versa.* At the nodes, there are maximum changes of pressure or density whereas of the anti-nodes, places of maximum amplitude of vibration, there are no variations of pressure.

7.5. Vibrations—It is a common experience that a vibration, once excited gradually has its amplitude diminished till finally it dies away. The vibration, unless it is maintained by external forces, must eventually cease and the body comes to the position of rest. Such a vibration is called a *damped vibration*, the damping being brought about by forces which resist the motion of the body. The damping may vary with the nature of the internal and external forces. Hence the amplitude of vibration may decrease (1) in arithmetic progression, (2) in geometric progression (or when the damping is very large), (3) to zero in one or half a period. The latter motion is called *aperiodic motion*.

A body vibrating with its natural frequency is said to execute its *free vibrations*. The vibrations of a body can be maintained by the action of the periodic vibrations of another body, even though their natural frequencies may be different. The driver forces its frequency on the driven which ultimately vibrates with the frequency of the driver. In such a case the driven is said to execute *forced vibrations*. If, however, the forced frequency coincides with the natural frequency of vibration of the driven, the latter takes up the vibration immediately and begins to perform sympathetic vibrations with a *large* amplitude. The driven is said to be in *resonance* with the driver. In all such cases it is implied that the driven or the resonator does not react upon the driver which would otherwise form a coupled system.

The phenomenon of resonance is of great practical importance.

7.6. Tone—Sound sensation perceived by the ear can be classified into two divisions—noise and musical sounds. The former is unpleasant whereas the latter are pleasant to the ear. Noise constitutes a combination of sounds producing irregular disturbances in air, a musical sound, on the other hand, is due to regular periodic vibrations. They are complex vibrations, which can be analysed into a combination of pure sine vibrations with the help of *Fourier's theorem*. A sound which is produced by a pure sine vibration is called a *tone*. It is loosely termed a note.

7.7. Pitch, Loudness and Quality—The *pitch* of a note depends upon the number of waves, comprising of compressions and rarefactions alternately, which reach the ear of the observer per second. When the source and observer have a *relative* motion the pitch is necessarily *altered* as the number of waves reaching the ear per second is changed depending upon the direction of relative motion. But when both are at rest or moving with the same velocity in the same direction, the number of waves perceived by the ear per second is essentially the same as the number of vibrations per second performed by the source. The pitch of a note depends on its frequency alone. A higher pitch corresponds to a higher frequency of vibration. The buzz of a mosquito is of a higher pitch than the roar of a lion since the frequency of vibration in the former case is much greater than that in the latter case.

Another feature that characterises a note is its loudness or intensity which depends on the amplitude of vibration of the source. Notes of the same frequency and intensity, sounded on two different musical instruments, may yet be distinguished from each other. The difference arises because of the presence of overtones or upper partials. The fundamental frequency of vibration of the notes is the same but each is accompanied by a different set of overtones, hence the resultant notes are different. Thus, as Helmholtz says :

"The quality of a musical sound is due to the presence of over-tones in addition to the fundamental."

Determination of Frequency

7.8. Tuning Forks—During recent years the importance of tuning forks has increased because they are now used as standards of frequency. Because of the great *purity* of tone and *constancy* of frequency, they are also used as sub-standards of time.

To understand the working of a fork, let us consider the different stages of bending of a bar as depicted in Fig. 7.3 *(a)*. Theoretical and experimental considerations reveal that the bending of the bar causes the *nodes* to approach nearer with the consequent lowering of its frequency. *In the final U-shaped form, the bar has anti-nodies at the two free ends and an anti-node in the middle.* The amplitude of vibration of this anti-node in the middle is much less compared to the amplitude of vibration of the free ends. It is further reduced by having a heavy stem attached to the U shaped bar which causes a further approach of the nodes towards the middle point.

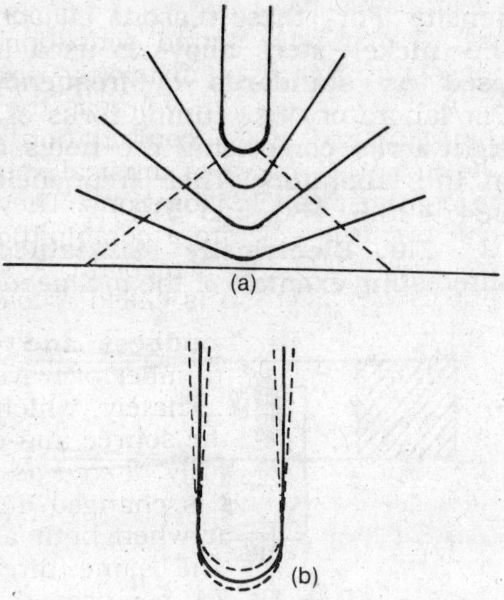

Fig. 7.3

When such a fork is sounded and placed on a sounding board, the energy for the maintenance of forced vibrations of the board is supplied by the anti-node at the middle point through the stem. As the energy is transmitted slowly the fork vibrates for a long period. If, however, we touch a vibrating prong of the fork, the vibrations suddenly stop because of the rapid withdrawal of energy. *It must be clearly borne in mind that an anti-node exists at the point where the stem of the fork is attached to the U-shaped portion and this only can explain why both the prongs should simultancausly close or open.*

The frequency of vibration of a fork varies inversely as the square of the length and directly as the thickness of the prongs and the velocity of sound in the material of which the prongs are made. The longer the length of the prongs, the lower is the frequency of vibration. Filing at the ends increases the frequency whereas loading at the ends lowers the frequency.

The velocity of sound in the material of the prong is $\sqrt{E/\rho}$ where E is the modulus of elasticity and ρ the density. To have a fork giving a pure note, E should be such as not to vary, specially with temperature. Also the material employed for the forks should have a high value for elasticity and low value for density. For these reasons *elinvar* (elasticity invariable), one of the nickel steel alloys is used in the construction of forks used as standards of frequency or sub-standards of time. For laboratory use tuning forks of steel serve well. A series of eight forks comprising the notes of the diatonic scale is used in the laboratory. Their frequencies are 256, 288, 320, 341.3, 384, 426.6, 480, and 512.

7.9. Electrically Maintained Tuning Fork—This is an interesting example of the *maintenance* of *vibrations* by electricity. The stem of the tuning fork is rigidly clamped to a heavy and stiff base. In between the two prongs and near the ends there is an electromagnet M, the current through which is maintained by an accumulator E. The circuit is made and broken by a pair of contacts, one being on the outside of one of the prongs and the other, on the tip of an adjustable screw (Fig. 7.4).

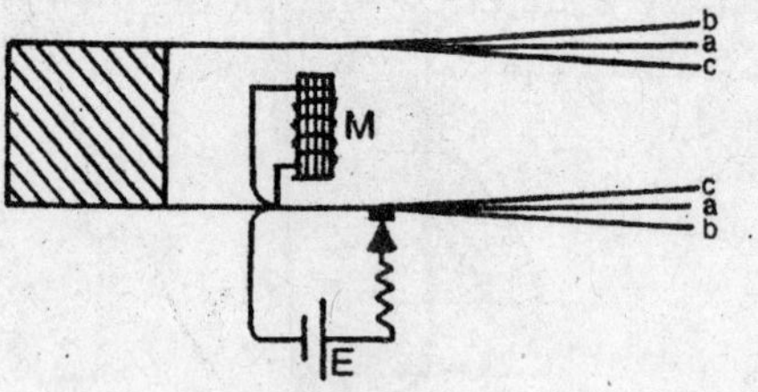

Fig. 7.4

The self-maintaining action of the fork is not so simple. *It is but for the presence of the inductance in the circuit that the fork continues to vibrate under the action of the electromagnet M.* This will be rendered evident from the following considerations :

As the prong moves from *a* to *b* and back again to *a*, the circuit remains closed ; as it advances from *a* to *c* and back again to *a*, the circuit remains open. Thus from the time t=0 to t=T/2 (where T is the *periodic time* of the fork), the current rises in value ; at t=T/2 the current suddenly drops to zero, assuming that there is no sparking at the contact points.

In the outward movement of the prong from *a* to *b* (t =0 to $t=T/4$) the electromagnet *opposes* its motion, whereas during its motion from *b* to *a* ($t=T/4$ to $t=T/2$) the electromagnet *assists* the closure of the prongs. Thus, if the quantity of electricity supplied in the two cases under consideration were the same, the vibrations of the prongs could not be maintained.

$$Q_1 = \int_0^{T/4} i\,dt \tag{7.3}$$

and

$$Q_2 = \int_{T/4}^{T/2} i\,dt$$

where Q_1 *and* Q_2 are the quantities of electricity supplied during the times $t = 0$ to $t = T/4$ and $t = T/4$ to $t = T/2$ respectively and

$$i = \frac{E}{R}\left\{1 - exp. - \left\{\frac{(R}{L}t\right\}\right\} \tag{7.4}$$

where E, R and L are the *e.m.f.*, resistance and inductance in the circuit. On account of its presence in the circuit the inductance prolongs the growth of current and hence Q_1 and Q_2 are essentially *different* and it is the energy $(Q_1 - Q_2)$ which is used up in the maintenance of the vibrations (Fig. 7.5) In the absence of inductance the current would quickly attain a maximum value, making Q_1 approach Q_2. Hence the efficiency would be lowered.

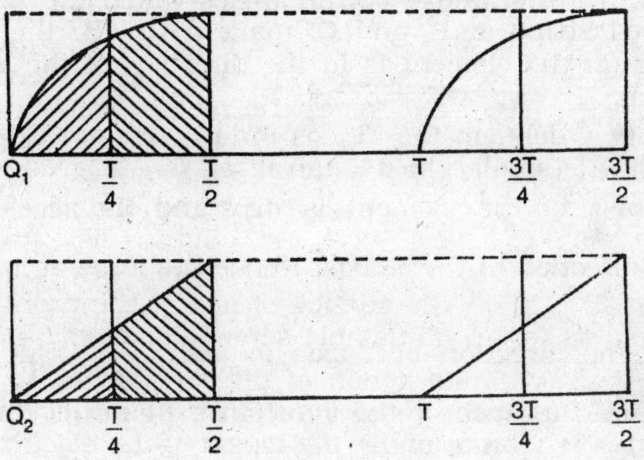

Fig. 7.5

These considerations show that inductance is necessary for the efficient maintenance of vibration but due to its presence arcing results at the points of contact at break. To quench this, a condenser is placed in parallel to the points of contact.

7.10. Transverse vibrations of strings—This is the type of vibration of a stretched string with which we are most familiar. The motion of the particles is *at right angles* to the line of string and the string is said to execute a *transverse vibration.*

A stretched string, sufficiently long and thin and held firmly between rigid supports, approximately fulfils the conditions of an ideal string which according to theoretical considerations (1) should have a *uniform* linear density, (2) should be flexible and (3) should not change in length during the state of vibration.

Let us consider such a string of mass per unit length *m* and stretched by a force T unit. Assume that the displacement of the string is *small*. Let X-axis be in the direction of the string

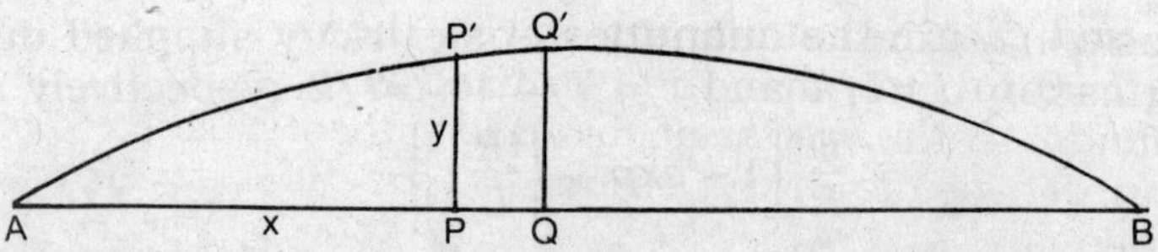

Fig. 7.6

APQB (Fig. 7.6) and Y-axis be in the direction of the displacement of the particles *at right angles* to the string. Let us study the motion of an element PQ (length=δx at a point distant x from A, the origin. At any instant let its displacement be Y. P'Q' is thus the *displaced* position of the element PQ and these we assume to be *equal.* If (θ+dθ) and θ be the angles which the tangents to the displaced string at P′ and Q′ make with AB, the resultant force acting on the element is in the direction of the Y-axis. Its magnitude is

T [sin (θ + dθ)—sin θ] = T cos θ dθ = Tδ_x (sin θ)′ = Tδ_x (tan θ)′ for, when θ is small, sin θ = tan θ.

The mass of the element is $m\delta x$ and its acceleration is $\frac{d^2y}{dt^2}$ in the direction of the Y-axis. Hence the force acting on this element in the direction of Y-axis is $m.\delta x.\ \frac{d^2y}{dt^2}$. But we have already reduced its value to be $T\delta x$ (tan θ). Equating the two, we have

$$m.\delta x\ \frac{d^2y}{dt^2} = T\delta x(\tan\ \theta)$$

but
$$\tan\theta = \frac{dy}{dx}$$

whence
$$m\frac{dy^2}{dt^2} = \mathrm{T}\ \frac{d^2y}{dx^2}$$

$$\therefore \qquad \frac{dy^2}{dt^2} = \frac{T}{m}\cdot\frac{d^2y}{dx^2}$$

This represents a wave motion travelling with a velocity

$$v = \sqrt{\frac{\mathrm{T}}{\mathrm{m}}} \qquad (7.5)$$

7.11. Formation of Stationary waves along strings— Consider the two ends of a string of length l to be fixed. A wave travelling along it is reflected at the two ends in succession : thus we have two waves travelling in opposite directions. They are in a position to form stationary waves, the condition imposed on this formation being that the two fixed ends must be nodes. *For the fundamental mode of vibration there should be an anti-node in the middle of the string. If λ be* the wavelength of these vibrations.

$$l = \lambda_1/2$$

$$\therefore \quad \lambda_1 = 2l \tag{7.6}$$

If n be the frequency of vibration of the string, the velocity of the transverse waves along it is

$$v = n\,\lambda_1 = 2nl \tag{7.7}$$

But we have from Eq. (7.5)

$$v = \sqrt{T/m}$$

whence

$$2nl = \sqrt{T/m}$$

$$\therefore \quad n = \frac{1}{2l}\sqrt{\frac{T}{m}} \tag{7.8}$$

For other possible modes of vibrations of the string we have still to impose the conditions of nodes at the two ends; in between these may be distributed *alternately* a number of *nodes* and anti-nodes. These are represented in Fig. 7.7.

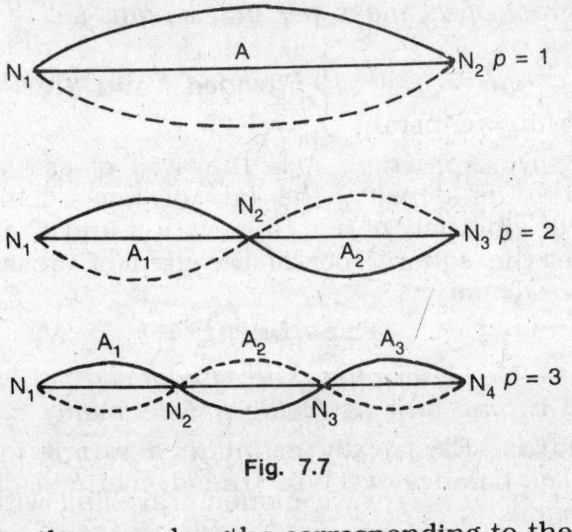

Fig. 7.7

Let λ_2, λ_3 be the wave-lengths corresponding to these modes of vibrations. Then,

$$l = \frac{2\lambda_2}{2} = \frac{3\lambda_3}{2} = \frac{4\lambda_4}{2} = \dots\dots = \frac{p\lambda_p}{2}$$

$$2l = 2\lambda_2 = 3\lambda_3\ 4\lambda_4 = \dots\dots = p\lambda_p. \tag{7.9}$$

where p is an integer. Combining (7.5). (7.7), and (7.9), we have

$$n_p = \frac{p}{2l}\sqrt{\frac{T}{m}} \tag{7.10}$$

The above equation embodies the laws of vibration of strings. These were experimentally determined in 1638 by Mersenne and may be stated as follows :

(1) Law of Length*—*The frequency of vibration of a stretched string varies inversely as the length, provided the tension and the linear density remain constant, i.e.*

$$n \propto \frac{1}{l} \quad \text{provided } T \text{ and } m \text{ are constant}$$

or $\quad nl$ = constant

(2) Law of Tension—*The frequency of vibration of a given length of a stretched string having a uniform mass per unit length, varies directly as the square root of the tension applied to it, i.e.,*

$$n \propto \sqrt{T} \text{ provided } l \text{ and } m \text{ remain constant.}$$

or $\quad \dfrac{n}{\sqrt{T}}$ = constant

(3) Law of mass—*The frequency of vibration of a given length of a string subjected to a constant tension, varies inversely as the sq. root of its mass per unit length, i.e.,*

$$\left.\begin{aligned} n &\propto \frac{1}{\sqrt{m}} \\ \text{or} \quad n\sqrt{m} &= \text{constant} \end{aligned}\right\} \text{provided } l \text{ and } T \text{ are kept the same.}$$

Since $m = A\rho$, where A is the area of cross-section of the string and ρ its density, the law of mass includes that the frequency of vibration of a string, with l and T constant, varies inversely as the square root of the area of cross-section of the wire and its density.

Experiment 7.1

Object—To determine the *frequency* of an electrically maintained tuning fork by *Melde's experiment.*

Apparatus—Electrically maintained tuning fork, stand with clamp, pulley, balance, wt. box, thread, metre scale, accumulator and a rheostat.

Theory—A thread, preferably fishing line (which, as a rule, is sufficiently uniform) is fixed to one of the prongs of an electrically maintained tuning fork. The other end of the string passes over a pulley and carries a known weight. Under the action of the fork this thread breaks up into *stationary waves,* when the length between the fork and the pulley has been properly adjusted. The experiment can be performed with two arrangements of the string with respect to the fork.

(1) Transverse arrangement—*When the thread has been spread out in a line with the length of the prong so that it executes a motion perpendicular to this line, the arrangement is called a* **transverse arrangement**. This is depicted in Fig. 7.9. The formation of *stationary vibrations* is due to the waves reflected at the pulley. The end of the string where it is fixed to the prong and the position where it touches the pulley, are the places of nodes. Hence for the formation of standing waves the length of the string should be a multiple of half the wavelength of the waves travailing along it. For the fundamental mode of vibration of the string there should be anti-node in the middle. Let l be the length of the string giving *fundamental note*, T the tension applied and m its linear density, then

n, the frequency of the driver (fork) = the frequency of the driven (string)

$$= \frac{1}{2l}\sqrt{\frac{T}{m}} \tag{7.11}$$

If the same length of the wire is to vibrate in two segments under the influence of the same fork, the tension applied must be reduced to a quarter of its previous value. The fork in this case excites, not the fundamental mode of vibration, but the second mode which it is capable of performing. In this manner the tension may be reduced and the fork then excites higher modes of vibration of the string. If the string vibrates in p segments under a tension T_p, the relation between this tension and the tension T for its prime mode of vibration, is

$$T_p = T/p^2 \tag{7.12}$$

(2) Longitudinal arrangement —As shown in Fig. 7.8, *In this mode of vibration the string is stretched in a line with the path described by the tip of the prong of the tuning fork.* This can very easily be secured by keeping the fork fixed and moving the string and the pulley through 90^0 from the position in the transverse arrangement. The *stationary waves* are formed, as explained above, due to the superposition of the direct and the reflected waves along the string. But the maintenance of vibrations in this case requires special consideration.

Suppose the prongs are in their *extreme* positions of *outward* displacements and the thread between the prong and the pulley is slackened. At this instant the prongs as well as the string have no velocity. This fact is represented in the diagram by small circles (Fig. 7.8a). Then the prongs begin receding till they reach

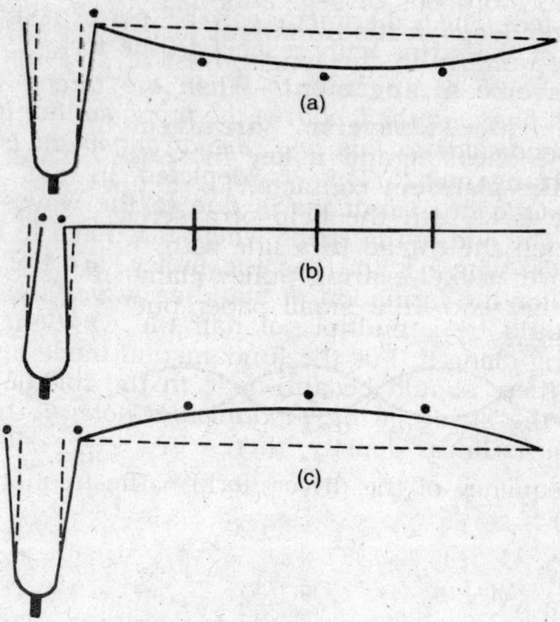

Fig. 7.8

the *extreme* positions of *inward* displacements, when the thread straightens up and is *horizontal*. The thread had no displacement from its mean position but possesses *maximum* velocity in the upward direction as indicated by arrows (Fig. 7.8b). Next the prongs move outward till they attain the same position as in (Fig. 7.8c), when it is noticed that the string, in virtue of its motion, reaches the position of maximum displacement upwards and has no velocity at this instant. Thus we observe that *the fork has executed one full vibration whereas the string has completed only one-half vibration.* The fork then possesses a *frequency double* the *frequency* of vibration of the string.

$\therefore$ frequency of the fork, n = 2 × frequency of the string

$$= 2 \times \frac{1}{2l}\sqrt{\frac{T}{m}}$$

$$\therefore \quad \frac{n}{2} = \frac{1}{2l}\sqrt{\frac{T}{m}} \qquad (7.13)$$

where l, T, and m have their usual significance. As for the *transverse arrangement*, we have a similar relation here.

$$Tp = T/p^2 \qquad (7.14)$$

From the Eqs. (7.11) and (7.13), it is evident that the same length of the string under the same tension will vibrate with different number of loops or segments in the two arrangements. If the number of loops in the transverse form is 2, 4, 6........, their corresponding number in the longitudinal arrangement reduces to 1, 2, 3,......

Method. **(*a*) Transverse arrangement**—Put in an accumulator, a rheostat and a key in series with the electro-magnet and the platinum contacts. Fix a thread to one of the prongs of the fork, with the help of a screw supplied for that purpose. Stretch the thread in a line with the prong of the fork and pass it over a ball-bearing pulley clamped to a stand (Fig. 7.9). At its other end tie a small paper bucket for the purpose of putting weights.

Fig. 7.9

Place suitable weights in the bucket. Adjust the gap between the platinum contacts and make the fork vibrate. The vibrations of the prong are at right angles to the line of the stretched thread.

Move the stand carrying the pulley so as to alter the length of the string. This is done in small steps when the thread will be found to break up into a number of segments. Well-defined nodes and anti-nodes appear. For finer adjustment add some sand in the bucket. This alters the tension and hence the frequency of vibration of the string. The nodes are found to reduce to points. When such an adjustment has been made and the plane of vibration of the string remains fixed, mark the position of nodes with ink. Measuring their distances apart, find the *average* distance between two nodes. This distance l, is half the wavelength of the stationary waves along the thread.

Weigh the bucket with its contents and calculate the tension applied to the thread. Cut out a metre or so of the thread and weigh it in an analytical balance correct to the fourth place of decimal. Calculate m, the mass per unit length of the thread. Then knowing the length l, the tension T and the mass per unit length m, evaluate the frequency of the fork with the help of Eq. (7.11).

Repeat the experiment with the same tension but with a

different number of segments. Again repeat by altering the tension applied.

From each set of observations calcutate n and then take the mean. This gives the average frequency of vibration of the fork as determined by the transverse arrangement of Melde's experiment.

(*b*) Longitudinal arrangement—Next stretch the thread at right angles to the length of the prong so that its vibrations are in a line with the thread (Fig. 7.10).

Fig. 7.10

Proceed with the adjustments and measurements as before and calculate the frequency of the fork with Eq. (7.13). Repeat with different lengths of the string and different tensions. It will be noticed that *with the same length and same tension, an even number of segments in the transverse arrangement reduces to its corresponding half number in the longitudinal arrangement.*

Sources of error and precautions—(1) The thread should be *uniform and inextensible.*

(2) The alignment of the thread and the prong of the fork is necessary. In the transverse arrangement, the thread should be stretched in a line with the length of the prong so that the vibrations of the tip of the prong are at right angles to it : in the longitudinal arrangement, the thread should be in a line with the line of vibration of the tip of the prong.

(3) Well-defined nodes should be obtained by fine adjustments of tension. This is secured by adding sand in the bucket in small quantities.

(4) The plane of vibration of the thread should not rotate. In the transverse arrangement, it should be horizontal whereas in the longitudinal arrangement, vertical.

(5) For the measurement of the distances between the nodes, those at the tip of the prong and at the pulley should be neglected as these have some motion. Intervening nodes should be used for the said purpose.

(6) The friction in the pulley should be as small as possible for it causes the tension to be less than that actually applied.

(7) To avoid sparking at the platinum points, a condenser of suitable capacity should be placed in parallel with the points of contact. This will increase the efficiency of the maintained vibrations in the fork.

Observations—[*A*] *Transverse arrangement.*

Set No.	S.No.	Tension (Constant for each set.)	No. of loops	Length of the corresponding thread	$l = \frac{\lambda}{2}$	Average of l for each set
A						

[*B*] *Longitudinal arrangement*

(N.B. Prepare a similar table)

[*C*] Determination of *m*

mass of the thread = gm.

length of the thread = cm.

Calculations —Mass of the thread per *unit* length= gm./cm

[*A*] *For transverse arrangement*

For set A. $n = \frac{1}{2l}\sqrt{\frac{T}{m}}$

=

= Hertz

Make similar calculations for other sets.

[*B*] *For longitudinal arrangement*

for set A, $n = \frac{1}{l}\sqrt{\frac{T}{m}}$

=

= Hertz

Make similar calculations for other sets.

Mean value of frequency of the fork = Hertz

Result —The *frequency* of the electrically maintained fork = Hertz.

Criticism of the method—Melde's experiment is one of the best methods of demonstrating the formation of stationary waves along strings. It gives a fairly satisfactory result for the value of the frequency of the fork. The arrangement, however, suffers from a number of defects, namely, (*i*) the friction at the pulley, (*ii*) the sparking at the platinum points which cannot be totally removed and (*iii*) the alteration in the frequency of the fork by the clamping screw. Some makers, however, supply forks, which along with the screw have definite frequencies 64 of 80. Such forks, if used, will eliminate the last error.

Experiment 7.2

Object To determine the *frequency* of A. C. mains by means of a sonometer.

Apparatus—Sonometer, half-kilogram weights, horse-shoe magnet, clamp stand, step-down transformer, metre scale and a screw gauge.

Theory—Let a sonometer wire stretched under a constant load be placed in a *uniform* magnetic field applied *at right angles* to the wire in the *horizontal* plane and let an alternating current of *low* voltage be passed through the wire. On account of *interaction* between the magnetic field and the current in the wire, the wire will be deflected, the direction of deflection being given by *Fleming's left-hand rule.* As the current is alternating, for half the cycle the wire will move *upwards* and for the next half cycle it will move *downwards*; in other words, the wire will receive impulses alternately in opposite directions at the frequency of the current. As a consequence *the wire will execute forced vibrations with a frequency* of the *A.C.*

If now the distance between the two bridges of the sonometer be so adjusted that the *natural frequency* of vibration n of the wire equals that of the current, there will be *resonance* and the wire will vibrate with a *large* amplitude. At *this* stage

$$f = n = \frac{I}{2l}\sqrt{\frac{T}{m}} \tag{7.15}$$

where l is the length of the wire between the bridges, m its mass per unit length and T is the tension in the wire.

Now $$m = \pi r^2 \rho \tag{7.16}$$

where r is the radius of the wire and ρ the density of its material. Thus, if m is evaluated from Eq. (7.16) the frequency of alternating current can be calculated from Eq. (7.15).

Method—Mount a small horse-shoe magnet N, (Fig. 7.11) *vertically* at the *centre* of a sonometer wire such that the wire passes *freely* in between the poles of the magnet and the face of the magnet is *normal* to the length of the wire. Apply a suitable tension to the wire, say, of a 4 kilograms weight. Reduce the voltage of the A. C. mains to about 6 or 8 volts by means of a step down transformer and then apply this *reduced* voltage between the two ends of the sonometer wire. As the current passes through the wire, it will begin to vibrate. The vibrations of the wire can be observed by naked eye or if very feeble they can be felt by touching the wire with a finger. Alter the distance between the two bridges so that the amplitude of vibration of the wire increases and finally adjust the positions of the bridges so as to get *maximum* amplitude of vibration of the wire. Measure the distance between the bridges. Repeat this determination several times and find out the mean value of l. Next alter the tension applied to the wire and take several sets of observations for l with different values of tension.

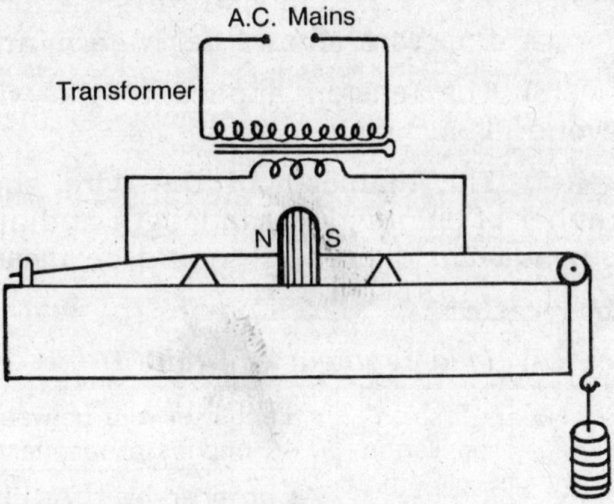

Fig. 7.11

Now, measure the diameter of the wire with a screw gauge and assuming a knowledge of the density of the material of the wire, calculate the mass per unit length m of the wire from Eq. (7.16).

Finally, calculate the frequency of A. C. mains from Eq. (7.15) for each set of observations separately and then take the mean.

Sources of error and precautions.—(1) The horse-shoe magnet should be placed *vertically* at the *centre* of the wire with its face *normal* to the length of the wire.

(2) The wire should be *horizantal* and pass *freely* in between the poles of the magnet.

(3) The A.C. should be stepped down to about 6 or 8 volts before passing through the wire.

(4) The sonometer wire and the clamp used to hold the magnet should be of *non-magnetic* material.

(5) The wire should be *uniform* and free from kinks.

(6) The tension applied to the wire should not stretch it beyond elastic limit.

(7) The diameter of the wire should be measured at a number of points along its length and at each point two mutually *perpendicular* diameters should be measured.

Observations

[*A*] *Measurement of l and T*

Set No.	Tension applied in kilo-gm.	Length of wire between the bridges at resonance		Mean length l *cm.*	Frequency of A. C. mains f Hz
		As obtained by increasing the length of the wire cm.	As obtained by decreasing the length of the wire cm.		
				Mean	

[*B*] *Measurement of diameter of wire*

Least count of the screw gauge = cm.

Zero error = cm.

S. No.	Reading along any diameter a cm.	Reading along perp. diameter b cm.	$\frac{a+b}{2}$ cm.
		Mean	

Calculations—Density of the wire (given) = gm/c.c.

Mean corrected diameter of the wire = cm.

radius of the wire = cm.

Mass per *unit* length of the wire

$m = \pi r^2 \rho$ = gm./cm.

Set I $f = \frac{1}{2l}\sqrt{\frac{Mg}{m}}$

= cycles/sec.

(Make similar calculations for other sets of observations)

Mean value of f = Hertz

Results—The frequency of A.C. mains = Hertz

Criticism of the method—This is a very simple, quick and convenient method of determination of the frequency of A.C. mains. It gives quite accurate results. The result is, however, affected by the defects inherent in a sonometer, *viz.*, stiffness of the wire, yielding of supports, friction at the pulley.

ORAL QUESTIONS

1. Define wave-length, frequency, amplitude and phase of vibration.
2. What conditions are essential for the production of sound?
3. What are transverse and longitudinal waves ?
4. Mention the characteristics of a progressive or advancing wave.
5. What do you mean by the principle of superposition?
6. What are beats ? Explain their production.
7. Why do you not hear beats when the difference of frequencies between two sound sources is large ?
8. Explain the formation of stationary waves. What are the chief characteristics of these waves ?
9. How do nodes differ from anti-nodes ?
10. What is the distance between two consecutive nodes or anti-nodes ? How would you demonstrate them experimentally ?
11. If a vibrating tuning fork held near the ear is rotated about a vertical axis containing the stem, how will the intensity of sound vary ?
12. Define pitch, loudness and quality.
13. Does any change in frequency occur with amplitude ? How is it associated with loudness ?

14. Does pitch always depend upon the frequency of vibration of the source of sound ?
15. What is Doppler effect ?
16. What is the difference between (1) a tone and note (2) harmonic and overtone ?
17. Differentiate between free, forced, sympathetic and maintained vibrations. Explain the part played by resonance in sound.

TUNING FORK

1. Explain the construction of a tuning fork. What is its importance in the laboratory ? To what uses is it put ?
2. Where are the nodes and antinodes situated in a vibrating fork ?
3. What purpose does the stem serve ? On what factors does the frequency of a fork depend ?
4. How does loading or filing at the prong affect its frequency.
5. How is the energy transferred from a vibrating fork placed on the sounding board ?
6. Why do the vibrations of a fork immediately stop when the prong is touched with a finger ?
7. What material is used in the manufacture of tuning forks?
8. What is elinvar and what is its practical importance ?
9. Why forks of frequencies 50, 100, 150...are not chosen for laboratory use ?
10. Explain the working of an electrically maintained tuning fork.
11. What part does induction in the circuit play ?
12. How may the efficiency of the maintenance of vibrations of the fork be increased ?
13. Why does sparking occur at platinum contacts ? How may it be minimised ?
14. How does the increase of current affect the vibrations of the fork?

VIBRATIONS OF STRINGS

1. What is the velocity of transverse progressive waves along a string under tension ? Will it be the same as of stationary waves along it ? How may these be experimentally demonstrated ?

2. What are the laws governing the vibrations in a stretched string ?
3. What characteristics must the string possess ?
4. Under constant tension if the length of the vibrating string is altered, how will its frequency change ?
5. How does the frequency change with tension, other factors remaining constant ? What will it be if the tension is quadrupled?
6. Which of the two wires of the same material and length and under the same tension, will have a higher frequency. (*i*) a thick one and (*ii*) a thin one ? What would be their ratio (*i*) if the area of cross-section of the thick wire is four times that of the thin wire (*ii*) if the radius of thick wire is four times that of the thin wire.
7. What do you mean by resonance ?
8. What is sharpness of resonance ? On what does this depend? What is the practical importance of this ?
9. Why is it better to have a thinner wire of lower Y and to load it with maximum load possible ? What is the principle of working in this experiment ? Why can you not wholly dispense with your ear in this experiment ?
10. Is the note given by the wire and the fork simple ?
11. Is it necessary to stretch the wire horizontal ? If not, which is better, horizontal or vertical, arrangement ?
12. Why is it better to stretch with a spring balance ?
13. How does friction at the pulley affect your result ?
14. What are the essentials of the string of the sonometer ? Can a rubber cord be used ?
15. What function does the sounding board serve? What type of vibrations are excited by it?
16. How is the wire tested for unison with the fork ?
17. Explain the principle of the rider. What precautions are to be taken during the performance of the experiment ?
18. What would be the effect of plucking the string at a place other than its middle point ?
19. What part do the bridges play? Explain their construction.
20. What are the sources of error in the experiment ? To what use can this sonometer be put ?

MELDE'S EXPERIMENT

1. What are the two arrangements of the string with respect to the fork in which the experiment can be performed ?
2. What is the essential difference between the two modes? What is the ratio of the frequency of the string in the two arrangements ?
3. Explain the working of the longitudinal form of the arrangement?
4. How do you proceed with the adjustments in the experiment? What precautions are to be taken ?
5. How do the number of loops alter as the string is changed over from transverse to longitudinal arrangement keeping all factors constant ?

FREQUENCY OF A C. MAINS

1. What do you understand by frequency of A.C. mains ?
2. Distinguish between A.C. and D.C.
3. How can you use a sonometer for determination of frequency of A.C. mains ?
4. What is the use of a magnet here ? How do you mount it and why ?
5. Why does the wire vibrate ? What is Fleming's left-hand rule? What is the principle of this method ?
6. What is resonance ? What precaution do you take in this experiment. ?
7. What is the material of sonometer wire ? Can it be of iron, if not, why ?
8. Discuss the accuracy of results obtainable by this method.

8

Determination of Velocity of Sound

8.1. Velocity of longitudinal vibrations in a medium— Referring to Fig. 8.1, consider a cylinder of a medium. Let a be its area of cross-section and let its axis be *parallel* to the axis of X. AB is a layer situated at a distance x from the origin O; CD, another layer under consideration at a distance $(x+dx)$ from O. Let AB be displaced through a distance y along the axis of X to the position A′B′ ; the layer CD accordingly occupies the position C′D′, the distance from the origin being $(x+y+dx+dy)$.

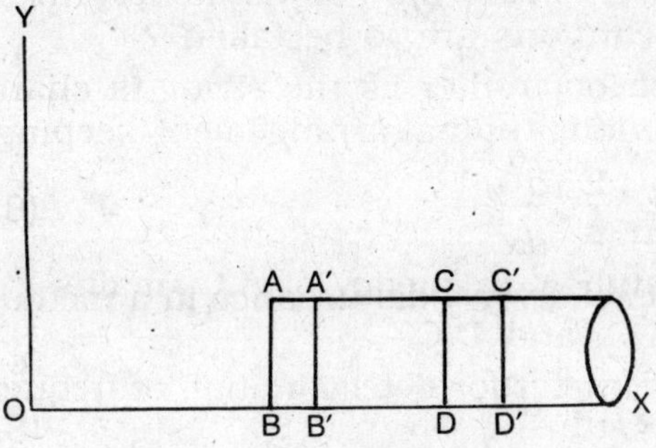

Fig. 8.1

The *initial* volume of the cylinder is $a.\ dx$ whereas the *final* volume becomes $a\ (dx+dy)$. The extension is then $a.dy$.

$$\therefore \quad \text{Strain produced by this displacement} = \frac{ady}{a.dx} = \frac{dy}{dx}$$

If E be the *elasticity* of the medium

$$\text{Stress} = \text{E} \times \text{strain}$$

$$= \text{E}\,\frac{dy}{dx} = \text{F}\ (x)$$

This is the *stress* at a distance x (or the restoring force for a displacement y). Applying *Taylor's theorem*, stress at $(x+dx)$ is

$$\text{F}\ (x + dx) = \text{E}\frac{dy}{dx} + \quad \text{E}\,\frac{d^2y}{dx^2}dx$$

Since *unequal* forces exist at the ends of the *dilated* cylinder, there is a force tending to move it. The magnitude of this force per unit area is $\text{E}\dfrac{d^2y}{dx^2}dx$, the difference of the two stresses.

∴ the *total* force over the entire cross-section $= E\frac{d^2y}{dx^2}dx.a$

Let ρ be the density of the medium and $\frac{d^2y}{dt^2}$ be its acceleration at x. The mass of the cylinder is $\rho.a.dx$; hence, the force acting on its cross-section is $\rho.dx\frac{d^2y}{dt^2}$. It must be equal to $E\frac{d^2y}{dx^2}\,dx.a$.

Hence,
$$\frac{d^2y}{dt^2} = \frac{E}{\rho}\frac{d^2y}{dx^2} \qquad (8.1)$$

The differential equations of a wave disturbance in a medium is

$$\frac{d^2y}{dt^2} = v^2\frac{d^2y}{dx^2} \qquad (8.2)$$

Comparing Eqs. (8.1) and (8.2), we get

$$v = \sqrt{E/\rho} \qquad (8.3)$$

Since sound is a *longitudinal* wave motion, the expression above gives the velocity of sound in the medium under consideration.

When waves are excited in bars, the modulus of elasticity concerned is *Young's modulus*. The velocity of sound is then given by

$$v = \sqrt{Y/\rho} \qquad (8.4)$$

When sound waves travel in a gas, the modulus of elasticity concerned is the adiabatic elasticity which is equal to γ P, where γ is the ratio of the two specific heats of the gas and P is its pressure. Hence.

$$v = \sqrt{\gamma P/\rho} \qquad (8.5)$$

The original formula $v = \sqrt{P/\rho}$ was given by Newton on the assumption that the modulus of elasticity concerned was the isothermal elasticity (= P.) From known values of ρ and P for air, the velocity calculated from Newton's equation gave a value of 280 metres/sec., a value much below the experimental value of 330 metres/sec. The discrepancy lay with the formula, as was suggested by Laplace. The changes associated with the production of compressions and rarefactions in the gas are adiabatic changes. The medium is compressed and relaxed so quickly that heat has no chance to enter the system or to leave it. As such the modulus of elasticity to be considered is the *adiabatic elasticity* (= γP). The modified formula (8.5) given above on calculation yields a value of 331 metres/sec., in close agreement with the experiment.

The velocity of sound in a gas is proportional to the square root of the absolute temperature. If α be the gas coefficient, and v_t and v_0 are the velocities of sound at t^o C and 0° C respectively.

$$v_t = v_0 \sqrt{1 + \alpha t} \tag{8.6}$$

For small values of t, we have, as an approximation

$$v_t = v_o \left(1 + \frac{1}{2}\alpha t\right)$$

The velocity of sound does not vary with change of pressure. The velocity of sound in air increases with humidity.

Experiment 8.1

Object—To determine the velocity of sound in air with a *Kundt's tube.*

Apparatus—Kundt's tube, brass rod with clamping device, resined leather, lycopodium powder (or cork dust).

Description of apparatus—As shown in Fig. 8.2, the apparatus consists of a tube about a metre or so in length and one inch in diameter. This is meant for the excitation of the *stationary vibrations* in the air enclosed in it. To limit the column of the enclosed air one of the ends of tube is corked and through the cork passes a thick wire carrying a closely-fitting disc. This wire can be moved in and out and thereby the position of the disc that limits the column of vibrating air on this end, can be altered. On the other end of the tube there is the end of the given rod (say of brass) carrying a disc just small enough to enter the tube wihout touching. It is this column of air between the two discs that is set into *stationary vibrations*. The driver, in the

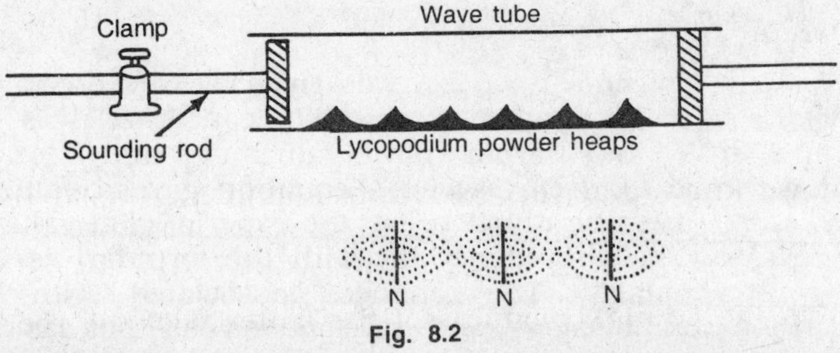

Fig. 8.2

present case, is the rod which is made to vibrate *longitudinally*. It is rigidly clamped in the middle, the clamping device and the supports of the tube being so fixed that the axis of the rod and the tube coincide. The free end of the rod is to be stroked with resined leather by pulling it along its length to set it in vibration. For a rod of glass as the driver, cloth moistened with alcohol may be used for the purpose of exciting vibrations. A paper scale may conveniently be pasted to the glass tube to enable the position of the nodes to be correctly observed.

Theory—The rod vibrates *longitudinally* and produces a note of, say, frequency n. The vibrations are such that *for the fundamental mode of vibration, there is a node in the middle and an anti-node at either end of the rod.* Let the length of the rod be l_r and the wavelenth of the *stationary waves* in the rod be λ_r then

$$\lambda_r = 2l_r$$

and

$$v_r = n\lambda_r = 2nl_r \tag{8.7}$$

where v_r is the velocity of sound waves in the material of the bar.

The vibrations of the bar so excited, in their turn, excite vibration of a stationary character in air enclosed between the two discs, but only when the two discs are a multiple of $\lambda_a/2$ apart, where λ_a is the wave-length of sound waves in air due to a frequency of vibration n, that of the *sounding rod. At the two discs we have nodes for the stationary vibrations in air* and others in between these two positions.

Let the average distance between the nodes as revealed by dust heaps be l_a. Then,

$$\lambda_a = 2l_a$$

and

$$v_a = n\lambda_a = 2nl_a \tag{8.8}$$

where v_a is the velocity of sound waves in air.

Dividing Eq. (8.8) by Eq. (8.7)

$$\frac{v_a}{v_r} = \frac{l_a}{l_r}$$

$$v_a = \frac{l_a}{l_r}\, v_r \tag{8.9}$$

But we know that the velocity of sound in a solid rod is given by

$$v_r = \sqrt{Y/\rho}$$

where Y is the Young's modulus of the material of the rod and ρ is its density. On substituting this value of v_r in Eq. (8.9), we get

$$v_r = \frac{l_a}{l_r}\sqrt{\frac{Y}{\rho}} \tag{8.10}$$

The factors Y and ρ are taken as known for the material of the rod; l_a and l_r are determined experimentally. On substituting the values so obtained, v_a can be calculated.

Method—Dry the tube. Spread some dry lycopodium powder inside it so as to form a strip about 2 mm wide. Then fix it in its proper supports. Clamp the given rod rigidly at its middle point and make sure that the axes of the tube and the rod coincide. Rotate the tube so that the powder is on the point of falling down. Make the rod 'speak' *i.e.*, emit a note by stroking it with resined leather along its length. The vibrations in the rod set the air in the tube into vibration; waves are sent along the tube and these, in their turn, are reflected from the disc at the other end. Thus the air in the tube is under the action of the incident wave and the reflected wave. Hence *stationary* waves can be produced in the air column provided its length is an integral multiple of $\lambda_a/2$. To attain such a position move the disc at the other end of the tube by very small amounts, each time the rod being stroked longitudinally and the air column tested for resonance. A position is finally reached when the air column is thrown into *resonant vibrations*. It is revealed by the agitation of the dust particles which settle down into prominent ridges at the anti-nodes. If the rod is continuously stroked with leather, the air vibrations cause the powder to move from the positions of the anti-nodes towards the nodes where it collects in heaps. But it is difficult to attain this result.

Mark out two prominent heaps at some considerable distance and count the number of heaps in between. Thence find the number of half-wavelengths in the air column between the positions already marked. Measure this distance and divide it by the number of half-wavelengths to get the value of $\lambda_a/2$ ($= l_a$) Repeat the observations and for each set get the value of l_a. Calculate the mean value. Record the room temperature.

Unclamp the rod and measure its length l_r. ($= \lambda_r/2$). From the tables of physical constants, record the values of Y and ρ for the material of the rod. From a knowledge of these four quantities, get the value of the velocity of sound in air at the recorded temperature with the help of Eq. (8.10).

Sources of error and precautions—(1) The tube should be

perfectly dry and so also the powder, as the presence of water vapour affects the velocity of sound.

(2) The rod should be clamped exactly and rigidly in the middle so that it may 'speak' freely.

(3) The alignment of the tube and the rod should be done before commencing the experiment.

(4) The presence of too much powder does not give the desired dust figures. A very thin strip of lycopodium powder should, therefore, be spread along the length of the tube. This may be done by first spreading such a thin strip of powder on a metre scale and then inverting it within the tube.

Observations

Room temperature = °C

[A] *Constants of the rod;*

(1) Length of the rod l_r. = cm.

(2) Young's modulus of material of the rod = dynes/cm^2

(3) Density of the material of the rod = gm./cc.

[B] *Determination of* l_a

S. No	Distance between two well-defined nodes d cm.	No. of nodes in between these $(n-1)$	No. of half-wavelengths in d n	$\lambda_a/2 = l_a = d/n$ cm.
1				
2				
3				

Calculations

$$v_a = \frac{l_a}{l_r}\sqrt{\frac{Y}{\rho}}$$

=

= cm./sec = m./sec.

Result —The velocity of sound in air at °C = m./sec.

Criticism of the method —The experiment is capable of modifications which are of great experimental value. The following are the determinations which can be made with it:

(1) Direct determination of the velocity of sound in a gas.

(2) Comparison of velocities of sound in two gases (using double tube method).

(3) Study of the variation of velocity of sound with temperature, pressure and humidity.

(4) Determination of γ, the ratio of the two sp. heats of gas. This is the only experimental method for finding the value of γ for gases available in small quantities.

The chief source of error lies in the fact that the velocity of sound in a limited column of gas is not the same as the velocity of sound in its extended atmosphere. It has been shown by Kirchhoff that

$$v_a = v\ (1 - k/a)$$

where v_a is the velocity of sound contained in a tube of radius a, v the velocity of sound in a large volume of air and k, a constant. Thus we see that the value of v_a obtained by experiment is less than the velocity of sound in a large volume of air.

This error can, however, be eliminated by using two tubes of different radii a_1 and a_2 and getting the values of v_{a1} and v_{a2} by experiment. For

$$v_{a1} = v\ (1 - k/a_1)$$

$$v_{a2} = v\ (1 - k/a_2)$$

Eliminating k, we have

$$v = \frac{a_2\, v_{a2} - a_1\ v_{a1}}{a_2 - a_1} \tag{8.11}$$

No such tube correction is necessary in the experiment for the comparison of velocities of sound in the gases provided the tubes used are of the same diameter.

The method can be adapted to determine the velocity of torsional vibrations along rods. The only change necessary in the above experiment is the mode of excitation of the rod. Instead of stroking it longitudinally with a resined cloth, the latter is kept near the free end and engaged so as to turn in a direction as may cause the rod to rotate about its axis. In such a case the note emitted by the rod is different from the note emitted when it vibrates longitudinally for this note corresponds to the torsional vibrations of the rod. Since the velocity of torsional vibrations is $\sqrt{n/\rho}$ where n is the modulus of rigidity and ρ the density of the material of the rod, the factor n can be calculated in the same manner as Y.

Exercise 1—To *determine Young's modulus of the material of the rod by Kundt's tube experiment.*

As deduced previously

$$v_a = \frac{l_a}{l_r}\sqrt{\frac{Y}{\rho}}$$

$$Y = v_a^2\ \frac{l_r^2}{l_a^2}\ \rho \tag{8.12}$$

If v_a, the velocity of sound in air at room temperature and ρ, the density of the material of the rod are taken as known, and l_a half the wavelength of stationary vibrations excited in air and l_r, half the wavelength of stationary vibration in the rod, are determined experimentally (as in the previous experiment), and these values substituted in Eq. (8.12), the value of, Y can be evaluated.

Exercise 2—*To determine the ratio of the two specific heats of air by Kundt's tube experiment.*

As already deduced $v_a = \frac{l_a}{l_r}\sqrt{\frac{Y}{\rho}}$

The velocity of sound in air can be calculated by taking Y and ρ for rod as known factors and determining l_a and l_r experi mentally (as in the previous experiment). But the velocity in air is also given by

$$v_a = \sqrt{\gamma\, P/\rho_a}$$

where Y is ratio of the two sp. heats C_p and C_v, P the pressure of air and ρ_a the density of air.

Equating the two values of v_a we have

$$\sqrt{\frac{\gamma P}{\rho_a}} = \frac{l_a}{l_r}\sqrt{\frac{Y}{\rho}}$$

whence squaring and rearranging we get

$$\gamma = \frac{l_a^2}{l_r^2}\,\frac{Y}{\rho}\,\frac{\rho_a}{P} \tag{8.13}$$

8.2. Frequency of a resonator—The air column contained in pipes can serve as examples of resonators but for analysis of sound the requirements of resonation should be sharpness of resonance. Helmboltz has described these, containing air and satisfying the necessary condition of *sharpness of resonance.* A resonator has two apertures, the *narrower* one is to be inserted into the ear and the wider one receives the sound. The internal capacity of a resonator is large compared with the orifice or neck of the resonator which remains in direct communication with the atmosphere. Two forms of resonators are depicted in Fig. 8.3.

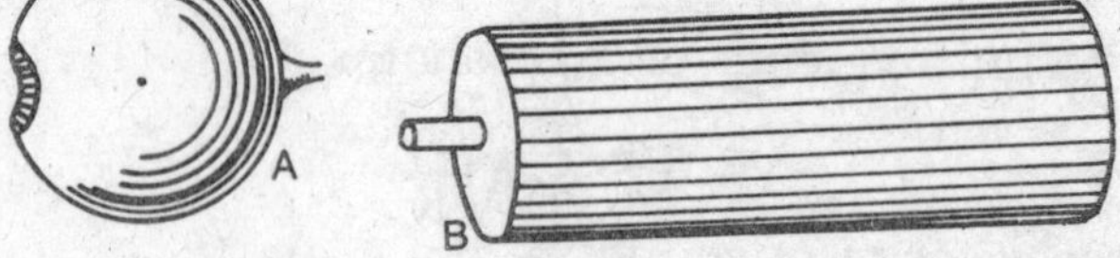

Fig. 8.3

For calculation of the natural frequency of the air column inside the resonator, we can imagine that the air in the neighbourhood of the neck alternately compresses and rarefies the air inside the resonator. Let A be the area of cross-section and *l*, the length of neck and ρ the density of air. The mass of this air which is responsible for periodic movements of air inside, is *Al*ρ. If the compression caused for a unit displacement of this mass is δ_p in the air inside the resonator, the restoring force per *unit* displacement is A. $\delta p = A\gamma\, p \dfrac{\delta\rho}{\rho}$ since the compression is adiabatic. Let v be the volume of the air column inside the resonator. Before compression it was (v + *A*). Since the mass of this air is constant,

$$v\,(p + \delta\rho) = (v + A)\,p$$

$$\therefore \qquad \delta\rho = A\,\rho/v$$

Hence the force per unit displacement is

$$f = A.\ \gamma p.\ \frac{A}{v} = \gamma p.\ \frac{A^2}{v}$$

Since the motion is *simple harmonic*, its period of oscillation T is given by

$$T = 2\pi\sqrt{\frac{A\rho l}{\gamma p A^2/v}} = 2\pi\sqrt{\frac{\rho l v}{y p A}}$$

The natural frequency of vibration of the air column inside the resonator is, therefore,

$$n = \sqrt{\frac{\gamma p/\rho}{2\pi}}\ \sqrt{\frac{A}{lv}} = \frac{v}{2\pi}\sqrt{\frac{A}{lv}} \tag{8.14}$$

where v is the velocity of sound.

The chief utility of resonators lies in the analysis of partials in a musical sound.

Experiment 8.2

Object—To show that the frequency of a resonator varies inversely as the square root of its volume and to estimate the *neck correction*.

Apparatus—A resonator fitted with a rubber tubing and pinch-cock at its lower end, a clamp stand, tuning fork set, and a graduated jar.

Theory—The frequency of a resonator is given by

$$n = \frac{v}{2\pi}\sqrt{\frac{A}{lv}}$$

where v is the velocity of sound in air. A and *l* are respectively the area of cross-section and length of the neck of the resonator and v its volume.

To establish that $n \propto 1/\sqrt{v}$, we shall prove that $n^2 v$ is constant and that the resulting curve between $1/n^2$ and v should be a *straight line* passing through the origin. In actual practice it will be found not to pass through the origin. The curve then obeys closely the relationship.

$$n^2(v + C) = \text{constant}$$

where C is the necessary correction to be applied to v. This is termed as the *neck correction*. Hence the intercept on the v-axis for $1/n^2 = 0$ is a measure of the neck correction.

Method—Fill the resonator with water up to the base of the neck and clamp it firmly on a clamp stand. Place a graduated jar of about 250 c.c. capacity below the rubber tubing (Fig. 8.4). Allow some water to run out by opening the pinch-cock. Hold a vibrating tuning fork of frequency, say, 512, close to the neck of the resonator and test for the resonance. By gradually running the water out of the resonator, a position will be attained when the resonator responds to the frequency of the vibrating fork. When *resonance* occurs, as tested by the response of the resonator, measure the volume of water that flowed out into the graduated jar.

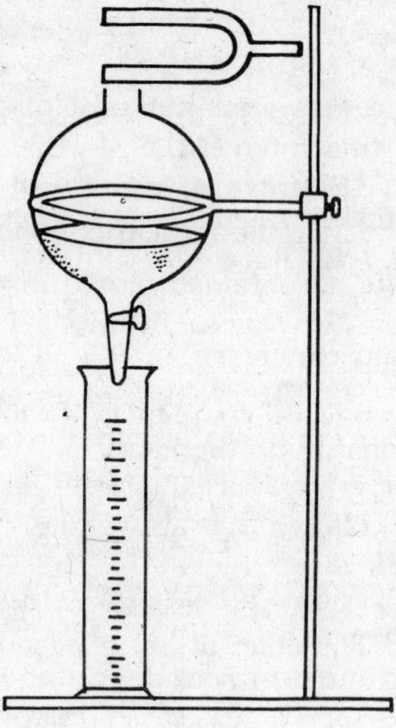

Fig. 8.4

Fill the resonator again to the original level, *i.e.*, up to the base of the neck and repeat the above observation. Hence calculate the mean value of the water flowed out from the resonator for the setting with the given fork. This is also the volume of the air column inside the resonator, which is in resonance with the fork.

Repeat the experiment with the forks of different frequencies and with each fork at least thrice. Calculate in each case the volume of the air column in the resonator that is in resonance with them. Plot a graph between v and $1/n_2$ (Fig. 8.5). It is found to be a *stralght line but not* passing through the origin.

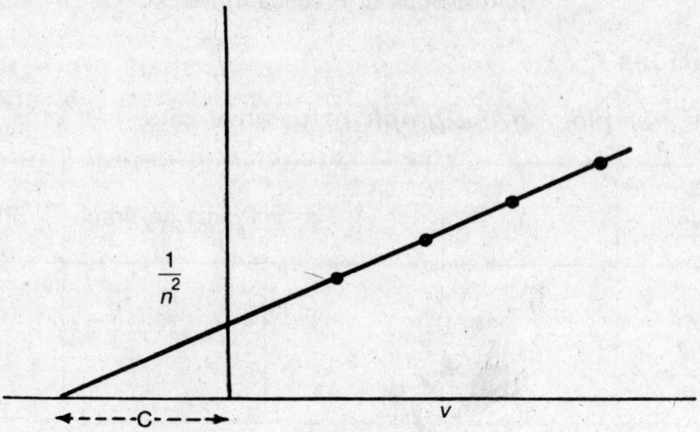

Fig. 8.5

Measure the intercept on the volume axis. The magnitude of this intercept, C, is the required *neck correction*.

Sources of error and precautions —(1) The agreement of pitch between the fork and the resonator should better be tested by blowing across the neck and noting the beats between the note so obtained and that of the vibrating fork.

(2) For each fork the observations should be repeated at least thrice.

(3) The volume of water that flows out of the resonator should be accurately measured, at least correct up to $\frac{1}{2}$ c.c. Accordingly the graduated jar should be chosen properly for this purpose.

(4) The graph between v and $1/n^2$ should be plotted on a *large* graph paper. $1/n^2$ being a small number, proper units must be chosen for its adequate representation on the graph.

Observations

Least count of graduated jar =   c.c.

S. No.	Frequency of the fork n	Volume of resonator V	Mean value of volume v for each set
1			
2			
3			

Calculations

[*A*] *For plotting the graph between* v *and* $1/n^2$.

S. No.	$1/n^2$	v	Points on graph	Product $n^2.v$

The graph is almost a straight line and the product $n^2 v$ is fairly constant.

[*B*] *For neck correction*

From the graph the intercept on the volume axis is....c.c.

∴ Neck correction, C = c.c.

Result—The graph is almost a straight line but does not pass through the origin. Neck correction = c.c.

ORAL QUESTIONS

KUNDT'S TUBE

1. On what factors does the velocity of sound in a medium depend ? What is Newton's expression for this ?

2. What correction was made by La place and why ? How does it vary with temperature, pressure and humidity ?

3. Describe the Kundt's tube arrangement for determining the velocity of sound in air.

4. Why do you clamp the rod in the middle ? Can you clamp it anywhere ? What other way of clamping do you suggest?

5. How does the 'note' emitted by the rod depend on the position of clamping ?

6. What kind of vibrations do you set up in the rod by so rubbing?

7. What other type of vibrations is it possible to set up and how?

8. How may vibrations be maintained by electricity ?

9. Why do you move the disc in the tube ?

10. When would stationary vibration be set up in air column inside the tube ?

11. What do you conclude when the dust in the tube is agitated ?

12. Why does the dust settle down at regular intervals ?

13. What assumption is made in the calculation of result for which there is not enough justification ? How will you check the justification of the assumption from your result ?

14. Why should the tube be made perfectly dry ?

15. What are important precautions in the experiment ?

16. What physical and chemical properties can you determine with this apparatus ?

17. What is γ and how is it related to sound ?

18. What light does the value of γ throw on the molecular complexity of a gas ?

RESONATORS

1. What is a resonator and to what use is it put ?
2. On what factors does the frequency of a resonator depend ? How does it vary with (*i*) A, the area of the neck (*ii*) *l*, the length of the neck and v, the volume of the air column in the resonator ?
3. What do you understand by the neck correction ? How do you determine it ?

4. What conclusions do you draw from the fact that the graph between v and l/n^2 does not pass through the origin ?
5. How do you test the resonance between the vibrating fork and the air column inside the resonator ?
6. What function is served by the air in the neck ?
7. What is mass reactance and how does it differ from stiffness reactance ?
8. When does resonance occur in a resonator ?
9. Can you use a resonator to determine the velocity of sound in air ? Do you expect to get a satisfactory result with it ?
10. What is the importance of resonator in Sound ?

SECTION III

OPTICS

9
Reflection of Light

9.1. Rotation of a plane mirror—It can be shown that if *a plane mirrior is rotated through a certain angle, the reflected ray is turned through twice this angle.* This result has been utilised in practice for the measurement of small angular deflections of rotating systems. We describe below two arrangements.

(*a*) *Telescope and scale arrangements*—A plane mirror (Fig. 9.1) about 1 cm. diameter is fixed to the rotating system parallel to its axis of rotation. In front of it and at a distance of about one metre is placed a well-illuminated graduated scale supported horizontally.

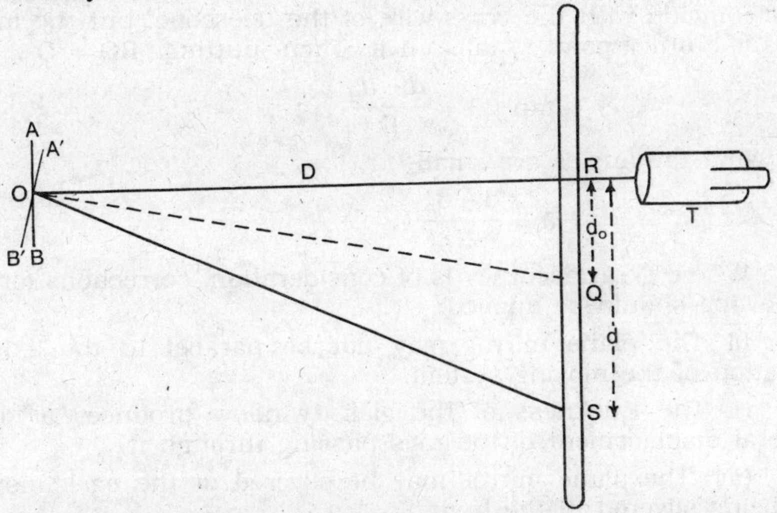

Fig. 9.1

In the position of test, the mirror AB is parallel to the graduated scale and the image of the scale division R will be seen coinciding with the vertical cross-wire of the telescope. When the mirror turns through an angle θ, attaining the position A′ B′, rays from some other point S now reach the telescope and form an image coincident with the vertical cross-wire. Since the angle through which the reflected ray is turned is twice the angle through which the mirror turns, ∠ ROS = 2θ. Hence, if RS = d and OR = D,

$$\tan 2\theta = \frac{RS}{OR} = \frac{d}{D}$$

From the table 2θ can be found. Then we can get the value of θ, the angle turned by the rotating system. Usually the angles are so small that we can fairly take tan 2θ=2θ.

$$\therefore \qquad \theta = \frac{d}{2D} \qquad (9.1)$$

Such a condition is easily attained by making observations at a distance of one metre and limiting the length of the scale to about 10 cms. Since θ α d, a comparison of the angles of deflection can be made by comparing the corresponding values of d.

If, in the initial position, the image of the foot of the perpendicular drawn from the axis of rotation to the scale, does not coincide with the cross-wire of the telescope, but the image to some other point Q falls on it. Then, putting RQ = D_0.

$$\tan 2\theta = \frac{d - d_0}{D}$$

or when the angles are small

$$\theta = \frac{d - d_0}{2D} \qquad (9.2)$$

Where great accuracy is of consideration, corrections for the following should be applied:

(*i*) The plane mirror may not be parallel to the axis of rotation of the moving system.

(*ii* The thickness of the glass window produces a slight lateral displacement of the rays passing through it.

(*iii*) The plane mirror may be silvered at the back instead of being silvered in the front.

(*b*) *Lamp and scale arrangement.*—Figure 9.2 illustrates the principle of this arrangement. By means of a short-focus convex lens light from a lamp is thrown on the plane mirror fixed parallel to the axis of rotation of the moving system. The rays of light are reflected from the mirror and fall on the scale. By suitably adjusting the distance of the lens from the bulb, the image of a straight filament of the bulb (or a wire or slit placed in the tube containing the lens) can be brought to a focus on a scale division R. *The function of the plane mirror is simply to reflect the pencil of light, emerging from the lens, that would have formed a real*

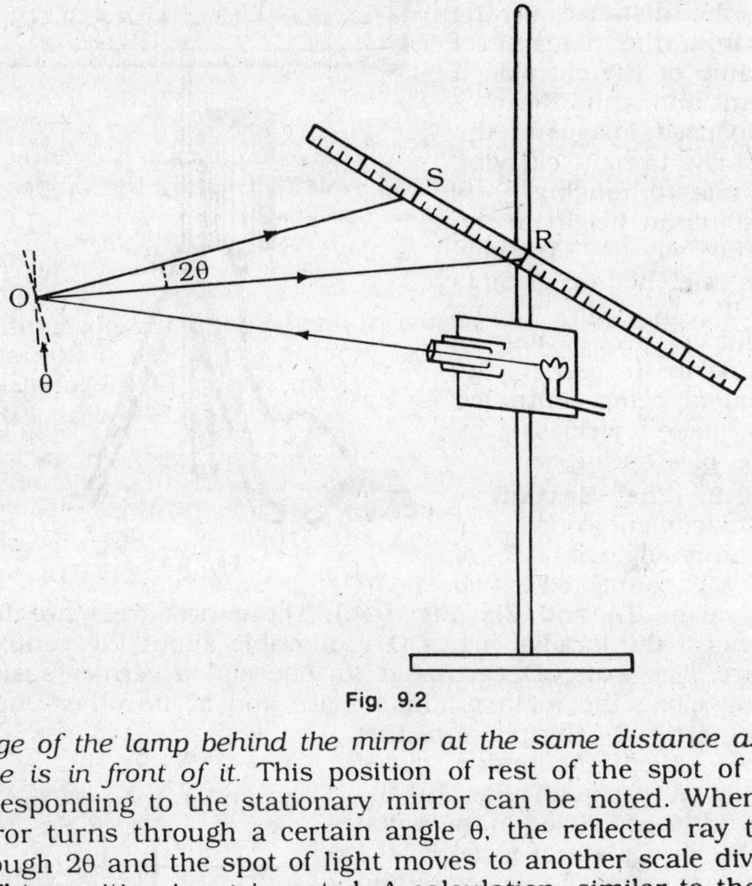

Fig. 9.2

image of the lamp behind the mirror at the same distance as the scale is in front of it. This position of rest of the spot of light corresponding to the stationary mirror can be noted. When the mirror turns through a certain angle θ, the reflected ray turns through 2θ and the spot of light moves to another scale division S. This position is again noted. A calculation, similar to the one applied in the previous arrangement, will show that

$$\tan 2\theta = 2\theta = \frac{SR}{OR} = \frac{d}{D}$$

or when angles are small

$$\theta = \frac{d}{2D}$$

The mechanical constructions of the apparatus are such as to permit great ease in handling it (Fig. 9.3). The lamp, operated on mains, is housed in a properly ventilated case. The housing is ball-jointed on a stand in such a manner that it can be rotated in any direction by working only one screw. Up and down, sideways and all angular motions are thus provided. The lamp-housing has a tube containing a short-focus lens. This tube can be moved in and out to alter the distance of the lens from the filament of the lamp or the slit placed therein and thus form a well-focused image on the scale. The upright carrying the lamp-housing is adjustable in height to the extent of, say 12″. The scale which is etched on glass is carried by the same upright. The scale is movable in its carrier, the movement being controlled and made jerkless by means of a spring.

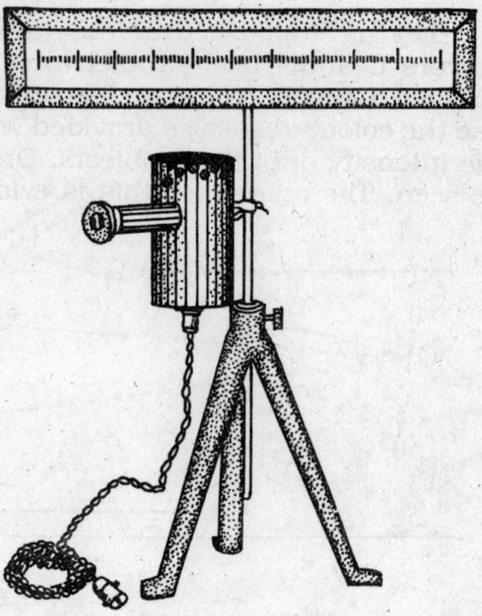

Fig. 9.3

9.2. The Sextant — The instrument consists of a graduated arc AB of about 60°, connected to two radial arms TA and EB (Fig. 9.4.). These two arms are fixed whereas a third radial arm CD is movable about the centre of the arc. The arm CD carries at its one end a vernier scale D moving along the main graduated arc and at its other end, a plane mirror C. The plane of this mirror, called the *index glass*, should be perpendicular to the plane of the arc. Fixed to the radial arm EB is a plate of glass E, the lower part of which is silvered and the upper part, transparent. The plane of this mirror, called the *horizon glass*, is also perpendicular to the plane of the arc. Pointing towards the horizon glass is fixed a telescope T to the arm A, to receive the direct rays through the transparent portion of E and the twice reflected rays from R and E. The axis of the telescope is parallel to the plane of the arc. The radial arm CD can be clamped in position with a clamping screw and then slow motion can be effected with the tangent screw.

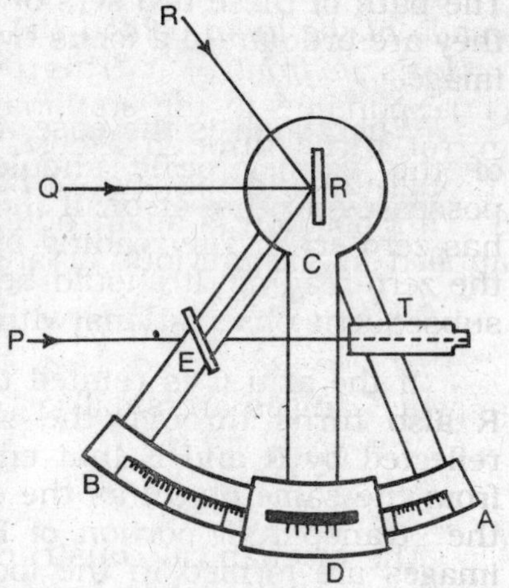

Fig. 9.4

Let us suppose that the radial arm CD is moved till the mirrors C and E are parallel to each other and the telescope is directed towards a distant object or the sun. In the latter case use the coloured glasses provided with the sextant for diminishing the intensity of brilliant objects. Only one image of the object will be seen. The reason for this is evident. One set of rays from the

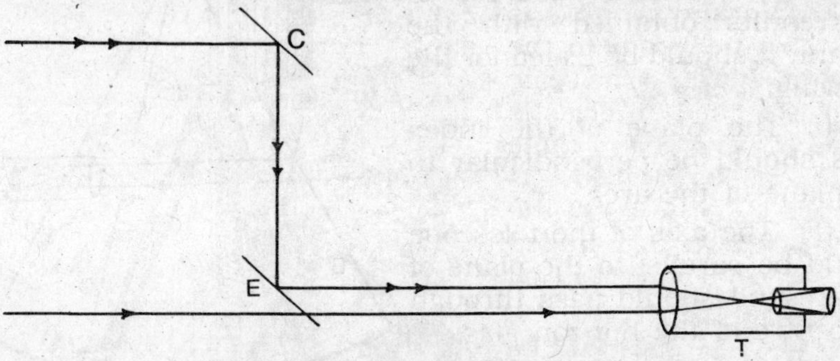

Fig. 9.5

object passes through the transparent portion of E and enters the telescope· the other set after reflection from C is reflected again at the silvered portion of E and enters the telescope. These two sets of rays are parallel since C and E are parallel and the incident pencil of light from the distant object is taken as parallel. The path of these two sets of rays is close together (Fig. 9.5) and they are brought to a focus by the telescope giving rise to a single image.

When such is the case, the zero of the main scale and that of the vernier scale should coincide. The instrument then possesses no zero-error. If the case be otherwise, the instrument has zero error. The reading on the vernier D is taken. This gives the zero-reading. It should be applied with its proper sign in the subsequent observations with the sextant.

If the arm C is rotated through an angle θ, the index glass R also turns through the same angle. But the set rays MR, reflected by R and E that enters the telescope no longer comes from the same object as the other set of rays S passing through the transparent portion of E (Fig. 9.6). Hence two superposed images are formed in the focal plane of the object glass of the telescope. The angle which the set of rays MR makes with the set SE is 2θ since the beam of light on reflection turns through twice the angle of rotation of the reflecting mirror. We note that the arm C has been rotated through an angle θ whereas the angle subtended by the two distant objects giving rise to the two sets of rays MR and SE, is 2θ. Thus, to make the instrument direct reading, the arc PQ is graduated such that each degree is numbered as two degrees on it.

To ensure the accuracy of the results obtained with the sextant, it should be tested for the following:

(*i*) The plane of the index glass should be perpendicular to the plane of the arc.

(*ii*) The axis of the telescope should be parallel to the plane of the arc and should pass through the centre of the horizon glass.

(*iii*) The plane of the horizon glass is also perpendicular to the plane of the arc and that the horizon glass and the index glass are parallel to each other when zero of the vernier falls against zero of the main scale.

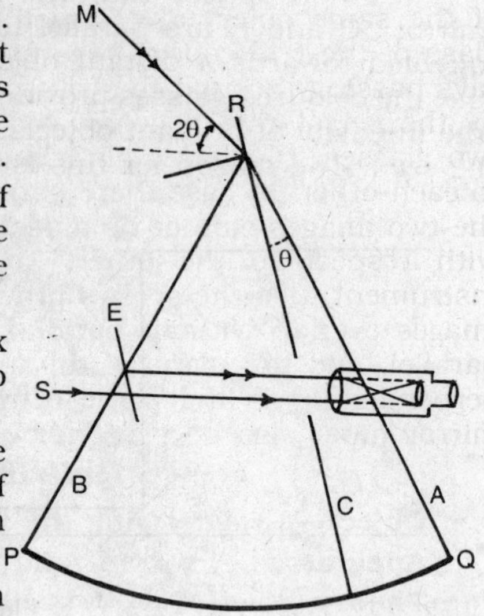

Fig. 9.6

To test for the first condition to be satisfied by the sextant, look at the image of the graduated arc in the index glass, the plane of which passes through the centre of the arc. If R is properly adjusted with its plane normal to the plane of the arc, the latter and its image will lie in the same plane. If the adjustment is faulty, use the screws at the back of this mirror till the correct position is secured.

The axis of the telescope is the line joining the centre of the object glass and the eye-piece. This line passes through the point of intersection of cross-wires, which evidently is the centre of the field of view. The axis of the telescope has to be adjusted parallel to the plane of the graduated arc. Observing two objects having an angular separation of about 10° to 90°, arrange for the coincidence of the two images in the centre of the field of view. Then tilt the instrument till the images lie towards the edge of the field of view and see if the images still coincide or not. Next tilt it till the images lie towards the other edge of the field of view and observe their coincidence. If in these positions the images remain coincident, the axis of the telescope is properly adjusted; otherwise, set the telescope in proper position by using the screw by which it is held to the sextant.

The third condition is tested by pointing the sextant towards distant object or the sun and turning the arm C till two images of the same object are formed in the focal plane of the object glass of the telescope. These two images, one obtained by set of rays passing through the transparent portion of E, and the other by the set of rays reflected by R and E, should coincide if the two glasses, the index glass and the horizon glass, are parallel to each other. In case there is non-parallelism of the two mirrors, the two images cannot be made to coincide, one will lie sideways with respect to the other. To correct for this defect in the instrument, use the screws provided at the back of E till the two images overlap without parallax. Now since the two mirrors are parallel, and the plane of the index mirror has already been set perpendicular to the plane of the arc, the plane of the horizon mirror must also be perpendicular to the plane of the arc.

Experiment 9.1

Object—To determine the height of a tower with a sextant.

Apparatus—A sextant and a 50′ measuring tape.

Theory— Let NX = h be the height of the tower and let α and β be the angular elevation of the tower from Y and Z respectively (Fig. 9.7), these being in a line normal to XN at N. If YZ be known and equal to x, say, we can calculate h in terms of x, α and β. Let NY = y, from the Δs XNY and XNZ, we have

$$h \cot \alpha = y$$

and $$h \cot \beta = y + x$$

whence $$h (\cot \beta - \cot \alpha) = x$$

or $$h = \frac{x}{\cot \beta - \cot \alpha} \tag{9.3}$$

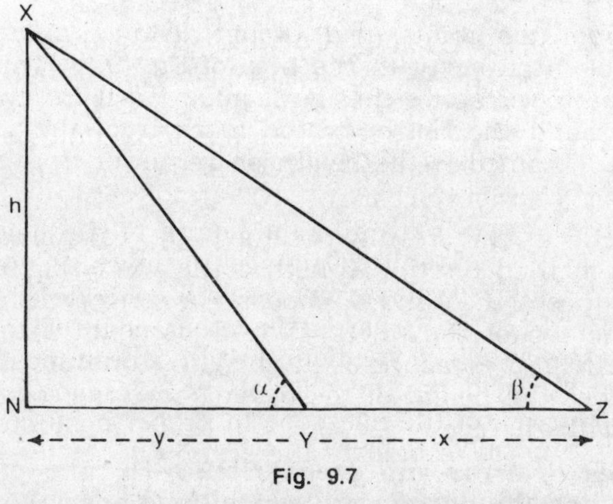

Fig. 9.7

Method—Having tested the sextant for its instrumental adjustments as described in Sec. 9.2, proceed as follows :

Determine the vernier constant of the sextant. Make a mark with a piece of chalk on the tower in level with the height of your eye. Move to a considerable distance from the tower and mark your position on the ground. Point the sextant towards the tower and move the arm C, which in its turn rotates the index mirror R, till the two images coincide. In this position take the reading on the main scale and the vernier scale. The computed reading then gives the zero reading of the sextant.

Now point the sextant such that the plane of the graduated arc is vertical and the axis of the telescope is directed towards the mark on the tower. See the image of the mark on the tower in the telescope. Rotate the arm C when it would be observed that the image of the tower descends. Continue the rotation of the arm C till the image on the top of the tower coincides with the image of the mark made on it. Clamp the arm C and make fine adjustment with the tangent screw to get the coincidence of the two images on the cross wires. With the help of a reading lens take the reading on the main scale and the vernier scale. The total reading gives the angular elevation of the tower with a horizontal line passing through the mark on it. Calculate the true angular elevation α by applying the zero-correction.

From this place of observation move through a distance, say, 5 metres or so towards or away from the mark, as may be found convenient. *Again find the zero error* and the angular elevation of the tower by the method described above. Then calculate the true angular elevation β of the tower.

Next from two points at a known distance apart, calculate the height of the tower with the help of Eq. (9.3). This gives the height of the tower above the mark made on it in level with the eye. To this add the height of the mark from the ground, the result thus obtained is the height of the tower from the level of the ground.

Now check the result obtained by calculation by the geometrical method. For this, construct the ΔXYZ. (Fig.9.7) with the observed values of x, α and β, choosing a convenient scale for it. From X, the top of the tower, drop a perpendicular XN to ZY produced. XN is a measure of the height. Compute the value of XN. This gives the height of the tower above the mark made on it in level with the eye.

Sources of error and precautions —(1) The plane of the index glass and the horizon glass should be normal to the plane of the graduated arc and that when the two are parallel, the zero of the vernier scale should coincide with the zero of the main scale.

(2) The axis of the telescope should be parallel to the plane of the graduated arc.

(3) The axis of rotation of the index mirror must coincide with the centre of the graduated arc.

(4) Determine the zero error for every distance of the object under examination and apply it properly in subsequent observations.

(5) While using the sextant, the plane of the graduated arc should be vertical. The axis of the telescope is then horizontal. It should pass through the mark on the tower.

(6) The mark on the tower and the two positions of observation for its angular elevation must be collinear.

Observations.—Value of one main scale division = min.

Total number of vernier divisions =

∴ Least count = min.

S. No.	From first place of observation					
	Zero Error			Angular elevation		
	Main Scale reading	Vernier Scale reading	Total reading	Main Scale reading	Vernier Scale reading	Total reading
1						
2						
3						
		Mean			Mean	

[*N.B.*.—Draw a similar table for second place of observation].

Distance between the two places of observation = cm.

Calculations

Mean corrected elevation from first place of observation =

Mean corrected elevation from second place of observation =

$$h = \frac{x}{\cot \beta - \cot \alpha}$$

=

= cm.

Height of the mark (on the tower) from the ground= cm.

∴ Height of the tower from the ground $= h +$ $=$ cm.

Also, from the Δ XNZ, XN $=$

$=$ cm.

∴ Height of the tower from the ground $=$ XN + $=$ cm.

∴ Mean height of the tower from the ground $= \dfrac{h + \ldots \text{ XN} + \ldots}{2}$

Result —The height of the tower = cm. = m.

Exercise 1 —To determine the altitude of the sun with a sextant.

Determine the zero error of the instrument by coinciding the image of the sun as seen through the clear portion of E with the image obtained by reflection from the two mirrors R and E. Let the reading be θ_1. Place a large dish containing mercury in a convenient position. The image S_1 of the sun in the mercury is observed. [Fig. 9.8 (*a*)]. Let O be the place of observation with the sextant. Rotate the radial arm C so that the mirrors R and E give rise to the image of the sun coinciding with s_1, the image formed by the Set of rays MD passing through the transparent portion of E. Let the reading on the graduated arc be θ_2. Then.

$$2\alpha = \theta_2 - \theta_1$$

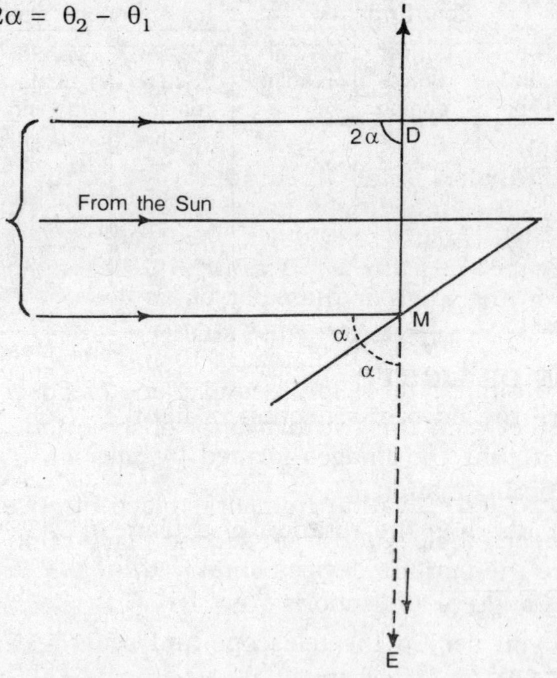

Fig. 9.8 (a)

or $\alpha = (\theta_2 - \theta_1)/2$

where α is the angular elevation or sun's altitude.

Exercise 2 —To determine the angular diameter of the sun with a sextant.

Hold the sextant vertically and look at the sun through the transparent portion of the horizon glass E, using sun-glasses to reduce the glare of the sun. Rotate the radial-arm C till the rim of the image of the sun S, obtained by reflection from the two mirrors R and E, just touches the rim of the direct image of the sun as illustrated in Fig. 9.8 (*b*). Note down the reading of the

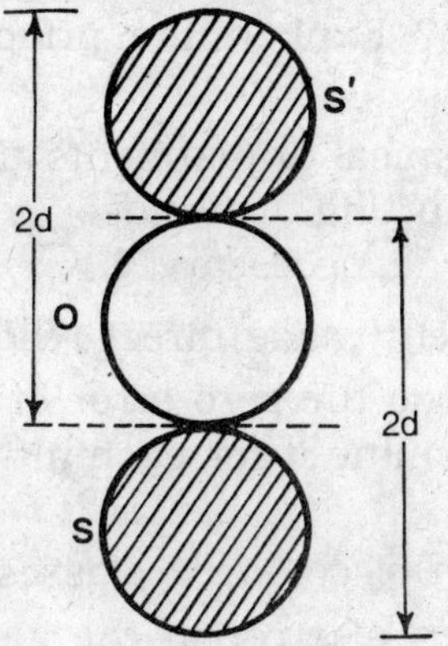

Fig. 9.8 (b)

sextant. Next further rotate the radial arm B till the reflected image of the sun shifts to the position S′ on the other side of the direct image. Again note down the reading of the sextant. The difference between the two readings of the instrument gives $2d$, where d is the angular diameter of the sun.

ORAL QUESTIONS

REFLECTION OF LIGHT

1. What are the laws of reflection of light ?
2. How is it that the images formed by mirrors are free from chromatic aberration ?
3. To what use has the rotation of a plane mirror been put?
4. What are the various devices employed for the measurement of small angular deflections ?
5. How do you set up the telescope and scale arrangement ?

6. What correction should be applied in measuring the actual values of angular deflections ?
7. How are the graduations printed on the scale ?
8. What is meant by lateral inversion ? Mention some practical application of this.
9. How do you set the lamp and scale arrangement ?
10. How is it that a plane mirror attached to the moving system forms a real image on the graduated scale?

THE SEXTANT

1. What is a sextant ? Explain the principle on which it is based.
2. What are the mechanical adjustments that are tested for in a sextant before putting it to use ? Explain why these adjustments are at all necessary ?
3. How is the instrument made direct reading?
4. How do you determine the zero error of the instrument ? Is it necessary to determine it for each altitude measured with it ?
5. Explain the function of coloured glasses provided with it.
6. What are the sources of error in the instrument ? How are they eliminated ?
7. To what other uses can the sextant be put ?
8. What is artificial horizon and what purpose does it serve ?

10

Refraction of Light
Focal Length Determinations

10.1. Cardinal points of an Optical system

(*a*) ***Focal points***—Consider a coaxial system having its axis AA′ and surrounded by media of refractive indices μ_1 and μ_2 as depicted in Fig. 10.1. Let I_1K_1 be an incident ray parallel to the axis of the system. After refraction through the system it takes up the direction K_2F_2 cutting the principal axis at F_2. This would

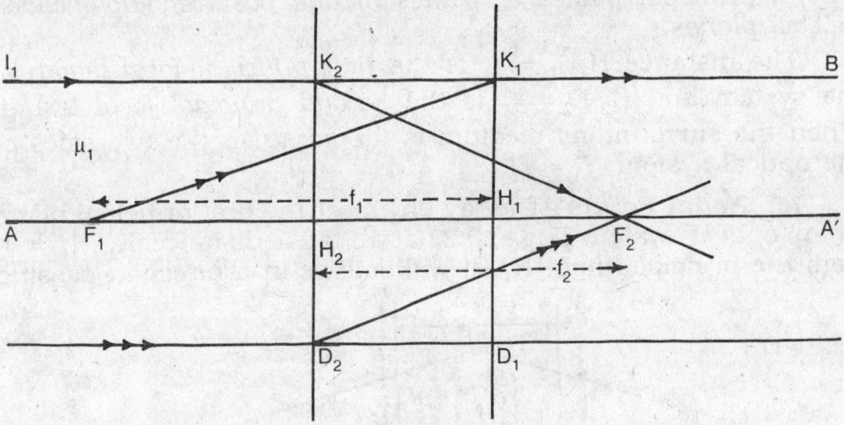

Fig. 10.1

be true for all the rays parallel to the principal axis, *i.e.*, all of them after refraction through the optical system would pass through F_2. (For divergent system, the refracted rays would appear to diverge from a point on the principal axis of the system.) *This point F_2 is termed as the second principal focus and corresponds to the object distance $u = \infty$.*

Similarly, if the rays starting from a point F_1 on the axis after refraction through the system become parallel to the axis, *this point is termed as the first principal focus of the system and corresponds to the image distance $v = \infty$.* The two planes passing through F_1 and F_2 and perpendicular to the axis are termed as the first and the second focal plane, respectively.

(*b*) ***Principal Points***—As shown in Fig. 10.1 the ray in the direction I_1K_1 after refraction through the system emerges in the direction K_2F_2 also the ray in the direction F_1K_1 after refraction through the system becomes parallel to the principal axis. The two rays I_1K_1 and F_1K_1 are directed towards the point K_1 and after refraction through the system appear to come from K_2. Thus K_2 is the image of the point K_1, *i.e.*, K_1 and K_2 are a pair of conjugate points. The plane K_1H_1 and K_2H_2 are a pair of conjugate planes. Since $K_1H_1 = K_2H_2$, it is evident that the lateral magnification for these two planes is +1. Thus, any ray CD_1 incident on the system parallel to the principal axis, cuts the plane K_1H_1 at a distance H_1D_1, therefore after refraction it must pass through the point D_2 in the plane K_2H_2 at a distance $D_2H_2=D_1H_1$ and also through F_2. Thus D_2F_2 determines the direction of the emergent ray from the system. The points H_1 and H_2 on the axis are called *principal points* or *unit points* and the planes through these points are spoken of as the first and the second principal planes or *planes of unit positive magnification* or *Unit planes*.

The distance $H_1F_1 = f_1$ is the *first principal focal length* of the system and $H_2 F_2 = f_2$ is the *second principal focal length.* When the surrounding medium is the same on the two sides of the optical system. $f_1 = f_2$.

(c) *Nodal Points*—The ray PK_1 meet the first principal plane of an optical system (Fig. 10.2) at K_1 at a distance $K_1 H_1 = h$ from the principal point H_1. It will emerge in a direction passing

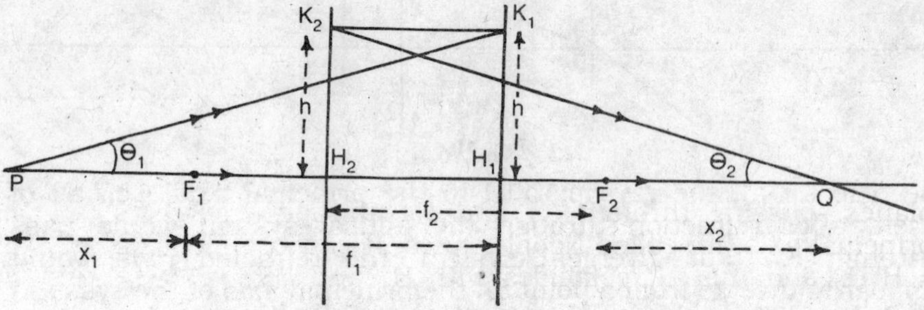

Fig. 10.2

through K_2 such that $K_2 H_2$ = h. Let the path taken by the ray be K_2 Q. Q is the image of the point P. Let f_1 and f_2 be the first and the second focal lengths of the system and let x_1 and x_2 be the distances of the object and the image respectively from F_1 and F_2. The angular magnification is given by

$$\gamma = \frac{\tan \theta_2}{\tan \theta_1}$$

Now $\tan\theta_2 = -\dfrac{h}{f_2 + x_2}$ and $\tan\theta_1 = \dfrac{h}{f_1 + x_1}$

$$\therefore \gamma = -\frac{f_1 + x_1}{f_2 + x_2} = -\frac{f_1}{x_2} = -\frac{x_1}{f_2} \qquad (\because x_1 x_2 = f_1 f_2)$$

For $\gamma = +1, x_2 = -f_1$ and $x_1 = -f_2$. Thus, these *two points* N_1 and N_2 on the axis of the system for which there is unit positive angular magnification, are termed as *Nodal Points*. Their positions can be marked on the axis as shown in Fig. 10.3. The property of these points is unit positive angular magnification. *i.e.*, a ray directed towards the point N_1 after refraction through the system, emerges parallel to itself through the point N_2. The

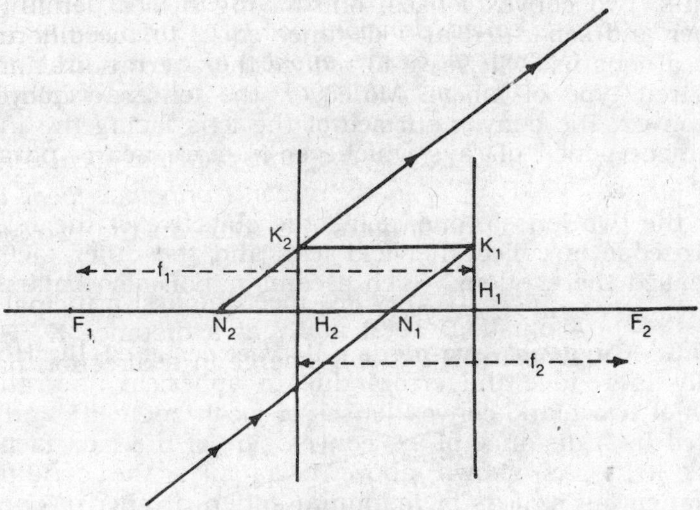

Fig. 10.3

planes, passing through N_1 and N_2 and perpendicular to the principal axis are called Nodal Planes. It can be shown that $N_1N_2 = H_1H_2$ and that N_1 coincides with H_1 and N_2 with H_2 when $f_1 = f_2$ (*i.e.*, when the incident and emergent rays lie in optically same media).

10.2. Eye-pieces—In optical instruments it is usual to employ combination of this lenses separated by a distance as an eye-piece. A single lens introduces defects in the image partly due to spherical aberration and partly due to chromatic aberration. Moreover, the field of view is narrow. In order to overcome these defects the single lens used as magnifier, is replaced by a combination of two lenses separated at a distance.

It can be shown that such a system is achromatic if their distance apart x is given by

$$x = \frac{f_1 + f_2}{2}$$

Also the spherical aberration is minimum when

$$x = f_1 - f_2$$

f_1 and f_2 being the focal lengths of the component lenses. Solving the two, we have

$$f_1 = 3f_2$$

and $$x = 3f_2 - f_2 = 2f_2$$

Thus, two convex lenses, one having a focal length thrice the other and separated at a distance equal to the difference of their focal lengths will serve to reduce the aberrations and form the desired type of image. Moreover, the lenses employed are plano-convex, the convex surface of the lens facing the incident or emergent pencil of rays whichever is more nearly parallel to the axis.

Of the two lenses, one facing the objective of the telescope or microscope is called the field lens and the other facing the eye is called the eye lens. Such a combination also enlarges the field of view.

10.3. Huygens' eye-piece —It was designed by Huygens primarily to reduce the errors due to spherical aberration. It consists of two plano-convex lenses of focal length $3f$ and f and separated by a distance of $2f$, convex side of the lens facing the incident light. As shown above it satisfies the condition of achromatism as well as of minimum spherical aberration.

The working of the eye-piece can be understood from the following considerations. The eye of the observer sees the final image at infinity with unaccommodated eye. Thus, the rays emerging out of the eye lens must be parallel and for that the image formed by the field lens must be situated at the first principal focus of the eye lens. As the distancc between the two lenses is $2f$, f being the focal length of the eye lens, the image formed by the field lens is at I_2 (Fig. 10.4), distant f from the field lens as well as from the eye lens. For the image to be formed at I_2 by the field lens, the image formed by the objective, which serves as the virtual object for the field lens should be situated at distance u given by the equation

$$-\frac{1}{f} - \frac{1}{u} = -\frac{1}{3f}$$

$$\therefore \qquad \frac{1}{u} = \frac{1}{3f} - \frac{1}{f} = -\frac{2}{3f}$$

$$\text{or} \qquad u = -3f/2$$

i.e., at I_2, midway between I_2 and the eye lens.

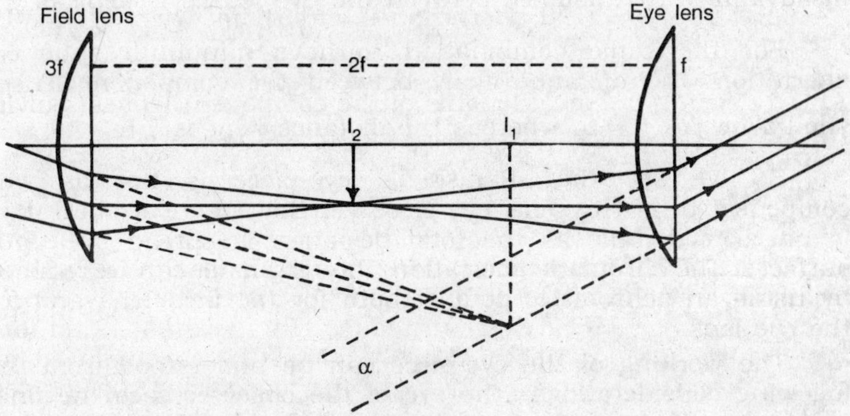

Fig. 10.4

As the image I_1 formed by the objective lies on the negative side of the field lens, the eye-piece is sometimes termed as a *Negative Eye-piece*. If a cross-wire or a micrometer scale is to be used with it, it must be inserted in between the lenses at a place where the field lens forms the real image I_2. But the micrometric measurement will not be reliable as the image of the virtual object is formed by both the lenses while that of the scale is formed by the eye lens alone and this is also curved and distorted. Moreover there is a practical difficulty also. Once the cross-wires are fixed in between the lenses, the eye-piece will be in adjustment for normal vision ; it cannot be adjusted for individuals of varying vision, as is practicable in a Ramsden's eye-piece.

The eye-piece is not truly achromatic ; it is achromatic in the same sense as an ordinary magnifying glass, for the various coloured images formed by it subtend the same angle at the eye.

10.4. Ramsden's Eye-piece—This eye-piece consists of two plan-convex lenses with their convex faces inwards and separated by a distance equal to $\frac{2}{3}f$, f being their common focal length.

For this combination to behave as achromatic the distance apart of the lenses should be $(f + f)/2 = f$. But if the field lens is at a distance f from the eye lens a serious disadvantage arises. The surface of the field lens and the image to be viewed both lie in the focal plane of the eye lens ; dust particles and any imperfection of the surface of the lens would be highly magnified and thus becomes a serious source of trouble. To overcome this disadvantage the distance between the two lenses is kept as $\frac{2}{3}f$.

For the same combination to have minimum spherical aberration, the distance apart between the component lenses should be $f - f = 0$, whereas the distance kept is $\frac{2}{3}f$.

So we see that Ramsden's eye-piece is not as well compensated as Huygens' eye-piece yet the spherical aberration is not considerable as the total deviation is spread over four surfaces. The chromatic aberration, though small, can be reduced by using an achromatic doublet both for the field lens and for the eye lens.

The working of the eye-piece can be understood from the following considerations. The eye of the observer sees the final image at infinity with unaccommodated eye. Thus, the rays emerging out of the eye lens must be parallel and for that the image formed by the field lens must be situated at the first principal focus of the eye. As the distance between the two lenses is $\frac{2}{3}f$, f being the focal length of either lens, the image formed by the field lens is at I_2 (Fig. 10.5), distant f from the eye lens or $f/4$ from the field lens and on the same side as the incident pencil. For the image to be formed at I_2 by the field lens, the

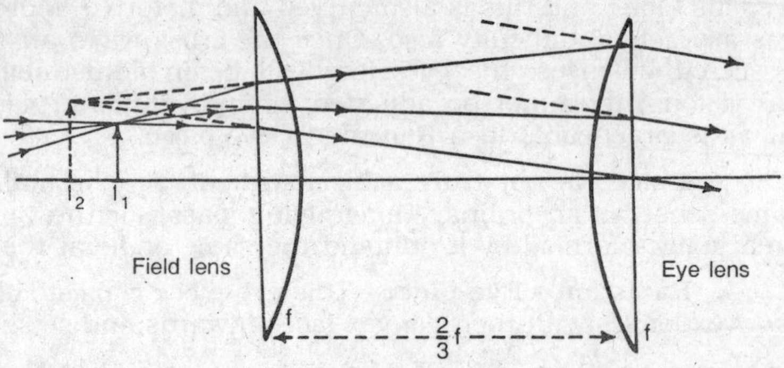

Fig. 10.5

image formed by the objective which serves as an object for the field lens should be at a distance u given by the equation :

$$\frac{1}{f/3} - \frac{1}{u} = -\frac{1}{f}$$

$$\therefore \qquad +\frac{1}{u} = +\frac{1}{f} + \frac{3}{f} = \frac{4}{f}$$

$$\text{or} \qquad u = f/4$$

Thus, the image I_1 formed by the objective lies at a distance $f/4$ in front of the field lens. As the image I_1 formed by the objective lies on the positive side of the field lens, the eye-piece is termed as *Positive Eye-piece.*

This eye-piece is best suited for micrometric measurements. A cross-wire or a micrometer scale can be introduced at a distance $f/4$ in front of the field lens, *i.e.*, at the place the image is formed by the objective. Both the cross-wire or scale and the image undergo equal magnification as the final image in each case is formed by both the lenses of the eye-piece and so it is practicable to make micrometric measurements with the eye-piece. As a matter of fact this eye-piece was mainly designed for this purpose. The eye-piece can be adjusted for normal vision as well as for distinct vision, only the combination has to be adjusted relative to the cross-wires.

10.5. Principle of Nodal Slide—Consider an optical system of which N_1 and N_2 are the nodal points. A ray AQN_1 of the incident parallel pencil of light, being directed towards the first nodal point N_1 will emerge out through the second nodal point parallel to itself along the direction N_2IF_2. Any other ray, say, DM of the parallel pencil takes the path PF_2. I is,

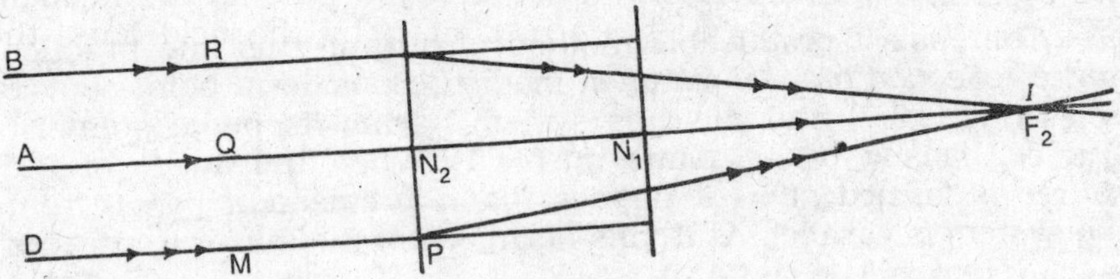

Fig. 10.6A

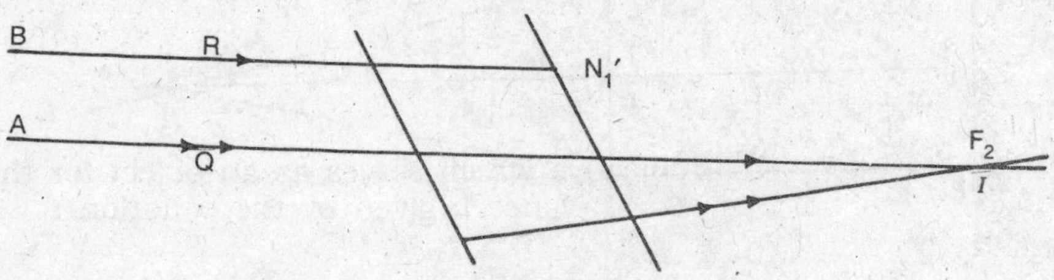

Fig. 10.6B

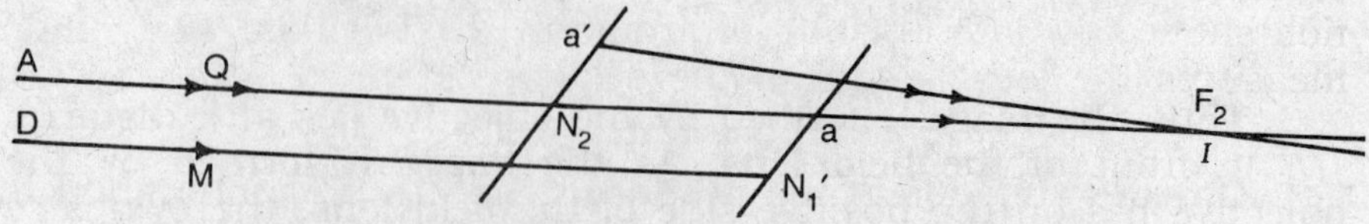

Fig. 10.6C

therefore, the image formed in the focal plane. Now suppose the system is such that it can be rotated about a vertical axis through the second nodal point N_2. As the system rotates the nodal point N_1 describes a short arc and now the ray BR begins to pass through N_1 giving rise to the corresponding ray N_2I. The ray AQ comes along PI, still all the emergent rays meet at I. Thus, the image I remains fixed on rotation of the system about a vertical axis through the second nodal point. So if such a point could be found on the axis of the optical system that a slight rotation of the system about a vertical axis through it would produce no shift of the image, then this point corresponds to the second nodal point and hence the second principal point (when the surrounding medium is the same on both sides of the system). Thus the distance between the screen which corresponds to the second focal plane and the axis of rotation which lies in the nodal plane (and hence the principal plane) gives the focal length of the system.

The easiest practicable method of putting this into practice is the *reflection auto-culmination method*. Consider a point source at F_2 (Fig. 10.7) and the lens system having its nodal point N_1 and N_2 backed by a plane mirror P. When the image of the source is formed at F_2 it implies that the emergent pencil from the system is parallel. As it falls normally on the mirror it retraces

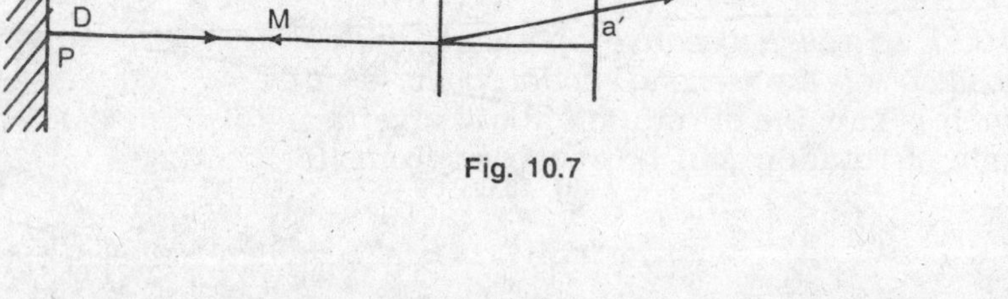

Fig. 10.7

its path and gives rise to the image at F_2. If the axis of rotation of the system is made to coincide with N_2, the image at F_2 will not show any lateral shift on rotation of the axis. N_2F_2 then measures the focal length of the system.

Experiment 10.1

Object —To determine the focal length of a combination of two convergent lenses separated by a distance and to verify the formula

$$\frac{1}{F} = \frac{1}{f_1} + \frac{1}{f_2} - \frac{x}{f_1 f_2}$$

where F represents the focal length of the combination of two convex lenses of focal lengths f_1 and f_2 when the two are separated by a distance x.

Apparatus —Nodal slide assembly and two convex lenses, usually of nearly the same focal length.

Description of apparatus —An optical bench, a metre and a half in length, is provided with four uprights. On one upright, placed at one extreme end of the bench, is mounted a lamp housing which has a circular opening for light from the frosted lamp to pass out and illuminate the cross-slits in a metal plate mounted on the adjacent upright. The heights of the lamp housing and the cross-slits can be adjusted by means of a screw. Next to the cross-slits is the upright which carries the *nodal slide*

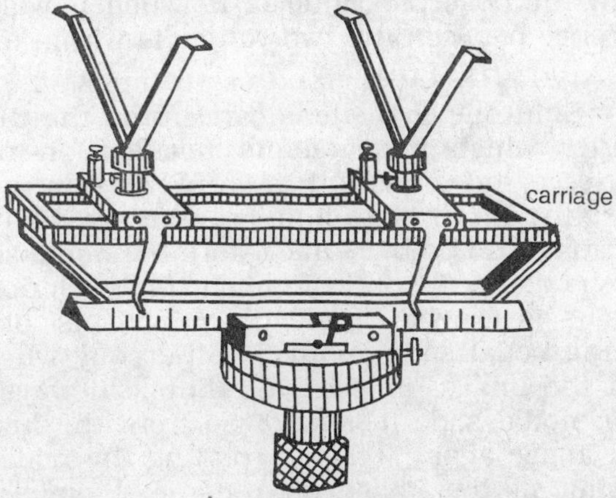

Fig. 10.8

(Fig. 10.8), so called because by sliding the carriage we can locate the position of the second nodal point of the lens or the lens system. It is capable of rotation about a vertical axis and if desired its angle of rotation can be recorded from its graduated circular

base. Lenses are placed in holders provided with the carriage of the slide, and the distance apart of the lenses can be directly read on the linear scale provided with this carriage. The carriage can be made to slide and as such the axis of rotation of the nodal slide, which remains fixed, can be made to pass through any point on the axis of the lens system. To enable one to find out through what point on the axis of the lens system the axis of rotation of the nodal slide is passing an index mark P is made on the slide past which the carriage scale moves. The index mark P marks the axis of rotation of the nodal slide.

The fourth upright carries a plane mirror which is capable of rotation about a horizontal axis perpendicular to the longer side of the bed of the bench.

The complete assembly is called the Nodal Slide Assembly.

Theory —If f_1 and f_2 be the focal lengths of the component lenses of a system and the lenses be situated at a distance x apart, the focal length of the combination is given by

$$\frac{1}{F} = \frac{1}{f_1} + \frac{1}{f_2} - \frac{x}{f_1 f_2}$$

Thus, to verify the above equation it is necessary to determine f_1, f_2 and F for a known distance x. Then the calculated value of F as obtained from a knowledge of f_1, f_2 and x should be compared with the observed value of F, when it will be noticed that the difference between the two values is within experimental errors.

Method —Shift one of the lens carriers on the carriage such that the pointer which indicates its position on the carriage coincides with the index mark P. Mount the given lens on it. Adjust the heights of the lamp-housing, the cross-slits, the lens and the plane mirror to be the same. Move the nodal slide upright carrying the lens along the optical bench till a well defined image of the cross-slits is formed beside the cross-slits, on the metal plate. Rotate the nodal slide about 5° either way when it will be observed that the image, in general, shifts sideways. Move the carriage of the nodal slide a mm. or so from the index mark P and focus the image again. Test on rotating the nodal slide that there is no shift of the image on the metal plate. When this position is achieved by trial, the plate screen with the cross-slits coincides with the focal plane of the lens and the axis of rotation of the nodal slide passes through the corresponding nodal point and hence through the corresponding principal point, for in this case, medium being air throughout, the nodal points are coincident with the principal points. The distance between the metal plate cross-slits and the axis of rotation of the nodal slide, therefore, gives the focal length of the given lens. Hence the position of the cross-slit upright and the nodal slide upright is noted on the optical bench and their difference computed to give the focal length.

Next turn the nodal slide through 180° so that the face of the lens which was away from the incident light now faces it. Adjust as before and determine its focal length. Take the mean of the two results. This gives the focal length f_1 of the first lens.

Replace the first lens by the second and repeat the above procedure to determine the mean focal length f_2 of the second lens.

Mount the two lenses some distance apart as recorded by the index marks of each lens on the linear scale of the carriage of the nodal slide. Move the carriage of the nodal slide so that the index mark P of the axis of rotation of the nodal slide is close to the first lens on which light from the cross-slits falls. Get a focused image of the cross-slits close to the cross-slits by shifting the nodal slide along the bench. Rotate the nodal slide about its axis when it is observed that the image is displaced. Note the direction of displacement of the image relative to direction of displacement of the axis of the system PP.' As depicted in Figs. 10.9 (*a*) and (*b*) this direction of displacement of the image is opposite to the direction of displacement of the axis PP′ of the system.

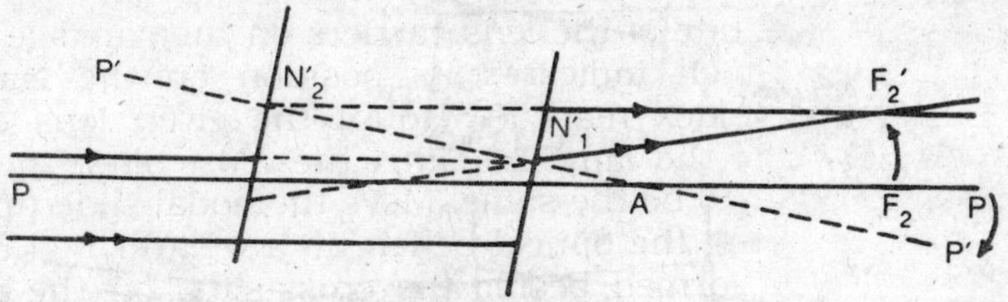

Fig. 10.9 (A)

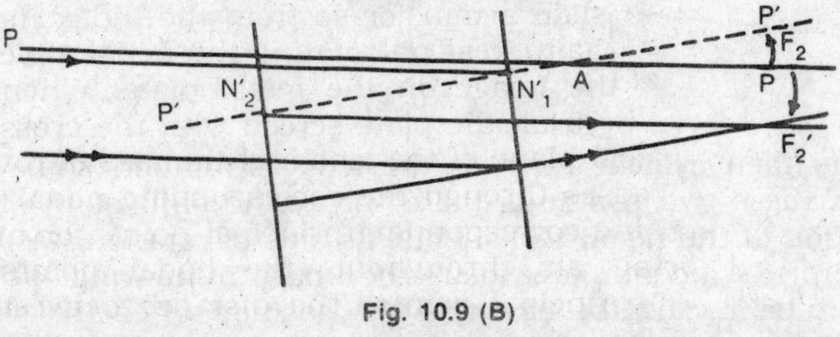

Fig. 10.9 (B)

This would be the observation even when the axis of rotation coincides with N_1 as shown in Fig. 10.9 (*c*).

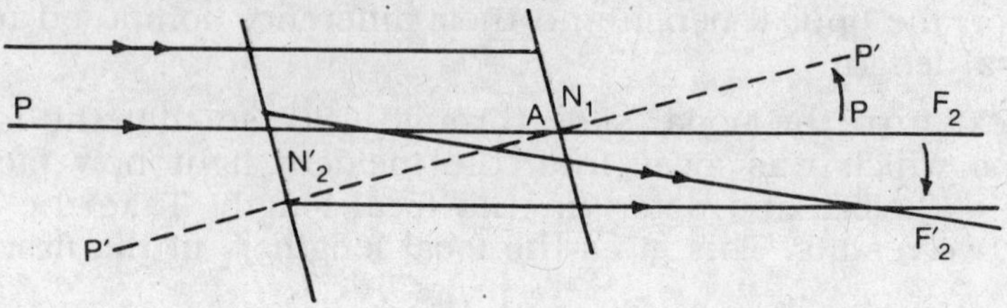

Fig. 10.9 (C)

Next shift the carriage of the nodal slide so that the index mark P is close to the second lens, *i.e.*, close to the lens nearer to the plane mirror. Focus the image by moving the nodal slide along the bench and observe the shift of the image by rotating the nodal slide on its axis. This displacement of the image relative to the displacement of the axis PP′ would be found to be in the same sense, *i.e.*, opposite to what was oserved when the axis of rotation was nearer to the lens close to the slits. This is depicted in Figs. 10.9 (*d*) and (*e*).

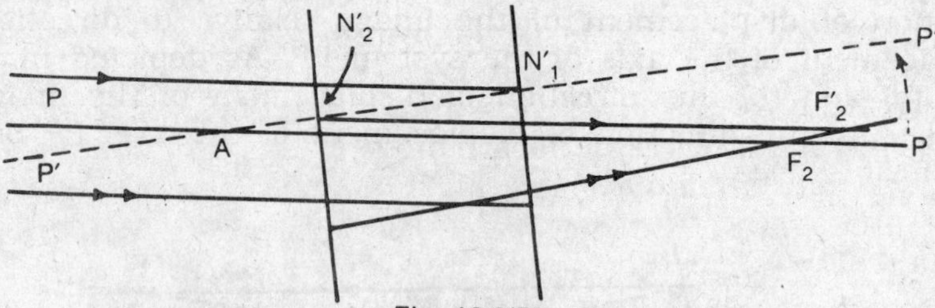

Fig. 10.9(D)

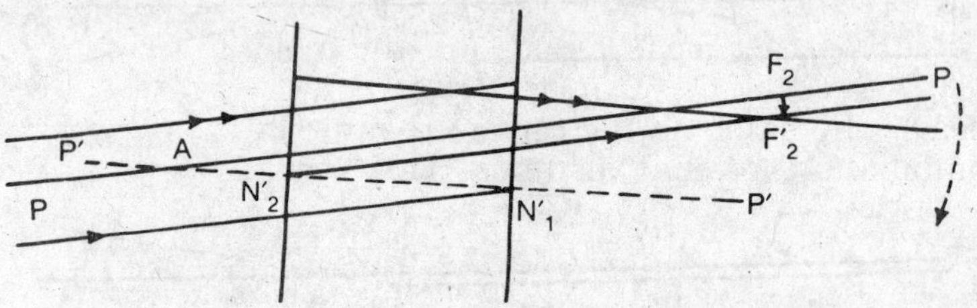

Fig. 10.9(E)

It is then evident that for the axis of rotation somewhere between these two positions there will be no shift of the image on rotation of the nodal slide about its axis of rotation. Move the carriage of the nodal slide, focus the image by moving the slide along the bench and by trial find the position when the shift of the image of the cross-slits ceases on slight rotation of the nodal slide about its axis. This is depicted in Fig. 10.9 (*f*).

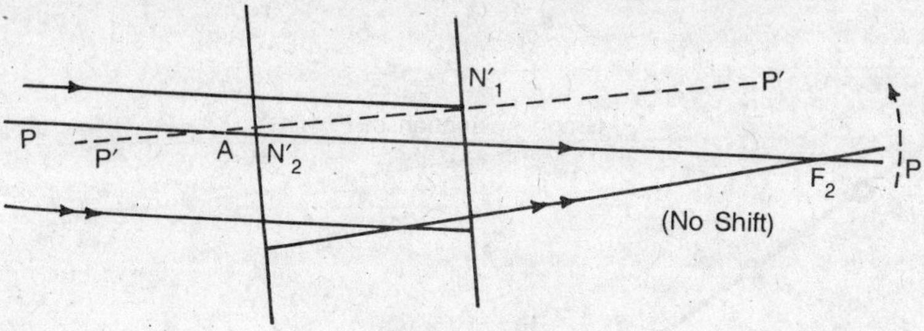

Fig. 10.9 (F)

It is evident that there is no shift of the image when and only when the axis of rotation A coincides with the second nodal point N_2.

At this stage note the positions of the cross-slit upright and the nodal slide upright on the optical bench. The difference of the two readings gives the focal length of the combination, for the plane of the cross-silts conincides with the focal plane of the combination and the axis of rotation of the nodal slide passes through the corresponding nodal point or the principal point.

Rotate the nodal slide through 180° and repeat the above adjustments. When the desired setting has been done, the distance between the cross-slits and the axis of rotation of the nodal slide again gives the focal length of the combination. This result will be found to be nearly the same as before. Take the mean value of the two results. Let this be F, the focal length of the combination.

Repeat the experiment with different distances between the component lenses.

Calculate the focal length of the combination from the observed values of f_1, f_2 and x. The result will be found to be very nearly the same as the observed value of the focal length of the combination for that distance. This verifies the equation

$$\frac{1}{F} = \frac{1}{f_1} + \frac{1}{f_2} - \frac{x}{f_1 f_2}$$

Plot a graph (Fig. 10.10) between x and I/F. The graph so obtained is a straight line.

Sources of error and precautions —(1) The cross-slits must be properly and intensely illuminated by the light coming from the lamp, which should preferably be a frosted one.

(2) The lens or the lens combination should be mounted

such that the principal axis of the system passes through the point of inter-section of the cross-slits.

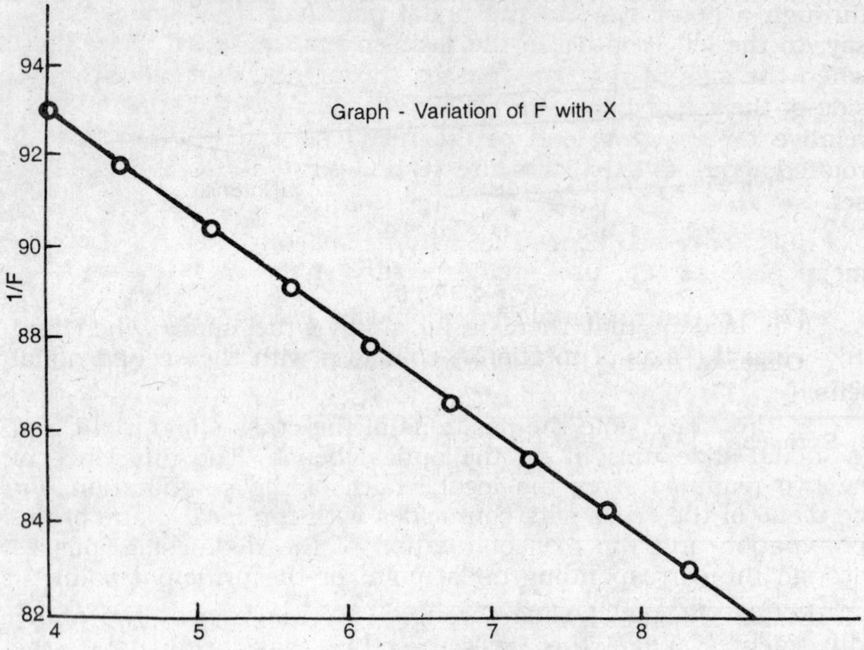

Fig. 10.10

(3) Sometimes a bright image is formed on the metal plate screen by reflection from the surface of the lens. This image should not be confused with the desired image which is formed by the rays of light which start from the cross-slits, pass through the lens or lens system, fall on the plane mirror and are reflected back from it through the lens or the lens system on to the metal screen to form an image of the cross-slits. If the plane mirror is slightly turned the genuine image will shift whereas the false image will remain stationary. This is the test which may be applied to differentiate the false image from the gennine image required for the experiment. Or else obstruct the light from falling on the plane mirror by placing your hand in front of the mirror when the genuine image will disappear whereas the false image will persist on the metal plate screen.

(4) The rotation of the nodal slide about the vertical axis while testing its coincidence with the nodal point should not exceed 5° or so.

(5) When the axis of rotation of the nodal slide passes

through the nodal point, the image of the cross-slits will not shift on rotating the nodal slide. When the axis of rotation passes through a point near to the nodal point, and the image shifts, say, to the left, relative to the near end of the nodal slide, then, when the axis of rotation passes through a point on the other side of the nodal point, the shift of the image will be to the right relative to the same end of the nodal slide when the latter is rotated about its axis. Thus, the desired position can be accurately set.

(6) For a well defined and sharp image to be formed on the metal plate screen, use small aperture of the lens or lenses.

(7) The mirror employed should be truly plane.

Observations —(A) Determination of focal length of the first lens.

S. Number	Light incident on	Position of Cross-slits a	Position of axis of Nodal slide b	Focal length of the lens $f_1 = a-b$
1	one face			
2	other face			
			Mean focal length	

(Make a similar table for the second lens).

(B) Determination of focal length of the combination.

Set No.	Distance between lenses	Light incident on	Position of cross-slits A	Position of axis of nodal slide B	Focal length of combination F = B − A
1		One lens			
		Other lens			
2		One lens			
		Other lens			

Calculations

$f_1 =$

$f_2 =$

$x =$

$$\frac{1}{F} = \frac{1}{f_1} + \frac{1}{f_2} - \frac{x}{f_1 f_2}$$

or $F =$

By experiment $F' =$

Results—(1) Focal length of the combination of the two given lenses at a distance of... cm. apart = cm.

(2) The calculated and the experimental values of the focal length of the combination being very approximately equal, the formula, $\frac{1}{F} = \frac{1}{f_1} + \frac{1}{f_2} - \frac{x}{f_1 f_2}$ is verified.

Exercise 1—To use $(x, 1/F)$ graph to find out the focal length of the combination :

(*i*) when the two lenses are in contact ; and

(*ii*) when the system would become telescopic.

Use the observations of the previous experiment and plot a graph between x along the axis of X and $1/F$ along the axis of Y, with the origin as (o, o). At least six points should be plotted. These points lie on a straight line. Produce the straight line to cut the axis of X and Y. The point of intersection on the axis of Y has co-ordinates $(0, 1/F)$ and thus for $x = 0$, the focal length of the combination is the reciprocal of this ordinate.

The point of intersection of the straight line on the axis of X has co-ordinates $(x, 0)$; hence $1/F$ being zero, the system is telescopic for this distance. This distance x should be nearly the same as $f_1 + f_2$. Check this from a knowledge of f_1 and f_2.

Exercise 2—To determine the focal length of a convergent lens by modification of Newton's equation $x_1 \, x_2 = f_1 \, f_2$.

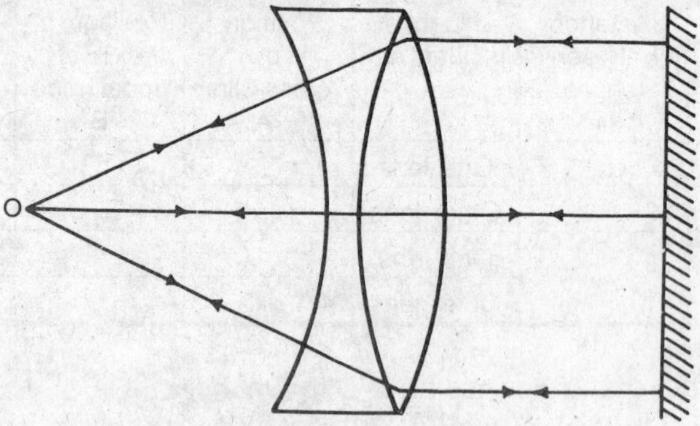

Fig. 10.11

Mount the test lens L on the Nodal slide. Use a point source of light O (Fig. 10.11) and set test tens for parallel light by *autocollimation method.*

Without disturbing O, shift the test lens L to the right (Fig. 10.12) through a measured distance x_1. The first principal focus F_1 of L has also then moved the same distance x_1, to the right.

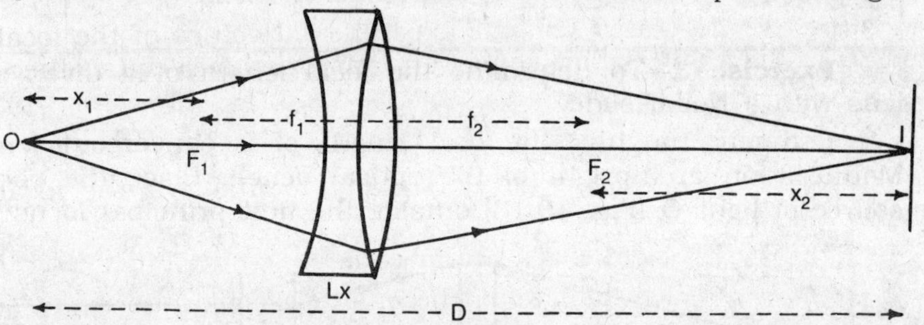

Fig. 10.12

By the use of a screen locate the position of the image I and take the reading on the optical bench. The difference of the readings for the positions of O and I is, say, D. If F_2 be the position of the second principal focus of the lens L_x, f_2 is its second focal length and the distance F_2 I is x_2

From Fig. 10.12, $D = x_1 + f_1 + f_2 + x_2$

or $$x_2 = (D - x_1) - f_1 - f_2 \quad (10.1)$$

Since the medium on both sides is air, $f_1 = f_2 = f$, say. Then from Newton's equation

$$x_1 . x_2 = f^2 \quad (10.2)$$

Combining Eqs. (10.1) and (10.2), we have

$$x_1 [(D - x_1) - 2f] = f^2$$

Let $(D - x_1)$ be d. Then

$$x_1 (d - 2f) = f^2$$

or $$f^2 + 2fx_1 - x_1 d = 0 \quad (10.3)$$

Since the distances x_1 and D have been measured, d can be evaluated. Substituting the values of x_1 and d in Eqs. (10.3) solve for f. Repeat the experiment for different values of x_1 and tabulate your observations as below :

S. No.	Position of O (To be kept constant) s	Position of L for parallel rays p	Position of L after Displacement g	x_1 $g - p$	Position of Image I r	D $(s - r)$	d $(s-r)$ minus $(g - p)$
1							
2							
3							

Exercise 3—To determine the focal length of a divergent lens with a Nodal slide.

Estimate roughly the focal length of a convergent lens. Mount it on an upright of the optical bench. Place the point source of light O (Fig. 10.13) outside the first principal focus of

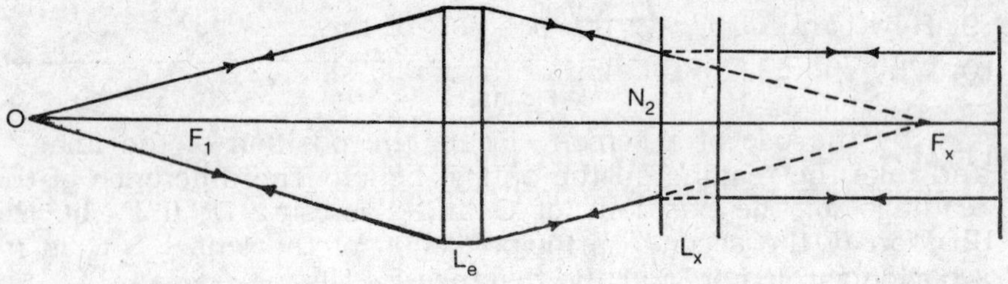

Fig. 10.13

the auxilliay divergent lens. Rays of light emerging from this auxilliary lens emerging towards the point F_x on its principal axis. Locate the position of F_x, with a screen which can be moved along the optical b :nch and note its position.

Mount the diverging lens L_x on the Nodal slide to the right of L_x and back it with a plane mirror M. The test lens L_x intercepts the rays of light which would have converged to F_x. Adjust the position of the test lens such that the image of O is formed coincident with O. This is possible only when the rays of light are rendered parallel by L_x. The mirror M being normal to the parallel pencil, the rays incident on it retrace their original path and form an image coincident with O. By inclining M slightly get the image beside O. Adjust the position of L_x on the nodal slide so that when it is rotated about a vertical axis through N_2, no motion of the image of O occurs. Determine the position of N_2 by means of the bench scale. Evaluate the distance N_2F_2 which is evidently the focal length of the test lens.

ORAL QUESTIONS

CARDINAL POINTS AND NODAL SLIDE

1. What are the cardinal points of an optical system?
2. Define focal points, principal points and nodal points.
3. Why are principal points called unit points ?
4. What are the properties of nodal points ?
5. How can you determine the cardinal points of an optical system ?
6. What is nodal slide and on what principle does it work ?
7. How can you use a nodal slide to determine focal length of a combination of two lenses separated by a distance ?
8. What precautions do you take in this experiment ?
9. How can you locate the position of the nodal points ?
10. Why should the rotation of the nodal slide about the vertical axis be small ?
11. How can you be sure that the image obtained is the desired image formed by refraction through the system ?
12. Why do you repeat your obserations by rotating the nodal slide through 180° ?

CHROMATIC AND SPHERICAL ABERRATIONS AND EYE-PIECES

1. What are the defects of images formed by a lens ?
2. What do you understand by chromatic and spherical aberrations ?
3. Distinguish between axial and lateral chromatism.
4. What is chromatic aberration due to ?
5. What do you understand by achromatic system ?
6. What conditions must be satisfied to make a system achromatic ?
7. Can an optical system be achromatised both for axial and lateral chromatism ?
8. How can spherical aberration be minimised ?
9. What is a crossed lens ?
10. What is an eye-piece ? Why is it made of two lenses separated at a distance ?
11. How many types of eye-pieces do you know ?

12. Describe Huygens' and Ramsden's eye-pieces.
13. Why is Huygens' eye-piece known as a negative eye-piece and Ramsden's eye-piece as a positive eye-piece ? Which of the two can be used for micrometric purposes and why ?
14. Are the images formed by the eye-pieces free from aberrations?
15. Why is Ramsden's eye-piece not well compensated for spherical aberration ?
16. Is Huygens' eye-piece achromatic ? If not in what sense can it be said to be achromatic ?
17. Give the uses of the two types of eye-pieces.

11
Dispersion of Light

11.1. Refraction through a prism—Consider a ray of light PQ incident on the face AB of the prism ABC (Fig. 11.1). At Q as it passes from a rarer into a denser medium, it is bent towards the normal and follows the path QR within the material of the prism. At R it passes from a denser into a rarer medium and consequently bends away from the normal along the direction RS. Originally the ray was travelling along the direction PQT but on emergence from the prism it takes up the path RS. From Fig. 11.1 it is evident that the ray has been deviated from its original path by an angle δ. This angle is termed as the angle of deviation. From the geometry of the fig., we can deduce that

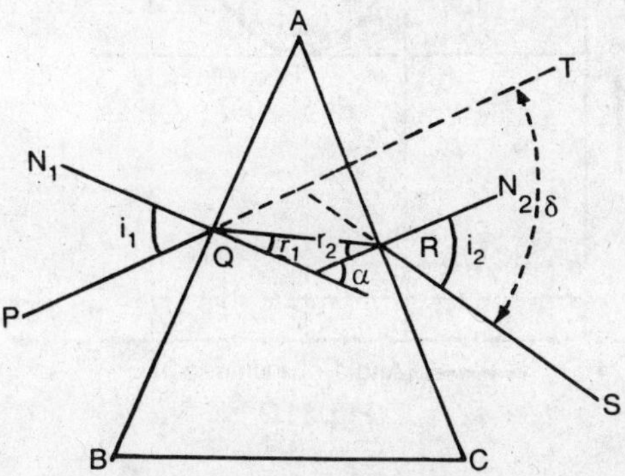

Fig. 11.1

$$\alpha = r_1 + r_2$$

and
$$\delta = (i_1 - r_1) + (i_2 - r_2)$$
$$= i_1 + i_2 - \alpha$$

where these angles have their usual significance. It is clear that the base of the prism plays no part in producing the deviation of the ray, it is only the angle of incidence i_1, the refracting angle α of the prism and its material (for i_2, the angle of emergence depends on it) which are responsible for deviating the ray through an angle from its original path.

It is of interest to examine the variation of deviation with the angle of incidence. Figure 11.2 represents the incidence-deviation curve for a prism. The variation is not uniform. For small angles of incidence the deviation is large. As the value of i increases, that of δ goes on diminishing till for a particular angle of incidence it attains minimum value. This angle of deviation is called the angle of *minimum deviation.* With further increase in the angle of incidence, the value of δ again increases. From the nature of the curve a few important conclusions can at once be drawn. Take PQ a line parallel to the axis representing the angles of incidence. It cuts the curve in two points P and Q. This means that for both the values of i ($i = i_1$ and $i = i_2$), the deviation produced by the prism is the same. It is also evident from the previous figure. If the ray is incident along SR at an angle i_2, it will emerge along QP at an angle i_1 and the deviation suffered will be same as in the first case when the ray was incident along PQ. Then from the $i-\delta$ curve we conclude that if i_1 is the angle of incidence, i_2 is the corresponding angle of emergence. Let PQ descend down towards the incidence axis but still parallel to it. The value of i_1 increases whereas that of i_2 decreases. This continues till PQ becomes tangential to the curve at B. Here $i_1 = i_2$; the angle of incidence has become equal to the angle of emergence hence the important result that a ray of light suffer minimum deviation when it passes through the prism symmetrically. The ray is *not necessarily parallel to the base except in the case of an isosceles or an equilateral prism.* When the ray passes symmetrically through the prism, it is said to be in the position of minimum deviation. Here

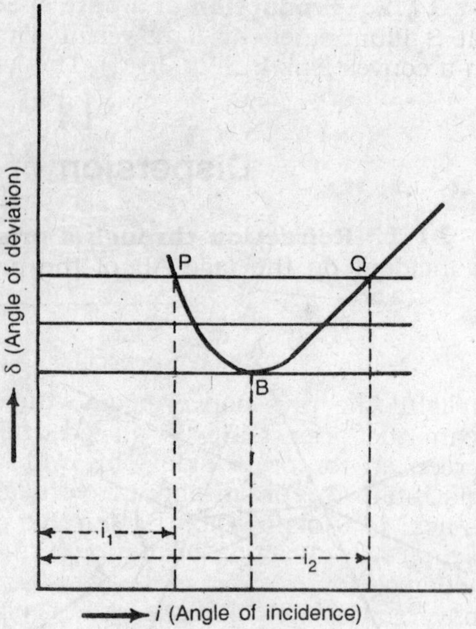

Fig. 11.2

$$\alpha = r_1 + r_2$$

and $$\delta = i_1 + i_2 - a = 2i_1 - \alpha$$

whence $$i_1 = \frac{\delta + \alpha}{2} \text{ and } r = \frac{\alpha}{2} \text{ (if } r_1 = r_2)$$

11.2. Production of a pure spectrum—Light from a narrow slit S illuminated by a powerful source of light is made to fall on a convex lens L_1 (Fig. 11.3). The lens gives a convergent pencil

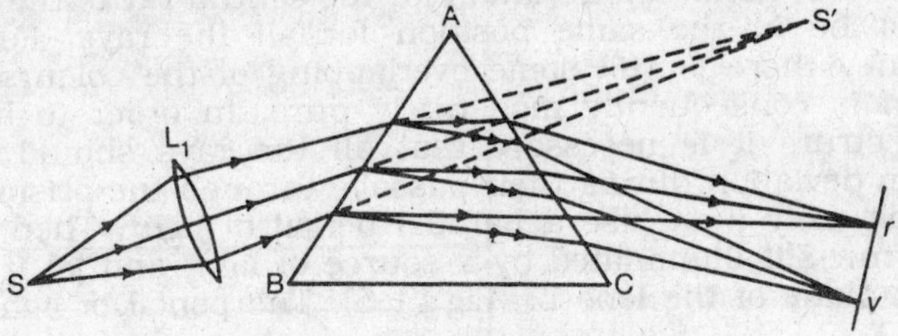

Fig. 11.3

of light, the rays converging to the point S. In this convergent beam the prism ABC is placed and the spectrum taken on a screen on the other side. This will not be quite pure. To attain this rotate the prism about a vertical axis and get the spectrum as near to S′ as possible. When this position is reached the prism occupies the position of minimum deviation and the spectrum is moderately pure, for in the position of minimum deviation, a convergent pencil of light will rise to a point image. Hence all the colours in the incident light will occupy distinct positions giving rise to a pure spectrum. In such an adjustment it will be noticed that the distance of S' and the spectrum from the prism is the same.

Another method employs a convex lens on the other side of the prism. A diverging pencil of light issues from a narrow slit and falls upon the prism ABC (Fig. 11.4). If looked at the prism on the side of the face AC, a virtual spectrum *vr* will be visible. If, at this stage, the rays are taken on the screen there is

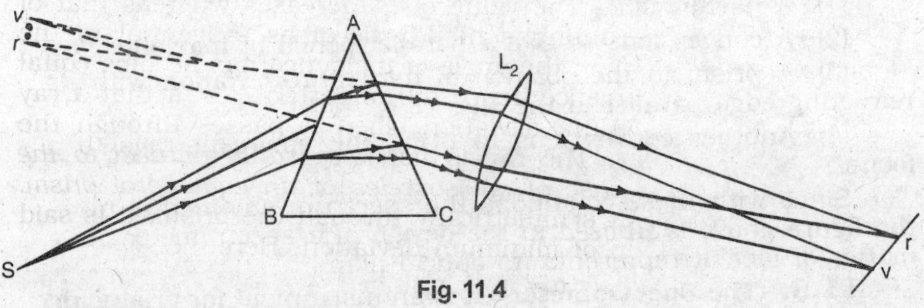

Fig. 11.4

considerable overlapping. To attain the required spectrum a lens L_2 is used to form a real image *rv* of *vr*. Again it would be necessary to adjust the prism in the position of minimum deviation for in that case alone the virtual spectrum will be pure. L_2 simply serves to throw it on the screen.

In the two methods described above, the prism may be in the position of minimum deviation for the central ray of the beam; it cannot be in the same position for all the rays. For this discrepancy there is still some overlapping of the colours ; the spectrum is consequently moderately pure. In order to have a pure spectrum it is necessary that all the rays should suffer minimum deviation during their passage through the prism. This is possible only if we use a parallel pencil of light. Then, let S be a narrow slit illuminated by a source of light and let it be in the focal plane of the lens L_1 (Fig 11.5). The pencil of light that

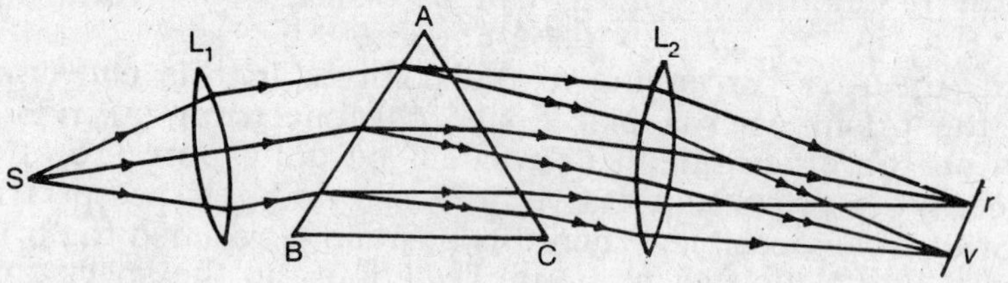

Fig. 11.5

issues from L_1 is a parallel one and falls on the face AB of the prism ABC placed in the position of minimum deviation for the incident rays. Dispersion occurs within the prism ; it results in various systems of parallel rays, the red rays parallel among themselves, the violet rays parallel among themselves and so on for the intermediate colours. On emergence from the prism, these systems of rays are made to fall on lens L_2. It brings each system of parallel rays to a different focus and we obtain a pure spectrum *rv*. Hence for the production of a pure spectrum the following conditions must be satisfled :

(1) A narrow slit.

(2) A convex lens to give a parallel pencil of incident light.

(3) A prism in the position of minimum deviation with its refracting edge parallel to the slit.

(4) Another convex lens to bring the emergent rays to a focus.

Since with these conditions the spectrum obtained is pure, the figure above is utilised in the design of a spectrometer. As a matter of fact, it represents its optical plan.

11.3. The Spectrometer—It is an instrument for producing the dispersion of the incident light so as to form its spectrum and provided with requisite arrangements for the measurement of the angles of deviation of the component ray. It essentially consists of the following parts :

(1) *The collimator* for producing an emergent pencil of parallel light which falls on the prism.

(2) *The revolving prism table* for mounting the prism used for producing the dispersion of the incident pencil of light.

(3) *The telescope* for observing the spectrum.

Below we give the description of each part in detail.

Figures 11.6 and 11.7 reveal the important parts of the spectrometer.

(1) *The collimator*—It consists of a main tube at one end of which there is a well-corrected convergent lens L_c and at the other a draw tube carrying a slit S. The draw tube can be moved inside the main tube by hand or by rack and pinion arrangement with the result that the slit S can be placed in the focal plane of L_c. The slit has one jaw fixed and the other jaw movable by

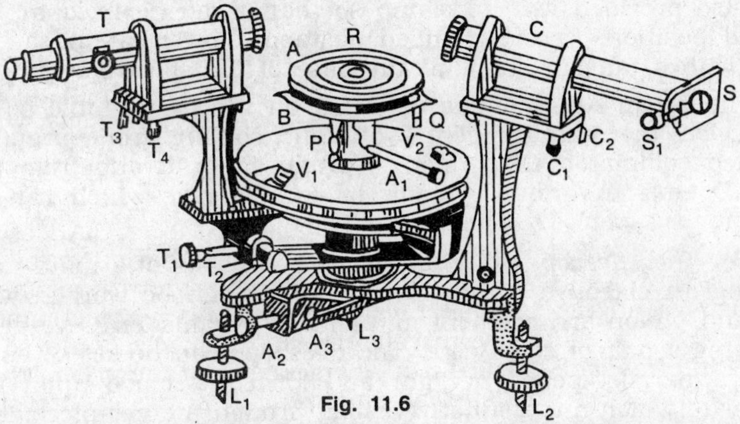

Fig. 11.6

a crew. In some slits both the jaws open out or close in simultaneously. The slit can be illuminated by the source of light the spectrum of which is to be examined. The collimator is rigidly fixed. In some instruments it can rotate about the axis of the spectrometer.

(2) *The revolving prism table*—The prism table P is a circular plane table which can be rotated about the axis of the instrument independently of the collimator C and the telescope

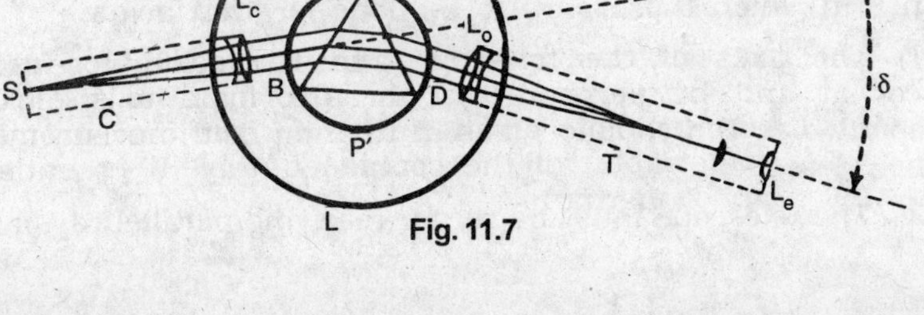

Fig. 11.7

T. It carries two verniers moving over the main graduated circular scale LMN, this being carried by the telescope. The height of the prism table can be adjusted by raising or lowering the rod carrying the table and then clamping it in position. The table is provided with three levelling screws, P.Q.R (Fig. 11.8). On the surface of the table are drawn concentric circles and also straight lines parallel to the line joining two of the levelling screws. It is also provided with of clamp so that it can be fixed in any desired position. In the clamped position, slow rotation can be given with a tangent screw at the base of the spectrometer.

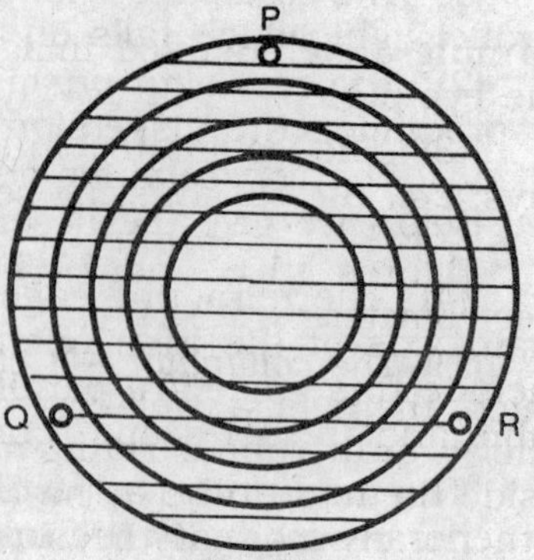

Fig. 11.8

The prism ABD can be mounted on the prism table so as to receive the parallel pencil of light coming from properly adjusted collimator. Dispersion is produced by it and from the face AD emerge various systems of parallel rays which can be brought to a focus by a telescope T.

(3) *The telescope* —The telescope T consists of a main tube carrying an objective L_0 and a tube which can be worked with rack and pinion arrangement and carries a Ramsden's eye-piece L_e having a pair of cross-wire. The telescope can be set so as to admit a parallel pencil of light and form an image upon the cross-wire which are distinctly visible through the eye-piece. The axis of rotation of the telescope coincides with the same axis as that of the prism table. With its rotation moves the main graduated circular scale. The number of degrees through which the telescope is rotated can be observed with verniers attached to the prism table. Like the latter it is also provided with a clamp and a tangent screw.

Some telescopes have the Gauss' eye-piece.

11.4. The adjustments of a spectrometer—Before proceeding to make any observations with a spectrometer it is necessary to test for its mechanical and optical adjustments. The sequence in which these have to be done is given below :

(1) The axes of the telescope and the collimator must intersect at and be perpendicular to, the main axis of the instrument, *i.e.*, the common axis of rotation of the prism table and the telescope.

(2) The telescope must be set for receiving parallel rays and

forming a well-defined image of the object on the cross-wires of the eye-piece.

(3) The collimator must be adjusted for emitting parallel rays. *i.e.*, the slit must be set in the focal plane of the collimating lens.

(4) The refracting edge of the prism must be parallel to the main axis of the instrument, *i.e.*, the faces of the prism which include this angle are parallel to the axis about which the prism table and the telescope rotate.

The mechanical adjustment embodied in condition (1) above is generally made by the instrument-maker and the student may not undertake it. However, we give the method by which this may be tested.

1. *Mechanical Adjustments* —(*a*) Mount a pin in the centre of the prism table and secure its proper illumination. Open the collimator slit as wide as permissible and look for the image of the pin. In case the axis of the collimator intersects the axis of the instrument the pin will be seen in the middle. If not, adjust the collimator by rotating it slightly about its vertical support till the requisite condition is satisfied. To adjust the telescope, remove the eye-piece and again look for the image of the pin. Arrange so that the image of the pin remains in the middle for at least three positions of the telescope on the circular scale. Here also it may be necessary to rotate the telescope about its vertical support. Replace the eye-piece. By this means we test that the axis of the collimator and the telescope intersect the main axis of the instrument.

Next, place a striding level[1] on the telescope tube. Turn the telescope so that it lies parallel to the line joining two of the levelling screws of the instrument. Adjust these screws to bring the bubble to the centre of the level. Now turn the telescope through 180° from this position and see if the bubble is still in the centre or not. If it lies there the adjustment is complete ; in case it is displaced, it should be brought back to the centre half way by adjusting the levelling screws and half way by rotating the telescope about its horizontal axis. The procedure may have to be gone through two or three times before the condition, that the bubble remains the centre for both positions of the telescope, is attained.

(*b*) For an instrument fitted with a Gauss' eye-piece the following method is adopted. Let a lamp be placed opposite the

1. A striding level is one provided with V's at both the ends so that these can rest on the upper surface of the telescope tube.

opening of the Gauss' eye-piece (Fig. 11.9). Light rays coming from the lamp are reflected along the axis of the telescope by a piece of clear glass fixed in the eye-piece inclined at 45° to the axis. These rays illuminate the cross-wires and proceed further till they emerge out of the objective of the telescope. If now they are reflected back into the telescope by a plane reflecting surface, a reflected image of the cross-wires will be visible beside the direct image of the cross-wires when viewed into the eye-piece. Mount a piece of plate glass with parallel sides on the prism table so that the faces are approximately normal to the table surface. Two images of the cross-wires will be visible, that formed by reflection being fainter than the direct image. Focus the telescope by moving the draw tube in and out till the two images coincide without parallax ; it is then adjusted for parallel light. For, in suh a case alone could the rays of light starting from the cross-wires and emerging out of the objective lens be reflected so as to retrace their own path and give rise to an image of the cross-wires coincident with them. Next turn the telescope through 180° so as to receive the reflected image from the other surface of the glass plate. If the coincidence of the cross-wire still continues the axis of the telescope is perpendicular to the main axis of the instrument about which it rotates. If, however, the reflected image and the direct image do not coincide without parallax, bring about the coincidence of the two by bringing the reflected image half way by adjusting the telescope and half way by adjusting the prism table on which is placed the plate of glass. This adjustment may have to be repeated till the coincidence remains unaltered for both the positions of the telescope.

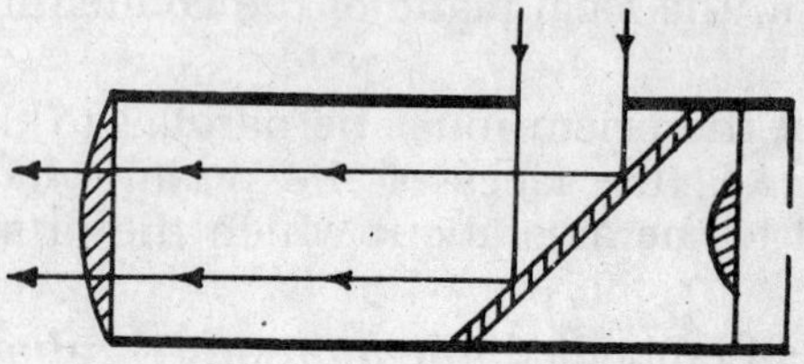

Fig. 11.9

(*c*) Having properly set the axis of the telescope, we proceed with the adjustment of the collimator axis. To adjust this, stretch a thin wire across the middle of the slit and view it through the telescope. The image of this thin wire should coincide with the intersection of the cross-wires. In case it is not so, use the screws attached to the collimator to secure the above setting. When this is done the collimator axis is also perpendicular to the main axis of the spectrometer.

II. Optical Adjustments —1. Adjustment of the Eye-piece

Before proceeding to make any tests or adjustments with a

spectrometer the primary setting that has to be done is to focus the eye-piece on the cross-wires. Turn the telescope towards a white wall a few feet away. With the help of the draw tube alter the distance of the eye-piece from the objective in such a manner that the field of view appears bright, although no image of the wall is formed. Now use one eye to view the wall directly and the other to look into the eye-piece. Alter the distance of the eye-piece from the cross-wires by drawing it in and out till the cross-wires are seen distinctly with one eye whereas the wall is also seen clearly with the other eye. This procedure ensures the fact that the eye-piece forms the image of the cross-wires at the same distance at which an object can be viewed by the person with an unaccommodated eye. For a short-sighted person the distance of the wall should be reduced while making this adjustment or else the wall may be replaced by a book and placed at such a distance that the print is distinctly visible with unaided eyes. This adjustment will vary from individual to individual in view of the varying power of accommodation of the eye.

2. *Adjustment of the Telescope.*

(*a*) A telescope fitted with a Gauss, eye-piece is automatically adjusted for parallel rays while testing for the mechanical adjustments of the spectrometer.

(*b*) For a telescope not provided with a Gauss, eye-piece, either of the two methods may be adopted.

(*i*) Point the telescope towards a distant object with a well-marked outline, *e.g.*, a telegraph post or clock-tower. A blurred image, in general, will be visible through the telescope. The reason is evident ; the image formed by the telescope objective does not lie in the plane of the cross-wires, for if it did, both the image and the cross-wires would be distinctly visible as the eye-piece had initially been adjusted on the cross-wires. So we have to adjust such that the focal plane of the objective coincides with the plane of the cross-wires. To attain this move the draw tube carrying the eye-piece with the cross-wires till the image of the distant object is seen distinctly on the cross-wires without parallax. This sets the telescope for admitting parallel rays of light and forming a distinct image of the distant object on the cross-wires.

(*ii*) **Schustea's method**—This method is suitable when the spectrometer cannot be moved out of the dark room for want of a distant object. Illuminate the slit with sodium light and adjust a prism in the position of minimum deviation. Turn the prism table so that the deviation of the refracted image is slightly geater than minimum deviation. From the nature of the i-δ curve it is

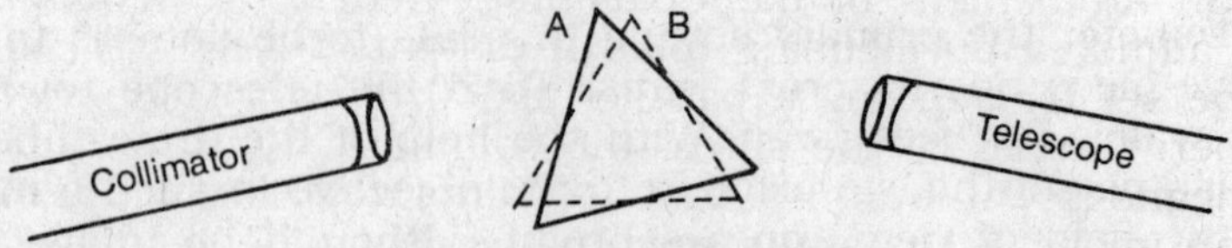

Fig. 11.10

evident that deviation corresponds to two values of the angle of incidence of the rays. In one position of the prism, shown by thick lines (Fig. 11.10) the angle of incidence of the rays is less than that for which we get minimum deviation ; in the other position of the prism, shown by dotted lines, the angle of incidence of the rays is greater than that which gives the position of minimum deviation. Let us suppose that the prism has been rotated so that the refracting edge is brought nearer towards the telescope, this is the position shown by dotted lines. Since the collimator and the telescope are out of adjustment the image seen in the field of view will be blurred. In this position of the prism when the rays are incident more obliquely than for minimum deviation, adjust the telescope so as to get a well-defined image of the slit in the field of view. Now rotate the prism to the other position shown by thick lines and adjust the distance of the slit from the collimating lens to get again a sharp image of the slit. Repeat this process of alternately focusing the telescope and the collimator till the image remains in good focus for both the positions of the prism. This adjustment is done usually in three to four alternate focusings. When this has been done both the telescope and the collimator have been adjusted, the former for receiving a parallel pencil of light and the latter for emitting it.

In case the prism is turned from the minimum deviation position so that the refracting edge moves towards the collimator, the first focusing should be made with the collimator. If the focusing has been made in the wrong order the image will rapidly become in-distinct and will serve as an indication that a mistake has been made in the procedure.

A simple rule, if remembered, *will always be found helpful to the student when making adjustments by this method*.

Turn the prism so that the refracting edge is brought
- towards the telescope—adjust the telescope.
- towards the collimator—adjust the collimator.

3. *Adjustment of the Collimator*—When the telescope has been adjusted by focusing it on a distant object, the collimator can be set for sending a parallel pencil of light in the following manner: Illuminate the slit of the collimator with the light from an incandescent bulb so that it points towards the collimator and an image of the slit is seen in the centre of the field of view. This image will, in general, be blurred. Alter the distance of the slit from the lens of the collimator till a sharp defined image is formed on the cross-wire and there is no parallax between them. When this condition is satisfied the required collimator adjustment has been done. For, the telescope is in adjustment to receive a parallel pencil of light and thereby form a defined image on the cross-wires without parallax and since the image of slit is so obtained, the incident light on the telescope, or, in other words, the emergent light from the collimator, must constitute a parallel pencil.

4. *Adjustment of the Prism Table*—(*a*) For a telescope fitted with a Gauss, eye-piece this adjustment is done in the following manner :

Mount the prism on the prism table so that the edge which we wish to use as the refracting edge in subsequent experiments, coincides with the centre of the table and one of the faces AB bounding this refracting angle A is perpendicular to the line joining two of the levelling screws Q and R (Fig. 11.11). Turn the telescope fitted with the Gauss, eye-piece so that it is normal to the face AB. The reflected image of the cross-wires and the direct image will lie along side and not, in general, coincident. Use the three levelling screws to get the coincidence of the two images. Next set the telescope perpendiular to the other face AD bounding the refracting angle A, and adjust as before but making use only of the levelling screw P. The motion of the levelling screw P will not disturb the previous adjustment with the face AB since it causes only a rotation of AB in its own plane. When these adjustments have been made, the refracting edge of the prism is parallel to the main axis of the instrument and we say that the prism table has been levelled.

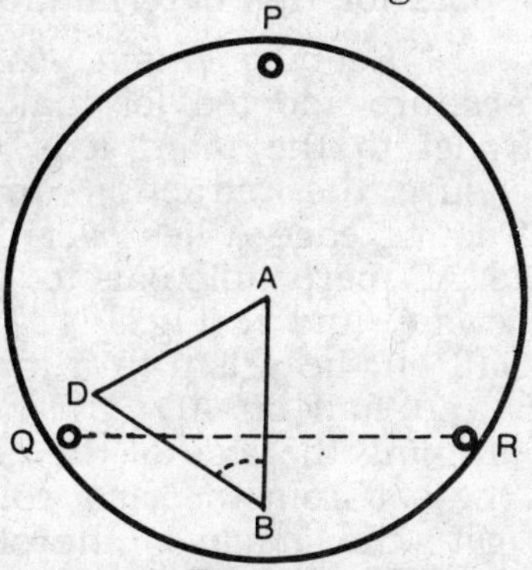

Fig. 11.11

(*b*) For a telescope not fitted with a Gauss' eye-piece the procedure that should be adopted is as follows :

Mount the prism on the table as described above. Illuminate the collimator with sodium light (or light from an incandescent bulb) and turn the table so as to receive the light on both the faces AB and AD, bounding the refracting angle A. Keep the table fixed : rotate the telescope so as to receive the rays of light reflected from the face AB. An image of the slit will be visible through the telescope but it may not lie in the centre of the field of view. Use the levelling screws of the prism table to get the image of the slit at the proper height in the field of view, *i.e.*, it should be bisected at the point of intersection of the cross-wires. Next turn the telescope to get the reflected image of the slit from the face AD and adjust it as before but using the levelling screw P alone. Hence the adjustment of the prism table is complete when the image lies in the middle of the field of view of the telescope in both its positions.

The spectrometer is now ready for making any measurements with it.

11.5. Measurement of the refracting angle of the prism—Either of the two methods described below may be adopted for this determination.

1. *Keeping prism fixed and moving the telescope.*—The procedure adopted for making the refracting edge of the prism parallel to the main axis of the instrument may be used to measure the refracting angle. Mount the prism so that the refracting edge A lies over the centre of the prism table and the face AB perpendicular to the line joining two of the levelling screws Q and R (Fig. 11.11). Allow the light from the collimator to fall on the prism in a manner that half of it falls on the face AB and half on AD. By moving the eye in a horizontal plane containing the axis of the collimator look for the reflected image of the slit from the face AB. Turn the telescope to this direction to get the image of the slit on the cross-wires and clamp it. Make the width of the slit as small as permissible and use the tangent screw of the telescope to get the image exactly on the cross-wires. With the help of a reading lens and a lamp, take the reading on both the verniers. Check that the difference between these two readings is very nearly 180°.

Turn the telescope to view the reflected image from the face AD and make a similar adjustment. Read the position of the telescope on both the verniers. Find the difference between the two readings of the same vernier. Take the mean of these two results. The angle so obtained is double the refracting angle of the prism. Half it to get the value of α the angle of the prism.

It sometimes happens that four images of the slit can be seen as we move the telescope to get the reflected images from the lace AB and AD. From Fig. 11.12 it is evident that the images formed by the set of rays 1 and 2 are due to reflection whereas those formed by 3 or 4 are due to refraction through the prism. This difficulty is likely to be felt more with a monochromatic source of light since the reflected and the refracted images present the same colour. To avoid this mistake either paste a piece of paper on the surface BD or scan the images first with the naked eye and then turn the telescope to point in their direction.

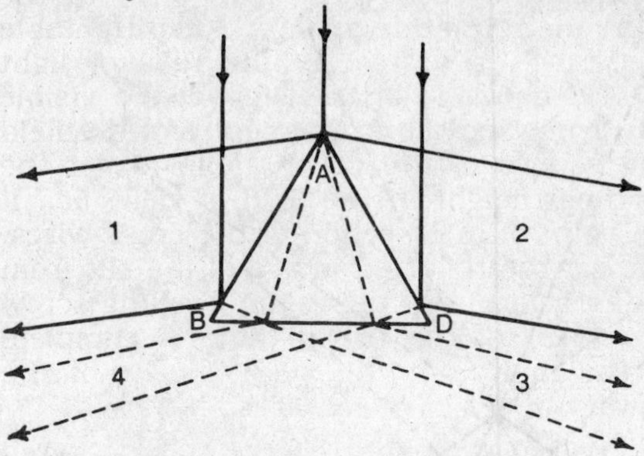

Fig. 11.12

The reason for the particular mode of placing the prism with its refracting edge coincident with the common axis of rotation of the telescope and the prism table, can be rendered clear by considering the effect of want of parallelism in the emergent beam from the collimator. Refer to Fig. 11.13 in which ABC is the prism so placed that A does not lie over O, the point through which passes the axis of rotation of telescope.

Let the incident beam on the prism be not parallel, *i.e.*, the image of the slit as formed by the collimating lens does not lie at infinity but at S. Then its images as seen through the faces AB and AC are respectively at S_1 and S_2. From the geometry of the figure it is clear that

$$\angle S_1AS_2 = \angle T_1'AT_2' = 2\alpha$$

where α is the angle of the prism.

Since the axis of rotation of the telescope is at O, the angle that we measure in the above method is T_1OT_2 which is less than $T_1'AT_2'$ (= 2a) by the sum of the angles OS_2A and OS_1A. To correct this error, either of the two conditions must be satisfied :

(*i*) S should be at infinity which means an exactly parallel pencil of light should emerge from the collimator. In that case AT_1' and OT_1' are parallel and so also AT_2' and OT_2. The two angles $T_1'AT_2$ and T_1OT_2 are each equal to 2a. But the condition that S may lie at infinity is difficult to be satisfied.

(*ii*) O should coincide with A which, in other words, means that the refracting edge of the prism should be placed at the

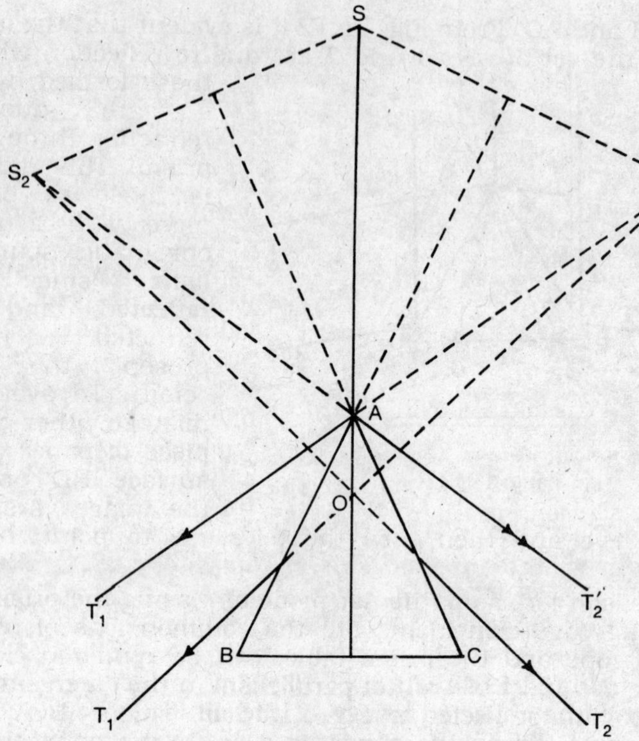

Fig. 11.13

centre of the prism table. For then the refracting edge and the axis of rotation of telescope coincide and the angle measured is T_1' AT_2' = $2a$.

2. *Keeping the telescope fixed and moving the prism.* Place the prism on the prism table with its centre over O, the centre of the prism table through which passes the main axis of the instrument. One face of the prism including the angle to be measured should be perpendicular to the line joining two of the levelling screws. Turn the prism table so that the light from the collimator may fall on the face AB. Move the telescope to get the reflected image of the slit from the face AB on the crosswires. Clamp the telescope and use the tangent screw for the final setting. Take the reading on the two verniers. Without disturbing the telescope, turn the prism table so that the reflected image of the slit from the face AC now comes into the field of view of the telescope. The new position of the prism (A)(B)(C) is shown by dotted lines. For setting the image on the cross-wires, clamp the prism table and use its tangent screw. When this is done, take

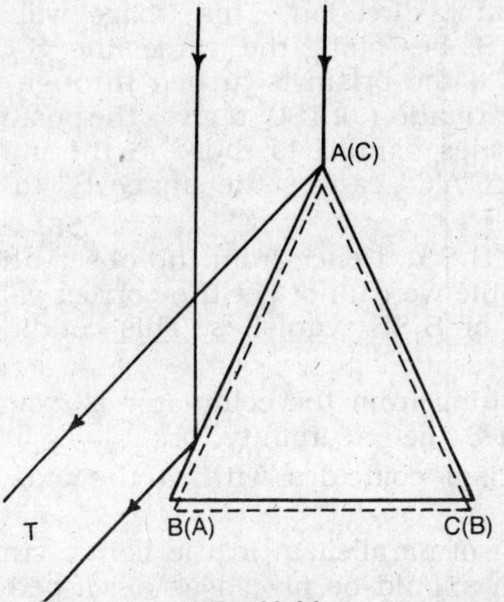

Fig. 11.14

the reading on the verniers. From Fig. 11.14 it is evident that the prism has been turned through an angle 180-α. Hence the difference of the readings on each vernier for the two settings is 180-α. Take the average and subtract it from 180 to get the value of α the refracting angle of the prism. Let us examine the effect of non-parallelism of the emergent pencil from the collimator. Let ABC be the prism and suppose its centre does not coincide with O, the common axis of rotation of the prism table and the telescope (Fig 11.15). Let the image of the slit formed by the collimating lens be at S. When the telescope is turned to receive the reflected image from the face AB, it is viewing the image at S′ along the direction S′OT. Now the prism is rotated so as to give the reflected image of the slit from face A′C′ (the new position of AC) on the cross-wires of the telescope which is

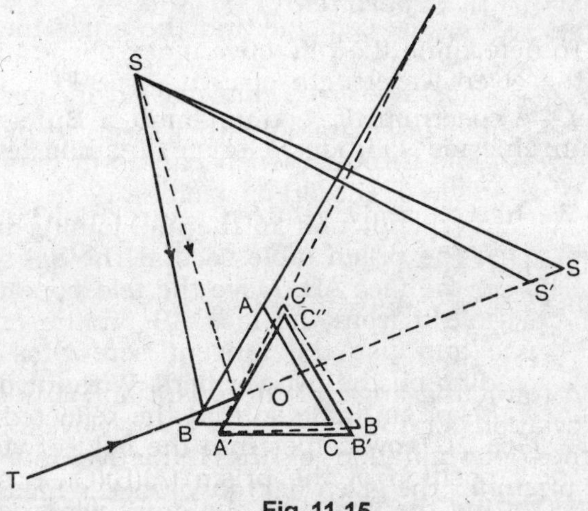

Fig. 11.15

kept fixed pointing in the same direction. This image will be visible along this direction if S_2 lies along the same line $S'OT$. This position can be attained if the prism is turned through an angle $180 - \alpha + C'AC'$; for the rotation of $180 - \alpha$ gives the position ABC. AA'C'B" with its three sides parallel to those in the initial position further rotation of $\angle C'A'C''$ causes the image S_2 to lie along the line S_1OT.

Now $\angle C'A'C'' = \angle BDA' = \angle S_1SS_2$. Hence from the observation of the rotation of the prism table we cannot get the correct value of α unless the angle $C'A'C''$ or S_1SS_2 vanishes. This condition can be attained when either.

(*i*) the pencil of light issuing from the collimator is exactly parallel in which case S_1 and S_2 lie at infinity, or

(*ii*) the centre of the prism coincides with O the axis of rotation of the prism table.

Hence to correct for want of parallelism in the light issuing from the collimator, the prism should be placed as described in the method above.

It is advisable to find the refracting angle by the two methods given above and compare the results. An error might creep in due to spherical aberrations of the objective lens while making determination with the first method in view of the fact that at the most half of the objective of the telescope is used in locating the image and that it is not the same half which is used in the two positions. However, this is not a very serious error for a good spectrometer.

Experiment 11.1

Object —To determine the refractive index of the material of the prism for the given wavelength of light ($\lambda = 5893 \times 10^{-8}$ cm.).

Apparatus —A spectrometer, sodium lamp or Bunsen burner fed with sodium chloride solution, a prism, reading lens and a lamp.

Theory —We have already deduced the relation.

$$\mu = \frac{\sin \frac{\alpha + \delta}{2}}{\sin \frac{\alpha}{2}} \tag{11.1}$$

where α is the refracting angle of the prism and δ is the angle of minimum deviation for the ray of the given wavelength. It is evident that the determination of the refractive index of the material of the prism for the given light involves the measurement of α and δ.

Method—The procedure to be adopted may be divided into three parts :

1. Mechanical and optical adjustment of the spectrometer as described in Sec. 11.4.

2. Measurement of the refracting angle of the prism by any of the two methods described in Sec. 11.5.

3. Measurement of the angle of minimum deviation of the prism for the given wavelength of light. For this proceed as detailed below.

Illuminate the slit of the collimator with sodium light. Mount the prism on the table such that its centre coincides with the main axis of the instrument and the face AC lies normal to the collimator axis. The angle A which has been measured should be used as the refracting angle. This method of mounting not only corrects for want of parallelism in the incident pencil on the face AC but also improves the resolving power of the prism since the entire pencil from the collimator falls on the face AC.

Turn the prism towards the left and look for the dispersed image of the slit through the face AB by moving the eye in a horizontal plane containing the collimator axis. Having obtained the image, turn the telescope to point in that direction to receive it in the centre of the field of view (Fig. 11.7). Turn the prism further and follow the image in the telescope. Since in the initial setting the face AC was almost normal to the incident pencil it means that the angle of incidence on it was small. The rotation of the prism table towards the left causes an increase in the angle of incidence which in its turn results in the decrease of the angle of deviation. To follow up the dispersed image the telescope has, therefore, to be moved towards the direction of the incident rays. As the prism table is turned further a position is reached when the rays suffer minimum deviation and the telescope is nearest to the line of the collimator axis produced. With further rotation of the prism table, the angle of deviation increases and the image begins to recede from the position of the previous adjustment. Hence the position of minimum deviation can be obtained by adjusting the telescope in such a position that the dispersed image moves up to the crosswires and then begins to retrace its path with the rotation of the prism. This position of the telescope gives the direction of the emergent rays from the prism. Read both the verniers.

Remove the prism without disturbing the prism table, which should be securely clamped. Unclamp the telescope and turn it so that its axis may be in line with the collimator axis. In such a case the direct image of the slit will be visible through the telescope and it can be finally set on the cross-wires by using the tangent screw of the telescope. Having done this, read the two verniers. This position of the telescope gives the direction of the incident rays. The angle of deviation is the angle between the direction of the incident and the emergent rays. To get this, take difference of the readings of the same vernier for the two positions of the telescope. Take the mean of the two results. It gives the value of δ.

Next turn the prism table so that the light from the collimator now falls on the face AB and the dispersed image has to be seen through AC. Repeat the experiment as above and get the value of δ. Finally get the mean value of δ by taking the average of the results obtained with the light incident first on the face AC and then on the face AB.

Substitute the values α and δ in the above formula and calculate μ.

Sources of error and precautions—(*i*) The mechanical adjustments of the spectrometer should be correct, *i.e.*,

(*a*) The axis of the collimator and the telescope must intersect the main axis of the spectrometer, the latter being defined as the common axis of rotation of the prism table and the telescope.

(*b*) The axis of the collimator and the telescope must be perpendicular to the main axis of the instrument.

(2) The optical adjustments of the spectrometer must be made correctly so as to ensure that

(*a*) the eye-piece is focussed on the cross-wires,

(*b*) the telescope is adjusted to receive a parallel pencil of light and form an image on the cross-wire without parallax.

(*c*) The collimator is in alignment with the source. For this, illuminate the slit of the collimator by placing the slit close to the window of the source of light. Open the slit wide enough and look through the collimating lens. The lens should be fully and symmetrically illuminated by the light coming out of the slit. If

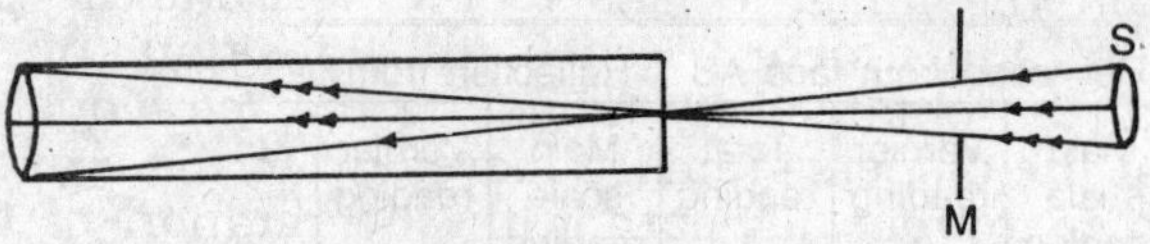

Fig. 11.15a Correct setting

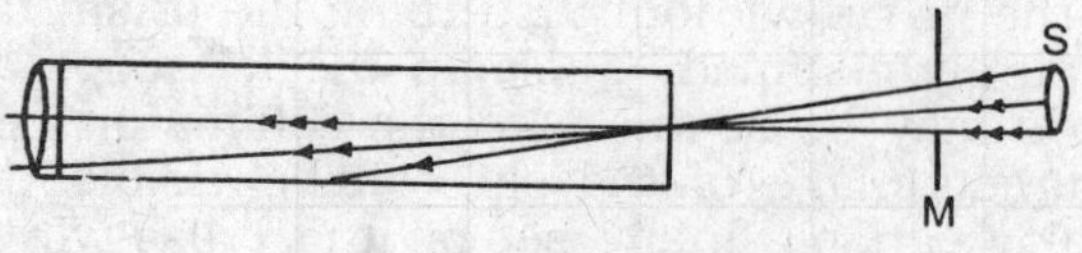

Fig. 11.15b Incorrect setting

this be not the case, the spectrometer as a whole should be displaced to achieve the adjustment. This adjustment is called alignment. Figures 11.15 (a) and 11.15 (b) illustrate the correct and the incorrect positions of alignment of the collimator.

(*d*) the collimator is set to emit a parallel pencil of light and

(*e*) the refracting edge of the prism is parallel to the main axis of the instrument and hence to the slit. Use of the levelling screws should be carefully noted.

(3) The slit used should be narrow : its width should be diminished so far as permissible.

(4) For an asymmetrical slit in which one jaw is fixed and the other moves out adjust the crosswires on the fixed jaw in all the settings ; for a symmetrical slit in which both the jaws are movable, adjust the crosswires in the middle of image seen in the field of view.

(5) Set the prism properly to correct for want of parallelism in the pencil of light coming from the collimator.

(6) At no time while taking observations should both the table and the telescope be unclamped. While reading the movements of the prism table, the telescope must be initially clamped and *vice versa*.

(7) Read both the verniers ; this eliminates the error due to non-coincidence of the center of the divided circle with the axis of rotation of the telescope or the table. Check that the difference between the readings of the two verniers is very nearly 180°.

Observations—[A] Measurement of the angle of the prism

Value of one main scale division =

Total number of divisions on the vernier scale =

∴ Least count = seconds

S. No.	Vernier	Reflection from face AB			Reflection from face AC			Difference of the two readings of the same vernier.	Mean value = 2α.
		Main scale reading	Vernier reading	Total reading	Main scale reading	Vernier reading	Total reading		
1	V_1								
2									
3									

1	V_2									
2										
3										

[A] Measurement of the angle of minimum deviation. Light incident on face AC.

S. No.	Vernier	For dispersed image			For direct image			Difference of the two readings of the same vernier.	Mean value = δ
		Main scale reading	Vernier reading	Total reading	Main scale reading	Vernier reading	Total reading		
	V_1								
	V_2								

[B] Measurement of the angle of minimum deviation. Light incident on face AC.

[C] Light incident on face AB.

Table as above.

Calculations

$$\therefore \qquad 2\alpha =$$

$$\alpha =$$

Mean value of $\delta =$

$$\mu = \frac{\sin \frac{\alpha + \delta}{2}}{\sin \frac{\alpha}{2}}$$

$$=$$

Result —The refractive index of the material of the prism for sodium light =

Criticism of the apparatus—The spectrometer when used for the determination of the refractive index of the material of the prism, gives a very satisfactory value for it. The determination involves the measurement of the angle of the prism and the angle of minimum deviation for the ray of given wavelength. The accuracy of the measurements depends on the correct way of placing the prism on the table and on the optically true nature of the faces of the prism.

A spectrometer is said to be a good one in its performance if it produces a bright spectrum, a large dispersion and has a large resolving power. The brightness of the spectrum depends upon the aperture of the collimator, *i.e.*, the ratio of the diameter of the collimating lens to its focal length. This is generally 1/12 and rarely exceeds 1/8, for then the lenses used for combination in the collimating lens have a large diameter and it becomes difficult to correct them. There are a few factors which cause the brightness of the spectrum to diminish : (1) the pencil of light emerging from the collimator exceeds the size of the prism and passes beyond its edges with a consequent loss of light, (2) reflection occurs at the face of the prism on which light falls and again internally when light passes out of the prism, and (3) absorption also occurs in the material of the prism, (4) incorrect alignment [Fig. 11.15(b)]. Yet in spite of all these factors the spectrum obtained is fairly bright and its brightness can be improved by illuminating the slit of collimator more strongly. For greater dispersion the prism used should have a large base and high refractive index. But the use of a large prism results in inhomogeneity of the material of the prism with a consequent large absorption in it. Hence in usual practice 1" equilateral prisms are used. To attain greater resolving power with a prism spectrometer use a dense flint glass prism having optically true surfaces.

It can be used for determining the dispersive power of the material of the prism, refractive index of a liquid and for studying Cauchy formula containing μ and λ.

Exercise 1—To determine the refractive index of a liquid for a given wavelength with a spectrometer.

For this we require a hollow prism to contain the experimental liquid. The two plates bounding the refracting angle of the prism should be optically true. The method to be adopted for the determination of the refractive index of liquid is the same as given in the previous experiment. Some points, however, are noteworthy.

While determining the angle of the hollow prism two images of the slit may be visible for each position of the telescope in case the surface of the plates of the prism are not plane and parallel. One of these is formed by reflection at the front surface whereas the other by reflection at the inside surface. For our purpose we require the image reflected by the front surface of the plate. To ascertain this, fill the hollow prism with experimental liquid ; this results in the loss of intensity of the image formed by reflection at the inside surface which can, therefore, be distinguished from the other. For measuring the angle of the prism, use the bright image.

Since the refractive index of a liquid varies rapidly with temperature it is necessary to record it with a thermometer at the time of setting for minimum deviation.

Having determined α and δ by the method given in the previous experiment, calculate μ for the liquid.

The method has this drawback that it is quite unsuitable for liquids available in small quantities.

Exercise 2—To determine the dispersive power of the material of the prism.

The experiment is but a repetition of finding out the refractive index of the material of the prism. Determine as described before the refractive index for three chosen lines in the spectrum, one in the violet, the other in the yellow and another in the red region. If the values obtained for these are μ_v, μ, and μ_R respectively, the dispersive power of the medium is given by

$$\omega = \frac{\mu_v - \mu_R}{\mu - 1}$$

The result may sometimes be expressed as

$$\gamma = \frac{1}{\omega} = \frac{\mu - 1}{\mu_v - \mu_R}$$

where γ may be called the "dispersion number."

Exercise 3—To study the variation of refractive index of the material of the prism with wavelength of light and to use the graph to calculate the minimum thickness of the base of the prism that can resolve the two lines of sodium of wavelengths 5890 and 5896 A.U.

The experiment is but a repetition of determining the refractive index of the material of the prism (*Experiment* 12.1) for a given wavelength. Here the refractive index has to be determined for a set of known wavelenghts. For this purpose the mercury vapour lamp is used and its principal visible lines indicated graphically below are employed. Their wavelengths are as follows :

R_1 6908 G 5461 V_1 4078
R_2 6234 BG 4916 V_2 4047
Y_1 5790 B 4358

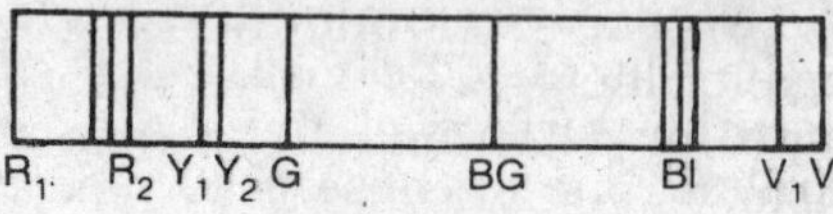

Fig. 11.16

The letters correspond to the lines in the graph and are also representative of the colour of the lines.

The angle of minimum deviation is obtained for each of these lines and the corresponding refractive index calculated by using Eq. 11.1.

Plot a graph between α and λ taking μ along the Y-axis and λ along the X-axis. The graph then represents the variation of the refractive index with wavelength of light.

From the graph Fig. 11.17 find the value of $d\mu/d\lambda$ in the region for the yellow light. Substitute this value of $d\mu/d\lambda$ in the equation

$$\frac{\lambda}{d\lambda} = t. \frac{d\mu}{d\lambda}$$

and knowing $\frac{\lambda}{d\lambda}$ = 5893/6, calculate the thickness of the base of the prism for the just resolution of the two sodium lines.

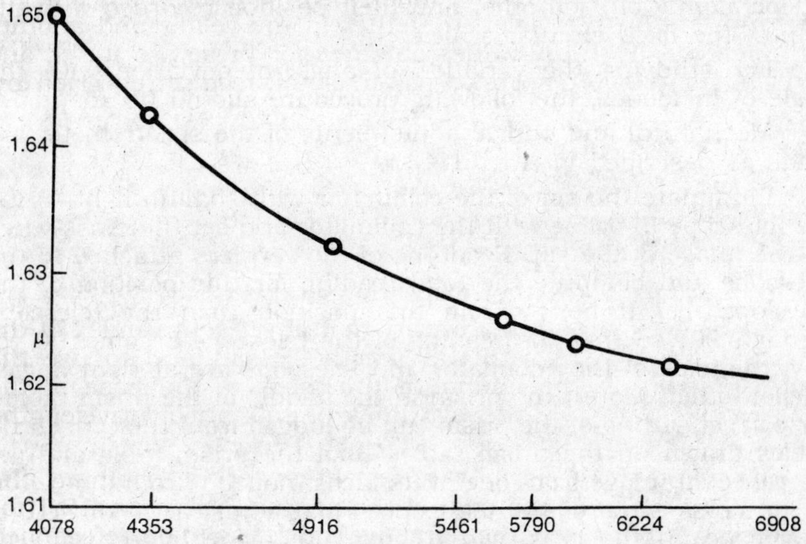

Fig. 11.17

Experiment 11.2

Object—To study the variation of angle of deviation with the angle of incidence for a prism and to use the incidence deviation graph for determining the refractive index of the material of the prism.

Apparatus—A spectrometer, a prism, a reading lens, a lamp and sodium lamp.

Theory—From the graph the angle of minimum deviations can be obtained. If this is δ and the corresponding angle of incidence is i, then

$$\alpha = 2i - \delta$$

where α is the angle of the prism. Knowing α and δ the refractive index of the material of the prism can be calculated using relation

$$\mu = \frac{\sin\frac{\alpha+\delta}{2}}{\sin\frac{\alpha}{2}}$$

Method—The spectrometer which is serviceable for this experiment is one having a fixed circular graduated scale and separate verniers for recording the positions of the telescope and prism table. Thus telescope and the prism table can be moved independently of each other and their position recorded with the help of the fixed circular scale.

For studying the variation of angle of deviation with the angle of incidence, the following procedure should be adopted.

Mechanical and optical adjustments of the spectrometer are made as described in Art. 11.4.

Illuminate the slit of the collimator with sodium light. Bring the telescope in a line with the collimator and set the cross-wires on the image of the slit. Read one of the verniers attached to the telescope and compute the total reading for this position of the telescope. Let it be ϕ. From this position turn the telescope through 90° so that the reading of the telescope becomes $90^o+\phi$. Now the axes of the collimator and telescope are at right angles to each other. Mount the prism in the middle of the prism table ; the correct setting of the prism can be judged from the concentric circles drawn on the prism table. Turn the prism table and get the reflected image from one of its faces (not the ground face) on to the cross-wires of the telescope. Final adjustment for setting the image on the cross-wires should be made by the tangent screw of the prism table. When this adjustment is complete, the face of the prism on which light from the collimator is incident is inclined at 45° with the collimator axis. Take the reading of the prism table and turn it through 45° so as to make this face normal to the collimator axis and hence to the incident rays. Take the reading of the prism table. Read both the verniers for the position which corresonds to zero angle of incidence.

From this position, turn the prism table through 10°, and look for the emergent rays from the other face of the prism bounding the refracting angle. In general no emergent rays can be obtained for angle less than 30° angle of incidence. The reason for the absence of the emergent rays that the refracted rays into the material of the prism meet this face (at which we are looking for emergent rays) at an angle greater than the critical angle for glass-air interface and are totally internally reflected.

Increase the angle of incidence by 5° ; let it now be 35°, Look for the emergent rays and set the cross-wires of the telescope on the yellow line of sodium. Record the reading of the telescope reading both the verniers. In this manner repeat the settings for various angles of incidence which are increased in steps of 5°.

The direct reading of the telescope ϕ, which had been previously recorded gives the direction of the incident rays and the various settings of the telescope on the yellow line of sodium

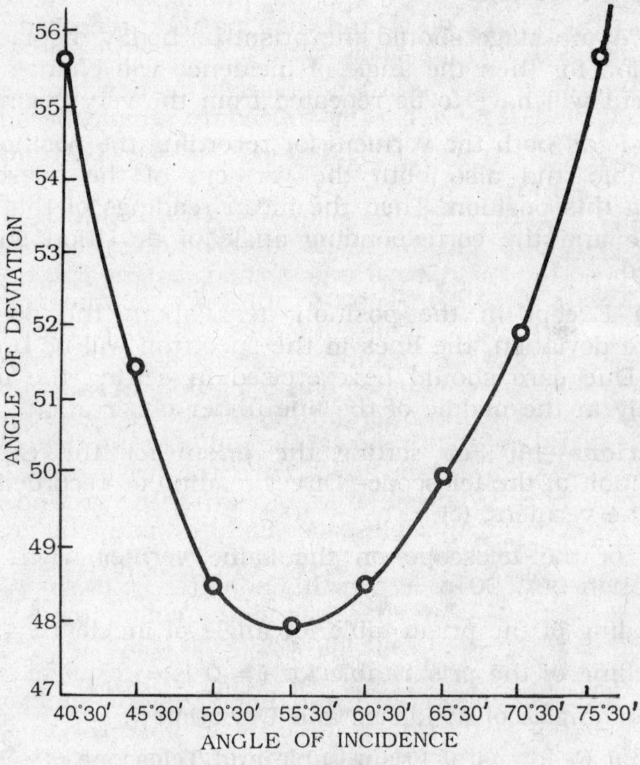

Fig. 11.18

give the direction of the emergent rays. The difference of the direct reading of the telescope and its reading in any setting gives the angle of deviation for that particular angle of incidence, for which the prism has been set. By finding these differences compute the angle of deviation for various angles of incidence for the given prism. Plot a graph between the angle of incidence as absissae and angle of deviation as ordinate. The nature of the graph is depicted in Fig. 11.18. From the graph find the angle of minimum deviation and find the corresponding angle of incidence. From these values of i and δ obtained from the graph, compute the angle of prism using the relation $\alpha = 2i - \delta$. Finally, calculate the refractive index of the material of the prism using Eq. 11.1.

Sources of error and precautions—All those enumerated in experiment 11.1 and

(8) At no stage should the prism be bodily displaced from its position for then the angle of incidence will change and the experiment will have to be repeated from the very beginning.

(9) Read both the verniers for recording the position of the prism table and also both the verniers of the telescope for recording this position. Then the mean readings of the angle of incidence and the corresponding angle of deviation should be computed.

(10) Except in the position near about the position of minimum deviation, the lines in the spectrum will be broad and diffuse. Due care should be exercised in setting the telescope accurately in the middle of the line under observation.

Observations—[A] For setting the prism for the experiment
Position of the telescope– Direct reading as recorded on one of the verniers, (ϕ) =

Reading of the telescope on the same vernier when turned, through 90°, $90+\phi$ =

Reading of the prism table for angle of incidence 45°=

Reading of the prism table for i = 0 is =

[B] For angles of Incidence and Deviation

Initial Readings of Prism Table and Telescope.

Prism Table for $i = 0$				Telescope for Direct image			
	Main Scale	Vernier Scale	Total reading		Main Scale	Vernier Scale	Total reading
Vernier A			a	Vernier C			c
Vernier C			b	Vernier D			d

POSITION OF

S. No.	PRISM TABLE for other angles of incidence						TELESCOPE for emergent rays					
	Vernier A			Vernier B			Vernier C			Vernier D		
	Main Scale	Vernier Scale.	Total	Main Scale	Vernier Scale.	Total.	Main Scale	Vernier Scale.	Total.	Main Scale	Vernier Scale	Total.
1			a'			b'			a'			b'
2			—			—			—			—
3			—			—			—			—
—			—			—			—			—

Calculations—For first reading $i = \dfrac{(a' - a) + (b' - b)}{2}$ and

$\delta = \dfrac{(c' - c) + (d' - d)}{2}$. Similarly, calculate δ for other values of i.

For plotting the graph tabulate the calculated values of i and δ as below :

Angle of		Point on Graph
Incidence	Deviation	
35	A	
—	—	
—	—	
—	—	
—	—	
70	H	

[B] From the graph,

Angle of minimum deviation =

Corresponding angle of incidence =

Angle of the prism = $2i-\delta$ =

$$\text{Therefore, } \mu = \frac{\sin \frac{1}{2}(\alpha + \delta)}{\sin \frac{1}{2}\alpha}$$

Result—The graph (Fig. 11.18) represents the variation of the angle of deviation with the angle of incidence.

The refractive index of the material of the prism for λ = 5895 A.U. =

Experiment 11.3

Object—To determine the temperature of sodium flame.

Apparatus—A 500 watt tungsten filament lamp, Bunsen burner asbestos loop for feeding sodium chloride solution, ammeter, a rheostat, voltmeter 0-250 volts, P.O.Box, direct vision spectroscope.

Description of Apparatus—*Direct Vision Spectroscope.* It consists of a narrow slit and a collimating lens which renders the light rays parallel. This followed by a train of prisms of crown

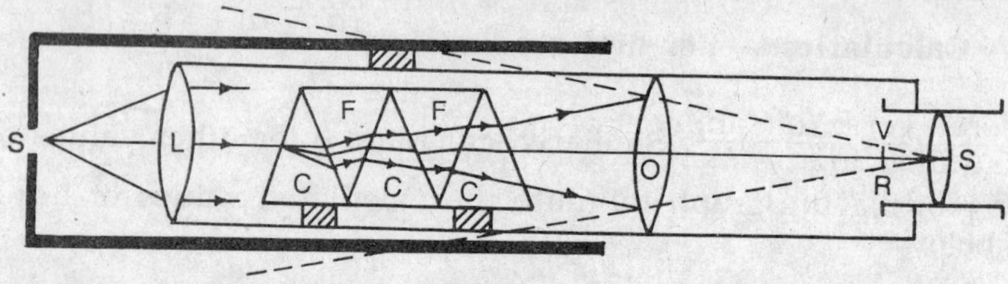

Fig. 11.19

and flint glass so combined as to produce dispersion without deviation. Usually three crown glass prisms are combined with two flint glass prisms and their angles are so chosen that the deviation of means ray produced by each set is the same whereas the flint glass prism produce greater dispersion than the crown glass ones. Thus when these are combined as shown in Fig. 11.19 the combination produces dispersion without deviation. To observe the spectrum produced by the combination use is made of a telescope which follows this train of prisms. As the spectrum is observed along the same direction in which the spectroscope is pointed towards the source, the instrument is termed direct vision spectroscope.

Theory—Kirchhoff's law states that at a given temperature the ratio of *the emissive power to the absorptive power is the same for all bodies and is equal to the emissive power of a perfectly back body*. Qualitatively it implies that if a body is capable of emitting certain radiations, it will absorb them when they fall on it. Thus the light from a heated tungsten filament is made to pass through sodium flame and viewed through a spectroscope. In the field of view is observed a faint continuous spectrum due to the filament and the bright yellow line of sodium. If the current through the filament is gradually increased it will be found that for a particular heating current, the yellow lines appear dark owing to absorption by sodium vapour. The temperature of the filament is then equal to the temperature of the flame. The temperature of the filament can be calculated from the relation,

$$R_t = R_o(1 + \alpha t)$$

where α the temperature coefficient of resistance of tungsten, is known. R_t can be calculated by observing the voltage across the lamp and the current flowing through it at the time of reversal of sodium lines. R_0 is found with a Post Office Box. The temperature so calculated gives the temperature of the heated filament and, therefore, also of the flame.

Method—Find the resistance of the filament when cold with a Post Office Box by connecting the former in the unknown arm.

Light the gas burner. Moisten the asbestos ring with sodium chloride solution and place it in the flame. Adjust its position accurately in the flame ; a yellow light is observed. Adjust the eye-piece of the spectroscope and get a well-defined image of the sodium lines on the crosswire. Usually only one line is visible because the dispersion produced by the prism train is not sufficient to resolve the two lines. Switch on tungsten lamp which is connected up in a series circuit consisting of an ammeter to measure the current and a rheostat to adjust the current flowing through the filament.

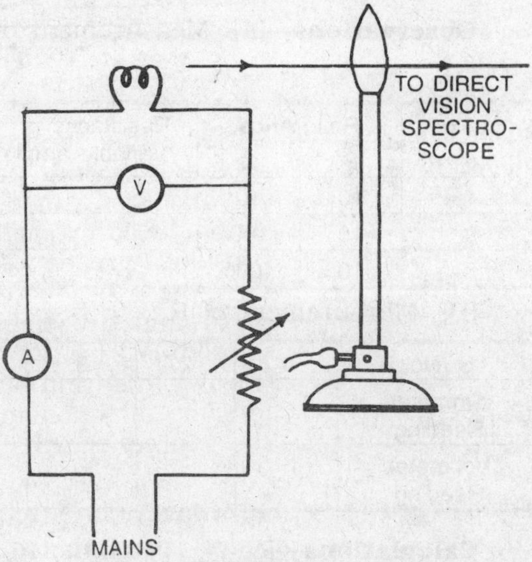

Fig. 11.20

The source of E.M.F. is the mains supply of 220 volts. A voltmeter is connected across the lamp to record the voltage applied to the filament. (Fig. 11.20) When the light from the filament is passed through sodium flame and examined with the spectroscope bright yellow line of sodium is seen against a faint continuous spectrum due to filament. Adjust the current in the filament circuit, which has to be increased, till the reversal of the sodium lines occurs. Now note the readings of the ammeter and the voltmeter. From these-values calculate R_t; R_o is already known as it has been determined with a P.O. Box. Taking the temperature coefficient of resistance of tungsten, to be known, calculate *t*, the temperature of the heated filament. This gives the temperature of the flame.

Sources of error and precautions—(1) The gas flame should be non-luminous.

(2) The asbestos loop should be well soaked in sodium chloride solution and should be fed with the solution when necessary.

(3) The current in the filament should be increased in small and regular steps; for this purpose an ammeter of range 0 to 0.5 amps, reading in steps of 0.01 amp. is essential.

(4) The position of reversal of lines should be observed very accurately.

Observations—[A] Measurement of R_0 with a P.O. Box.

S. No.	Ratio Arms		Resistance in variable arm	Direction of deflection	Inference
	P	Q			
1	10	10			
2	10	100			
3	10	1000			

[B] Measurement of R_t

S. No.	1	2	3	4	5	Mean value
Ammeter Reading						i
Voltmeter Reading						V

Calculations—$R_t = \frac{V}{i}$ = ohms; R_o = ohms.

$$R_t = R_0 (1 + \alpha t)$$

$$t = \frac{R_t - R_0}{R_o.\alpha}$$

$$= {}^oC$$

Result—The temperature of the flame = °C.

Criticism of the method—The method is not accurate for determining the temperature of flame. The results obtained are low. Kirchhoff's law applies to black body radiation and the radiation from the lamp and the flame cannot be treated as black body radiation corresponding to their temperatures. Hence the results obtained are low.

12.6. Sources of light—(*a*) The Bunsen Flame—The simplest method of producing spectra is to feed the inorganic salts into the Bunsen burner flame. However, this method is suitable for salts of sodium, potassium, lithium, calcium, strontium and barium. The salts are fed into the flame by a small loop of platinum wire. The loop is first dipped in concentrated HCl, and heated over the flame till it ceases to give off the yellow colour. Then it is dipped in the given salt and fed into the flame of the Bunsen burner. Owing to the high temperature the salt gets volatilised and gives its characteristic flame. If the light from the flame is now examined with a spectrometer, the spectrum obtained will be the characteristic line spectrum of the salt. If the salt need be fed for a longer period some asbestos is impregnated with the salt solution and held in the flame. This impregnation on the salt can be renewed when necessary. Some devices of continuous automatic feeding have also been employed.

(*b*) The Electric Spark—If the terminals of the secondary coil of an induction coil are connected to two metal electrodes and they are brought close to each other when the induction coil is working, a stream of sparks passes between them. The colour of the spark depends upon the gas or vapour present between them. With air the colour of the spark is violet and the light analysed gives the spark spectrum of air.

If the condenser C is connected in parallel with the spark gap S, the spark is stopped and the current in the secondary is used in charging the condenser (Fig. 11.21). When the potential of the condenser is raised to such a value that it is able to break the dielectric resistance of the gap, the condenser discharges through the gap which is kept extremely small. The spectrum

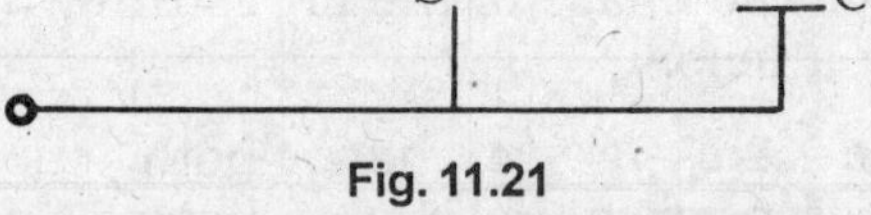

Fig. 11.21

obtained under such conditions is the spark spectrum of the metal points of the electrodes between which the spark-passes.

To obtain the spark spectrum of a solution, it should be placed in a test-tube through the bottom of which a platinum wire is fused (Fig. 11.22). The tube is then corked and glass rod, having a fused platinum wire throughout its length, is passed through it. The lower end of the wire is adjusted to be about a mm. or so above the surface of the solution. The two platinum electrodues, one at bottom of the test tube and the other in the glass rod are then connected to the secondary of an induction coil. The spark passes between the wire and the surface of the liquid when the induction coil is worked and the light emitted gives the spark spectrum of the solution.

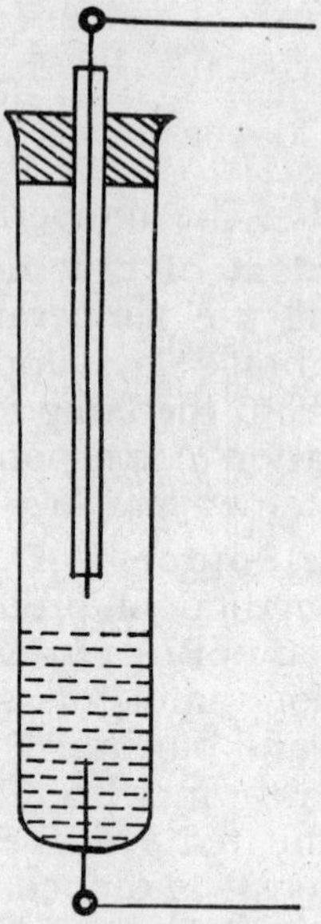

Fig. 11.22

(c) The Electric Arc—(*i*) The Carbon Arc—The arc lamp can be started by bringing the carbon electrodes into contact suddenly and then drawing them apart. A bright discharge passes between the electrodes. As the arc continues to run, a crater is formed in the positive electrode and the end of the negative electrode gets tampered. Most of the light comes from the crater and very little (about 5%) from the luminous discharge. The spectrum of the arc is continuous superposed on which is the line and band spectra of the luminous discharge. It requires about 70 to 80 volts to start the arc.

The arc spectra of salts can be examined by feeding them in the crater and exciting the arc. The image of the arc is thrown on the slit of a spectrometer by a lens.

(*ii*) The Metal Electrode Arc—If we have both the electrodes made of copper, or iron or aluminium, instead of carbon, and the electrode arc started, the spectrum so obtained by examining the light of this arc will be the arc spectrum of the metal electrode. Due care must be taken that the current is not large otherwise the electrodes will melt.

Peter Cooper-Hewitt in 1901 invented a type of arc entirely different from those described above. One electrode is a metal wire and the other a pool of mercury. The entire unit is sealed

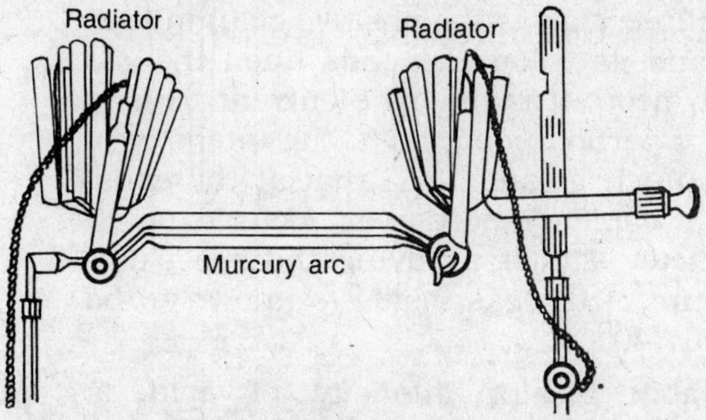

Fig. 11.23

in an evacuated tube. To start the arc the tube is tilted until the stream of mercury from the cathode touches the metallic anode. This completes the circuit and the current passes. It is then tilted the other way back so that the tube returns to its normal position, the stream breaks and the arc forms across the gap thus created. The mercury which vaporises maintains the light arc. The electrodes are kept comparatively cool and the light is due almost entirely to the glowing vapour. The brightness of the mercury arc depends upon its design and condition. The lines obtained in the mercury arc spectrum are very bright and they serve the purpose of monochromatic light when used alongwith certain filters.

The Cooper-Hewitt arc Lamp made in quartz is a powerful source of ultra-violet light.

(*d*) Discharge Tubes—Luminous discharge tubes can be classified under two categories : (*i*) Cold cathode lamps and (*ii*) Hot cathode lamps.

(*i*) Cold cathode lamp—These tubes contain gas at low pressure which ranges between 5 mm. and 10 mm. depending upon the gas enclosed. The electrodes are of aluminium welded to the ends of platinum or tungsten wires, the latter being sealed through the glass. Aluminium is used because it sputters very little. Normally these tubes require a high voltage supply. So when worked with an induction coil or a high voltage transformer, a diffuse glow fills the tube which for spectroscopic work has the shape and form depicted in Fig. 11.24. This diffuse glow is the positive column type of discharge and its colour depends upon the gas enclosed; with neon, it is red. The current density in the capillary portion being high, the intensity of light there is much greater than that at the ends. Light from the capillary portion is focused on to the spectrometer slit for studying the spectrum. The temperature of the gas inside is known to be very low (about 40°C).

Fig. 11.24

Considerably greater intensity of light is obtained with the end on type of tube shown in Fig. 11.25. The electrodes are made of sheet aluminium which is rolled up and fitted inside the loosely-fitting glass tubes G, G. They are connected to platinum or tungsten wires, the latter being sealed through glass. When worked with a transformer, the tube lights up and the light

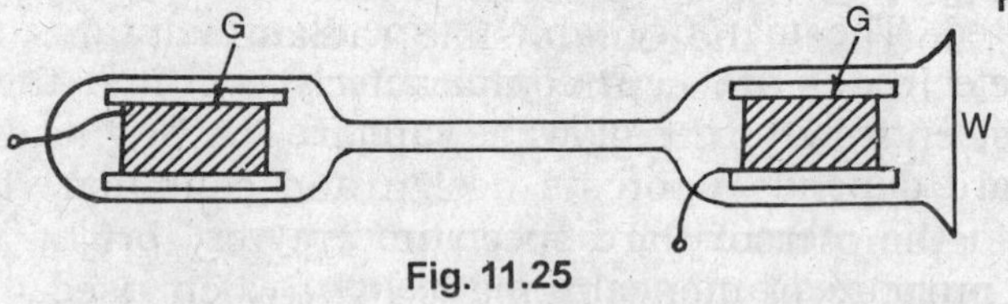

Fig. 11.25

through the plane glass window W is examined through a spectrometer. The larger area of the electrodes permits the use of greater currents; hence the intensity obtained in this type of tube is much greater than that obtained with the other type.

(ii) Sodium Discharge Lamp—In the sodium discharge lamp as illustrated in Fig. 11.26, neon gas at a pressure of about 10 mm of mercury is filled in the discharge tube and some specks of metallic sodium are deposited on the sides of tubes. For the discharge to start voltage of about 470 volts is required which is supplied by a single-wound leak transformer. Initially the discharge passes between the electrodes E through the neon gas and the colour of the light is red. During this neon discharge the metallic sodium evaporates and the colour of the light is tinged yellow. As time elapses more of sodium vapour is produced which ultimately maintains the discharge giving its characteristic yellow light. The pressure of the sodium vapour is about 0.01 mm. and temperature of the tube is sufficiently high. To reach the necessary temperature quickly and to maintain it with the smallest possible heat loss, the lamp is surrounded by a *double* walled evacuated glass vessel V.

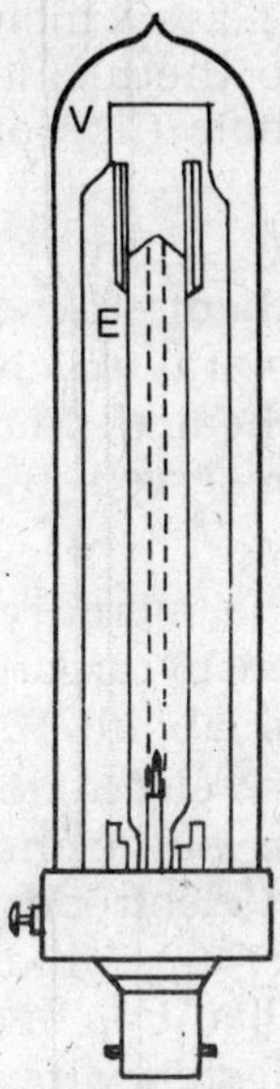

Fig. 11.26

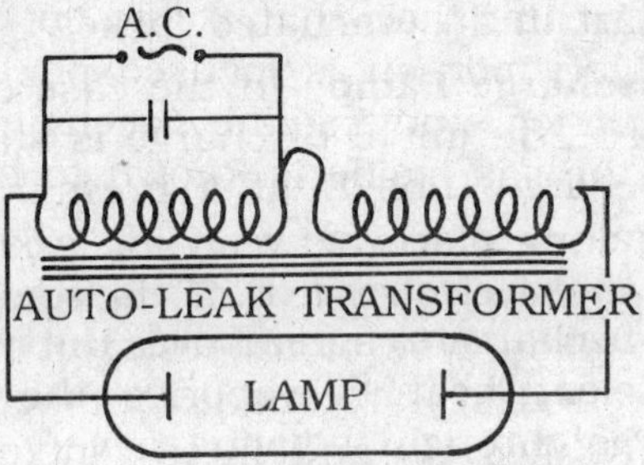

Fig. 11.27

The sodium lamp is essentially a low pressure vapour lamp because an increase in pressure above a few mm. of mercury does not increase the only visible radiant energy, namely, the yellow D lines. Since the characteristic of a low pressure discharge is a low intensity, the lamp is, therefore, in the form of a long tube.

The lamp is of the cold cathode type, *i.e.*, the electrodes are not preheated and hence as the ionisation voltage of neon is high, 21.5 Volts, a high voltage is necessary to start the discharge. The electrical connections are shown in Fig. 11.27. At first sight it would appear that the lamp's voltage will always be greater than the supply voltage, and also that the circuit has no ballast. However, the auto-leak transformer is one in which the leakage reactance is enormously increased by the use of magnetic shunts and hence the necessary characteristics can be obtained from a single transformer instead of using a transformer of the ordinary constant voltage type in conjunction with a separate choke. Take the case of a 140 W sodium lamp. When running at a discharge of voltage 165 volts and the lamp current is 0.9 amps; the starting voltage is 410 volts. The condenser in the circuit is used for power factor improvement. There is another constructional feature of the lamp which needs to be noted and it is the glass used for the lamp tube. A thin coating of this special glass which does not blacken under sodium discharge, is given to the inside of the lamp tube. It is not necessary to make the whole tube of this glass as it is very expensive.

***(iii)* Hot cathode Lamps**—These lamps employ electrodes in the form of cylinders with heating filament inside them. These electrodes are made of barium oxide coated tungsten spiral which readily provides a copious supply of electrons when heated. For the efficient running of the lamp, the heat is conserved by enclosing the lamp part in an evacuated jacket.

***(iv)* Mercury Discharge Lamp**—In the case of the mercury discharge lamp (Fig. 11.28), initial discharge is started by argon contained in the discharge tube D at a pressure of about 10 mm., the discharge taking place between an auxiliary electrode A and the adjacent main electrode E. This sets up sufficient ionisation A and the discharge to spread over the main electrodes B resulting in sufficient heat to vaporise the mercury and maintain the discharge through it giving a very brilliant light. The main electrodes are in the form of spirals of tungsten wire,

Fig. 11.28

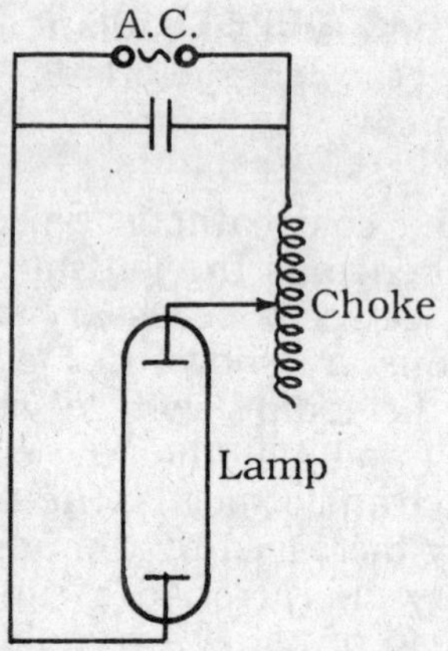

Fig. 11.29

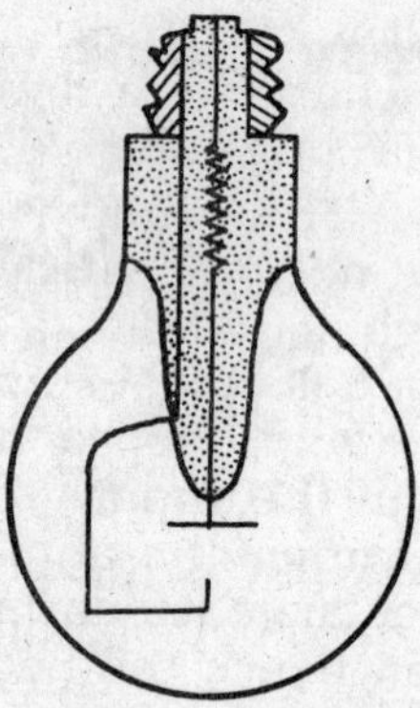

Fig. 11.30

each holding picket of electron-emitting barium compound. There is no preheating as in the case of low pressure lamp. The discharge tube contains just sufficient mercury to give a pressure of one atmosphere at the temperature of the tube which is very high about 600°C. To limit the current at the auxiliary electrode a ballast B of about 50,000 ohms is placed in series with it. The efficiency of the lamp is very high. The radiation from lamp is chiefly green, violet and ultra-violet. It works at a lower voltage and a proper choke is provided for use with it. The electrical connections are shown in Fig. 11.29.

(e) Neon Lamps—The neon spectrum tubes make use of the positive column of the discharge and hence require high voltages to operate them. But the neon lamp makes use of the different parts of the discharge. When a P. D. of about 70 volts is applied, the positive column is extremely small, but the cathode becomes covered with the floating luminous cathode glow of orange colour. For D.C. operated lamps the cathode consists of an iron disc with a coating of potassium and the anode consists of a point. Using the light from the edge of the disc we get a good linear source of light. As the voltage required to operate it is about 70, a resistance in series with the cathode is put in the cap of the lamp which is then used direct on mains. (Fig. 11.30).

When operated on A.C. there is a glow on both the electrodes.

ORAL QUESTIONS

DISPERSION OF LIGHT

1. Explain the action of a prism in producing dispersion of light.
2. What is meant by angular dispersion and the dispersive power of a medium ?
3. What is the angle of deviation and on what factors does it depend ? When is it a minimum ?
4. What is the practical importance of the minimum deviation position of a prism ? What part does it play in the production of a pure spectrum ?
5. Explain the different methods used for the production of a pure spectrum ? Which method is the best and why ?
6. What conditions must be satisfied when a pure spectrum is to be produced with a prism ?

THE SPECTROMETER

1. What is a spectrometer ?
2. What are the requisites of a good spectrometer ? Explain the working of its essential mechanical and optical parts ?
3. How and why is the collimating lens achromatized ?
4. Why is the slit not permanently fixed at the focus of the collimating lens ?
5. Has the aperture of the collimating lens anything to do with the brightness of the spectrum ?
6. Explain the working of the slit.
7. What is a symmetrical slit and how does it differ from an asymmetrical slit?
8. Why are lines drawn on the prism table ? Why is it provided with levelling screws and why is the spectrometer not provided with these ?
9. Explain the use of the three levelling screws in the adjustment of the prism table.
10. What is the function of the clamping and the tangent screw attached to the prism table ?
11. What are the essential parts of the spectrometer telescope ?
12. Why is there a weight attached to the telescope frame ?
13. What type of eye-piece is used ? Why does it consist of two lenses ?
14. What is meant by achromatic combination of lenses ?
15. What is chromatic aberration ? How may it be minimised for two lenses separated at a distance ?
16. What is spherical aberration ?
17. Explain the importance of planoconvex lenses used in eye-pieces.
18. Explain the difference between the construction and working of Ramsden's eye-piece and Huygen's eye-piece.
19. Why are cross-wires not fitted in Huygens' eye-piece ?
20. How do you focus the eye-piece on the cross-wires ?
21. How will the focusing of an eye-piece on the cross-wires vary with the sight of the observer ?
22. How do you focus the telescope for parallel rays ? Explain

the various methods used for this adjustment.

23. Why is the telescope not permanently focused for infinity ?
24. How is the collimator set to emit a parallel pencil of light ?
25. What is auto-collimating method ?
26. How do you level the prism table ? Can a spirit level be used for this adjustment ?
27. What is the meaning of levelling prism table ?
28. Why should the refracting edge of the prism be parallel to the axis of rotation of the telescope and also to the slit ?
29. What mechanical adjustments should the spectrometer be tested for before making any optical adjustments ?
30. How do you measure the angle of a prism with a spectrometer ? Explain the two methods in frequent use for this measurement. Bring out the importance of the particular mode of placing the prism over the prism table in each of the above methods.
31. How do you set the prism in the position of minimum deviation ? How does this angle vary with wave-length of light ? Why is the spectrum best defined in this position ? Is dispersion also more in this position of the prism ? If not why ?
32. How would you find out the refractive index of a liquid with a spectrometer ? How may it be employed to determine the dispersive power of the material of the prism ?
33. What are the precautions that should be observed while using a spectrometer ? Why are two verniers provided with it ? What is the difference between the readings of the two verniers ?
34. Explain the difference between a spectrometer, a spectroscope and a spectrograph.

12

Interference of Light

12.1. Introduction—From the principle of superposition of waves, as enunciated by Thomas Young, we know that when two wave trains arrive simultaneously at a point, it is thrown into vibration having an amplitude equal to the algebraic sum of the amplitudes of the constituent vibrations. If the two waves are in the same phase of vibration, the resultant amplitude is large ; if the two waves reach the point in opposite phases, the resultant amplitude is very small. In the latter case, if the constituent vibrations are of the same amplitude, the resultant amplitude of vibration reduces to zero. Since the two vibrations completely destroy each other, this is usually spoken of as *destructive interference.* There is sufficient experimental evidence on this subject in the branch of sound where a sound wave superposed on a sound wave produces silence just in a manner two sets of water waves when superposed may produce a dead level. Such a phenomenon, experimentally demonstrated with light, simply bears out the validity of the undulatory theory of light. Thus, two light waves of the same amplitude and period superposed on each other may produce a maximum intensity or complete darkness depending upon the phase of vibration in which the waves arrive at a point. It must, however, be borne in mind that there is *no real destruction of energy at places of complete darkness ; only a redistribution occurs, the energy reappearing at places of maximum intensity.*

12.2. Composition of two vibrations of the same period and amplitude and having a certain phase difference—Let a be the common amplitude of vibration and δ be the phase difference between the two when these reach a certain point. From the Δ of amplitudes, (Fig. 12.1) the resultant amplitude A, represented by the side DC, is given by

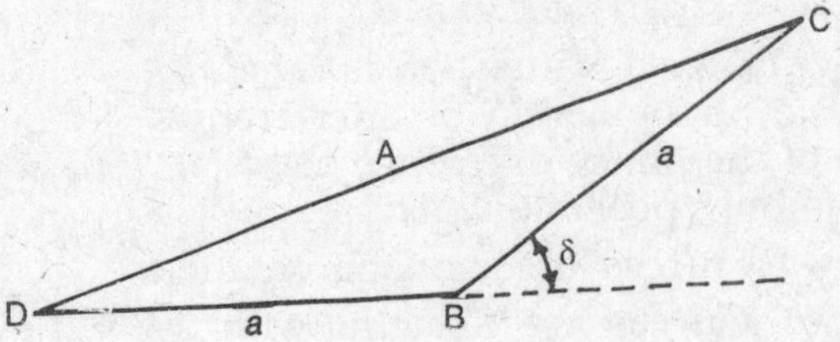

Fig. 12.1

$$A^2 = DB^2 + BC^2 + 2DB.\,BC \cos \delta$$
$$= a^2 + a^2 + 2a^2 \cos \delta$$
$$= 2a^2 + 2a^2 \cos \delta$$

As δ goes on changing, the resultant amplitude varies

For $\delta = o$, $A^2 = 4a^2$, a maximum value

$\delta = \pi$, $A^2 = o$, a minimum value

$\delta = 2\pi$, $A^2 = 4a^2$, a maximum value

$\delta = 3\pi$, $A^2 = o$, a minimum value

Generalizing the results, whenever

$$\delta = 2n\pi$$

we have a maximum amplitude of the vibration and whenever

$$\delta = (2n + 1)\,\pi$$

we have zero amplitude of resultant vibration.

These results may also be expressed in terms of path difference. Then, the condition of maximum amplitude, *i.e.*, maximum intensity at a point becomes

$$e = \frac{\lambda}{2\pi}.\delta = 2n.\frac{\lambda}{2}$$

where e is the path difference between the two vibrations reaching the point. For destructive interference

$$e = (2n + 1)\,\lambda/2$$

Thus, if there are two sources of light emitting into space waves of the same wavelength and amplitude, at points where these arrive having a path difference an even multiple of $\lambda/2$, we get maximum intensity of light whereas at points where the path difference between the waves reaching there is an odd multiple of $\lambda/2$ we get complete darkness.

12.3. Conditions of Interference—From the discussion above it is clear that we have a constant illumination at a certain point on the screen so long as the resultant amplitude remains constant.

$$\therefore \qquad A^2 = 2a^2 + 2a^2 \cos \delta,$$

A will have a fixed value if δ is fixed. Hence the essential condition for the phenomenon of interference to be observed is this constancy of the phase difference between the waves arriving at a point. It is only possible when

(*i*) the two sources are monochromatic,

(*ii*) the two sources are vibrating with the same amplitude, (*iii*) the two sources are vibrating in phase or with a constant phase difference.

Such two sources are said to be *coherent*. The only way to realise these vibrating in phase or with a constant phase difference or vibrating such that every change of phase which occurs in one is accompanied by the same change in the other, is to make one source the image of the other or to divide a pencil of rays into two either by reflection or refraction and re-unite them by superposing one over the other.

12.4. Width of Interference Bands—Let S_1 and S_2 denote two sources of monochromatic light vibrating parallel to each other in the same phase and with the same amplitude. Thus, let these be the two coherent sources of light. The waves originating from the two sources travel with the same velocity. At the point M, (Fig. 12.2) equidistant from S_1 and S_2 the displacements will be in the same phase which will thus unite increasing the

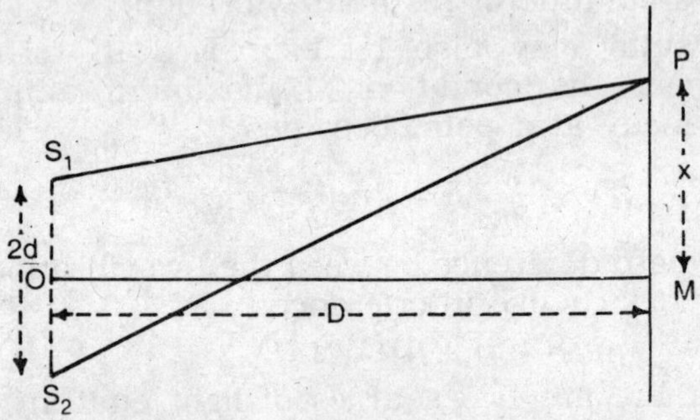

Fig. 12.2

intensity of light there. Let us now proceed to determine the illumination at different points on the screen. For this consider a point P, distant x from M. The waves of light from S_1 travel a distance S_1P whereas those from S_2, a distance S_2P. Thus the waves at P will not, in general, arrive in the same phase. This difference of phase between the two waves will depend upon the path difference between them. If the path difference is an even number of half wavelengths, the waves reaching P meet in the same phase; if the path difference is an odd number of half wave lengths, the waves at P arrive in opposite phases. Accordingly the point P is of maximum intensity when reinforcement of waves occurs and of minimum intensity when the waves reach it in opposite phase resulting in destructive interference. The conditions are, therefore,

$$S_2P - S_1P = 2n\,\lambda/2 \qquad (12.1)$$

and $$S_2P - S_1P = (2n + 1)\ \lambda/2 \tag{12.2}$$

for maximum brightness and complete darkness respectively, n being any integer. To evaluate this path difference, let $S_1S_2=2d$ and OM=D. Then,

$$S_2P^2 = D^2 + (x + d)^2$$

$$S_1P^2 = D^2 + (x - d)^2$$

$$S_2P^2 - S_1P^2 = 4xd$$

$$\therefore \qquad S_2P - S_1P = \frac{4xd}{S_2P + S_1P}$$

Let d and x be very small in comparison with D. Then,

$$S_2P = S_1P = D \text{ approximately}$$

$$\therefore \qquad S_2P - S_1P = \frac{4xd}{2D} = \frac{2xd}{D}$$

Substituting this value in Eqs. (12.1) and (12.2) we have, for maximum brightness,

$$\frac{2xd}{D} = 2n\ \frac{\lambda}{2}$$

or $$x = n.\ \frac{D\lambda}{2d}$$

and for complete darkness

$$\frac{2xd}{D} = (2n + 1)\ \frac{\lambda}{2}$$

of $$x = (n + \frac{1}{2})\ \frac{D\lambda}{2d}$$

These two equations reveal that the intensity of light on the screen varies from a maximum to a minimum and so on as the point P moves away from M on either side of it. Thus, we get a number of alternately dark and bright bands or fringes on the screen. Such fringes are termed interference fringes.

Now, the distance between two consecutive fringes can be obtained by giving to n any two consecutive integral values and obtaining the difference in x; this distance is termed as the *fringe width*. If this be denoted by β, then,

$$\beta = \frac{D\lambda}{2d} \tag{12.3}$$

This is the value obtained whether we use the equation for maximum intensity or for complete darkness on the screen. A direct conclusion of this is the fact that the fringes are equally spaced. As the fringe width depends on λ, the fringes for blue rays will have lesser spacing than those for red rays. As a consequence of this only a few coloured fringes can be seen when the sources S_1 and S_2 are of white light. The central fringe at M will, however, be achromatic in view of the fact that waves of all wavelengths reach M in the same phase of vibration.

From the expression for the fringe width it is also clear that fringes having a measurable width can be obtained only if D is large and $2d$ small. Hence, the plane of observation should be at a large distance from the two coherent sources whereas the distance between the sources should be very small.

$$\lambda = \beta \frac{2d}{D} \tag{12.4}$$

This equation provides us with a method of finding the wavelength of light if the fringe width, the distance between the two sources and the distance between the plane containing the sources and the plane of observation, can be measured. Methods based on this theory have been employed for the measurement of λ. We shall, however, deal only with Fresnel's biprism method in detail.

12.5. Displacement of fringes—In Fig. 12.2 the central fringe lies at M as this point is equidistant from the two coherent sources S_1 and S_2. If a thin plate of mica be interposed in the path of one of the interfering beams, the central fringe of the system is shifted towards the path of the beam in which the plate is interposed. Suppose it is displaced to the point P. Then the time taken by the waves to travel from S_2 to P in air must be the same as the time taken by waves to travel (S_1 P–t) in air and a thickness t in mica ; for, then alone the two optical path lengths would be equal. Let υ_1 and υ_2 be velocities of light in air and in mica, then the equation of time is

$$\frac{S_2P}{\upsilon_1} = \frac{S_1P - t}{\upsilon_2} + \frac{t}{\upsilon_2}$$

If the shift takes place to the nth fringe of the original system,

or $$S_2P - S_1P = n\lambda$$

$$(\mu - 1)\ t = n\lambda$$

$$= \frac{x.\lambda}{\beta}$$

or $$t = \frac{x.\lambda}{\beta\ (\mu - 1)}$$

Thus, the equation can be used to determine the thickness of a mica sheet provided the fringe width β, the displacement x

on the screen are experimentally found out (λ and μ being known factors).

12.6 Formation of Newton's Rings—Let us consider a plano-convex lens placed with its convex face in contact with a plane plate of glass. This provides us with a film of air of gradually increasing thickness enclosed between the convex surface of the lens and the plane surface of the plate. The curves of constant thickness are then circles having their centre at C (Fig. 13.3), the point of contact of the surfaces of the lens and the plate. Consequently, when mono-chromatic light is made to fall on the lens from above, a series of concentric rings are formed with their centre at C. These rings are due to interference of light waves reflected from the upper and the lower surface of the film enclosed between the two surfaces. The phenomenon is essentially the same as the colour of thin films of varying thickness and the results deduced for that can be applied. Let us suppose that the thickness of the film at a certain point P is $PQ = t$. This thickness remains constant along a circle of radius CP and centre C. If the wavelength of light used is λ and μ the corresponding refractive index of the film, the point P, and therefore, the circle with radius CP, is dark or bright, according as

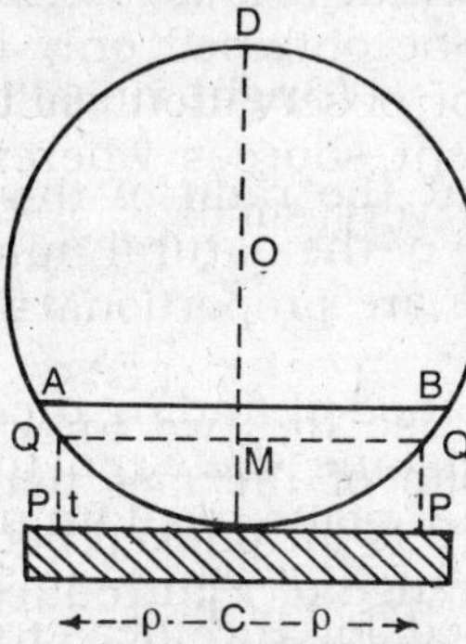

Fig. 12.3

$$2\ \mu t \cos (r + \theta) = 2n\ \lambda/2 \qquad \text{(Dark rings)} \qquad (12.5)$$

$$2\ \mu t \cos (r + \theta) = (2n + 1)\ \lambda/2 \qquad \text{(Bright rings)} \qquad (12.6)$$

where r is the angle of refraction for the ray into the film, θ the angle of the film and n, any integer.

Let CP be ρ and the radius of curvature of the convex surface of the lens be R. Then

$$CP^2 = CM.\ MD.$$

$$\therefore \qquad \rho = t\ (2R - t)$$

or $\qquad i = \rho^2/2R$ to a first order of approximation.

Substituting the value of t in Eqs. (12.5) and (12.6)

$$2\ \mu\ \frac{\rho^2}{2R} \cos\ (r + \theta) = (2n + dt)\ \frac{\lambda}{2} \qquad \text{(Dark rings)}$$

and $$2\,\mu\,\frac{\rho^2}{2R}\cos(r+\theta) = (2n+1)\frac{\lambda}{2} \qquad \text{(Bright rings)}$$

Whence we can express ρ_n the radius of the nth dark or bright ring as

$$\rho_n^2 = \frac{n\lambda R}{\mu\cos(r+\theta)} \qquad \text{(Dark rings) (12.7)}$$

and $$\rho_n^2 = \frac{(2n+1)\lambda R}{2\,\mu\cos(r+\theta)} \qquad \text{(Bright rings) (12.8)}$$

The above two equations reveal that the radii of the dark rings are proportional to the square roots of the natural numbers 1,2,3....while the radii of the bright rings are proportional to the square roots of the odd natural number.

The central spot at C is dark because the two interfering rays have a parth difference $\lambda/2$ in spite of the fact that the thickness $t = 0$. This path difference of $\lambda/2$ is introduced between the two rays as a result of the phase change of π introduced in the ray which suffers reflection at the surface of the plate of glass. No phase change occurs in the other ray.

The Eqs. (12.7) and (12.8) can be simplified to a great extent by putting $\mu = 1$ (for air film), $r = 0$ (considering normal incidence of the rays) and $\theta = 0$ (using a lens of a large radius of curvature). We then have

$$\rho_n^2 = n\lambda R \qquad \text{(Dark rings)}$$

and $$\rho_n^2 = \frac{(2n+1)\lambda R}{2} \qquad \text{Bright rings)}$$

whence their diameters D_n are given by

$$D_n^2 = 4n\lambda R \qquad \text{(Dark rings)} \qquad (12.9)$$

and $$D_n^2 = 2(2n+1)\lambda R \qquad \text{(Bright rings)} \qquad (12.10)$$

Equations (13.9) and (13.10) provide us with a ready means of measuring the wavelength of light if D_n, n and R can be determined experimentally. But it is not advisable to use these results for reasons to be given below. Instead of these, a modified formula may be used. The diameter of ($n+p$)th ring is given by

$$D_{n+p}^2 = 4(n+p)\lambda R \qquad \text{(Dark ring)} \qquad (12.11)$$

and $$D_{n+p}^2 = 4(2n+2p+1)\lambda R \qquad \text{(Bright ring)} \qquad (12.12)$$

Subtracting either (12.9) from (12.11) or (12.10) from (12.12) we get

$$D_{n+p}^2 - D_n^2 = 4p\lambda R$$

$$\therefore \qquad \lambda = \frac{D_{n+p}^2 - D_n^2}{4pR} \qquad (12.13)$$

This equation has an advantage over Eqs. (12.9) and (12.10) which are deduced on the assumption that theoretically at C. $t = 0$ and that if the lens and the plate are in contact, there is no deformation of the two contacting surfaces due to the pressure employed to make contact between them. Theoretically, therefore, the rings are to be numbered with the central dark spot as the zero order dark ring. But it is just possible that due to encroachment of dust particles, other imperfections and deformations of two surfaces, the thickness at the point of contact may not be zero. As a consequence of this, though the central spot may be dark, it may be the second, the third or the fourth........order dark ring occupying the centre and as such the nth order ring as counted from the centre is as a matter of fact the $(n+1)$th or $(n+2)$th or...Therefore, Eqs. (12.9) and (12.10) do not correctly represent the diameter of the so counted nth order ring. Equation (12.13), on the other hand, involves the difference of the squares of the diameters of two rings and this is independent of the thickness at C. For this reason Eq. (12.13) is recommended for the determination of the wavelength of light with Newton's rings.

The equations for the diameters of the rings reveal that with the change of the wavelength of light, the diameters of the rings also change. Therefore, the rings for the violet rays will have a smaller diameter than those for the red rays. When white light is used, a series of coloured rings are exhibited, violet at the inner and red at the outer edge. Each wavelength gives rise to its own system of concentric rings. The diameters of the rings of these various systems are different. As the coloured rings soon become superposed and overlapped, this results in almost uniform illumination after a few rings. It is for this reason that the rings observed with white light are few in number. The central spot is, however, dark.

The central spot can be changed into a bright one. We have only to interpose a film of refractive index less than that of the material of the lens and greater than that of the material of the plate. Then the two interfering streams of light are reflected under identical conditions, *i.e.*, during a passage from a denser to a rarer medium. Hence no phase difference is introduced between the interfering rays. It follows then that the central spot for $t = 0$ shall be bright under these conditions.

It is of importance to bear in mind that no rings can be observed at all in one case, *viz.* when the film is backed by a polished reflector simply because the reflection is complete and the reflected and the transmitted systems of rings get superposed. These being of a complimentary character, no rings are exhibited.

12.7 Interference in Wedge-shaped Film

A wedge shaped air film can be created by putting a thin sheet of paper at the one end between two optically plane glass plate. When a monochromatic light is made to fall on it normally, interference of light occurs between the light reflected from the upper and lower surfaces (A and A′ in Fig. 12.4) of the film respectively. If t is thickness of the film at the point of consideration (i.e. AA′ = t), then the path difference between the two rays, reflected from the top and bottom of the film is equal to $2\mu t$.

As for air $\mu = 1$

$$\therefore \quad \text{Path difference} = 2t \tag{12.14}$$

An addition path difference of $\frac{\lambda}{2}$ is introduced due to the fact that the reflection at the bottom end is from an optically denser medium.

$$\therefore \quad \text{Total path difference} = 2t + \frac{\lambda}{2} \tag{12.15}$$

Thus, a bright fringe is formed where

$$2t + \frac{\lambda}{2} = 2n.\frac{\lambda}{2}$$

$$2t = (2n-1)\frac{\lambda}{2}, \; n = 1, 2, 3... \tag{12.16}$$

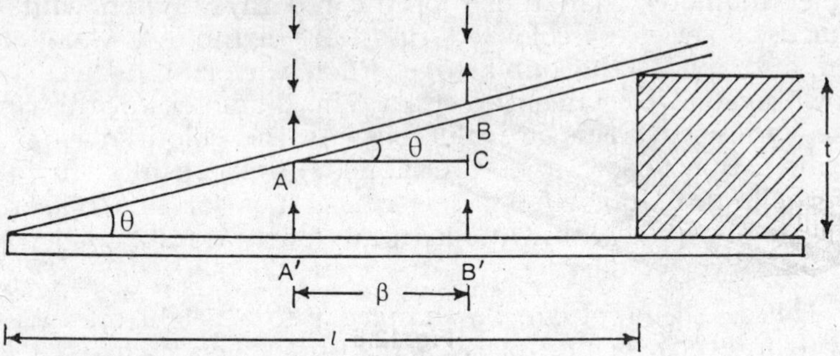

Fig. 12.4

A dark fringe is formed where

$$2t + \frac{\lambda}{2} = (2n+1)\frac{\lambda}{2}$$

or $$2t = n\,\lambda, \; n = 0, 1, 2,.... \tag{12.17}$$

Straight fringes of equal thickness are formed. Each fringe is a locus of equal thickness of the film. Fringes formed are known as *fringes of equal thickness.* This method is used to find the thickness of a thin sheet or paper (see experiment No. 13.3) and test the flatness of a surface.

Experiment 12.1

Object—To determine the wavelength of sodium of light with Fresnel's biprism.

Apparatus—Good quality optical bench, slit, biprism, sodium lamp, lens for focusing light on the slit, a short focus lens provided with circular stop of variable aperture.

Description—(1) *Fresnel's Biprism.*—As the name implies, a biprism may be regarded as consisting of two prisms of very small refracting angle placed base to base. As a matter of fact it is ground from a single plate of glass and then polished. It has one of its angles only slightly less than 180°. The remaining two angles are of the order of 30′ or less.

(2) *The optical bench* (Fig. 12.5).—The bench consists of a solid cast iron base about 175 cms, in length supported by four levelling screws. The scale on one of its arms is engine divided

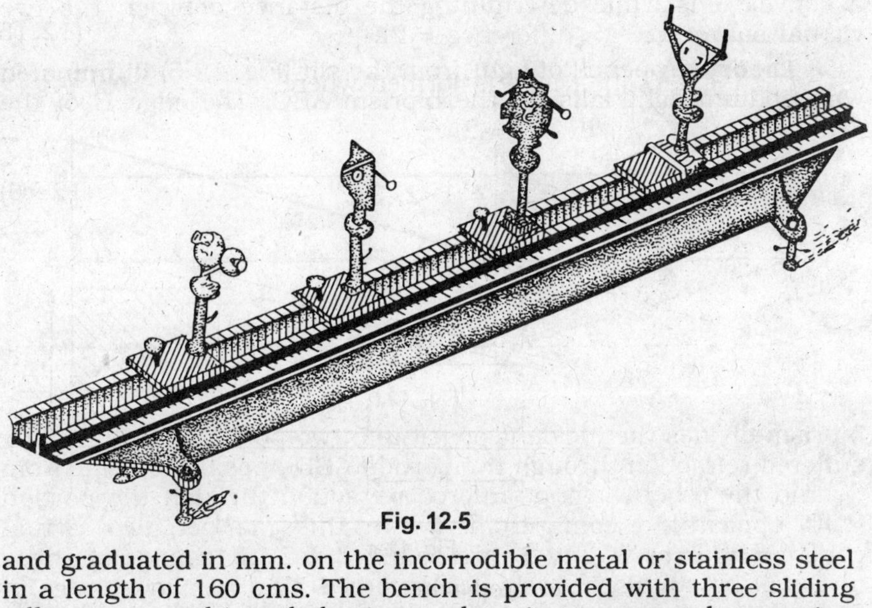

Fig. 12.5

and graduated in mm. on the incorrodible metal or stainless steel in a length of 160 cms. The bench is provided with three sliding pillars or upright each having a clamping screw and a vernier. These can be clamped in position and the reading taken on the scale correct to 0.1 mm with the help of the vernier. By means of rack and pinion arrangement the pillars can be moved verticaliy and thus adjusted to a suitable height. Suitable arrangements are provided for the following motions of each sliding pillar :

(1) Transverse motion at right angles to the bed of the optical bench—the position of the sliding pillar is indicated by an index along a scale graduated in $\frac{1}{2}$ mm.

(2) Rotation about the axis of the sliding pillar.

(3) Two sliding pillars carry an attachment for the necessary mountings of the biprism and the slit. With the help of the tangent screw, each of these attachments is capable of rotation in its own plane.

The remaining pillar carries a micrometer eye-piece. It consists of a Ramsden's eye-piece fitted with cross-wires and so mounted that it can be moved across the field of view and at right angles to the bed of the bench by means of a micrometer screw with a divided head reading to 0.001 cm.

The bench is provided with other accessories depending upon the use to which it is being put. However, for the biprism experiment a spare pillar of simpler construction without vernier and rack motion is also supplied with it. It is used for holding a convex lens while determining the distance between the two virtual sources.

Theory—A pencil of light from the slit (Fig. 12.6) illuminated with sodium light falls on the biprism ABC. The edge B of the

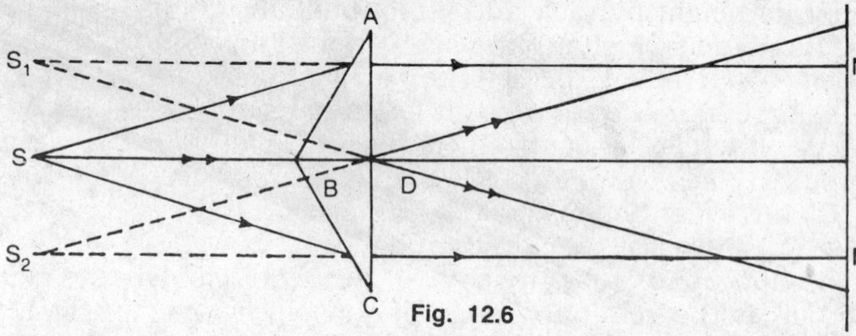

Fig. 12.6

biprism divides the incident pencil into two portions. One, having suffered refraction through the portion ABD appears to come from S_1 and the other, having suffered refraction through the portion BDC appears to come from S_2. In this manner two virtual coherent sources S_1 and S_2 are obtained, for any change of phase in S will be simultaneously accompanied by a similar change in S_1 and S_2 which will consequently be in the same phase of vibration at all times. Interference bands can be observed on the screen MN which coincides with the plane of crosswires of an eye-piece actuated by a micrometer screw. Thus a measurement of the fringe width with the help of the eye-piece, and the determination of the distances SM and S_1S_2 can be used to evaluate the wave length of sodium light for

$$\lambda = \beta \frac{2d}{D}$$

where β = fringe width

$2d = S_1S_2$

and $D = SM$

Method—The following adjustments are necessary before any measurements connected with the experiment can be made :

Level the bed of the optical bench with the help of levelling screws and a spirit level.

Focus the micrometer eye-piece on the cross-wires and mount it at a certain height on the upright or pillar meant to carry it. Turn it so that one of the cross-wires becomes vertical. Move it up to the upright carrying the slit and adjust the two to the same height. Next make the slit vertical with the help of the tangent screw which provides rotation of the slit in its own plane about an axis parallel to the length of the bed. To test for the correctness of this setting mount a convex lens of short focal length on the spare upright and get the image of the slit on the vertical cross-wire. The slit is then parallel to the vertical cross-wire.

Mount the biprism on its proper upright following the slit and adjust its height by rack and pinion motion provided for this purpose. Its plane face should be set at right angles to the length of the bed. Then make the slit narrow and illuminate it with sodium light. Use a convex lens to focus it on the slit. This will increase the intensity of light emerging out of the slit and yield better results. Place the eye on the side of the biprism away from the slit. Owing to refraction through the prism two virtual images will be seen. Move the eye across the bench when one of the images will appear to cross the edge of the biprism from one side to the other. If the refracting edge of the biprism is parallel to the slit, the images as a whole will appear to cross the edge ; otherwise, when the adjustment is faulty, either the top or the bottom of the image will cross the edge first. Adjust the biprism by rotating it in its own plane to effect this sudden transition of the full image. When this is done, the slit, the cross-wire and the edge of the biprism are all parallel.

But the line joining the slit and the edge of the biprism should be parallel to the length of the bed of the optical bench. To make this adjustment adopt the following procedure :

Look through the eye-piece when a set of interference fringes will be visible. In case they are not visible move the biprism upright at right angles to the bed of the optical bench till the fringes appear. These can be made more distinct by narrowing the slit within permissible limits. Set the cross-wire on the central bright fringe and move the eye-piece away from the biprism and note if there is any lateral shift of the fringes. If the cross-wire remains fixed on the fringe on which it was set, the above adjustment is complete. In case there is any lateral shift move the biprism at right angles to the bed of the bench till the shift ceases. This is illustrated in Fig. 12.7.

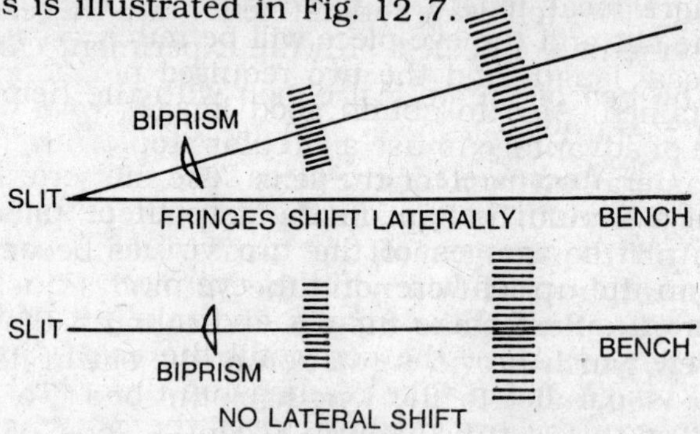

Fig. 12.7

At this stage the necessary measurements can be made.

1. *Measurement of S.M (= D).*—It is the distance between the slit and the focal plane of the eye-piece, *i.e.*, the plane of cross-wires. The difference between the readings of the position of the upright carrying the slit and that carrying the eye-piece should give the distance S.M. But there may be bench error which must be taken into consideration. To find it, adopt the following method. Remove the upright carrying the biprism taking due care that its adjustments are not disturbed. Also remove the cap carrying the cross-wire of the eye-piece. Then take a rod of glass about 25 cm. long and clamp it parallel to the length of the bed with one end just touching the end of the slit. Move the eye-piece slowly till the other end of the rod comes into focus. Read the position of the two uprights carrying the slit and the eye-piece. The difference of these two readings is, say, a. Measure the length of the rod. Let it be b. The conclusion is that the actual distance between the slit and the eye-piece is b whereas that recorded by the difference in the readings of the index of the two uprights is a. Hence the index error of the bench is $(b-a)$. Now mount the upright carrying the biprism and set the eye-piece, with its cross-wires refitted, at a convenient position so that a good set of interference fringes is observed. The distance between the slit and the eye-piece should be large, for then alone widely spaced fringes can be obtained. Having fixed the position of the eye-piece clamp the upright with the clamping screw and read its position on the bench. Also take the reading of the

position of the upright carrying the slit. The difference of the two readings gives the distance between them as measured on the bench. To this distance add the index error algebraically to get the actual distance between the slit and the place of the cross-wires. This is $SM = D$.

2. *Measurement of S_1S_2 (= 2d).*—Introduce a convex lens of short focal length between the biprism and the eye-piece. It is mounted on the spare upright provided for this purpose. This lens of short focal length is necessary for then the distance between the slit and the eye-piece will be much greater than four times its focal length and the two required positions of the lens can be obtained. Also to obtain good images with a single lens it would be of advantage to use a circular stop of variable aperture (as in a camera) in front of the lens. The slit, the biprism and the eye-piece should be kept fixed. Move the convex lens along the bench till the images of the two virtual sources are seen distinctly on the cross-wire of the eye-piece. Set the vertical cross-wires on one of these images and take the reading on the micrometer screw. Move the screw till the cross-wire comes on the second image. Again take readings on the micrometer screw. The difference of the two readings gives the distance d_1 between the two images formed on the cross-wire. Next move the lens to another position when the images are again seen distinctly through the eye-piece. As before, measure the distance apart between the two images. Let it be d_2. The distance, $2d$, between S_1 and S_2 can be calculated from the formula.

$$S_1S_2 = 2d = \sqrt{d_1\, d_2}$$

This is evident, for, if the distance of the slit from the lens be a and of the eye-piece from the lens be b, we get for the first position of the lens

$$2d/d_1 = a/b$$

For the second position of the lens

$$2d/d_2 = b/a$$

hence multiphying two equations

$$4d^2/d_1d_2 = 1$$

$$2d = \sqrt{d_1 \cdot d_2}$$

Repeat the experiment by altering the position of the upright carrying the eye-piece. However, do not alter the relative positions of the slit and the biprism, for this would result in a change in the value of S_1S_2. Then get the mean value of $2d$.

Finally, set the upright carrying the eye-piece at the mark for which the distance SM (= D) has been measured and remove the convex lens.

3. *Measurement of fringe width* β —Move the screw so that

the cross-wire falls in the middle of one of the bright fringes. Take the reading of the micrometer screw. Move the screw such that the cross-wire falls in succession on the fringes following the first and take their corresponding readings. Measure a set of twenty fringes in this manner and evaluate the fringe width as detailed below in the observations.

Repeat the experiment by moving the eye-piece to another position. Measure SM (= D) for this position and also β, this having changed with D. However, $S_1S_2 = 2d$ remains unaltered, for the slit and the biprism are not disturbed. For each set of D and β calculate λ. Finally calculate the mean wavelength.

Sources of error and precautions—(1) For the success of the experiment it is very essential that the slit should be as narrow as permissible. It should be parallel to the vertical cross-wire of the eye-piece.

(2) The biprism edge should be parallel to the slit. A faulty setting should be corrected by rotating the biprism in its own plane by a tangent screw provided for this purpose.

(3) The line joining the slit and the edge of the biprism should be paralled to the length of the bed of the optical bench. Test it by seeing that there is no lateral shift of the fringes as the eye-piece is shifted along the bench.

(4) Take into consideration the bench error while measuring SM (= D).

(5) While using the micrometer screw avoid back-lash error.

(6) Always set the cross-wire in the middle of the bright fringe. The readings can be taken with dark ones also but the setting of the cross-wire against a dark background is more difficult than its setting on a bright background. The former setting is liable to introduce some error in the measurement.

(7) While determining $2d$ use a convex lens of a short focal length. For, then alone will the distance between the slit and the cross-wires be more than four times its focal length and the two required positions of the lens can be obtained.

(8) It would be found of advantage to use a circular stop of variable aperture. When the aperture is small, the central part of the lens is used in image formation and hence the images formed are fairly good.

(9) Care must be taken to see that the axis of the lens is parallel to the bed of the optical bench.

(10) The mean separation for one fringe should be calculated as given in the observation table.

Observations—[A] Measurement of SM (= D)

Least count of the vernier = cm

S. No.	Position of the slit				Position of the eye-piece				Observed value of SM
	Main Scale reading	Vernier reading	Total reading	Mean	Main Scale reading	Vernier reading	Total reading	Mean	
1									
2				A				B	A–B
3									

Actual length of the glass rod = cm

Corresponding distance as measured by the vernier indices of the two uprights = cm

∴ Bench error = cm

∴ Corrected value of SM (= D) = cm

[B] Measurement of S_1S_2 (= $2d$).

Pitch of the screw = cm

No. of divisions on circular head = cm

Least count of the screw = cm

S. No.	Micrometer reading		2d	Micrometer reading		2d
	First image	Second image		First image	Second image	
1.						
2.						

[C] Measurement of β, the fringe width, for a distance SM (= D).

No. of fringe	Micrometer reading *a*	No. of fringe	Micrometer reading b	Separation of 10 fringes b – a
1		11		
2		12		
3		13		
4		14		
5		15		
6		16		
7		17		
8		18		
9		19		
10		20		
Mean separation for 10 fringes			=	
Mean separation for 1 fringe			=	

Similar other sets for different values of D to be recorded as above.

Calculations

Corrected value of SM=

$$S_1S_2 = 2d = \sqrt{d_1\, d_2}$$

$=$ from first set

Mean value of $2d$ $=$

Mean value of β $=$

$$\therefore \quad \lambda = \beta \frac{2d}{D}$$

$=$

$=$ cms.

Similar calculations from other sets of observations of β and corresponding values of D should be made.

$\therefore$ Mean value of the wavelength of sodium light $=$ cm

Result—The wavelength of sodium light $=$ A.U.

Criticism of the Method—The success of the experiment depends upon the preliminary adjustments. The illumination of the slit should be adequate and the bed of the bench should be in the diverging patch of light emerging from the opening in front of the source. The slit, the biprism, edge the cross-wire should be parallel to each other and the line joining the first two should be parallel to the bed of the optical bench. If this adjustment is faulty the actual width of the fringes cannot be measured because the plane of the fringes will not be at right angles to the bed but inclined to it at a certain angles whereas the cross-wire carried by the eye-piece moves at right angles to the bed of the bench. The distance thus measured is not actually β but $\beta \cos \theta$, whereas θ the inclination of the plane of the fringes with the plane normal to the length of the optical bench. Hence the measured width will always be less than β except when $\cos \theta = 1$ or $\theta = 0$ which is the case when the line joining the slit and the biprism edge is parallel to the bed of the bench. Herein lies the importance of the adjustment.

The measurement of $2d$ and β should be done accurately since these quantities are of a small order of magnitude. A slight error in the measurement of these has its full effect on the values of λ. However, the results obtained with this arrangement are not so accurate as those obtained with a diffraction grating.

Where the number of fringes obtained with the biprism is large it would be preferable to measure the fringes at intervals of 5 fringes and then compute the fringe separation for 20 or 30

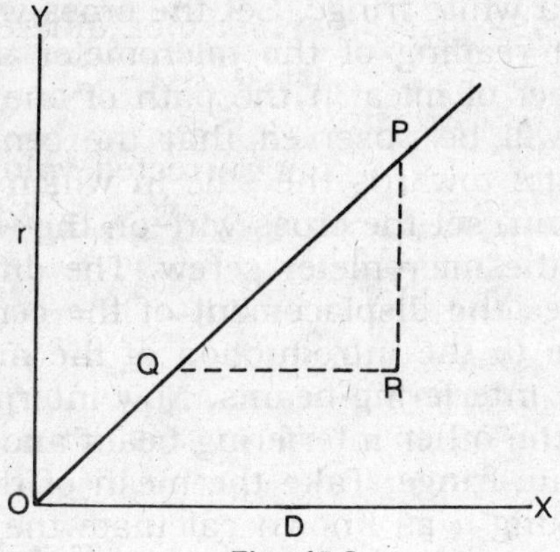

Fig. 12.8

fringes. From the mean value of this, calculate the fringe width β.

For more accurate results, measure the fringe width for different values of D keeping $2d$ constant. Plot a graph between D along the X-axis and β, the fringe width, along the Y-axis. The graph is a straight line (Fig. 12.8). Then if b be the bench error.

$$\beta_1 = \frac{(D_1 \pm b)\,\lambda}{2d} \text{ and } \beta_2 = \frac{(D_2 \pm b)\,\lambda}{2d}$$

whence $$\beta_1 - \beta_2 = \frac{\lambda}{2d}(D_1 - D_2)$$

or $$\lambda = \frac{2d.\ (\beta_1 - \beta_2)}{(D_1 - D_2)}$$

This equation eliminates the uncertainty of the bench correction and the results obtained therefrom are more accurate.

The same graph may also be used to estimate the bench error. Take the origin of the graph (o,o) produce the straight line to cut the X-axis. The value of D corresponding to this of intersection is the bench error, for at this point β is zero. This is evident from the equation

$$\beta = \frac{(D \pm b)\,\lambda}{2d} = 0$$

or $$b = \pm D$$

Exercise.—To determine the thickness of a given mica sheet by biprism method.

Make all the preliminary adjustments of the biprism as described in Expt. 12.1, and measure β, the fringe width for sodium light. Replace the sodium light by an incandescent filament lamp, the source of white light, and keep all the uprights fixed at their previous positions. On looking through the micro-meter eye-piece, the field of view will be covered with a few coloured fringes having a central white fringe. This is so because the waves of all wavelengths arrive in the same phase and the overlapping of the maxima of intensity for each wavelength results in a white fringe. Set the cross-wire on the white fringe and take the reading of the micrometer screw. Introduce the given thin sheet of mica in the path of one of the interfering beams, when it will be observed that the central white fringe of the system shifts towards the side in which the plate had been interposed. Again, set the cross-wire on the white fringe and take the reading of the micrometer serew. The difference of the two observations gives the displacement of the central white fringe of the system due to the introduction of the mica sheet in the path of one of the interfering beams. Now interpose the same sheet in the path of the other interfering beam and observe the shift of the central white fringe. Take the mean of the two results ; this gives us x. Taking μ as known calculate the thickness of the mica sheet in terms of λ from the equation.

$$t = \frac{x.\lambda}{\beta\ (\mu - 1)}$$

If the value of λ be substituted in the above, we get the thickness in cms.

It should be borne in mind that β and x are to be determined for the fixed values of D and $2d$.

Experiment 12.2

Object—To determine the wavelength of sodium light by Newton's rings method.

Apparatus—Optical arrangement for Newton's rings, travelling microscope, sodium lamp, short focus convex lens, reading lens and a spherometer.

Description—The experimental arrangement for Newton's rings consists of heavy seasoned wood stand for supporting the travelling microscope. This stand has a projection 3" square in the middle of one of its longer edges. On this projection is placed a clean old photographic plate. It is used since it provides us with nearly an optically true plane surface. Over the plate rests a biconvex spectacle lens of about 200 cm. radius of curvature. Slightly above this combination is an arrangement for mounting another thin plate of glass, the inclination of which with the vertical can be altered with a screw motion. It is set inclined at 45° to the vertical. The objective of the microscope is placed above this and adjusted to get the rings in to focus. Figure 13.9 shows the optical arrangement only.

Theory—We have already proved that

$$\lambda = \frac{D_{n+p}^2 - D_n^2}{4pR}$$

where the letters have their usual significance. Hence the determination of λ involves the measurement of radius of curvature of the surface of the lens in contact with the plate of glass and the diameters of two rings, the difference of their number being p.

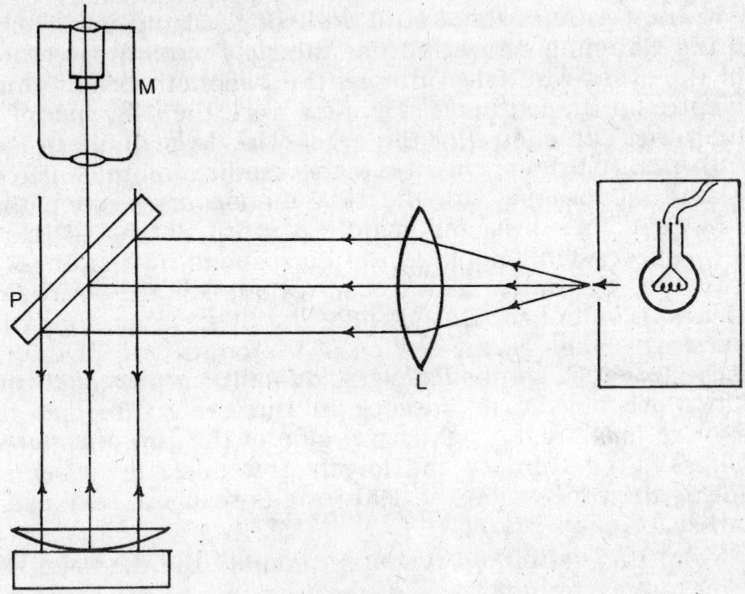

Fig. 12.9

Method—Very carefully clean the surface of the convex lens of a large focal length and place it on a clean plate of glass. Switch on the sodium lamp and wait till it attains full brilliancy. Use a short focus lens L. (Fig. 12.9) suitably mounted on a stand to allow the rays to fall on the glass plate P inclined at 45° to the vertical. The lens L should be so placed with respect to the source of light that the emergent pencil of light is approximately parallel. The glass plate P sends this pencil of light vertically downwards and thus the angle of refraction into the air film is practically zero. This is what the theory of the experiment also demands. The rays of light are partly reflected at the two surfaces bounding the air film and give rise to the interference bands in the form of concentric circular rings. The student will find no difficulty in observing the rings with naked eye but some difficulty may be experienced in getting the rings into focus in the microscope M. *These rings are localised fringes, being formed in the air film.* Therefore, for quick setting the following procedure may be adopted. Remove the convex lens and focus the microscope on a mark on the plate. On replacing the lens and adjusting the microscope over the point of contact and on slight re-focusing a set of bright rings becomes quite distinct.

Having obtained the bright rings, set microscope properly so that the point of intersection of the two perpendicular cross-wires lies on the centre of the certral dark ring. Clamp the microscope with the clamping screw and use the slow motion screw to move it till the cross-wire falls on, say, the twentieth bright ring from the centre on the left hand side. Now work the slow motion screw in the other direction till the cross-wire falls in the middle of, say, the 16th bright ring. Take the reading of the microscope. Move the microscope with the slow motion screw such that the cross-wire comes over the middle of 15th, 14th, 13th....bright ring in succession and take the corresponding readings of the microscope. Continue taking observations till the fourth or the third ring is obtained. Do not take the observations for the first two or three rings as the setting of the cross-wire in the middle of these bright rings is somewhat difficult in view of their large width. Move on the microscope in the same direction till the cross-wire falls on the right hand side of the ring measured last of all, say, the third or the fourth and take the observations. Continue the process till the 16th ring is reached, readings being recorded for each setting.

From this set of observations calculate the diameter of each ring by taking the difference of the readings of the same ring on the left hand side and on the right hand side. Then calculate the value of the squares of the diameters of these rings and evaluate the expression $D^2_{n+p} - D^2_n$ choosing such a value of p as to make use of all the observations. Then get the mean value of $D^2_{n+p} - D^2_n$ for the known value of p.

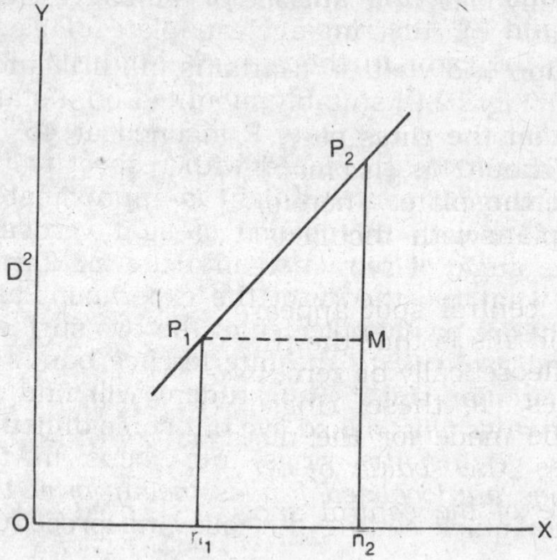

Fig. 12.10

Determine the radius of curvature R of that surface of the lens which was in contact with the plate of glass. For this determination use either a spherometer or an optical lever.

Thus, knowing the values of $D^2_{n+p} - D^2_n$ and R, calculate λ.

An alternative method of evaluating $D^2_{n+p} - D^2_n$ is to plot the squares of diameters along the Y-axis and the number of ring along the X-axis (Fig. 12.10). Take two points P_1 and P_2 on this curve which should be a straight line. From P_1 and P_2 drop perpendicular P_1n_1 and P_2n_2 on X-axis. Then $P_2n_2 = D^2_{n2}$; $P_1n_1 = D^2_{n1}$, where n_1 and n_2 are the corresponding number of the rings.

$\therefore P_2n_2 - P_1n_1 = P_2M$ for a value of $p = (n_2 - n_1)$.

$$\lambda = \frac{P_2M}{4(n_2 - n_1)\,R}$$

Sources of error and precautions—(1) The light from the sodium lamp should be rendered approximately parallel by means of a short focus lens. Its height should be so adjusted that the emergent pencil from it falls on the plate of glass almost horizontally. Since the plate is inclined at 45° to the vertical, the pencil of light is reflected vertically downwards. The angle of refraction into the air film is then practically zero. This setting is essential in view of the fact that theory demands it.

(2) The spectacle lens used for the production of Newton's rings should have surfaces having large radius of curvature. This is neessary for two reasons : (*i*) the angle θ enclosing the air film becomes very small and (*ii*) the rings observed have than a large diameter and hence the error in the measurement of their diameters is minimised.

(3) It may happen that the rings seen are not circular but possess an elliptical shape. This is mainly due to unclean surfaces of the lens and the plate. Therefore, in such a case, clean the lens and the plate with rectified spirit and then set them to get the rings.

(4) Sometimes, the central spot appears bright instead of a dark one. The reason for this is that the thickness at the point of contact which should theoretically be zero, is not zero, probably due to unclean surfaces. If these rings are circular, the measurements can still be made for the method of calculation described above involves the *value of* $D^2_{n+p} - D^2_n$ *which is independent of the nature of the central spot, whether dark or bright.*

(5) Having obtained the rings set the microscope properly so that the point of intersection of the cross-wires lies on the

centre of the central ring, for, then alone can the diameters of the rings be measured. For a faulty setting the measurement will give the length of a chord and not the diameter of the ring.

(6) When proceeding to measure the diameter of the rings use bright rings and not dark ones. The cross-wire can be easily set in the middle of bright ring because of the contrast between the cross-wire and the bright background of the ring. It is not possible to set the cross-wire correctly on a dark ring. First, two or three rings should not be measured as the accurate setting of the cross-wire in the middle of these rings is rather difficult in view of their large width.

(7) Since the slow motion screw of the vernier microscope is used in the measurements, it should be initially seen that its motion covers the diameters of the rings to be measured. Usually the working length of the screw is about 2 cm, and it may be, if the above precaution is not taken, that by the time the measurements are completed, the working length of the screw is finished. As a consequence the whole set of observations has to be neglected and a fresh one taken after setting such that the slow motion screw covers the diameters of the rings under measurement. Such a precaution is not essential in a microscope having slow motion screw for its entire working length of nearly 15 cm.

(8) The value of $D^2_{n+p} - D^2_n$ should be calculated as given in the table below.

Observations—[A] Determination of $D^2_{n+p} - D^2_n$ and *p*.

No. of Rings	Microscope reading		Diameter $(L-R)$ cm.	Diameter^2 $(L-R)^2$ cm^2.	$D^2_{n+p} - D^2_n$ cm^2.	p
	Left hand side (L) cm.	Right hand side (R) cm.				
16 15 14 . . . 10 9 . . . 5					$D^2_{16} - D^2_9$ $D^2_{15} - D^2_8$ — $D^2_{10} - D^2_3$	7
				Mean value =		

[B] Measurement of R using a spherometer

Pitch of the screw = cm

No. of divisions on circular head =

Least count of the spherometer = cm.

S. No.	Reading on Plane glass plate cm.	Convex surface of the lens cm.	b cm.	a cm.
1. 2. 3.				
		Mean =		

Calculations

$$R = \frac{a^2}{6h} + \frac{h}{2}$$

$$= \quad = \quad \text{cm.}$$

$$\therefore \quad \lambda = \frac{D_{n+p}^2 - D_n^2}{4pR}$$

$$= \quad \text{cm.}$$

From the graph (Fig. 13.10)

$$P_2M \quad = \quad \text{cm}^2.$$

and $p = n_2 - n_1 \quad =$

Also $R = \quad$ cm.

$$\therefore \quad \lambda = \frac{p_2 M}{4(n_2 - n_1) R}$$

$$= \quad \text{cm.}$$

∴ Mean wavelength of sodium light = cm.

Result—The wavelength of sodium light = cm.

Criticism of the method—The condition demanded by the theory of the experiment that $\mu=1$, $\theta = 0$ and $r = 0$ are satisfied by using a thin film of air enclosed between the plate and the lens of a large radius of curvature and by using an approximately parallel pencil of light. The rings are localised fringes and formed in the air film enclosed between the lens and the plate of glass. They are not seen directly but after refraction through the lens. For this reason the observed diameters do not correctly represent the diameters of the rings formed in the air film. To overcome this difficulty, the plate may be placed over the given lens; in this case the rings so formed are viewed after refraction through a thin parallel plate of glass. The observed diameters here represent the true diameters of the rings. However, this is inconvenient because the plate of glass cannot be kept steady over the lens during the course of measurement of diameters of the rings. It is because of this inconvenience that the usual method of placing the lens over the plate is adopted. The error arising out of this method is avoided appreciably by using a very thin lens of a large radius of curvature. The rings formed are nearly coincident with the surface of the lens and hence lie at its principal plane. Their image due to the lens lies at the second principal plane ; it is, therefore, of the same size as the object—the rings formed in the air film. Hence, in spite of the fact that the rings are viewed through the lens, their observed diameters are the same as those of the actual rings. In actual practice, however, the magnification factor only slightly differs from unity and is left out of consideration. In the case of a thick lens employed in the experiment the true diameters of the rings should be obtained by dividing the observed diameters with the magnification factor of the lens.

The radius of curvature of the lens surface in contact with the plane plate of glass may also be found out by Boys' method, details of which are given below :

Mount the lens on an optical bench. Illuminate the cross-slits in the metal plate by the light coming from a frosted lamp. Back the lens by a plane mirror. Adjust the position of the lens by moving it along the bench so that the image of the cross-slits is formed beside the cross-slits. Take the readings of the slider carrying the cross-slits and the slider carrying the lens. The difference of the two readings gives the focal length F of the lens.

To determine the radius of curvature of the lens surface B, (Fig. 12.11) the plane mirror which backed the lens is removed, and the lens is moved along the bench until a faint image of the cross-slits is formed adjacent to the cross-slits in the metal plate screen. It should be kept in view that two images will be seen by reflection in the two faces of the lens, one erect from the front surface of the lens which behaves like a convex mirror and other inverted by the concave back surface. It is the latter image that

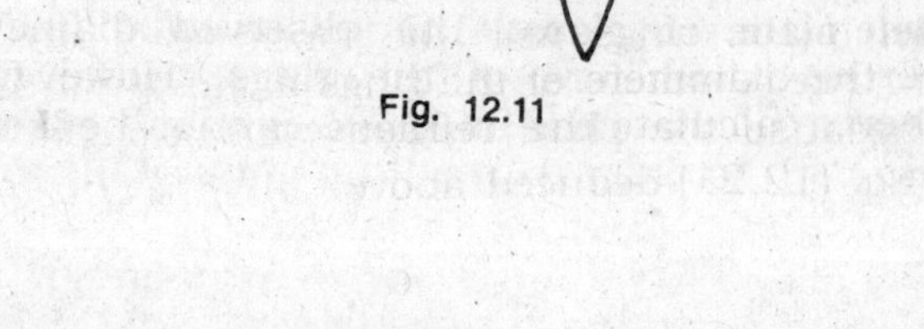

Fig. 12.11

is required for experiment. To test this image stretch a cross-wire in the upper end of the vertical slit ; in its image the cross-wire will appear at the lower end. This makes us certain that it is this image which is required to be formed adjacent to the cross-slits. When this adjustment has been made, it is apparent that the rays after refraction at A must have taken the path AB to strike the back surface normally so as to return along their incident path.

If one were to look from the side of the lens remote from P, the transmitted ray ABC will be received by the eye of the observer and it would appear to come from Q, along the path QB, which is the radius of curvature of the lens surface B. Since the rays starting from P after refraction through the lens appear to come from Q, Q is the virtual image of P formed by the lens. Hence, writing $u = OP$ and $v = OQ = R$, we have

$$\frac{1}{R} - \frac{1}{u} = -\frac{1}{F} \qquad (12.18)$$

where F has been initially determined and hence Eq. (12.18) gives the value of R, the radius of curvature of face B.

The method can be adapted for the determination of the refractive index, of a transparent liquid. For, with the film of air enclosed betwen the two surfaces

$$(D^2_{n+p} - D^2_n)\text{air} = 4\, p\, \lambda\, R \qquad (12.19)$$

If a thin film of liquid is interposed between the surface of the lens and the plate, a contraction in the diameter of the rings is observed. A similar expression for the liquid has a value given by the equation

$$(D^2_{n+p} - D^2_n)\, liq = \frac{4\, p\, \lambda\, R}{\mu} \qquad (12.20)$$

Dividing Eq. (13.18) by Eq. (13.20).

$$\mu = \frac{(D^2_{n+p} - D^2_n)\, air}{(D^2_{n+p} - D^2_n)\, liq} \qquad (12.21)$$

Exercise—To determine the refractive index of the given liquid by Newton's Rings method.

Place the lens over the plate of glass so as to enclose a thin air film. Proceed as described in the previous experiment to measure the diameters of the rings. Then calculate the value of $(D^2_{n+p} - D^2_n)$ air. Now, introduce a film of liquid between the lens and the plate, when the rings will be found to contract in diameter. Measure the diameters of the rings again and evaluate $(D^2_{n+p} - D^2_n)$ liq. Then, calculate the refractive index of the liquid with the help of Eq. (12.21) deduced above.

Experiment 12.3

Aim—To determine the thickness of a thin paper by interference of light in a wedge shaped film.

Appartus—A sodium lamp, an optically plane glass plate, a microscope slide, a travelling microscope, an arrangement having an optically plane glass plate inclined at an angle of 45°, a thin (sharply cut) piece of paper (e.g. cigarette paper).

Theory—A wedge-shaped air film is obtained by inserting a thin piece of paper between two optically plane glass plates, as shown in Fig. 13.12. The fringe at A (Fig. 12.4) is produced by the light beam reflected at A, the bottom of microscope slide (the upper surface of air film) and A′ the upper surface of the glass plate. The next fringe will be located at a point (say, at B). Where the thickness of air film increases by $\frac{\lambda}{2}$ (i.e., BC = $\frac{\lambda}{2}$). If θ is the angle of the air wedge (Fig. 12.4) then

$$\tan\theta = \frac{BC}{AB} = \frac{BC}{A'B'}$$

$$= \frac{\lambda}{2} \times \frac{1}{\beta}$$

where, β (= A′B′) is the fringe width. If t is the thickness of the paper, then from Fig. 12.4, we have

$$\tan\theta = \frac{t}{l}$$

where l is separation between the points of contact of the glass plates, and location of paper. From the above two equations, we get

$$\frac{t}{l} = \frac{\lambda}{2\beta}$$

or

$$t = \frac{l\lambda}{2\beta}$$

Thus, knowing λ and calculating l and β we can find the thickness of the paper.

Procedure—1. Level the microscope table and adjust the microscope tube in a vertical position and find the vernier constant of the horizontal scale of the microscope.

2. Arrange the appartus as shown in Fig. 12.12, with one edge of the thin plate (microscope slide) placed on the glass plate and the other on the paper. Make sure light falls on the air film normally.

3. Adjust the source and film so that straight parallel fringes are formed and are seen through the microscope.

4. Put the cross-wire at the centre of one of the straight fringes and note down the microscope verinerir on the scale in

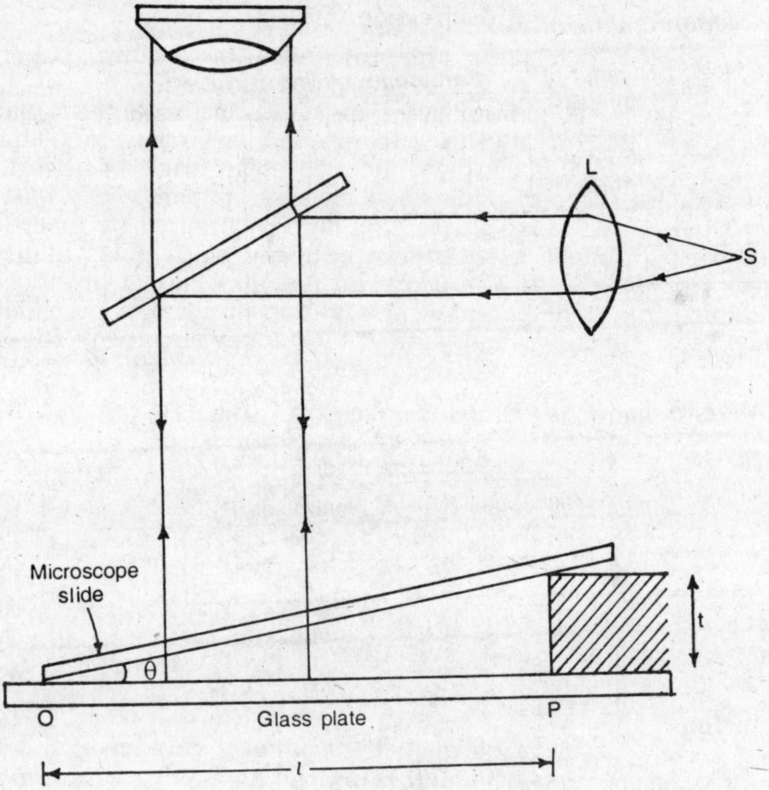

Fig. 12.12

this position. Now shift the cross-wire and start counting the number of fringes crossed. Adjust the cross wire at the centre of a particular, say, 20th fringe, and note down the microscope vernier on the scale in this position. Calculate the distance traversed in crossing 20 fringes and calculate the fringe width by dividing the distance traversed by 20.

5. Repeat the above process at least 3 times and find the mean fringe width β.

6. Now measure the distance *l*, i.e., distance from the line of contact of the microscope slide (upper thin plate) with the glass plate at O to P. For this purpose bring the cross-wire of the travelling microscope at the point of contact, note down the microscope vernier on the scale and then bring the cross-wire at the location of paper tip at P, and find the distance traversed.

Observations : Vernier constant = cm

Wavelength of sodium light λ = 5893×10^{-10} m

Measurement of β

Sr. No.	Number of fringes crossed (n)	Microscope reading		Distance traversed (d) (cm)	Fringe width $\beta = \frac{d}{n} =$ (cm)
		Initial point	Final point		
1					
2					
3					

Mean β = cm
= m

Measurement of l

Sr. No.	Microscope reading		l (cm)
	Initial point (O)	Final point (P)	
1			
2			
3			

Mean l = cm
= m

Calculation

Thickness of paper, $t = \frac{\lambda l}{2\beta} =$ m

Log error : Taking logarithm and differentiating, we get

$$\frac{\Delta t}{t} = \frac{\Delta\lambda}{\lambda} + \frac{\Delta l}{l} + \frac{\Delta\beta}{\beta}$$

In this experiment $\delta\lambda = 3A^{o} = 3 \times 10^{-10}$ m $\Delta t = 0.001$ cm $= 10^{-5}$ m.

$\Delta\beta = 0.001$ cm $= 10^{-5}$ m

Substitute the value of λ and measured values of t, l and β and calculate the maximum possible error in m.

Result—The thickness of given piece of paper is = ...m ($t \pm \Delta t$).

Precautions : 1. The glass plate and the microscope slide should be neat and clean so that both are optically plane.

2. The glass plate should be perfectly horizontal.

3. The end of paper placed between the plates should be sharply cut and not torn one.

4. The light incident on the arrangement should be incident perfectly normal.

5. The field of view should be in the central region of the air film.

ORAL QUESTIONS

INTERFERENCE OF LIGHT

1. What do you mean by interference of light?
2. What conditions must be satisfied if we want to observe interference of light in the laboratory?
3. Is it possible to observe interference with light coming from two candles? If not, Why?
4. Explain the terms—coherent sources and monochromatic light.
5. Why is it essential that monochromatic source of light should be employed in experiments on interference of light?
6. Why should the two sources vibrate in phase or with a constant phase difference?
7. On what factors does the fringe width depend? To have a measurable fringe width how should these factors be altered?
8. Is there any loss of energy at places where destructive interference occurs?

BIPRISM—DETERMINATION OF λ

1. Explain the formation of fringes with a Fresnel's biprism. What is a biprism and how is it made?
2. How do you obtain the two coherent sources here ? Do you know other methods of obtaining such sources?
3. Describe the mechanical parts of the optical bench you are using.
4. Explain the functions of the various motions which are provided with each upright.
5. What conditions are necessary for the success of the experiment?
6. How do you adjust to satisfy each of the conditions enumerated by you? What are the chief precautions that should be observed while performing the experiment?
7. Why do you not make measurements with dark fringes ?
8. How is bench correction determined and applied in the measurement of D ?

9. How do you measure 2d ? Why should a short focus lens be employed in the above measurement? What part does a circular stop of variable aperture placed before the lens play in the formation of images ?
10. Why do you suppose that the two virtual sources are formed in the plane of the slit?

NEWTON'S RINGS—DETERMINATION OF λ

1. How are these rings formed ?
2. Why are the fringes circular?
3. Why is the central ring dark ?
4. Can you arrange to get a bright spot at the centre? If so how ?
5. Sometimes you see elliptical rings or circular rings with a bright centre. Explain your observation.
6. Why do the rings get fine as we move out from the central ring ?
7. What is the difference between the rings seen by reflected light and those seen by transmitted light ?
8. Will any ring be visible if the lens is placed over a mirror instead of a plate?
9. On what factors does the diameter of a ring depend ?
10. Why should a lens of a large radius of curvature be used in the experiment ? How can you determine the wavelength of light from the determination of the diameter of the *n*th ring ? What disadvantage has this method?
11. Why do you use the equation $\lambda = \frac{D_{n+p}^2 - D_n^2}{4pR}$ for evaluating λ?
12. How are the conditions $\mu = 1$, $\theta = 0$, $r = 0$ attained in your experiment ?
13. Where are the rings formed ?
14. Suggest any easy method of focusing the microscope on the rings?
15. What are the sources of error in the experiment and how are they eliminated? Bright circular rings with a bright central spot are visible : can they be used for the measurements required for the determination of λ.
16. Should you make measurements with dark rings or bright rings ?

17. Explain how you can use this method for the determination of refractive index of a liquid.
18. Why do the rings contract on the introduction of the liquid between the lens and the plate?
19. Explain why only a few rings are seen with white light.

WEDGE SHAPED FILM

1. How can a wedge shaped film be formed?
2. Why should the plates be optically plane?
3. Why the fringes formed are known as 'fringes of equal thickness'?
4. How are the conditions $\mu = 1$, $r = 0$ obtained in this experiment?
5. Why should θ not be large?

13

Diffraction of Light

13.1. Introduction—Those phenomena which show that light, when it passes through a very narrow aperture or close to the sharp edge of an obstacle, deviates from its rectilinear path and bends round the edges, are classified as *diffraction* phenomena. In the case of sound waves which have a sufficiently large wave-length the existence of diffraction phenomena may often be noticed. But in the case of light where the wavelength is extremely small it requires a careful setting and closer examination. The obstacle or the aperture should be of dimension comparable with the wave-length of light if the phenomena of bending of light waves is to be observed.

The essential difference between interference and diffraction of light can be understood from considerations given below. In interference the resultant intensity at a point is obtained by considering the super-position of two wave-fronts coming from two coherent sources ; whereas in the diffraction phenomena the resultant intensity at a point is due to the interference between the component parts of a single wave front.

13.2. Action of a Diffraction Grating—Let us consider a number of parallel slits N, each of width *e* and separated by a distance *d* (Fig. 13.1). Consider a plane wave-front incident on these slits. Such a piece of apparatus providing a number of parallel equidistant slits in an opaque screen is called a *diffraction grating*. All the points within the slits become the sources of secondary wavelets and at each slit their resultant can be supposed to act at its middle point, S_1, S_2.....S_n, each vibration having an amplitude A $\frac{\sin \alpha}{\alpha}$, where $\alpha = \frac{\pi}{\lambda} e \sin \theta$.

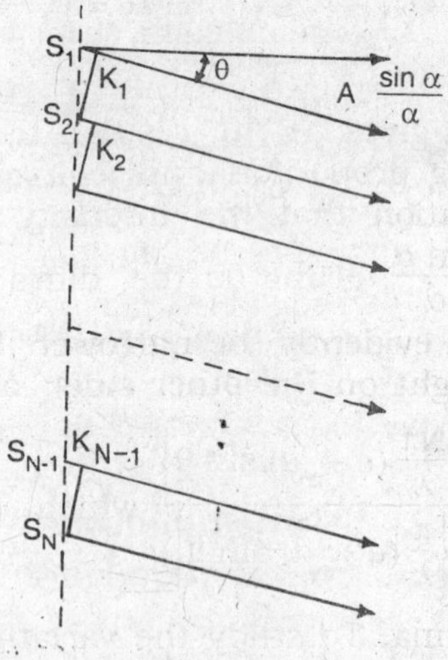

Fig. 13.1

Let us now proceed to study the effect of these

vibrations in a direction inclined at an angle θ to the direction of the incident rays. It is evident that the path difference between the two corresponding rays from S_1 and S_2 is $S_1K_1 = (e + d) \sin\theta$.

This is equivalent to a phase difference $\frac{2\pi}{\lambda}(e + d) \sin\theta$. Thus, as we pass on from one vibration to another, the phase difference goes on changing by the same amount $\frac{2\pi}{\lambda}(e + d) \sin\theta$. The problem of calculating the intensity along a particular direction θ involves the summation of a number of vibrations N, each of the same amplitude $A \frac{\sin\alpha}{\alpha}$ and having a common phase difference $\frac{2\pi}{\lambda}(e + d) \sin\theta$.

The resultant amplitude R of a number of vibrations N, each of the same amplitude a and having phase difference d increasing in A.P. is

$$R = a \frac{\sin Nd/2}{\sin d/2}$$

Here $\alpha = \frac{A \sin\alpha}{\alpha}$, $Nd = N \frac{2\pi}{\lambda}(e+d) \sin\theta$;

$$d = \frac{2\pi}{\lambda}, (e + d) \sin\theta$$

$$\therefore \qquad R = A \frac{\sin\alpha}{\alpha} \frac{\sin\left\{\frac{N\pi}{\lambda}(e + d)\sin\theta\right\}}{\sin\left\{\frac{\pi}{\lambda}(e + d)\sin\theta\right\}}$$

This gives us the resultant amplitude, the intensity of light being proportional to its square. It is clear from the above equation that the intensity depends upon two factors (1) $A \frac{\sin\alpha}{\alpha}$, giving us the diffraction pattern due to a single slit (and evidently the narrower the slit the greater the spreading of light on the other side) and (2)

$\frac{\sin\left\{\frac{N\pi}{\lambda}(e + d)\sin\theta\right\}}{\sin\left\{\frac{\pi}{\lambda}(e + d)\sin\theta\right\}}$ which gives us a series of maxima and minima. To study the variation of intensity with this factor, put

$\frac{\pi}{\lambda}$ (e+d) sin θ = x ; then this factor becomes $\frac{\sin Nx}{\sin x}$.

Minima of intensity.—The minima of intensity are given by

$$\sin Nx = 0 \text{ but } \sin x \neq 0$$

or

$$Nx = \pm n\pi$$

$$N \frac{\pi}{\lambda} (e+d) \sin \theta = \pm n\pi$$

$$\sin \theta = \pm \frac{n\lambda}{N(e+d)} \quad (13.1)$$

n having integral values with the exception of 0, N, 2N, 3N... because in that case the same condition will make sin x = 0. The positive and the negative signs indicate the positions of the minima on the right and on the left of the central maximum.

From the above equation it is clear that minima occur along directions for which the path difference between the extreme rays is $n\lambda$, n having integral values with the exception of 0, N, 2N,........

Principal Maxima.—The condition for principal maxima is

$$\sin x = 0$$

$$\therefore \quad \frac{\pi}{\lambda} (e+d) \sin \theta = \pm n\pi$$

$$\therefore \quad \sin \theta = \pm \frac{n\lambda}{(e+d)} \quad (13.2)$$

n having all integral values. For $n = 0$, sin θ = 0 or θ = 0, the angle of diffraction is zero. In this direction all the rays arrive in the same phase of vibration and so we get a *bright central image.* This image is termed the *image of zero order.* The diffracted images corresponding to values of $n = 1$ and $n = 2$.....are called respectively the *principal maxima* of the first order and that of the second order, etc., Putting $n = 1$ gives us $\sin \theta = + \frac{\lambda}{(e+d)}$. It implies that $S_1K_1 = \lambda$ or $S_2K_2 = \lambda$. Thus in that direction we get the principal maximum of the first order for which the path difference between any two consecutive corresponding rays is equal to λ. Similarly, we get the second order image along the direction for which

$$\sin \theta = \pm \frac{2\lambda}{(e+d)}$$

It is clear from Eq. (14.2) that for fixed values of n and $(e + d)$, the angle of diffraction θ varies with λ. If heterogeneous light be made to fall on the slit, the diffracted images of the given order n for the constituent wavelength will be formed in different positions. The grating thus produces a spectrum which can be observed with a properly adjusted telescope. However, along the direction $\theta = o$, all the waves arrive in the same phase and hence the direct image has the same colour as the source. No dispersion is produced along this direction. For images of other order, the grating produces dispersion and gives rise to spectra of the incident light. The significance of positive and negative signs in the equation above lies in this that there are two spectra of the same order lying equidistant on either side of the zero order image.

Secondary Maxima.—The position of the secondary maxima is given by the equation obtained by differentiating $\frac{\sin Nx}{\sin x}$ and equating it to zero. Thus,

$$\frac{d}{dx}(\sin Nx \operatorname{cosec} x) = o$$

or

$$N \tan x = \tan Nx$$

The roots of this equation other than $x = \pm n\pi$ give set of maxima called secondary maxima. As regards their number between any two principal maxima, let us consider the equation for the minima deduced before,

$$\sin \theta = \frac{n\lambda}{N(e+d)}$$

$n = 0$ and $n = N$ change the condition for that of the principal maxima hence these values of n have to be excluded. That is, n can have values 1, 2, 3...$(N - 1)$ giving $(N-1)$ minima. The number of intervening maxima is then $(N - 2)$. It can be shown that the larger the value of N, the weaker are these secondary maxima. They are, therefore, not visible in the grating spectrum and are of no practical importance.

13.3. Diffraction Gratings—A diffraction grating consists of a large number of slits arranged very close together at regular intervals. The distance between the centres of the two successive slits, *i.e.*, the quantity $(e + d)$ in the discussion above, is termed the *grating element*. Gratings are either of the transmission type or of the reflection type. They are produced by ruling a large number of fine parallel lines at regular spacing on a piece of plate glass or speculum metal by means of a sharp diamond point of a dividing engine. Very good gratings were made on glass by Nobert and on speculum metal by Rowland. The number of lines ruled varies from 12000 to 30000 per inch and the ruled surface from 2" to 6." Since it requires a great amount of labour to produce a grating, copies of these are made available in the form of celluloid casting for usual work. These are called replica gratings. The transmission gratings are essentially plane whereas reflection gratings may be plane or concave. The latter is successfully used in the investigation of the ultra-violet region as it produces a sharp image of the spectrum directly by reflection. There is no question of any absorption here as in the case of transmission gratings which employ lenses for focusing the spectrum.

13.4. Light incident obliquely—minimum deviation—The most important adjustment in the diffraction grating experiment for the determination of λ is to set the grating perpendicular to the path of the rays from the collimator if the grating is not normal to the incident rays but the latter fall upon it at an angle i with the normal path retardation, as shown it Fig. 13.2, is

$$K_1C+CK_2 = (e+d)\ (\sin i + \sin \theta).$$

The position of the nth order spectrum in this case is given by

$$(e + d)\ (\sin i + \sin \theta) = n\lambda$$

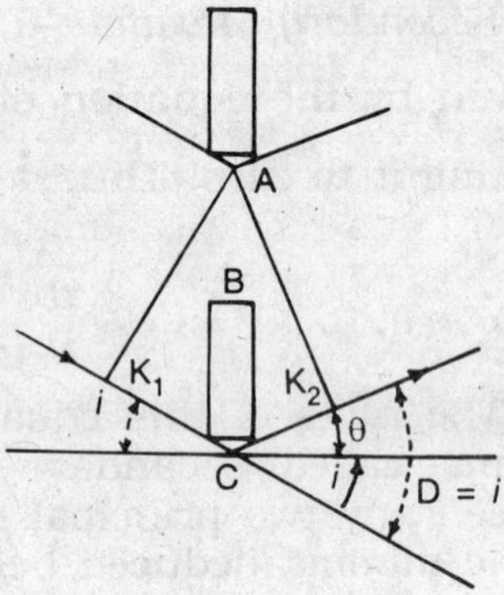

Fig. 13.2

Now, let a small change in the angle of incidence produce a change $\delta\theta$ in the angle of diffraction. On differentiating the above equation

$$\cos i\ \delta i + \cos \theta \delta\theta = 0$$

$$\delta\theta = -\cos i \sec \theta\ \delta i$$

If $i = 0$, *i.e.*, the grating is normal to the incident rays $\cos i = 1$ and $\delta\theta = \sec \theta \delta i$. Since the angle of diffraction is usually large for a grating containing a large number of lines and for a higher order image, small error in θ will cause a greater error in sec θ, the latter being much greater than unity. Hence, for the position of the normal setting of the grating with respect to the incident rays, a small deviation from this position causes a correspondingly large error in the angle of diffraction. To overcome this difficulty the grating may be used in the position of minimum deviation.

The value of the deviation is given by $D = i + \theta$. We have the following condition for its minimum value ;

$$\delta D = \delta i + \delta\theta = 0$$

But $\cos i\,\delta i + \cos\theta\,\delta\theta = 0$

whence $\cos i = \cos\theta$

or $i = \theta$

for each is less than 90°. Hence the angle of minimum deviation $D = i + \theta = 2i = 2\theta$. We have then the condition for the nth order spectrum modified to

$$2\,(e + d)\sin\frac{1}{2}D = n\lambda \qquad (13.3)$$

Using the grating in the position is more advantageous as the definition of the spectrum is considerably improved and a slight error of adjustment would not alter the minimum angle of diffraction when measured as half the angle of deviation.

13.5. Dispersive power of grating—If λ_1 and λ_2 be two close wavelengths and θ_1 and θ_2 their corresponding angles of diffraction for the nth order spectrum obtained with a grating, $(\theta_1 - \theta_2)$ is a measure for the angular separation for a wavelength difference $(\lambda_1 - \lambda_2)$. The expression $\dfrac{\theta_1 - \theta_2}{\lambda_1 - \lambda_2}$ is taken to be a measure of the dispersion produced by the grating in the region λ_1 to λ_2. When, however, these wavelengths are very nearly the same, $\dfrac{\theta_1 - \theta_2}{\lambda_1 - \lambda_2} = \dfrac{d\theta}{d\lambda}$. This is called the *dispersive power* of the grating. It is the *change in the angle of diffraction corresponding to a unit change in wave length.* We can deduce the expression for this with the help of the equation for the nth order spectrum, which is

$$(e + d)\sin\theta = n\lambda.$$

$$\therefore \quad (e + d)\cos\theta\, d\theta = n d\lambda.$$

$$\therefore \quad \frac{d\theta}{d\lambda} = \frac{n}{(e + d)\cos\theta} \qquad (13.4)$$

Hence, for a given order spectrum, the dispersive power varies inversely as the grating element. In other words, it depends on the closeness of the rulings on the grating. For a given grating it increases with the order of the spectrum.

13.6. Width of a Principal maximum—The angular separation between the first two minima lying on either side of the principal maximum of the nth order, is called the width of the principal maximum in that order. If θ_n be the angle of diffraction of wave-length λ in the nth order, it is given by

$$(e + d)\sin\theta_n = n\lambda. \qquad (13.5)$$

Here, $n\lambda$ is evidently the path difference between any two consecutive corresponding points on the grating and as such the path difference between the extreme rays of the grating of N lines is $Nn\lambda$. As we pass on to either side of the direction θ_n, the direction giving the principal maximum of the nth order, we come to the directions $(\theta_n \pm d\theta_n)$ for which the path difference between the extreme rays has changed by λ *i.e.*, from $nN\lambda$ to $Nn\lambda \pm \lambda$. It is along these directions that we get the minima surrounding this principal maximum. Thus, $2d\theta_n$ is the width of the principal maximum and half of this *i.e.*, $d\theta_n$ is usually spoken of as half width of the principal maximum. Writing the condition for adjacent minima to the principal maximum, we have

$$N(e + d)\sin(\theta_n \pm d\theta_n) = Nn\lambda \pm \lambda \tag{13.6}$$

Eliminating $(e + d)$ between Eqs. (13.5) and (13.6) and expanding $\sin(\theta_n + d\theta_n)$, taking into consideration that $d\theta_n$ is small, we have

$$d\theta_n = \frac{1}{Nn \cot \theta_n}$$

Thus, the principal maxima are very sharp in view of that fact that N is very large and so also θ_n for the order n.

13.7. Resolving power of a Grating—As shown in the previous article the principal maxima have a finite width and are never geometrical line. As such the distribution of intensity of light of nth principal maximum of the wave length λ, is somewhat as shown by the curve P in Fig. 13.3. It is maximum in the middle of the line and sharply fades away on both sides.

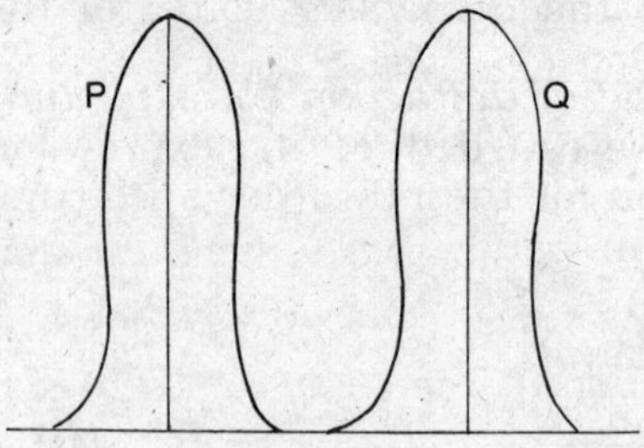

Fig. 13.3

Now, let the incident light consist of two wave lengths λ and λ_1. Their principal maxima for the order n will be separated as shown in Fig. 13.3, P being that of λ and Q of λ_1. However, if λ_1 happens to lie close to λ, this angular separation between P and Q will become less and for one particular value of λ_1 the

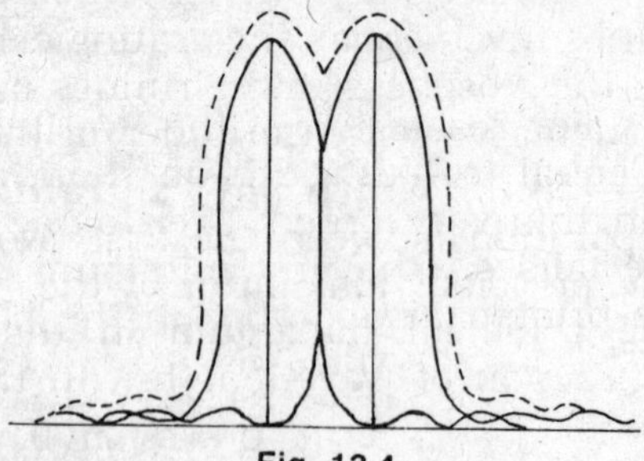
Fig. 13.4

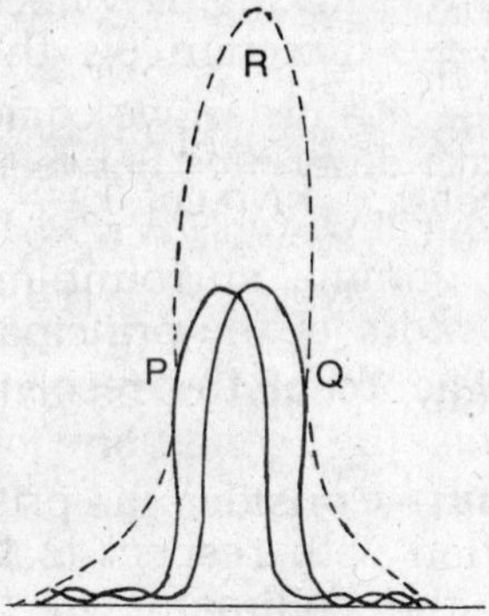

Fig. 13.5

central maximum of each pattern lies over the first minimum of the other and we have the condition represented in Fig. 13.4; the resultant illumination is obtained by summing up the two curves. It is represented by the dotted curve and reveals that there is a distinct dip in the middle. It is this finite fall in the middle which permits us to see the two maxima corresponding to two near wavelengths close together. If λ_1 has a value still closer to λ, there is overlapping of the two principal maxima. The resultant curve, R in Fig. 13.5 indicates that the eye fails to distinguish the two wave-lengths. Hence for the two wavelengths to be resolved, the degree of closeness which their principal maxima can have, is that the principal maximum of one should lie on the first minimum of the other. This is regarded as the criterion which establishes the *limit of resolution*. For, if the maxima lie at this distance or further apart, they can be seen separately, if they lie close together they will be seen as one image. We are indebted to Lord Rayleigh for this clarifying discussion.

Let us now proceed to find an expression for the resolving power of the grating. If λ and λ_1 $(= \lambda + d\lambda)$ are the two wave-lengths, the angular dispersion between them in the nth order spectrum is given by

$$\frac{d\theta_n}{d\lambda} = \frac{n}{(e + d) \cos \theta_n}$$

$$\therefore \qquad d\theta_n = \frac{n.d\lambda}{(e + d) \cos \theta_n} \qquad (13.8)$$

For resolution of the two lines this angular dispersion must be equal to half-width of the principal maximum of wave-length λ in this very order. Then alone the principal maximum of one line falls on the first minimum of the other. Now, half-width of the principal maximum in the nth order is given by

$$d\theta_n = \frac{1}{Nn \cot \theta_n} \qquad (13.9)$$

Equating (13.8) and (13.9) we have the desired condition for the resolution of the two lines.

$$\frac{n\,d\lambda}{(e+d)\cos\theta_n} = \frac{1}{Nn\cot\theta_n}$$

But $$\frac{n\lambda}{(e+d)} = \sin\theta_n, \quad \therefore \quad \frac{d\lambda}{\lambda\cot\theta_n} = \frac{1}{Nn\cot\theta_n}$$

or $$\frac{\lambda}{d\lambda} = Nn \tag{13.10}$$

The product Nn is taken to be a measure of the resolving power of a grating.

13.8. Resolving Power of a prism—Consider a prism placed in the position of minimum deviation with respect to the incident parallel light *ab* coming from the collimator of the spectrometer. M_0N_0 represents the incident wave-front just striking the prism at N_0. Dispersion will take place as light traverses the prism. N_0D is the path followed by a ray of wave-length λ. On emergence from the prism we have a parallel pencil of light AM, DN. The plane wave-front MN is received by the telescope objective and the image of the slit corresponding to this wave-length is formed in its focal plane. The first minimum of the image will occur along a direction for which the wave-front is, say, represented by ME such that NE=λ. If in addition ME is

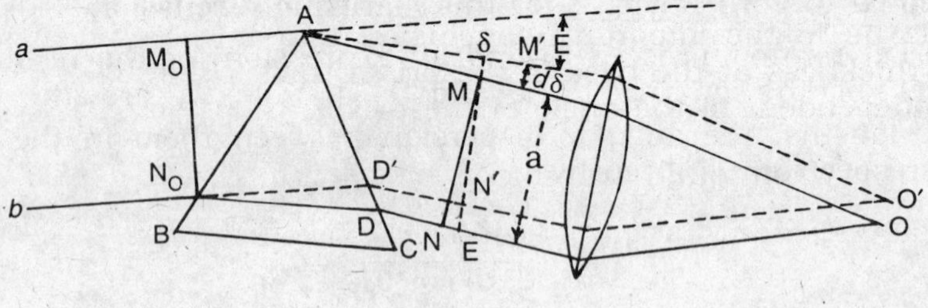

Fig. 13.6

the actual wave-front for another wave-length $\lambda_1 = \lambda + d\lambda$, the two wave-lengths will be just resolved since then the maximum of λ will fall over the minimum of λ_1. The direction of the ray for the wave-length λ is represented by N_0D' in the prism and $D'N'$ parallel to AM' outside the prism.

For the sake of simplicity let us consider the paths M_0AM and M_0AM' to be entirely in air. Since AM and AM' are equal to within a quantity small compared with λ, we may take the paths M_0AM and M_0AM' to be almost equal. Hence, the corresponding path lengths N_0D (in glass)+DN (in air) and N_0D' (in glass)+$D'N'$ (in air) are also equal. If μ is the refractive index of the material of the prism, for the wave-length λ and $\mu + d\mu$ for $\lambda + d\lambda$, then

$$\mu N_0 D + DN = (\mu + d\mu)\ N_0 D' + D'N'$$

But $D'N' = DE = DN + NE, N_0 D' = N_0 D$ approximately.

$$\mu N_0 D + DN = (\mu + d\mu)\ N_0\ D + DN + NE$$

$$\therefore \quad 0 = N_0 D.\ d\mu + NE$$

But $NE = \lambda$, for this is the condition of resolution of the two lines,

$$\therefore \quad \lambda = -N_0 D.\ d\mu$$

where $N_0 D = t$, practically the thickness of the base of the prism. Dividing by $d\lambda$,

$$\frac{\lambda}{d\lambda} = -t.\ \frac{d\mu}{d\lambda} \tag{13.10a}$$

The negative sign will eventually vanish since $\frac{d\mu}{d\lambda}$ is a negative quantity as an increase in wave-length results in a decrease in the value of the corresponding refractive index. The equation gives the theoretical resolving power of a prism and leads us to the important conclusion that it is proportional to the thickness at the base of the prism traversed by the ray and is independent of the angle of the prism.

It is often found convenient to evaluate $\frac{d\mu}{d\lambda}$ in terms of other constants. From Cauchy's formula connecting μ and λ we have

$$\mu = A + \frac{B}{\lambda^2}$$

$$\therefore \quad \frac{d\mu}{d\lambda} = -\frac{2\,B\,t}{\lambda^3}$$

$$\therefore \quad \frac{\lambda}{d\lambda} = \frac{2\,B\,t}{\lambda^3} \tag{13.11}$$

The dispersion D of the prism is the ratio of the change in the angle of deviation with wave-length. Thus, if $d\delta$ is the change in the angle of deviation with a change $d\lambda$ in wave-length.

$$D = \frac{d\delta}{d\lambda}$$

Now, in Fig. 13.6, $d\delta = \angle M'AM = \angle NME = \lambda/a$, if a is the width of the emergent beam,

$$D = \frac{1}{a}\frac{\lambda}{d\lambda} = -\frac{t.}{a}\frac{d\mu}{d\lambda} \tag{13.12}$$

Thus, the dispersion varies inversely as the width of the beam. In the position of minimum deviation, the width of the beam is a maximum and the lines are consequently well defined and sharp. Using the prism in any other position results in greater dispersion since a is reduced. But this greater dispersion is of little practical use as the lines are then broadened and do not remain well defined.

13.9. Prism and grating spectra—From Cauchy's formula we know the relation between wavelength of light and the corresponding refractive index for the material of the prism. According to this relation,

$$\mu = A + \frac{B}{\lambda^2}$$

whence

$$\frac{d\mu}{d\lambda} = -\frac{2B}{\lambda^2}$$

This shows that it is not a constant but increases with diminishing wave-length. So the prismatic spectra are drawn out at the violet end. They are not normal ; a normal spectrum being one for which $\frac{d\mu}{d\lambda}$ is a constant. On account of this irrationality of dispersion and the dependence on the material of the prism, the prismatic spectra, obtained by two experimenters working at different places, cannot be compared.

On the other hand, for a grating $\frac{d\theta}{d\lambda} = \frac{n}{(e+d)\cos\theta}$ the grating dispersion is independent of the material on which the lines are ruled and depends on n, $(e+d)$ and θ. For a given grating in a given order it depends only on θ ; $\frac{d\theta}{d\lambda}$ being large if $\cos\theta$ is small or θ is large. We know from the grating equation, $(e+d)\sin\theta = n\lambda$, that θ is large for longer wavelengths. Hence $\frac{d\theta}{d\lambda}$ increases with wavelength.

The grating spectrum is consequently drawn out at the red end in contrast to the prismatic spectrum which is drawn out at the violet end. Like the prismatic spectra it is also not normal. However, in a concave grating it is normal when $\theta = 0$, for then $d\theta/d\lambda$ becomes constant and equal to $\frac{n}{(e+d)}$. So grating spectra obtained by two experimenters at different places can be compared with each other the only difference being that their lengths are in the inverse ratio of the grating elements.

In the prismatic spectra deviation for the violet ray is the greatest and for the red, the least. Just reverse is the case in grating spectra. Hence when placed side by side and examined it will be noticed that the order of the colours in the two spectra are reversed. Prismatic spectrum is much brighter than diffraction spectrum because in the former, the light is concentrated only in one spectrum whereas in the latter it is thrown into spectra of different orders.

So far as the resolution of lines in the spectrum is concerned it is much greater in the grating spectrum than in the prismatic spectrum. Prismatic spectrum is free from spurious lines or 'ghosts' which are observed in a grating spectrum. Another point of difference between the two is that the spectral lines in the prismatic spectra are curved, the convex side of each line being turned towards the red end of the spectrum.

13.10. Resolving Power of a Telescope—Let O be a monochromatic point source of light emitting light of wavelength λ. Let G be the objective of a telescope (Fig. 13.7). In usual practice the telescope objective alone limits the beam of light, entering it but for the sake of clarity of discussion, let us suppose a rectangular aperture AA kept in front of it. The incident wave-front MN is brought to a focus at P, the objective, and is seen magnified with the help of the eye-piece E. The image P is not a point image but its intensity curve is of the form *p*. Consider another point object O′ lying to the right of O. Its image will also be formed in the focal plane but displaced to the left of P. However, as O′ approaches O, its image P′ *in the focal plane approaches* P and for one position of O, P′ will be so near to P that they will be just resolved. This condition is obtained

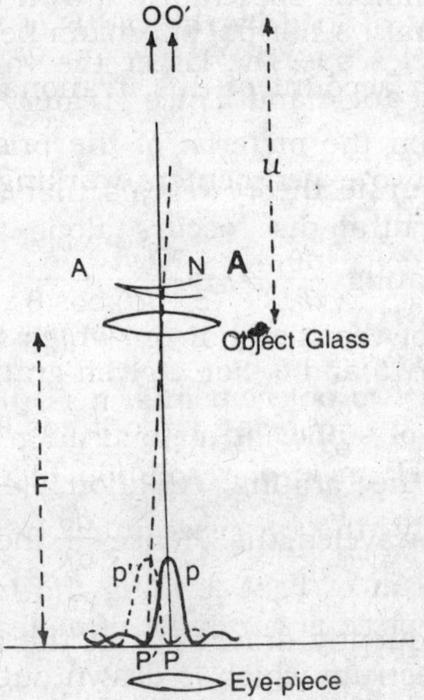

Fig. 13.7

when the intensity patterns P′ and p are so close that the maximum of one falls over the first minimum of the other. It is this criterion which establishes the limit of resolution ; if O′ approaches O any further, the images overlap and cannot be distinguished as two.

As the aperture AA is rectangular, the first minimum of the patten p occurs when the distance NE = λ. Hence, for the maximum of the pattern P′ to fall on the first minimum of p, ME, the diffracted wave-front for O must be the actual wave-front from the object O. The angle between the two wave-fronts is $\frac{NE}{MN} = \frac{\lambda}{d}$ where d is the width of the aperture. Since the rays are normal to the wave-front, the angle which the incident rays from O and O′ subtend at the objective is also λ/d. Incidentally, this is also the angle subtended by images at the objective. If F be the focal length of the objective,

$$\frac{OO'}{u} = \frac{\lambda}{d} = \frac{PP'}{F} \qquad (13.13)$$

The relation above reveals that the theoretical least angle of resolution is the wavelength of light divided by the width of the aperture admitting light into the telescope. The larger the value of d the smaller is the value of this angle and hence greater the resolving power of the telescope.

If the aperture is circular, accurate theory reveals that the first minimum surrounding the central disc occurs along the direction for which NE=1.22λ. As such,

$$\frac{OO'}{u} = 1.22\,\frac{\lambda}{d}$$

Supposing a telescope resolves two object points, it is then also necessary to use an eye-piece of sufficient magnification to see the images separately. The least angle of resolution of the normal human eye is $1\frac{1'}{2}$. Therefore, the eye-piece must be of sufficient power to separate the beam of light forming the two images so that these leave the eye-piece at an angle of at least $1\frac{1'}{2}$.

Experiment 13.1

Object— To determine the wavelength of sodium light with a diffraction grating.

Apparatus—A spectrometer, a diffraction grating of known grating element, sodium lamp and a reading lens.

Theory—If light from a narrow slit, rendered parallel by a lens, is made to fall on the grating and another lens employed to converge the rays issuing from the grating, we obtain an image of the slit along the same direction as the incident rays. This is the image of the zero order and has the same colour as the source of light. Surrounding this image on either side are the images of the first order, the second order and so on. For heterogeneous light we have the first order spectrum, the second order spectrum, etc., on either side of the central image. If θ be the angle of diffraction in the order n for any wavelength λ then

$$\sin\theta = \frac{n\lambda}{(e+d)}$$

$$\therefore \qquad \lambda = \frac{(e+d)\sin\theta}{n} \qquad (13.14)$$

Thus, the necessary measurements with the grating for the determination of λ are the angle of diffraction θ and the order of the spectrum n.

Method—Make the preliminary mechanical and optical adjustments of the spectrometer. Then proceed as follows :

Set the telescope and the collimator in a line so that the direct image of the slit falls on the intersection of the cross-wires of the eye-piece of the telescope and lies in the centre of the field of view. Note the reading for this position of the telescope on both the verniers. Let it be ϕ on one of them. Next mount the grating with the help of its proper clamping device so that its ruled surface lies at the centre of the turn-table and also perpendicular to the line joining two of the levelling screws Q and R. Now turn the telescope so that the reading of the vernier becomes $\phi + 90^\circ$ and clamp it. It means that the axes of the collimator and the telescope are now at right angles to each other. Turn the grating with the help of the turn-table so that the reflected image from its surface falls on the inter-section of the cross-wires of the eye-piece of the telescope. Adjust the levelling screws Q and R, if necessary, to get the reflected image in the centre of the field of view of the telescope. In case the preliminary adjustments of the spectrometer are correct it will be noticed that when the telescope is rotated through 180° from this position and the turn-table carrying the grating also rotated to send the reflected image of the slit on to the intersection of the cross-wires, the image occupies the centre of the field of view.

So when the collimator and the telescope axes are at right angles to each other and the grating has been adjusted to send the reflected image on to the intersection of the cross-wires, the incident rays on the grating are inclined at 45° to the plane of the grating. Take the reading on the verniers. From this position rotate the turn-table through 45° or 135°, as the case may be, *so that the grating surface is normal to the incident rays and faces the telescope.*

Now it remains for us to see that the rulings of the grating are parallel to the main axis of the instrument. When the axis of the telescope is parallel to that of the collimator, an image of the slit will be seen lying with its centre on the intersection of the cross-wires of the telescope. Turn the telescope about the axis of the instrument until the diffracted image of the first order is visible on the cross-wires of the telescope. If necessary, adjust the levelling screw P so as to get the middle of the diffracted image on the intersection of the cross-wires of the telescope. When this adjustment has been carefully attended to the centre of all diffracted images on either side of the direct image will be found to lie on the intersection of the cross-wires of the telescope. Then the rulings of the grating are parallel to the main axis of the instrument.

Finally, the slit should be parallel to the rulings of the grating. To carry out this setting, adjust the telescope on a diffracted image and rotate the slit in its own plane till the image observed becomes well defined and bright. Due care is necessary not to disturb the collimating arrangement, *i.e.*, the distance between the slit and the collimating lens should remain unaltered as the slit is rotated.

For all these adjustments light from an incandescent filament lamp may be used. Illuminate the slit with sodium light. Turn the telescope to get the first order image on the left to the intersection of the cross-wires, final setting being done with the help of the tangent screw or the telescope. Read the two verniers. Next turn the telescope to get the first order image to the right on the intersection of the cross-wires and setting as before, read the two verniers. The difference in the readings of the same vernier for the two settings gives twice the angle of diffraction. Calculate its mean value. Use this and the known grating element for the calculation of the wavelength of sodium light.

Repeat the experiment for the second order. Finally, calculate the mean wavelength by taking the average of those obtained with the first order image and the second order image.

Sources of error and precautions—(1) The mechanical adjustments of the telescope should be correct (Prec. 1 Expt. 11.1).

(2) The optical adjustment of the spectrometer must be made correctly.

(3) The slit used should be as narrow as permissible.

(4) In handling the grating do not touch the faces of the glass; hold it between the thumb and the fingers by the edges.

(5) The grating should be so set that its ruled surface faces the telescope and is normal to the incident pencil. In this setting when the telescope and the collimator axes are perpendicular, the reflected image brought on to the intersection of the cross-wires should be one obtained by reflection from the ruled surface. Care must be taken to select this, for two or more images are sometimes observed. A mistake made here will be evident when the grating has been rotated through 45° or 135° for, now it will be observed that the grating does not lie normal to the incident pencil coming from the collimator. This faulty adjustment is apparent to the eye.

The turn-table should be properly rotated through 45° or 135° so that the grating faces the telescope. It is not to face the collimator.

(6) At no time while taking observations should both the table and the telescope be unclamped, *i.e.*, while reading the movement of the table, the telescope must be clamped and vice versa.

(7) Read both the verniers; this eliminates any error due to non-coincidence of the centre of the divided circle with the axis of rotation of the telescope or the table.

(8) Since the two lines of sodium will be resolved with the grating, the telescope should be set on the same line in all the observations.

Observations—(A) Measurements of the angle of diffraction.

Value of one main scale division = min.

Total number of vernier divisions =

Least count = sec.

Order of Spectrum	Vernier	Spectrum left of direct image			Spectrum right of direct image			Difference of the two readings of the same vernier θ	Mean value of θ
		Main scale reading	Vernier reading	Total reading.	Main scale reading	Vernier reading	Total reading.		
1	V_1								
	V_2								
2	V_3								
	V_4								

(B) Determination of the grating element.

No. of lines = per inch

∴ Grating element $e + d$ =

Calculations— $\lambda = \dfrac{(e+d)\sin\theta}{n}$

= A°

Similar calculations for the second order spectrum.

∴ Mean value of λ = A°.

Result—The mean wavelength of sodium light = A°.

Criticism of the method—Of the various methods for the determination of the wavelength of light, this one based on the diffraction grating is by far the most accurate. When the experiment is performed and the necessary precautions taken, the value of the wavelength of light obtained by this method is very satisfactory. One of the essential precautions for the success of the experiment is to set the grating normal to the incident rays, for a small variation δi in the angle of incidence causes correspondingly large error in the angle of diffraction. In using the method, therefore, one may find that the angles of diffraction measured to left and right are not exactly equal. This evidently shows that the grating has not been set exactly normal to the incident rays. When this occurs and the difference between the angles of diffraction is small, rotate the grating through half the difference and the new readings will be found approximately equal to allow the method to be used without significant error.

When the difference is large, the method of two deviations should be used. (Exercise 2).

The wavelength of light is given by the expression.

$$\lambda = \frac{(e+d)}{n} \cos\phi \sin\frac{1}{2}(U_1 + U_2)$$

where $\phi = i + \frac{1}{2}(U_1 - U_2)$

It is evident from this expression that if U_1 and U_2 differ by a small amount so as to keep the error of determination below a certain limit, cos ϕ can be taken as unity and the simple equation

$$\lambda = \frac{(e+d)}{n} \sin\frac{1}{2}(U_1 + U_2)$$

employed[1].

Reese[2] has pointed out that if U_1 and U_2 are approximately 20° they may differ by 3/4° before the error exceeds 1 part in 1000. He, therefore, suggests that the energy spent in setting the grating normal to the incident pencil achieves nothing.

Hence, the grating may either be used inclined at a small angle to the incident pencil and λ determined by measuring two deviations or may be employed in the position of minimum deviation. Though the latter method improves the definition of the images and considerably minimises the error in the measurement of the angle of diffraction, it is rather very laborious and "can claim only the interest of any optical curiosity"[1]

The grating may conveniently be employed in the constant deviation method[3] for determining λ. The method was reported by Einsporn in 1927.

It has been mentioned in the method that when properly mounted the grating surface should be adjusted to be normal to the incident pencil and that it should face the telescope. It will be clear from the following discussion that it is necessary to set the surface of the grating facing the telescope and not the collimator.

Let AB be the grating surface facing the collimator and let IP be a representative ray of the parallel pencil of light falling on it. It suffers diffraction at P. The angle of diffraction, θ_g in glass, is given by the reaction.

$$(e+d) \sin\theta_g = n\lambda_g \qquad (13.15)$$

At Q, the ray PQ suffers refraction. We then have

1. Candler—Modern Interferometers—Higher Publication, 1951.
2. Reese—Jour. Ot. Soc. Amer 22, 496, 1932
3. Prakash & Singh—Optics.

$$\frac{\sin\theta}{\sin\theta_g} = \mu = \frac{v_a}{v_g} = \frac{\lambda}{\lambda_g}$$

$\therefore$ $$\text{Sin } \theta_g = \frac{\lambda_g}{\lambda} \sin\theta$$

whence, from (14.15)

$$(e+d)\sin\theta \frac{\lambda_g}{\lambda} = n\lambda_g$$

or $$(e+d)\sin\theta = n\lambda$$

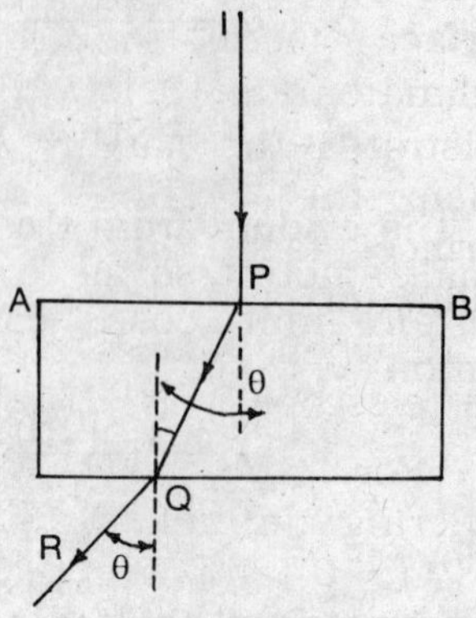

Fig. 13.8

This is the usual grating equation and therefore θ corresponds to the angle of diffraction in air. Thus the decrease in the angle of diffraction at P to θ_g is compensated by the refraction effects at Q and the angle of emergence θ is the same as if diffraction at the grating surface had occurred into air. The glass plate, therefore, simply serves as a mount for the grating and does not play any part in the diffraction effects. From this it appears that so far as the measurement of the angle of diffraction is concerned, it is immaterial which way the grating surface faces—the collimator or the telescope. But this statement is subject to one condition—that the incident pencil of light is parallel. For only in this case the angles measured are the same. Figures. 13.9 and 13.10 illustrate this point.

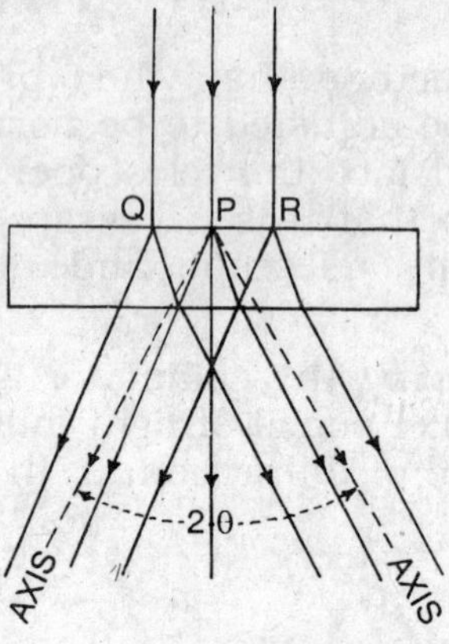

Fig. 13.9

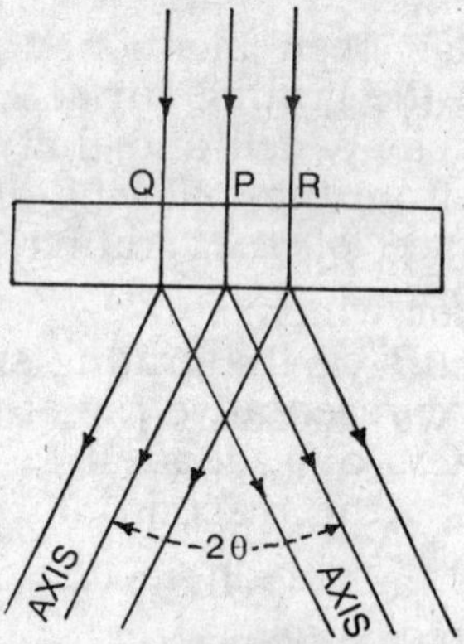

Fig. 13.10

But when the incident pencil is not parallel, as is normally the case, we cannot measure the angle of diffraction correctly, when the grating surface faces the collimator. Fig. 13.11 illustrates the case when the grating surface faces the collimator and Fig. 13.12 when it faces the telescope.

For the ray suffering diffraction at P, $i = 0$, and the corresponding angle of diffration is θ. The angle usually measured is $2\theta = \angle AOB$ in Fig. 13.11 and $= \angle T_1PT_2$ in Fig. 13.12. The axis of rotation of the telescope lies in the grating surface at P and not at O, the point whose position varies with wave-length of light. It is, therefore, apparent that in Fig. 13.12 we measure 2θ correctly for any wavelength and for a ray for which $i = 0$, while in Fig. 13.11 we measure $T'_1 PT'_2$ which is greater than angle AOB and the measured angle in this case is certainly not for the ray for which $i=0$ but for the rays Q and R for which $i \neq 0$.

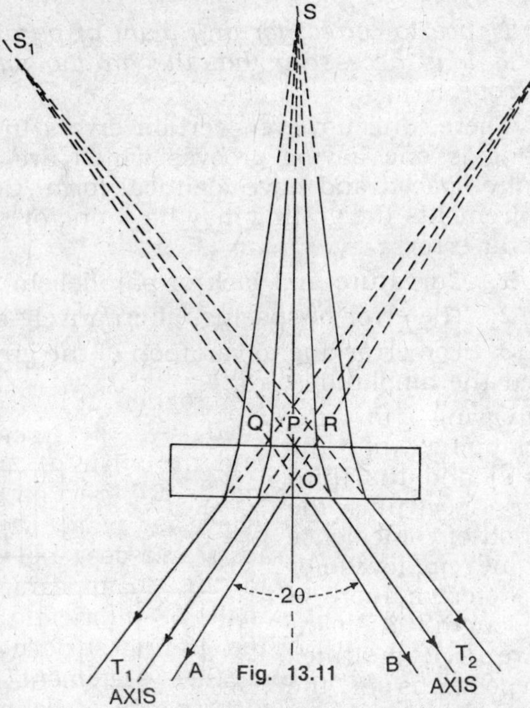

Fig. 13.11

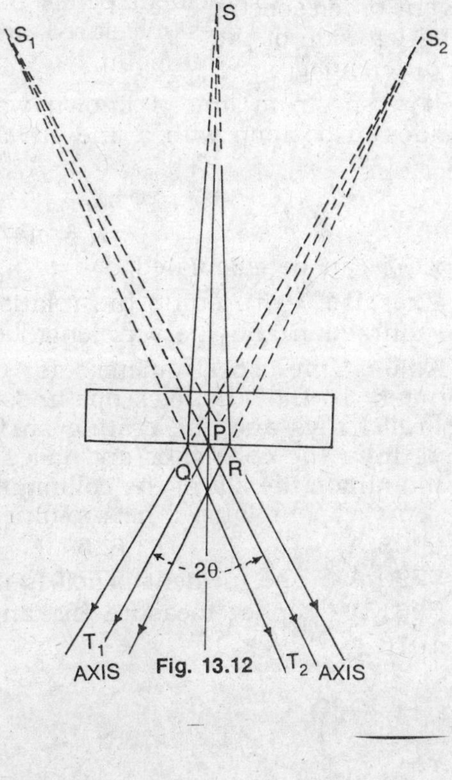

Fig. 13.12

Hence to correct for any want of parallelism in the incident pencil, it is necessary that the grating surface should face the telescope.

There, are, however, certain errors in the grating. An ideal grating is one having grooves which are straight and parallel, equally spaced and have identical form. Corresponding to these requirements the grating may have one or several of the following general errors:

1. Curvature and lack of parallelism of the grooves.
2. The error of spacing, alternatively called the phase error.
3. The error due to variation of the groove form alternatively called the amplitude error.

Owing to the overlapping of the second and the third order spectrum, the method of coincidence may be employed for the determination of wavelength of light. Figure 13.13 indicates the positions of the different order spectra on the left side of the zero order image.

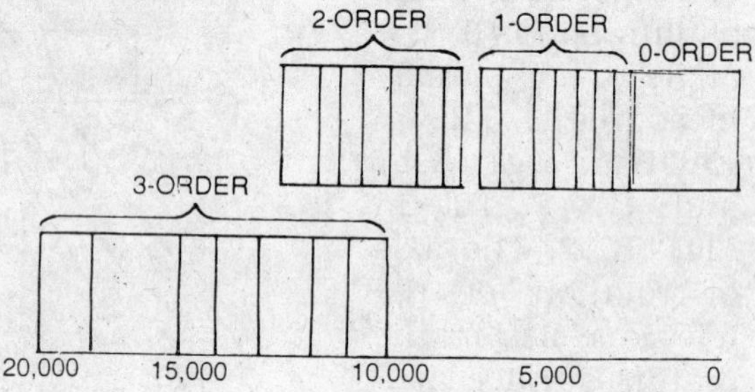

Fig. 13.13

If a spectrum line of known wavelength λ in the order n coincides in position with a line of wavelength λ' in the order n, then

$$n\lambda = n'\lambda'$$

or

$$\lambda' = n\lambda / n'$$

whence λ can be calculated.

Exercise 1—To study the relation between the sine of the angle diffraction and the wavelength of light.

Adjust the spectrometer as described in the grating experiment so that the telescope and the collimator are adjusted for parallel rays and the grating surface is normal to the light coming from the collimator and faces the telescope.

Illuminate the slit of the collimator by the light coming from a mercury vapour lamp. The prominent lines of mercury are.

6908 A°,	6234 A°,	red
5790 A°,	5770 A°,	yellow
5461 A°		yellowish green
4916 A°		deep green

4385 A° violet
4047 A° deep violet

Determine the angle of diffraction in the first order for each line.

Plot a graph between wavelength λ, taken along the X-axis, and the sine of the angle of diffraction, taken along Y-axis. The graph (Fig. 13.14) is a straight line indicating linear relationship between sin θ and λ

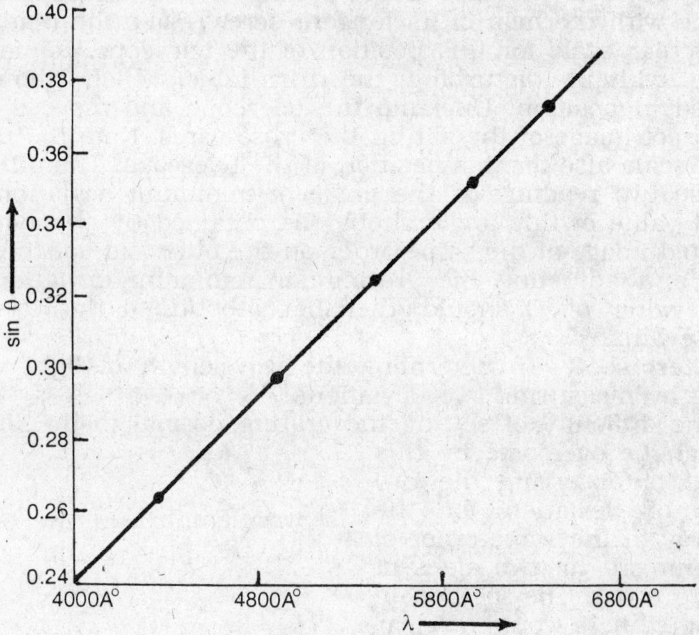

Fig. 13.14

The graph can be said to be the calibration graph of the grating in any subsequent determination of wavelength of a spectral line. Measure the angle of diffraction with this grating and determine from the graph the wavelength corresponding to the sine of the observed angle of diffraction.

From the graph one can also compute the grating element. For ease of calculation, find the value of λ corresponding to the value of sin θ = 0.25. The value of the grating element is evidently given by

$$(e+d) \sin \theta = \lambda \quad \text{(for first order)}$$

or $$(e+d) \times 1/4 = \lambda$$

or $$(e+d) = 4\lambda$$

Exercise 2—To use the grating in the position of minimum deviation for the determination of wavelength of sodium light.

Level the grating as in the previous experiment. From the position of the direct image received in the telescope, turn it so as to receive the diffracted image in the field of view. Then rotate the table carrying the grating in a direction such that the deviation decreases ; follow the diffracted image in the telescope. When the deviation appears to have reached a minimum value, set the cross-wire of the telescope on the image. From this position, whatever direction the grating is turned the deviation will be found to increase. Final setting of the telescope should be done with the help of its tangent screw. Take the reading on the circular scale for this position of the telescope. Remove the grating without disturbing the turn-table which should be clamped in position. Unclamp the telescope and move it to get the direct image of the slit on the cross-wires. Note the reading on the scale also for this position of the telescope. The difference of these two readings is the angle of minimum deviation D. A second value of this angle should be obtained by observing the diffracted image of the same order on the other side of the direct image and adjusting the grating for minimum deviation. The mean value of D should then be substituted in Eq. (13.3) and λ evaluated.

Exercise 3—To determine the wavelength of light with a grating by measuring two deviations[1].

The difficulty of setting the grating normal to the incident rays can be overcome by this method of observing the two angles of deviation for the same ray in the same order of spectrum on the two sides of the zero order. The method to be adopted here would be clear from the discussion below.

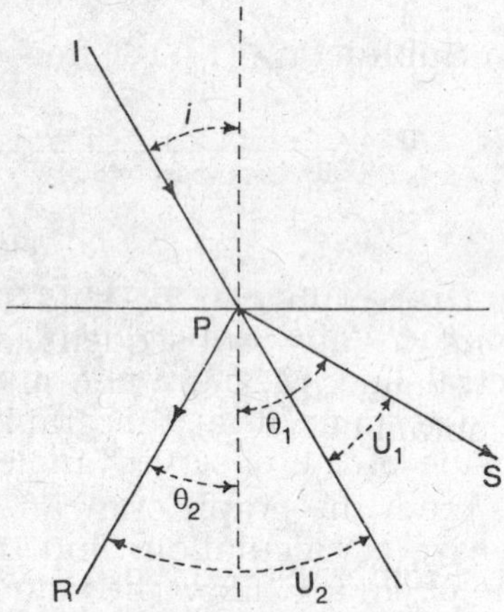

Fig. 13.15

Let the light coming from the collimator be incident on the grating, say, at an angle i (Fig. 13.15). Suppose that a ray of wavelength λ is deviated in the nth order through an angle U_1 on the right and U_2 on the left.

For the ray deviated to the right, the angle of diffraction θ_1 is given by $\theta_1 = U_1 + i$. Hence the necessary condition satisfied by this ray is

1. Uhler—Jour, Opt. Soc. Amer, 21,410 ; 1931. Details taken from "OPTICS" by I. Prakash & L. Singh.

$$(e + d)\ (\sin\ \theta_1 - \ \sin i) = n\lambda$$

$$\therefore \quad (e + d)\{\sin\ (U_1 + i - \sin i\} = n\lambda$$

$$\frac{n\lambda}{(e + d)} = 2 \cos \left(i + \frac{U_1}{2}\right) \sin \frac{U_1}{2}$$

$$= 2 \cos \left[\left(i + \frac{U_1}{2} - \frac{U_2}{2}\right) + \frac{U_2}{2}\right] \sin \frac{U_1}{2}$$

$$= 2 \cos \left[\phi + \frac{U_2}{2}\right] \sin \frac{U_1}{2}$$

$$= 2 \sin \frac{U_1}{2} \cos \frac{U_1}{2} \cos \phi - 2 \sin \frac{U_1}{2}$$

$$= 2 \sin \frac{U_1}{2} \cos \frac{U_1}{2} \cos \phi - 2 \sin \frac{U_1}{2}$$

$$\sin \frac{U_2}{2} \sin \phi \quad (13.16)$$

where $\phi = i + \frac{1}{2} (U_1 - U_2)$ (13.17)

Similarly, it can be shown for the ray deviated in the same order to the left that

$$\frac{n\lambda}{(e + d)} = 2 \cos \frac{U_1}{2} \sin \frac{U_2}{2} \cos \phi + 2 \sin \frac{U_1}{2} \sin \frac{U_2}{2} \sin \phi \quad (13.18)$$

Subtracting (14.16) from (14.18) we have

$$\tan \phi = \frac{\sin \frac{1}{2} (U_1 - U_2)}{2 \sin \frac{1}{2} U_1 \sin \frac{1}{2} U_2}$$

This equation clearly reveals that for purposes of calculation of ϕ, the angle of incidence i on the grating need not be known.

From Eqs. (13.16) and (13.18) on adding and rearranging we obtain

$$(e + d) \cos \phi \sin \frac{1}{2} (U_1 + U_2) = n\lambda \quad (13.19)$$

From this equation it is evident that a calculation of the wavelength of light can be made by simply measuring the two deviations U_1 and U_2.

Hence the following procedure should be adopted. Having set the spectrometer and levelled the grating-table as described in Expt. 13.1, set the grating, say, inclined at a small angle to the pencil of light coming from the collimator. View the first order spectrum to the right and turn the telescope to recieve the required line into the field of view. Adjust the cross-wires on this line and take the readings on the two verniers for this position of the telescope. Turn the telescope now towards the left and set it on the zero order image. Take the reading on the two verniers. Compute the mean deviation U_1 to the right. Then turn the telescope more towards the left and set the crosswires on the same line in the same order. Take the readings of the two verniers and compute as before the angle of deviation U_2 to the left.

From these values of U_1 and U_2 calculate the value of tan ϕ and finally get the value of cos ϕ. Then from Eq. (13.19) calculate the wavelength λ

The experiment can be repeated for the same wavelength in the second order.

Similarly, readings can be taken with other lines in the spectrum and their wavelengths calculated.

Exercise 4—To determine the value of $\frac{d\theta}{d\lambda}$ for the grating.

Having mounted and adjusted the grating as in Expt. 13.1, turn the telescope to bring the D_1 line in the first order on the intersection of the cross-wires of its eye-piece. Take the reading on the circular scale. Then set the telescope on the same line in the first order but now on the other side of the direct image. Again take the reading on the circular scale. The angle between these two positions of the telescope gives twice the angle of diffraction for this line. In a similar manner make observations with the D_2 line in the first order spectrum and get the value of the angle of diffraction for it. The difference of these two angles of diffraction gives us the angular separation of the two wavelengths corresponding to D_1 and D_2 lines. If these angles of diffraction are θ_1 and θ_2.

$$\text{Dispersive power, } \frac{d\theta}{d\lambda} = \frac{\theta_1 - \theta_2}{6 \times 10^{-8}}$$

since $d\lambda = 6 \times 10^{-8}$ cm. for the two sodium lines.

Compare the result so obtained with the value of $d\theta/d\lambda$ calculated from the equation.

$$\frac{d\theta}{d\lambda} = \frac{n}{(e + d) \cos \theta}$$

Repeat the experiment for the second order spectrum and make similar calculations as above. It will be noticed that the dispersion increases with the order of the spectrum.

Exercise 5—To determine the resolving power of a grating.

Mount the grating on the turn-table of the spectrometer and

having set it as in Expt. 13.1, proceed as follows. Illuminate the slit with sodium light. Turn the telescope to get the first order images of the D_1 and D_2 lines of sodium within its field of view. Now reduce the part of the grating producing the spectrum with the help of a slit having both jaws movable. The two lines will be found to approach each other. Continue reducing the width of the grating till the two lines just cease to be resolved. Then with the help of a microscope measure this width of the grating between the jaws of the slit. Since the number of lines to an inch in the grating is known, calculate how many lines are included in the measured width of the grating. The product of this calculated number of lines and the order of the spectrum gives the resolving power of the grating. Repeat the experiment for the same order on the other side of the direct image. Again repeat the experiment for the second order when it will be found that the width of the grating has to be further reduced to get to the position where the two lines cease to be resolved. Make a calculation for the resolving power for the second order also.

Compare your results with the value of $\frac{\lambda}{d\lambda}$ for the sodium lines.

Exercise 6—To determine the resolving power of a prism.

Referring to Eq. (13.11) we note that the evaluation of the resolving power of a prism involves the determination of the constant B, the thickness *t* tranversed by rays of wave-length λ the wafe-length difference being $d\lambda$. Proceed as follows :

(*a*) *Determination of B.*—Illuminate the slit of the spectrometer with a light of known wave-length. Place the prism on the table of the spectrometer already tested for mechanical and optical adjustments. Measure the angle of the prism and also that of minimum deviation. Use these to calculate the refractive index of the material of the prism for this wave-length. Next use light of another known wave-length and determine, as before, the corresponding refractive index. If λ_1 and λ_2 be these wave-lengths and μ_1 and μ_2 corresponding refractive indices, then, by Cauchy's formula.

$$\mu_1 = A + \frac{B}{\lambda_1^2}$$

and

$$\mu_2 = A + \frac{B}{\lambda_2^2}$$

$$\mu_1 - \mu_2 = B + \left(\frac{1}{\lambda_1^2} - \frac{1}{\lambda_2^2}\right)$$

$$B = \frac{\lambda_1^2 \lambda_2^2 (\mu_1 - \mu_2)}{\lambda_2^2 - \lambda_1^2} \tag{13.20}$$

(*b*) *Determination of t.*—Illuminate slit with sodium light and adjust the prism in the position of minimum deviation. Measure the angle for the D_1 line. Then place an adjustable slit in front of the objective of the telescope to cut down its aperture. The separation of the two lines will diminish and for a certain width of the aperture the two lines will cease to be resolved. When this has been attained measure the width of the aperture with a microscope. Let it be *a*. This is then the width of the emergent pencil of light used for the just resolution of these lines. To calculate *t* let us see what other factors need be known. Refer to Fig. 13.16. Here

$$AE = a/\cos i \text{ and } \frac{1}{2} DE = t/2 = AE \sin(\theta/2)$$

$$t = 2\, AE. \sin\theta/2 = \frac{2a\sin(\theta/2)}{\cos i} \tag{13.21}$$

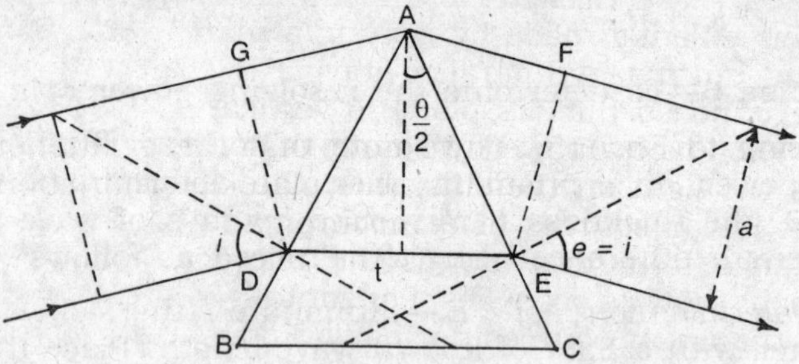

Fig. 13.16

where θ is the angle of the prism and *i* the angle of incidence. The latter can be found from the relation $\delta = 2i - \theta$, where δ is the angle of minimum deviation already determined experimentally. Since all the factors θ, *i* and *d* are known, *i* can be calculated.

(*c*) *Culculation of the resolving power of the prism.*—Having thus obtained the value of the constant *B*, and the thickness *t* for the just resolution of the two lines, calculate the resolving power with the help of Eq. (13.11). Compare the results so obtained with the value of $\lambda/d\lambda$ for the sodium lines.

Experiment 13.2

Object—To verify the expression for the resolving power of a telescope fitted with a variable rectangular aperture.

Apparatus—A telescope, a powerful source of monochromatic light, variable rectangular aperture, a sheet of plate glass coated with a tin foil and containing two fine parallel lines distant 2 to 3 mm, and micrometer microscope.

Theory—For a telescope fitted with a rectangular aperture the limiting angle of resolution θ, is given by

$$\theta = \frac{OO'}{u} = \frac{\lambda}{d}$$

where OO' = the distance between the two objects,
u = the distance of the object from the objective of the telescope.
λ = wavelength of light coming from the objects,
d = width of the aperture (placed before the objective of the telescope) for just resolution of the two objects.

The reciprocal of the limiting angle of resolution is the resolving power.

Method—A serviceable object for this experiment is one that may be conveniently made by coating a plane glass sheet with a tin foil and cutting two parallel lines 2 to 3 mm. apart. Mount this on a stand so that its plane is vertical and so also the two parallel lines. Place a powerful source of sodium light behind this and focus the light on the slits; this plate then provides us with two bright linear sources of monochromatic light close together. At a distance from this mount the telescope on a stand and adjust such that its axis lies horizontally and meets the plate providing the linear sources normally. Focus the telescope on the two sources. By a horizontal movement of the telescope at right angles to its axis adjust that the images of the two slits lie symmetrically with respect to the intersection of the cross-wires in the field of view. Mount an adjustable slit, a rectangular aperture, as close to the objective of the telescope as possible. Now diminish in small steps the width of this aperture and thus adjust its minimum width for which the two images just cease to be resolved. Measure this minimum width of the slit with a micrometer microscope. Also measure the distance between the plate providing the two linear sources and the object glass of the telescope.

Repeat the experiment for different distances of the plate providing the sources from the objective of the telescope. Finally, measure the distance between the two linear sources with the help of a micrometer microscope.

For each distance, calculate the value of $\frac{OO'}{u}$ and $\frac{\lambda}{d}$ and compare the two results.

Sources of error and precautions—(1) The plane of the target consisting of two bright linear sources should be vertical and so also these sources.

(2) The axis of the telescope should be hoizontal and should meet the target normally.

(3) For each distance of the target from the telescope, the latter should be focused on the two sources. The images of these two should lie symmetrically with respect to the intersection of the cross-wires.

(4) Due care must be taken while measuring the width of the aperture and distance between the two linear sources with a microscope. Avoid back lash error during the course of measurements made with it if it is a micrometer microscope.

(5) The minimum width of the aperture for resolution of the two objects should also be found by opening the slit.

Observations and Calculations

No.	Minimum width of slit for resolution d	Wave-length of light λ	Theoretical resolving power d/λ	Distance between two linear sources OO'	Distance between target and objective u	Practical resolving power u/OO'	$\frac{u/OO'}{d/\lambda}$
1.							
2.							
3.							
4.							
5.							

Result—The expression for the resolving power of the telescope fitted with a rectangular aperture has been verified.

Criticism of the method—When we compare (last column of the observation table) the theoretical and practical resolving powers, it is noticed that the value of the expression OO'/u is about 10 to 20 per cent higher than that of λ/d. There is no occasion for surprise when the theoretical resolving power has been exceeded, for, as we know there is a certain degree of arbitrariness in the establishment of what we call the limiting value of the resolution of the two objects. For cheap telescopes where the object glass is imperfect, the resolving power obtained practically does not reach the theoretical value. Such is also the case when the telescope is imperfectly focused on the target.

II. An alternative method of the measurement of OO' and d is with a magnifying Lantern of known magnification factor.

Exercise 7—To determine the diameter of lycopodium particles by Corona rings method.

A small hole 0.25 mm. or 0.5 mm, in diameter is made in a metal plate. Along two perpendicular lines passing through the centre of the hole, pin holes are made at distances of 1 cm. apart. These holes simply serve the purpose of marking equal distances. The plate is mounted on the upright of an optical bench and light from sodium lamp is focused on the central hole. Its position on the bench is recorded.

Some lycopodium powder is dusted on a glass plate. The dusted plate is also mounted on one of the uprights of the optical bench. On viewing through the glass plate at the hole illuminated by sodium light, a series of circular fringes of yellow colour would be visible. Alter the distance of the glass plate and adjust in such a way that the first dark ring of the pattern coincides with first set of pin holes, distant *r* from the centre of the central hole. Take the reading of the position of the glass plate on the optical bench.

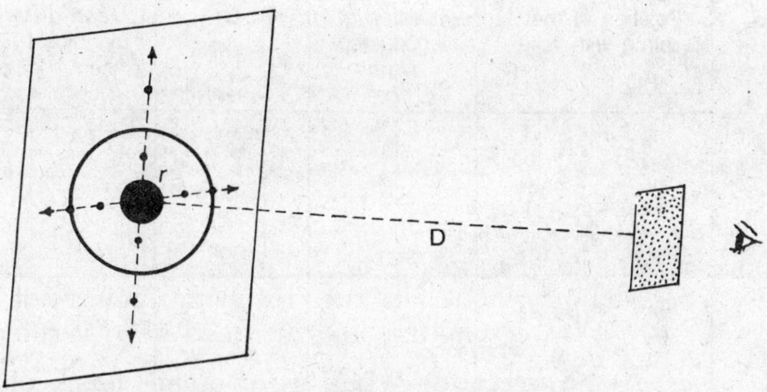

Fig. 13.17

Let the distance of the glass plate from the hole in the metal plate be D. Then, the angle of diffraction giving the first dark ring of the diffraction pattern due to the lycopodium powder particle of diameter *d*, is evidently given by

$$\theta_1 = r/D$$

But the angle of diffraction is also 1.22 λ/d. Hence

$$1.22\ \lambda/d = r/D$$

$$d = \frac{1.22\ \lambda.\ D}{r} \qquad (13.24)$$

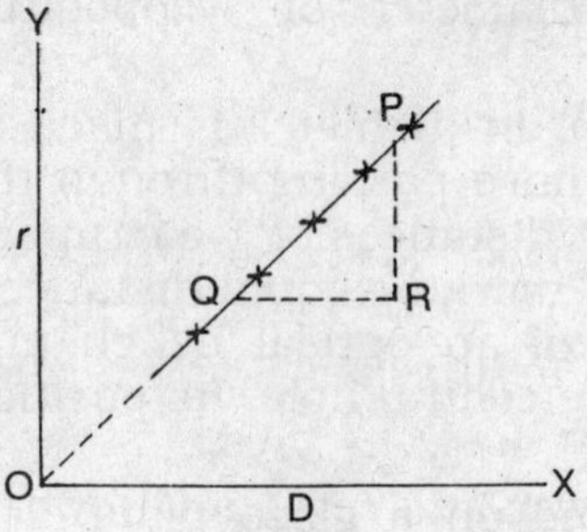

Fig. 13.18

The distance of the glass plate is again altered so that the first dark ring now falls on the second set of equidistant points from the centre of the hole. Similarly, other sets are taken and in each case the values of r/D is found. The mean value is substituted in the equation above to get the diameter of the particles.

Alternatively, a graph is plotted taking D along the X-axis and r along the Y-axis. The slope PR/QR of the straight line thus obtained (Fig. 14.18) will give r/D which may be substituted in Eq. (13.22) to calculate d.

When the second dark ring is used in the experiment the equation becomes

$$\theta_2 = 2.27\ \lambda/d = r/D$$

Tabulate as below.

S. No.	Position of metal plate with hole	Position of the dusted plate b	D $(b-a)$	Mean distance of pinholes from the centre of plate r

ORAL QUESTIONS

THE DIFFRACTION GRATING

1. What do you mean by diffraction of light ? How does it differ from interference of light ?
2. What light does the phenomena of diffraction throw on the nature of light ?
3. What is a diffraction grating ? Name the various types you know.
4. What are the essentials of good grating ?
5. How is grating made and what is a replica casting ?

6. What are the general errors of a grating and what effect do these produce on the spectrum observed with it ?
7. Explain how a grating produces dispersion. Compare its action with a prism.
8. What do you mean by its dispersive power. On what factors does it depend ?
9. How does ruling additional lines in the same space affect dispersion and dispersive power ?
10. Distinguish between dispersive power and resolving power of a grating.
11. What is meant by resolving power of a grating and on what factors does it depend ?
12. Compare a prism and a grating spectrum. Bring out the essential points of difference between the two.
13. How do you use a transmission grating for determining the wave-length of light ? What adjustments are necessary with it ?
14. Explain how and why you set the grating normal to the incident rays with its ruled surface facing the telescope ?
15. Why should the rulings of the grating be parallel to the axis of rotation of the prism table ?
16. How would the slit be arranged parallel to the rulings of the grating ?
17. What advantages has the minimum deviation position setting of the grating over the normal setting ?
18. How do you determine experimentally the dispersive power and the resolving power of (*a*) a grating and (*b*) a prism ?

RESOLVING POWER OF A TELESCOPE

1. What do you mean by the resolving power of a telescope ? On what factors does it depend ?
2. Why are telescopes fitted with objective of large diameter?
3. How do you measure the resolving power of a telescope experimentally? Does it depend upon the distance between the objects under examination and the telescope ?
4. Explain with what relative accuracy should the different distances used in the formula be measured. Suggest an alternative method of their measurements.

5. Explain why sometimes the theoretical limit is surpassed by the value obtained practically? What precautions should be observed while performing the experiment ?

MAGNIFYING POWER OF A TELESCOPE AND A MICROSCOPE

1. Define the magnifying power of (*a*) a telescope, (*b*) a microscope. On what factors does it depend in each case?
2. Distinguish between the resolving power and magnifying power of (*a*) a telescope and (*b*) a microscope.
3. Where should object be placed in determining the magnifying power of (*a*) a telescope, (*b*) a microscope, and why ?
4. What is the focal length of the telescope objective ?
5. What is the focal length of the telescope adjusted for normal vision?
6. How does the magnifying power of a telescope vary with distance?
7. What precaution do you observe in determining the magnifying power of the two ?
8. When it is inconvenient to place the scale at a large distance from the objective, what method would you adopt to find its magnifying power ? How would you verify the results in both the cases ?
9. Does the magnifying power of a telescope depend on the aperture of the objective ?
10. What is the practical utility of a large magnification ?

DIAMETER OF LYCOPODIUM PARTICLES

1. What is the principle on which the experiment is based?
2. What type of diffraction fringes do you get and why?
3. Why should the hole in the middle of the metal plate be small?
4. What is the function of the pinhole along two perpendicular lines drawn on the metal plate?
5. Can you use this method for finding the diameter of french chalk particles?
6. Why is this method suited for lycopodium particles?
7. What other method can be employed to measure it accurately?

8. Can the diameter of the particles be measured directly by a microscope?
9. How would the two results agree ?
10. What would be the nature of *r-D* curve from the observations?
11. Can white light be used at the source in this experiment?
12. What would be the nature of the fringes in this case?
13. What precautions are necessary for the success of the experiment ?

14

Polarisation of Light

14.1. Introduction—A cord or string is capable of both longitudinal and transverse vibrations. In the former case the appearance of the string is the same in all planes drawn through it whereas in the latter it has definite sides on it with respect to the surrounding space. It appears flat just like a ribbon. Since it has acquired a two-sidedness, it is said to be polarised. If a narrow slit be placed in the path of the longitudinal vibrations, these will pass through even when the slit is rotated. But such is not the case with transverse vibrations. For when the slit is parallel to the plane in which the vibrations of the string are restricted these pass on freely, but in a position when the slit lies perpendicular to the above plane of vibration, the vibrations are completely checked. Thus this slit on rotation detects the presence of polarisation in the vibrations of the string. A similar experiment in the case of light throws considerable light regarding the nature of vibrations of a light wave. When a beam of light is allowed to pass through a tourmaline crystal, the transmitted light is found to show a two-sidedness of vibrations. For when the transmitted light is examined by another tourmaline crystal, then in one position, when the two are parallel, maximum amount of light is transmitted. But on rotation of the second plate the intensity of the transmitted light falls till when the two are at right angles to each other, light is completely refused transmission through it. On further rotation light reappears till it attains maximum intensity once again; this is when the rotation of 180° has been completed from the initial position. The first plate is said to polarise the light vibrations and the second plate analyses this light, *i.e.*, detects the state of plane-polarisation in the light vibrations. The former is called the *polariser* and the latter *analyser.*

14.2. Methods of production of polarised light—(1) By Reflection—If a pencil of light rays *I* falls on a plate glass *P* having its rear side blackened, a part is transmitted and absorbed at the back of the plate and the reflected part *M* is left over for observations. If a similar plate A is placed parallel to P (Fig. 14.1), the rays are finally reflected along the diretion N. If A is rotated variations of intensity in the reflected portion N are observed. And for one particular angle of incidence of the pencil, I, these variations in the intensity of the portion N are most

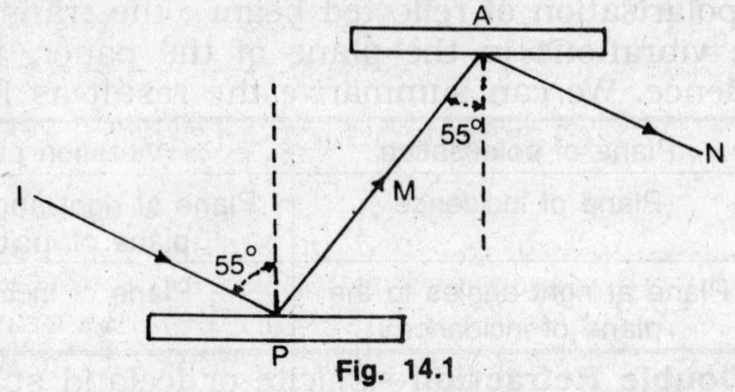

Fig. 14.1

pronounced. This angle of incidence of the pencil I is called the *polarising angle*. This experiment shows that the plate P polarises the light whereas A analyses it. The pencil M obtained by reflection at the surface of P is almost completely plane polarised when the angle of incidence for the pencil I is 55°. It must be clearly borne in mind that the light here is said to be polarised in the plane of incidence. This plane of incidence is called the plane of polarisation. But the vibrations are taking place in a plane at right angles to the plane of incidence, *i.e.*, the plane of polarisation.

(2) By Refraction—If a parallel pencil of light falls on a glass plate at a polarising, angle, the reflected portion, as described above, is polarised in the plane of incidence. If the transmitted portion be examined with an analyser (Fig. 14.2) it is also found to be plane polarised. The difference between the light polarised by reflection and that by refraction is that in the latter case the polarisation is incomplete. There is no complete extinction of light on the rotation of the analyser A_2. To achieve complete polarisation the light is made to pass through a number of plates in succession, called pile of plates. The plane of polarisation of the

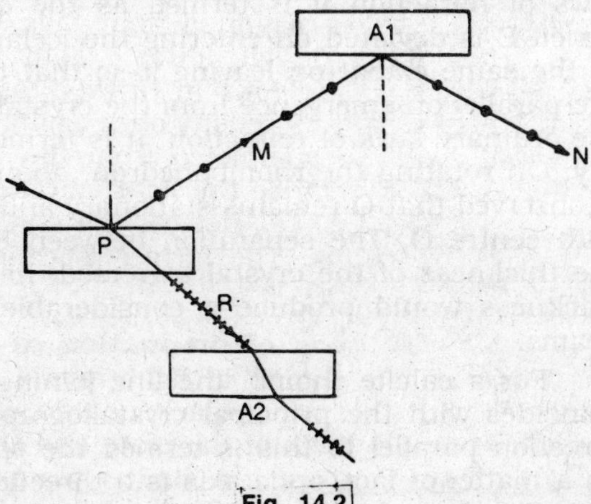

Fig. 14.2

transmitted beam is at right angles to that of the reflected beam. In the figure the dots represent the vibrations at right angles to the plane of polarisation of reflected beam ; the transverse lines represent the vibrations in the plane of the paper, *i.e.*, in the plane of incidence. We can summarize the result as follows :

Light	Plane of polarisation	Vibration plane
Reflected	Plane of incidence	Plane at right angles to the plane of incidence
Transmitted	Plane at right angles to the plane of incidence	Plane of incidence

(3) By Double Refraction—Calcite or Iceland spar has the peculiar property of exhibiting the phenomenon of double refraction. If a rhombohedron of calcite is placed in the path of ray in such a way that the latter fall normally on the face of the former the pencil of light is broken up into two pencils O an E (Fig. 14.3). The pencil O goes through the crystal along the same direction as the incident pencil for it obeys the laws of refraction. Since the angle of incidence on the crystal is zero, the angle of refraction for the pencil O is also zero. As this pencil of light is obtained in accordance with the laws of refraction it is termed as the *ordinary ray*. The other pencil E is deviated on entering the iceland spar and is deviated to the same extent on leaving it so that the two beams E and O are parallel on emergence from the crystal. Since E does not obey the ordinary laws of refraction, it is termed as the *extraordinary* ray. On rotating the rhombohedron, about the incident pencil, it is observed that O remains stationary and E moves round a circle with centre O. The separation between E and O depends upon the thickness of the crystal traversed; hence a crystal of a large thickness would produce a considerable separation of the two beams.

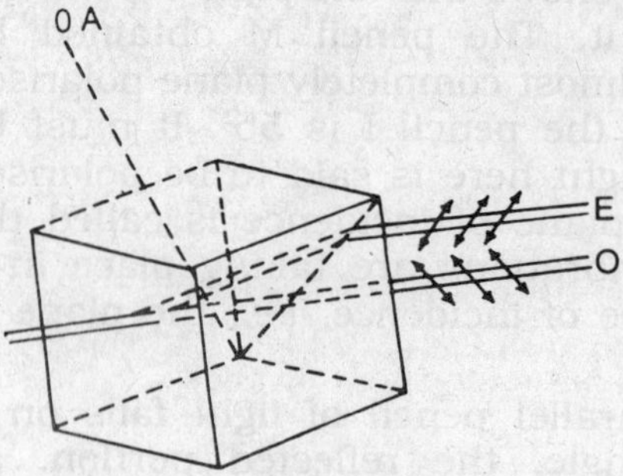

Fig. 14.3

For a calcite rhomb, the line joining the two blunt corners coincides with the principal crystallographic axis. This line or a direction parallel to this is termed the *optic axis of the crystal*. As a matter of fact, optic axis is a direction making equal angles with the three faces at the blunt corner. When the crystal is a rhomb, this line also passes through the other blunt corner, hence the first definition given above. A plane normal to the end faces and containing the optic axis is termed as the *principal plane or section*. The optic axis is also characterised by the fact that it is the line along which the crystal is singly refracting, *i.e.*, the incident rays do not break up into two beams O and E (Fig. 14.4).

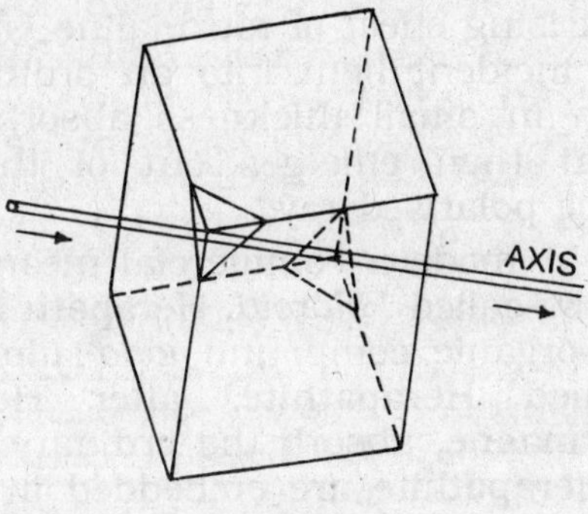

Fig. 14.4

Thus, in general, the calcite crystal exhibits the phenomenon of double refraction and gives rise to the O and E beams. If the two beams are analysed, each is found to be plane polarised, but in planes at right angles to each other. The ordinary ray is polarised in the principal section of the crystal and the extra-ordinary ray in a plane at right angles to the principal section.

Let us now study what methods are adopted in practice for the production of plane-polarised light using this property of double refraction. It is clear that the methods should be based on the following :

(*a*) Separation of the two beams *E* and *O*.

(*b*) natural absorption of one beam by crystal.

(*c*) Suppression of the *O* beam by artificial device.

Let us examine each in detail.

(*a*) *Separation of the two beams F and O*—We have already stated that the separation of the two beams depends upon the thickness of the calcite crystal traversed. So if we use a calcite crystal of sufficient thickness, we shall get the two beams of light *F* and *O* quite wide apart. One of these may be checked by placing an opaque object in its path when the other beam would be left over for experimental work. But this method suffers from one serious drawback. In a natural crystal it is very difficult to get a piece of sufficient length and transparency. If it is not transparent there is a considerable loss in the intensity of the emergent beams and no useful purpose is served. For this reason this method is seldom employed in practice.

(*b*) *Natural absorption of one beam by crystal*—Many minerals exhibit the property of polarising the light they transmit. This is shown in a striking manner by a tourmaline crystal. The polarising effect of tourmaline is due to the fact that it splits up the incident light into an ordinary and extraordinary ray and even in small thickness absorbs the ordinary ray completely. What then emerges out of the crystal is the extraordinary plane-polarised ray.

A modern commercial means of obtaining polarised light is the so-called *Polaroid.* Herapath investigated upon the crystals of the organic compound *iodo-sulphate of quinine.* It was later on termed Herapathite, after Herapath. These crystals, like tourmaline, absorb the ordinary ray. Ultra-microscopic crystals of Herapathite are embedded in nitro-cellulose films, their axis being parallel to each other. These sheets are sold in the form of discs called Polaroid discs. Polaroid has been used for the analysis of crystals, control of light intensity, conversion of instruments into polarising instruments, *etc.*

(c) *Suppression of the O beam by artificial device*—The end faces of a rhombohedron of Iceland spar having a length nearly thrice its width, are ground away until they make an angle of 68° with the long edges. It is then cut through diagonally by a plane at right angles to the new end faces and the principal section of the crystal. These new diagonal surfaces obtained are ground flat, polished and then cemented together by a thin layer of Canada Balsam so as to occupy the same position as before. Such a prism is called a *Nicol prism.*

Consider its working with reference to Fig. 14.5 showing the diagonal principal section of the Nicol prism. An incident pencil of light I is split up into the ordinary and the extraordinary ray, the former being strongly refracted along the path *HG* whereas the latter takes up the path *HF* till these rays meet the surface of Canada Balsam. Now the refractive index of Canada Balsam is 1.55 whereas the values of the index of refraction of the crystal for the ordinary and the extraordinary ray are 1.66 and 1.48 respectively. As such the ordinary ray, *HG,* as it is passing from a denser into a rarer medium and the angle of incidence at *G* is greater than the critical angle for the two media, is in a position to suffer total internal reflection and take up the path *GO.* The ray *HF,* on the other hand, passes through and ultimately emerges along *E,* parallel to I. It suffers only a trifling lateral

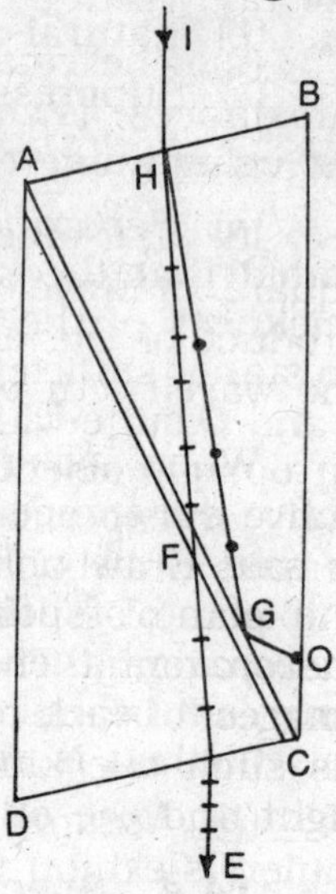

Fig. 14.5

displacement. The intensity of the emergent beam is less than that of incident light in view of the fact that the ordinary beam has been removed. A Nicol prism is the most perfect form of polariser and provides us with a plane-polarised beam, polarised in a plane at right angles to the principal section of the crystal. The vibrations in the beam are in the principal section of the crystal. This is shown in the figure by short lines drawn perpendicular to the E ray.

14.3. Rotation of the plane of polarisation—When two nicols are crossed the field of view remains dark. If some canesugar solution dissolved in water is interposed between the two, the field of view becomes lighted and to restore the extinction of light the analyser has to be rotated through a certain angle. From such an experiment it is clear that it is the sugar solution which rotates the plane of polarisation of the incident light coming from the polarising nicol. If, instead of the sugar solution, the experiment is repeated with a thin quartz plate cut at right angles to the optic axis, a similar conclusion can be drawn. Thus some solids, liquids and vapours exhibit the phenomenon of rotation of the plane of polarisation of light. Such substances are said to be *optically active.* Boit successfully investigated the phenomena and enunciated the following laws :

(1) The amount of rotation produced is directly proportional to the thickness of the optically active substance traversed by the ray.

(2) The rotation produced by two or more optically active substances is the algebraic sum of the rotation produced individually.

(3) The amount of rotation produced varies inversely as the square of wavelength. This is only approximately true for the product of the amount of rotation produced and the square of the wavelength of light increases with refrangibility of light.

When looking towards the source of light if the optically active substance rotates the plane of polarisation to the right, it is said to be *right handed or dextrorotatory*; others which rotate the plane of polarisation to the left are termed *left-handed or laevorotatory*. There exist two kinds of quartz which are mirror images of each other even in their crystalline structure and form. As such one kind rotates the plane of polarisation towards the right and the other towards the left.

14.4. Specific and Molecular Rotation—The specific rotation of a substance, *e.g.*, turpentine, is the rotation produced by a decimeter of the pure substance divided by its density. This definition applies to optically active liquids. When substances like sugar are to be experimented upon, we require to dissolve them in an inactive solvent. Then the *rotation produced per decimeter length of the solution divided by the number of gms. of the active substance per c.c. of the solution is called the specific rotation of the dissolved substance.* For example, if

Mass of sugar dissolved	= 6 gm.
Volume of solution	= 60 c.c.
Length of tube	= 22 cm.
Rotation	= 5.50°

∴ Specific rotation at the temperature of observation.

$$(\alpha)_t = \frac{5.5 \times 160}{2.2. \times 6} = 66.7^\circ.$$

The specific rotation is not a constant. However, the variations are small and may be neglected. It is found to vary with the wavelength of light, the nature of the inactive solvent, the concentration of the solution and its temperature. The specific rotation of cane-sugar dissolved in water is given by the expression.

$$(\alpha)_t = 66.5^\circ - 0.0184\ (t - 20)$$

where t is the temperature of the solution. Evidently at 20°C its value is 66.5°.

The product of the specific rotation and the molecular weight of the substance is called *molecular rotation.*

For solid crystalline transparent substances the specific rotation is merely expressed in degrees per mm thickness. For quartz it is 21.72° per mm. at 20°C. Its value at different temparetures in case of quartz is given by the equation.

$$(\alpha)_t = 21.72^\circ\ [1+0.000147\ (t - 20)]$$

The wavelength of light for which these rotations are referred to in the equations above is for the D-line of sodium.

14.5. Polarimeters—The essential parts of a simple polarimeter are two Nicol prisms, one of which serves to polarise a beam of light passing through it and the other serves to detect the state of polarisation in the light coming from the polariser. The latter prism is consequently called the analyser. The light vibrations in the transmitted light through the polariser take place in its principal section and when the principal section of the analyser is parallel to that of the polariser, all the light is completely transmitted by the analyser. If the analyser is gradually rotated, a position is reached when no light is allowed transmission by the analyser; in this position the two Nicols are said to be *crossed* and the principal sections of the two are at right angles to each other.

In between the two Nicols a glass tube containing an optically active substance is placed. Initially the two Nicols were in the crossed position. On introduction of the optically active substance, the plane of vibration of the light coming from the polariser is rotated and as such the analyser transmits some light through it. To get to the crossed position the analyser has to be rotated through a certain angle and this angle must correspond to the angle of rotation of the plane of vibration produced by the optically active substance. We therefore ought to be able to measure the rotation of the plane of vibration by measuring the angle turned through by the analysing Nicol.

The method has, however, one essential drawback. The accurate setting of the analyser is practically uncertain in view of the fact that in the region of darkness it is very difficult to judge when the field is completely dark. This uncertainty can be slightly minimised by the following procedure. Rotate the nicol till the position when darkness just sets in is reached. Carefully mark the brightness of the field of view. Note the reading on the circular scale. Then continue rotation of the nicol when the field of view grows darker and the position of complete darkness is reached. Still continue the rotation of the nicol but now slowly till the field of view has the same brightness as in the initial setting. The average of these two readings, giving the position of equal brightness on either side of the position of complete darkness, will give a more accurate value than the one obtained by complete extinction of light. The method is based on the quality of the eye to judge two positions of equal intensity more accurately than one of maximum brighthness or darkness. Even this method has the handicap that the fields of view for comparison are presented one after the other and not side by side at the same time. For this reason various other additional devices have been introduced working on the line of presenting two halves of the field of view of equal brightness for judgment of this equality. These devices enable a sharp adjustment to be made.

(1) Laurent plate—(Glass-Quartz Combination).—It consists of a quartz plate cut parallel to the optic axis and of such a thickness that a phase difference of 180 or path difference of $\frac{\lambda}{2}$ is introduced between the ordinary and the extraordinary ray during their passage through it. The quartz plate covers half the field of view and the other half is covered by glass. Its thickness is such that it provides the transmitted light of the same intensity as coming out of quartz. Such a combination is called a *Laurent Plate.* When in use, it is placed immediately after the polarising nicol.

In Fig. 14.6 A.B.C., shaded portion, represents the quartz plate having the optic axis parallel to BC; the clear portion represents the side glass plate. Light coming from the polariser falls on the combination. Let the incident light be polarised in a plane at right angles to OP, *i.e.*, the vibrations of the incident light are taking place parallel to OP. In quartz the incident vibrations are broken up into two perpendicular vibrations, one parallel to the optic axis, *i.e.*, along OB and the other perpendicular to the optic axis, *i.e.* along OA. However, when the two vibrations emerge from the quartz plate, they have a phase difference of 180°, those parallel to OB being retarded by this amount. Hence on emergence from the quartz plate, these vibrations recombine to produce a linear vibration parallel to OQ, the inclination of OQ with OB being the same as of OP with OB. Thus the plane of polarisation of light emerging from quartz is inclined to the plane of polarisation of light emerging from glass at an angle POQ = 2POB = 2ϕ. The value of ϕ can be altered at will, this being the angle between the principal section of the polarising nicol and the optic axis of the quartz plate.

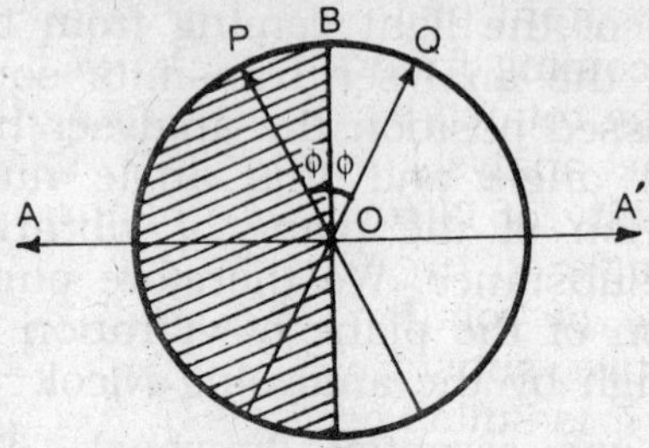

Fig. 14.6

Thus, on emergence from the combination we have now two plane polarised beams, one having vibrations parallel to OP and the other parallel to OQ. If the analyser be set so that its principal section lies along AOA′, (Fig. 14.6*a*) the two fields of view will appear to be equally bright since the portion ON and ON′ transmitted by the analyser are equal. A slight rotation of AA′ from this position causes a considerable change in intensity since the transmitted portion for one rapidly increases and for the other rapidly decreases; this causes the two portions of the field of view to be unequally illuminated. So

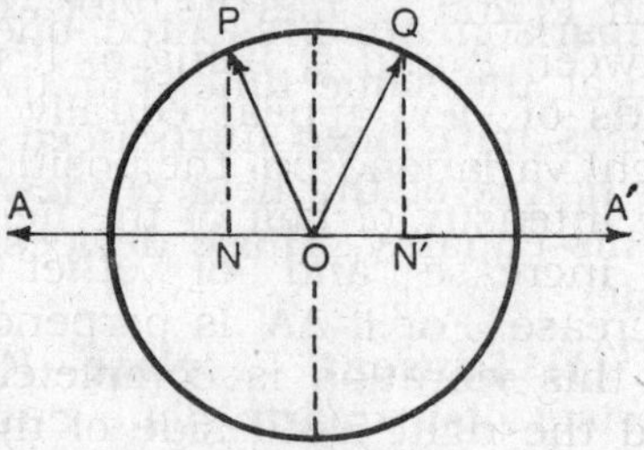

Fig. 14.6(a)

the purpose of this device is clear. The eye is by far a better judge of equality of illumination of two portions displayed side by side than the portion of complete darkness, or even of maximum brightness, in the field of view.

Having set the analyser as discussed above record the position on the graduated circular scale. Now if an optically active liquid is introduced between the Laurent plate and the analyser, the plane of polarisation of light coming from quartz as well as that coming from glass plate, will be rotated by the same amount in the same direction. Hence the analyser has to be turned through again to restore the equality of illumination of the two halves of the field of view. The angle so turned through by the analyser gives the required angle of rotation of the plane of polarisation due to optically active substance. The essential drawback of such a device is that it is suitable for one particular wavelength for which the path retardation in E and O beams is $\lambda/2$.

(2) Lippich tow prism arrangement—The optical plan is given in Fig. 14.7. The field of view is divided into two portions. One half of it is illuminated by the light coming through P alone and the other half by light passing through both the nicols P and Q. As the principal sections of P and Q are slightly inclined at a small angle ϕ to each other the light that falls on the analyser consists of two plane polarised beams. One coming through P has its vibrations parallel to P (Fig. 14.8) and the other coming through both P and Q has its vibrations parallel to Q, the inclination between P and Q being ϕ. If the analyser is set at AA′ the two fields of view appear equally bright. A slight variation from the position causes the intensity of half of the field of view to increase and of other half to decrease. For if AA′ is perpendicular to Q, this vibration is completely cut off and the right hand side of the field of view is dark; AA′ is set perpendicular to P, this vibration is completely cut off and the left hand side of the field of view becomes completely dark. Hence this arrangement, quite similar

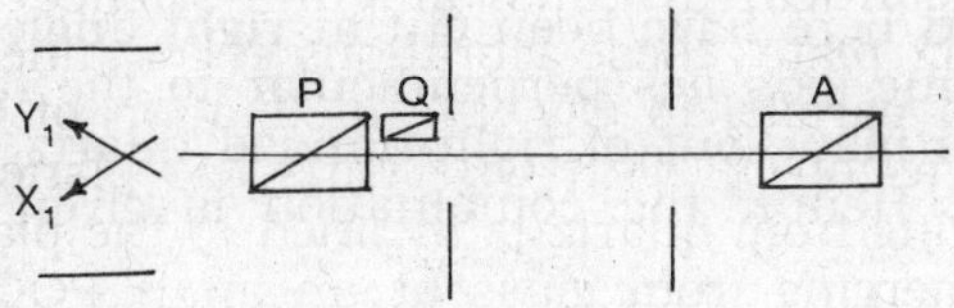

Fig. 14.7

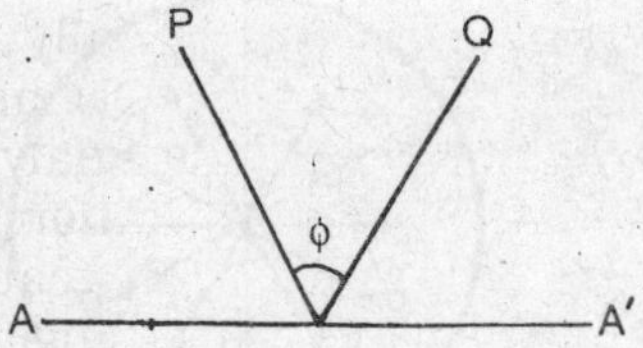

Fig. 14.8

to Laurent plate, is very sensitive, for a slight rotation of the analyser from the position AA′ causes a sharp diminution in the intensity of one-half of the field of view. This greatly helps in setting the analyser to the position AA′ when the two portions of the field of view are equally bright. It is also clear from the above discussion that the sharpness of the setting depends on ϕ. If we reduce ϕ the arrangement becomes more sensitive but there is loss of intensity of light transmitted by the analyser in the position of AA′ and thus the illumination of the field of view becomes poor. Hence, there is a certain value of ϕ which we have to set by trial so that the arrangement may remain sensitive and the field of view may not be poorly illuminated.

The principle of working of this is the same as of Laurent plate. However, this has an advantage that it can be used with light of any wavelength and hence with white light also.

(3) Bi-quartz—This is another device of increasing the sensitiveness of the arrangement of two nicols for studying the rotation of plane of polarisation. It consists of two semi-circular plates of quartz cemented together so as to form a disc. The thickness of each of the plates is the same and equal to about 3.75 mm. The plates used here have been cut at right angles to the optic axis, *i.e.* the optic axis lies perpendicular to the faces of the plate. One piece is made out of right-handed quartz and the other of left-handed. Hence, the combination is given the name bi-quartz.

When white light plane polarised by a nicol falls on the bi-quartz, rotatory dispersion occurs in each plate. The plane of polarisation in the two halves is rotated in opposite directions. The thickness of each plate has been chosen to be nearly 3.75 mm, so that the plane of polarisation of the yellow light may be rotated through nearly 90°. It is thus the brightest part of the spectrum that gets rotated through 90°. For the red the rotation is the least and for the violet it is the greatest. In Fig. 14.9, PQ represents the vibration plane of the incident light; *r, t, o, y, g, v,* the vibration planes for the red, orange....violet rays. Since the rotation of the plane of polarisation by quartz for yellow is 90°, yy is one straight line and hence both the vibrations can be simultaneously quenched for one setting of the analyser. Hence, if the principal plane of the analyser is set so

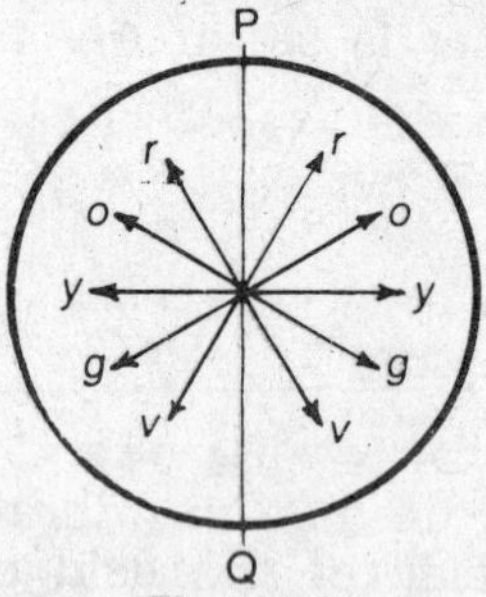

Fig. 14.9

as to quench the yellow light, the common colour of both the halves of the field of view will be a mixture of red and blue which produces grey colour. This is called the *tint of passage*. If the analyser be rotated slightly from this position, one-half of the field of view is predominantly blue and the other red and vice versa on rotation in the reverse direction. Hence, the position of the tint of passage is a very sensitive one and the analyser can be very quickly and accurately set on this.

To use it obtain the tint of passage without the experimental solution and note the position of the analyser. Then introduce the given solution in a tube placed between the bi-quartz and the analyser when the field of view will be found to present different colours in the two halves owing to the rotations of the planes of polarisation of the incident beam but all in the same sense. Again obtain the tint of passage with the analyser. Note its position. The difference of the two readings gives the rotation of the plane of polarisation.

Experiment 14.1

Object—To determine the polarising angle for the glass prism surface and to determine the refractive index of the material of the prism using Brewester's law $\tan \phi = \mu$.

Apparatus—A spectrometer, sodium lamp, a polaroid attachment for the objective of the telescope, a prism, a reading lamp and a reading lens.

Theory—If the angle of incidence on a surface of, say, a glass prism is varied and the reflected light analysed with a rotating polaroid there will be variations of intensity from a maximum to minimum for each angle of incidence. It implies that there will always be a minimum of intensity but this minimum of intensity will be zero for only one particular angle of incidence on the glass surface. This critical angle of incidence which gives a zero minimum when analysed by a rotating polaroid or nicol is called the *polarising angle*. For this angle of incidence on the glass surface in question we have the maximum polarisation in the reflected beam. According to Brewester, maximum polarisation occurs in the reflected beam when the reflected and the refracted rays are mutually at right angles to each other. Thus, from Fig. 15.10 which depicts this state, we have

$$\mu = \frac{\sin \phi}{\sin r} = \frac{\sin \phi}{\sin (90 - \phi)} = \frac{\sin \phi}{\cos \phi} = \tan \phi$$

This law is known as Brewester's law. If the polarising angle ϕ be determined, the refractive index can be calculated.

Method—Make the preliminary adjustments of the

spectrometer as outlined in Expt. 14.1. Mount the prism on the prism table so that the centre of the circumcircle of its base coincides with the axis of rotation of the prism table. Mount the polaroid attachment (or a nicol attachment) on the telescope of the spectrometer. Turn the prism to receive the light from the collimator on one of its faces (not the ground face). Turn the telescope to receive the reflected image on to the cross wires. Rotate the polaroid slowly and observe the position of minimum intensity. In general this will not be zero. Turn the prism table to increase the angle of incidence on the face of the prism table and again test for zero intensity of the reflected beam with the polaroid. On gradually increasing the angle of incidence and testing as above, a position will be attained when the intensity of the reflected beam shows variations of intensity with the rotating polaroid and in one position of the latter, there is complete darkness in the field of view. The image of the slit completely disappears and the field of view presents a totally dark appearance. For this position take the readings of the two verniers to record the position of the telescope. Remove the prism and set the telescope on the direct image and record the readings on the two verniers. From these observations, find the difference θ between them. Then the polarising angle is given by

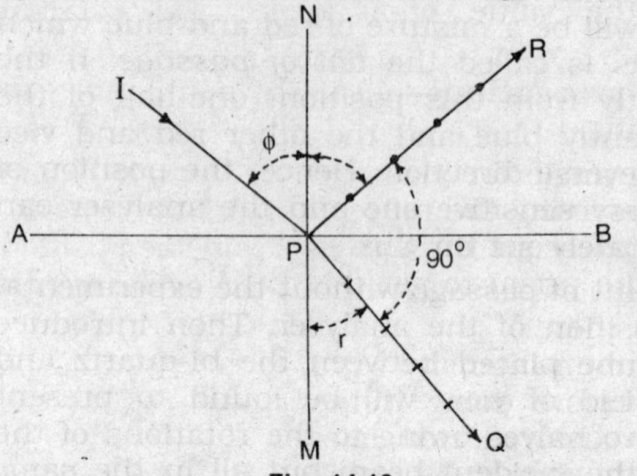

Fig. 14.10

$$\phi = (90 - \theta/2)$$

This is clear from the geometry of Fig. 14.11. Since the reflected ray enters the telescope along the direction PR, angle IPA is equal to angle RPB. Also angle IPA = angle PBQ. Hence, PB bisects the angle QPR, which we have measured as θ. ∠ RPB = θ/2 and the angle of reflection and hence the angle of incidence IPN is (90 − θ/2).

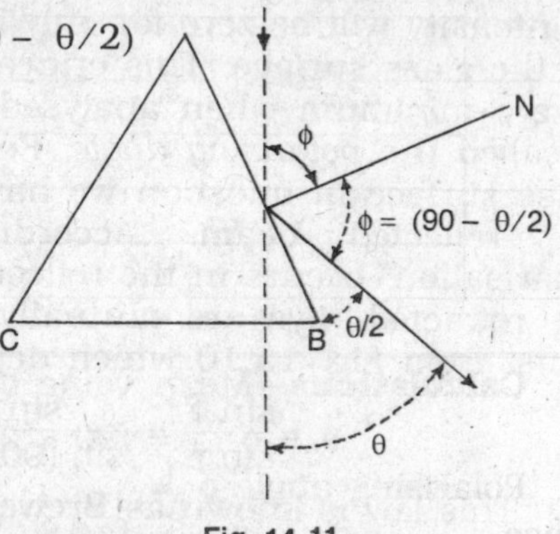

Fig. 14.11

Repeat the experiment a number of times and get the mean value of θ. Calculate the value of μ using the relation tan $\phi = \mu$.

Source of error and precautions—All those enumerated in Expt. 12.1 and the following.

(8) As the polarising angle for glass lies between 50° and 60°, it is better to start observing the reflected image for $i = 50°$. For this purpose turn the telescope through 80° from the position for direct image and set the prism to get the reflected image on the cross-wires.

(9) The test for minimum intensity equal to zero should be done very carefully and before doing it the crosswires of the telescope must be set on the reflected image. The position of complete extinction of light should be checked by rotating the polaroid in clockwise and then in anti-clockwise direction.

(10) In practice it will be found that the light is not quite all cut off but proceeding in the manner described in precaution 8, a position will always be found when it is vey nearly true.

(11) The width of the slit can be kept wide here to allow greater amount of light to enter the telescope after reflection from the prism surface but due care must be exercised in setting the cross-wires on it. If the slit is symmetrical, the crosswires must be set in the middle; in an asymmetrical slit, the crosswires should be fixed on the image of the fixed jaw of the slit.

Observations—[A] For polarising angle.

S. No.	Vernier	Position of telescope for complete extinction of reflected ray	Position of direct image	Difference of the two readings of the same vernier
1.	V_1			
	V_2			
2.				
3.				
			Mean	

Calculations—Mean value of θ =

∴ ϕ = $(90 - \theta/2)$ =

∴ Polarising angle ϕ =

Hence μ = $\tan\phi$

=

Result—(*i*) The polarising angle =

(*ii*) The refractive index of the material of the prism, as calculated from Brewester's law =

Criticism of the method—The spectrometer method of determining the polarising angle is superior to the method of Noremberg's polariscope because here the accuracy of measurements of the angle of incidence is much greater than in the latter. The use of a polaroid is better than analysing nicol or a plane mirror because the two appreciably alter the direction of the transmitted or reflected ray, as the case may be; and this has to be followed by the eye which is to examine the complete extinction of light. With a polaroid, the transmitted ray is fixed and the observations are made to greater precision as the telescope is used as an aid to vision.

Experiment 14.2

Object—To determine the specific rotation of cane-sugar dissolved in water.

Apparatus—Polarimeter with a sensitive device (Half-shade or Bi-quartz), sodium lamp (for use with half-shade) or white light source (for use with Bi-quartz). 100 c.c. flask, beaker, funnel, pipette, analytical balance and a wt.-box.

Description of Polarimeter—Polarimeter, in general, consists of a source of light, a polariser and an analyser provided with a graduated circular scale. Figure 14.12 below represents the general optical arrangement of most polarimeters. S is a

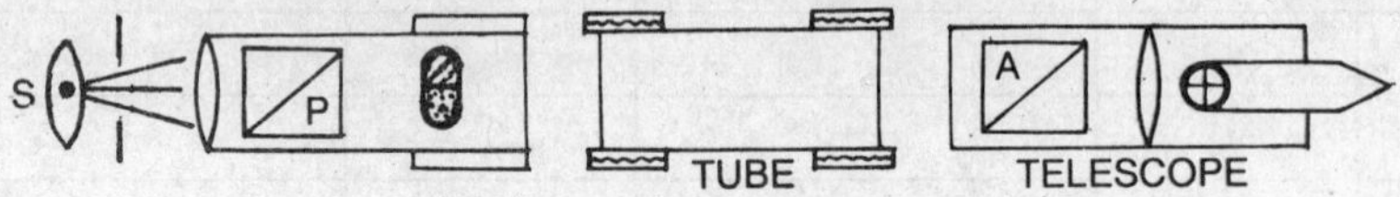

Fig. 14.12

source of light so placed that it is nearly at the focus of the lens L, thus providing a parallel pencil of light. This falls on the polarising nicol P which produces plane polarised beam. The polarising nicol is immediately followed by either of the sensitive arrangements, *e.g.*, Laurent plate bi-quartz, *etc.* The analysing nicol A is in front of a low-power telescope. In between the polarising arrangement and A is placed the tube containing the liquid under investigation. The tube is closed on both sides by plates which are optically worked and held at the ends of tube with metal caps.

Theory—Let

Mass of sugar dissolved $= x$ gms.

Volume of solution $= v$ c.c.

Length of tube $= l$ decimeter.

Rotation produced $= \theta^\circ$

Temperature of solution $= t^\circ C$

$$[\alpha] = \frac{\theta / l}{x / v} = \frac{\theta v}{l x} \quad (14.1)$$

Method—Weigh a 100 c.c. flask. Introduce about 10-15 gms of cane-sugar into this and weigh again. Fill some distilled water in the flask and dissolve the sugar. Finally fill the flask upto the 100 c.c. mark on it. The solution is now ready for use.

Clean the polarimeter tube and fill it with distilled water. See that there is no air bubble in the tube when the end caps have been screwed. Place the polarimeter tube in its proper position and secure the proper illumination depending upon the sensitive device used therein (sodium light for half-shade, and white light for bi-quartz). Look through the analyser when it will be observed in general that the two portions of the field of view are of unequal intensity (half-shade) or present different colours, red and blue (bi-quartz). Rotate the analyser till the two portions of the field of view are of same intensity (half-shade) or acquire the same tint, called the tint of passage (bi-quartz). Take the reading of the analyser on the main scale and the vernier scale and compute the total reading.

Remove the distilled water from the tube and rinse with the prepared sugar solution. Fill it completely with this solution and replace it at its proper position. On looking through the analyser, the previous setting would be disturbed. Adjust again till the two portions of the field of view are of the same intensity (half-shade) or acquire the grey tint (bi-quartz). Take the reading of the analyser as before.

These settings of the analyser should be done by rotating the analyser in the clockwise direction as well as by rotating it in the anti-clock-wise direction 180° apart from the first setting. From these observations, calculate the rotation of the plane of vibration produced by the sugar solution.

Repeat the experiment with solutions of different strengths. Plot a graph between the rotation produced and the strength of the solution. The graph is a straight line; it shows that the rotation of plane of vibration produced by an optically active substance is directly proportional to the amount of the active substance in the same volume of solution.

Measure the length of the tube. Convert this to decimeters. From the observed data, calculate the specific rotaion for sugar. State the temperature at which the determination is made.

Sources of error and precautions—(1) The cap of the tube containing the solution should not be trightly screwed so as to strain the glass that allows the passage of polarised light into the solution. Glass in a strained condition is likely to produce elliptic polarisation which will interfere with the settings of the nicol analyser.

(2) There should be no air bubble in the solution contained in the polarimeter tube.

(3) The temperature of the solution must be recorded because specific rotation varies, though slightly, with temperature.

(4) Having set the analyser in correct position with respect to the polariser, turn the former through 180° and again make a similar setting. Under no circumstances should the polariser be touched during one complete set of observations for this will change the plane of polarisation of the incident beam and the readings obtained for rotation later will not tally with those taken before.

Observations

Value of one main scale division =

Vernier div. coincide with... main scale div.

Least count of the vernier of the analyser =,

Mass of the flask	=	gm.
Mass of flask+sugar	=	gm.
Mass of sugar (x)	=	gm.
Volume of solution (v)	=	c.c.
Length of the tube (l)	=	decimeter
Temperature of the solution	=	t° C.

[A] Observations with distilled water in the tube.

S. No.	First position of analyser						Position of analyser 180° apart from the first setting						Mean value $\frac{a+b+c+d}{4}$
	Equal intensity of the two portions of the field of view obtained by						Equal intensity of the two portions of the field of view obtained by						
	Clockwise rotation			Anti-clockwise rotation			Clockwise rotation			Anti-clockwise rotation			
	Main scale reading	Vernier reading	Total reading a	Main scale reading	Vernier reading	Total reading b	Main scale reading	Vernier reading	Total reading c	Main scale reading	Vernier reading	Total reading d	

[B] Observations with solution in the tube.

S. No.	First position of analyser						Position of analyser 180° apart from the first setting						Mean value $\frac{a'+b'+c'+d'}{4}$
	Equal intensity of the two portions of the field of view obtained by						Equal intensity of the two portions of the field of view obtained by						
	Clockwise rotation			Anti-clockwise rotation			Clockwise rotation			Anti-clockwise rotation			
	Main scale reading	Vernier reading	Total reading a'	Main scale reading	Vernier reading	Total reading b'	Main scale reading	Vernier reading	Total reading c'	Main scale reading	Vernier reading	Total reading d'	

Calculations

$$\text{Angle of rotation } \theta = \frac{a+b+c+d}{4} - \frac{a'+b'+c'+d'}{4}$$

$$\therefore \quad (\alpha)_t = \frac{\theta}{l} \cdot \frac{v}{x} \text{ degrees.}$$

Standard value of α at t°C = ...

% Error = ...

Result— The specific rotation of cane-sugar solution at °C for sodium light is......degrees.

ORAL QUESTIONS

POLARISATION OF LIGHT

1. What is plane-polarised light ? How does it differ from natural light ?
2. What are the methods adopted for the production of plane polarised light ?
3. What is the relation between the plane of polarisation and plane of vibration ?
4. What is polarising angle and what is its importance?
5. Why is the polarisation of transmitted light incomplete ?
6. How does a pile of plates help in rendering the polarisation of transmitted light complete ?
7. What is double refraction of light ?
8. Explain the polarisation of the ordinary and the extraordinary ray.
9. What are the blunt corners of a crystal ?
10. Define optic axis and principal section of a doubly refracting crystal.
11. How is it that a crystal is singly refracting along the optic axis ?
12. Explain the construction of a nicol prism.
13. What functions does Canada Balsam used for cementing the two portions of the prism serve ?
14. In what plane is the ray emerging of the nicol prism polarised ?
15. In what plane are the vibrations taking place ?
16. Explain the working of a tourmaline crystal.

17. What is polaroid and what are its practical applications ?
18. What is circularly and elliptically polarised light ? How do you produce and detect them ?
19. What is rotatory polarisation ? What are the laws relating to this ?
20. What is an optically active substance ?
21. What are dextero-rotatory and laevo-rotatory substances ?
22. Define specific and molecular rotation.
23. Is specific rotation a constant quantity for pure substances ? On what factors does it depend ?

SPECIFIC ROTATION AND ITS DETERMINATION

1. Describe the construction of the simple Polarimeter.
2. Can the analyser be correctly set in the position of complete extinction of light ? If not, why ? How can you improve upon this method of obtaining the position of the analyser without any additional aid ?
3. What are the sources of error in the experiment ?
4. What precautions do you observe while performing the experiment ?
5. What for are the Laurent plate, be quartz, *etc.*, employed in polarimeters ? Explain the action of each such device.
6. What are polarimeters ? To what use are they put ?

15

Laser and Optical Fiber

15.1. Laser

The term LASER stands for Light Amplification by Stimulated Emission of Radiation. A laser device produces a light beam which has the following characteristics :

(*i*) The laser beam is highly intense ; much more than an ordinary light beam.

(*ii*) The laser light is coherent, i.e., all the waves are exactly in phase with each other.

(*iii*) As a laser beam propagates there is almost no divergence. Therefore, it can be sent from Earth to Moon and the reflected beam can be detected back at the Earth.

(*iv*) A laser light is nearly monochromatic.

We try to understand below the basic laser action.

15.2. Absorption and Emission (Spontaneous and Stimulated) of Radiation

According to quantum theory, atoms exist only in certain discrete energy states and absorption and emission of light causes them to make a transition from one discrete energy state to another. These energy states for the atom correspond to electrons occurring in particular energy levels relative to nucleus. Hence, different energy states for the atom correspond to different electronic configurations and a single electron transition between two energy levels within the atom will provide a change in energy suitable for the absorption or emission of a photon.

Fig. 15.1 (a) illustrates a two energy state (or level) atom, in which the atom is initially in the lower energy state E_1. When a photon with energy $E_2 - E_1$ is incident on the atom, it may be excited into the higher energy state E_2 due to absorption of the photon. This process is known as *stimulated* (or *induced*) *absorption*. Usually, absorption takes place in a radiation field, *i.e.*, a radiation is incident on the atom. However, only photons of frequency υ_{21} (such that $h\upsilon_{21} = E_2 - E_1$) can be absorbed by the atom, out of the incident photons of different frequencies.

In an atom, absorption process strongly depends on the

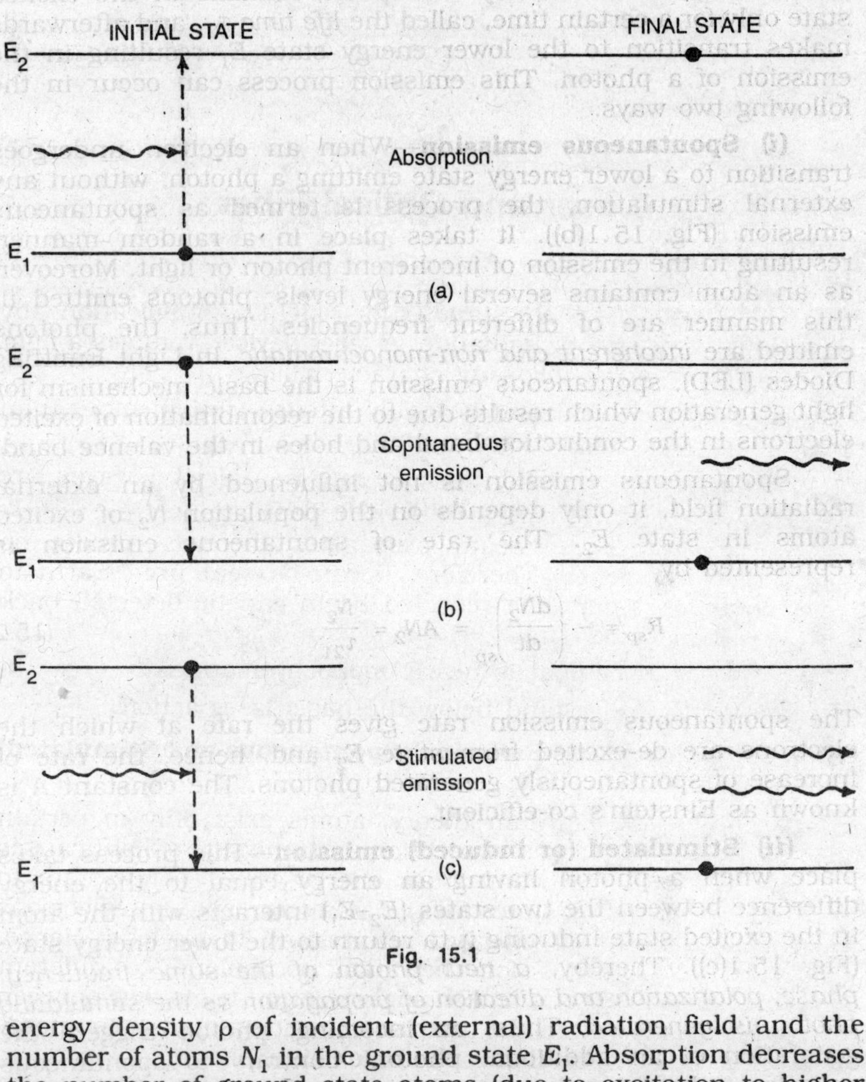

Fig. 15.1

energy density ρ of incident (external) radiation field and the number of atoms N_1 in the ground state E_1. Absorption decreases the number of ground state atoms (due to excitation to higher energy state) with a rate of

$$R_{ab} = -\frac{dN_1}{dt} = B_{12}\ \rho N_1 \tag{15.1}$$

where, B_{12} is known as the *Einstein's absorption co-efficient.* R_{ab} also represents the *upward transition rate* of atoms.

Emission—The excited state is an unstable state. An atom which has been excited by absorption, remains in the excited state only for a certain time, called the *life time* τ_{21} and afterwards makes transition to the lower energy state E_1 resulting in the emission of a photon. This emission process can occur in the following two ways.

(i) Spontaneous emission—When an electron undergoes transition to a lower energy state emitting a photon, without any external stimulation, the process is termed as spontaneous emission (Fig. 15.1(b)). It takes place in a random manner, resulting in the emission of incoherent photon or light. Moreover, as an atom contains several energy levels, photons emitted in this manner are of different frequencies. Thus, the photons emitted are *incoherent and non-monochromatic*. In Light Emitting Diodes (LED), spontaneous emission is the basic mechanism for light generation which results due to the recombination of excited electrons in the conduction band and holes in the valence band.

Spontaneous emission is not influenced by an external radiation field, it only depends on the population N_2 of excited atoms in state E_2. The rate of spontaneous emission is represented by

$$R_{sp} = -\left(\frac{dN_2}{dt}\right)_{sp} = AN_2 = \frac{N_2}{\tau_{21}} \qquad (15.2)$$

The spontaneous emission rate gives the rate at which the electrons are de-excited from state E_2 and, hence, the rate of increase of spontaneously generated photons. The constant A is known as Einstein's co-efficient.

(ii) Stimulated (or induced) emission—This process takes place when a photon having an energy equal to the energy difference between the two states (E_2–E_1) interacts with the atom in the excited state inducing it to return to the lower energy state (Fig. 15.1(c)). Thereby, *a new photon of the same frequency, phase, polarization and direction of propagation as the stimulating photon is generated.* Thus, an incoming, photon triggers the generation of an additional photon. Contrary to spontaneous emission, *coherent radiation* is obtained in this mode of emission. This implies that when an atom is stimulated to emit light energy by an incident wave, the liberated light energy add upto the wave in a constructive manner, providing *amplification of light.*

This process of light amplification is the basic principle of the device LASER which stands for Light Amplification by Stimulated Emission of Radiation. Unlike spontanous emission, stimulated emission is directly proportional to the energy density

ρ of the external radiation field. Also, stimulated emission rate R_{st} increases with the increase in number N_2 of excited atoms. It is given by

$$R_{st} = -\left(\frac{dN_2}{dt}\right)_{st} = B_{21}\,\rho N_2 \tag{15.3}$$

where, B_{21} is known as the *Einstein's stimulated coefficient.*

15.3 Population Inversion

Absorption and stimulated emission are mutually reverse processes. Moreover, they are absolutely homogeneous, that is

$$B_{12} = B_{21} = B \tag{15.4}$$

The relationship between coefficients A and B, as shown by Einstein, is given by

$$A = \frac{8\pi\, hn^3}{\lambda_{21}^3}\, B = 8\pi h \left(\frac{\nu_{21}}{c}\right)^3 n^3\, B \tag{15.5}$$

Where λ_{21} is wavelength of the emitted light wave, n the refractive index and c the speed of light in free space.

Consider a system composed of large number of atoms, in which some atoms are in ground state and others in the excited state. At any given time numerous absorption and emission processes occur. Each absorption process attenuates the incident light wave whereas each stimulated emission amplify it. Unlike light waves generated by stimulated emission, light waves generated by spontaneous emission exhibit a random phase. The small number of spontaneously emitted photons which accidentally have the same phase as that of the photons generated by stimulated emission, aid the lasing action.

In a closed system, absorption and emission processes are in state of thermal equilibrium. This means that

$$R_a = R_{st} + R_{sp} \tag{15.6}$$

Using Eqs. (15.1), (15.2) and (15.3), we get

$$B_{12}\ \rho N_1 = B_{21}\ \rho N_2 + AN_2$$

Laser operation, *i.e.*, amplification of incident light wave requires that stimulated emission is dominant compared to absorption. To determine which of these processes is actually dominant, the ratio

$$\frac{R_{st}}{R_{ab}} = \frac{B_{21}\,\rho N_2}{B_{12}\,\rho N_1} = \frac{N_2}{N_1} \tag{15.7}$$

of stimulated emission rate to absorption rate is a well-suited parameter. This is evident from Eq. (15.7) that this ratio depends only on the population of electrons in the excited state and ground state. On the basis of the above ratio, following classification can be made.

$N_1 > N_2$: That is, population of the atoms in the ground state is higher than that in the excited state. In this case, *absorption is dominant* and, hence, the intensity of light wave exponentially decreases.

$N_1 = N_2$: That is, population of atoms in the ground state is equal to that in the excited state. In this case, *absorption and stimulated emission are in equilibrium* and, hence, the intensity of light wave remains constant.

$N_1 < N_2$: That is, population of atoms in the excited state is more than that in the ground state. This condition, known as *population inversion, is the most important case and must be achieved for the laser action.* In this case, stimulated emission is dominant and hence, the intensity of the incident light increases, *i.e., the incident light wave is amplified.*

Under the conditions of thermal equilibrium, the lower energy level E_1 of the two level atomic system contains more atoms than the upper energy level ; represented by the Boltzmann distribution

$$\frac{N_1}{N_2} = \frac{g_1 \exp\left(-\frac{E_1}{kT}\right)}{g_2 \exp\left(-\frac{E_2}{kT}\right)} = \frac{g_1}{g_2}\, exp\left(\frac{E_2 - E_1}{kT}\right) \tag{15.8}$$

$$= \frac{g_1}{g_2}\, exp\left(\frac{h\nu}{kT}\right) \tag{15.9}$$

where g_1 and g_2 are degeneracies of the levels. Sometimes, an atom has several sublevels of equal energy within an energy level, the energy level is then said to be degenerate. The degeneracy parameters g_1 and g_2 indicate the number of sublevels within the energy levels E_1 and E_2 respectively. If the system is non-degenerate, g_1 and g_2 may be equal to unity. k is Boltzmann's constant and T is the temperature in Kelvin.

This normal situation ($N_1 > N_2$) is depicted in Fig. 15.2 (a). However, to achieve optical amplification, *i.e.*, laser action, it is essential to create a non-equilibrium distribution of atoms in

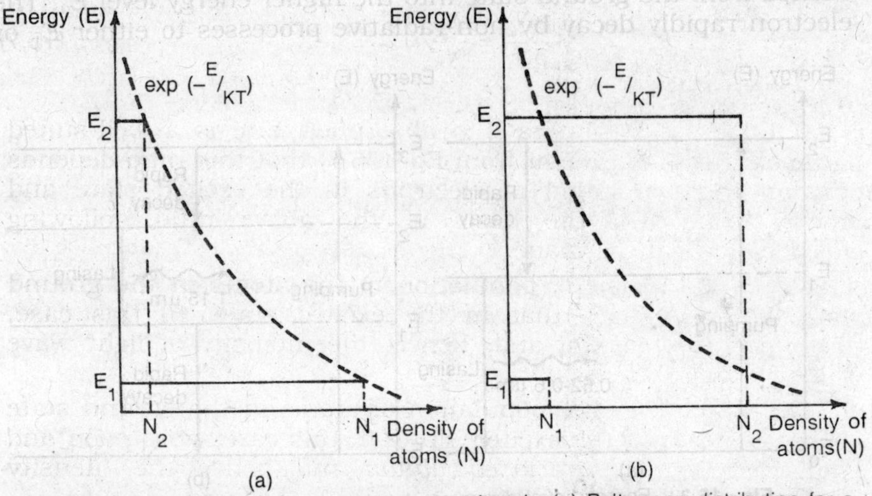

Fig. 15.2 : Polulation of atoms in two energy Level : (a) Boltzmann distribution for a system in thermal equilibrium, (b) A non-equilibrium distribution with population inversion.

such a way that the population of the upper energy level is greater than that of the lower energy level ($N_2 > N_1$). This is depicted in Fig. 15.2 (b).

Pumping—The process of achieving population inversion is known as *pumping*. A common method used for pumping involves the application of intense radiation, *e.g.*, from an optical flash tube, due to which atoms are excited into higher energy state through stimulated absorption.

The two level atomic systems are, however, not suitable for population inversion. Because, if $B_{12} = B_{21}$, the probabilities of absorption and stimulated emission are equal, making the population in the two levels equal.

For achieving population inversion, three or four energy level atomic systems are used. The energy level diagrams of such systems are shown in Fig. 15.3. These atomic systems are characterized by the presence of a central *metastable state* in which the atoms spend an unusually long time. It is the transition from this metastable level that the stimulated emission or lasing action takes place.

Three-level system—As shown in Fig. 15.3(a), a three level atomic system consists of a ground level E_0, a metastable level E_1 and a third level E_2 above the metastable state. In equilibrium, the atomic distribution follow the Boltzmann's law. However, with suitable pumping the electrons in some of the atoms may be excited from the ground state into the higher energy level E_2. The electron rapidly decay by non-radiative processes to either E_1 or

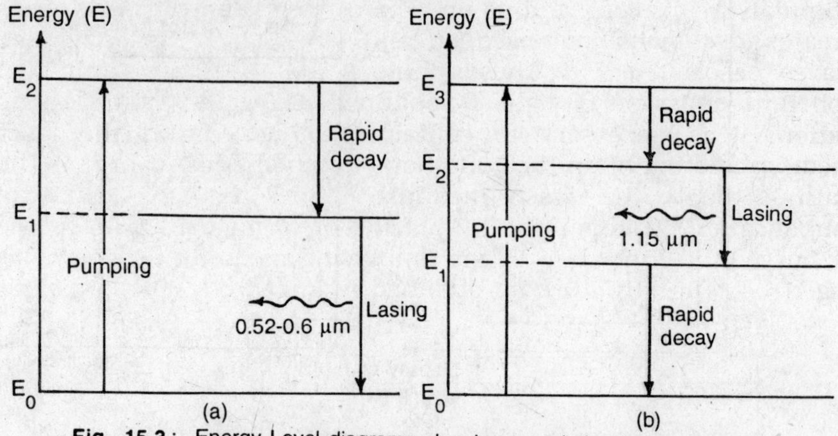

Fig. 15.3: Energy Level diagrams showing population inversion and lasing for (a) three level system. (b) four level system

directly to E_o. Hence empty states are always available in E_2. The life time of electrons in the metastable state E_1 is longer than that in E_2 which results in the accumulation of larger number of atoms in E_1. Over a period, the number of atoms N_1 in the metastable state E_1 becomes more than that in the ground state E_o, thereby, achieving population inversion between the levels E_o and E_1. Stimulated emission and lasing action can then occur by radiative transition of electrons from level E_1 to E_o. A ruby laser is a three level laser. The draw back with three level laser is that it generally requires very high pump power. It is so because ground state being the terminal state, more than half the ground state atoms must be pumped into the metastable state to achieve population inversion.

Four-level System—A four level system, depicted in Fig. 15.3(b) (*e.g.*, Helium-Neon laser) is characterized by much lower pumping requirement. In such systems, pumping excites the atoms from the ground state E_o to the level E_3. The excited atoms then decay to the metastable level E_2. Since the population of E_3 and E_1 remains essentially unchanged, even a small increase in the number of atoms in energy level E_2 creates population inversion. Thus, stimulated emission takes place between level E_2 and level E_1.

15.4 Optical Cavity : Optical feedback and laser oscillation

Light amplification in laser occurs when a photon colliding

with an atom in the excited state causes the stimulated emission of a second photon and then both these photons cause two more stimulated emissions and so on. The continuation of this process creates *avalanche multiplication* and when the electromagnetic waves associated with these photons are in phase, amplified coherent emission of light is obtained. To achieve this lasing action, it is necessary to contain photons, within the laser medium and maintain the conditions for coherence. Longer is the path through the laser medium, higher is the degree of amplification. This is done by placing or forming mirrors (plane or curved) at either end of the amplifying medium, as shown in Fig. 15.4. The structure so obtained is known as *optical cavity*.

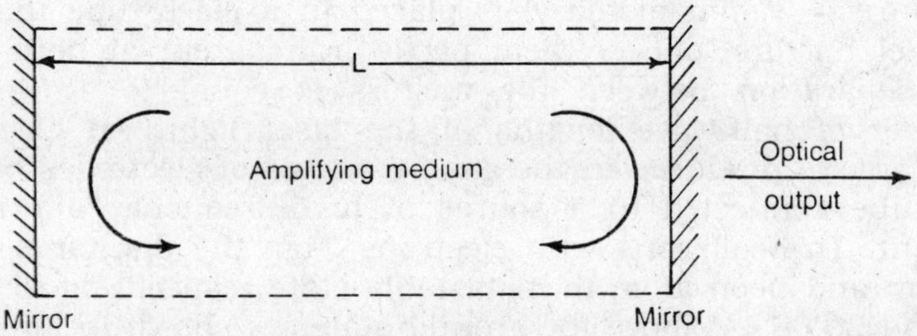

Fig 15.4 : Optical Cavity

The optical cavity is analogous to an oscillator as it provides positive feedback of the photons by reflection at the mirrors at either end of the cavity. Therefore, the area enclosed inside the optical cavity is also called an *optical resonator*. Amplification of light due to a single pass through medium is small. However, after multiple passes due to reflection the net optical path becomes a multiple of cavity length and light amplification becomes very large. One of the two mirrors is made partially transmitting and a part of laser light is obtained from the cavity for external application.

15.5 Types of Lasers

Lasers are principally classified on the basis of either the material used for the active area or the pumping process employed. Gases, crystal and semiconductors are some of the active materials used for lasers. Optical pumping (using another highly intense light (flash) source) and injection current pumping are the employed pumping methods. A broad classification of laser on the basis of active material used is given below.

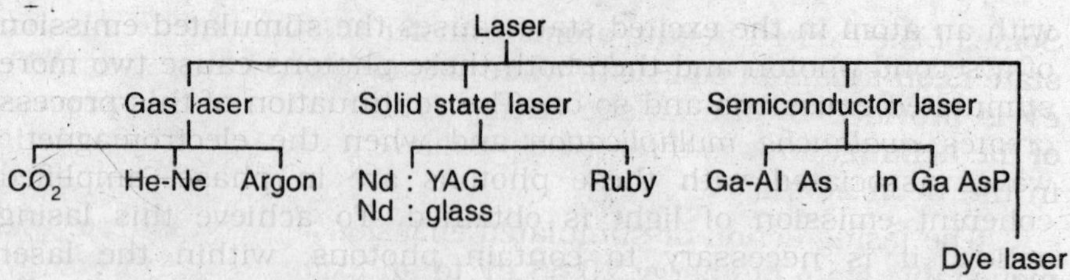

In the college laboratories, Helium-Neon laser is the most commonly used laser and is, therefore, discussed below.

15.6. Helium-Neon laser

In this laser 7 parts of helium and 1 part of neon at a low pressure (≈ 1 mm mercury) is placed in a glass tube that has parallel mirrors, one of them partly transparent at both ends. The separation between the mirrors is equal to an integral number of half-wave lengths of the laser light. An electronic discharge is produced in the gas by means of electrodes outside the tube connected to a source of high frequency alternating current. The collisions with electrons from the discharge excite helium and neon atom to metastable state respectively to 20.61 *eV* and 20.66 *eV* above the ground state, as shown in Fig. 15.5.

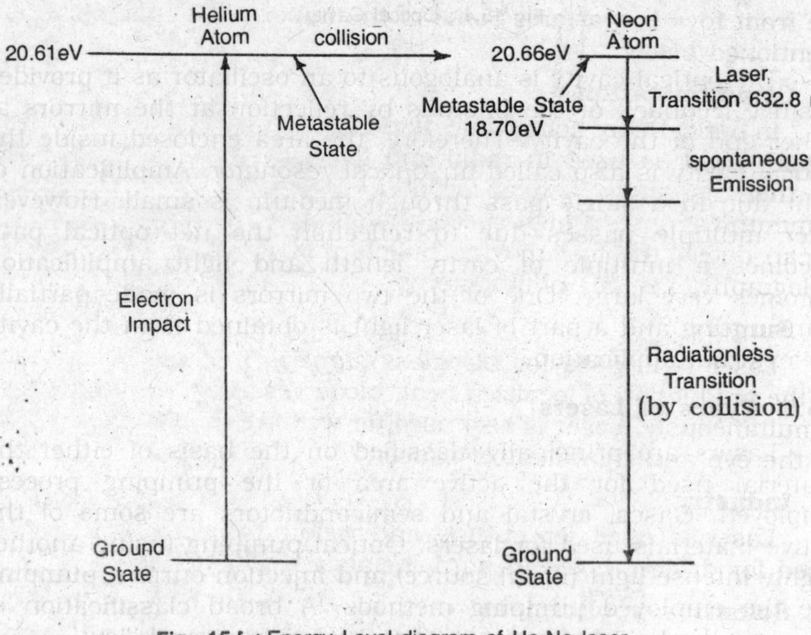

Fig. 15.5 : Energy Level diagram of He-Ne laser.

Some of the excited helium atoms transfer their energy to ground state neon atoms in collisions while the additional energy of 0.05 *eV* is provided by the kinetic energy of the atoms. The purpose of the helium atoms is thus to help achieve a population inversion in the neon atoms.

The lasing action or stimulated emission in neon is from the metastable state of energy 20.66 *eV* to an excited state of energy 18.70 *eV*, with the emission of a photon of wavelength 6328 A°. This is followed by spontaneous emission of a photon due to the transition of the electron to a metastable state which yields incoherent light. The remaining excitation energy is lost in collisions with the tube walls. Unlike a ruby laser in which pulsed laser light is generated due to pulsed excitation from xenon lamp, a He-Ne laser operates continuously due to the excitation of He and Ne atoms Continuously. The power output of He-Ne laser usually lies between 1 mw to 50 mw.

A He-Ne laser being a four-level laser, population inversion is easily achieved compared to more conventional ruby laser which is a three level laser.

15.7 Applications of lasers

Laser is being used extensively nowadays in every field of life from toys to warfare. Some of the important applications are mentioned below.

1. Communication

In optical fibre communication laser is used as optical carrier signal. Laser is used in radio and television transmission. Also, as there is almost no absorption, it is used for under-water communications. In high speed photography, laser light is useful, it can detect the fast moving bullets and missiles. It is used in holography, i.e., 3D photography.

2. Surgery

Laser is is used for *bloodless surgery*. In this process, due to the production of localized heat, blood vessel are cut and sealed simultaneously. Laser is also used for welding the detached retina of the eye and treatment of cancer.

3. Industry

Laser is used for cutting and welding purposes. It is also used for drilling holes in hard metals and diamonds.

4. Scientific applications

Laser is used for determination of chemical and crystalline structures of molecules, *i.e.*, laser spectroscopy. In Raman spectroscopy and various other related phenomenon Laser is being extensively used.

5. Warfare

High power laser is used to destroy enemy's aeroplane, tanks, etc. The concept of *starwar* is based on this application.

15.8. Optical fiber

In early sixties of 20th century, experiments were carried out to find the viability of propagating information carrying light beam through the open atmosphere. It was realised that besides huge loss of intensity, variable factors like rain, fog, dust, etc., are hindrances and a guiding medium is a must through which light can propagate. After intensive investigation optical fibre was found to be the most viable option.

Using optical fiber information carrying light can be transmitted over long distance without any significant loss. Optical fibers are increasingly replacing the transmission line in the communication systems.

An optical fiber is a very thin flexible fiber (as thin as a human hair) normally having a cylindrical shape, as shown in Fig. 15.6. It consists of the following three layers :

(*i*) **Core**—Core is transparent innermost section of refractive index μ_1 and is usually made up of glass or plastic. It is the core through which light actually propagates.

(*ii*) **Cladding**—The core is surrounded by cladding. It is also made up of glass or plastic. The refractive index of cladding μ_2 is always lower than that of core μ_1.

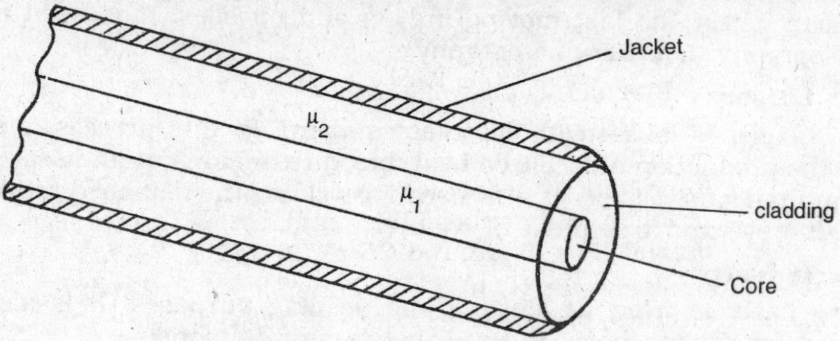

Fig. 15.6 : Structure of Optical fiber.

(*iii*) **Jacket**—To protect the fibre from abrasion, crushing, moisture, etc. an outer layer made up of plastic or polymer is formed which is known as jacket or *protective jacket.* The light propagating through the fiber does not interact with jacket.

Plastic fibres have the advantage of higher flexibility than glass fibres and preferred in systems that require higher flexibility. However, attenuation of signal is higher in the plastic fiber than that in the glass fibre. Also core and cladding should be free from microscopic cracks or flaws in the interior.

Advantages of optical fibres. There are many advantages of optical fiber, some of which are mentioned below.

(*i*) **Extremely large bandwidth**—More than 100 GHz of bandwidth is available in optical fiber. Such a large bandwidth makes it possible to transmit thousands of voice conversations or dozens of video signals over the same fiber simultaneously.

(*ii*) **Smaller diameter and light weight**—Compared to conventional copper cables, optical fibres are smaller in diameter and lighter in weight. This makes optical fiber specially suitable for use in satellites, ships, missiles, high rise buildings, etc.

(*iii*) **Immunity to inductive interference**—Because optical fiber is made up of dielectric materials it is free from induced interference (in the form of electrical noises), unlike in the metallic cables. This makes it possible to use optical fibre even in noisy electrical environment.

(*iv*) **No emission of signal**—As optical fiber cables do not radiate signals, the optic fiber transmission does not interfere with other services. In other words, signal confinement is excellent.

(*v*) **Long life**—Matallic cables corrode and has a life span of around 12 to 15 years on the average. As glass (or plastic) does not corrode, there is no corrosion problem with optical fiber and estimated life of an optical fiber is around 25 to 30 years.

(*vi*) **Greater safety**—In a conventional cable fear of short circuit, prevents its use through areas such as chemical plants and coal mines with volatile gas. There is no such problem with optical fiber and fibre cable can be repaired even when transmission is on.

(*vii*) **High tolerance to environmental factor**—As optical fiber cables are made of glass, these have a high tolerance to temperature extremes as well as to liquids and corrosive gases.

(*viii*) **Conservation of natural resources**—As commonly available sand and plastic, rather scarce resources like copper and iron, is used to make optical fiber, it helps conserve the depleting natural resources.

(*ix*) **Low cost**—The glass, of which generally optical fiber is made, is much cheaper than conventional conductors made from copper.

15.9 Propagation of light in optical fiber

The propagation of light through an optical fiber is depicted in Fig. 17.7. It is the phenomenon of *total internal reflection* at core-cladding interface which guides the light from one end of

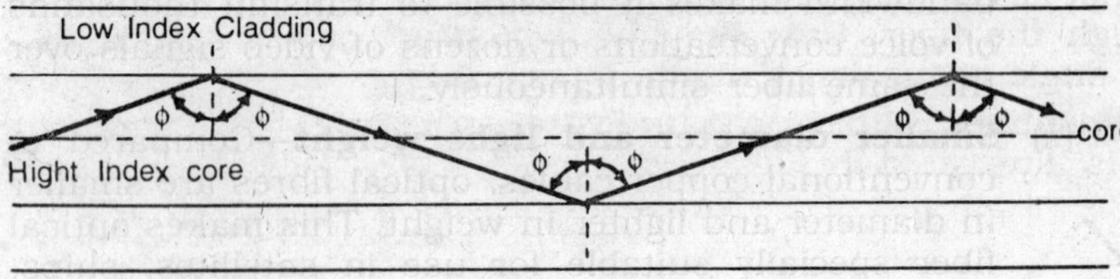

Fig. 15.7 : Propagation of a Light ray in an optical fiber.

the fiber to the other end through the core of the fiber. We know that the two necessary conditions for the total internal reflection to take place are :

(*i*) The light wave should be propagating from an optically denser medium to rarer medium.

(*ii*) The angle of incidence should be greater than critical angle of incidence.

As the refractive index μ_1 of core is higher than that of cladding μ_2 the first condition is satisfied for the light incident from the core on core-cladding interface. If the angle of incidence is greater than critical angle of incidence for the core-cladding interface ($\phi > \phi_c = \sin^{-1}\frac{\mu_2}{\mu_1}$) the light rays undergo repeated total internal refelection at the core-cladding interface.

For a core of refractive index μ_1 = 1.50 and a cladding of refractive index μ_2 = 1.47, the critical angle of incidence ϕ_c is equal to

$$\phi_c = \sin^{-1}\left(\frac{1.47}{1.50}\right) = 78.5^o$$

In Fig. 15.7 it is assumed that the angle of incidence ϕ at the interface is greater than the critical angle of incidence ϕ_c for the core-cladding interface and is, therefore, totally internally reflected according to the laws of reflection. A light ray such as the one shown in Fig. 15.7 is known as *meridional ray*. A meridional ray is one which crosses the axis of the fiber core.

Here it is important to mention that we have assumed a perfect fiber, *i.e.*, without any imperfection. Any imperfection and discontinuity at the core-cladding interface can cause refraction instead of total internal reflection causing loss of light ray into the cladding.

Acceptance angle—From the above discussion it is clear that not all the light rays incident on the core shall be transmitted through the fiber. Only the rays which are incident such that their angle of incidence at the interface is greater than ϕ_c shall be transmitted. Other rays are lost even upon entering the fiber. This is illustrated in Fig. 15.8.

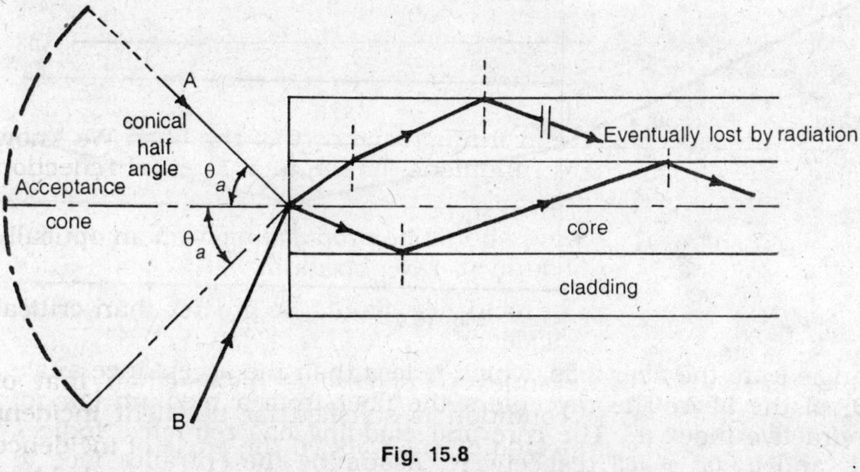

Fig. 15.8

As shown in Fig. 15.8, any ray of light incident on the fiber core at an angle greater than θ_a (such as the ray *B*) will not be transmitted through the fiber because such a ray will be incident at the interface at an angle less than ϕ_c. Such a ray is refracted into the cladding and is eventually lost. In other words, only the rays incident on the core at an angle less than a specific angle called *acceptance angle* θ_a is transmitted through the fiber.

The maximum (conical) angle (to the fiber axis) at which light entering the core is transmitted is called the *acceptance angle*. All the rays incident at the fibre at an angle greater than θ_a is eventually lost. The acceptance angle in unique for a particular fiber but different for different fiber and depends on the materials of the fiber.

From symmetry considerations it can be seen that the emerging angle of the outgoing ray from the fiber is equal to the incidence of the entering ray into the fiber, provided the medium at both the ends are the same or they have the same refractive index.

15.10. Numerical aperture

It is not always practical to talk about acceptance angle and refractive indices of core and cladding. Using ray theory of light, a parameter, called the *numerical aperture* (NA) of the fibre, is more useful and is introduced below.

Fig. 15.9 shows a light ray incident on the fiber core at an

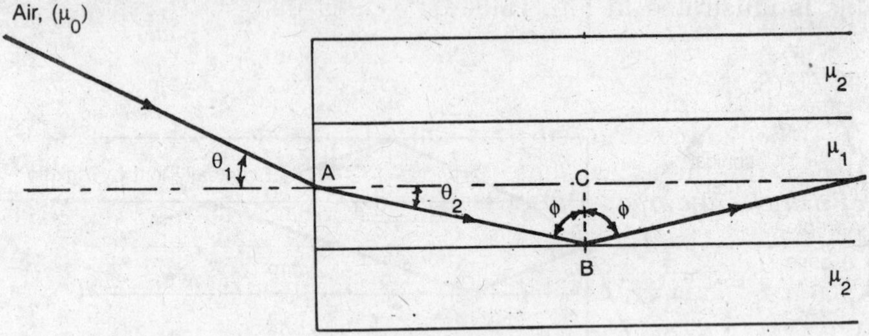

Fig. 15.9

angle θ_1 to the fiber axis, which is less than the acceptance angle θ_a of the fiber. The ray enters the fiber from a medium (air) of refractive index μ_o. The core and cladding has refractive indices μ_1 and μ_2 $(\mu_1 > \mu_2)$ respectively. Assuming the entrance face to be normal to the axis and considering the refraction at the air-core interface at A, we have from Snell's law,

$$\frac{\sin\theta_1}{\sin\theta_2} = \frac{\mu_1}{\mu_0} \tag{15.10}$$

From the right angled Δ ABC, we have

$$\phi = \frac{\pi}{2} - \theta_2$$

or $\theta_2 = \frac{\pi}{2} - \phi$

$\therefore \quad \sin \theta_2 = \sin(\pi/2 - \phi) = \cos \phi$

Substituting for $\sin \theta_2$ in Eq. (15.10), we get

$$\frac{\sin \theta_1}{\cos \phi} = \frac{\mu_1}{\mu_0} \tag{15.11}$$

Using the identity $\sin^2 \phi + \cos^2\phi = 1$, we have $\cos \phi = \sqrt{(1 - \sin^2 \phi)}$

$\therefore$ Eq. (15.11) , can be written as

$$\frac{\sin \theta_1}{\sqrt{1 - \sin^2 \phi}} = \frac{\mu_1}{\mu_0} \tag{15.12}$$

For the limiting case of total internal reflection, $\phi = \sin^{-1}\left(\frac{\mu_2}{\mu_1}\right)$. Also, in the limiting case θ_1 becomes equal to acceptance angle for the fiber, $\theta = \theta_a$. Putting these limiting values in Eq. (15.12), we get

$$\mu_0 \sin \theta_a = \sqrt{\mu_1^2 - \mu_2^2}$$

The parameter $\mu_0 \sin \theta_a$ is called the *numerical aperture* (NA). Hence, the numerical aperture, sometimes also called the *figure of merit* of the optical fibre is given by

$$\text{NA} = \mu_0 \sin \theta_a = \sqrt{\mu_1^2 - \mu_2^2}$$

If the end of the fiber is in air, then $\mu_0 = 1$. In that case

$$\text{NA} = \sin \theta_a = \sqrt{\mu_1^2 - \mu_2^2} \tag{15.13}$$

Normally, μ_2 is only a few percentage lower than μ_1 and, therefore, the following approximation can be made in order to express NA in terms of the relative refractive index difference $\Delta \left(= \frac{\mu_1 - \mu_2}{\mu_1}\right)$. From Eq. (15.13), we have

$$\text{NA} = \sqrt{(\mu_1 + \mu_2)(\mu_1 - \mu_2)} = \sqrt{2\mu_1 (\mu_1 - \mu_2)}$$

$$= \sqrt{2\mu_1^2 \frac{(\mu_1 - \mu_2)}{\mu_1}} = \mu_1 \sqrt{2\Delta} \tag{15.14}$$

where, $\mu_1 + \mu_2 = 2\mu_1$ and $\Delta = \frac{\mu_1 - \mu_2}{\mu_1}$

Fig. 15.10) shows the variation of numerical aperture with acceptance angle θ_a.

As obvious from the above equations, a larger value of NA implies a larger acceptance angle. NA is a measure of the light gathering ability of a fiber. It is independent of the fiber core diameter and depends upon the refractive indices of the core and cladding.

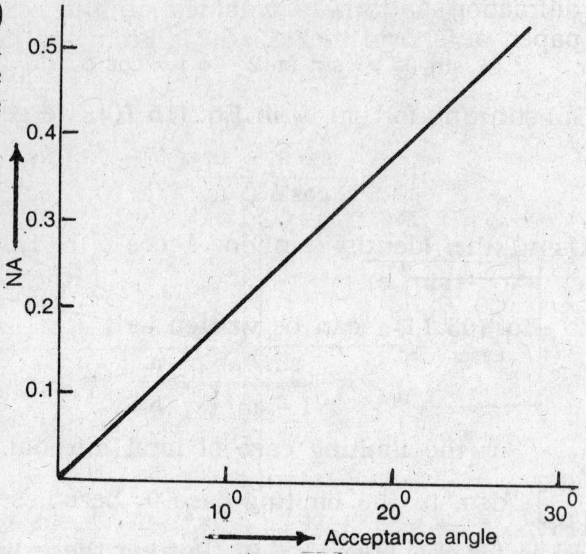

Fig. 15.10

Normally, the numerical aperature for fibres used in short distance communications are in the range of 0.4 to 0.5 while for the fibres used in long distance communications it is kept in between 0.1 to 0.3.

Experiment 15.1

Object—To find (*i*) wavelength and (*ii*) angular divergence of a He-Ne laser.

Apparatus—He-Ne laser, diffraction grating, screen (graph paper), scale.

Theory—(*i*) **Wavelength** : When monochromatic radiation of wavelength λ is diffracted by a diffraction grating, the *n*th order principal maxima is formed at an angle θ_n given by

$$(a + b) \sin \theta_n = n \lambda$$

where, $(a + b)$ is grating constant which is equal to

$$(a + b) = \frac{1}{N} \quad (N = \text{No. of lines per inch})$$

$$\therefore \quad \lambda = \frac{\sin \theta_n}{n.\,N} \text{ inch}$$

$$= \frac{\sin \theta_n}{n.\,N} \times 2.54 \text{ cm.} \quad (15.17)$$

Laser light is incident on the diffraction grating and

diffraction pattern is obtained on the screen usually a graph paper, as shown in Fig. 15.11. From the figure, we have

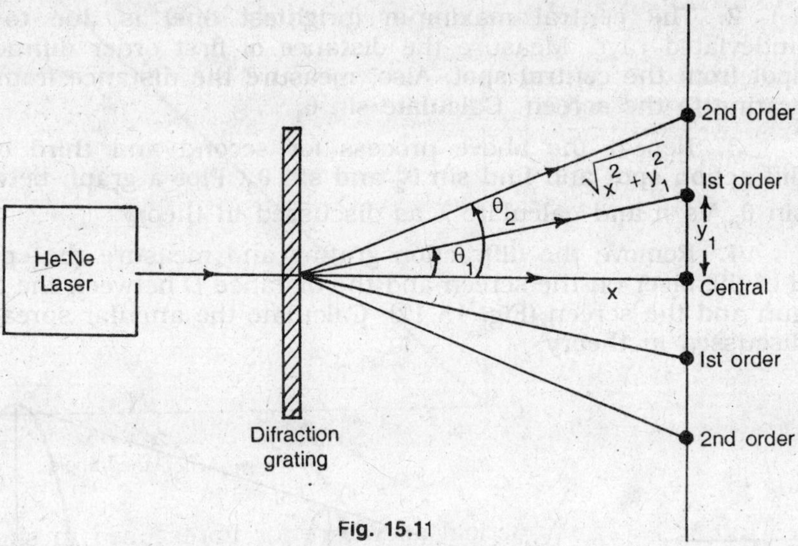

Fig. 15.11

$$\sin \theta_n = \frac{y_n}{\sqrt{x^2 + y_n^2}}.$$

$$\therefore \qquad \lambda = \frac{y_n}{\sqrt{x^2 + y_n^2}} \cdot \frac{1}{nN} .2.54 \text{ cm.}$$

A graph is plotted between $\sin \theta_n$ Vs. n for n = 1, 2, 3, 4,.. and from the slope of the graph $\frac{\Delta \sin \theta_n}{\Delta n} \left(= \frac{\lambda}{a+b} = \frac{\lambda N}{2.54} \right)$, wavelength of the laser light is calculated.

(*ii*) **Angular divergence (spread)**—Angular spread θ is given by

$$\theta = \frac{d}{D}$$

where d = diameter of spread, and

D = distance between grating and screen.

It is assumed that θ is extremely small as actually the case is.

Procedure—1. Place a diffraction grating in front of the laser beam so that a diffraction pattern is obtained on a graph paper which serves as screen placed at a distance of around 2m from the grating.

2. The central maximum (brightest one) is due to the undeviated rays. Measure the distance of first order diffraction spot from the central spot. Also, measure the distance from the grating to the screen. Calculate sin θ_1.

3. Repeat the above process for second and third order diffraction spot and find sin θ_2 and sin θ_3. Plot a graph between sin θ_n Vs n and calculate λ as discussed in theory.

4. Remove the diffraction grating and measure the spread d of the laser on the screen and the distance D between the laser gun and the screen (Fig. 15.12). Calculate the angular spread as discussed in theory.

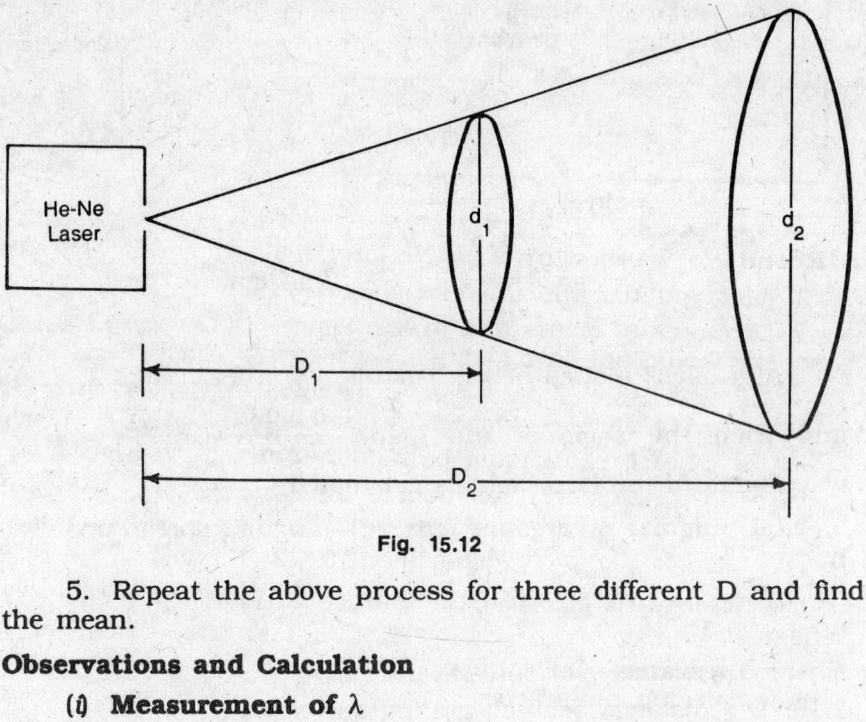

Fig. 15.12

5. Repeat the above process for three different D and find the mean.

Observations and Calculation

(i) Measurement of λ

Grating constant (a + b) =

$$N = \frac{1}{a + b}$$

Distance of screen from grating =

Sr. No.	Order of diffraction	Position of screen x	Distance from Central Spot y	$\sin\theta$
1	0		—	
2	1			
3	2			
4	3			

Slope of $\sin\theta$ Vs n, $\dfrac{\Delta \sin\theta}{\Delta n}$ = ...

$$\lambda = \frac{\Delta \sin\theta}{\Delta n} \cdot \frac{1}{N} \cdot 2.54 \text{ cm} = \text{.......... m} = \text{..... A}^{\circ}$$

Actual λ of a He-Ne laser light = ... A°

% Error =

(*ii*) **Measurement of angular spread**

Sr. No.	Distance from laser to screen (D)	Linear spread d	Angular Spread $\theta = \dfrac{d}{D}$
1			
2			
3			

Mean θ = radian
= degree

Result—(*i*) Wavelength of He-Ne laser light =

(*ii*) Angular spread of laser light =

Sources of errors and precautions—1. The graph used as screen should not have foldings. If present it may introduce error.

2. The laser should neither be too close nor too far from the screen. Keep a distance of a few metres.

3. The spread should be obtained on a wide screen.

4. Do not stare laser source directly, it may damage your eye.

Experiment 15.2

Object—To measure the numerical aperture of an optical fiber.

Apparatus—Optical fiber, He-Ne laser, microscopic objective, photodetector, multimeter.

Theory—Numerical aperture (NA) is a basic parameter of an optical fiber. It is a measure of *light gathering power*, or *degree of openness* of the fiber. The sine of acceptance angle θ_a of the fiber as discussed earlier, is equal to the numerical aperture of the fiber. That is,

$$NA = \sin \theta_a$$

Thus, to determine *NA*, acceptance angle of the fiber should be known.

The laser light is made to pass through the fiber and the emergent beam is incident on a photodetector. The photodetector

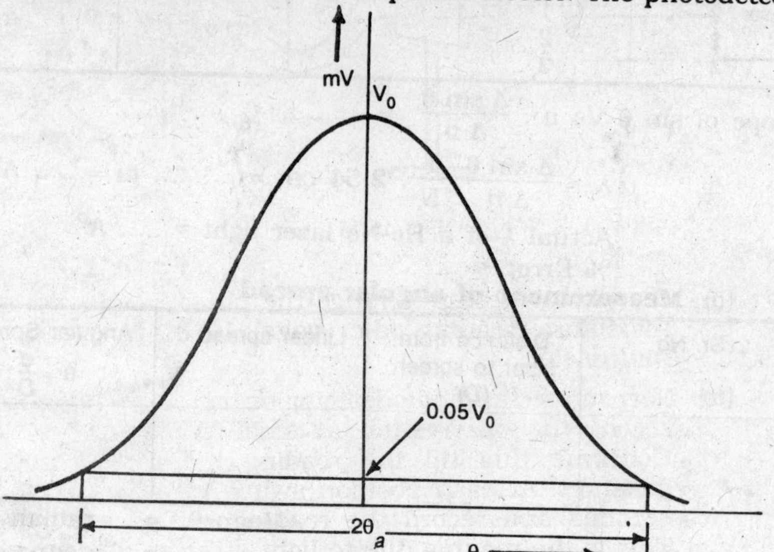

Fig. 15.13

produces a current the magnitude of which is proportional to the intensity of the light incident on it. A multimeter, set on d.c. volt, is connected to the photometer, which measures the developed d.c. voltage in the photometer. The variation of amplitude of d.c. volt with angle of incidence on either side of the maximum value angle is plotted, as shown in Fig. 15.13. Determine 5% of peak d.c. voltage V_o and mark this value on the voltage (y -) axis. Draw a horizontal line passing through this point meeting the curve on both the ends. The angular separation between two ends gives $2\theta_a$. Find θ_a and $\sin \theta_a$ gives the NA of the fiber.

Procedure

(*i*) Arrange, laser source, microscopic objecuve and input end of optical fiber at the same horizontal level (Fig. 15.14.)

(*ii*) Switch on the laser and make the laser light pass through the optical fiber using the microscopic objective.

(*iii*) The laser light from emerging end is made to fall on

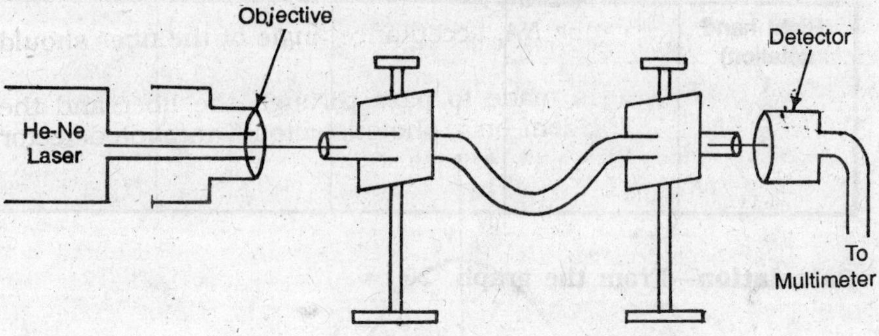

Fig. 15.14

the photodetector. Arrange the photodetector so that maximum reading (d.c. volts) is recorded in the multimeter.

(*iv*) Now rotate the photodetector on one side slowly and record the observation at small intervals, say, 2°. Continue this till the reading becomes almost a constant. At each position switch off laser for a few seconds and record the reading in the multimeter. This is the reading due to light other than laser light and is called background reading.

(*v*) Repeat the above process on the other side of the maximum reading position.

(*vi*) Plot a graph between d.c. volt Vs θ and determine θ_a and hence sin θ_a or numerical aperture of the fiber.

Observations

L.C. of photodetector micrometer =

L.C. of multimeter (d.c. volt) =

Maximum reading in multimeter =

Sr. No. (Left hand rotation)	Angle w.r.t. the central position θ	Multimeter reading (mV)	Background multimeter reading (mV)	Net d.c. volt (mV)
1 2 3				

(Contd.)

(Right hand notation) 1 2 3 . .				

Calculation—From the graph, $2\theta_a$ =

$\therefore \quad \theta_a = \ldots\ldots$

NA = $\sin \theta_a$ =

Result—Numerical aperture of the optical fiber is =

Sources of errors and Precautions

1. The receiving end of fiber should be close to the laser source and the emerging end should be closely placed to the detector.
2. The ends of the fiber should be polished and clean so that no incident light is reflected back from the core.
3. The plane of the receiving end of the fiber should be normal to the incident laser.

ORAL QUESTIONS

LASER

1. What for LASER stands ?
2. Explain the basic mechanism of lasing action.
3. What is meant by population inversion and how is it achieved ?
4. What is an optical cavity ?
5. Why are mirrors placed on the two sides of an optical cavity ?
6. Explain why it is easier to achieve population inversion in a four-level laser compared to that in a three-level laser.
7. Give one example each of a three-level laser and a four-level laser.
8. Name various types of laser.
9. What is the wavelength of laser light from (*i*) a Ruby laser and (*ii*) a He-Ne laser ?
10. Mention five useful applications of laser.

OPTICAL FIBER

1. What is an optical fiber ? Briefly explain its structure ?
2. Why the refractive index of cladding must always be lower than that of core ?
3. What is the need of 'jacket' in an optical fiber ?
4. On what factors the critical angle of incidence of core-cladding interface depend ?
5. Define acceptance angle.
6. Why light from a laser source and not from a LED is preferred for optical fiber ?
7. Define numerical aperture (NA). On what factor(s) does its value depend ?
8. Mention advantages of optical communication over other modes of communications.

16
Miscellaneous Experiments in Optics

Experiment 16.1

Object—To determine the compressibility of a liquid (*e.g.* Kerosene oil) by ultrasonic diffraction of light.

Apparatus—Radio frequency oscillator, glass cell with piezoelectric crystal attached on its side, spectrometer, monochromatic light source, liquid.

Theory—Sound waves having frequencies above the limit of human audibility, *i.e.*, frequency greater than 20,000 Hz, are called *ultrasonic waves* or *ultrasonics*. The piezoelectric crystal, attached on a side in the glass cell, filled with liquid, is connected to a R.F. oscillator. When the frequency of the oscillator is equal to the natural frequency of the piezoelectric crystal, resonance occurs and ultrasonic waves of the same frequency is produced in the liquid. The ultrasonic wave passing through a liquid is an elastic wave in which compressions and rarefactions travel one behind the other regularly spaced equally apart. The distance between two successive compressions or rarefactions is the wavelength λ_l of ultrasonic wave in the liquid.

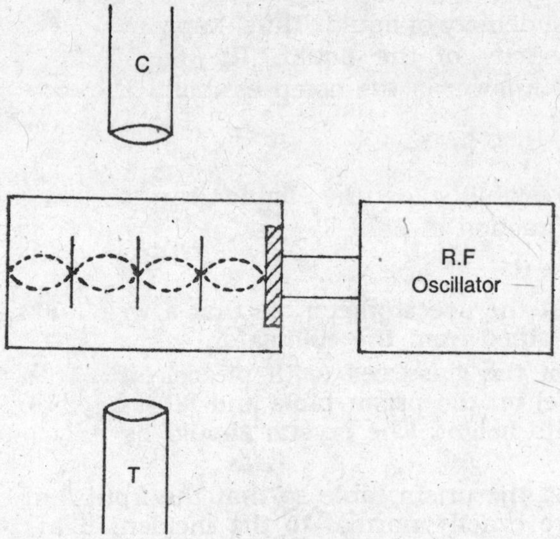

Fig. 16.1

Due to reflections at the opposite sides of the glass cell a stationary wave pattern is obtained with nodes and antinodes formed at equally apart (Fig. 16.1). Now the liquid serves as an accoustic diffraction grating to the light incident normally on it and produces diffraction of light according to the grating rule.

If λ is the wavelength of ultrasonic wave in air, λ_l that in the liquid and θ_n is the angle of diffraction of the nth order, then

$$(a + b) \sin \theta_n = n\lambda$$

In the present case, $a + b = \lambda_l$, therefore,

$$\lambda_l \sin \theta_n = n\lambda \tag{16.1}$$

If ν is the frequency of the ultrasonic wave and υ its velocity in the liquid, then

$$\nu\, \upsilon = \lambda_l \tag{16.2}$$

Thus, knowing λ and n and measuring the angle of diffraction, the wavelength of ultrasonic wave in the liquid can be determined by using Eq. (16.1). Also, knowing the frequency of ultrasonic wave υ and having calculated λ_l we can determine its velocity ν in the liquid using Eq. (16.2).

It is known that the velocity of sound wave in a liquid is related to its bulk elasticity E by the relation.

$$\upsilon = \sqrt{\frac{E}{\rho}} \tag{16.3}$$

where, ρ is the density of liquid. Thus, knowing ν we can calculate the bulk elasticity of the liquid. The reciprocal of this bulk elasticity E is known as the compressibility K of the liquid, *i.e.*,

$$K = \frac{1}{E} \tag{16.4}$$

Thus, compressibility of the liquid can be determined by ultrasonic diffraction of light.

Procedure

1. Adjust the spectrometer, so that a well collimated beam of light is obtained from the collimator.

2. Mount the glass cell (with piezoelectric crystal fixed on one of its side) on the prism table and fill it by the given liquid upto 3/4 of its height. The crystal should be fully immersed in the liquid.

3. Adjust the prism table so that the front and back faces of the cell are exactly normal to the incident light. Also make sure that the piezoelectric crystal in the liquid is located in such a way that the ultrasonic waves generated by it travel in a

direction perpendicular to the direction of light.

4. Switch on the radio frequency oscillator and adjust its frequency to match with the natural frequency of the crystal. When this happens, diffraction pattern is observed in the spectrometer telescope.

5. Measure the angle $2\theta_1$ between the first order spectrum lines on either side of the undeviated image of the slit. Also, measure $2\theta_2$, $2\theta_3$, ...similarly.

6. Note down the frequency of r.f. oscillator at which diffraction pattern is obtained. This is also the frequency of the ultrasonic wave in the liquid.

Observation and Calculations

Wavelength of light λ =

Frequency of ultrasonic wave ν =

Density of liquid ρ =

No.	Order of diffraction	Vernier I	Vernier II	Mean θ_n	$\lambda_l = \frac{n\lambda}{\sin\theta_n}$
1	2				
2	1				
3	0				
4	1				
5	2				
				Mean λ_l =	m

Velocity of ultrasonic wave in liquid $\upsilon = \nu\lambda_l$ =

Bulk Elasticity $E = \upsilon^2\rho$

Compressibility $K = \frac{1}{E}$ = ... N/m^2

Result

(*i*) Velocity of ultrasonic wave in the liquid = ...m/s

(*ii*) Compressibility of the liquid = ...N/m^2

Sources of errors and Precautions

1. Care should be taken to make sure that the liquid is completely stationary before experiment is started.

2. The piezoelectric crystal should be completely immersed in the liquid.

3. The incident light should be perfectly normal to the direction of propagation of ultrasonic wave in the liquid.

4. The light used should be monochromatic.

5. As the angles measured is extremely small, angles should be observed very carefully.

Experiment 16.2

Object—To determine the angle of the biprism by measuring the straight fringes in wedge-shaped films formed by placing the biprism edge on an optical flat.

Apparatus—An optical flat, a biprism, optical arrangement to illuminate the wedge shaped films, sodium lamp, travelling microscope, short focus lens and a reading lens.

Description—The experimental set up is similar to the one used in Newton's ring method for measurement of R. The biprism edge is placed over an optical flat, a little wax may be used to let it rest in a steady fixed position. Slightly above this combination is an arrangement for mounting another thin plate of glass, the inclination of which with the vertical could be altered. It is set inclined at 45° with the vertical so as to direct the slightly convergent pencil of sodium light downwards on to the two wedge-shaped films of air formed between the flat and the surface of the biprism. A vernier microscope with an objective of about 7.5 cm. focal length is usually employed for the purpose of measurment of fringe width of the fringes formed in the wedge-shaped films. This is depicted in Fig. 16.2.

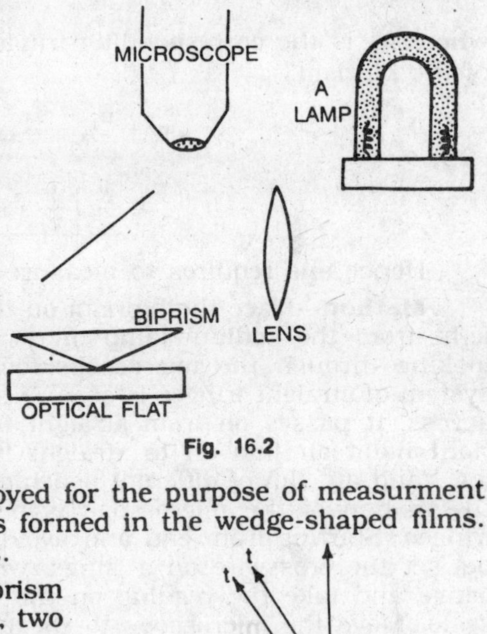

Fig. 16.2

Theory—When the biprism is placed over an optical flat, two wedge-shaped films of air are formed between the upper surface of the optical flat and the lower surface of the biprism. The angles of the two wedge-shaped films will, in general, be unequal, say, θ_1 and θ_2 unless, of course, the biprism is symmetrically placed when the two will be equal. In view of this we get two fringe systems : one corresponding to the right wedge-shaped film of angle θ_1 and

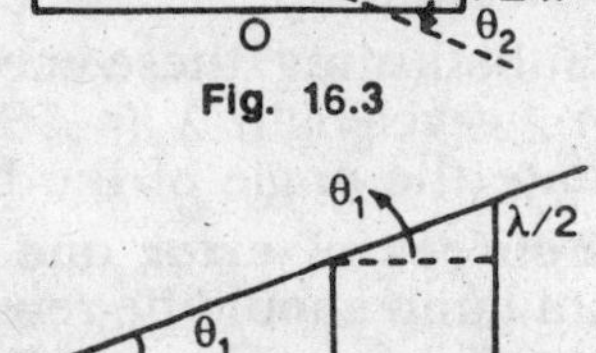

Fig. 16.3

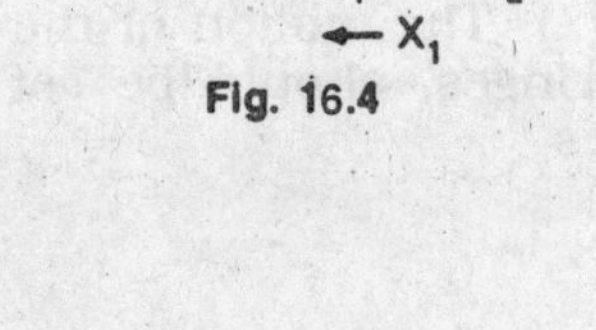

Fig. 16.4

the other corresponding to the left wedge-shaped film of angle θ_2. The two fringe systems evidently differ in their fringe spacing. It can be shown that if B_1 and B_2 are the places on the wedge-shaped film corresponding to consecutive bright fringes, then,

$$\theta_1 = \frac{\lambda/2}{X_1}$$

where X_1 is the fringe width.

Similarly,

$$\theta_2 = \frac{\lambda/2}{X_2}$$

where X_2 is the corresponding fringe width. Evidently the angle of the biprism,

$$\theta = \frac{\theta_1 + \theta_2}{2} = \frac{1}{2}\left(\frac{\lambda/2}{X_1} + \frac{\lambda/2}{X_2}\right)$$

$$= \frac{\lambda}{4}\left\{\frac{1}{X_1} + \frac{1}{X_2}\right\} \qquad (16.5)$$

Hence one requires to measure X_1 and X_2.

Method—Place the biprism on the optical flat and direct the light from the sodium lamp on to it in such a way that, on looking through the microscope focussed on the flat, a fringe system of straight fringes is seen. When the microscope is moved across, it passes on from straight fringes corresponding to the right hand air film on to straight fringes corresponding to the left hand air film of different spacing. Adjust in such a way that the motion of the microscope is perpendicular to the straight fringes. Starting from end and avoiding back lash error (if need be) set the crosswire on a fringe, which may be called the first fringe and take the reading on the main scale and the vernier scale. Move the microscope to the fourth fringe and again take the reading. In this manner take the readings for the seventh, tenth... fringes. From this compute X_1. Repeat the observations with the straight fringes corresponding to the left hand air film and calculate X_2.

Substituting these measured values of X_1 and X_2 and the known wavelength λ (= 5893 Å) for sodium light in Eq. (16.5) calculate the angle of the biprism.

Sources of error and precautions—(1) The light from the sodium lamp should be rendered approximately parallel by means of the short focus lens, an exactly parallel pencil is not required.

(2) The motion of the microscope, focussed in the plane of the fringes, should be set in a direction perpendicular to the fringes.

(3) Since the slow motion screw of the vernier microscope is used in the measurements, it should be seen initially that its motion covers the entire set of fringes to be measured.

Observations and Calculations—(A) Measurement of X_1.

No. of fringe	Microscope Reading	No. of fringe	Microscope reading	12 X_1	Mean X_1
1		13			
4		16			
7		19			
10		22			

(B) Similar table for measurement of X_2

From equation $\theta = \frac{\lambda}{4}\left\{\frac{1}{X_1} + \frac{1}{X_2}\right\}$ =

Result—The angle of the biprism =

16.1. Diffraction at a narrow wire—We can apply the results of diffraction at a straight edge to the study of the diffraction effects at a narrow wire. Let S be a narrow slit illuminated with monochromatic slight. WW′ is the cylindrical wave front from S incident on the narrow wire of diameter AB. The extremities of this wire serve as two straight edges and, therefore, the shadow on the screen is bounded by straight, unequally spaced diffraction fringes on both sides and their origin can be traced to the same cause as the diffraction fringes of a straight edge. The system of fringes produced on either side of the shadow is evidently produced by the light that passed that side of the obstacle and is independent of that which passed the other side. Thus, the upper system of fringes is due to wave front portion AW and the lower system due to the portion BW′.

Apart from these two systems of diffraction fringes bounding the shadow region, there is another system of bands exhibited within the shadow region CD. Unlike the diffraction fringes, this system consists of fine fringes which are qually spaced and the maxima and minima of intensity which are exhibited here show a better contrast than in the diffraction system, because the minima of intensity correspond to complete darkness, *i.e.*, zero intensity. They are essentially a set of interference bands and we must now explain the cause of their existence in the shadow region.

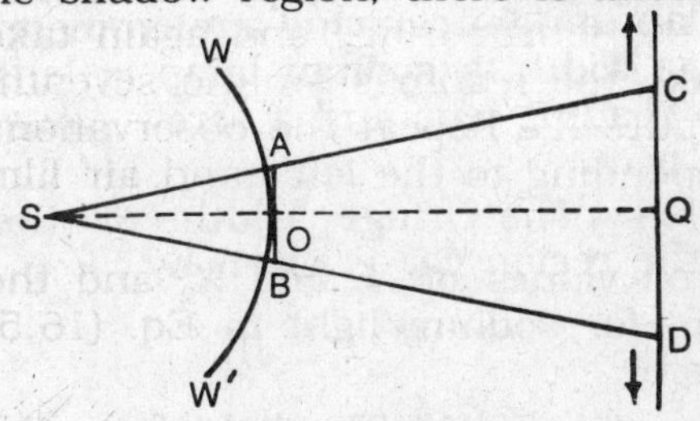

Fig. 16.5

In straight edge diffraction we have seen that the intensity of light gradually fades in the shadow region and that this is due to the half period strips exposed in the region AW. Thus, the intensity at any point Q inside the shadow region is determined entirely by the few effective half period strips in AW, the others of higher order nullifying each other's effect. The portion of the wave front AW can, therefore, be considered as a luminous source at A; likewise the lower portion BW′ can be replaced by a similar source at B. They behave as two coherent sources, and therefore, produce interference effects in the region CD. The bands are interference bands, characterised by equal spacing and sharp contrast between maxima and minima.

The fringe spacing is given by

$$X = \frac{D\lambda}{d}$$

where D is the distance OQ, d the diameter of the wire AB and λ the wavelength of light used in the set up.

All these results were verified by Young. He showed that when the light on the side of A was obstructed, not only did the diffraction fringes corresponding to this edge disappear but also the interference fringes in the shadow region. This was a conclusive evidence of the fact that interference bands exist as a result of the joint effect of the portions of the wavefront, AW and BW′ and that the diffraction fringes exist on either side of the shadow region independent of each other.

Experiment 16.3

Object—To determine the wavelength of light by measuring the interference fringes obtained by diffraction at a wire.

Apparatus—An optical bench carrying (*i*) a slit of variable width capable of rotation in its own plane, (*ii*) an upright on which the experimental wire can be fixed vertically, also capable of rotation in its own plane, (*iii*) an upright carrying a micrometer eyepiece for measuring the fringe width, a sodium lamp, a lens for focussing light on the slit, a reading lens and a screw gauge to measure the diameter of the wire.

Theory—Refer to Fig. 16.4. The fringe width of the interference system existing in the region CD is given by

$$X = \frac{D\lambda}{d}$$

and hence

$$\lambda = \frac{Xd}{D}$$

where D is the distance OQ and d is the diameter of the wire.

Method—(Note—For the use of the optical bench for this experiment, refer to the biprism experiment 16.1)

1. Adjust the slit, the wire fixed vertically (approximately) and the micrometer eyepiece at the same height on the optical bench.

2. Use a convex lens of short focal length to focus the light from the sodium lamp on to the slit.

3. Move the upright carrying the wire along the optical bench and fix its position relative to the slit at a distance of, say, 15 cm., and at a distance of about 50 cm. from the wire, fix the micrometer eyepiece.

4. With the light focussed on the slit, look through the micrometer eyepiece and move it across the bench to obtain the patch of light corresponding to the region CD (may be the fringes are seen) in the field of view. Move the upright carrying the wire in its own plane when it will be noticed that fringes with good contrast appear in the field of view. This will be the case when the slit and the wire are parallel.

5. Remove the lateral shift as discussed in the biprism experiment.

6. With the help of the micrometer eyepiece measure these six to ten fringes seen in the field of view and find the average fringe width.

7. Measure the diameter of the wire at a number of places with the help of a screw gauge and find the mean diameter.

8. Calculate λ using the relation, $\lambda = \dfrac{X d}{D}$

Sources of error and precautions—As in biprism experiment.

Observations and Calculations.—(*i*) Position of wire on the bench = cm.

(*ii*) Position of the micrometer eyepiece on the bench = cm.

Fringe no.	Micrometer reading	Fringe no.	Micrometer reading	4*X*	Mean *X*
1	a	5	e	e—a	
2	b	6	f	f—b	
3	c	7	g	g—c	
4	d	8	h	h—d	

Measurement of diameter of wire—Take at least 10 sets and find their mean.

Hence, $$\lambda = \frac{X d}{D} =$$

$$= \text{Å}$$

Result—The wavelength of sodium light = Å

***Experiment* 16.4**

Object—To verify the law of Malus.

Apparatus—An optical bench carrying a lamp with its housing, a lens to render the light issuing from the lamp approximately parallel, a polaroid to produce plane polarised light, a polaroid provided with a graduated circular scale to give its orientation, a photovoltaic cell, a galvanometer with lamp and scale arrangement.

Theory—Let the transmission plane of the analyser be inclined at an angle θ with the transmission plane of the polariser. The amplitude A of the light emerging from the polariser is resolved into two components,

$$a = A \cos \theta$$

and $$b = A \sin \theta$$

the former having vibrations in the transmission plane of the analyser and the latter having vibrations perpendicular to this. Evidently, only the former component is transmitted. The intensity of the transmitted component of the beam is

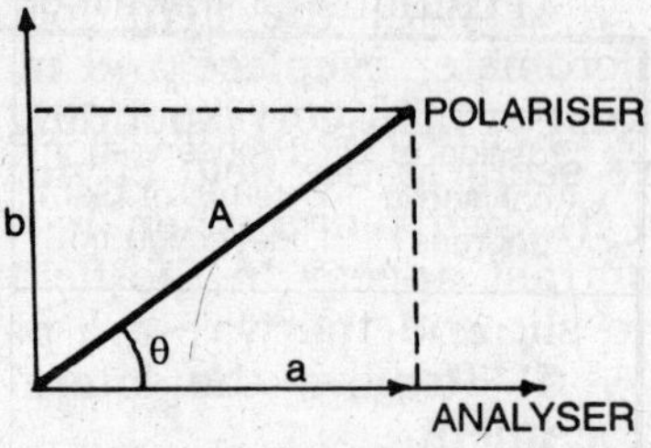

Fig. 16.6

$$I_0 = a^2 = A^2 \cos^2 \theta = I_0 \cos^2 \theta$$

where I_0 ($= A^2$) is the intensity of the incident polarised beam. This relation between I_1 and I_0 is called the law of Malus.

Method—1. Arrange lens, polariser, analyser and photovoltaic cell on the optical bench.

2. Note the position of the analyser on its graduated scale and observe the deflection of the galvanometer.

3. Alter the position of the analyser (which can be rotated about incident beam as axis) in steps of 10° as measured on its graduated circular scale and observe the corresponding deflections of the galvanometer. This may be done for the rotation of the analyser through about 150° or so from its initial setting.

This completes the observations.

4. Plot a graph between the position of the analyser (as recorded on its graduated circular scale) on X-axis and the deflection of the galvanometer (which is proportional to I) along Y-axis.

5. From the graph read the position of the analyser for the maximum deflection of the galvanometer. This evidently corresponds to the position when the polariser and the analyser are parallel, *i.e.*, when θ = 0.

6. From the above position determined graphically, compute θ for the different positions of the analyser and then calculate $\cos^2\theta$ and $I_0 \cos^2\theta$ (*i.e.* $d \cos^2\theta$) and compare the values of $d \cos^2\theta$

with the corresponding observed I, *i.e.*, with the observed deflection of galvanometer.

7. Plot a graph between $\cos^2\theta$ along X-axis and galvanometer deflection along Y-axis. This graph is a straight line and this proves the validity of the law of Malus.

Sources of error and precautions—1. The position of the polariser is not to be disturbed through the experiment.

2. Rest of the precautions as in Experiment 16.2 with a photo-voltaic cell.

Observations and Calculations

Positions of Polariser fixed throughout the experiment.

1	2	3	4	5
Position of Analyser in degrees	I (observed) in terms of Gal. deflection cm.	Angle for Max. intensity I_o	$\cos^2\theta$	$I_0 \cos^2\theta$
141	7	—50	0.4132	6.56
131	8.5	—40	0.5867	9.33
121	12.0	—30	0.7499	11.92
111	14.0	—20	0.8830	14.04
101	15.2	—10	0.9698	15.41
91	15.9	ZERO	1.0000	15.9
81	15.2	+10	0.9698	15.41
71	14.1	20	0.8830	14.04
61	12.0	30	0.7499	11.92
51	9.5	40	0.5867	9.33
41	7.0	50	0.4132	6.56
31	3.9	60	0.2500	3.97
21	2.1	70	0.1169	1.85
11	1.2	80	0.0301	0.48
01	0.6	90	0.0000	0.00
351	1.0	100	0.0301	0.48
341	2.0	110	0.1169	1.85

Result—The graph between $\cos^2$ and *d*, the deflection of the galvanometer (proportional to I) is a straight line. This proves the law of Malus. Comparison of columns 2 and 5 in the table above also proves the validity of the law of Malus.

2. Interference of polarised light—Consider a crystal plate of calcite cut parallel to the optic axis inserted between two crossed nicols (Fig. 16.9). The observation now is that light does get through. The physical explanation of this is quite evident. On emergence from the crystal plate, the light, in general, is elliptically polarised, because, on passage through the crystal plate a certain phase difference is introduced between the two rectangular vibrations O and E, both plane polarised. These rays do not interfere at this stage in the ordinary sense but they give

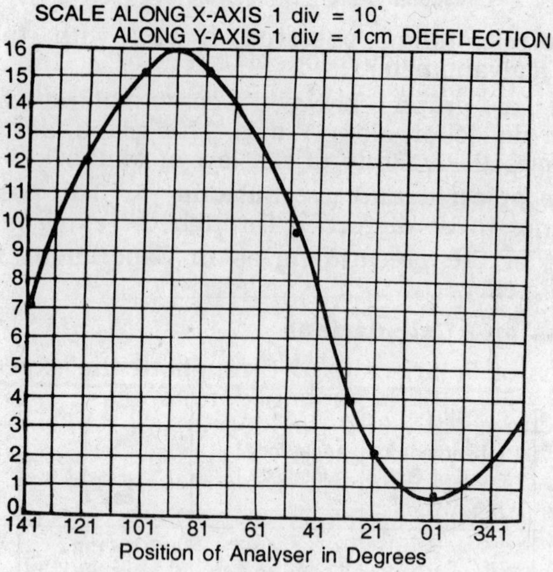

Fig. 16.7

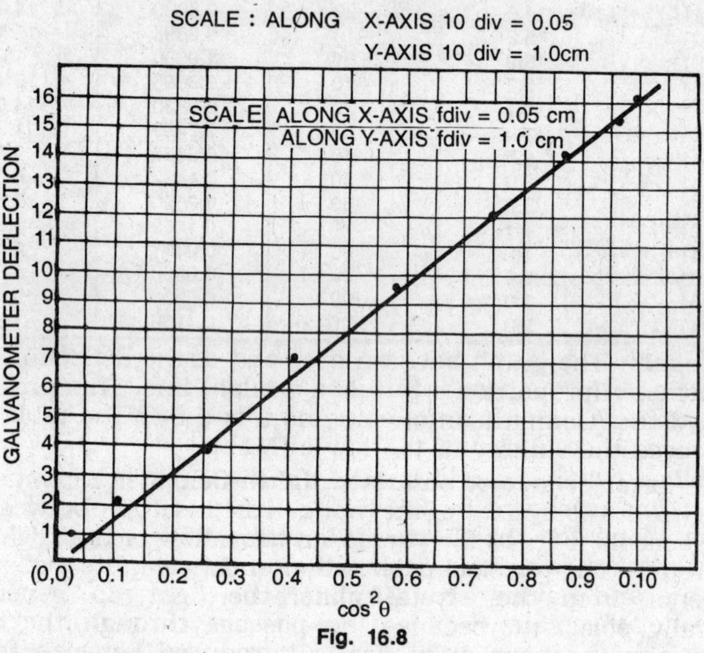

Fig. 16.8

rise to elliptically polarised light. The emergent elliptically polarised light falls on the Nicol which shall not be able to extinguish this light and will transmit some component of this.

For the purpose of calculation of intensity of the transmitted beam through the analyser, we can consider the phenomenon as one of interference between the two component vibrations of the elliptically polarised light when part of each has been transmitted by the analyser with vibration in its principal section. This satisfies the condition for interference of polarised light—two rays

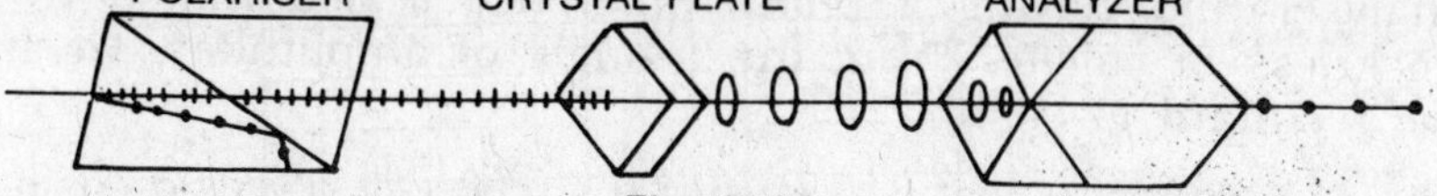

Fig. 16.9

polarised at right angles (obtained from plane polarised light) interfere when brought into the same plane of polarisation. The component vibrations of elliptically polarised light are derived from plane polarised light and the analyser brings components of these into the same plane of polarisation.

The whole question may now be conveniently discussed analytically and in a more general way. Let OP represents the vibration plane of the plane polarised light from the polariser and let this be inclined to the optic axis of the crystal plate at angle α. OQ represents the principal section of the plate and OR, a plane at right angles to this. Let *a* be the amplitude of the incident vibration. It is broken up into O and E rays within the calcite plate and their amplitudes are respectively a sin α and a cos α represented by OB and OE. The E ray has vibrations in the plane of incidence (which coincides with the principal section of the crystal) ; this is shown by OE in Fig. 18.10. OB represents the vibration plane for the O ray. What then emerges out of the crystal plate are two plane polarised vibrations at right angles to each other and having a phase difference δ determined by the thickness of the plate.

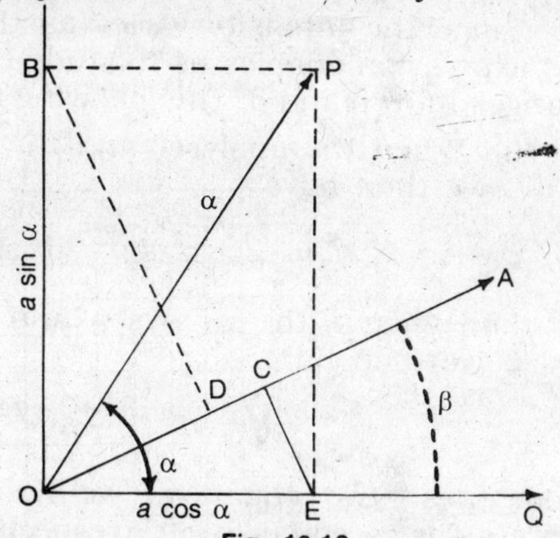

Fig. 16.10

When these components strike the analyser with its transmission plane OA inclined at angle β with the E component of the emergent light from the crystal, it will transmit the component OC and OD with amplitudes a cos α cos β and a sin α sin β whereas the components a cos α sin β and a sin α cos β being normal to OA will not be transmitted. The transmission through the analysing Nicol does not add anything to the existing phase difference. Hence, the problem of finding the resultant intensity due to interference of these components is simply this—

To compound two vibrations of amplitudes $a \cos \alpha \cos \beta$ and $\alpha \sin \alpha \sin \beta$ taking place in the transmission plane of the analyser and having a phase difference δ determined by the thickness of calcite. From the triangle of amplitudes, we have, the resultant given by

$$A^2 = a^2 \cos^2 \alpha \cos^2 \beta + a^2 \sin^2 \alpha \sin^2 \beta + 2a^2 \cos \alpha \cos \beta \sin \alpha \sin \beta \cos \delta$$

$$= a^2 \left\{ (\cos \alpha \cos \beta + \sin \alpha \sin \beta)^2 - 4 \cos \alpha \cos \beta \sin \alpha \sin \beta \sin^2 \frac{\delta}{2} \right\}$$

$$= a^2 \{\cos^2 (\alpha - \beta) - \sin 2 \alpha \sin 2 \alpha \sin 2 \beta \sin^2 \frac{\delta}{2}\}$$

Since the intensity varies as the square of the amplitude, this expression enables us to study the variation of intensity with changes in α, β and δ. The following cases are of special interest :

(i) When the analyser and the polariser are crossed, $\alpha - \beta = 90°$, we then have

$$A^2 = I_0 = a^2 \sin^2 2 \alpha \sin^2 \frac{\delta}{2} \tag{16.6}$$

(ii) When the analyser and the polariser are parallel, $\alpha = \beta$, we then have

$$A^2 = I_p = a^2 \{1 - \sin^2 2 \alpha \sin^2 \frac{\delta}{2}\} \tag{16.7}$$

From these two equations we find that the sum of the two expressions is equal to a^2 irrespective of α and δ. This is to be expected because the sum of the two beams represents the total intensity which should be the same as the intensity of the incident polarised light provided there is no absorption anywhere in the system.

When the analyser and the polariser are parallel, for a given value of δ if the crystal plate is rotated in its own plane, for $\alpha = 0, \pi/2, \pi$ and $3\pi/2$, the resultant intensity is $I_p = a^2$, *i.e.*, the transmitted intensity is equal to the intensity of the incident beam, neglecting reflection and absorption losses. In all these positions there is no double refraction. For $\alpha = 0$ and π the vibration plane of the incident light coincides with the principal section of calcite, the incident light is, therefore, transmitted through the plate as the extraordinary beam and since the transmission plane OA of the analyser coincides with OQ the beam is freely transmitted by analyser. For $\alpha = \pi/2$ and $3\pi/2$ the vibration plane of the incident light OP is inclined at 90° with the principal section OQ of the calcite plate, hence it is transmitted as the ordinary beam with vibrations perpendicular to OQ. As the analyser is parallel to the polariser, it will again freely transmit this beam. Hence, we can say that these four positions occur when the E and O transmission directions in the crystal coincide with the transmission planes of the analyser and polariser which are parallel.

A similar argument will show that when the polariser and the analyser are crossed, there would be no transmission of light through the analyser in these positions.

If the incident light is white and not monochromatic, the phase difference at any point on the crystal plate will vary from wavelength to wavelength. Hence, in any position of the arrangement under reference, the intensity will be zero for some wavelengths but not for others. The transmitted light is, therefore, coloured. When the analyser and the polariser are crossed, the resultant intensity is given by

$$I_c = a^2 \sin^2 2\alpha \sin^2 \frac{\delta}{2}$$

the colours obtained in this case are very rich and pure. However, when the analyser and the polariser are parallel, the colours obtained in the transmitted beam are complimentary because the resultant intensity in this case is

$$I_p = a^2 - a^2 \sin^2 2\alpha \sin^2 \frac{\delta}{2}$$

But these complimentary colours are mixed with a good deal of white light simply because the intensity in this case also depends upon a^2, a factor independent of the wavelength of light. The colours are, therefore, impure in this case.

Experiment 16.5

Object—To determine the value of $(n_e - n_o)$ for quartz.

Apparatus—A parrallel plate of Quartz cut parallel to its optic axis in such a way that it lies in its surface, a pair of polaroids, a strong source of white light, a pair of short focus lens, a spectrometer, a plane transmission grating, a reading lens.

Theory—When a parallel quartz plate with its optic axis lying in its surface is introduced between two crossed nicols, the intensity or the beam transmitted by the analyser, as has been shown earlier, is given by

$$I_e = a^2 \sin^2 2\alpha \sin^2 \frac{\delta}{2}$$

where a is the angle between the vibration plane of the polariser and the optic axis of the crystal plate and δ is the phase difference between O and E rays and this is determined by the thickness of the quartz plate.

The above equation may be written as

$$I_e = \frac{1}{2} a^2 \sin 2\alpha \{1 - \cos \delta\}$$

where $$\delta = \frac{2\pi}{\lambda}(n_e - n_0)\, t,$$

t being the thickness of the quartz plate and $(n_e - n_o)$, the difference between the refractive indices of quartz for the E and O rays, being the birefringence.

Since the retardance R is given by

R = birefringence × thickness = $(n_e - n_o) \times t$

the expression for the intensity of the transmitted beam is

$$I_e = \frac{1}{2} a^2 \sin 2a \{1 - \cos \frac{2\pi R}{\lambda}\}$$

With the white light source the phase difference will be different for different wavelength and if the spectrum of the light emerging from the analyser is viewed through a spectrometer, it will be crossed by dark bands corresponding to zeroes of $\{1 - \cos \frac{2\pi R}{\lambda}\}$ and occur for values of λ in such a way that

$$\frac{R}{\lambda} = 1, 2, 3 \ldots$$

When the observed values of $1/\lambda$ are plotted vrs. integers, the slope of the resulting line gives I/R. Thus its reciprocal gives R. The birefringence may then be calculated from the relation

$$(n_e - n_o) = \frac{R}{t}$$

where the thickness t may be determined with a screw gauge or be taken as a known factor supplied by the manufacturer.

Method—1. Make the preliminary adjustments of a grating spectrometer as described in Experiment 16.1.

2. Take a white light source and place a lens L_1 in such a way that the source is approximately in its focal plane.

3. Allow the light to pass through a polariser (Polaroid) and at some distance from it place the analyser (Polaroid) which can be rotated about the incident beam as axis.

4. Allow the light coming out of the analyser to fall on a

lens L_2 which can be adjusted to focus this beam on to the slit of the collimator of the grating spectrometer.

5. Having focussed the light on the spectrometer slit rotate the analyser in such a way that it is crossed with the polariser and no light is, therefore, received on the slit of the collimator. The field of view of the telescope is completely dark.

6. Introduce the quartz plate between the polariser and the analyser. It will be observed that some light is transmitted by the analyser and the direct image of the slit is seen in the field of the telescope.

7. Move the telescope to the right of the direct image when a continuous spectrum crossed by dark lines will be observed in the field of view of the telescope. This is the first order spectrum towards the right, a similar one can be seen to the left of the position for direct image.

8. Measure the angles of diffraction corresponding to the dark lines seen in the first order and compute their wavelengths.

9. Calculate $1/\lambda$ for each of the above wavelengths and plot a graph between $1/\lambda$ on the Y-axis and the integers 1, 2, 3... on the X-axis.

10. Determine the slope of the resulting straight line, which gives I/R.

11. Compute R and knowing t, calculate $(n_e - n_o)$ for quartz.

Sources of Error and precautions—1. As in the grating experiment 16.1.

2. For better intensity, it is desirable to adjust the quartz plate by trial in such a way that $\alpha = 45°$.

Observations and Calculations.—$n = 1$ (First order.)

S. No.	Vernier	SPECTRUM left	right	Difference 2θ	Mean θ
1	V_1				
2	V				
13					

Calculate the wavelengths corresponding to these measured lines by using the relation

$$\lambda = \frac{(e + d) \sin \theta}{n}$$

and tabulate the results as given in the following Table.

Serial Order	Wavelength in Angstrom	$\frac{1}{\lambda}$ cm^{-1}
1	6430	1.555×10^4
2	6290	1.590

Serial Order	Wavelength in Angstrom	$\frac{1}{\lambda}$ cm^{-1}
3	6190	1.616
4	5986	1.670
5	5846	1.710
2	5062	1.975

GRAPH

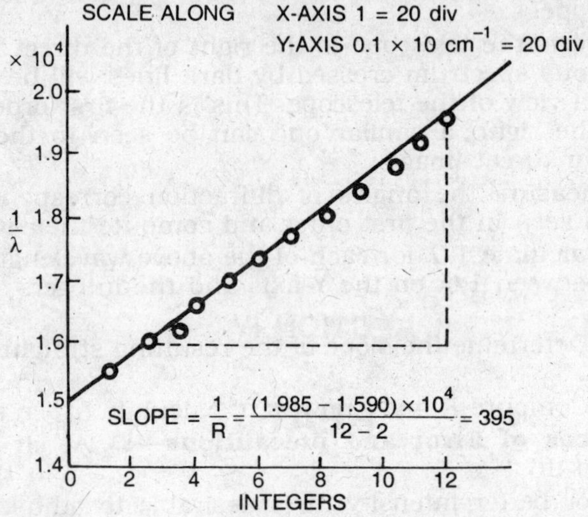

Fig. 16.11

From the graph,

$$\text{Slope} = \frac{1}{R} = \frac{(1.985 - 1.590) \times 10^4}{12-2} = 395$$

whence

$$R = \frac{1}{395}$$

Since the thickness of the quartz plate (plate t) = 0.2 mm = 0.02 cm

$$(n_e - n_o) = \frac{R}{t} = \frac{1}{395 \times 0.02} = \frac{1}{7.9} = 0.12.$$

Result—The birefringence of quartz is = 0.12.

Exercise—To determine the birefringence of Cellulose tape :

The quartz plate in the above experiment is replaced by a glass plate on which as many as 24 to 32 layers of cellulose tape are placed. The thickness of a layer of the tape may be separately measured with a screw gauge after the glue has been removed from it with alcohol.

SECTION IV

HEAT

17

Expansion of Gases

17.1. Gaseous Laws—Since solids and liquids have a size of their own, the effect of external pressure on their size (volume) is quite *inappreciable* when compared with that on a gas which is distinguished from the other two states of matter in having no size of its own and so tending to fill up the whole of the vessel intended to contain it. For this reason, the problem of volume-variation of a gas is a bit more complicated than that of similar variation on a solid or liquid. In the case of gases, therefore, we study the mutual variations of volume, pressure and temperature. With a view to know how each of these influences the other, it has been found more convenient to study the mutual variation of any two of them, keeping the third constant. This subject was studied in this way by the two well-known scientists, Robert Boyle and Charles who enunciated the results of their experiments in the form of two well-known laws called after them the Boyle's and the Charles' laws.

Boyle's law refers to variations between volume and pressure, the temperature remaining constant. It states that *the product of the pressure and volume of a constant mass of the gas is constant provided the temperature remains the same,* or if *P* and *V* denote the pressure and the volume of the gas,

PV = constant (at constant temperature).

Charles' law refers to variations between volume and temperature, pressure remaining constant, and says that *all gases have the same coefficient of expansion.*

The third alternative, *viz.*, the variation of pressure with temperature, volume remaining constant was not simultaneously studied by Charles. But later when this study was undertaken the results of experiments showed that not only all gases have the same coefficient of expansion, but also that *all gases have the same coefficient of increase of pressure with increase of temperature at constant volume,* and further that these two coefficients are the same. Hence, we have,

$$\alpha_p = \alpha_v$$

if α_p and α_v are respectively the pressure coefficient and volume coefficient of a gas.

If P_0 and V_0 are the pressure and volume of a gas at 0°C, and P_t and V_t are their values at t°C we have,

$$V_t = V_0 (1 + \alpha_v t) \tag{17.1}$$

or

$$P_t = P_0 (1 + \alpha_p t) \tag{17.2}$$

$$\alpha_v = \frac{V_t - V_0}{V_0 + t} \tag{17.3}$$

and

$$\alpha_p = \frac{P_t - P_0}{P_0 \ t} \tag{17.4}$$

19.2. Absolute Zero and Absolute Scale of Temperature—If we suppose[1] that a gas continues to behave in accordance with these laws at all temperatures, we see from Eq. (19.3) and (19.4) that at a temperature $1/\alpha_p$ below the melting point of ice V_t or P_t will vanish.

This leads to a startling conclusion, *viz.*, that there is a temperature at which the volume and pressure of a gas should vanish. The vanishing of volume means vanishing of matter, and so an absurd conclusion. Vanishing of pressure, in terms of dynamical theory of gases, means vanishing of kinetic energy; a result having a deeper significance given by Nernst Heat Theorem.

Though an actual gas does not obey Charles' law at very low temperatures, yet the above conclusion of a temperature at which the pressure vanishes has a theoretical importance in giving us a new scale of temperature called the *Absolute Scale of Temperature* Measurement.

At this temperature $1/\alpha_p$ below 0°C the gas ceases to exert any presure on the walls of the containing vessel. On the basis of kinetic theory of gases, it means that the gas has lost all motions, molecular and atomic. As heat content of a body is supposed to be due to such motion, then, when no such motion exists the body must be devoid of all heat. It is unimaginable to have a body at a temperature lower than the one at which it is devoid of all heat. Therefore, this temperature $I/\alpha_p{}^o$ is called *absolute zero*. Taking the value of α_p as 0.0036, the absoute zero is

$$-\frac{1}{.00366} = -273°C$$

i.e. 273°C below the Zero of the Centigade scale.

A scale of temperative, having–273°C as Zero, is called the

1. Though Boyle and Charles gave us these simple laws, subsequent experiment with more refined instruments showed that actual gases obey these laws only at very low pressures and high temperatures far above the critical temperature.

absolute scale of termperature. Temperatures reckoned on this scale are called *absolute temperature.* On this scale, the melting point of ice, 0°C, is 273°K, whereas the boiling point of water, 100°C, is 373° K. Thus, any temperature t°C can be expressed on the absolute scale by adding 273° to it. Hence, we have.

$$t^{o}\ C = (273 + t)^{o}\ K = T^{o}\ K \tag{17.5}$$

17.3. Perfect Gas—Most of the gases are found to obey Charles' law and Boyle's law approximately within fairly wide limits of pressure and temperature; we are, therefore, led to the conclusion that an *ideal gas be one which obeys both the laws accurately.* Such a gas deserves the name a *perfect gas.* No actual gas can claim to be called perfect. However, permanent gases like air, hydrogen, nitrogen, oxygen and helium may be treated as approximately perfect. Other gases show greater deviation from these laws.

On the basis of the laws of Charles and Boyle *the equation of state of a gas* connecting its pressure *P*, volume *V*, and absolute temperature *T*, has been deduced. The equation,

$$PV = RT$$

where *R* is the gas constant, brings out clearly the fact that of the three quantities *P*, *V* and *T*, only two are variable, the third being a quantity dependent on the other two and connected by the above equation.

Experiment 17.1

Object—To determine the *pressure coefficient* of air by constant *volume* air thermometer.

Apparatus—Constant volume air thermometer, ice-bath and arrangement for boiling water and mercury thermometer.

Description of apparatus—The convenient form of the apparatus for laboratory use is due to Jolly (Fig. 17.1). It essentially consists of a glass globe A which can be immersed in a bath. The globe is in communication with a mercury manometer through a glass tubing of a fine bore. The manometer tube is quite wide and consists of two limbs BC and DE containing mercury and connected by a thick pressure tubing. The level of mercury in BC can be adjusted by raising or lowering the other limb till the meniscus of the mercury level in BC just touches a glass index placed at the junction of the limb BC and the narrow tube. The function of the index is to ensure constancy of the enclosed volume in the globe when the mercury meniscus just touches it. The pressure exerted by the enclosed gas can be determined by noting the difference in the levels of the two limbs and adding this to or subtracting it from, the atmospheric pressure as the case may be. The reading for atmospheric pressure should be taken with a Fortin's barometer at the time of recording observations. For accurate measurement of the difference in levels of the mercury surfaces, the apparatus is so arranged that the two limbs lie close together beside a paper scale attached to the frame.

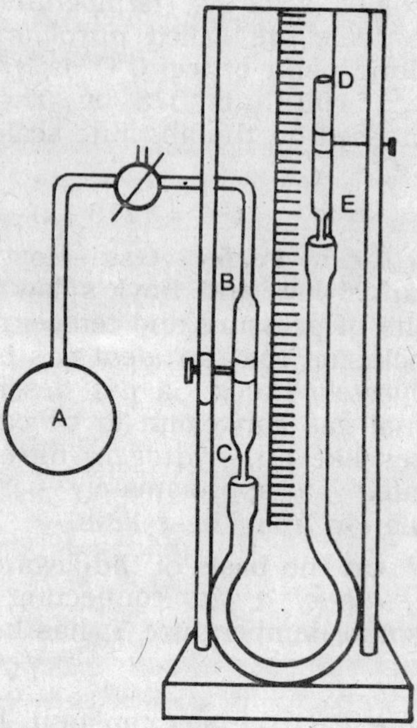

Fig. 17.1

Theory—Let P_1 be the pressure of the enclosed air at t_1°C and P_2 be the pressure of the same mass of air at t_2^oC, the volume of the gas being kept constant. Then, if α_p be the pressure coefficient of expansion for air, and P_o be the pressure exerted by the same mass of the gas at constant volume at 0°C,

$$P_1 = P_o(1 + \alpha_p t_1)$$

and

$$P_2 = P_o(1 + \alpha_p t_2)$$

$$\frac{P_1}{P_2} = \frac{1 + \alpha_p t_1}{1 + \alpha_p t_2}$$

Solving for α_p we get,

$$\alpha_p = \frac{P_2 - P_1}{P_1 t_2 - P_2 t_1} \qquad (17.6)$$

Thus, the determination of the pressure coefficient of a gas involves the measurement of the pressures exerted by the given mass of the gas under constant volume at two known temperatures t_1 and t_2.

Method—Take a beaker sufficiently large to accommodate the glass globe A. Fill it with water so that A is completely submerged. Place a thermometer near A to record the temperature of the enclosed air in it. Heat the water to its boiling point when the level of mercury in the limb BC will be found to be lowered due to the expanding gas. Raise the limb DE upwards so that the meniscus of mercury in BC just touches the tip of the glass index placed therein. When the two levels remain stationary, read the difference between the two on the paper scale fixed in between the two limbs. Also record the atmospheric pressure from a Fortin's barometer. To this, add the observed difference between the two levels; this gives the pressure exerted by the enclosed gas at the temperature of boiling water indicated by the mercury thermometer.

Allow the liquid to cool, taking due care that the mercury in the limb BC is not drawn over into the globe A. To prevent this the limb DE is lowered so that the level in BC is well below the junction of the wide and the narrow tubes. Keep the water well stirred as the temperature falls. When it has fallen to the desired value (say, to 90°C) the adjustment of the limbs is again effected so that the mercury level in BC remains just in contact with the tip of the index, thus ensuring a fixed volume of the enclosed gas. But before taking final readings, the temperature should be made steady for 4 to 5 minutes because the air inside the glass takes time to attain the temperature of the bath, glass being a bad conductor of heat. The temperature is maintained steady by adjusting the flame and constant stirring of the water in the beaker. Again read the difference in the levels of mercury in the two limbs and add to the atmospheric pressure, thus obtaining the pressure of the same amount of enclosed air but at a lower temperature (90°C).

In this manner take 5 or 6 readings at regular intervals of 10°C and tabulate the observations. Then plot a graph between the temperature of the gas as abscissae and the pressure exerted by it in mm. of mercury as ordinates. The graph will be a straight line. From the graph observe the pressure P_1 and P_2 corresponding to two chosen temperatures t_1 and t_2 and then calculate a_p from the Eq. (17.6).

The graph can be employed to get the value of P_o, the pressure of the enclosed gas at 0°C, by extrapolation. Produce the straight line to cut the line $t = 0$; the point of intersection gives the value of P_o. The value of a_p may then be evaluated from the formula:

$$\alpha_p = \frac{P_t - P_0}{P_0\, t}$$

If the straight line is produced further towards the side of lower temperature, it will cut the temperature axis at a point where the pressure becomes zero. This temperature is the absolute zero.

Sources of error and precautions—(1) The mercury employed in the limbs of the thermometer should be pure for, if impure, it might stick to the sides. For this reason the limb should be tapped before taking the final observation.

(2) The air enclosed within the bulb should be completely dry. The presence of even a small quantity of moisture will affect the result since vapours do not obey the gas laws.

(3) The globe containing air should be completely immersed in the liquid and the latter should be constantly stirred.

(4) Since glass is a bad conductor of heat, it takes time for the air inside to acquire the temperature of the bath. The readings should, therefore, be taken when the temperature of the bath has remained steady for some time; for this reason it is recommended that the observation should be taken while the bath is cooling and not while it is being heated.

(5) While the bath is cooling, the mercury in the limb BC should stand at a low level to avoid its being withdrawn inside the bulb due to the reduction of the pressure of the cooling gas.

Observations—Atmospheric pressure (H) = cm.

No. of readings	Temperature of the bath $t^{o}C$	Constant level of mercury in the closed limb b cm.	Level of mercury in the open limb X cm.	Difference in levels b-x cm.	Pressure of air in the bulb. H± (b − x) cm.

Calculations—From the graph, Fig. (17.2)

(*a*) For the point A,

Pressure = cm.

Temperature= °C

(*b*) For the point B,

Pressure = cm.

Temperature = °C

$$\alpha_p = \frac{P_2 - P_1}{P_1 - P_2 t_1}$$

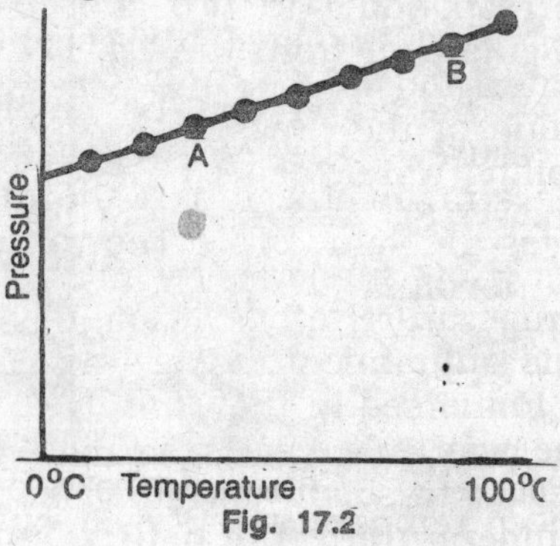

Fig. 17.2

Result—The mean pressure coefficient of air = per °C

Experiment 17.2

Object—To determine the melting point of a substance with a constant *volume* air thermometer.

Theory—Let P_t and P_0 be the pressure of given mass of air at t°C and 0°C respectively, the volume of air being kept constant. Then,

$$P_t = P_o (1 + \alpha_p t)$$

or
$$t = \frac{P_t - P_o}{P_o \alpha_p} \tag{17.7}$$

This equation provides us with a ready means of measuring temperature if the values of P_t, P_0 and α_p are experimentally determined. For a quick determination of α_p, the bulb may be enclosed in a steam bath and ice bath, thus giving the values of P_{100} and P_0 at 100°C and 0°C. Then,

$$\alpha_p = \frac{P_{100} - P_0}{P_0 \times 100} \tag{17.8}$$

Then recording P_t at an unknown temperature, the value of the latter can be found.

Method—Immerse the globe of the air thermometer in a mixture of ice and water at 0°C. When the bulb has remained in the container for a few minutes, adjust the mercury in the limb BC with the help of the other limb DE so that the level of mercury in the former touches the tip of the glass index. It will be found that the levels then remain stationary indicating that the air inside the globe has acquired the steady temperature of ice. Read the difference in the levels in the two limbs. Record the barometric height also and from a knowledge of these two, calculate the pressure exerted by the enclosed air at 0°C. Let it be P_o.

Next immerse the globe of the air thermometer in boiling water and making the necessary adjustments as before, get the value of the pressure exerted by the gas at the temperature of boiling water. Let it be P_{100}. Then calculate the value of α_p from Eq. (17.8).

Now immerse the bulb in a water bath and maintain its temperature to be the same as the melting point of wax. To make this adjustment, some wax contained in a thin-walled test tube is immersed in the water bath which is heated from below. When the wax just begins to melt withdraw the flame : when it just begins to solidify, again heat the bath. In this manner maintain temperature of the bath at the melting point of wax and record the value of the Pressure P_t. Then calculate t from Eq. (17.7).

Next taking temperature along the axis of X and pressure along the axis of Y, plot the points $(0, P_0)$ and $(100, P_{100})$ on a graph paper. Join the two points by a straight line. Find out the point P_m on this straight line (Fig. 17.3) whose ordinate is P. The obscissa of the point will give the melting point of wax.

Observations—Atmospheric pressure (H) = cm.

No. of readings	Temperature of the bath $t^{o}C$	Constant level of mercury in the closed limb b cm.	Level of mercury in the open Limb x cm.	Difference in levels $(b - x)$ cm.	Pressure of air in the bulb $H x \pm (b - x)$ cm.
	0°C 100°C Melting pt. of wax				

Calculations $t = \dfrac{P_t - P_o}{P_0\, \alpha_p}$

$= \quad ^{o}C$

From the graph (Fig. 19.3) the temperature corresponding to the pressure...... = °C

∴ Melting point of wax = °C

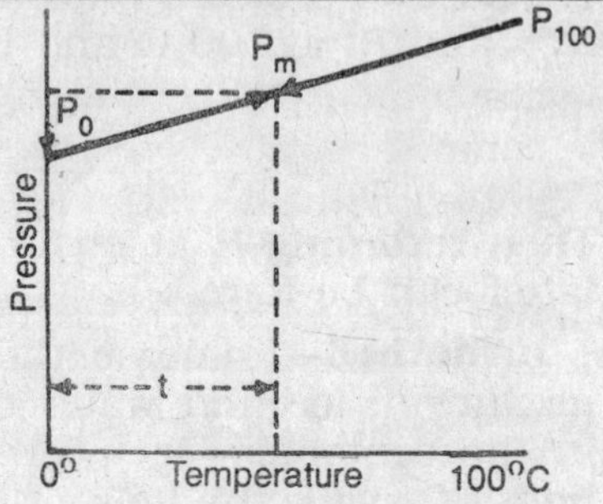

Fig. 17.3

Result—The melting point of wax, as determined by the constant volume air thermometer = °C

Criticism of the method—The method yields a satisfactory value of α_p and can be successfully employed for the determination of the melting point of wax. However, there are certain sources of error which limit the accuracy of the results obtained with this apparatus. We do not take into consideration the expansion of the glass bulb since it is negligible on comparison with the expansion of the gas contained in it. If the coefficient of expansion of the glass bulb be taken as known and duly applied even then the accuracy of the result is limited, for, it is doubtful to treat the material of the vessel to be isotropic when it has been blown into the shape of the bulb. To eliminate such a doubtful error, use may be made of a globe made of fused silica which has a very small coefficient of cubical expansion. The result then obtained is independent of the material of the containing vessel. Moreover, corrections should be applied (1) for the "dead space" gas which does not acquire the same temperature or pressure as the gas contained in the bulb and (2) for the density of mercury in the two limbs being different.

The constant volume air thermometer as used in the above experiments, has been modified to the form of the *Standard Gas Thermometer*. The latter has certain advantages over the former. For the measurement of pressure in the former, four readings of mercury surfaces are necessary, *viz.*, those in the two limbs of the air thermometer and the zero and upper surface of the Fortin's barometer. In the standard gas thermometer these readings are reduced to two by incorporating a barometer within the apparatus, the possible error in the measurement is thus halved. Moreover, hydrogen is employed which shows a uniform increase in pressure with temperature, volume remaining constant. Hence the constant volume hydrogen scale of temperature is uniform and can give accurate results of unknown temperatures that may be determined with this thermometer.

17.4. Constant Pressure Air Thermometer—This instrument (Fig. 17.4) consists of a glass bulb A in communication with the graduated limb BC through a capillary tubing M and a three way tap T which can be so turned that both A and BC can be put in communication with atmospheric air. When desired the tap can also be turned to cut both A and BC from the atmosphere but at the same time establishing communication between them.

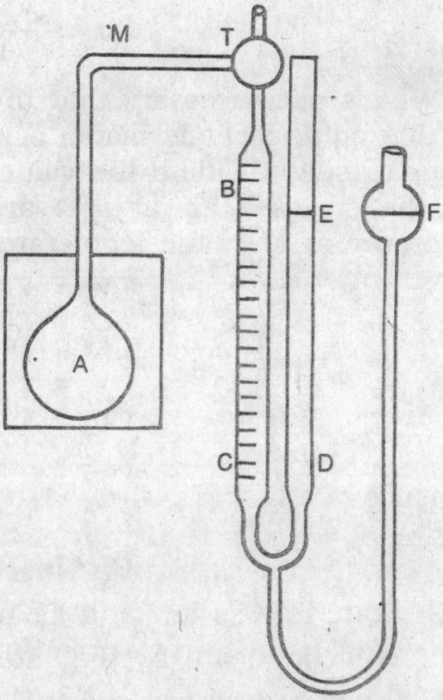

Fig. 17.4

To use the apparatus for the determination of the volume coefficient of expansion of air, the following procedure is adopted.

The bulb A is immersed in an ice bath containing a mixture of ice and cold water, the tap T being so turned as to put both A and BC in communication with the atmosphere. After some time when air inside the bulb has acquired the temperature of the melting ice, the reservoir F is so adjusted that the meniscus of mercury in BC and ED stands at the same level. The tap T is then closed and turned so as to put A in communication with BC. The volume of the air filling the bulb at 0°C is then observed. Let it be V_0.

Next immerse the bulb in a bath of boiling water, the temperature being recorded with a thermometer. Owing to the rise of temperature the volume and pressure of the enclosed air will change, but our object is to measure the volume change due to change of temperature, keeping the pressure constant. For this reason, the reservoir *F* is lowered till the mercury in the limb BC and ED again stands at the same level. The reading in the limb BC is then taken. The difference between this reading and the previous reading gives the change in volume at constant pressure brought about by the change in temperature from 0°C to 100°C.

The value of α_v may now be calculated with the help of the relation,

$$\alpha_v = \frac{V_t - V_0}{V_0\, t}.$$

It is not necessary that the initial reading be taken at 0°C. If this adjustment is made, say, at the room temperature, t_1°C, we can still calculate the value of α_v. For, let V_1 be the volume of the enclosed air at t_1°C and V_2 be its volume at constant pressure at t_2°C, the temperature of boiling water. Then,

$$V_1 = V_0 (1 + \alpha_v . t_1)$$

and

$$V_2 = V_0 (1 + \alpha_v . t_2)$$

∴

$$\frac{V_1}{V_2} = \frac{1 + \alpha_v . t_1}{1 + \alpha_v . t_2}$$

whence

$$\alpha_v = \frac{V_2 - V_1}{V_1 t_2 - V_2 t_1} \qquad (17.9)$$

ORAL QUESTIONS

1. State Boyle's law and Charles' law.
2. What is meant by the equation of state of a gas ?
3. What is a perfect gas ? Does such a gas exist in nature ?
4. What are permanent gases ?
5. Define volume coefficient and pressure coefficient of expansion of a gas. Use these to get the idea of Absolute zero.

6. What is absolute scale of temperature ? What other scales of temperature do you know ?
7. What is a thermometer ? Mention the various types of thermometers and their ranges. Point out their merits and demerits.
8. What are the advantages of gas thermometers ? Of the two gas thermometers : the constant volume and constant pressure—which do you prefer for experimental work and why ?
9. What are the sources of error and precautions in the experiment with constant volume thermometer ? Should the readings be taken when the temperature of the bulb is rising or when it is falling ? Give reasons for your answer.
10. What corrections are applied in the case of this thermometer ? Can it be used for determining the melting point of wax ?
11. What is meant by the dead space?

18

Measurement of Specific Heat

18.1. Calorimetric Definitions—(*a*) *Unit of heat—The unit of heat is defined as the quantity of heat required to raise the temperature of unit mass of water by one degree.* This has given rise so three different units of heat according to the system of units chosen.

(*i*) *Calorie*—It is the unit of heat on the C.G.S. system and is the quantity of heat required to raise the temperature of one gramme of water through one degree centigrade (or more precisely from 14.5°C to 15.5°C).

(*ii*) *British Thermal Unit (B. Th. U.)*—It is the unit of heat on the F.P.S. system and is defined as the quantity of heat required to raise the temperature of 1 *lb.* of water through 1°F. A bigger unit frequently used for commercial purposes equals 1,00,000 (B. Th. U.) and is called the *therm*.

Centigrade Heat Unit (C.H.U.) or Centigrade Thermal Unit (C. Th. U.) This is the quantity of heat required to raise the temperature of 1 *lb.* of water through 1°C. It is also known as *lb. calories*.

These units of heat are related as follows :

1 B. Th. U. = 252 calories, 1 calorie = 0.003968 B. Th.U.

1 C. H. U. = 453.6 calories, 1 calorie = 0.002205 C.H.U.

(*b*) *Specific Heat*—Specific heat of a substance is defined as the quantity of heat required to raise the temperature of unit mass of the substance by one degree. Its unit on the C.G.S. system is cal. per gm. per °C.

Specific heat of a substance changes slightly with temperature, *e.g.*, sp. heat of mercury decreases slowly with temperature and that of water at first diminishes and then increases.

(*c*) *Thermal Capacity*—Thermal capacity of a body is defined as the quantity of heat required to raise the temperature of the body by 1°C. Thus, if m is the mass of a body and s its specific heat, its thermal capacity is equal to ms. Unit of thermal capacity is cal. per °C.

(*d*) *Water Equivalent*—The water equivalent of a body is the amount of water in grammes which would take the same amount of heat to raise its temperature by 1°C as the body. Thus, if m is the mass of the body and s its sp. heat, the heat required to raise the temperature of the body by 1°C is ms calories. This quantity of heat will raise ms gm. of water through 1°C. Hence the water equivalent of the body is ms gm.

It may be noted that the numerical values of water equivalent and thermal capacity of a body are equal, but they are expressed in different units.

18.2. Laws of Cooling—When a heated body is placed in an enclosure maintained at a temperature lower than that of the body, the body gradually cools till its temperature becomes equal to that of the enclosure, the modes of transference of heat being conduction, convection and radiation. *Newton's law of cooling states that the rate of loss of heat (expressed in calories per unit time) from a body is directly proportional to the excess of temperature of the body over that of the surroundings, provided other conditions remain constant.* Expressed mathematically,

$$-\frac{dH}{dt} = K(\theta - \theta_0)$$

where dH is the heat lost in a time dt, θ is the temperature of the body, θ_0 that of the enclosure and K a constant of proportionality depending upon the area and nature of the radiating surface.

Let m be the mass of the body and s its specific heat, then, if $\frac{d\theta}{dt}$ be the rate of cooling, the above equation gives,

$$-ms\frac{d\theta}{dt} = K(\theta - \theta_0) \quad (18.1)$$

whence

$$-\frac{d\theta}{dt} \propto (\theta - \theta_0)$$

Thus, if we plot the rate of cooling as abscissae against the excess of temperature as ordinates, the curve will be found to be a straight line. This straight line relation between the rate of cooling and the excess of temperature proves the validity of Newton's law of cooling.

Since the law was found to be valid for small temperature differences, say, 30° to 40°C, elaborate experiments were carried out by Dulong and Petit to extend the result to large variations of temperature and their conclusions are embodied in the equation,

$$-\frac{d(\theta - \theta_0)}{dt} = k\,(\alpha^{\theta} - \alpha^{\theta^o}) \qquad (18.2)$$

where α is a constant and k is a factor depending on the nature of the surface radiating heat to the enclosure.

Later on, Stefan gave a simple relation which expresses the *amount of radiation emitted by a body varies as the fourth power of its absolute temperature*. The relation leads us to the conclusion that,

$$-\frac{dT}{dt} = a(T_1^4 - T_0^4) \qquad (18.3)$$

where a is a universal constant and T_1 the T_0 are the absolute temperatures of the body and the surroundings. It has been proved to be valid for black bodies over a very wide range of temperatures.

Newton's law of cooling is included in Stefan's law. For, consider a black body at a temperature T_1^o placed in an enclosure, also black at a lower temperature T_2^o. Then, the rate of cooling is

$$a\,(T_1^4 - T_2^4)$$

Let us suppose that T_1 is slightly greater than T_2, *i.e.*, let $T_1 = T_2 + t$ where t is small compared with either of the absolute temperatures T_1 or T_2. Then, the rate of cooling is,

$$= a[\,(T_2 + t)^4 - T_2^4]$$

$$= aT_2^4\,\{1 + \frac{t}{T_2})^4 - 1\}$$

$$= aT_2^4\,\{1 + \frac{4t}{T_2} + \text{..... higher powers of } \frac{4t}{T_2}\text{.......} - 1)$$

$$= 4a\,t\,T_2^3$$

$$= Kt$$

where $K = 4\,aT_2^3$

This shows that the rate of cooling and hence the rate of loss of heat is proportional to the excess of temperature of the body over the surroundings. This is Newton's law of cooling.

Experiment 18.1

Object—To determine the specific heat of a liquid (......) by the method of cooling.

Apparatus—A double walled vessel with lid containing two holes to carry two rubber corks, two similar calorimeters blackened from outside, two thermometers, stop-watch, balance and wt. box.

Description of apparatus—The double-walled enclosure is made of two rectangular chambers B and C made of brass or copper (Fig. 18.1). In the chambers, cold water is placed to

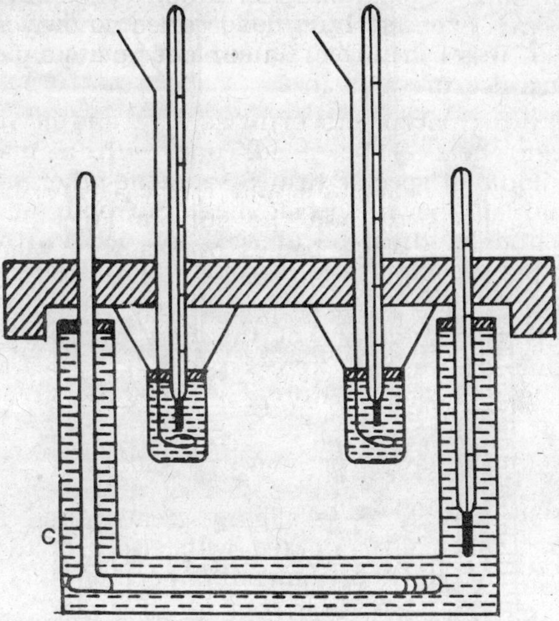

Fig. 18.1

maintain the temperature of the surroundings constant. The wooden lid covering the chambers is cork lined below and has two holes for admitting rubber corks through which pass the two thermometers for recording the temperatures of the two liquids. The lower stems of the thermometers engage to other corks which tightly fit the calorimeters and thereby support it. The calorimeters have nearly the same mass and are made of the same metal, usually aluminium. These are blackened from outside so that the loss of heat may be primarily due to radiation. Into one of these calorimeters the experimental liquid is placed and into the other a liquid of known specific heat. From a knowledge of their cooling curves, it is possible to calculate the specific heat of the liquid.

Theory—When a body cools while suspended in an enclosure, the rate of loss of heat from the body depends on the excess of temperature of the body above that of the enclosure, the area of the body exposed and the nature of its surface. Thus, if dQ is the loss of heat from a body in time dt at any instant t when the excess of its temperature above that of the enclosure is θ, we have,

$$dQ = a.f(\theta)dt \qquad (18.4)$$

where a is constant depending upon nature of the surface of the body and its area exposed and $f(\theta)$ is an unknown function of θ and only of θ. This function of θ need not be a simple algebraic or trigonometrical expression in θ.

Now, let two calorimeters of masses m and m' respectively and of specific heat S be filled, one with a mass m_1 of experimental liquid of specific heat S_1 and the other with a mass m_2 of a liquid of known specific heat S_2, both at the same temperature, and let them be allowed to cool. Further, let the times taken by the first and second calorimeters and their contents to cool from θ_1 to θ_2 be t_1 and t_2 respectively. Then, if $d\theta$ is the fall of temperature of the first calorimeter and its contents in time dt, we have from Eq. (18.4),

$$(mS + m_1S_1)\ d\theta = dQ = a_1 f(\theta)\ dt$$

whence $$(mS + m_1S_1) \cdot \frac{d\theta}{f(\theta)} = a_1\ .\ dt$$

Integrating from θ_1 to θ_2 we have,

$$(mS + m_1S_1) \int_{\theta_1}^{\theta_2} \frac{d\theta}{f(\theta)} = a_1\ t_1 \qquad (18.5)$$

Similarly, for the second calorimeter and its contents, we have,

$$(m'S + m_2S_2) \int_{\theta_1}^{\theta_2} \frac{d\theta}{f(\theta)} = a_2\ t_2 \qquad (18.6)$$

If the two calorimeters have identical surfaces and are filled with equal volumes of the two liquids, *i.e.*, area exposed for loss of heat is the same, we have $a_1 = a_2$. In that case dividing Eq. (18.5) by Eq. (18.6), we get,

$$\frac{mS + m_1\ S_1}{m'S + m_2\ S_2} = \frac{t_1}{t_2}$$

whence on rearranging

$$\frac{mS + m_1S_1}{t_1} = \frac{m'S + m_2S_2}{t_2} \qquad (18.7)$$

This equation can be used to calculate the specific heat S_1 of the experimental liquid, if other quantities are determined.

It should be noted that at the instant when the temperature of the two calorimeters and their contents is the same, it is their *rate of loss of heat* which *is the same and not their rates of fall of temperature.*

Method—Weigh the two calorimeters empty. Heat separately the given liquid and water to the same high temperature, say, 60°C. Into one of the calorimeters introduce the given liquid and into the other, water, taking care that the two liquids are nearly equal in volume. Then support the calorimeters from the lid by engaging them in rubber corks through which the bulbs of the thermometers pass so as to dip completely in the liquids. Next, place the lid over the double-walled enclosure, taking due care that neither of the two calorimeters touches the sides of the inner enclosure. When the lid has been securely placed and some time has elapsed, start taking observations as follows :

When the second's hand of the stop-watch is at 60, take the reading of the thermometer immersed in the experimental liquid and when the second's hand is at 30, take the reading of the other thermometer immersed in water. Proceeding in this manner, get the temperatures of the liquid and the water at regular intervals of one minute. Continue taking observation till the temperatures of the two liquids are a few degrees above the room temperature.

Next, plot the cooling curves for the two liquids and from the curves find the times taken by the liquid and the water to cool through the same range of temperatures.

Weigh the calorimeters after the observations. Take the specific heat of the material of the calorimeters as known and calculate the specific heat of the liquid from Eq. (18.7).

Sources of error and precautions—(1) The space between the two chambers is to be filled with cold water to ensure a uniform temperature of the surroundings. No water is to be placed in the inner chamber.

(2) The determination of the specific heat of a liquid depends on the assumption that the rate of loss of heat is the same for the liquid as for water. This is valid only when losses of heat take place under identical conditions. To procure this, it is necessary that,

(*a*) two calorimeters should have a good coating of lamp-black.

(*b*) The two liquids should be used equal in volume and not equal in masses.

(*c*) The liquid should not be heated in the calorimeters as the flame is likely to spoil the coated surface :

they are first heated in beakers and then transferred into the calorimeters.

(d) The temperatures of the two liquids when transferred to the calorimeters must be nearly the same.

(3) None of the two calorimeters should touch the inside chamber.

(4) Taking of reading should not be commenced unless the two thermometers indicate nearly the same temperature. In the mean time, the regular state of affairs shall have set in.

(5) The calorimeters must be made of a good conductor of heat and the walls should be very thin. When these conditions are satisfied, the temperature of the liquid can be considered to be the same as that of the calorimeter radiating heat to the surroundings.

Observations—[A] *For cooling of water and liquid.*

Least count of thermometers = °C

No. of readings	Time in minutes	Temperature of liquid in °C	Temperature of water in °C
1	0	60.5	—
2	1/2	—	60.0
3	1	59.5	—
4	$1\frac{1}{2}$	—	59.0
5	2	58.0	—
6	$2\frac{1}{2}$	—	57.5

[B] Mass of first empty calorimeter (m) = gm.

Mass of first calorimeter + liquid = gm.

Mass of second empty calorimeter (m') = gm.

Mass of second calorimeter + water = gm.

Calculations—Mass of liquid (m_1) = gm.

Mass of water (m_2) = gm.

From the graph (Fig. 18.2),

Time taken by water to cool from...to...°C = sec.

Time taken by liquid to cool through the *same* range = sec.

From equation

$$\frac{mS + m_1S_1}{t_1} = \frac{m'S + m_2S_2}{t_2}$$

we have

$$S_1 = \frac{t_1}{m_1t_2}(m'S + m_2S_2) - \frac{mS}{m_1}$$

Substituting for the quantities on the right hand side of this equation and solving for S_1, we have

S_1=

Result—The sp. heat of the given liquid (....) as determined by the method of cooling = cal./gm./°C.

Criticism of the method—It is a very convenient and accurate method for determination of specific heat of liquids, but is a failure in the case of powders and solids for they cannot be stirred to keep the temperature uniform and the cooling conditions constant. There is no mixing of two substances in this method and hence no heat due to chemical action is evolved or absorbed and so the method is superior to the method of mixtures in this respect.

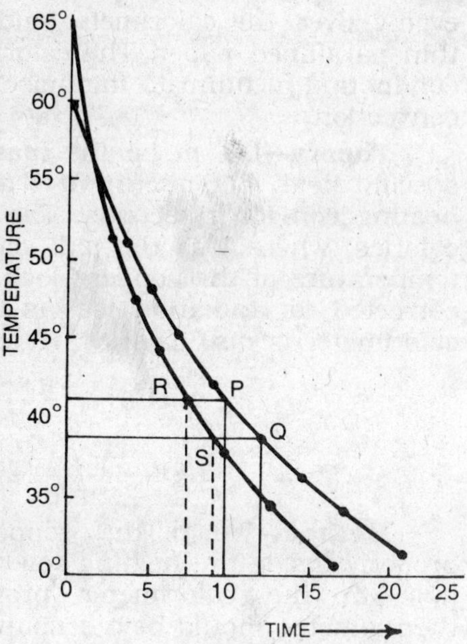

Fig. 18.2

The theory assumes that the rate of loss of heat is the same in the two cases under conditions apparently similar. This may, however, not be the case in practice and any variation or uncertainty in this respect will have its full effect on the result. Moreover, observations with a rapidly falling mercury-in-glass thermometers are difficult and introduce some error.

It should be noted that *this method* of determination of specific heat of a liquid *does not depend upon Newton's law of cooling* and so the accuracy of result is not affected by the validity of that law.

Experiment 18.2

Object—To determine the specific heat of copper by copper block calorimeter.

Apparatus—Copper block calorimeter, a sensitive thermometer, 6 volt battery, ammeter, voltmeter, rheostat, a heating coil and stop watch.

Description of apparatus—The copper block calorimeter is a block of copper having a hole drilled in the middle for inserting the thermometer to record its temperature. Heat is supplied by a heating coil carrying a current of about 1.5 amperes and wound evenly over the calorimeter and insulated from it electrically by thin paraffined paper. The calorimeter is surrounded by a non-conducting medium to minimise loss of heat by conduction and convection.

Theory—Let m be the mass of the calorimeter and s its specific heat. Let a current i amperes be passed through the heating coil for t seconds. The heat so generated $Vit \times 10^7/J$ calories, where V is the $p.d.$ across the heating coil, raises the temperature of the copper block. If the rise of temperature duly corrected for radiation loss is θ°C, the heat taken up by the calorimeter is $ms\theta$ calories. Hence,

$$ms\,\theta = \frac{Vit \times 10^7}{J}$$

or

$$s = \frac{Vit \times 10^7}{m\theta J} \qquad (18.8)$$

Method—Weigh the copper block calorimeter. Place it properly inside the heating coil and insert a thermometer in the hole in the calorimeter provided for this purpose. The thermometer should be a sensitive one reading to at least 1/5°C. Electrical connections are made as shown in Fig. 18.3. The current i and voltage V are suitably adjusted prior to fitting in the calorimeter and their values recorded.

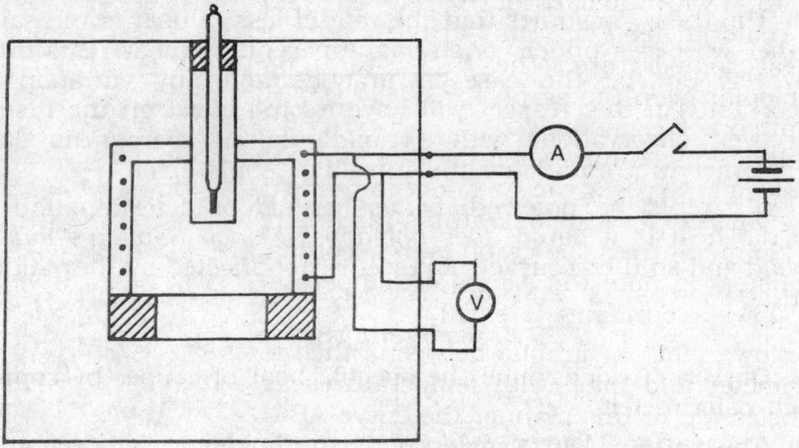

Fig. 18.3

Switch on the current and start the stop watch. At regular intervals of time, say, 1 minute, record the temperature of the calorimeter. Observations of temperature of the calorimeter when the temperature is rising should be taken very carefully at the requisite intervals and there should be no lag in observations of the time and temperature. When there has been an appreciable rise in the temperature of the calorimeter (usually about 6^0 to 7^0C), switch off the current and record the time for which the current had been passed.

Allow the calorimeter to cool and record the fall in temperature of the calorimeter, again, at regular intervals of time. The duration for recording the fall in temperature may be about 10 minutes.

From the observations obtained for the cooling of the calorimeter, calculate its rate of cooling when the calorimeter has its highest temperature. For evaluating this, it is best to consider the fall of temperature during four or five minutes and divide this by the number of seconds in the interval considered. Then on the graph paper, represent the temperature of the calorimeter along the *X*-axis and the rate of cooling along *Y*-axis. Plot the point *M* on the graph (Fig. 18.4) with the data for the calorimeter at its highest temperature. The other point *N* on the graph is plotted corresponding to its initial temperature when the rate of cooling is essentially zero. The straight line *MN* can be taken to represent the rate of cooling of the calorimeter at temperatures ranging between its initial and final temperatures. The graph is in accordance with Newton's law of cooling, as here the excess of temperature of the calorimeter over the surroundings is small.

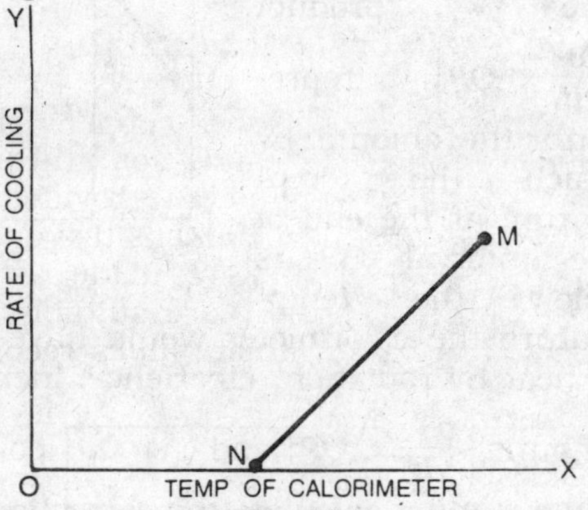

Fig. 18.4

Now, plot a graph between the temperatures of the calorimeter and time, with temperatures as ordinates and time as abscissae. In this manner the curve ABCDEF is obtained (Fig. 18.5). The portion ABCD corresponds to the rise in temperature of the calorimeter during the stage heat is being supplied to it and the portion DEF to the fall in temperature after heating has been discontinued and the calorimeter left to itself to cool.

Divide the curve ABCDEF into a number of portions which are nearly straight, *e.g.*, AB, BC, CD....Let the temperatures corresponding to the points A and B be θ_1 and θ_2 respectively. Hence, the mean temperature during the time interval *OP* is $\left(\frac{\theta_1 + \theta_2}{2}\right)$. Corresponding to this mean temperature obtain the rate of cooling, say, $\frac{d\theta'}{dt}$ from the graph (Fig. 18.4). Multiply this rate of cooling by the time interval *OP*. The product $\left(\frac{d\theta'}{dt} \times OP\right)$ represents the amount by which the temperature at the end of the interval OP is below the temperature the calorimeter would have attained in absence of loss of heat by radiation, etc. Hence, increase the ordinate PB by the amount $\left(\frac{d\theta'}{dt} \times OP\right)$ and get the point *B′* on the graph. In the same manner calculate the correction necessary for the next time interval *PQ*. Let this be $\left(\frac{d\theta''}{dt} \times PQ\right)$. Then the corrected temperature for the point C is

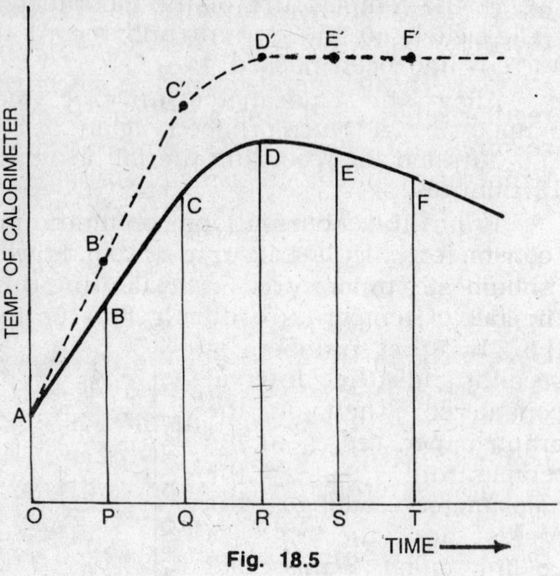

Fig. 18.5

$$QC\left(\frac{d\theta'}{dt} \times OP\right) + \left(\frac{d\theta''}{dt} \times PQ\right).$$

This gives the point C′.

In this manner obtain the graph A′ B′ C′ D′ E′ F′ which reveals that the portion D′ E′ F′ is a straight line parallel to the time axis. This implies that if there were no loss of heat from the calorimeter by radiation, etc., its temperature would have remained constant with time. Hence, the corrected temperature of the calorimeter is the ordinate corresponding to the portion D′ E′ F′. Take the reading from the graph. Let it be θ.

Having then determined all the factors occurring in equation (18.8), calculate the specific heat of copper.

Sources of error and precautions—(1) The temperatures of the calorimeter should be measured with a sensitive thermometer reading to $1/5^{o}C$.

(2) The temperature of the calorimeter should be noted at regular intervals of time and there should be no lag between the readings of time and temperature, especially when the temperature of the calorimeter is rising.

(3) The readings of ammeter and voltmeter should remain constant during the course of the experiment and their values should be properly adjusted prior to placing the calorimeter inside.

Observations—(A) Mass of the calorimeter = gms.
P.D. across the heating coil = volts.
Current through the coil = amperes
Time for which the current is passed = seconds

[B] *Determination for applying Radiation Correction*
Least count of thermometer = °C

Calculations—Corrected rise of temperature as obtained from the graph = °C

Hence,
$$s = \frac{Vit \times 10^7}{m\theta J}$$
= cal./gm./°C

Criticism of the method—The results obtained are fairly satisfactory. The method of correcting for the loss of heat by radiation is rather unsatisfactory in one respect. It is the question of the lag in the thermometer in recording rapidly changing temperatures. The temperature of the thermometer bulb will not necessarily be the same as that of the calorimeter when its temperature is rising. Hence, this factor introduces error and the portion of the graph ABCD has lower ordinates than what they would have been in absence of the thermometer lag. Necessary correction[1] for the thermometer lag should be applied.

ORAL QUESTIONS

SPECIFIC HEAT BY METHOD OF COOLING

1. Define sp. heat, calorie and thermal capacity.
2. Distinguish between rate of cooling and rate of loss of heat of a body. On what factors does the loss of heat from a body depend ?

---

1. Consult Waston, *A Text Book of Practical Physics*, Third Edition, p.220-12

3. For the determination of sp. heat by cooling which do you equate—the rates of cooling of the two liquids or the rates of loss of heat from the two liquids? Why ?
4. Why is the outer surface of the calorimeter containing the cooling liquid blackened? What are the essentials of a good calorimeter? Are these the same for the one used in the method of mixtures and the other used here in this method of cooling ?
5. Why should the liquids be taken in equal volumes and not in equal masses? Why should the two liquids have the same temperature nearly?
6. What are the sources of error and precautions in the experiment ? How is the temperature of the sorroundings kept constant?
7. Does the experiment yield good values of sp. heat of liquids? Is the method applicable to solid? Criticise the method pointing out its demerits.
8. What is the shape of the cooling curves you have obtained? If Newton's law of cooling is obeyed, What is the shape of the curve geometrically? What would be the shape of $(d\theta/dt.\theta)$ curve?
9. Why is the cooling curve called logarithmic curve? What law does the rate of loss of heat obey when effects of conduction and convection are entirely eliminated?
10. What is Stefan's law of radiation? How do you apply radiation correction in an experiment ?

COPPER BLOCK CALORIMETER

1. Describe the copper block calorimeter and explain how it can be used to determine the sp. heat of copper.
2. How is the heat produced in the coil measured?
3. How is the final temperature of the calorimeter corrected for loss of heat by radiation ?
4. What precautions do you take in this experiment ?

19

Thermal Conductivity

19.1. Modes of Transference of Heat—When one end of a metal bar is heated, the flow of heat from this end to the other can readily be observed by the change in temperature along the length of the bar. To a varying degree, this property of transmitting heat is possessed by all material substances, and the phenomenon is known as *conduction*. In this process the particles take up the heat from the neighbouring ones and pass on the same to those next to them. There are two other modes of propagation of heat, namely, *convection* and *radiation*. In the process of convection the transference of heat takes place by the actual movement of the heated particles, a fact that distinguishes it from conduction in which no transfer of the particles is apparent. In the case of fluids both conduction and convection appear to be processes of diffusion. In conduction, the action is molecular whereas in convection it is molar, a group of molecules move about in the form of convection currents, thus imparting some of their kinetic energy to others. Both conduction and convection are rather slow processes. Radiation, on the other hand, is a quick process. It consists in the transfer of energy, the presence of the material particles being unnecessary. The velocity of propagation of radiant heat is identical with the velocity of light.

19.2. Definitions—(1) *Thermal Conductivity, Steady State*—Consider a thin plate of a material with parallel faces. Let one of its faces be maintained at a temperature θ_1 and the other at a temperature θ_2. When the amount of heat flowing into the thin plate at one end is equal to the amount of heat flowing out through the other face, no part of this being used up in raising the temperature of the plate, it is then said to have attained a *steady state*. In this state the quantity of heat Q passing through the plate is found to be directly proportional to (1) the temperature difference between the two faces $(\theta_1 - \theta_2)$, (2) the area *A* of the plate and (3) the time *t* for which the flow of heat is considered, and inversely proportional to (4) the thickness *x* of the plate. Hence

$$Q \propto A \frac{(\theta_1 - \theta_2)\, t}{x}$$

$$\text{or} \qquad Q \propto K.\ A.\ \frac{(\theta_1 - \theta_2)\ t}{x} \qquad (19.1)$$

where K is a constant of proportionality, called the *coefficient of thermal conductivity* of the body.

This equation may be used to define the constant K. Suppose the area A of the plate = 1 sq. cm. the temperature difference $(\theta_1 - \theta_2) = 1^\circ C$, the time for which heat flows = 1 sec. and the Thickness of the plate = 1 cm., then $Q = K$

Thus, *the coefficient of conductivity on the C.G.S. system may be defined as the number of calories per second that would flow across every sq. cm. of the surface of a material of one cm. thickness when the temperature difference between the two faces is* 1^0C*, the flow of heat being considered normal to the surfaces.*

Similarly, we can define K in M.K.S. units. In M.K.S., the unit of K is J/s.m.k. or W/m.k., while that in C.G.S. it is Cal/s.cm.k.

In the above discussion it has been stated that the flow of heat should be normal to the faces of the material. In the simple case of a parallel plate the isothermal surfaces are parallel to the faces of the plate and the lines of heat flow are consequently normal to these isothermal surfaces. For this reason the temperature gradient $(\theta_1 - \theta_2)/x$ in this case, is the same everywhere along its length. When, however, isothermal surfaces are not planes parallel to the end faces of the material, but are curved and so the lines of flow are also curved, the problem of calculating the flow of heat becomes complicated. The equation may then be conveniently written as

$$Q = K(\theta_1 - \theta_2) \times t \times \text{shape factor} \qquad (19.2)$$

shape factor, being dependent on the nature of the isothermal surfaces. This factor has been evaluated for the simple case of radial flow of heat from the centre of a sphere, and from the axis of a cylinder.

(2) *"Diffusivity" or "Thermometric Conductivity." Variable State*—Consider a long cylindrical bar heated at one end. It takes sufficient time before the bar attains the steady state when flow of heat is primarily governed by the thermal conductivity of the material of the bar. Let us study what conditions influence the temperature of a small portion of the bar before the steady state is reached. The bar is said to be in a *variable state*. The quantity of heat flowing across any isothermal surface of the bar is partly

retained by the bar in raising its temperature and partly conducted away to the colder end. Thus, in the variable state it is the thermal capacity combined with the thermal conductivity of the material of the bar that govern the flow of heat. For this reason, in the equation governing the variable state, the factor $K/\rho c$ is found to be a constant, (where K is the thermal conductivity, ρ is the density and c, the specific heat of the material of the bar). This constant has been named *diffusivity* by Lord Kelvin and *thermometric conductivity* by Maxwell. It measures the change of temperature which would be produced in a unit volume of the substance when the quantity of heat that flows is equal to K. *i.e.*, the quantity that flows in unit time through unit area of the bar of unit thickness having a temperature difference of unity between its faces.

19.3. Classification of Conductors—Solids may be classified as poor, medium and good conductors.

(1) *Poor Conductors*—Such materials as are used for purposes of heat-insulation come under this head of poor conductors. They include asbestos, wool, cork, kieselguhr, etc. The value of K ranges from 0.00008 to 0.00003 C.G.S. units. They do not form homogeneous solids, their structure being either fibrous or cellular. Consequently, conduction is not the only process of heat transmission for such materials. The best heat insulators have nearly twice the conductivity of air (0.00005).

(2) *Medium Conductors*—The refractory materials, bricks various kinds of wood, chalk, sawdust, etc., fall under the head of medium conductors. The value of conductivity varies from 0.0003 to 0.01 C.G.S. units. The division between poor and medium conductors is only a matter of convenience.

(3) *Good Conductors*—All metals come under the class of good conductors. The value of conductivity for these ranges from 0.015 to 1.00 C.G.S. units. Metals are good conductors both of heat and of electricity. This may be explained on the supposition that the two phenomena of heat conduction and electric conduction depend on the diffusion of the free electrons. Wiedemann and Franz have given the empirical law that the ratio of the thermal and electrical conductivities at a particular temperature is the same for all metals. This has been partially confirmed by experiments.

All pure metals have a negative temperature coefficient of conductivity whereas the alloys have a positive coefficient. Thus, with the rise of temperature the value of thermal conductivity for metal decreases.

Experiment 19.1

Object—To determine the coefficient of thermal conductivity of copper by Searle's apparatus.

Apparatus—Searle's apparatus for conductivity of metals, two thermometers reading $\frac{1}{2}$°C, two thermometers reading $\frac{1}{5}$°C, boiler for generating steam, constant-head water flow arrangement, beakers and balance.

Description of apparatus—Cylindrical copper bar about a foot in length has a copper steam heater incorporated at one end and a copper spiral at the other through which cold water circulates (Fig. 19.1). At a known distance apart on the copper bar, two thermometers are inserted to record the temperatures there so that value of the temperature gradient may be calculated. For efficient thermal contact between the bar and thermometers, the latter are placed in the cups in the bar containing mercury. These cups should be well above the axis of the bar otherwise the rate of flow of heat is likely to be affected. The temperatures of the inflowing and the outflowing water can be observed by the thermometers inserted in the cups for that purpose at ends of the spiral. The whole piece of apparatus is felt-lined and covered in a wooden case with a removable front. All this is to ensure that the flow of heat is along the axis of the bar and normal to the isothermal surfaces.

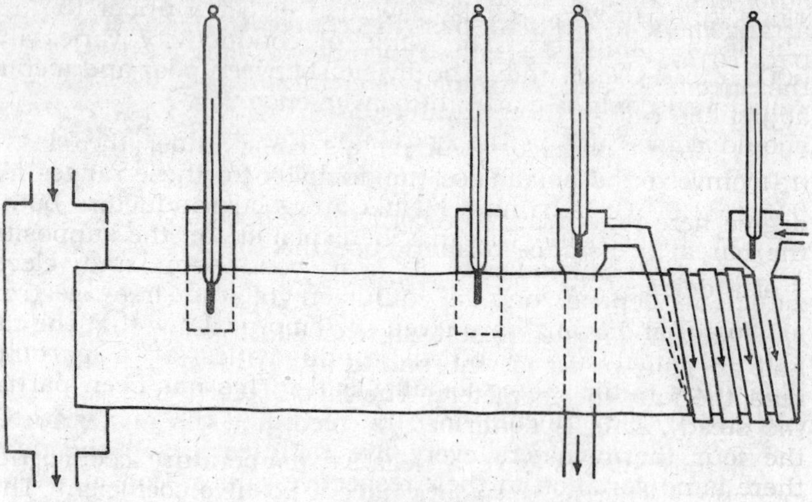

Fig. 19.1

Theory—If m be the mass of water flowing through the copper spiral per second and t_3 and t_4 are the steady temperatures of the inflowing and the outflowing water as indicated by the two thermometers, the quantity of heat taken up by water per second is $m\ (t_4-t_3)$ calories.

Now, in the steady state of the bar, the quantity of heat flowing across any normal section per second is,

$$Q = K\,A\,\frac{t_2 - t_1}{d}$$

where K is the coefficient of thermal conductivity of the material of the bar. A is the area of cross-section of the circular bar of radius r and t_1 and t_2 are the steady temperatures at two points, a distance d apart along the length of the bar. In the steady state, this heat flowing across a cross-section of the bar is taken up by the water circulating past the copper spiral. Hence,

$$K.\ \pi r^2\,\frac{t_2 - t_1}{d} = m\ (t_4 - t_3)$$

$$\therefore \qquad K = \frac{m\,(t_4 - t_3)\,d}{\pi r^2\,(t_2 - t_1)} \qquad (19.3)$$

The value of K can be evaluated by recording all the factors occurring on the right hand side of the above equation.

Method—Pass a steady current of steam from the boiler into the steam chamber to heat the end of the copper bar. At the other end, allow a steady current of water from a constant level arrangement to circulate past the copper spiral surrounding the bar. To have a measurable difference of temperatures between the incoming and outgoing water through the copper spiral, adjust the rate of flow of water to be low, *i.e.*, a trickle of water should come out of the spiral. As the temperature difference (t_4-t_3) involved is small, use the thermometers reading to $\frac{1}{5}$ of a degree here ; whereas, to record the temperatures t_1 and t_2 on the bar at a distance d apart, use the thermometer reading to $\frac{1}{2}$ of a degree.

Wait till the steady state is reached; it generally takes about half an hour or so for the bar to attain it. In the steady state read the four thermometers. The fact that the bar has attained the steady state is confirmed by recording the temperatures of the four thermometers every five minutes and observing that there is no variation in their respective values. Consider the last observations as giving the steady temperatures of the four thermometers.

To determine the rate of flow of water, collect it in a graduated jar and note the time for which the collection is made. From the known volume of water, calculate the mass of water collected. Instead of a graduated jar the collection of water may be made in a weighed beaker which is then re-weighed. From the difference of the two weights calculate the mass *m* of water collected. Since we also know the time for which the collection is made, the rate of flow of water is easily determined. Make the determination of the rate of flow of water a number of times and find the mean value.

Then measure the distance *d* between the two thermometers on the bar. Also determine the diameter of the bar at a number of places with the help of vernier callipers. At every place make the diameter along two mutually perpendicular diameters. Having thus determined all the factors occurring in Eq. (19.3), calculate the value of K for the material of the bar.

Repeat the experiment by interchanging the thermometers and also by altering the rate of flow of water and calculate the value of K. Then get the mean value of K from such determinations.

Sources of error and precautions—(1) Care should be taken to have the flow of water from a constant level arrangement. Water should not be allowed to flow direct from the tap in which case the rate of flow of water will not be constant.

(2) A number of determination of the rate of flow of water should be made and their mean value should be substituted in the formula.

(3) To have a measurable difference of temperature between the inflowing and outflowing water, the rate of flow should be low but continuous.

(4) To confirm that the bar has attained a steady state, begin recording the temperatures of the four thermometers after about half an hour or so, when the readings do not show any apparent change. Then record these every five minutes and when practical no variation in their respective values is observed with time, the layer can be supposed to have attained the steady state.

(5) The experiment should be repeated by altering the rate of flow of water.

(6) For each value of the rate of flow of water, the thermometers in each pair should be interchanged. *i.e.*, interchange T_1 and T_2 and also T_3 and T_4. This will eliminate any error due to defects in the thermometers used.

(7) The diameter should be measured at a number of places;

at every place the determination should be made along two mutually perpendicular diameters.

Observations—[A] Length of the bar between the two fixed thermometers T and T_2 = cm.

[B] *Determination of the diameter of the bar*

Vernier constant = cm.

Zero error = cm.

S. No.	Reading along one diameter *a* cm.	Reading along a perpendicular diameter *b* cm.	$\frac{a+b}{2}$ cm.
1. 2. 3.			
		Mean	

[C] *Determination of* m, t_1, t_2, t_3 and t_4.

S.No.	Mass of water flowed gm.	Time taken min.	*m* gm/sec	L.C.= °C		LC= °C	
				t_1 °C	t_2 °C	t_3 °C	t_4 °C
1. 2. 3.							

ON INTERCHANGE OF THERMOMETERS

1. 2. 3.							

Calculations—Mean corrected value of the diameter = cm.

∴ Mean value of the radius = cm.

Value of (t_2-t_1) before interchange = °C

Value of (t_2-t_1) after interchange = °C

Value of (t_4-t_3) before interchange = °C

Value of (t_4-t_3) after interchange = °C

∴ Mean value of (t_2-t_1) = °C

and Mean value of (t_4-t_3) = °C

$$K = \frac{m\,(t_4 - t_3)\,d}{\pi r^2\,(t_2 - t_1)} =$$

Result—The coefficient of thermal conductivity of copper = calories per sec. per sq. cm. per unit temp. gradient.

Criticism of the method—The method is straightforward although it takes some time to carry out in practice. It is suitable only for materials which are good conductors of heat and can be made available in the form of a bar or a cylinder. The mercury-in-glass thermometers are in good thermal contact with the bar since they are dipped in mercury contained in the cups made in the bar for the said purpose. To eliminate any errors due to defects in the thermometers employed, the thermometers in each pair are interchanged and the readings repeated. For more accurate result the heating should be done by an electrical arrangement using platinoid resistance wire wrapped round the end of the bar, and the temperatures should be measured by platinum resistance thermometers.

19.4. Radial Flow of Heat in a Cylinder—Let a hollow cylinder of length l and radii r_1 and r_2 as shown in Fig. 19.2, be heated axially at a constant rate. Since there is a radial flow of heat from the axis of the cylinder to its outside surface, the isothermal surfaces are cylindrical. Considering an isothermal surface at a distance r from the axis of the cylinder where the temperature gradient is: $d\theta/dr$, the amount of heat flowing per second across the surface is, from Eq. (19.1) given by

$$Q = KA\left(-\frac{d\theta}{dr}\right)$$

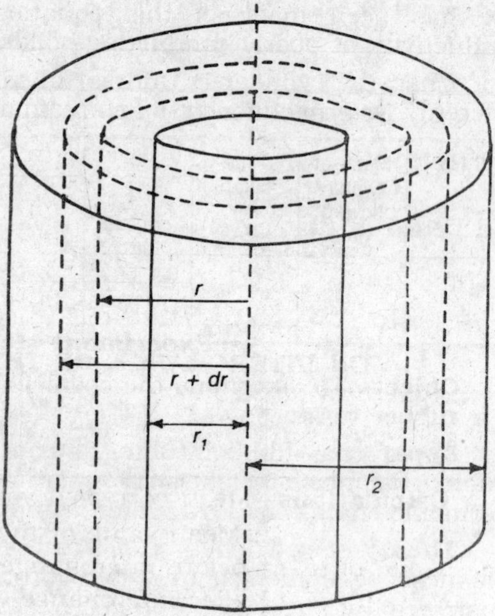

Fig. 19.2

where A is the area of the cylindrical surface and K is the thermal conductivity of the material of the cylinder.

$$A = 2\pi rl$$

$$Q = -2\pi rl\, K\frac{d\theta}{dr}$$

whence
$$-d\theta = \frac{Q}{2\pi\, l\, K}\cdot\frac{dr}{r} \qquad (19.4)$$

Since the heat is supplied along the axis of the tube at a constant rate, Q must remain constant. Hence, if θ_1 and θ_2 are the temperatures of the inner and outer surfaces of the cylinder in the steady state, we have by integrating Eq. (19.4),

$$-\int_{\theta_1}^{\theta_2} d\theta = \frac{Q}{2\pi lK}\int_{r_1}^{r_2}\frac{dr}{r}$$

or $$(\theta_1 - \theta_2) = \frac{Q}{2\pi lK} \log_e \frac{r_2}{r_1}$$

whence $$Q = \frac{2\pi lK(\theta_1 - \theta_2)}{\log_e (r_2/r_1)} \quad (19.5)$$

$$Q = \frac{2\pi lK(\theta_1 - \theta_2)}{2.3026 \log_{10} (r_2/r_1)} \quad (19.6)$$

Use has been made of this equation to determine thermal conductivity of poorly conducting solids.

When the cylinder is thin so that r_1 and r_2 are only slightly different, we have as a first approximation,

$$\log_e \frac{r_1}{r_2} = 2\,\frac{r_2 - r_1}{r_1 + r_2}$$

and hence, in that case,

$$Q = \frac{\pi lK(r_1 + r_2)(\theta_1 - \theta_2)}{(r_1 - r_2)}$$

Experiment 19.2

Object—To determine the coefficient of thermal conductivity of a rubber tubing.

Apparatus—Rubber tubing, steam generator, burner, stop-watch, balance, weight-box, calorimeter of large capacity, thermometer, tin can, graduated cylinder and callipers.

Theory—Let a length l of a rubber tubing of uniform wall thickness be immersed in water contained in a calorimeter as illustrated in Fig. 19.3, and let a steady stream of steam be allowed to pass through the rubber tubing for a time t. Then if θ_1 and θ_2 are the temperatures of the inside and outside surfaces of the rubber tubing, the amount of heat passed on to the calorimeter and its contents from the inside of the tube, is from Eq. (19.6) given by

$$Q = \frac{2\pi lK(\theta_1 - \theta_2)t}{2.3026 \log_{10} (r_2/r_1)}$$

where r_1 and r_2 are the internal and external radii of the rubber tubing and K its thermal conductivity.

If θ_3 and θ_4 are the initial and final temperatures of the calorimeter and its contents, the amount of heat absorbed by them is given by

$$Q = (Ms+m)\ (\theta_4-\theta_3)$$

when M is the mass of the calorimeter and *s* its specific heat and *m* is the mass of the water taken. Hence, equating the two expressions for Q, we have,

$$\frac{2\pi lK\,(\theta_1 - \theta_2)t}{2.326 \log_0 (r_2/r_1)} = (Ms + m)\ (\theta_4 - \theta_3)$$

whence

$$K = \frac{2.36 \log_{10} (r_2/r_1)}{2\pi l\ (\theta_1 - \theta_2)t} \times\ (Ms+ m)\ (\theta_4 - \theta_3)$$

θ_1 is equal to the temperature of the steam and θ_2 may be taken equal to $(\theta_3+\theta_4)/2$, the average temperature of the calorimeter. Hence,

$$K = \frac{2.3026 \log_{10} (r_2/r_1)\,(Ms + m)\,(\theta_4 - \theta_3)}{2\,\pi lt\left[\theta_1 - \dfrac{\theta_3 + \theta_4}{2}\right]} \qquad (19.7)$$

This equation can be used to calculate the value of K for rubber.

Method—Take a calorimeter of large capacity, say, 1000 c.c. and weigh it in a balance. Fill about two-thirds of it with water measured with a graduated cylinder. Coil a large length of rubber tubing and immerse it into the water contained in the calorimeter,

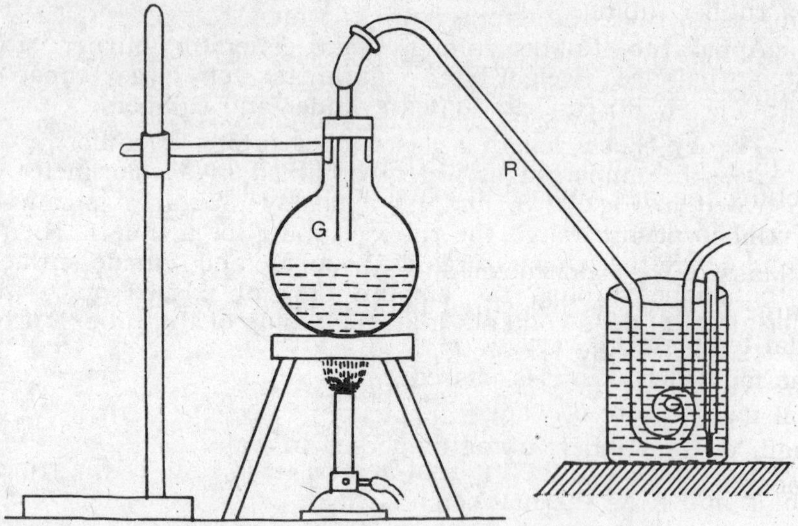

Fig. 19.3

allowing its both ends to project a little outside the calorimeter. Tie two cotton threads round the rubber tubing, one where it enters the water and the other where it leaves the water in the calorimeter. Note down the initial temperature of water.

Next connect one end of the rubber tubing R to the delivery tube of the steam generator G as shown in Fig. 19.3, and put its other end inside a tin can C to collect the condensed steam. Allow a steady stream of steam to pass through the rubber tubing for a known interval t till the temperature of the water rises by about 15° or 20^oC. Stir the water continuously and note down its final temperature.

Measure the length of the rubber tubing between the cotton loops and determine the internal and external diameters of the tubing with callipers. Note down the atmospheric pressure as measured by a Fortin's barometer and find out from tables the temperature of steam corresponding to that pressure. Finally, calculate the coefficient of conductivity of rubber from Eq. (19.7)

Sources of error and precautions—(1) The calorimeter should be of large capacity so that a large length of the rubber tubing may be placed in it.

(2) The rubber tubing should be of uniform thickness throughout.

(3) The steam should be passed through the tubing for such a time as to raise the temperature of water by at least 15^oC.

(4) The internal and external diameters of the rubber tubing should be measured very accurately with callipers, readings being taken along two mutually perpendicular diameters.

(5) The cotton loops tied round the tubing should be kept touching the *free* surface of water in the calorimeter.

Observations

[A] Mass of the calorimeter and stirrer $M =$ gm.
Volume of water as measured by graduated cylinder $V =$ c.c.
Initial temperature of water $\theta_3 =$ ^oC
Time for which steam is passed $t =$ sec.
Final temperature of water $\theta_4 =$ ^oC
Length of the rubber tubing immersed in water $l =$ cm.
Atmospheric pressure $p =$ cm of Hg

[B] Measurement of internal and external diameter of the tubing vernier constant of callipers. = cm
Zero error = cm.

(i) *Internal diameter*

S. No.	Upper end of tubing		Lower end of tubing		Mean diameter cm.
	Reading along any diameter cm.	Reading along perp. diameter cm.	Reading along any diameter cm.	Reading along perp. diameter cm.	

(ii) *External diameter*

S. No.	Reading along any diameter a	Reading along perp. diameter b	Mean $\frac{a+b}{2}$
		Mean	

Calculations—Mass of water $M = V =$ gm.

Mean internal radius of the tubing $r_1 =$ cm.

Mean external radius of the tubing $r_2 =$ cm.

Temperature of steam from tables $\theta_1 =$ °C

Sp. heat of the material of calorimeter given= cal./gm./°C

$$K = \frac{2.3026 \log_{10}(r_2/r_1)(Ms + m)(\theta_4 - \theta_3)}{2\pi l t \left[\theta_1 - \frac{\theta_3 + \theta_4}{2}\right]}$$

Cal. per sec. per sq. cm. per unit temp. gradient.

Result—The coefficient of conductivity of rubber.

= Cal. per sec. per sq. cm. per unit temp. gradient.

Criticism of the Method—The method gives satisfactory results. Besides Lee's method this is the only method which can be used to determine the conductivity of rubber. The temperature of outside surface of the rubber tubing does not remain constant as stipulated in theory but changes continuously as the steam is passed through it. This is a serious source of error in the method.

Experiment 19.3

Object—To determine the coefficient of thermal conductivity of glass in the form of a tube.

Apparatus—Glass tube enclosed in a wider tube which forms the steam jacket, constant head water flow arrangement, boiler for generating steam, graduated jar, two thermometers reading to 1/5°C.

Description of Apparatus—As illustrated in Fig. 19.4 a glass tube *PQ* of bore about 1.5 cm. is enclosed for a greater part of

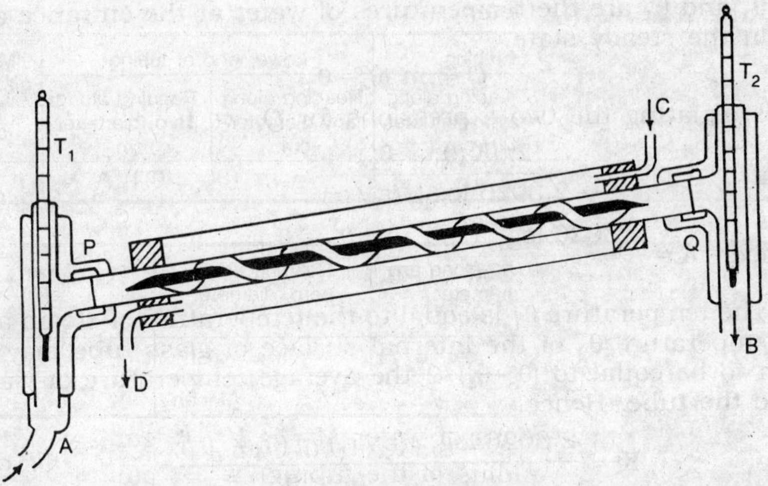

Fig. 19.4

its length in a wider tube J which forms the steam jacket. Steam is made to enter the jacket at C and leave it at D. Two T-pieces are connected to the tube PQ one at each end by short rubber tubing and each encloses a thermometer T_1 and T_2. With the help of a constant water-level tank (not shown in the Figure) a steady stream of water is made to flow past the tube and the temperatures at entrance and exit are measured by T_1 and T_2 respectively.

In order to maintain the temperature of water constant over any cross-section of the tube the water is made to traverse the tube along a spiral path. This is achieved by winding a rubber spiral of a wide spacing over a thin brass rod and placing this within the experimental tube PQ.

Theory—When the steam is passed through the jacket there is a radial flow of heat from the outside surface of glass to its inner surface and the isothermal surfaces are coaxial cylinders. If θ_1 and θ_2 are the temperatures of the outer and inner surfaces of the glass tube in the steady state, the amount of heat crossing its walls per sec. is from Eq. (19.6) given by

$$Q = \frac{2\pi l K(\theta_1 - \theta_2)}{2.3026 \log_{10}(r_2/r_1)}$$

where r_1 and r_2 are the internal and external radii of the glass tube, l its length and K the coefficient of conductivity of glass.

This heat Q is removed by the steady flow of water inside the tube. If the water flows at a constant rate of m gm, per sec., and θ_3 and θ_4 are the temperatures of water at the entrance and exit in the steady state,

$$Q = m(\theta_4 - \theta_3)$$

Hence, equating the two expressions for Q, we have,

$$\frac{2\pi l K(\theta_1 - \theta_2)}{2.3026 \log_{10}(r_2/r_1)} = m(\theta_4 - \theta_3)$$

whence $$K = \frac{2.3026 \log_{10}(r_2/r_1)\, m(\theta_4 - \theta_3)}{2\pi l(\theta_1 - \theta_2)}$$

The temperature θ_1 is equal to the temperature of steam and the temperature θ_2 of the internal surface of glass tube may be taken to be equal to $(\theta_3+\theta_4)/2$ the average temperature of water inside the tube. Hence,

$$K = \frac{2.3026 \log_{10}(r_2/r_1)\, m(\theta_4 - \theta_3)}{2\pi l\left[\theta_1 - \dfrac{\theta_3 + \theta_4}{2}\right]} \qquad (19.8)$$

Method—Set up the apparatus as shown in Fig. 19.4 and pass a steady current of steam from the boiler into the steam jacket J to heat the outer surface of the glass tube PQ. Allow a steady stream of water to flow past the tube spirally. It enters the inlet end at A and leaves the exit end at B. Keep the glass tube inclined so that the exit end is at a higher level than the inlet end. There are two reasons for doing so. Firstly, the tube will not be empty at any time and secondly, any air evolved from the water owing to its rise in temperature is easily swept out. To have a measurable difference of temperature between the incoming and outflowing water through the tube, adjust the rate of flow of water to be low. As the difference of temperature θ_4 and θ_3 is small, use the thermometers reading to 1/10 of a degree C.

Wait till the steady state is reached. The fact that the steady state has been attained can be confirmed by recording the temperatures θ_2 and θ_4 every five minutes and observing that there is no variation in their respective values. Consider the last observations as giving the steady temperatures θ_3 and θ_4.

To determine the rate of flow of water, collect it in a graduated jar and note the time for which the collection is made. Taking the density of water as unity, calculate the rate of flow of water. Make the determination of the rate of flow of water a number of times and calculate the mean value of m.

Note the barometric height H and evaluate the temperature of steam θ_1 from the tables.

Measure the length of the glass tube, *l*, exposed to steam.

Measure the external diameter of the tube at a number of places with a vernier callipers. At every point make the determiation along two mutually perpendicular diameters. From the mean value of the external diameter, calculate the mean external radius r_2.

To otain the mean internal radius r_1, fit a cork tightly at one end of the tube and clamp it vertically. From a burette fill a known volume of water into the tube. Let this volume be V. Measure the height of water column in the tube. Let it be h. Then

$$\pi r_1^2 h = V$$

or

$$r_1 = \sqrt{\frac{V}{\pi h}} \tag{19.9}$$

Having thus determined all the factors occurring in Eq. (19.8), calculate K.

Repeat the experiment by altering the rate of flow of water. Finally, calculate the mean value of K.

Sources of error and precautions—(1) Water should be made to flow from a constant level arrangement and not directly from the tap.

(2) The rate of flow of water should be low but continuous so as to get a measurable difference of temperature $(\theta_4 - \theta_3)$.

(3) The water should traverse the tube spirally and along an up-gradient of height.

(4) Observations of θ_3, θ_4 and *m* should be made when steady state has been attained.

(5) External diameter should be measured at a number of places; at every place the determination should be made along two mutually perpendicular diameters.

(6) The length of the tube exposed to steam should be measured and used in calculations. This length is the length inside the jacket.

Observations—[A] Barometric Height, H = cm.

[B] *Determination of m*, θ_3 and θ_4.

S. No.	Mass of water flowed gm.	Time Taken sec.	m gm./sec.	θ_3 °C	θ_4 °C
1.					
2.					
3.					

[C] Length of the glass tube exposed to steam, (l) = cm.

[D] *Determination of external diameter of the tube*

Vernier Constant = cm.

Zero Error = cm.

S.No.	Reading along one diameter *a* cm.	Reading along a perpendicular diameter b cm.	$\frac{a+b}{2}$ cm.
1. 2. 3.			

[E] *Determination of internal radius of the tube.*

S.No.	Height of water column h cm.	Volume of water taken in V c.c.
1. 2. 3.		

Calculations

From the tables,

Temperature of steam = ^{o}C

Mean corrected external diameter = cm.

Mean corrected external radius, r_2 = cm.

Internal radius $r_1 = \sqrt{\frac{V}{\pi h}}$ (From Eq. 21.9)

Mean internal radius, r_1 = cm

$$K = \frac{2.3026 \log_{10} (r_2/r_1) . m (\theta_4 - \theta_3)}{2 \pi l \left[\theta_1 - \frac{\theta_3 + \theta_4}{2}\right]}$$

=

=

Result—The coefficient of thermal conductivity of glass = cal per sec. per sq. cm. per unit temperature gradient.

Criticism of the method—The method is capable of giving result accurate to within about 5 per cent. The theoretical assumption that the isothermal surfaces across which heat is conducted in glass are coaxial cylinders is far from true. At the exit end of water, the temperature of the glass surface inside is essentially greater than surface temperature at the entrance end of water. To eliminate any errors due to defects in the thermometers employed, these should be interchanged and the observation repeated.

Experiment 19.4

Object—To determine the conductivity of glass by the method of Lees and Chorlton.

Apparatus—Lees and Chorlton's apparatus, steam generator, two thermometers preferably reading to 1/10th of a degree Celsius, vernier callipers and screw gauge.

Description of apparatus—The apparatus consists of a heavy retort stand with a clamp and a metal ring R from which is suspended a cylindrical slab *C* of brass or copper of diameter about 12.5 cm. by means of threads so that its top surface is horizontal. On this rests the steam chamber *S* the bottom of which is a thick brass or copper block *B* in which a hole is bored for inserting a thermometer T_1. The experimental substance *G* in the form of a circular disc is sandwiched between the block *B* and the block *C*, the latter carrying the second thermometer T_2.

Theory—Let *A* denote the area of cross-section of the experimental slab *G* and *d* its thickness, and let the thermometers T_1 and T_2 record temperatures θ_1 and θ_2 in the steady state. Then the heat transmitted through the slab per second is

$$Q = KA. \frac{\theta_1 - \theta_2}{d}$$

where *K* is its thermal conductivity, neglecting, of course, the heat lost by it (G) radially.

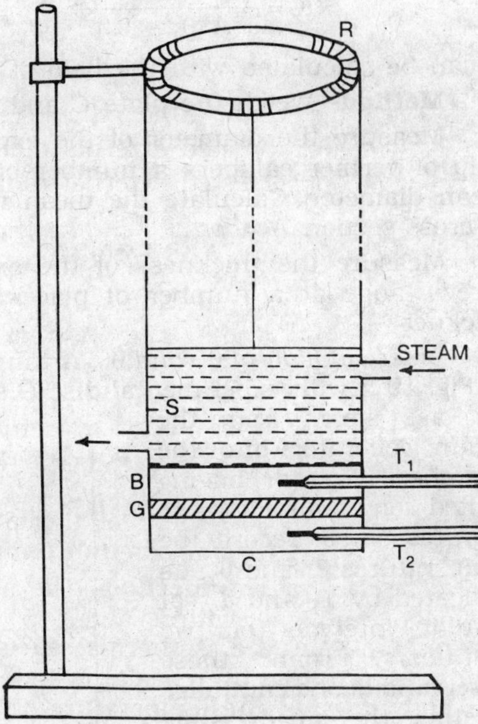

Fig. 19.5

In the steady state this rate of flow of heat *Q* is equal to the rate at which heat is radiated from the curved surface and the bottom surface of the disc *C*. Hence, if the plate *C* is heated to a temperature about

10°C above θ_2 and then allowed to cool with the specimen disc G above it, a cooling curve of the plate may be plotted by noting its temperature at regular intervals of time until the temperature falls to about 10°C below θ_2. From the cooling curve, $\left[\frac{d\theta}{dt}\right]_{\theta=\theta_2}$ is computed by drawing a tangent to the curve at the point corresponding to $\theta = \theta_2$. Hence, if m be the mass of the disc C and s its specific heat, the heat lost per second by the disc C by radiation when at a temperature θ_2 is

$$Q = ms\left[\frac{d\theta}{dt}\right]_{\theta=\theta_2}$$

Hence, we have

$$\frac{KA(\theta_1 - \theta_2)}{d} = ms\left[\frac{d\theta}{dt}\right]_{\theta=\theta_2}$$

$$K = \frac{dms\left[\frac{d\theta}{dt}\right]_{\theta=\theta_2}}{A(\theta_1 - \theta_2)} \qquad (19.10)$$

K can be calculated with the help of this equation.

Method—Weigh the plate C and determine its mass m.

Measure the diameter of the experimental disc G with the help of vernier callipers a number of times and determine the mean diameter. Calculate the mean radius r and then its area of cross-section $A = \pi r^2$.

Measure the thickness of the experimental disc G with a screw gauge at a number of places and determine the mean thickness d.

Determination of θ_1 *and* θ_2. Arrange the apparatus as shown in Fig. 19.5 with experimental disc G sandwiched between B and C. Pass steam from the steam generator into the chamber S and having waited for about forty-five minutes or so, record the temperatures θ_1 and θ_2, as indicated by T_1 and T_2, at regular intervals of two minutes. From these observations one can judge whether the steady state has been reached or not.

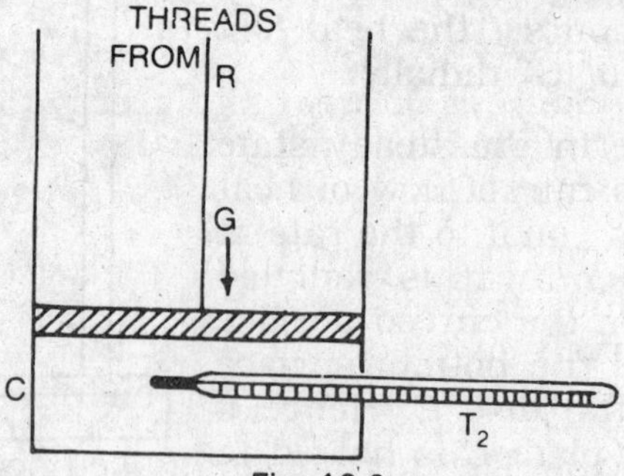

Fig. 19.6

When θ_1 and θ_2 become steady, *i.e.*, do not change with time, record these.

The first part of the experiment ends here.

Determination of $\left[\frac{d\theta}{dt}\right]_{\theta = \theta_2}$: Heat the disc C alone separately by allowing a Bunsen flame to play on it until T_2 registers a temperature about 10°C above θ_2 as recorded during the steady state. Suspend C by means of threads attached to the ring R with the specimen G placed on top of it and allow it to cool. Record the temperature of the disc C at regular intervals of time (may be at 30 second intervals if the fall is rapid) until the temperature falls to about 10°C below θ_2, the temperature of C as recorded during the steady state.

Plot a graph between θ and t, the time. This gives the cooling curve of the disc C. At a point on the curve corresponding to $\theta = \theta_2$ draw the tangent to the curve and determine the value of $\left(\frac{d\theta}{dt}\right)_{\theta = \theta_2}$.

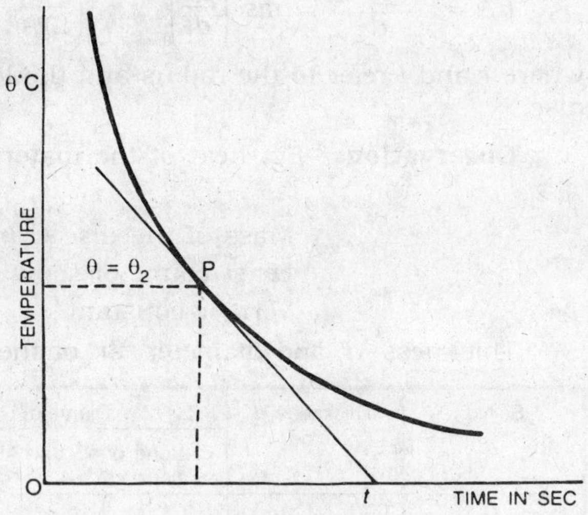

Fig. 19.7

Taking s the specific heat of the material of the disc C as known and having now determined all the other factors in Eq. (19.10), calculate K.

Sources of error and precautions—(1) The constants of the experimental disc must be carefully determined with the help of vernier callipers and a screw gauge.

(2) The fact that a steady state has been reached should be judged by recording the temperatures θ_1 and θ_2 at regular intervals of time, say, every two minutes, and then noting that these temperatures do not change with time.

(3) The thermometers used should be sensitive reading upto at least 1/5 of a degree Celsius.

(4) A thin disc of the specimen must be employed in this experiment because in the discussion of the theory we have assumed no heat is lost from the curved surface of the specimen.

(5) For the validity of the equation,

$$KA.\frac{(\theta_1 - \theta_2)}{2} = ms\left(\frac{d\theta}{dt}\right)_{\theta = \theta_2}$$

it is necessary that the *disc C is cooled with the specimen G above it.*

If the specimen G is not placed over the disc C and the disc C is allowed to cool, then the exposed surface is increased. *The equation under such circumstances becomes,*

$$KA = \frac{(\theta_1 - \theta_2)}{d} = ms\left(\frac{d\theta}{dt}\right)_{\theta = \theta_2} \times \left(\frac{r + 2l}{2r + 2l}\right)$$

where r and l refer to the radius and the length of the cylindrical disc.

Observations—Sp. heat of the material of the disc.

= cal/gm./°C

Mass of the disc C (m) = gm.

Least count of screw guage = cm.

Vernier constant = cm.

Thickness 'd' and diameter '$2r$' of the specimen disc.

S. No.	Thickness d cm	Diameter $2r$		Mean cm.
		Along one diam. cm.	Along perp. diam. cm.	
1. 2. 3. 4.				
Mean		Mean		

Table 2. Steady State Temperatures

S. No.	Time in minutes	Reading of the thermometer	
		T_1 °C	T_2 °C
1	0		
2	2		
3	4		
4	6		
5	8		

Table 3. Observations for cooling curve of Disc C.

S. No.	Time in seconds	Temp. of C θ°C	S. No.	Time in seconds	Temp. of C θ°C

Calculations—Area of cross-section of the disc (A) = πr^2

θ_1 = °C = sq. cm.

θ_2 = °C

From the graph,

$$\left[\frac{d\theta}{dt}\right]_{\theta = \theta_2} = \quad {}^\circ\text{C per sec.}$$

$$K = \frac{msd}{A(\theta_1 - \theta_2)} \times \left[\frac{d\theta}{dt}\right]_{\theta = \theta_2}$$

= Cal. per sec. per sq. cm. per unit temperature gradient.

or

= Cal. per sec. per cm. per °C.

Result—The thermal conductivity of glass = cal. per sec. per cm. per °C.

ORAL QUESTIONS

THERMAL CONDUCTIVITY

1. What are the various modes of transference of heat? Point out the difference between them ?
2. Mention some examples of poor, medium and good conductors of heat. Does any relation exist between thermal and electrical conductivities of metals?
3. Define thermal conductivity and diffusivity. What are their units?
4. Explain the flow of heat in the variable and the steady state of a heated body?
5. What is meant by temperature gradient ? Why does it take some time for the four thermometers to attain steady temperatures ?
6. How is the temperature gradient determined?
7. Do the thermometers come in contact with the bar? How

is the efficient thermal contact established between bar and the thermometer bulb?

8. How are we to ascertain whether the four thermometers indicate steady temperature? What arrangement do you employ to maintain steady flow of water past the copper spiral surrounding the bar?
9. Mention the chief precautions and sources of error in the experiment ? Is this method suitable for bad conductor of heat ?

CONDUCTIVITY OF RUBBER

1. How do you determine conductivity of rubber? What precaution do you take in this experiment ?
2. Why should the rubber tubing be of uniform wall thickness ?
3. Why should the calorimeter be of large capacity?
4. How do you measure length of the rubber tubing ?
5. What should be the rise of temperature of water and why ? What are the defects in this method?
6. Do you know of any other method for determination of K of rubber.

CONDUCTIVITY OF GLASS

1. Why should the tube be kept inclined and not horizontal?
2. Why should water be made to traverse the tube spirally? When is the steady state achieved ? How would you verify that the steady state has been attained? Why should the rate of flow of water be low?
3. Why should the temperature of the inner surface of the tubes be taken as the mean of the temperatures of water at the inlet and outlet ends?
4. What is the nature of the isothermal surface in conduction of heat from the outer to the inner surface of the tube?
5. Is it quite true to assume a cylindrical isothermal surface? How do you determine the mean internal radius?
6. What length of the tube should be measured for the calculation of K and why? Can this method be employed for determining K of rubber in the form of a tube?

20

The Mechanical Equivalent of Heat

20.1. Definition of J—It was Dr. Joule (1818-1889) who for the first time experimentally established the equivalence of work and heat. This result is embodied in the First Law of Thermodynamics which, after Maxwell, may be stated as follows:—

Whenever work is transformed into heat, or heat into work, the quantity of work is mechanically equivalent to the quantity of heat.

The law then includes two facts.—

(1) The conversion of work into heat and *vice versa.*

(2) The existence of a definite relationship of direct proportionality between the two.

Thus, if the quantity of heat produced is H when an amount of work performed is W, we have

$$\frac{W}{H} = \text{a constant, J}$$

where J is called the *mechanical equivalent of heat.* From the relation we can arrive at the definition of J. For, when the heat produced is a unit quantity of heat, W = J. Hence the mechanical equivalent of heat *is the amount of work done to produce unit quantity of heat* (1 caloric on C.G.S. system). The present accepted value of J in the C.G.S. system is 4.18×10^7 ergs per calorie and in S.I. system 4.18 Joule per calorie.

Experiment 20.1

Object—To determine the value of J by Searle's Friction Cone method.

Apparatus—Searle's apparatus, wt. box, balance and a thermometer.

Description of apparatus—A truncated cone of brass is carried by a spindle which can be rotated by belt and wheel drive (Fig. 20.1), the number of revolutions performed by the cone being recorded by a counter. A second truncated brass cone closely fits into the first, but is prevented from revolving with the outer cone by means of two pins engaging a wooden disc which

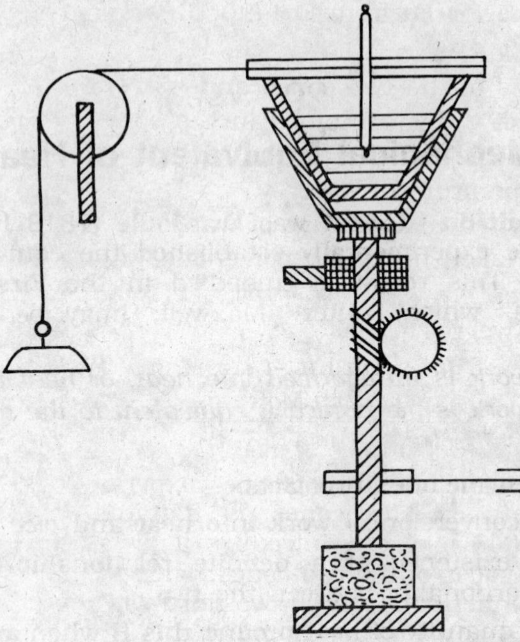

Fig. 20.1

is kept stationary by a suitable force, applied tangentially to the disc through a weight carried by a string passing over a pulley. When the outer cone moves, the inner cone remaining stationary, the friction between the two cones imparts heat to the cones and their contents.

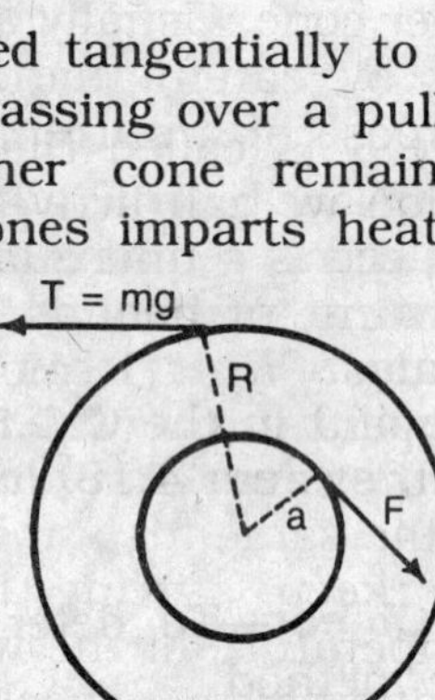

Fig. 20.2

Theory—Let a be the mean radius of the surfaces of contact of the two cones and R the radius of the disc. If F is the mean value of the force of friction between the two cones, then, for steady suspension of the mass M (Fig. 20.2), we have :

$$F \times a = Mg \times R \qquad (20.1)$$

When the outer cone rotates, the inner remaining at rest, the work done in one complete revolution of the outer cone is $F \times 2\pi a$.

Consequently, the work done in n revolutions is :

$$W = F2\pi a \times n$$
$$= Fa \times 2\pi n$$

Substituting for *Fa* from Eq. (20.1), we get :

$$W = MgR \times 2\pi n \tag{20.2}$$

If the mass of the two cones and stirrer $= m_1$
The mass of water contained in the cone $= m_2$
Sp. heat of the material of the cones $= s_1$
Initial temperature of water $= t_1$
Final corrected temperature of water $= t_2$
Then heat taken up by the cones and water $= (m_1s_1+m_2)(t_2-t_1)$

But $W = JH$

Therefore, we get :

$$J.\ (m_1s_1+m_2)(t_2-t_1) = MgR \times 2\pi n$$

or

$$J = \frac{2\pi n\ MgR}{(m_1s_1 + m_2)(t_2 - t_1)} \tag{20.3}$$

Method—Before beginning the experiment, test whether the friction between the cones is suitable in amount or not. If the friction is great, introduce a drop of lubricating oil between the surfaces of the two cones. Then weigh the two cones empty. Fill the inner cone nearly two third with water and weigh the two cones again. Replace them in the apparatus. Introduce a thermometer reading to $1/5°C$ into the water contained in the inner cone. Usually a stirrer is also provided for stirring the water but by careful handling the thermometer itself can serve the purpose of a stirrer which is then dispensed with.

Now, by trial adjust the weight in the pan in such a way that the string remains *tangential* to the disc and the mass suspended remains stationary at the same height when the other cone rotates. When this adjustment is completed, note the initial temperature of water in the cone and also the number indicated by the revolution counter. Then begin rotating the driving wheel, at the same instant starting the stop watch. As the rotation goes on, keep stirring the water from time to time. When the temperature of the water has risen by about 8 or 10°C, stop the rotation of the wheel at the same time noting the time of rotation of the wheel with the stopwatch. Then note down the final temperature of water and also the reading of the revolution counter.

To make the radiation correction, leave the apparatus for the same length of time as that taken for the experiment and note the fall of temperature during this period. Let it be δt. The mean excess of temperature over the surroundings during the experiment is half the final excess, and therefore, the mean rate of loss of heat during the experiment would be half the rate at the end of the experiment. Hence, to correct for radiation loss during the experiment add $\delta t/2$ to the observed rise of temperature.

To determine the quantity $2\pi R$ occurring in the expression for J, pass a thread once around the circumference of the disc and measure the length of the thread with a scale. This gives the value of $2\pi R$ for the disc.

Sources of error and precautions—(1) Initially it must be tested that the two cones are properly lubricated, otherwise the surfaces in contact will "seize".

(2) The initial temperature of water should be nearly the same as the room temperature. This is attained in preliminary adjustment of the correct weight for steady suspension.

(3) The string should leave the disc tangentially and the mass suspended should remain at a constant level. As the temperature rises, it will be found necessary to increase the speed of rotation of the wheel to keep the weight suspended at the same level.

(4) Correction for radiation loss should be applied in the manner indicated in the method.

(5) The friction in the pulley should be at a minimum, as this would affect the value of the tension applied and thereby cause an error in the result.

(6) The thermometer itself should be used as a stirrer in the experiment.

Observations

Least count of thermometer	=	°C
Sp. heat of the material of the cones	=	(given)
Mass of the two cones	=	gm.
Mass of the two cones + water	=	gm.
Initial temperature of water	=	°C
Initial reading of the counter	=	
Final temperature of water	=	°C
Final reading of the counter	=	
Temperature of water after cooling for the same time during which the experiment is performed	=	°C
Load at the end of the string	=	gm.
Circumference of the disc	=	cm.

Calculations—Mass of water = gm.

Rise of temperature = °C

Radiation correction = °C

∴ Corrected rise in temp. = °C

No. of revolutions performed by the cone =

$$J = \frac{Mg.\ 2\pi\ R.\ n}{(m_1 s_1 + m_2)(t_2 - t_1)}$$

= ergs/cal. =joule/calorie

Result—The mechanical equivalent of heat = joule per calorie.

Criticism of the method—The value of J obtained by this method is usually slightly higher than its accepted value. This is due to the fact that a portion of the heat generated is lost by radiation. Although a correction is applied for this radiation loss, yet the value of the amount of heat measured by this method is always less than its actual value. To obtain greater accuracy in the value of J, therefore, any method which does not require correction for radiation loss such as Callendar and Barne's continuous flow calorimeter method, may be used.

Experiment 20.2

Object—To determine the value of mechanical equivalent of heat with Callendar and Barne's continuous flow calorimeter.

Apparatus—Callendar and Barne's continuous flow calorimeter, a constant water level tank, a storage battery, a rheostat, an ammeter, a voltmeter, a beaker, a stop-watch, two thermometers and connecting wires.

Description of apparatus—It consists essentially of a narrow glass tube along the axis of which is stretched a heating coil of nichrome wire (Fig. 20.3). The ends of the heating coil are

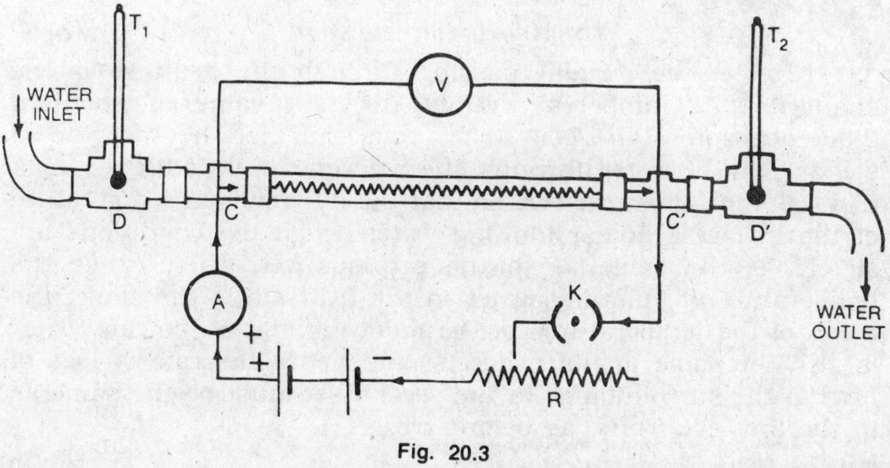

Fig. 20.3

soldered to binding screws fixed on small metal tubes C and C′, one at either end of the glass tube, the metal tubes being connected to the glass tube by means of short pieces of rubber tubing. The metal tubes C and C′ are connected, by means of other piece of rubber tubing, to wider metal tubes D and D′ mounted vertically and provided with an inlet or outlet side tube for water flowing through the calorimeter. The tubes D and D′ are closed with corks bored to carry thermometers for recording the temperature of the inflowing and outflowing water.

Theory—Let a current of I amperes be allowed to pass through the heating coil and let V volts be the potential difference across it, then the electrical energy transformed into heat in one second is $VI \times 10^7$ ergs. If this be equivalent to H calories of heat, we have :

$$H = VI \times 10^7/J$$

where J is the mechanical equivalent of heat.

Let the heat produced in the heating coil be removed by a continuous flow of water and let m grams per second be the steady rate of flow of water through the calorimeter when the conditions are steady, *i.e.*, the temperatures of the inflowing and outflowing water are constant. Then, if θ be the difference in temperatures of the inflowing and outflowing water, the heat removed by water per second is $m\theta$ cal. Now, if the loss of heat to the surroundings be X cal, per sec., we have, by equating the rate of production of heat in the heating coil to the sum of the rate of removal of heat by water and the rate of loss of heat to the surroundings.

$$VI \times 10^7/J = m\theta + X \qquad (20.4)$$

Now let the current passing through the heating coil be changed to I' amperes and let the new value of potential difference across the coil be V' volts. Further, let the rate of steady flow of water through the calorimeter be adjusted to a new value m' gms. per sec. so that the difference in temperature of the inflowing and outflowing water, when the conditions are steady, is the same as in the previous set. Then, since the temperature of inflowing water in the two sets is the same, the *mean* of the temperatures of the inflowing and outflowing water is also the same in the two sets; and hence the rate of loss of heat to the surroundings in the second set must be the same as in the first set. Thus, as before, we get :

$$V'I' \times 10^7/J = m\theta + X \qquad (20.5)$$

Subtracting Eq. (20.5) from Eq. (20.4), we get

$$\frac{VI - V'I'}{J} \times 10^7 = (m - m')\ \theta$$

whence
$$J = \frac{VI - V'I'}{(m-m')\theta} \times 10^7 \qquad (20.6)$$

From this equation the value of J can be calculated, if other quantities are determined.

Method—Connect the inlet tube of the constant-level water tank to a water tap and its outlet tube to the inlet tube of the Callendar and Barne's continuous flow calorimeter. Allow tap water to run into the tank and adjust its flow so that the level of water in the tank remains constant. This ensures a continuous and steady flow of water through the calorimeter. Adjust the height of the water tank so that the flow of water through the calorimeter is small but continuous. Next make connections as shown in Fig. 20.3, connecting the heating coil to a storage battery in series with a rheostat and an ammeter, and placing a voltmeter in parallel with the coil. In case of A.C. supply, current may be taken from the main with the help of a step-down transformer. In such a case both the ammeter and the voltmeter should be A.C. measuring instruments.

Then start the current and by means of the rheostal adjust its value to be about 2 amperes. The resistance coil inside the calorimeter will get heated and the heat produced in it will be continuously removed by water except for small heat loss to the surroundings. The temperature recorded by the thermometer at the outlet end of the calorimeter will gradually rise until steady conditions are reached. Now adjust, if necessary, the flow of water through the calorimeter by raising or lowering the water tank until the difference in temperatures of the inflowing and outflowing water when conditions are steady, remains constant at about at least 5°C. When this stage is reached, note down at regular intervals the temperatures registered by the thermometer at the inlet and outlet ends of the calorimeter correct upto 1/10 of a degree C. When three constant consecutive readings for T_1 and T_2 the temperatures of the inflowing and outflowing water, have been obtained, commence collecting the outflowing water from the calorimeter in a graduated jar, noting the time of commencement by a stop-watch. When sufficient quantity of water has been collected and at the same time the stop-watch records an integral number of minutes, stop the stop-watch simultaneously removing the graduated jar with the collected water and determine the rate of flow of water m through the calorimeter. Take at least three sets of observations for m and determine the mean value of m. Note down the readings of the ammeter and the voltmeter which give the values of the current I in the circuit and the potential difference V across the heating coil.

Next alter the current in the heating coil by means of the rheostat (or in the case of A.C. supply by altering the tapping of the transformer) and, by raising or lowering the constant level water tank according as the current has been increased or decreased, and adjust the rate of flow of water through the calorimeter until the difference in temperatures registered by the two thermometers, when the conditions are steady, is the same as in the previous set. Then, determine, as in the first set, the mean value of m' of flow of water through the calorimeter. Note down the current I' and the potential difference V'. Now stop the current and after it stop the flow of water. Calculate the value of J from Eq. (20.6).

Source of error and precautions—(1) The heating coil should have a resistance of about 10 ohms. It should be taken in the form of a twisted strip which serves the double purpose of making the surface exposed as great as possible and also of causing the water to circulate in its flow and then get thoroughly mixed.

(2) If the current is taken from a storage battery, see that it enters the ammeter or the voltmeter by the +ve marked terminal. In case of A.C. supply, the ammeter and the voltmeter should be A.C. measuring instrument.

(3) A constant level water tank should be used for controlling the rate of flow of water through the calorimeter, for with it the rate of flow may be easily varied and yet always remains steady.

(4) The rate of flow of water through the calorimeter should be steady and hence the pressure difference between the inlet and outlet ends of the calorimeter should be maintained constant by adjusting the supply of water from the tap into the tank so that level of water in the tank is maintained constant.

(5) When beginning the experiment, the water should be allowed to flow through the calorimeter *before the current is* started but at the end of the experiment reverse order should be followed, *i.e.*, the current should be stopped before the flow of water is stopped.

(6) The rate of flow of water through the calorimeter and the strength of the current in the heating coil should be so

adjusted as to give, when conditions are steady, a difference of temperatures of the inflowing and outflowing water of at least 5°C. For the slow flow of water the current should be adjusted to about 1.5 amperes and for the quick flow the current should be about 2.0 amperes.

(7) As the rise in temperature of the water flowing through the calorimeter is small, the temperatures of the inflowing and outflowing water should be noted correct upto 1/10 of a degree C.

Observations—Set I.

Least Count of ammeter = amp.
Least Count of voltmeter = volts
Least Count of thermometers = °C
Current through the heating coil = amperes
Potential difference across the coil = volts.

S. No.	Temperature of inflowing water T_1	Temperature of outflowing water T_2	Rise in Temperature θ	Volume of water collected in c. c.	Time taken		Rate of flow water gm./sec.
					Min.	Sec.	
1.							
2.							
3.							
4.							

Mean

Note—Similarly, record the observations for second set.

Calculations— $J = \dfrac{VI - V'I'}{(M - M')\,\theta} \times 10^7$

$= \times 10^7$ ergs/cal. = Joule/cal.

Result—The value of mechanical equivalent of heat

$J = \quad \times 10^7$ ergs/cal. = Joule/cal.

Criticism of the Apparatus—The apparatus is capable of yielding very good results. The water equivalent of the calorimeter, the heating coil and thermometers has been eliminated since the observations for temperature are taken when the temperature distribution becomes constant and no more heat is taken up by the calorimeter. The temperatures of the inflowing and outflowing water are taken when they are steady and hence can be noted down very accurately. The error due to loss of heat to the surroundings has been eliminated by taking two sets of observations with different current strenghts and rates of flow of water but keeping the mean of the temperatures of the inflowing and outflowing water, when the conditions are steady, in each set the same. The loss of heat to the surroundings may be reduced by enclosing the flow tube in a hermetically sealed glass vaccum-jacket which may be silvered and surrounded by a second jacket containing water at the temperature of the inflowing water through the calorimeter. There is no variation in the resistance of the heating coil after the readings of the two thermometers become steady and hence the current through the heating coil and the potential difference across it can be kept constant.

The accuracy of the result is limited only by that of the measuring instruments used in conjunction with the calorimeter, *i.e.*, the thermometers, the ammeter and the voltmeter. For very accurate results the temperatures may be measured with a differential pair of platinum thermometers, the bulb of each thermometer being surrounded with a thick copper tube of negligible resistance attached to the end of the heating coil. The copper tube on account of good conductivity keeps the bulb at the temperature of the adjacent water and due to its low resistance prevents the generation of any appreciable amount of heat by the current near the thermometer. The potential difference across the heating coil may be measured in terms of a standard Clark cell by means of an accurately standardised potentiometer. This potentiometer also serves to determine the current through the heating coil by measuring the potential difference across a standard resistance included in the same circuit.

The continuous flow in calorimeter is equally suitable for the determination of J or the specific heats of liquids or for studying the variation of specific heat of liquids with temperature. When the specific heat of mercury is to be determined, the heating coil is dispensed with for the mercury itself serves as the conductor. The calorimeter can also be used for the determination of specific heats of gases at constant pressure if the current strength is kept small and cotton-wool plug placed in the outlet just in front of the thermometer to steady the flow of the gas over the bulb.

ORAL QUESTIONS

MECHANICAL EQUIVALENT OF HEAT

1. Enunciate the first law of thermodynamics. Define J. What are its units ?
2. How is heat generated in your experiment ? Is it possible to convert work into heat without friction?
3. How are work, heat and friction related to each other?

4. What are reversible and irreversible changes ?
5. What is the moment of the couple applied to the inner cone ? Which of the two cones rotates ? Can we not use two cylinders instead of two cones ?
6. Why has the wheel to be turned faster as the temperature of the calorimeter rises?
7. What do you mean by radiation correction? How is it applied in this experiment ? Is heat lost in any way other than radiation?
8. Why is the vessel containing cones made of ebonite ?
9. Why should the cord carrying the suspended mass leave the disc tangentially ? Why should the mass suspended be kept at the same height ? What does its slight rise and fall indicate ? How can you calculate the work done even if the mass does not remain perfectly at rest ?
10. How does the counter register the revolutions of the outer cone ? Is the number of rotations of the cone and the hand wheel the same ? If not, what is their ratio ?
11. What are the precautions and sources of error in the experiment ?
12. If the diameter of the disc is reduced to half, by what amount should the mass be increased or decreased for its steady suspension ?
13. What other methods of determining J do you know ? Discuss their relative merits and demerits ?

21

Miscellaneous Experiments in Heat

Platinum Resistance Thermometer—It is a practical application of variation of resistance with temperature for measurement of temperature. As shown in Fig. 21.1, a platinum resistance thermometer consists of a *bare* spiral of fine platinum wire usually about 0.2 mm, in diameter and of resistance about one ohm, wound *non-inductively* on a mica frame and placed in a glass tube for temperatures up to 300°C and in glazed porcelain or fused silica-tube for higher temperatures. The ends of the wire are soldered on two stout copper rods to which are connected the leads to the bridge which measures the resistance of the thermometer.

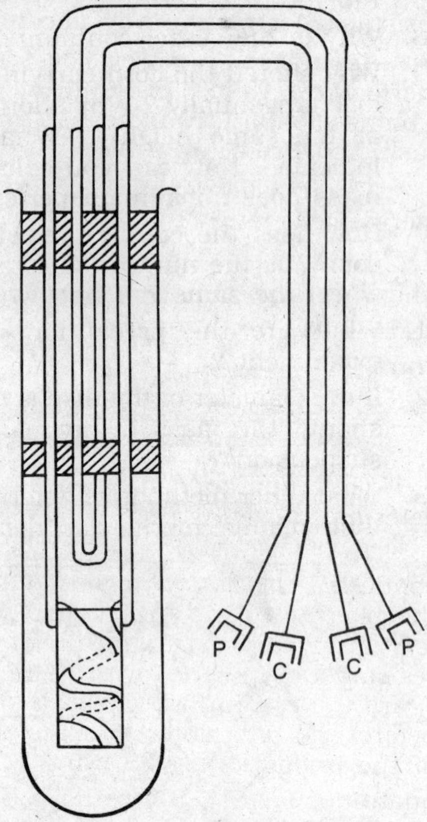

Fig. 21.1

Since the measuring bridge is often situated at a considerable distance from the body whose temperature is to be measured, the leads are sufficiently long and hence any change in their resistance with temperature will introduce errors in the measurement, unless *compensating leads are used. The compensating leads or dummy leads are exactly similar in size and material to the leads of the thermometer and are soldered to similar rods placed within the thermometer tube* the lower ends of the rods being connected together as shown. The ends of the leads are soldered to copper connecting strips, those connected to the thermometers leads being lettered PP and those connected to the compensating leads being lettered CC. The compensating leads running along with the thermometer leads are connected in the *opposite* arm of the bridge to that in which the thermometer itself is connected. Any change of temperature affects both the pair of leads *equally* and hence the balance of the bridge is not affected by them.

Experiment 21.1

Object—To determine the *temperature co-efficient of resistance* for platinum by means of *Carey Foster's bridge*.

Apparatus—A Carey Foster's bridge, a platinum resistance thermometer, a decimal-ohm box, two resistance coils of about 2 ohms each, a Leclanche cell, a weston galvanometer, a thermos flask, a hypsometer, a barometer and a mercury thermometer.

Theory—Let R_0 be the resistance of the platinum resistance thermometer at 0°C and R_t that at t °C. Then,

$$R_t = R_0 \quad (1 + \alpha\, t)$$

where α stands for *temperature coefficient of resistance* for platinum.

Thus,

$$\alpha = \frac{R_t - R_0}{R_0 t} \qquad (21.3)$$

This equation can be used to calculate α, if R_t and R_0 are determined.

To determine R_t and R_0 let two *nearly equal* resistances be connected in the *inner* gaps of a Carey Foster's bridge and let the platinum resistance thermometer be connected in the *outer left* gap of the bridge and the *compensating leads* with a known resistance S in series with them in the *right* gap. Then, if l_1 and l_2 are the reading on the scale of the positions of the null point before and after interchanging the resistance in the *outer* gaps of the bridge for $S = S_1$, and l'_1 and l'_2 are the corresponding quantities for $S = S_2$, we have :

$$R - S_1 = \rho\,(l_2 - l_1) \qquad (21.4)$$

and

$$R - S_2 = \rho\,(l_2' - l_1') \qquad (21.5)$$

where R is the resistance of the platinum resistance thermometer. Dividing Eq. (21.4) by Eq. (21.5), we have,

$$\frac{R - S_1}{R - S_2} = \frac{l_2 - l_1}{l_2' - l_1'}$$

or $R(l_2' - l_1') - S_1(l_1' - l_1') = R(l_2 - l_1) - S_2(l_2 - l_1)$

which gives

$$R = \frac{S_1(l_2' - l_1') - S_2(l_2 - l_1)}{(l_2' - l_1') - (l_2 - l_1)} \qquad (21.6)$$

The resistance *R* of the platinum resistance thermometer can also be determined *graphically* without requiring the knowledge of ρ. It is evident that if the known resistance S in the *outer* gap of the Carey Foster's bridge is given different value S_1, S_2, S_3, etc., and the corresponding values of $(l_2 - l_1)$ determined, then *the plot of* $(l_2 - l_1)$ against S will come out to be a straight line as shown in Fig. 21.2. *The intercept of this straight line on the resistance-axis* equals R, the resistance of the thermometer, for if we put $(l_2 - l_1) = 0$ in Equ. (21.4) evidently S = R.

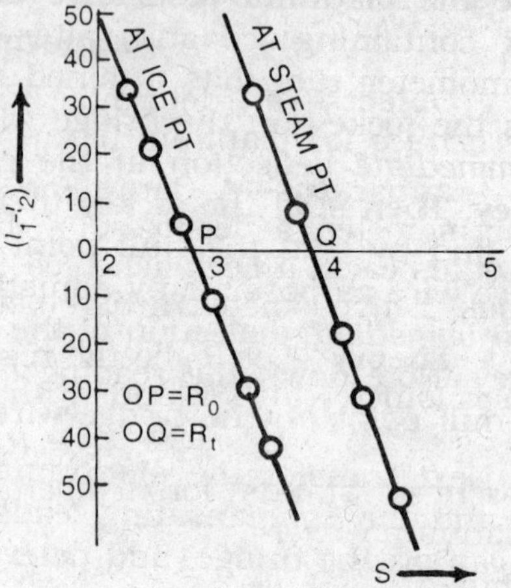

Fig. 21.2

By placing the platinum resistance thermometer first in melting ice and then in steam and determining the values of R_0 and R_t from Eq. (21.6) or graphically, α can be calculated from Eq. (21.3), *t* being the temperature of steam.

Method—Connect the leads from the platinum resistance thermometer lettered PP to the *outer left gap* of the Carey Foster's bridge and those lettered CC in series with a decimal-ohm box

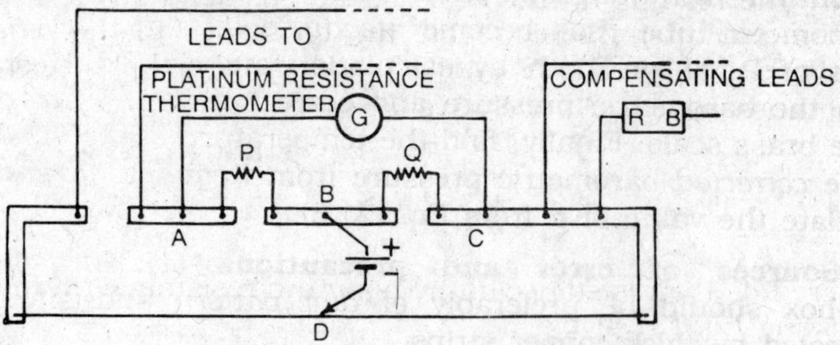

Fig. 21.3

to the *outer right gap* of the bridge. Next connect two approximately *equal* resistance coils in the *inner* gaps of the bridge and the weston galvanometer between A and C. Finally, connect the cell between B and D.

Adjust the decimal-ohm box to S_1 equal to about one ohm. Place the platinum resistance thermometer tube in a thermos flask containing ice and, after about half-an hour when the thermometer tube has attained the *steady* temperature of 0°C, pass the jockey on the bridge wire and note down the direction of *immediate* deflection in the galvanometer on depressing the jockey. Then shift the jockey to another point on the bridge wire and find by trial the null point, as close to the *middle* of the bridge wire as possible by adjusting S_1, if necessary, where there is no *immediate* deflection of the galvanometer on depressing the jockey. Note down the reading l_1 on the scale of the position of the null point on the bridge wire.

Next transfer the platinum thermometer to the outer right gap and the compensating leads in series with S_1 to the outer left gap of the bridge, and determine as before the reading l_2 on the scale of the position of the null point on the bridge wire.

Now adjust the decimal-ohm box to S_2 and determine the values of l_1' and l_2' as for l_1 and l_2 above and calculate the value of R_0 from Eq. (21.6). Take at least three sets of observations with slightly different values of S_1 and S_2 and find the mean value of R_0. Also plot a graph between (l_2-l_1) and the various values of S. This will be a straight line as shown in Fig. 21.2. Find its intercept *OP* on the resistance-axis. This will give R_0.

Next, place the platinum resistance thermometer in a hypsometer and after some time when the temperature of the thermometer tube has become *steady*, determine the value of R_t, as of R_0 above, both by calculation and graph. Then note down the barometric pressure and correct it for the expansion of the brass scale. Finally, find the temperature of the steam t°C at the *corrected* barometric pressure from *Regnault's tables* and calculate the value of α from Eq. (21.3).

Sources of error and precautions—(1) The decimal ohm-box should be preferably of *dial pattern* and should be connected by *thick* copper strips.

(2) The usual positions of the cell and the galvanometer should be *interchanged*, *i.e.*, the galvanometer should be

connected between A and C and the cell between B and the jockey. The contact of the jockey with the bridge wire should not be more than momentary. This will prevent heating of the platinum spiral due to prolonged flow of current and consequent change in its resistance.

(3) As the usual positions of the cell and the galvanometer have been interchanged, the latter will remain *permanently* connected in the circuit and hence unless all the resistances used are non-inductive, a throw due to induction effects will be produced in the galvanometer when the jockey is pressed over the bridge wire. Accordingly both the resistances P and Q in the inner gaps of the bridge should be *non-inductive* and the connecting wires should never be coiled.

(4) In order that the bridge may have high sensitiveness, the resistance of the four arms of the bridge should be of the *same order* of magnitude.

(5) The balance point should be determined by noting the *immediate* deflection of the galvanometer when the jockey is passed over the bridge wire, for when the current continues to flow a deflection will be produced in the galvanometer on account of change of resistance of the platinum spiral due to heating by the current, even when the bridge was balanced before pressing the jockey over the bridge wire.

(6) The null point should be as close to the *middle* of the bridge wire as possible.

(7) The galvanometer should be shunted by a low resistance wire to avoid *excessive* deflections in it when the bridge is out of balance, the shunt being removed when the exact position of the null point is being determined.

(8) The jockey should be pressed gently over the bridge wire and the contact between the jockey and the bridge wire should not be made while the jockey is being moved along.

(9) While the positions of the null points for the measurement of R_0 and R_t are being determined, the temperature of the platinum thermometer should remain *steady*.

(10) The graph between $(l_2 - l_1)$ and S is a *straight* line and must be *smoothly* drawn. While plotting the graph the *sign* of $(l_2 - l_1)$ should be carefully taken into account.

Observations—[A] *Determination of R_0 and R_t.*

Temp. of platinum resistance thermometer	S. No.	Resistance in series with compensating leads S ohms	Position of null pt. with platinum resistance thermometer in the		For graph (l_2-l_1) cm.	Resistance of platinum resistance thermometer R ohms	Mean R ohms
			Left gap cm.	Right gap cm.			
°C	1	S_1 =	l_1	l_2			
		S_2 =	l_1'	l_2'			
	2	S_3 =					
		S_4 =					
	3						
t°C	1						
	2						
	3						

[B] *For determination of t*

Barometric pressure H = cm.

Room temperature θ = °C

Coefficient of linear expansion for brass = 0.0000189 per °C (given).

Calculations—R_0 $= \dfrac{S_1 (l_2' - l_1') - S_2 (l_2 - l_1)}{l_2' - l_1' - l_2 + l_1}$

= ohms

Similarly, calculate R_t.

Barometric pressure corrected for expansion of the scale

= H (1+0.0000189

= cm.

Temperature t of steam at the above corrected barometric pressure from *Regnault's tables* = °C.

$$\alpha = \frac{R_t - R_0}{R_0\, t}$$

= per °C

From the graph (27.30)

Intercept OP on resistance axis =

R_0 = ohms

Intercept OQ on resistance axis =

∴ R_t = ohms

$$\therefore \quad \alpha = \frac{R_t - R_0}{R_0 t} = \quad {}^{\circ}C$$

Result—The *temperature co-efficient of resistance α for platinum* = per °C

Accurate value of *a* =

Percentage error =

Criticism of the method—The charge in resistance (R_t–R_0) of the platinum resistance thermometer is a very *small* quantity, hence the accuracy in the determination of α depends upon the accuracy with which this change in resistance is measured. By means of Carey Foster's bridge the value of R_0 and R_t can be measured sufficiently accurate. The use of compensating leads running along the thermometer leads has eliminated errors due to any variations of temperature and consequent alteration in the resistance of the leads.

The result is slightly affected by a certain rise in temperature of the platinum spiral which occurs owing to the passage of current through it, quite apart from that due to the heat supply in which the thermometer tube is placed. The Eq. (21.3) used to calculate α is true for small ranges of temperature only and hence the value of α obtained by this method will represent only the mean value of α between the melting point of ice and the boiling point of water. For more accurate value of α exact expression $R_t = R_0 (l + \alpha t + \beta t^2)$ should be used and the resistance R_0 and R_t measured more accurately by Callendar and Griffith's bridge.

Experiment 21.2

Object—To determine the temperature coefficient of resistance of platinum by means of Callendar and Griffith's pattern bridge.

Apparatus—Platinum resistance thermometer, Callendar and Griffith's pattern bridge, a Leclanche cell, a galvanometer, beaker (600 c.c. capacity), electric stove, a decimal-ohm dial box, a mercury thermometer and connection wires.

Description—Callendar and Griffith's pattern bridge. The circuit diagram of the bridge is given in Fig. 21.4.

The ratio coils P and Q have equal resistance, so that the

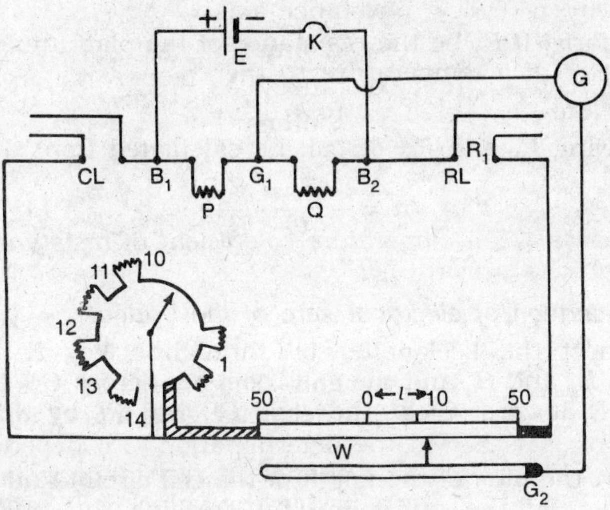

Fig. 21.4

ratio is 1 : 1. CL and RL are the gaps for connecting the compensating and the resistance leads of the thermometer. R is a dial pattern resistance, each coil having a resistance of one ohm. In series with this is a stretched wire W, also of resistance one ohm. The scale for the reading of the length of the wire is graduated with zero in the middle and 50 division marks at either end. Thus, the value ρ, the resistance per unit length of the bridge wire is 1/100 ohms per unit length.

Theory—Suppose with connections as shown in the figure above, the null point is at a length l measured from middle points of the wire. Let R_1 be the required resistance of the platinum wire, say, at room temperature t_1 and R the resistance introduced in the dial box. Then

$$\frac{P}{Q} = 1 = \frac{y + R + (50 + l)\rho}{y + R_1 + (50 - l)\rho}$$

where ρ is the resistance per unit length of the bridge wire and y is the resistance of the leads. From the above equation we have

$$y + R_1 + (50 - l)\rho = y + R + (50 + l)\rho$$

$$R_1 = R + 2l\rho \qquad (21.6a)$$

If ρ is known, R_1 can be calculated from the observed values of R.

Similarly, if R_2 be the resistance of the platinum wire of the thermometer at a temperature t_2, then

$$R_2 = R' + 2\, l'\rho \tag{21.6b}$$

Knowing R_1 and R_2, α, can be calculated from the relation

$$\alpha = \frac{R_2 - R_1}{R_1 t_2 - R_2 t_1} \tag{21.7}$$

Method

I. *Determination of electrical zero of the bridge.*

Connect the Leclanche cell through a key K across the terminals B_1 and B_2 and the galvanometer across G_1, G_2. Put the dial box R at zero. Close the gaps CL and RL by thick copper strips.

Insert the plug of the key K of the cell circuit and determine the position of the null point by pressing the jockey in the neighbourhood of the middle of the wire. Note this position. This gives the position of the electrical zero of the bridge wire. For an accurate bridge it should be at zero of the scale—this zero of the scale being at the middle point of the wire and 50 division marks on either side being at the ends of the wire. The difference between the zero of the scale and the electrical zero may be termed as the zero error of the bridge. Due correction must be applied in subsequent observations.

II. *Determination of* ρ, *the resistance per unit length of the bridge wire.*

Remove the copper strip from the gap CL. Connect a decimal ohm resistance box across CL. The dial box R stands at zero. Insert a resistance r of, say, 0.3 ohm in the gap CL, by rotating the dial of the decimal ohm box and locate the null point on the bridgewire. Measure the length from the zero of the scale. Correct this length for the zero error of the bridge by subtracting from this length the zero error of the bridge with its appropriate sign. Let this corrected length be L_1 units. Then evidently, since $R = 0$.

$$r = 0.3 = 2L_1\rho$$

or $$\rho = \frac{r}{2L_1} = \frac{0.3}{2L_1} \text{ ohm per unit length.}$$

Repeat the observations by introducing resistances of value 0.4, 0.5, 0.6 ohm from the decimal—ohm dial box across CL. Calculate ρ from each set and find its mean value.

III. *Determination of R_1 and R_2*

Disconnect the decimal ohm dial box from the gap *CL* and the copper strip from the gap *RL*.

Connect the compensation leads of the platinum resistance thermometer across *CL* and its resistance leads across *RL*.

Introduce a resistance of 2 ohms in the dial box *R* and locate the balance point on the bridge wire. (Usually the resistance of the thermometer lies between 2 and 3 ohms). Correct this length for the zero error of the bridge wire and knowing the values of *R*, *l* and ρ, calculate R_1 from Eq. (21.6a).

By changing the value of *R* to 3 ohms, redetermine the value of R_1. It may be noticed here that with *R* equal to 2 ohms, the balance point was on the positive side of the scale (*i.e.*, on to the right of zero of the scale) whereas with *R* equal to three ohms, the balance point is obtained on the left-hand side of the zero, *i.e.*, on the negative side of the scale. In the latter case R_1 will be calculated from the relation,

$$R_1 = R - 2l\rho$$

Find the mean value of R_1. Record the room temperature t_1 °C. Thus the resistance R_1 of the platinum coil is known at a temperature t_1 °C.

Next, place the platinum resistance thermometer in a large beaker or container with water. Allow the water to boil by heating the container over an electric stove. Also put a mercury thermometer to record the temperature t_2 of boiling water.

When water in the container has been boiling for about ten minutes or so, introduce a resistance of 3 ohms in the dial box *R* and locate the new balance point on the bridge wire, which will now be again on the positive side of scale. Repeat this observation after every two minutes and when the reading does not change it can be assumed that the platinum coil has attained the temperature t_2 °C of boiling water, as recorded by the mercury thermometer. Note this length for the balance point and correct it for zero error. From a knowledge of the resistance *R* (= 3 ohms), the corrected length l' for balance and ρ, calculate R_2 from Eq. (21.6b).

Put *R* at 4 ohms and redetermine the value of R_2. Compute the mean value of R_2 from those two determinations.

Knowing R_1, R_2 at temperatures t_1 and t_2, calculate the temperature coefficient of resistance α for platinum.

Sources of error and precautions

1. The decimal ohm box should be preferably of dial pattern

and should be connected by short, thick wires.

2. The cell circuit should be made on by the plug key only when observations are being taken; for the rest of the time, the plug should be out of the key. This would prevent unnecessary heating of the resistance coils due to the passage of current.

3. The galvanometer should be shunted by a low resistance wire to avoid excessive deflections in it when the bridge is out of balance the shunt should be removed when the exact position of the null point is being determined.

4. The jockey should be pressed gently over the bridge wire and the contact between the jockey and the bridge wire should not be made while the jockey is being moved along.

5. To ascertain that the platinum coil has attained the temperature of boiling water, the position of the balance point must be found out every minute or every two minutes; if there is no shift in the balance point it can be assumed that the platinum coil has attained the temperature of boiling water. If there is any shift, it should be of a negligible order, a mm. or so. The mean of such close lengths should be taken to obtain the length for balance of the bridge.

Observations

TABLE 1—*For Zero Error*

S. No.	Resistance in R	Balance point in the bridge wire + or − length units of length	Mean Position of Electrical Zero units of length	Zero Error (Difference between Scale Zero and Electrical zero) ± Z units of length.
1.	Zero	}		
2.	Zero			
3.	Zero			
4.	Zero			

TABLE 2—*For Determination of* ρ

S. No.	Resistance used in Decimal ohm Box r ohm	Position of Balance point units of length L_1	Corrected length L_1 for zero error	$\rho = \frac{r}{2L_1}$ ohm unit length
1.	0.3			}
2.	0.4			
3.	...			
4.	...			

Mean

TABLE 3—*For determination of R_1 and R_2 at temperatures t_1°C and t_2°C.*

Least count of the temperature = °C.

S. No.	Resistance used in dial box R ohm	Lenght for balance point l units of length	Corrected length for Zero error $l \pm Z$ units of length	Resistance of the platinum coil R_1 $R_1 = R \pm 2(l \pm Z)\rho$	Temperature t_1 °C
1.	2	. .	. .	$R_1 = R + 2(l \pm Z)\rho$ =	
2.	3	. .	. .	$R_1 = R - 2(l \pm Z)\rho$ =	

Make similar Table for R_2 with values of R as 3 and 4 ohms.

Calculations

(1) For ρ

$$= \frac{x}{2L_1} = \quad = \text{ohm/unit length}$$

Similar calculation for other sets.

Mean value of ρ = ohm/unit length

(2) Calculation for R_1 and R_2

From Set 1, $R_1 = R + 2(l \pm Z)\rho$

=

= ohm

From Set 2, $R_1 = R - 2(l \pm Z)\rho$

=

= ohm.

Mean value of R_1 = ohm.

Similarly,

mean value of R_2 = ohm.

Hence,

$$\alpha = \frac{R_2 - R_1}{R_1 t_2 - R_2 t_1}$$

=

= per °C

Result—The temperature coefficient of resistance for platinum = per °C.

Standard value of a = per °C.

Percentage error =

Experiment 21.3

Object—To determine the value of Stefan's constant.

Apparatus—Stefan's constant apparatus, thermometers, suspended coil galvanometer, lamp and scale arrangement, resistance box 0-10,000 ohm, calorimeter, beakers, stopwatch, steam generator and connection wires.

Description of apparatus—The Stefan's constant apparatus used in the laboratory has been shown below (Fig. 21.5). It consists of a blackened hollow metal hemisphere B about 25 cm.

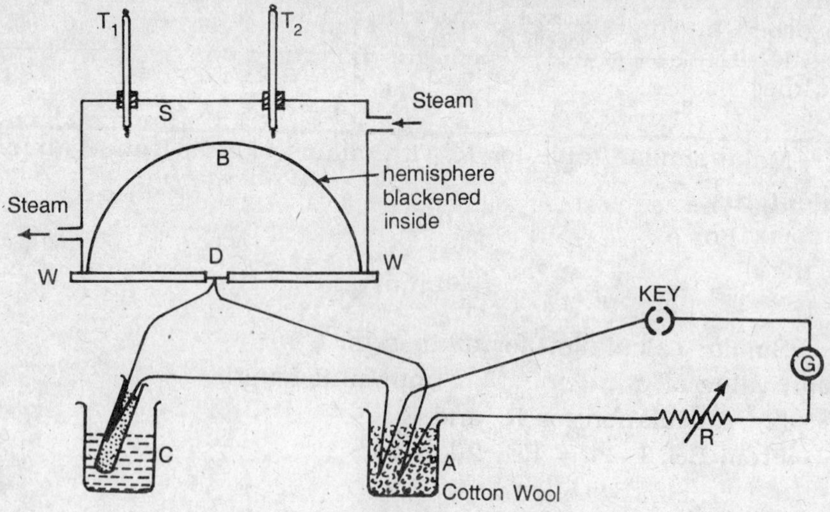

Fig. 21.5

in diameter fitted into a wooden board *WW* lined with tin. *B* is heated to a uniform temperature measured by the thermometer T_1 and T_2 by passing steam through the steam chamber *S* above the hemisphere *B*. In the wooden board at the centre of the hemisphere is a hole where a small silver disc *D* can be supported. The upper surface of the disc *D* is blackened to prevent reflection. It is better to fit the disc in a vulcanite frame rather than to allow it to touch the tin-lined wooden board directly. Thus, the hemisphere *B* with the tin lined wooden board and disc *D* constitute a black body enclosure. The black surface of *B* is the radiator and heat is received by the silver disc *D*. A thermocouple junction, usually of copper and constantan, is soldered to the bottom of the disc *D*, the other junction being placed in a tube containing oil, standing in a calorimeter *C* containing water. The junctions to the galvanometer are kept in a vessel *A* packed with cotton wool to prevent any electrical effect due to difference of temperature at these junctions. To keep the deflection on the scale, in case the galvanometer is sensitive, it may be necessary to introduce a resisance R in series with the galvanometers.

Theory.—The hemisphere B, the wooden frame W at the bottom along with the silver disc D behave as a black body encolsure. Suppose the temperature of the whole enclosure including disc D is T_1°K. Then there would be equilibrium, and the disc would both emit and absorb the same amount of radiation per unit times. Let R_1 be the amount of radiation absorbed by the silver disc per unit area per second, and let A be the area of the disc. Then the amount of heat absorbed by the disc per unit time is :

$$R_1 A = \sigma A\ T_1^4$$

The area of the disc D is negligible compared to the surface area of the enclosure. Hence, even if the disc D is at a temperature T^0K, less than T_1°K, the amount of energy absorbed by the disc is still $R_1 A = \sigma\ AT_1^4$. But the amount of heat radiated by the silver disc at this temperature T will no longer be $R_1 A$ but

$$RA = \sigma AT^4$$

where R is the amount of radiation emitted per unit area per second by the disc when at temperature T^0K. Hence, the gain of heat energy per recond is :

$$(R_1 - R)A = \sigma\ A\ (T_1^4 - T^4)$$

Owing to this gain of heat energy by the disc, its temperature gradually rises. At any instant let the temperature of the disc be T and $\frac{dT}{dt}$ the rate of rise of temperature with time. We then have,

$$m.s.\frac{dT}{dt} = \sigma A(T_1^4 - T^4) \qquad (21.8)$$

where m is the mass of the disc and s its specific heat.

In the equation above, the units of the quantity on the left hand side are calories per second. Hence, as such, will be expressed in units as Cal sec^{-1} cm^{-2}. °C^4. If σ is to be expressed in ergs $sec.^{-1}$ $cm.^{-2}$ °C^{-4}, the equation would be :

$$J.m.s.\ \frac{dT}{dt} = \sigma\ A\ [T_1^4 - T^4]$$

Whence
$$\sigma = \frac{Jms\ \dfrac{dT}{dt}}{A\ [T_1^4 - T^4]} \qquad (21.9)$$

In this equation A, m and s are known constants of the disc supplied by the manufacturer and J = 4.2 × 10^7 ergs per calorie. By experimentally observing T_1, the temperature of the enclosure and $\frac{dT}{dt}$, the rise of temperature per unit time when radiation is falling on it and it has a temperature T, σ can be calculated from Eq. (21.9).

For these determinations it would first be necessary to ascertain the relation between the deflection of the galvanometer and the difference in temperature between the two junctions. For this purpose the disc is surrounded with cotton wool and the cold reservoir placed over it. The calorimeter C is then heated to a certain temperature and then allowed to cool. As it cools, the temperature of the oil bath is read with a thermometer and the corresponding deflection of the galvanometer is noted. The difference in temperature between the hot junction and the cold disc is recorded on a graph against the readings of the galvanometer deflection. This plot is used as a calibration curve of the thermocouple and is shown in Fig. (21.6).

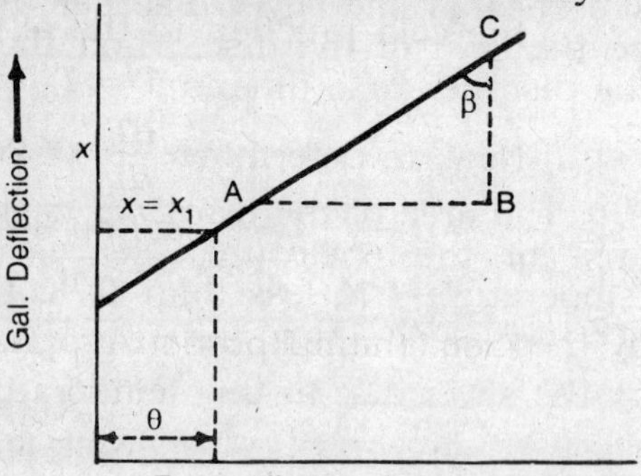

Fig. 21.6

Now we require $\frac{dT}{dt}$ when the disc is at a temperature T. To determine T and $\frac{dT}{dt}$ the following procedure is to be adopted. As soon as the temperature T_1 of the enclosure is steady, the disc is brought at the centre of the hemisphere and as the temperature of the disc rises, the galvanometer deflection is noted at regular intervals of time the junction in the oil bath being kept at room temperature. A plot between deflection x and time t is obtained as

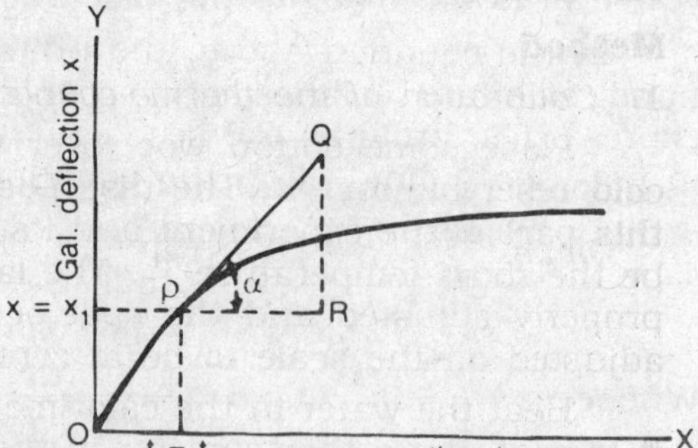

Fig. 21.7

depicted in Fig. 21.7. Let us choose a point P on this plot corresponding to any instant of time t. The corresponding deflection of the galvanometer is x_1. The temperature difference between the disc D and the cold junction in C corresponding to the deflection x_1 is read from the calibration graph of the thermocouple. Suppose this is θ°C. Hence,

$$\theta = T - T_0$$

where T is the required temperature of the disc and T_0 that of the cold junction in the oil bath. Hence,

$$T = T_0 + \theta \tag{21.10}$$

Now, to determine $\frac{dT}{dt}$ we proceed as follows :

$$\frac{dT}{dt} = \frac{d(T_0 + \theta)}{dt} = \frac{d\theta}{dt} = \frac{d\theta}{dx} \cdot \frac{dx}{dt}$$

From the calibration graph,

$$\frac{d\theta}{dx} = \tan \beta = \frac{AB}{BC}$$

and from the second graph, with a tangent drawn to the curve at the point P whose temperature we have determined as $T = T_0 + \theta$,

$$\frac{dx}{dt} = \tan \alpha = \frac{QR}{PR}$$

Hence,

$$\left[\frac{dT}{dt}\right]_P = \tan \beta . \tan \alpha = \frac{AB}{BC} \times \frac{QR}{PR}.$$

Thus having determined T and $\left(\frac{dT}{dt}\right)_{atT}$ we can substitute these values in Eq. 21.9 and determine the value of Stefan's constant.

Method

1. *Calibration of the thermo-couple*

Place some cotton wool over the disc D and then put the cold reservoir over it. The disc D serves as the cold junction in this part of the experiment and its temperature may be taken to be the room temperature T_0. The lamp and scale arrangement is properly adjusted and the spot of light with the index mark is adjusted on the scale to be at zero mark.

Heat the water in the calorimeter C to about 90°C and allow it to cool. Put a thermometer in the oil bath which contains now the hot junction of the thermocouple. Adjust the resistance R in the circuit so that the deflection, with the maximum temperature difference between the junctions, remains on the scale. As the oil cools, record the temperature of the hot junction and the corresponding deflection of the galvanometer on the scale. Continue the observations upto a temperature of the hot junction which is about 5° above the cold junction.

Plot a graph between the difference of temperature of the two junctions of the thermocouple, $(T - T_o) = \theta$, against the deflections of the galvanometer. The graph is a straight line as shown below :

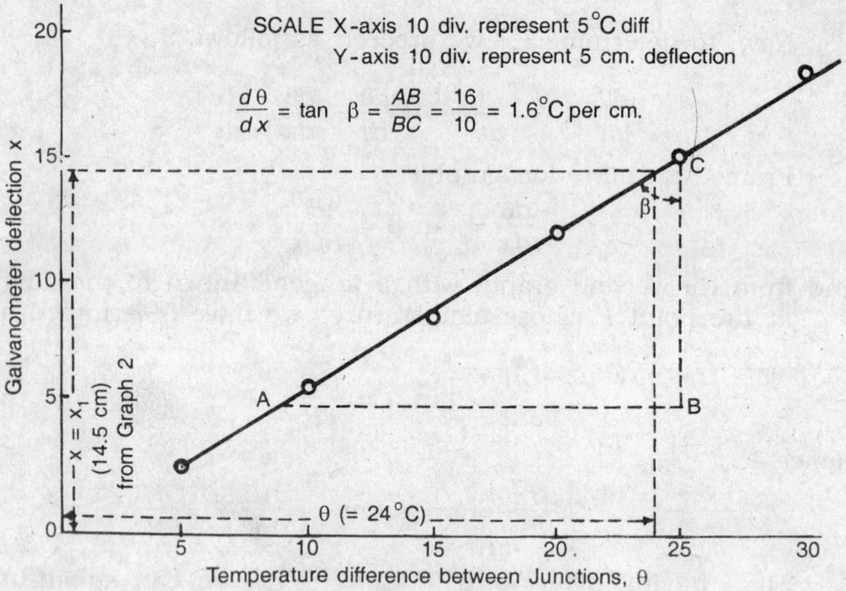

Graph 1 between θ and x

From the graph find:

$$\frac{d\theta}{dx} = \tan\beta = \frac{AB}{BC}$$

(From this graphical data $\dfrac{d\theta}{dx} = 1.6°\text{C per cm.}$)

2. *Determination of* T_1, T *and* $\dfrac{dT}{dt}$ *at* T.

Withdraw the disc *D* from the centre of the enclosure. Change the water in the calorimeter so that the junction in the oil now stands at the room temperature. If it is different from room temperature, record its actual temperature. Let it be T°*K*.

pass steam into the steam chamber. When the temperature recorded by the two thermometers becomes steady for more than five minutes or so, record these observations and take their mean. This is the temperature in Celsius degrees. Add 273 to this and get the mean value of T_1^oK. This gives the temperature of the enclosure in Kelvin degrees.

When this steady state has been reached, the disc D is brought at the centre of the enclosure and now the junction soldered to D acts as the hot junction. At regular intervals of 10 or 15 seconds, read the deflection of the galvanometer. As time elapses, the deflection at first decreases rather rapidly but later on, very slowly. Continue the observations for three or four minutes.

Plot a graph between time t along the axis of X and the corresponding deflection of the galvanometer x along the axis of Y. Choose a point P on the graph close to the origin and draw a tangent to the curve at this point P.

From the graph (given on the next page) find the value of

$$\frac{dx}{dt} = \tan \alpha = \frac{QR}{PR}$$

[From this graphical data $\frac{dx}{dt}$ = 0.125 cm. per sec.]

Calculate $\frac{dT}{dt}$ for the point P from the relation

$$\frac{dT}{dt} = \frac{d\theta}{dx} \cdot \frac{dx}{dt} = \tan \beta \tan \alpha$$

[From the graph drawn, $\frac{dT}{dt} = 1.6 \times 0.125 = 0.2$ °C per second].

To determine the temperature of the silver disc corresponding to the point P at which we have determined its rate of rise of temperature with time, $\frac{dT}{dt}$, find the value of x, the galvanometer deflection for this point. This is x_1. (In our graph this is 14.5 cm.). From the calibration graph of the thermocouple for this value of $x = x_0$, find the temperature difference θ between the two junctions. If T_0 be the temperature of the cold junction, then evidently, the temperature of the disc D corresponding to the point P is :

$$T = (T_o + \theta)^o K$$

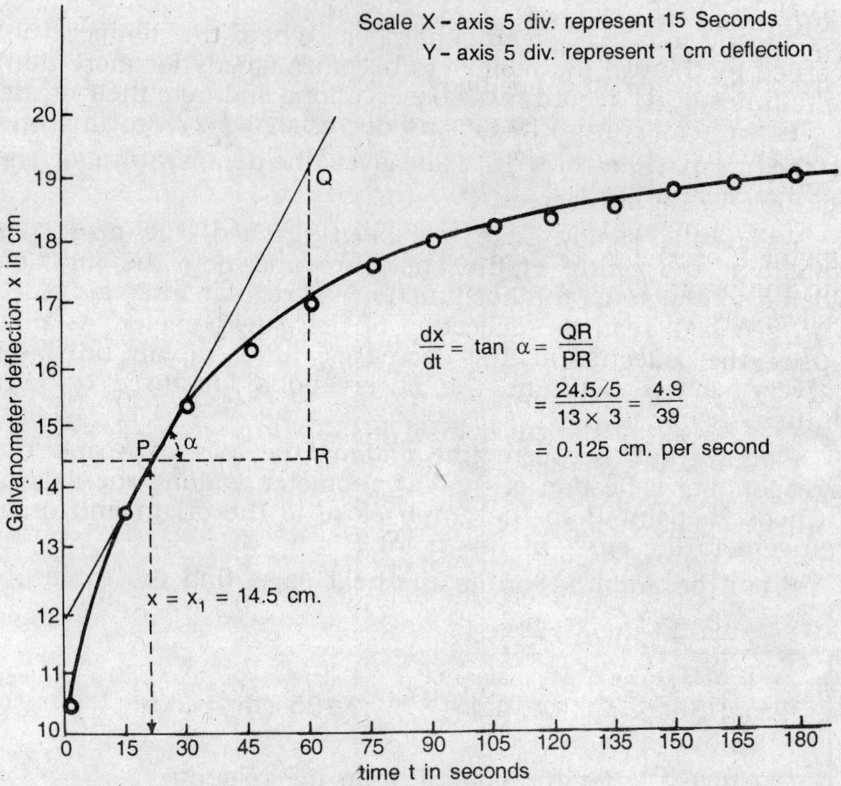

Graph 2 between t and x

[From our graphs, for $x_1 = 14.5$, $\theta = 24°C$, hence if $T_o = (273+20)°K$, $T = (273+20+24)°K = 317°K$.]

m, s and *A* are constants of the disc supplied by the manufacturer. Knowing all the factors in Eq. (24.9) calculate σ.

Sources of error and precautions

1. If the galvanometer used is very sensitive it may be found necessary to use a high resistance from a four dial resistance box in series with the galvanometer so that the deflection remains with the scale for a maximum temperature difference between the junctions.

2. In the calibration part of the experiment, the junction soldered to the disc *D* is the cold junction. As such the disc *D* should be maintained at the constant temperature of the room ; it must not receive radiation from any source. So it is desirable to cover it with cotton wool.

3. The leads from the two junctions leading to the galvanometer are kept in a container *A* packed with cotton wool

to prevent any disturbing effect due to difference of temperature in the leads. The container *A* is kept at a considerable distance from the heater and hot body.

4. Care must be taken to see that the wires constituting the thermocouple are in contact only at the junction and do not touch each other elsewhere.

5. Care must be exercised in drawing the tangent at the point P on the (*x*, *t*) graph. The point P should be chosen close to the origin O, since errors soon arise by conduction from the silver disc.

Observations

1. Known constants

Mass of the silver disc, (*m*) = gm.
Radius of the silver disc (*r*) = cm.
Sp. heat of silver(*s*) = Cal/gm. °C
Value of. J = 4.2×10^7 ergs/calorie.

2. For the Calibration of the thermocouple.

Room temperature = θ_R °C

Initial position of the spot of light on Scale = 0 cm.

S. No.	Temperature of hot junction	Difference in temperature between the two junctions	Deflection of Galvanometer.
1. 2. 3. 4.		5°C	

3. Measurement of T_1.

Temperature as recorded by thermometer T_1 = °C
Temperature as recorded by thermometer T_2 = °C
Mean temperature θ_1 = °C
∴ Temp. of the enclosure (θ_1 °C) = °C

4. Table for determination of T and $\frac{d[illegible]}{dt}$ at T.

Tem. of cold junction = T_0°K

S. No.	Time in Seconds	Deflection	S. No.	Time in Seconds	Deflection

Temperature of the enclosure $T_1 = (273 + \theta_1)^oK$
$= \qquad {}^oK$

From the calibration graph,

$$\frac{d\theta}{dx} = \tan\beta = \frac{AB}{BC} = \qquad \text{per cm.}$$

From the (x, t) graph, at the point P

$$\frac{dx}{dt} = \tan\alpha = \frac{QR}{PR} = \text{cm. per sec.}$$

Hence, $$\frac{dT}{dt} = \frac{d\theta}{dt} = \frac{d\theta}{dx}\,\frac{dx}{dt} = \frac{AB}{BC} \times \frac{QR}{PR}$$

$$= \qquad {}^oK \text{ per sec.}$$

For the point P, $x = x_1$.

For this value of x_1, the corresponding difference in the temperature of the two junctions is θ, as determined from the calibration graph. Hence the temperature of the silver disc corresponding to the point P is

$$T = (T_0 + \theta)^oK$$
$$= \qquad {}^oK.$$

Hence, $$\sigma = \frac{J\,m\,s\left(\dfrac{dT}{dt}\right)_{at\,P}}{A(T_1^4 - T^4)}$$

$=$

$= \qquad$ ergs, sec.$^{-1}$ C^{-2} $^{\circ}$C^{-4}.

Result—The value of Stefan's constant = ergs, sec^{-1} cm.$^{-2}$ $^oC^{-4}$.

Standard $\sigma = 5.67 \times 10^{-10}\ J\,s^{-1}\,m^{-2}\,K^{-4}$

% Error = ...

SECTION IV

MAGNETISM AND ELECTRICITY

22
Magnetic Measurements

22.1. Magnetic Field due to Current in a Conductor

When an electric current is started in a conductor, immediately a magnetic field is established in the space around the conductor. The conductor itself does not become a magnet for it does not possess any magnetic poles, neither can it attract iron filings. The nature or form of the field depends upon the shape of the conductor and can be explored with iron filings or a compass needle in the usual manner. The intensity B of the *magnetic field* at a point P due to a *short* element *dl* of the conductor as shown in Fig. 26.1, is (*i*) directly proportional to the length of the element *dl* as seen from the point P, (*ii*) directly proportional to the strength *I* of the carrent and (*iii*) inversely proportional to the square of the distance *r* of the element from the point P. Thus,

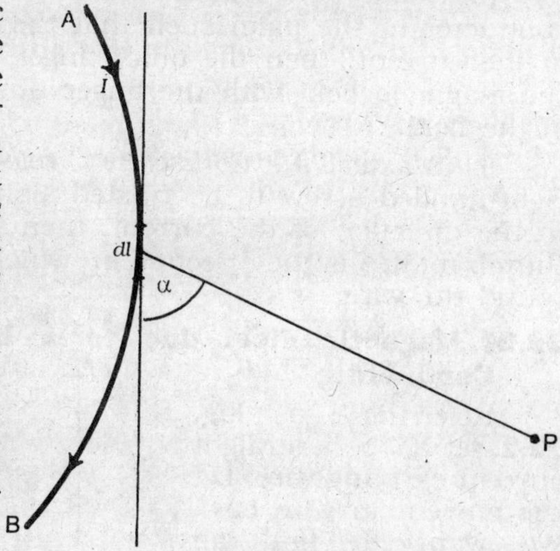

Fig. 22.1

$$dB = K = \frac{I\, dl \sin \alpha}{r^2}$$

In SI units, $K = \frac{\mu_0}{4\pi}$ where μ_0 is the permeability of free space. Its value is $4\pi \times 10^{-7}$ Tm A^{-1} or, Wb m^{-1} A^{-1}.

$$dB = \frac{\mu_0}{4\pi} \cdot \frac{I\, dl \sin \alpha}{r^2}$$

The above rule for calculating the intensity of the magnetic field at a point due to a small element of a conductor carrying current is known as *Biot-Savart law*. The rule is applicable to each of the small elements of a current carrying conductor and hence can be extended to the whole conductor. The direction of the field is dependent upon the direction of the current and can be obtained from any of the following rules :

(*a*) Look along the conductor in the direction of the current, the magnetic lines of force will go round in the direction in which the hands of a clock move, *i.e.*, the north pole of the magnetic needle will be deflected in the clockwise direction.

(*b*) *Right hand Thamb rule*—If we grasp the current carrying conductor in the palm such that the thumb is in the direction of the current, then the other fingers curve in the direction of the magnetic field with the finger tips pointing in the direction of the field.

(*c*) *Maxwell's Corkscrew rule*—Imagine an ordinary right-handed screw to be twisted along the wire so as to move in the direction of the current, then the direction in which the thumb rotates is the direction in which north pole tends to move round the wire.

22.2. Magnetic Field due to a Linear Current Carrying Conductor

Referring to Fig. 22.2, let XY be a straight current carrying wire. Let the current in wire be *I*. We want to find an expression for the magnetic field B at a point P at a distance *r* from the wire. Let us consider an infinitesimally small current element CD of Length *dl*. Let *s* be the distance of P from the mid-point of the current element; let θ be the angle which OP makes with the direction of current.

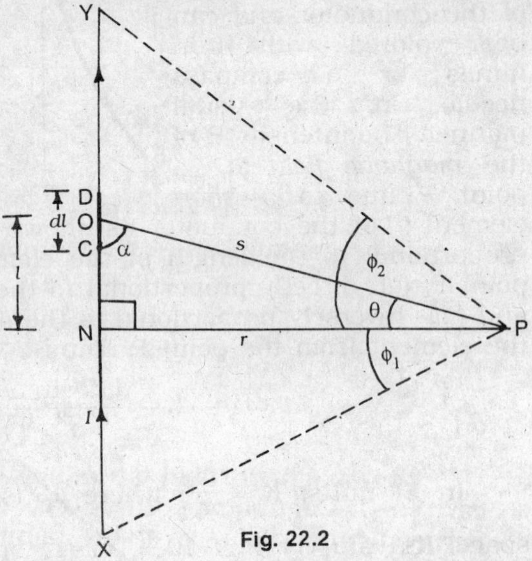

Fig. 22.2

Using Biot-Savart law, the magnetic field at P due to the current element CD is given by

$$dB = \frac{\mu_0}{4\pi} \cdot \frac{I\, dl \sin \theta}{s^2} \qquad (22.1)$$

In right angled Δ ONP, $\theta + \alpha = \pi/2, \Rightarrow \theta = \pi/2 - \alpha$

$$\therefore \qquad \sin\theta = \sin(\pi/2 - \alpha) = \cos\alpha$$

Also, $\cos\alpha = \dfrac{r}{s}$ or $s = \dfrac{r}{\cos\alpha} = r\sec\alpha$

Again, $\tan\alpha = \dfrac{l}{r}$ or $l = r\tan\alpha$

On differentiation we get :

$$dl = r\sec^2\alpha\, d\alpha$$

Substituting dl in Eq. (22.1), we get

$$dB = \frac{\mu_0}{4\pi}\frac{I(r\sec^2\alpha\, d\alpha)\cos\alpha}{r^2\sec^2\alpha}$$

or

$$dB = \frac{\mu_0 I}{4\pi r}\cos\alpha\, d\alpha$$

The magnetic field at P due to the entire wire XY is given by integration of above expression within the limits $-\phi_1$ to ϕ_2,

$$B = \int dB = \int_{-\phi_1}^{\phi_2}\frac{\mu_0 I}{4\pi r}\cos\alpha\, d\alpha$$

$$= \frac{\mu_0 I}{4\pi r}\int_{-\phi_1}^{\phi}\cos\alpha\, d\alpha = \frac{\mu_0 I}{4\pi r}\left|\sin\alpha\right|_{-\phi_1}^{\phi_2}$$

$$= \frac{\mu_0 I}{4\pi r}[\sin\phi_2 - \sin(-\phi_1)]$$

or,

$$B = \frac{\mu_0 I}{4\pi r}(\sin\phi_1 + \sin\phi_2)$$

If the conductor is infinitely long, then,

$$\phi_1 = \phi_2 = \pi/2$$

$$B = \frac{\mu_0 I}{4\pi r}(\sin\pi/2 + \sin\pi/2) = \frac{\mu_0}{4\pi}\cdot\frac{2I}{r}$$

or,

$$B = \frac{\mu_0 I}{2\pi r} \qquad (22.2)$$

The magnetic field at a point due to a straight current carrying Conductor, therefore, varies *inversely* as the perpendicular distance of the point from the conductor and is the same at all points at the same distance from it. The direction of field at P is at *right angles* to the plane through P and the axis of the wire and hence the field has no component in that plane. The magnetic lines of force are, therefore, circles in planes perpendicular to the axis of the conductor with their centres on its axis.

If the conductor has n turns, then,

$$B = \frac{\mu_0 n I}{2\pi r} \qquad (22.3)$$

Experiment 22.1

Object—(*i*) To plot the resultant magnetic field of a vertical *straight* current and the earth in a horizontal plane,

(*ii*) to show that the field due to the straight current varies *inversely* as the distance, and

(*iii*) to calculate the value of the horizontal component of earth's magnetic field from the positionof the neutral point.

Apparatus—A large rectangular framework covered with several turns of copper wire and mounted vertically as shown in Fig. 22.3, with one of its vertical sides passing through a drawing board centrally, a bulb of resistance of about 140 ohms, a suitable

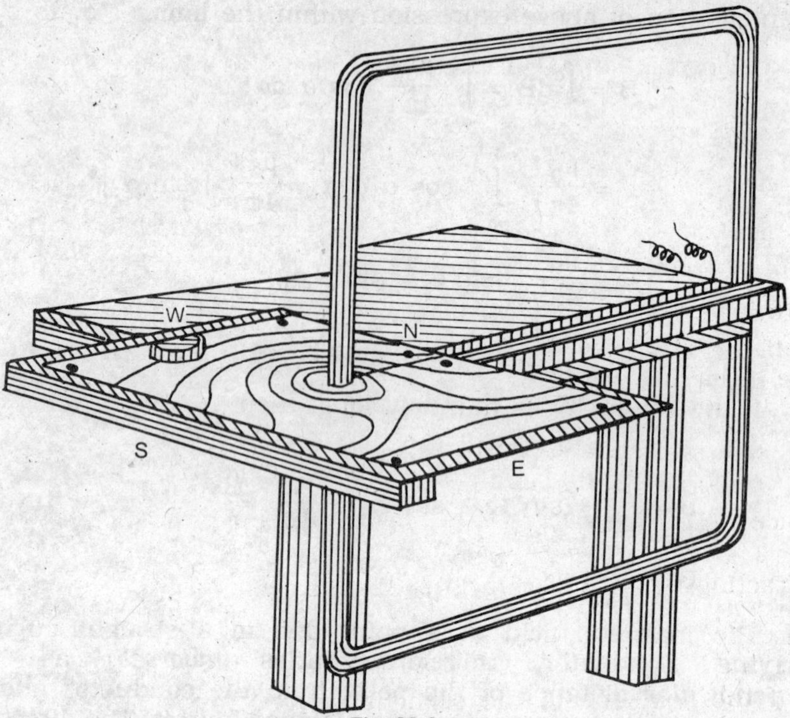

Fig. 22.3

rheostat, an ammeter, a compass needle, a Searle's oscillating needle, a stop-watch, a sheet of drawing paper and wax.

Theory—Referring to Fig. 22.4, let *AB* be the *magnetic east-west* line passing through the axis of the straight conductor which cuts the *lines of force* in the *resultant* field of the straight current and the earth at points x_1, x_2, x_3, etc., on the *opposite* side of the *neutral point.* If tangent is drawn at any of these points, say, the tangent will lie in the *magnetic north-south* direction. This is because the lines of force in the field of the straight current *alone* are concentric circles.

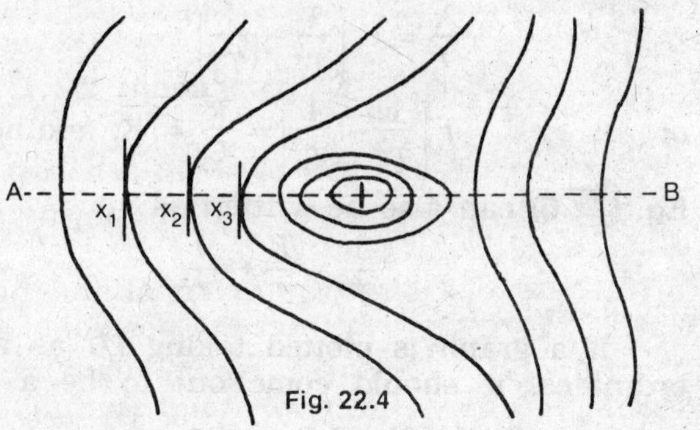

Fig. 22.4

Evidently, the resultant magnetic field at x_1 is $(B+H)$, where, B is the field due to the straight current and H the horizontal component of the earth's field.

Now, if a Searle's needle be allowed to oscillate at x_1, its *period* of oscillation is given by :

$$T = 2\pi\sqrt{\frac{I}{M(B + H)}}$$

This gives :

$$B + H = \frac{4\pi^2 I}{MT^2} = \frac{C}{T^2} \quad (22.4)$$

where $C = 4\pi^2 I/M$ is a constant.

If the same needle were allowed to oscillate in earth's field *alone* its period would be :

$$T_0 = 2\pi\sqrt{\frac{I}{MH}}$$

whence $$H = \frac{4\pi^2 I}{MT_0^2} = \frac{C}{T_0^2} \quad (22.5)$$

Subtracting Eq. (22.5) from Eq. (22.4), we have :

$$B = C\left(\frac{1}{T^2} - \frac{1}{T_0^2}\right)$$

If the field due to the straight current varies *inversely* as the distance

$$B = \frac{k}{r}$$

where r is the distance of the point x_1 from the conductor and k is a constant of proportionality, hence, we have :

$$\frac{k}{r} = C\left(\frac{1}{T^2} - \frac{1}{T_0^2}\right)$$

or $$r\left(\frac{1}{T^2} - \frac{1}{T_0^2}\right) = \frac{k}{C} = \text{K, another constant.} \qquad (22.6)$$

Eq. (22.6) can also be written as :

$$\frac{1}{T^2} = \frac{K}{r} + \frac{1}{T_0^2}$$

If a graph is plotted taking I/r as abscissae and I/T^2 as ordinates, it should come out to be a *straight line* with an intercept on the Y-axis equal to $\frac{1}{T_0^2}$.

At the *neutral point*, the field due to the straight current is $\frac{\mu_0 nI}{2\pi r}$, where r is the distance of the neutral point from the conductor, I the current in *amperes* flowing through it and n is the number of turns of wire in the conductor. Since at the neutral point *resultant* field intensity in the horizontal plane is *zero*, the field due to the straight current is exactly *equal* and *opposite* to the horizontal component H of the earth's field. Hence,

$$H = \frac{\mu_0 nI}{2\pi r} \qquad (22.7)$$

Method—Fix with wax a sheet of drawing paper on the drawing board in such a manner that the vertical side of the rectangular framework carrying the straight conductor passes through the paper *centrally*. With the help of the compass needle draw the magnetic north and south line, on the drawing paper by a fine arrow. Connect the two open ends of the straight conductor to the mains through a bulb resistance, an ammeter and a suitable rheostat all *in series*, the value of the bulb resistance being so chosen as to produce a current of strength between one and two amperes.

After the electric connections have been completed and thoroughly checked, start the current in the conductor when a magnetic field due to the current in the conductor will be produced in the space around it. With the help of the compass needle, trace on the drawing paper, in the usual manner, the magnetic lines of force in the combined field of the straight current and the earth and locate the *neutral point*. The neutral point will lie due magnetic east or west of the conductor according as the current is flowing downwards or upwards through the conductor. The neutral point will be found to be enclosed by four sets of lines of force forming a curvi-linear quadrilateral. Diminish the area of this quadrilateral by tracing lines of force within it. When the area of the quadrilateral has been considerably reduced, its centre will give the position of the neutral point. Around the neutral point, mark a circle of needle's size and measure the distance r of the neutral point from the straight conductor. Then note down the reading of the ammeter and calculate the value of H from Eq. (22.7).

Now, with the help of the compass needle draw the magnetic east-west line through the axis of the conductor, and on the *opposite* side of the neutral point, mark on this line a number of points x_1, x_2, x_3, etc., at known distances r_1, r_2, r_3, *etc.* from the conductor. Allow a *Searle's needle* to oscillate at any one of these points, say, x_1 in the *resultant* field $(B+H)$ and note down the time taken by it to complete 20 oscillations and thus find out the period of oscillation of the needle at x_1. Then allow the needle to oscillate at other points and determine as before the period of oscillation T at the various points. Now, switch off the current and determine the period of oscillation T_0 of the needle at any point in the earth's field *alone*. For each point x_1, x_2, etc., calculate the value of the expression $r\left(\frac{1}{T^2}-\frac{1}{T_0^2}\right)$. This will be the *same* for *each* point. Plot a graph taking $1/r$ along the X-axis and $1/T^2$ along the Y-axis. This will come out to be a straight line with an intercept on the Y-axis equal to $1/T_0^2$.

Sources of error and precautions

(1) All magnetic materials, current-bearing conductors and the mains should be at a considerable distance from the straight conductor otherwise the field produced in the space around the conductor will be more complex in character than that which is intended to be produced. Connections of the straight conductor with the rest of the circuit should be made by twin *flexible wires* twisted together so that the passage of current through them may not affect the field of the straight conductor.

(2) The magnetic meridian should be represented on the drawing paper by a fine arrow and once this magnetic north and south lines have been drawn, the paper should never be disturbed throughout the experiment.

(3) Unless it is a hot-wire instrument, the ammeter should

be kept at a *sufficient* distance from the conductor as most ammeters contain a strong permanent magnet.

(4) The current flowing through the straight conductor should be of one to two amperes so that a *strong* magnetic field is produced in the space around the conductor and hence the *neutral point* may be at a considerable *distance* from it. Further, the magnetic field produced should be *steady* and hence the reading of the ammeter should remain constant during the experiment, the rheostat being adjusted from time to time, if necessary, to keep the current constant.

(5) After tracing a magnetic line of force, its direction should be clearly marked by an arrow head.

(6) For determination of *T Searle's needle* should be allowed to oscillate only at points lying on the magnetic *east-west* line passing through the conductor and on *opposite* side of the *neutral point.*

(7) The amplitude of oscillation of the needle should be small, say, 3 or 4.

(8) The graph between $1/r$ and $1/T^2$ should be a *straight line* and should be *smoothly* drawn.

Observations

[A] *Determination of H*

(i) Distance of the *neutral point* from the conductor $r =$ cm. = m.

(ii) Current through the circuit $I =$ amp.

(iii) Number of turns in the conductor $n =$

[B] *Determination of T*

Least count of stop watch = sec.

Set No.	Distance of point x from conductor m.	No. of oscillations	Time taken min.	Time taken sec.	Period *T* sec.	*Mean* period *T* sec.
A						

[C] *Determination of* T_0

S. No.	No. of oscillations	Time taken		Period T_0
		min.	sec.	
1. 2. 3.				

Mean

Calculations

Point on graph	$\frac{1}{r}$	$\frac{1}{T^2}$	$r\left[\frac{1}{T^2}-\frac{1}{T_0^2}\right]$

For point x_1, $r =$ cm = m, $T =$ sec.

$$r\left[\frac{1}{T^2}-\frac{1}{T_0^2}\right] = \quad =$$

(*N.B.*—Similar calculations should be done for other points.)

$$H = \frac{\mu_0 \, nI}{2\,\pi\, r}$$

$$=$$

$$= \quad \text{Tesla}$$

Result—(*i*) The graph between $1/r$ and $1/T^2$ is a *straight line* showing that the field due to a straight current varies *inversely* as the distance.

(*ii*) The value of horizontal component of earth's magnetic field H in the laboratory at...... = Tesla

Standard value of H at....... = Tesla

∴ Percentage error =

Criticism of the method—It is not an accurate method of determining the value of the horizontal component of earth's field for the resultant field due to the linear conductor and the earth is liable to be affected by the presence of neighbouring apparatus carrying current. The magnetic field in the neighbourhood of the neutral point is very *weak* and the behaviour of the needle in this region becomes almost *uncertain.* The position of the neutral point cannot, therefore, be *correctly* located and, hence, the accuracy of result cannot be expected to be great.

22.3. Magnetic Field at Centre of a Circular Current Carrying Coil

Consider a circular coil of radius r through which a current

I is flowing. The wires leading to and from (also called *leads*) are twisted so that the magnetic fields due to them cancel out. Consider an infinitesimally small element of length *dl* of the wire (Fig. 22.5) According to Biot-Savart's law, the magnetic field at the centre is given by :

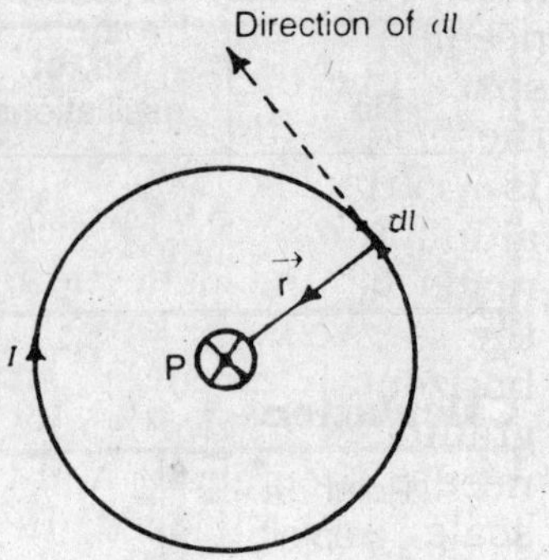

Fig. 22.5

$$dB = \frac{\mu_0}{4\pi} \frac{I\,dl \sin\theta}{r^2}$$

where θ is the angle between $\vec{dl}$ and $\vec{r}$, which is a right angle in this case. Therefore,

$$dB = \frac{\mu_0}{4\pi} \frac{I\,dl \sin 90^\circ}{r^2} = \frac{\mu_0}{4\pi} \frac{I\,dl}{r^2}$$

The loop can be considered to be made up of large number of such small elements. The magnetic field due to all the elements will be in the same direction. Thus, the net magnetic field at the centre P is given by

$$B = \int dB = \int \frac{\mu_0}{4\pi} \frac{I\,dl}{r^2} = \frac{\mu_0}{4\pi} \frac{I}{r^2} \int dl$$

or $$B = \frac{\mu_0}{4\pi} \frac{I}{r^2} 2\pi r$$

or $$B = \frac{\mu_0 I}{2r} \qquad (22.8)$$

If the coil is of *n* turns,

$$B = \frac{\mu_0 nI}{2r} \qquad (22.9)$$

The direction of the field at P is at right angles to the plane of the coil and may be determined according to the following rule : Looking at the face of the coil, if the current is clockwise, the direction of the field inside the coil is away from the observer; if the current is anti-clockwise the direction of the field inside the coil is towards the observer.

22.4. Tangent Galvanometer

As depicted in Fig. 22.6 the tangent galvanometer consists essentially of a magnetic needle pivoted or suspended at the centre of a circular coil consisting of many turns of insulated copper wire, the plane of the coil being *vertical*. The magnetic needle is *small* and the radius of the coil *large* compared to the needle so that the field due to the current in the coil in entire space in which the needle moves may be *uniform* and equal to the field at the centre of the coils. At right angles to the needle is attached a light aluminium pointer which reads off the deflection of the needle on the horizontal circular scale graduated in degrees. The needle, the pointer and the scale are enclosed in a metal box with glass cover to shield them from air daughts. Below the needle at the base of the box is fixed a plane mirror with the help of which *errors* due *to parallax* in reading deflections are avoided. To level the instrument the base is provided with three levelling screws.

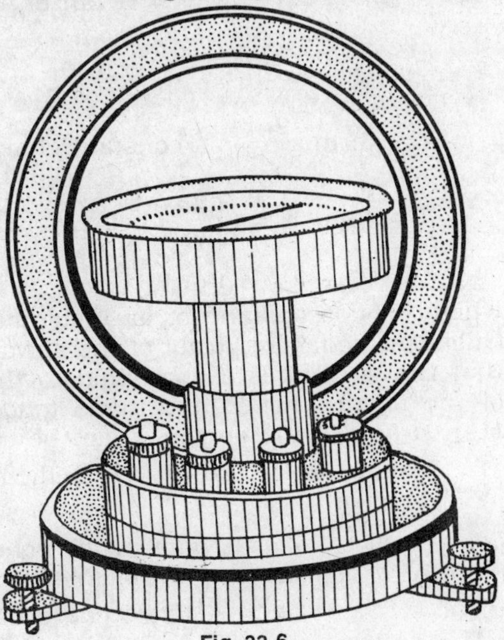

Fig. 22.6

The tangent galvanometer usually employed in laboratories has 500 turns of copper wire wound on the circular frame which terminate in four binding screws. Between the first and second terminals there are only two turns of wire which have a resistance of about 0.02 ohms. This is known as *ammeter coil* and is used with *strong* currents. Between the first and the third terminals there are 50 turns of wire which have a resistance of about 1.1 ohms and are used with *moderate* currents. Between the first and the fourth terminals there are 500 turns of wire having a resistance of about 226 ohms. This is called *voltmeter coil* and is used with *weak* currents.

To use the instrument the plane of the coil is set in the *magnetic meridian* when the needle and the coil will lie in the *same* vertical plane. Then the current to be measured is passed through the coil and the deflection θ produced in the needle is read on the scale. Since the field B at the centre of the coil is perpendicular to earth's horizontal field *H*, we have from Eq. (22.5).

$$B = H \tan \theta$$

If the coil has *n* turns of radius *r*, then magnetic field prodduced at the centre of the coil when a current I is passed is given by :

$$B = \frac{\mu_0}{4\pi} \cdot \frac{2\pi nI}{r}$$

$$\frac{\mu_0}{4\pi} \cdot \frac{2\pi nI}{r} = H \tan\theta$$

or

$$I = H. \frac{2r}{\mu_0 n} \tan\theta \qquad (22.10)$$

The term $\frac{2r}{\mu_0 n}$ is a constant characteristic of the instrument.

If we put $K = \frac{H\,2r}{\mu_0 n}$, then Eq. (22.10) can be expressed as :

$$I = K \tan\theta \qquad (22.11)$$

where K is a constant for an instrument for a particular location called the *reduction factor* of the galvanometer. Eq. (22.11) shows that the current is proportional to the tangent of the deflection of the needle, hence, the name *tangent galvanometer*. Knowing H, r, n and θ, the current I can be calculated.

When $\theta = 45°$, $\tan\theta = 1$, and, hence, from Eq. (22.11) :

$$K = I$$

i.e. reduction factor is numerically equal to the current required to produce a deflection of 45° in the galvanometer.

Experiment 22.2

Object—To determine the *reduction factor* of the two turns coil of a tangent galvanometer using a copper voltameter and then to calculate the value of H, the *horizontal component of the earth's magnetic field.*

Apparatus—The tangent galvanometer, a battery, a rheostat, a commutator, a key, a copper voltameter, a test plate of copper, a weight box, a chemical balance and a spirit level.

Theory—Let the plane of the coil of the tangent glavanometer be placed in the *magnetic meridian* and let a current of strength I *amperes* be allowed to pass through the coil. If the needle is deflected through an angle θ from the *magnetic meridian* :

$$I = K \tan\theta \qquad (22.12)$$

where K is the *reduction factor* of the tangent galvanometer. If the *same* current passes through a copper voltameter placed *in series* with the tangent galvanometer and if the mass of copper *deposited* on the cathode plate in t seconds be m gms., then from Eq. (22.1) :

$$m = Z\,I\,t \qquad (22.13)$$

where Z gms. per coulomb is the *electro-chemical equivalent of* (cupric) copper.

Eliminating I between Eqs. (22.12) and (22.13), we get

$$K = \frac{m}{Zt \cdot \tan\theta} \quad (22.14)$$

This equation gives the value of K, the value of Z being already known from table of constants.

Now, if n be the number of turns in the galvanometer coil and r be its *mean* radius :

$$K = \frac{2Hr}{\mu_0 n}$$

Hence,
$$H = \frac{K \mu_0 n}{2r} \text{ Tesla} \quad (22.15)$$

Method—Clean the copper *cathode plate* of the copper voltameter *thoroughly* with sand paper and a rug or emery cloth until it is quite bright and weigh it in an accurate *chemical balance* correct up to 1/10 of a milligram. Then level the compass box and also the base of the tangent galvanometer by means of levelling screws testing the levelling with a spirit level. Now rotate the coil if it is movable, or the instrument as a whole if the coil is fixed, about a *vertical* axis till *the coil, the needle and its image in the plane mirror, all lie in the same vertical plane.* In this position of the coil, if the instrument is perfect, the aluminium pointer in the compass box, if the latter is fixed, will point 0-0, if the box is, however, not fixed, rotate it *without disturbing the position of the coil,* till at least one of the ends of the pointer stands at zero. Now using the test plate of copper as the cathode in the copper voltameter connect the tangent galvanometer (coil of 2 turns) through a commutator to a battery *in series* with a suitable rheostat, key and the copper voltameter as shown in Fig. 22.7. Start the current

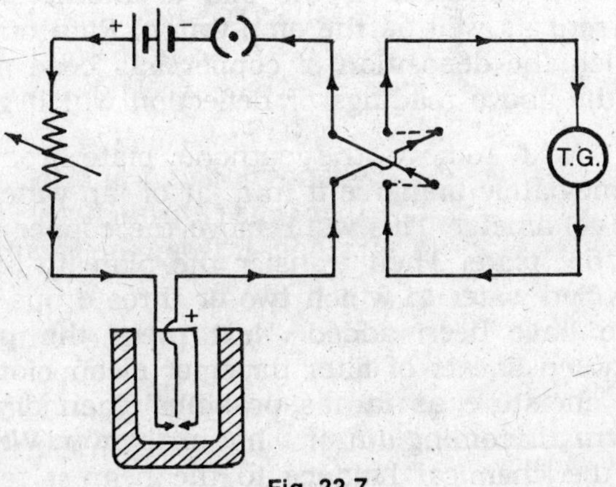

Fig. 22.7

and after gently tapping the compass box with figure, read the deflections at both ends of the pointer. Then *reverse* the current through the tangent galvanometer by means of the commutator and again read the deflection at the two ends of the pointer. If the *mean* values of readings for deflection before and after reversing the current in the tangent galvanometer differ, if at all, by more than 1°, slightly turn the coil until they agree as closely as possible. *This sets the plane of the coil of the tangent galvanometer parallel to the magnetic meridian.* Again level the galvanometer if necessary and then mark the positions of the levelling screws of the tangent galvanometer on the table. Adjust the rheostat so that a deflection in the neighbourhood 45° or failing which between 25° and 65° is produced in the tangent galvanometer.

Now, stop the current and take out the *test plate* of copper from the copper voltameter. If the connections of the voltameter are correct the plate should be covered with a smooth salmon coloured deposit of copper. Now replace this plate in the voltameter by the *clean* copper plate over which the deposit of copper is to be made, taking care not to touch it with fingers but holding it from the top screw or between a double strip of paper. Then start the current and *immediately* start stop-watch. Note down the deflection in the tangent galvanometer at the *two* ends of the pointer after *every* 5 minutes. Keep the deflection constant by adjusting the rheostat, if necessary. When the current has passed for ten minutes, quickly reverse the direction of the current in the tangent galvanometer by means of the commutator, and again note down the deflection at the two ends of the pointer after every 5 minutes. As the end of another ten minutes stop the current as well as the stop-watch. Find out the exact time t for which the deposition of copper has been made. Take the *mean* of the above readings or deflection which gives the value of θ.

Next remove the cathode plate from the voltameter and immediately immerse it in a jar of tap water already placed near the voltameter. This will remove the copper sulphate solution left on the plate. Then transfer the plate to another jar containing distilled water to which two or three drops of sulphuric acid per litre have been added. Next press the plate *without rubbing* between sheets of filter paper or clean blotting paper to remove the moisture as far as possible. Then dry the plate finally in warm air coming out of a hot-air blower. Weigh the *cool dry* plate in the chemical balance to the nearest tenth of the milligram with the help of a rider and thus find out the mass m of the copper deposited. Then taking Z = 0.0003295 *gm. per coulomb*, calculate the value of K from the Eq. (22.14). Finally, putting n = 2 and using the value of r for two turns as specified by the maker, calculate the value of H from the Eq. (22.15).

Sources of error and precautions

(1) The experiment should be performed on a *rigid* table preferably of stone. The base of the tangent galvanometer and the compass box should be carefully levelled so that the plane of the coil is vertical and the needle is *free* to swing in a *horizontal* plane.

(2) The plane of the galvanometer coil should be set in the *magnetic meridian* otherwise the *tangent law* will not hold good.

(3) When the tangent galvanometer has been carefully levelled and its coil placed in the magnetic meridian and the setting tested, the position of the levelling screws on the table should be lightly marked as a precaution against accidental displacement of the instrument.

(4) All magnetic materials, current-bearing conductors and the mains should be at a considerable distance from the tangent galvanometer. The connections between the galvanometer and the commutator should be made by two *flexible wires* twisted together and they should lead away from the galvanometer so that *the magnetic field produced by the passage* of current through them has no appreciable effect on the needle. As there are only two turns of copper wire in the galvanometer coil, the effect of current in any wire near the galvanometer on the needle is easily *comparable* with the effect due to the current in the galvanometer coil itself.

(5) The copper sulphate solution in the copper voltameter should have a density of about 1.18 gms./c.c. and should be made slightly more acidic than the acqueous solution of the salt by the addition of 0.1% by volume of concentrated sulphuric acid which increases the conductivity of the solution. The total area of the cathode surface immersed in the solution should be about 50 sq. cm. per ampere passing otherwise the deposit of copper on the plate will not be smooth and firm and hence will be liable to scale off. The area for deposition of copper can be doubled by depositing it on both sides of the cathode plate for which two anode plates connected together should be immersed in the solution one on each side of the cathode plate.

(6) The cathode plate over which deposit of copper is made should be clean and free from any trace of brown oxide or grease, otherwise the deposits will not adhere properly and evenly. The

plate should be thoroughly cleaned with sand and moist rug, or with emery cloth or with sand paper or pumice until the surface on either side becomes quite bright. It should never be touched with fingers otherwise the parts touched will be rendered greasy.

(7) The cathode plate of the voltameter, *i.e.*, the plate over which the deposit of copper is to be made should be connected to the –ve pole of the battery or to a lower potential point in the circuit. The plate should always be weighed before and after deposit in an accurate chemical balance upto 1/10 of a milligram with the help of a rider.

(8) The strength of the current should be between 1.5 and 2 amperes so that in about 20 minutes a deposit of about 1/2 gm. of copper may be obtained. With weaker current a very long time is required to obtain a deposit sufficient to be weighed accurately. On the other hand, if the current is too strong, a very large area of the plate has to be immersed in the solution (see prec. 5), otherwise the copper will adhere very poorly.

(9) The strength of the current should be kept *constant* when the deposition of copper on the plate is being made. The deflection in the tangent glvanometer should, therefore, be kept constant by adjusting the rheostat from time to time, if necessary.

(10) The deflection in the tangent galvanometer should be about 45° for then the readings of deflection would be *least liable* to error. In no case the deflection should be less than 25° and greater than 65°.

(11) While reading the deflection in the tangent galvanometer, avoid *error due to parallax.*

(12) Both the ends of the pointer in the compass box should be read to avoid *error due to eccentricity* of the pivot on which the needle rotates with respect to the circular scale.

(13) Readings of deflection in the galvanometer should be taken first with current passing in one direction and then by reversing it to avoid errors due to thermo-electric effects and due to any want of accurate setting of the galvanometer coil in the magnetic meridian. A commutator should be used to reverse the current in the galvanometer without changing its direction in the rest of the circuit.

(14) When the cathode plate is taken out of the voltameter after the deposition of copper has been made, it should be immediately immersed in a beaker or jar of *tap* water otherwise the delay will help in the oxidation of fine deposit of copper to copper-oxide.

Observation—[A] *Determination of m and t.*

Mass of cathode plate *before* deposition of copper = gm.

Mass of cathode plate *after* deposition of copper = gm.

Time for which current was passed = min. = sec.

[*B*] *Determination of* θ

Time min.	Deflection of pointer with current				Mean
	Direct		Reverse		
	Rt. end	Left end	Rt. end	Left end	
0	45°	45°	...	...	
5	45°	45°	...	...	
10	45°	45°	45°	45°	45°
15	...	...	45°	45°	
20	...	...	45°	45°	

[*C*] *Constants of the coil*

Number of turns used =

Circumference of the coil =

= cm. = m.

Calculations—Mass of copper deposited on the cathode plate

= gm.

Electro-chemical equivalent of copper from tables

= gm. per coulomb.

$$K = \frac{m}{Zt \cdot \tan\theta}$$

= amperes

Mean radius of the coil = cm. = m.

$$H = \frac{K \mu_0 n}{2r}$$

= Tesla.

Results—(*i*) The reduction factor of the tangent galvanometer for two turns coil = amperes

(*ii*) The value of horizontal component of earth's magnetic field in the laboratory at......= Tesla

Standard value of H = Tesla

∴ Percentage error =

Criticism of the method—The method suffers from a number of defects :

(*i*) The plane of the coil cannot be set exactly in the magnetic meridian and there is always an uncertainty in this adjustment.

(*ii*) The needle is not always exactly at the centre of the coil.

(*iii*) Unless the coil has a single and exactly circular layer, the value of r is somewhat uncertain. The magnetic field is uniform only in a small space at the centre of the needle and since the needle has a *finite* size, its movement can never be in an absolutely uniform field.

The last two defects have been removed in Helmholtz modification of tangent galvanometer in which the needle is placed between two coils in the *uniform* field produced by them.

The friction resistance to the movement of the needle is not negligible for it is pivoted and saddled with a long pointer. The pointer over scale method of measuring deflections has a very limited accuracy. This can, however, be considerably increased by fixing a plane mirror vertically at the centre of the needle and measuring the deflection by lamp and scale or telescope and scale method.

22.5. Magnetic Field at an Axial Point of a Circular Current Carrying Coil

Consider a circular loop of radius r carrying a current I. We intend to calculate the magnetic field at a point P on the axis of

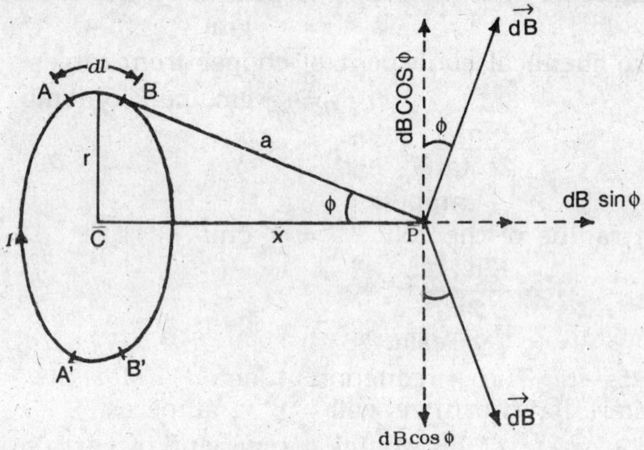

Fig. 22.8

the loop at a distance x from its centre C. In Fig. 22.8, the loop is perpendicular to the plane of the paper and its axis is in the plane of the paper. We consider an infinitesimally small element AB of the coil of length dl. Let 'a' be the distance of P from the midpoint of this element. Using Biot-Savart's law, magnetic field at P due to this element is given by :

$$dB = \frac{\mu_0}{4\pi} \frac{I\,dl \sin\theta}{a^2}$$

As r is small, therefore, $\theta \cong 90^\circ$.

$$\therefore \qquad dB = \frac{\mu_0}{4\pi} \frac{I\,dl}{a^2} \sin 90^\circ$$

$$= \frac{\mu_0}{4\pi} \frac{I\,dl}{a^2}$$

As $\vec{dl}$ is perpendicular to the plane of the figure, $\vec{dl} \times \vec{r}$, *i.e.*, the direction of dB must be in the plane of the figure. The magnetic field at P due to this element is in the direction of PQ. This field can be resolved into two components : the vertical component $dB \cos\phi$ and the horizontal component $dB \sin\phi$. The vertical component is exactly cancelled by the vertical component of the field due to a current element of the same size, radially opposite A′ B′. Thus, only the axial component survives and add up due to such small current elements. So, the total magnetic field at P due to the entire circular loop is given by :

$$B = \int dB \sin\phi = \int \frac{\mu_0}{4\pi} \frac{I\,dl}{a^2} \sin\phi$$

$$= \frac{\mu_0 I}{4\pi} \frac{1}{a^2} \sin\phi \int dl = \frac{\mu_0 I}{4\pi a^2} \sin\phi .\, 2\pi r$$

$$= \frac{\mu_0 Ir}{2a^2} \sin\phi = \frac{\mu_0 Ir}{2a^2} . \frac{r}{a} \qquad \left(\because \sin\phi = \frac{r}{a}\right)$$

$$= \frac{\mu_0 Ir^2}{2a^3} = \frac{\mu_0 Ir^2}{2\,(r^2 + x^2)^{3/2}} \qquad (\because a^2 = r^2 + x^2) \qquad (22.16)$$

If the number of turns of the coil is n, then,

$$B = \frac{\mu_0 Ir^2 n}{2\,(r^2 + x^2)^{3/2}} \qquad (22.17)$$

At the centre of the loop $x = 0$

$$\therefore \qquad B = \frac{\mu_0 I}{2r}$$

which is the same expression as obtained previously for the field at the centre of the coil.

If the values of field B and the corresponding values of x for various points on the axis of a circular coil carrying current be plotted on a graph, a curve as shown in Fig. 22.9 is obtained.

The curve is at first concave towards *O*, the point correspondig to the centre of the coil, but the curvature goes on decreasing as x increases and soon *changes* its sign, the curve becoming convex towards *O*. At the *point of inflection*, *i.e.*, where the curvature changes its sign $d^2B/dx^2 = 0$.

Now, $$B = \frac{\mu_0 Ir^2}{2(r^2 + x^2)^{3/2}} = \frac{k}{(x^2 + r^2)^{3/2}}$$

$$\therefore \quad \frac{dB}{dx} = k\frac{d}{dx}(x^2 + r^2)^{-3/2} = -3k\,x(x^2 + r^2)^{-5/2}$$

and $$\frac{dB^2}{dx^2} = 3\,k\,[x^2 + r^2)^{-5/2} - 5x^2\,(x^2 + r^2)^{-7/2}]$$

Now, if $$\frac{d^2B}{dx^2} = 0$$

$$-3k\,[x^2 + r^2)^{-5/2} - 5x^2\,(x + r^2)^{-7/2}] = 0$$

or $$5x^2\,(x^2 + r^2)^{-1} = 1$$

or $$5x^2 = x^2 + r^2$$

whence, $$x = \frac{1}{2}r \qquad (22.18)$$

Thus, at the *point of inflection* $x = \frac{1}{2}r$.

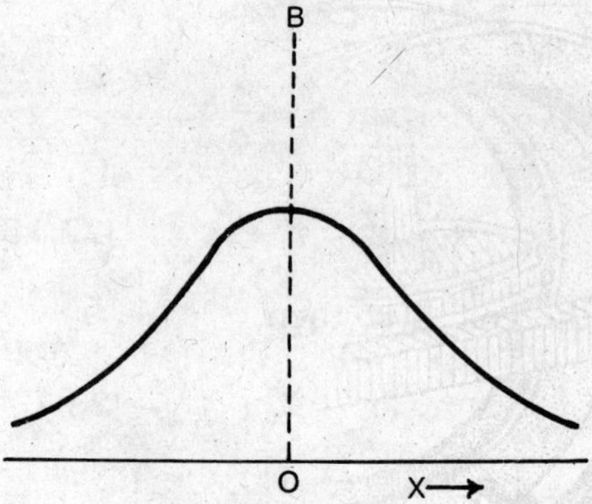

Fig. 22.9

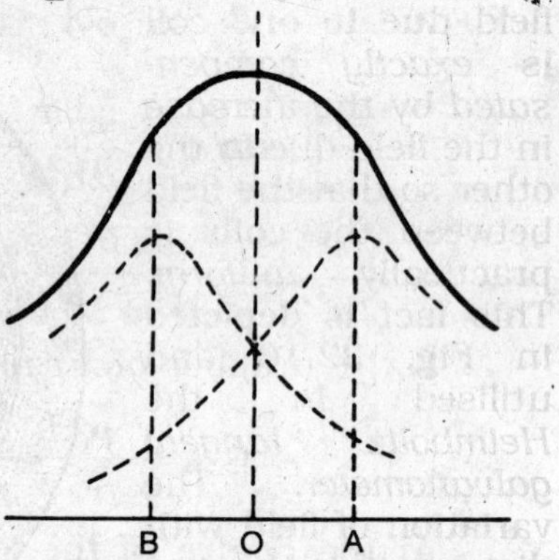

Fig. 22.10

At this point the rate of change of field dB/dx is *constant*. Hence, if two similar coils be placed with their axis coincident and separated by a distance equal to the radius of *either*, and if the *same* current is passed through them in the *same* direction, *the rate of increase of field due to one coil at midpoint between the coils is equal to the rate of decrease of field due to the other*

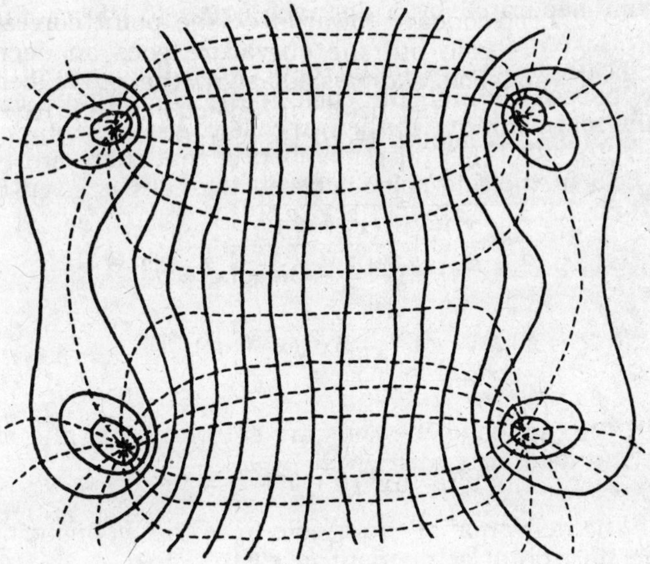

Fig. 22.11

at the same point, and as we move along the axis from the mid-point any decrease in the intensity of the field due to one coil is *exactly compensated* by the *increase* in the field due to the other so that the field between the coils is practically *uniform*. This fact is depicted in Fig. 22.10 and utilised in the *Helmboltz tangent galvanometer*. The variation of field with distance for the two coils is shown graphically in Fig. 22.11.

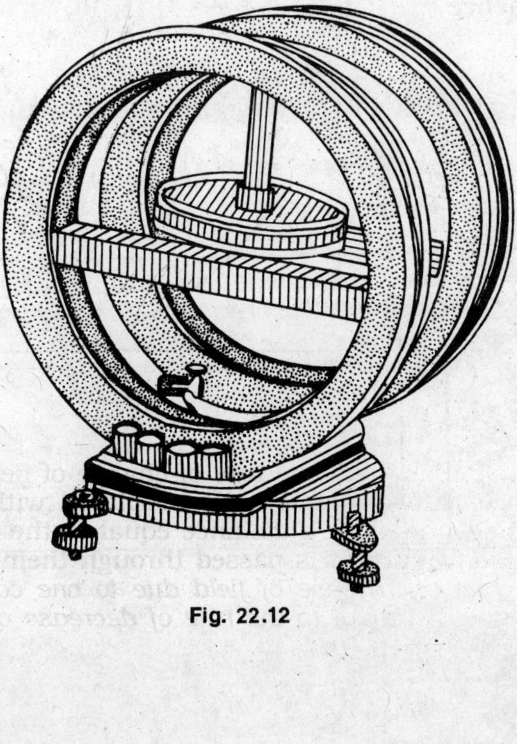

Fig. 22.12

As depicted in Fig. 22.12, the *Helmboltz tangent galvanometer* consists of two *equal coaxial* coils separated by a distance equal to the radius of the coils.

The needle is pivoted *mid-way* between them ; and the *same* current is passed through the coils in the *same* direction so that fields at the mid-point due to the two coils *assist* each other. Putting $x = r/2$ in Eq. (22.17) and remembering that there are two coils the total field at the mid-point between the coils is given by :

$$B = 2\left(\frac{\mu_0 I r^2 n}{2\left(\left(\frac{r}{2}\right)^2 + r^2\right)^{3/2}}\right)$$

or $$B = \frac{\mu_0 n I}{1.3975\, r}$$

When the planes of the coils are set *parallel* to the *magnetic meridian*, the field B is also given by :

$$B = H \tan \theta$$

where θ is the deflection of the needle from the *magnetic meridian* and H the horizontal component of earth's field.

Thus, from the above two equations, we get :

$$\frac{\mu_0 n I}{1.3975\, r} = H \tan \theta$$

whence $$I = \frac{1.3975\, r}{\mu_0 n} H \tan \theta \qquad (22.19)$$

Experiment 22.3

Object—To plot graph showing the variation of magnetic

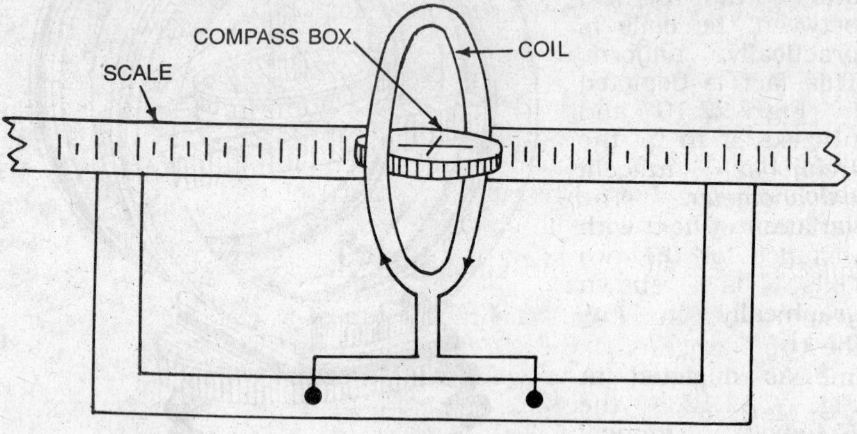

Fig. 22.13

field with distance along the axis of a circular coil carrying current and to estimate from it the radius of the coil.

Apparatus—Tangent galvanometer of the Stewart and Gee type (Fig. 22.13), and a current source with a switch to reverse the current direction.

Description of Apparatus—The apparatus called *Stewart and Gee tangent galvanometer* consists of a circular coil of thin copper wire fixed with its plane vertical on a suitable horizontal bench, and a magnetometer compass box which can slide on the bench so that the centre of the needle always lies on the axis of the coil. Along the bench on both sides of the coil is fixed a scale graduated in millimetres on which can be read off the distance of the centre of the needle from the centre of coil.

Theory—Let the plane of the coil be placed *parallel* to the *magnetic meridian* and let a current of I amperes be allowed to pass through the coil. The magnetic field B at a point situated on the axis of the coil and a distance x from its centre, due to the current in the coil is given by :

$$B = \frac{\mu_0 I r^2 n}{2 (r^2 + x^2)^{3/2}}$$

where r is the radius of the coil and n the number of turns in it. The field B is *perpendicular* to the direction of the horizontal component H of the earth's field and hence, if θ be the deflection from the *magnetic meridian* of a compass needle placed at that point, then,

$$B = H \tan \theta$$

Equating the above two expressions for B, we get :

$$\frac{\mu_0 n I r^2}{2 (r^2 + x^2)^{3/2}} = H \tan \theta$$

If a graph be plotted taking the distances of various points along the axis of the coil from *one end* of the instrument as abscissae and the corresponding values of tan θ as ordinates, the curve obtained will be as in Fig. 22.15. At the *point of inflection, i.e.*, where the curvature changes sign, $x = \frac{1}{2} r$ and hence the distance between the points of inflection P and Q on the two branches of the curve situated on the two sides of the line XC will be equal to r, the radius of the coil.

Method—Place the magnetometer compass box on the sliding bench so that its magnetic needle is at the centre of the coil. Rotate the bench in a horizontal plane until *the coil, the* *meedle and its image all lie in the same vertical plane.* This sets the coil roughly in the *magnetic meridian* and hence its axis in the magnetic east and west direction. In this position of the bench the pointer in the compass box will read 0-0 on the circular scale.

The current source contains one *current set knob*, the output *terminals* and display indicating the magnitude of the current flowing through the outer circuit. There is reverse/normal switch, called polarity switch, with which the direction of the current through the coil can be reversed.

Without disturbing the adjustment of magnetometer, connect the two coil terminals to the current source. Polarity of the connection does not matter. Keep the set current knob at the

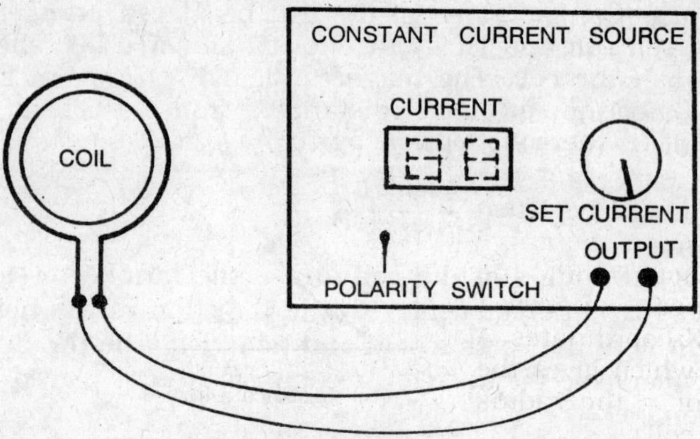

Fig. 22.14

minimum and switch on the current source. Increase the current with set current knob to produce a large deflection of about 75° in the compass box. Gently tapping the compass box with finger, record the deflection at the two ends of the pointer. Then reverse the current in the coil using the polarity switch and again measure the deflection. If the mean value of readings for deflection before and after reversing the current in the coil differ by more than 1°, slightly turn the coil with the help of the bench until *they agree as closely as possible.* This sets the coil in the magnetic meridian.

Now, place the compass box on east or west arm at its *end* and note down the readings of deflection before and after reversing the current in the coil, and from the four readings of deflection thus obtained, calculate the *mean* value of θ. Measure the distance of the needle from the *end* of the arm. Then, shift the compass box along the bench by 2 cm. and determine the value of θ for another point measuring its distance again from the end of the arm. Continue to move the compass box on the bench in steps of 2 cm. and determine the *mean* value of θ for each position of the needle on the axis of the coil till the compass box passing through the centre of the coil reaches the *other* end of the bench or the deflection is reduced to about 5°, *taking special care in determining deflection when x is nearly equal to r/2.*

Next, plot a graph taking the various values of distances of the centre of the needle from the *end* of the bench as abscissae and the corresponding values of tan θ, the tangent of the *mean* value of deflection, as ordinates. This will consist of two symmetrical branches, as shown in Fig. 22.15. Find out the two points of *inflection P and Q* on the curve and determine the difference between their X-coordinates *A* and *B* which gives the value of *r*, the radius of the coil.

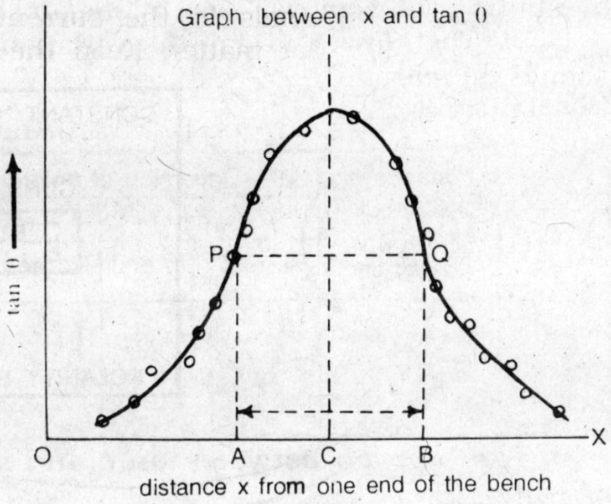

Fig. 22.15

Sources of error and precautions

(1) All magnetic materials and current-bearing conductors should be at a considerable distance from the apparatus.

(2) The plane of the coil should be *parallel* to the *magnetic meridian*, otherwise the *tangent law* will not hold good.

(3) The current through the coil should be sufficient to give a *large* deflection, say, of about 75° when the needle is at the centre of the coil and its strength should be *constant.* The current through the circuit should, therefore, be sent from a constant current source.

(4) The *error due to parallax* in reading deflection of the needle should always be avoided.

(5) *Both* ends of the pointer in the compass box should be read to avoid *error due to eccentricity* of the pivot on which the needle rotates with respect to the circular scale.

(6) Readings of deflection in the compass box should be taken first with current passing in one direction and then by *reversing* it.

(7) Since the *exact* position of the centre of the coil is not accurately known, the distances of the centre of the needle in all positions of the compass box should be measured from one of the ends of the bench.

(8) Since the curve becomes almost *vertical* in the neighbourhood of $x = r/2$, observations for deflection when the distance of the needle from the centre of the coil is approximately equal to *half the radius* of the coil should be very carefully made by shifting the compass' box in steps of about a cm.

(9) The curve showing the variation of tan θ with distance should be *smooth*.

Observations

Current I = ... Ampere

S.No.	Distance of the needle from *one end* of bench	Deflection of pointer with current					Tan θ
		Direct		reverse		Mean θ	
		Rt. end	Lft. end	Rt. end	Lft. end		
1.							
2.							
...							
...							

Plot a graph between tan q and x and locate the point of inflection.

Result—The radius of the coil as estimated from the *distance* tan θ curve = cm.

Radius of coil by *actual* measurement = cm.

∴ Percentage error =

Criticism of the method—The theory underlying the method assumes (*a*) that the plane of the coil lies exactly in the *magnetic meridian* and is exactly *vertical*, (*b*) that the needle is *infinitely small*, its axis is exactly *horizontal* and its centre always lies exactly on the axis of the coil, and (*c*) that the coil is exactly circular. These assumptions are obviously not all justifiable in practice and hence the result obtained by this method cannot be very accurate. Further, the friction resistance to the motion of the needle is not negligible for it is pivoted and saddled with a long pointer. The accuracy of the pointer-over-scale method of measuring deflections is very limited and consequently the position of inflection points on the curve cannot be determined very accurately.

ORAL QUESTIONS

STRAIGHT CURRENT CARRYING CONDUCTOR

1. State Biot-Savart rule for calculating the magnetic field at a point due to small current-element.
2. How can you determine the direction of the magnetic field at a point due to a current in a conductor ? State the rules.
3. How does the magnetic field at a point due to a straight current vary with the distance of the point from the conductor ?
4. What is the direction of magnetic field at a point due to straight current ?
5. What type of lines of force do you get in the field of a straight current ?
6. What will be the character of the resultant field of straight current and the earth ? How many neutral points do you get in this case ? Where will it lie ?
7. How can you determine the value of H by plotting the resultant field of the straight current and the earth's horizontal field ? What precautions do you observe in this experiment ?
8. Why should the straight conductor be connected with the rest of the apparatus by flexible wires ?
9. Why should the ammeter be kept at a distance from the straight conductor ?
10. Why do you take current from mains ? Can an accumulator of E.M.F. 2 volts not do ?
11. How can you accurately locate the position of the neutral point ? Why does the behaviour of the needle in the neighbourhood of the neutral point become uncertain ?
12. What is the accuracy of the result obtainable by this method? Do you know any other method of determining H ?

TANGENT GALVANOMETER

1. Describe the construction of tangent galvanometer. Why is it so called ?
2. What is the tangent law ? Why is only a short needle used in the compass box ?
3. What is the relation between current and the angle of deflection ?
4. How is the needle of the T.G. deflected ?
5. What are the two fields ? Point out their directions.

6. How is the direction of magnetic field at the centre of the coil determined ?
7. What are the factors on which the field at the centre of the coil depends ?
8. What is meant by reduction factor of tangent galvanometer ? Explain its physical significance. Why is it so called ?
9. What is galvanometer constant or coil constant of a T.G. ? Is it the same as R.F. of a T.G. Do their values remain constant at all places ?
10. How does R.F. vary with (*a*) the number of turns, (*b*) the radius of the coil and (*c*) the place of working ? How will you determine R.F. of a T.G. ? How can your result be verified ?
11. What is the use of the pointer ? Why is it made of aluminium? Can you replace it by an iron pointer or by a pointer made of straw ?
12. Why is the pointer attached at right angles to the needle ? Can you not do without the pointer ?
13. What is the purpose of the mirror fixed at the base of the box? What is the harm if this is not avoided ?
14. Why do you place the plane of the coil in the magnetic meridian ?
15. How will the deflection change if the coil be displaced from the magnetic meridian ? How do you set the coil in the magnetic meridian and how is the accuracy of the adjustments tested ?
16. What will be the deflection if the plane of the coil be placed at right angles to the magnetic meridian ?
17. Why are both ends of the pointer read ?
18. Why do you take readings of deflection with current direct and reverse ?
19. What is the accuracy of observations with a T.G. ? How can it be increased ?
20. Why should the deflection be in the neighbourhood of 45° ?
21. What is the harm if the deflections are very small or very large?
22. What is the range in which deflections should lie ?
23. What is the approximate resistance of your instrument ?
24. Of which wire is the coil of the tangent galvanometer made ?
25. What is meant by sensitivity of a galvanometer ? How can a T.G. be made more sensitive ?

26. Is tangent galvanometer a dead-beat galvanometer ? If not, how can you make it so ?
27. Why are three levelling screws provided at the base of the instrument ?
28. How are the readings affected if the base of the galvanometer is not perfectly levelled ?
29. How do you level a T.G. ?
30. Having levelled the instrument why is it necessary to mark the positions of the levelling screws ?
31. What are the uses of a T.G.?
32. Why in the T.G. there are three coils of turns 2, 50 and 500 ?
33. Why is copper valtameter preferred to an ammeter to measure current ?
34. Describe a copper voltameter and explain the reactions which take place when the current is passed through it.
35. What should be the strength of the copper sulphate solution ? Why is some H_2SO_4 added to it ?
36. Why do you connect the plate on which deposition of copper is made to the -ve of the battery or to a lower potential point in the circuit ?
37. From where does copper come to the cathode ?
38. When the copper liberated at the cathode comes from the anode, why don't you weigh the anode plate before and after deposition of copper to get the mass of copper liberated ?
39. Why are two plates used for the anode ?
40. Is it necessary to clean the anode plates also ? What will be the harm if they remained dirty ?
41. How do you clean the cathode plate ? How should a clean plate be handled and why ?
42. Why do you weigh the cthode plate to such a high degree of accuracy ?
43. While removing the cathode plate from the voltameter after deposition of copper, why do you immediately immerse it into water ?
44. How is the plate dried ?
45. How do you measure H with a tangent galvanometer ? Discuss the accuracy in the value of H obtainable by this method.

CIRCULAR COIL

1. How does the field due to a circular current vary along its axis ? How does it vary with (*a*) the distance and (*b*) the number of turns in the coil ?
2. Where is field maximum ?
3. How do you find out the radius of the coil from the *x*-tan θ graph ?
4. What sort of graph do you get in this case ?
5. Why is the needle in the compass box very small ?
6. Do you get uniform field around any point on the axis of the coil ? If not, how can you modify the apparatus to get uniform field ?
7. What is an Helmholtz tangent galvanometer ?
8. What advantages does it possess over an ordinary tangent galvanometer ?
9. What precautions do you take in this experiment ?
10. Why a constant current source is must in this experiment? Discuss the accuracy of the results obtained in this experiment.

MAGNETIC SUSCEPTIBILITY

1. Define magnetic susceptibility.
2. Give typical values of magnetic susceptibility for different types of magnetic materials.
3. How is the magnetic susceptibility dependent on temperature ? Explain.

23

Measurement of Resistance and Potential Differences

23.1. Potential Difference and its Unit

When a current flows through a conductor, it gets heated. This is due to the fact that when electricity moves from one end of the conductor to the other, it does so at the expense of a part of its electrical energy which is transformed into heat energy in the conductor. This *energy transformed into heat or work done when unit quantity of electricity passes between any two points equals the potential difference (P.D.) between those two points.* If the work done or the energy transformed into heat between the two points is unity when unit quantity of electricity passes between them, then *P.D.* between the two points is also unity. *Volt is the unit of P.D. and is equal to the P.D. between the two points which are such that the amount of work done or energy transformed into heat is one joule or 10^7 ergs when one coulomb of charge passes between them.*

23.2. Ohm's Law and Unit of Resistance

We know that when the flow of liquid in a tube is *streamlined*, the quantity of liquid flowing through the tube per second is directly proportional to the pressure difference between the ends of the tube. In exact analogy with this *when a current is allowed to pass through a conductor, the strength of the current through it is directly proportional to the P.D. between the ends of the conductor, provided the physical state of the conductor remains the same*. Thus, mathematically,

$$I = kV$$

where V and I are the *P.D.* and current respectively and k is a quantity characteristic of the particular conductor, which measures the capacity of the conductor to allow electricity to pass through it and is called its *electrical conductivity*. This important relation between I and V is called *Ohm's Law*.

Now, $$V/I = 1/k = R$$

where $R = 1/k$ is called the *electrical resistance* of the conductor for the greater the value of R, greater must be the *P.D.* established for the same current through the conductor.

Now, if V and I each be equal to unity, R is also unity. The unit of resistance is Ohm (Ω). *The resistance of a conductor is one ohm when a P.D. of one volt applied at its ends causes a current of one ampere to flow through it.*

Since conductance is the reciprocal of resistance, the practical unit of conductance is reciprocal ohm or *mho.*

Since the resistance of a conductor remains constant so long as its physical state remains the same, attempts have been made to construct a standard ohm. The alloys chosen for the construction of standard ohm are *manganin* and *constantan.* This is because temperature co-efficient of resistance of these alloys are very small.

23.3. Electromotive Force (E.M.F.) and Potential Difference (P.D.)

We have seen above that whenever electricity passes between two points of a conductor there is an expenditure of energy which appears as heat between those two points. This implies that in order to maintain the current in the circuit, energy must be continuously supplied to it by some source situated somewhere in the circuit. This source may be a cell in which case the necessary energy is supplied by the chemical action inside it. But whichever may be the source of supply of energy in the circuit, we say, from analogy of flow of water in a closed circuit where the flow is maintained by a mechanical effect of a pump, which may be called the water-motive-force, that the current in an electrical circuit is maintained by an electrical effect called the *electromotive force (E.M.F.).* This *E.M.F. of a source of supply of energy to an electrical circuit is measured by the rate of working or the energy supplied per second when it is maintaining unit current in the circuit* and is thus clearly equal to the work done or energy transformed when unit quantity of electricity passes completely round the circuit containing the source. This work is equal to the sum of the work done to drive unit quantity of electricity in the *external* circuit, *i.e.*, the potential difference V across the external resistance and the work done in driving unit quantity of electricity inside the source, *i.e.*, the potential difference v across the *internal* resistance of the source ; and hence if E be the E.M.F. of the source of electrical energy in the circuit.

$$E = V + v$$

But from *ohm's* law $V = IR$ (23.1)

and $v = I\,r$

where I is the current flowing through the circuit and R and r are the external resistance of the circuit and the internal resistance of the source respectively.

Thus, $$E = IR + Ir \tag{23.2}$$

Dividing Eq. (23.1) by Eq. (23.2) we get

$$\frac{V}{E} = \frac{R}{R+r} = 1 - \frac{r}{R+r}$$

whence, $$V = E\left(1 - \frac{r}{R+r}\right) \tag{23.3}$$

Thus, the *P.D.* between the terminals of a battery is always *less* than its E.M.F. when it is supplying a current. If, however, no current is being supplied by the source, *i.e.*, it is *on open circuit, the P.D. between its terminals equals the E.M.F. of the source*. The *E.M.F.* is measured in the same units as the *P.D.*

An *E.M.F.* like a mechanical force is a *directional* quantity and determines the direction of the current in the circuit. If there are more than one *E.M.F* in the circuit the direction of current, depends upon the *resultant* or effective *E.M.F.* in the circuit. An *E.M.F.* corresponds to energy transformation in a *reversible* process for, if an *E.M.F.* opposes the flow of current which is maintained by a bigger *E.M.F.* in the circuit then energy will be absorbed from the circuit at the smaller *E.M.F.* and not supplied to it and hence the energy of the source of *E.M.F.* opposing the flow of current will increase at the expense of the energy of the source of *E.M.F.* maintaining the current in the circuit. On the other hand, *P.D.* has a direction which depends upon the direction of the current in the circuit and, hence, upon that of the resultant or effective *E.M.F.* in the circuit. *P.D.* corresponds to energy transformation in an *irreversible* process for whichever may be the direction of current in the circuit, *P.D.* will always correspond to the dissipation of electrical energy from the circuit in the form of heat.

23.4. Combination of Conductors

When several conductors are connected in a circuit, the following general relationships hold :

(1) **Law of Resistors in Series**—Referring to Fig. 23.1, when several conductors are connected in *series, i.e.*, one after another so that the *same current* passes throug them, *the total resistance is equal to the sum of the individual resistances, i.e.*,

$$R = r_1 + r_2 + r_3 = \ldots = \Sigma r_n$$

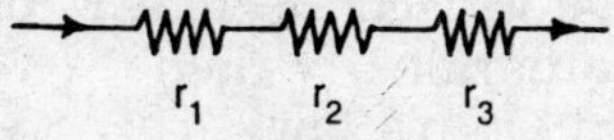

Fig. 23.1

(2) **Law of Resistors in Parallel**—Referring to Fig. 23.2 when several conductors are connected in *parallel, i.e.*, one end of all conductors is joined to one point of the circuit and the other end of all conductors is joined to a second point of the circuit so that the *same P.D.* exists across each conductor, *the equivalent resistance is such that its reciprocal is equal to the sum of reciprocals of the individual resistances, i.e.*,

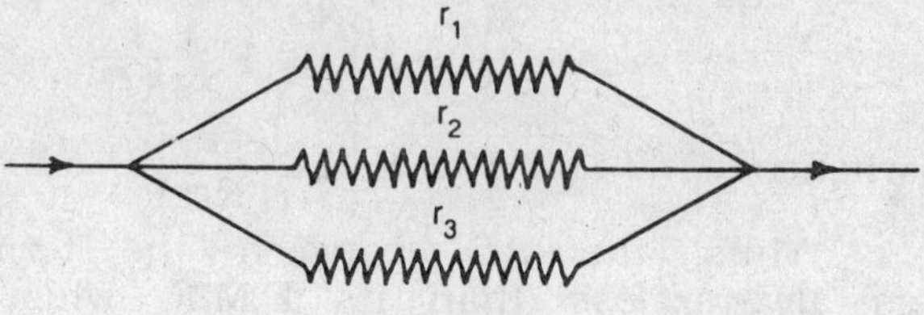

Fig. 23.2

$$\frac{1}{R} = \frac{1}{r_1} + \frac{1}{r_2} + \frac{1}{r_3} + \ldots = \Sigma\frac{1}{r_n}$$

Since $1/R = k$, the *electrical conductivity*, it follows from above that *when several conductors are connected in parallel, their total conductivity is equal to the sum of the individual conductivities, i.e.*,

$$k = k_1 + k_2 + k_3 + \ldots = \Sigma k_n$$

Note that *when the resistors are connected in parallel, the total resistance is always less than the least of the component resistances.* If $r_1 = r_2 = \ldots = r$, $R = r/n$, *i.e.*, when n equal resistances are connected in parallel, the total resistance is equal to nth part of one of the resistances.

23.5. Current in Divided Circuits and Use of Shunt

Let Fig. 23.3 represent a position of circuit in which the main current I divides itself among a number of resistors connected in parallel. Let V be the *P.D.* across each resistor and let R be the equivalent resistance of these resistors connected in parallel, then from Ohm's law,

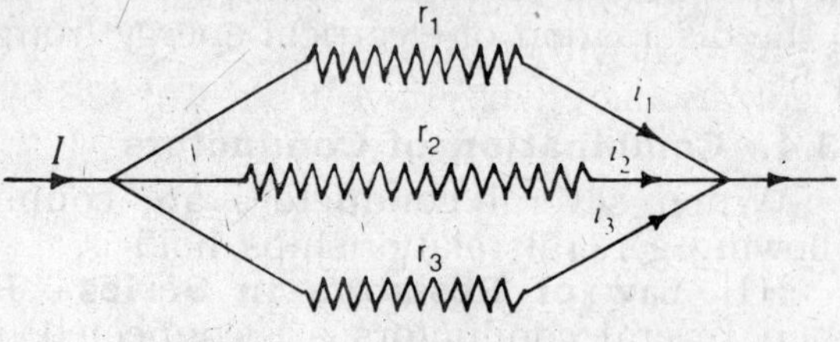

Fig. 23.3

$$V = IR = i_1 r_1 = i_2 r_2 = i_3 r_3 \qquad \ldots\ldots \qquad (23.4)$$

and since the *total* current is unchanged

$$I = i_1 + i_2 + i_3 + \ldots\ldots$$

From Eq. (23.4), we get :

$i_1 = IR/r_1$; $i_2 = IR/r_2$ and so on, where R is given by

$$\frac{1}{R} = \frac{1}{r_1} + \frac{1}{r_2} + \frac{1}{r_3}$$

If there are only two conductors, r_1 and r_2 in parallel, then,

$$R = \frac{r_1 r_2}{r_1 + r_2}$$

and hence

$$i_1 = \frac{Ir_2}{r_1 + r_2} \text{ and } i_2 = \frac{Ir_1}{r_1 + r_2} \quad \ldots \quad (23.5)$$

or

$$\frac{i_1}{i_2} = \frac{r_2}{r_1}$$

i.e., the current divides inversely as the resistances.

The above principle is used in sending only a *small* portion of the main current through an instrument, *e.g.*, a galvanometer by connecting a low resistance wire in *parallel* with it. This is illustrated in Fig. 23.4. The wire connected across the instrument is called a 'shunt' and provides a by pass for the current and hence reduces the sensitiveness of the instrument as well as prevents it from being damaged when the total current is large. Let I be the total current and I_g the current through the galvanometer after it has shunted with a resistance S. Then, if G be the resistance of the galvanometer, we have from Eq. (23.5).

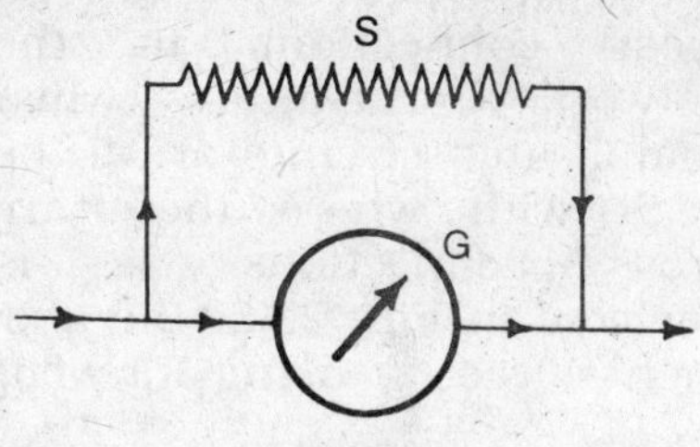

Fig. 23.4

$$I_g = \frac{S}{S+G} I$$

or

$$I = \frac{S+G}{S} I_g$$

The factor $(S+G)/S$ by which the galvanometer current has to be supplied to get the total current is called the *multiplying power* of the shunt. Now, if only 1/10 of the total current is to be passed through the galvanometer, then clearly the multiplying power of the shunt must be 10, *i.e.*,

$$\frac{S+G}{S} = 10$$

whence, $S = G/9$

Similarly, if 1/100 or 1/1000 of the total current is to be passed through the galvanometer, the resistance of the shunt must be 1/99 or 1/999 respectively of the galvanometer resistance.

Sometimes, galvanometers are provided with special shunt boxes containing shunts of different multiplying powers. These are known as *range multipliers.* This system has the disadvantage that a shunt box can be used only with the instrument for which it has been made and hence each galvanometer requires its own special shunt box. This difficulty is overcome in Ayrton and Mather Universal shunt which will serve for any galvanometer.

23.6. Kirchhoff's Laws

The total resistance and the current in various branches of a divided circuit can be readily calculated by the application of laws of resistances in series and in parallel and *Ohm's law* when the circuit is a *simple* one, *i.e.*, when there are no cross connections in the network. But when complicated circuits are to be dealt with where there are cross connections as is depicted in Fig. 23.5, the above method fails and the problem is then solved by using *Kirchhoff*'s following two *laws:*

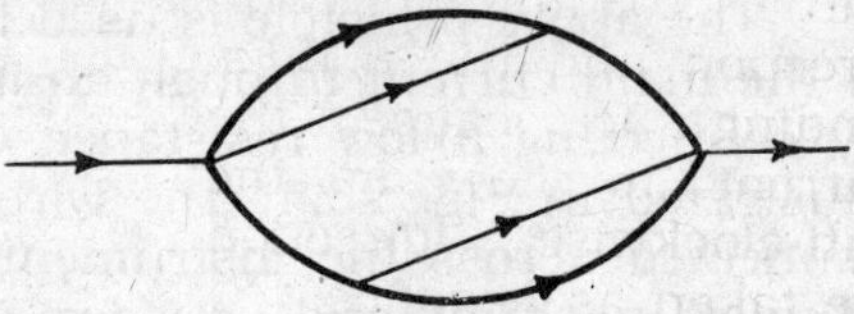

Fig. 23.5

I Law

The sum of currents arriving at any point in a circuit equals the sum of the currents leaving that point, *i.e.*, *the algebraic sum of the currents at any point in a circuit is zero* or mathematically,

$$\Sigma i = 0$$

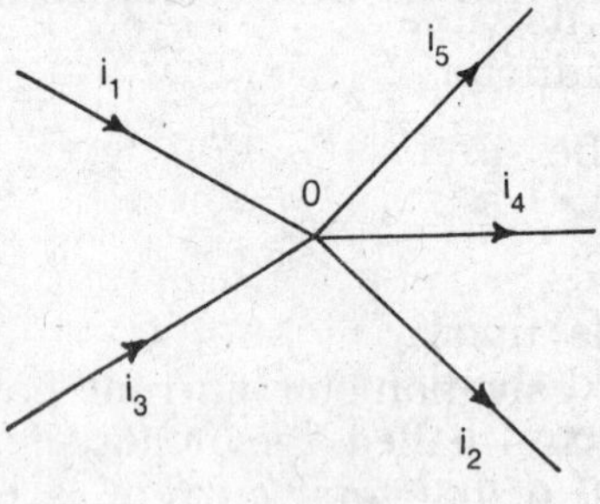

Fig. 23.6

Thus, referring to Fig. 23.6, if two currents i_1 and i_3 flow *towards* a point, O in a circuit and the currents i_2, i_4 and i_5 flow *away* from it, then applying the law to the point O, we have :

$$i_1 + i_3 = i_2 + i_4 + i_5$$

or $$i_1+i_2+i_3+i_4+i_5 = 0$$

This law is simply an expression of the fact that for a steady current in a circuit, *there can be no accumulation of charge at any point in the circuit*, and corresponds exactly to the *equation of continuity* of a perfectly incompressible fluid.

II Law

The algebraic sum of the electromotive forces acting in a closed circuit (or mesh) in a network of conductors is equal to the algebraic sum of the products of current and resistance of each part of the circuit or mathematically,

$$\Sigma E = \Sigma(ir)$$

The currents flowing in the clockwise direction and the *E.M.F.'s* tending to cause the current to flow in the clockwise direction are reckoned as *positive* while currents flowing in the anti-clockwise direction and the *E.M.Fs.* tending to cause the current to flow in the anti-clockwise direction are taken as *negative*.

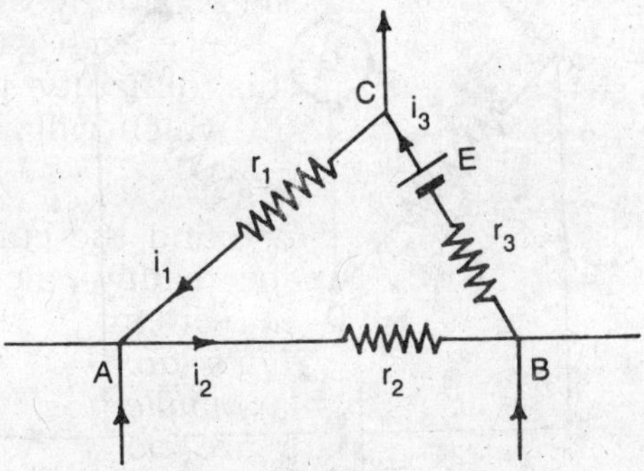

Fig. 23.7

Applying the law to the circuit *ABC* in the network of Fig. 23.7, we have,

$$i_1r_1+i_2r_2+i_3r_3 = E$$

This law is merely a generalisation of *Ohm's* law.

23.7. Specific Resistance or Resistivity

The resistance of a conductor is directly proportional to its length and inversely proportional to its area of cross-section. Thus, if R be the resistance of a conductor of length l and of area of cross-section S assumed to be uniform, then $R \propto \frac{l}{S}$

or $$R = \rho l/S$$

where ρ is a constant which depends upon the nature of the material of the conductor, and is called the *specific resistance* or *resistivity* of the material of the conductor. If $l=1$ and $S=1$, then $R = \rho$. Thus, *the specific resistance of a material is numerically equal to the resistance of a piece of it of unit length and uniform unit cross-sectional area.* It is also *numerically equal to the resistance offered by a cubical block of the material with edge of unit length when the flow of current is normal to one of its faces.* The unit of resistivity is Ωm. The reciprocal of resistivity is called *specific conductance* and is expressed in *mho per m.*

23.8. Wheatstone Bridge

Wheatstone bridge consists of four resistances r_1, r_2, r_3 *and* r_4 arranged as shown in Fig. 23.8 the one pair of *opposite*

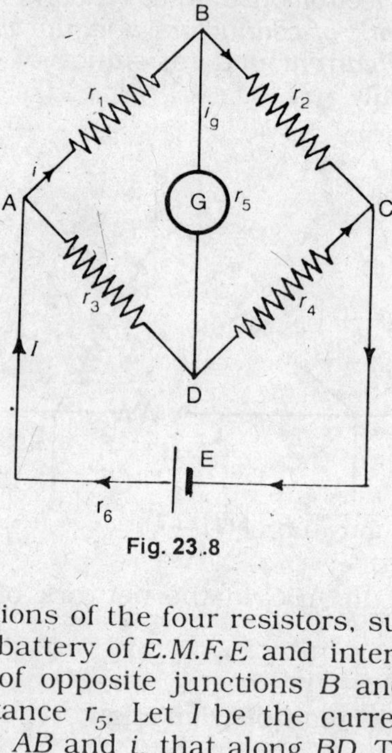

Fig. 23.8

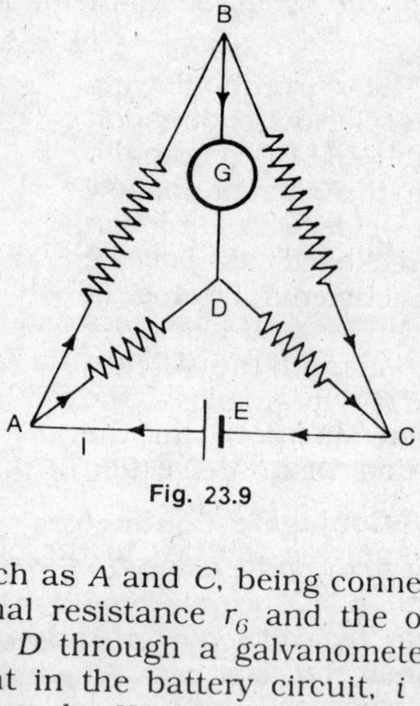

Fig. 23.9

junctions of the four resistors, such as *A* and *C*, being connected by a battery of *E.M.F. E* and internal resistance r_6 and the other pair of opposite junctions *B* and *D* through a galvanometer of resistance r_5. Let *I* be the current in the battery circuit, *i* that along *AB* and i_g that along *BD*. Then, by *Kirchhoff's first law*, the currents in *BC*, *AD*, and *DC* will be $(i - i_g)$ $(I - i)$, $(I - i + i_g)$ respectively. Applying *Kirchhoff's second law* to the meshes *ABDA* and *DBCD*, we get :

$$ir_1 + i_g r_5 - (I - i)\, r_3 = 0$$

and

$$(i - i_g)\, r_2 - (I - i + i_g)\, r_4 - i_g r_5 = 0$$

Rearranging these equations, we get :

$$i\,(r_1 + r_3) + i_g r_5 - Ir_3 = 0 \qquad (23.6)$$

$$i\,(r_2 + r_4) - i_g\,(r_2 + r_4 + r_5) - Ir_4 = 0 \qquad (23.7)$$

When there is no current flowing through the galvanometer, *i.e.*, $i_g = 0$, the bridge is said to be *balanced* and in that case Eqs. (23.6) and (23.7), reduce to :

$$i\,(r_1 + r_2) = Ir_3 \tag{23.8}$$

$$i\,(r_2 + r_4) = Ir_4 \tag{23.9}$$

Dividing Eq. (23.8) by Eq. (23.9), we have, since i and I are not equal to zero :

$$\frac{r_1 + r_3}{r_2 + r_4} = \frac{r_3}{r_4}$$

or
$$\frac{r_1 + r_3}{r_3} = \frac{r_2 + r_4}{r_4}$$

whence,
$$\frac{r_1}{r_3} = \frac{r_2}{r_4}$$

or
$$\frac{r_1}{r_2} = \frac{r_3}{r_4} \tag{23.10}$$

Thus, knowing the values of any three resistances, say, r_1, r_2, and r_3, when the bridge is balanced, the value of the unknown resistance r_4 can be calculated from the above relation. The relation is independent of the current in the battery circuit. An examination of Fig. 23.10 will show that when the positions of the galvanometer and battery are interchanged, the above condition for no deflection of the galvanometer will still hold.

23.9. Conjugate Conductors

If two conductors of any network of conductors are so related that an E.M.F. in one produces no current in the other, they are said to be conjugate to each other. The 'battery arm' and the 'galvanometer arm' in a Wheatstone bridge are *conjugate* to each other when the bridge is balanced, *i.e.*, when the resistance of other arms are so adjusted as to satisfy the usual Wheatstone bridge relationship of Eq. 23.10. It can be shown that *the current in one of the two conjugate conductors due to an E.M.F. in any part of the network, is independent of the resistance of the conjugate conductor.*

23.10. Sensitiveness of Wheatstone Bridge

In the practical use of Wheatstone bridge for the measurement of a wide range of resistances, the accuracy in the determination of the bridge depends upon the *sensitiveness* of the bridge. The sensitiveness of the bridge depends upon the current in the galvanometer, for the greater the current in the galvanometer for a given *small* want of balance of the bridge, greater is the sensitiveness of the bridge. Hence, to make the bridge most sensitive, the positions of the galvanometer and the battery and the values of the resistances in the various arms of the bridge should be so adjusted as to produce *maximum* current in the galvanometer for a given *small* want of balance.

(a) Positions of galvanometer and battery—The galvanometer can be connected in *BD* and the battery in *AC* as shown in Fig. 23.10 or their positions can be interchanged, but the two arrangements are, in general, not equally sensitive. It can be shown that if, *of the two resistances, that of battery and that of galvanometer, the greater connects the junction of the two greatest with that of the two least of the four arms of the bridge*, the current in the galvanometer and, hence, the deflection in it is greatest for a given small want of balance and, therefore, the bridge most sensitive for the given resistances.

(b) Galvanometer resistance—The sensitiveness of the galvanometer should be maximum, for the greater the sensitiveness of the galvanometer, greater will be the accuracy in the determination of the proper balance of the bridge. The sensitiveness of a galvanometer is proportional to the deflection for a given small current, and this deflection is proportional to the number of turns in the coil, if the current is the same. Now, as the channel of the bobbin on which the coil is wound is fixed with respect to its cross-section and the mean radius, the number of turns can be increased only by decreasing the cross-section of the wire. Let the cross-section of the wire be decreased to $1/n$ of its previous value, then the number of turns can be increased to n times their original value. But this also increases the length of the wire n times and hence the total resistance of the coil is increased to n^2 times its original value, *i.e.*, $r_5 \propto n^2$ where r_5 is the resistance of the galvanometer and n is the number of turns in the coil. Hence, for a given value of current the deflection in the galvanometer is proportional to $\sqrt{r_5}$. The deflection also varies with the current. Hence the sensitiveness of the galvanometer is proportional to the quantity $i_g \sqrt{r_5}$. It can be shown that this quantity is *maximum* when :

$$r_5 = \frac{r_4 (r_1 + r_3)}{r_3 + r_4} = \frac{r_1 (r_2 + r_4)}{r_1 + r_2}$$

This gives the value of the resistance of the galvanometer in order to make the bridge most sensitive for the given values of r_1, r_2, r_3, r_4 and r_5. The tangent galvanometer is very unsuitable for Wheatstone bridge for its sensitivity is very small, *i.e.*, it requires comparatively large current to produce an appreciable deflection.

(c) Battery Resistance—When the resistance of the battery is at the disposal of the experimenter, the bridge is most sensitive for the given values of r_1, r_2, r_3, r_4 and r_5 when the current given by the battery is maximum, and this is the case *when the external resistance equals the internal resistance of the battery*. Thus, for nearly complete balance,

$$r_6 = \frac{(r_1 + r_2)(r_3 + r_4)}{r_1 + r_2 + r_3 + r_4} = \frac{r_4(r_1 + r_2)}{r_2 + r_4}$$

This gives the battery resistance to make the bridge most sensitive for the given values of r_1, r_2, r_3, r_4 and r_5.

(d) Galvanometer and battery resistances fixed and other variables —Usually the resistance r_5 of the galvanometer and the resistance r_6 of the battery are not at the disposal of the experimenter and hence, if r_4 be the unknown resistance, the only resistances which can be varied to make the bridge most sensitive are r_1, r_2, and r_3. For a particular value of r_3, there will be a pair of values of r_1, and r_2 which will constitute the best arrangement for that value of r_3; and there will be a *particular* value of r_3 which, with the corresponding values of r_1 *and* r_2, will be the best arrangement for the given values of r_4, r_5 and r_6. It can be shown that in this case the values of r_1, r_2 and r_3 which will make the bridge most sensitive should be as follows :

$$r_1 = \sqrt{r_5 r_6},\ r_2 = \sqrt{r_4 r_6 \frac{r_4 + r_5}{r_4 + r_6}} \text{ and } r_3 = \sqrt{r_4 r_5 \frac{r_4 + r_6}{r_4 + r_5}}$$

The resistance r_1 is independent of the value of r_4 and has a definite value $\sqrt{r_5 r_6}$. Having given this value to r_1, the value of r_6 is roughly determined and then the values of r_2 and r_3 calculated.

(e) All resistances variable—When all resistances except r_4 the unknown resistance, can be varied, *i.e.*, the resistances of the battery, the galvanometer and other arms of bridge except the unknown resistance is at the disposal of the experimenter, the most sensitive arrangement is that in which *each of the resistances is equal to the unknown resistance*, i.e., $r_1 = r_2 = r_3 = r_5 = r_6 = r_4$.

In this case it is immaterial whether the galvanometer is connected in *BD* and the battery in *AC* or *vice versa*. The heating effects will also be the same in all branches. This arrangement gives the most accurate value of the unknown resistance but such an arrangement is practically impossible for every unknown resistance requires the setting up of a special bridge and the employment of special galvanometer and battery.

23.11. Metre Bridge

It is one of the practical application of the Wheatstone bridge and, as shown in Fig. 23.10, consists of a straight and uniform wire of manganin, constantan or some other suitable alloy of *high* specific resistance, stretched along a metre scale, with its ends soldered to two copper strips of negligible resistance. The wire is one metre long and has a resistance of about 2 to 3 ohms. Between these two copper strips and a third one fixed along the panel there are two gaps, G_1 and G_2, in which are connected with the help of binding screws, the unknown resistance X and a suitable resistance box. A jockey serves to make contact with the stretched wire and a mark on it or a pointer attached to it indicates the position of the point of contact on the measuring scale. The battery is joint between terminals on the two end strips, a key being included in the circuit so that the current may be passed when required. The galvanometer is connected between a terminal on the middle copper strip and the jockey.

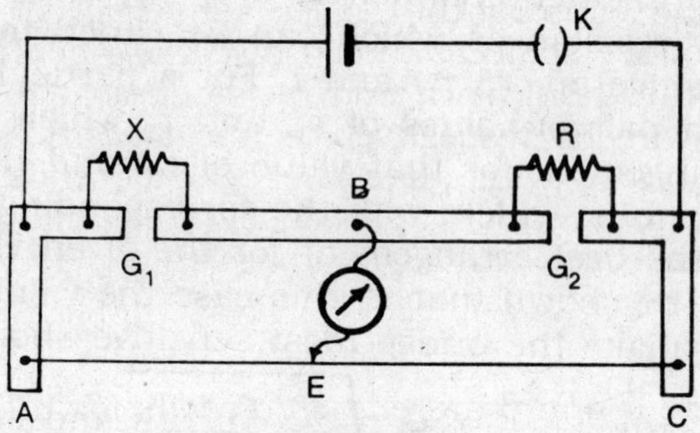

Fig. 23.10

To determine the *unknown* resistance X, the connections are made as shown. A suitable resistance R is introduced in the gap G_2 by means of the resistance box and a point of contact E is found on the wire by sliding the jockey at which there is no deflection in the galvanometer. If the two resistances into which the bridge wire is then divided be R_1 (AE), and R_2 (CE), we have by Eq. (23.10) since the bridge is *balanced* :

$$\frac{X}{R} = \frac{R_1}{R_2}$$

Since the wire is assumed to be of *uniform* cross-section, the resistance of any portion of it is *proportional* to the length. Hence, if $AE = l$, $CE = (100 - l)$ and $R_1 \propto l$ and $R_2 \propto (100 - l)$. Thus,

$$\frac{X}{R} = \frac{l}{100 - l} \qquad \dots \qquad (23.11)$$

whence
$$X = \frac{l}{100 - l} R$$

This gives the value of the resistance to be measured.

23.12. End Corrections for Metre Bridge

We have assumed above that in a metre bridge the resistance of the copper pieces and the soldering at the ends of the bridge wire are negligible, but *actually they are not so small as to be neglected*. Whenever the soldering is not done carefully, the solder spreads over the bridge wire and the cross-section of the wire no longer remains uniform. The solder forms an alloy with the material of the wire, the specific resistance of which is not equal to that of the material of the wire. Further, the bridge wire may not be exactly one metre long and the scale may not be accurately placed with respect to the wire, either because the wire is slightly longer or shorter or because of the difficulty of making the connections between the ends of the wire and the strips exactly at the ends of the scale. The effect of all these factors is to introduce some *unknown resistance* at the two ends of the wire. Let these unknown end resistances at the left and right ends of the wire be respectively equivalent to α and β cm. of length of the bridge wire, then the *effective* ends of the wire will be at distances α and β respectively from the 0 and 100 cm. readings of this scale and hence the Eq. (23.11) will be modified to :

$$\frac{X}{R} = \frac{l + \alpha}{100 - l + \beta}$$

where α and β are *end corrections* for the bridge wire. If the end of the bridge wire is beyond the *extreme* graduation at that end, the end correction is *positive* ; while if the extreme graduation is beyond the end of the wire, and correction may be positive or negative according to the total end resistance at that end is greater or less than the resistance of the length of the wire by which it falls short of the extreme graduation of the scale.

23.13. Interchanging Commutator

This commutator is very helpful in interchanging the resistances P and Q connected in the two gaps of a metre bridge. As shown in Fig. 23.11, it consists of a circular ebonite disc 4 to 5 cm. in diameter, drilled and fitted with 8 terminals marked 1, 2, 3.....up to 8, *equally* spaced near its edge and projecting about 1/2 beneath the disc. Below the ebonite disc is a circular

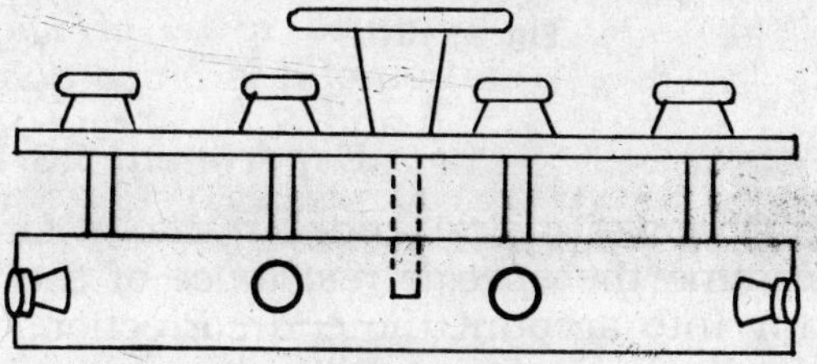

Fig. 23.11

block of ebonite (Fig. 23.12) about 1 cm. thick and of equal diameter. The ebonite block is drilled with 8 holes, about 3/4 cm. deep corresponding in position with the terminals of the disc, and eight terminals screwed into the side of the block and marked 1, 2, 3.....up to 8, pass into the holes. The holes are filled with mercury which secures contact between the block terminals and the upper disc terminals. The ebonite

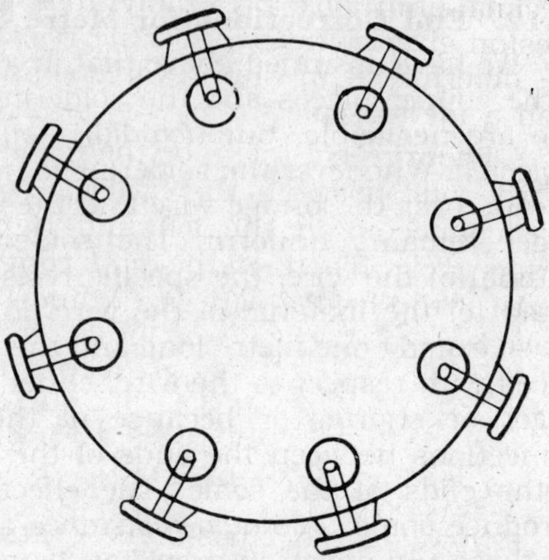

Fig. 23.12

disc can be rotated about a screw passing through the centre of the disc into a hole drilled *centrally* into the ebonite block. By lifting the disc and turning it round different terminals can be made to fit into the holes in the ebonite block. The terminals on the ebonite disc marked 2 and 3, 1 and 4, 5 and 8, and 6 and 7 are connected together by means of thick copper wires (Fig. 23.13). One of the resistances to be interchanged is connected to the terminals of the ebonite block marked 1 and 2 and the other resistance between those marked 5 and 6. The terminals marked 3 and 4 are connected to one gap of the bridge and the terminals marked 7 and 8 to the other. A rotation of the ebonite disc through a *right angle* will affect the interchange of the resistances.

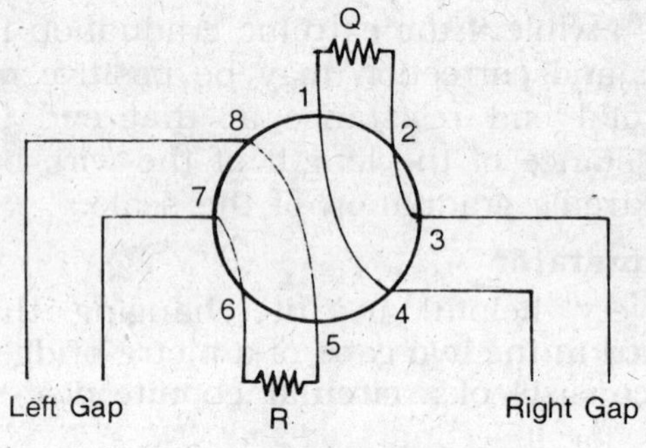

Fig. 23.13

Experiment 23.1

Object—To find *end corrections* for a metre bridge and to determine the *specific resistance* of the material of a given wire taking into account the end corrections for the bridge.

Apparatus—Metre bridge, two ordinary resistance boxes, one decimal-ohm box preferably *dial* pattern, a Leclanche cell, a weston glavanometer, wire the resistivity of whose material is to be determined, a shunt wire, connecting wires, a screw gauge and a metre scale.

Theory—Let l_1 cm. be the reading on the scale of the position of the *null point* on the bridge wire when resistances P and Q are connected in the left and right gaps of the metre bridge respectively. Then, if the *end corrections* for the left and right ends of the bridge wire are α and β m of length of the wire, we have :

$$\frac{P}{Q} = \frac{l_1 + \alpha}{100 - l_1 + \beta}$$

or $$Q\alpha - P\beta + Ql_1 - P(100 - l_1) = 0 \quad \ldots \quad (23.12)$$

Now, let the positions of the resistances P and Q be *interchanged*. Then, if l_2 cm. be the reading on the scale of the position of the null point of the bridge wire, we have :

$$\frac{P}{Q} = \frac{l_2 + \alpha}{100 - l_2 + \beta}$$

or $$P\alpha - P\beta + Ql_2 - P(100 - l_2) = 0 \quad \ldots \quad (23.13)$$

Multiplying Eq. (23.12) by Q and Eq. (23.13) by P and then subtracting Eq. (23.12) from Eq. (23.13), we get :

$$(P^2 - Q^2)\alpha + P^2 l_2 - Q^2 l_1 + PQ(l_2 - l_1) = 0$$

or $$(P^2 - Q^2)\alpha + Pl_2(P + Q) - Ql_1 + l_1(P + Q) = 0$$

whence, $$\alpha = \frac{Ql_1 - Pl_2}{P - Q} \quad \ldots \quad (23.14)$$

Similarly, $$\beta = \frac{Ql_1 - Pl_2}{P - Q} - 100 \quad \ldots \quad (23.15)$$

Interchanging Commutator-Plug type—This is depicted in Fig. 23.14. It consists of brass blocks arranged as shown in the figure with tapering holes for inserting the plugs; *a, b, c*.......k are the terminals for connections. When in use, only eight of these will be connected. A resistance *X*, as shown, is connected across *h* and *g* while the other resistance *Y* across *e* and *f*. The terminals *i* and *l* are connected to the left gap of a Carey Foster Bridge while *g* and *k* to the right gap.

With the plugs inserted in the blocks as shown by positions 1, 1, 1, 1, the resistance *X* lies in the right gap while the resistance *Y* lies in the left gap. It can be seen that when the

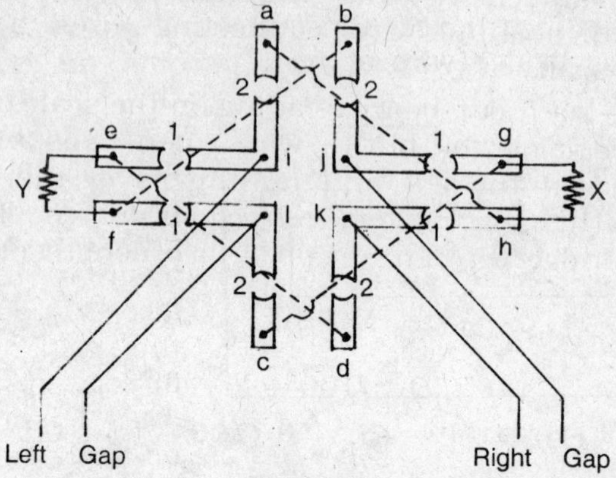

Fig. 23.14

plugs are removed from the positions 1, 1, 1, 1, and inserted at the positions 2, 2, 2, 2, X is transferred to the left gap and Y to the right gap.

From Eqs. (23.14) and (23.15) the values of end corrections α and β can be calculated, having determined l_1 and l_2 for known values of P and Q.

Now, if X be the resistance of the wire the specific resistance of whose material is to be determined :

$$X = \rho \frac{L}{\pi r^2}$$

or

$$\rho = \frac{\pi r^2}{L} X \quad \ldots \quad (23.16)$$

where L is the length of the wire and r its radius.

If the wire be connected in the left gap of the bridge and a known resistance R in the right gap and the reading on the scale of the position of null point on the bridge wire is l cm., we have :

$$\frac{X}{R} = \frac{1 + \alpha}{100 - l + \beta}$$

or

$$X = \frac{1 + \alpha}{100 - l + \beta} R \quad \ldots \quad (23.17)$$

The value of specific resistance ρ can be calculated from Eq. (23.16) having determined the value of X from Eq. (23.17).

Method—Make the connections as shown in Fig. 23.14. Shunt the galvanometer with a low resistance wire. Introduce a resistance P = 10 ohms, say, in the left gap of the metre bridge and resistance Q = 1 ohm in the right gap by adjusting the respective resistance boxes. Close the cell circuit and press the jockey over the bridge wire at near its one end and note down the direction of deflection in the galvanometer. Move the jockey to near the other end of the wire and again note down the direction of deflection in the galvanometer. The directions of deflection in the two cases must be *opposite* otherwise there is a faulty contact or wrong connection in some part of the circuit.

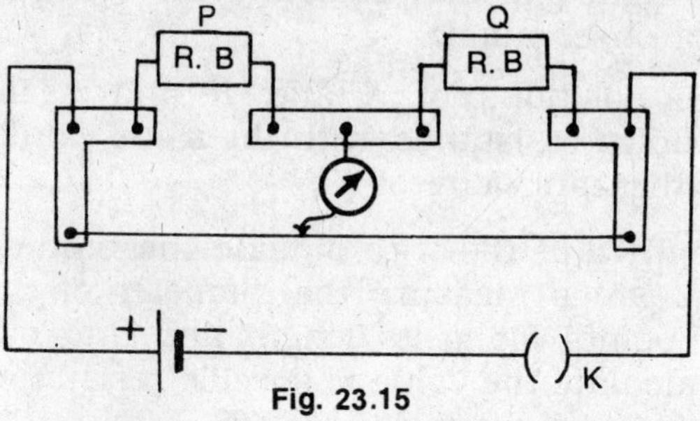

Fig. 23.15

Having tested the connections of the bridge, find by trial the approximate position of the null point. This will be near one end of the bridge wire. Then, remove the shunt from the galvanometer so that it may have its full sensitivity possible and determine the accurate position l_1 of the null point.

Now, interchange P and Q, and repeat the above determinations to find the value of l_2. This time the null point will be near the other end of the wire. Calculate the values of α and β from Eqs. (23.14) and (23.15).

Keeping Q = 1 ohm, alter the value of P to 15 ohms and 20 ohms. Calculate the values of α and β for each set of observations separately and then take their mean.

Now replace the resistance box in the left gap of the bridge by the wire the specific resistance of whose material is to be determined and that connected in the right gap by a decimal-ohm box.

Adjust the decimal-ohm box to a suitable resistance, say, R = 2 ohms. Close the battery circuit and with the galvanometer shunted with a low resistance wire, find out the position of the null point on the bridge wire and from it calculate roughly the value of the unknown resistance. Adjust the decimal-ohm box to this value and then find out the approximate position of the null point which will be in the 'middle third' of the bridge wire. Remove the shunt from the galvanometer and determine the exact position of the null point as accurately as possible with the full galvanometer sensitivity. Calculate the value of X, the unknown resistance, from Eq. (23.17). Take at least two more sets of observations by slightly decreasing and increasing the value of known resistance R in the right gap and calculate the value of X separately for each set of observation.

Next, interchange the positions of X and R and again determine the value of unknown resistance, with the same values of known resistance R. Find mean value of X.

Now, measure the length L of the wire outside the binding screws. Then with a screw gauge measure the diameter of the *bare* wire at a number of points along its length and find out the mean radius. Finally, calculate the value of specific resistance ρ of the material of the given wire from Eq. (23.16).

Sources of Error and Precautions

(1) A plug key should be included in the cell circuit which should be closed only when observations are being made. This prevents prolonged flow of current and hence consequent alteration in the values of the various resistances due to appreciable heating.

(2) The galvanometer should be shunted by a low resistance wire to avoid excessive deflection in it when the bridge is out of balance. The shunt must be removed when the null point has been almost obtained.

(3) The battery circuit should be closed before depressing the jockey on the bridge wire, but when breaking reverse order should be followed. Otherwise, if the resistance to be measured is inductive, *e.g.*, that of a coil wire, a momentary kick will be produced in the galvanometer due to self-induced E.M.F. This momentary kick is obtained even if the exact balance exists and may sometimes be excessive enough to damage the galvanometer.

(4) The jockey should always be pressed gently so that good contact may be made with the bridge wire without flattening the jockey. The contact between the jockey and the bridge wire should not be made while the former is being moved along, otherwise the wire will be worn unevenly in various parts and hence will no longer remain uniform in cross-section.

For determination of end corrections only

(5) The resistances P and Q should differ as much as possible. If $P = Q$ and so $l_1 = l_2$, the terms $\dfrac{Q\,l_1 - P\,l_2}{P - Q}$ and

$\frac{P l_1 - P l_2}{P - Q}$ of Eqs. (23.14) and (23.15) will reduce to *0/0, i.e.*, become *indeterminate* and so become the values of α and β. Hence, the difference between *P* and *Q* should be as *large* as possible. Convenient values of *P* and *Q* are *Q* = 1 ohm and *P* = 10 to 20 ohms. With higher values of *P/Q*, the balance point may lie on the copper strip.

For determination of specific resistance only

(6) The null point should be as near the *middle* of the bridge wire as possible and hence the known resistance *R* should be of the same order of magnitude as unknown. *When the null point is in the middle of the wire, a small error in determining its position produces the least error in the value of the unknown resistance, i.e., the accuracy in the result is greatest.* For if *dl* be a small error introduced in the determination of *l* and *dX* be the corresponding error produced in the result, we have, since

$$X = \frac{l}{L-l} R$$

$$dX = \frac{L\,dl}{(L-l)^2} R$$

$$\therefore \quad \frac{dX}{X} = \frac{L\,dl}{l(L-l)}$$

Now, dX/X is minimum when $l\,(L - l)$ is maximum. But $l + (L - l) = L$, a constant. Hence, the product $l\,(L - l)$ is maximum when $l = L - l$ or $l = L/2$.

(7) To eliminate error due to non-coincidence of the jockey with the metallic edge on underside of it which comes in contact with the bridge wire, two separate measurements of the unknown resistance must be made, one by introducing the unknown resistance in one gap and known resistance in the other and the second by interchanging them.

(8) The length of only that portion of the wire which is *outside* the binding screws should be measured for specific resistance calculations. It is advisable to make sharp marks on the wire at the points where it *leaves* the binding screws *before it is unclamped.*

(9) The radius of the wire is a *small* quantity and occur raised to the *second* power in the expression for ρ. Hence, the diameter of the *bare* wire should be measured at a number of points along its length and at each point readings along two *mutually perpendicular* diameters should be taken.

Observations

[*A*] *Determination of* α *and* β

S.No.	P ohms	Q ohms	Position of balance point on the scale with *P* in the left gap l_1 cm.	Position of balance point on the scale with *P* in the Right gap l_2 cm.	α cm.	β cm.
1.						
2.						
3.						
				Mean		

[*B*] *Determination of unknown resistance X of the wire*

S. No.	Known resistance *R* ohms	*X* in the left gap: Position of balance point *l* cm.	(100–*l*) cm.	*X* ohms.	*X* in the right gap: Position of balance point *l′* cm.	(100–*l′*) cm.	*X* ohms
			Mean			Mean	

[*C*] *Constants of the wire.*

(*i*) Length of the wire = cm.

(*ii*) Measurement of diameter of the wire

Least count of the screw gauge = cm.

Zero error = cm.

S. No.	Reading along any diameter *a* cm.	Reading along *perpendicular* diameter *b* cm.	$\frac{a+b}{2}$ cm.
1.			
2.			
3.			
4.			
5.			
		Mean	

Corrected diameter of the wire = cm.

Calculations

$$\alpha = \frac{Ql_1 - Pl_2}{P - Q}$$

$$= \quad \text{cm.}$$

$$\beta = \frac{Pl_1 - Ql_2}{P - Q} - 100 = \quad \text{cm}$$

When X in left gap

$$X = \frac{l + \alpha}{100 - l + \beta} R$$

$$= \quad \text{ohms.}$$

When X in right gap

$$X = \frac{100 - l' + \beta}{l' + \alpha} . R$$

$$= \quad \text{ohms}$$

Mean value of resistance of wire = ohms.

Mean radius of the wire = cm.

$$\rho = \frac{\pi r^2 X}{l}$$

$$= \quad \text{ohm} \times \text{cm.}$$

$$= \quad \text{ohm} \times \text{m.}$$

and the specific resistance of the material (.......) of the given wire = micro-ohms × m.

Standard value = micro-ohms × m.

error = %

Criticism of the method

The method yields satisfactory results, but the accuracy is limited due to the length of the bridge wire being one metre only, and can be increased by increasing the length of the bridge wire. While determining the end corrections for the bridge wire the balance points are obtained near the ends of the bridge wire and so the accuracy in the measurement of lengths of the ratio arms is not great and hence the value of α and β obtained are not very accurate. Further, the bridge wire may not be of uniform cross-section throughout and as such it would be incorrect to take the resistances of the ratio arms proportional to their lengths. Lastly, unless the resistances of the various arms of the bridge are equal and made of the same material, they will be heated up differently and the consequent alternation in their values will not be the same.

23.14. Post Office Box

It is a compact form of Wheatstone bridge and was originally designed to measure the resistances of electric cables and telegraph wires. As shown in Fig. 23.16, it consists of three separate sets of coils of known resistances arranged to form the three arms of a Wheatstone bridge, the fourth arm being the unknown resistance whose value is to be determined. The two sets of known resistance coils which are exactly *alike* and *similarly* placed form

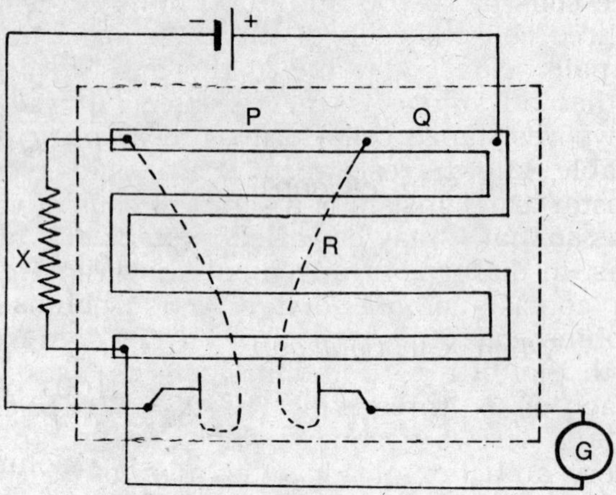

Fig. 23.16

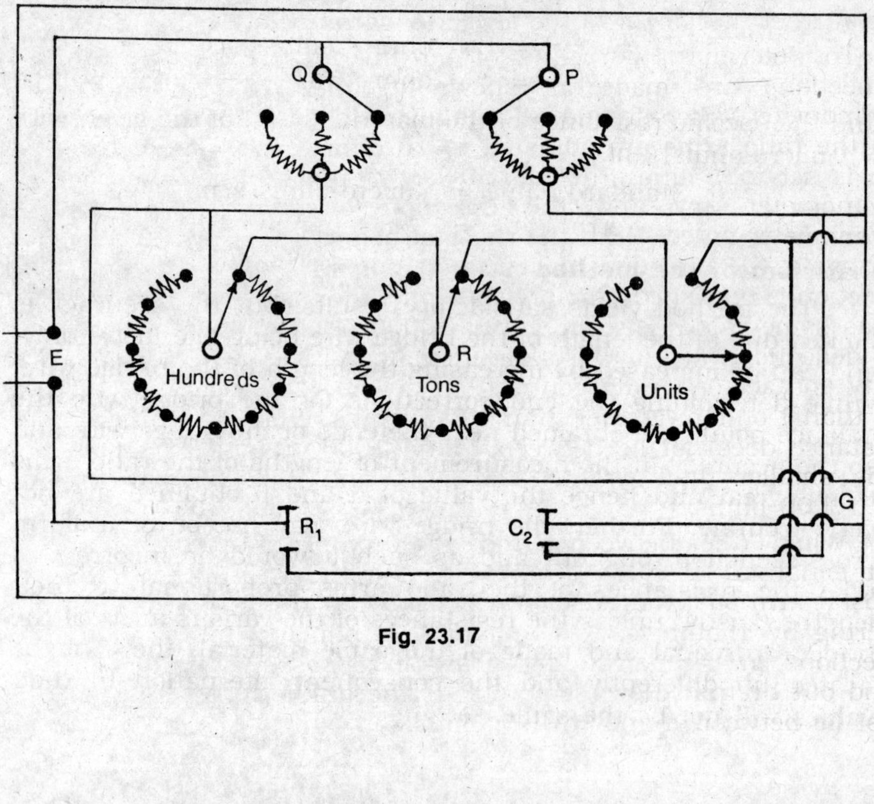

Fig. 23.17

the two *ratio arms* P and Q of the bridge. Each ratio arm contains three coils of 10, 100 and 1000 ohms resistance. One advantage of three coils in each of the ratio arms is that we can choose the pair nearest in value to the resistance to be measured and another advantage is the extension of range. The third set of known resistance coils forms the *rheostat arm, i.e.*, the arm of variable known resistance R. In plug type arrangement, the rheostat arm consists of a series of coils of values from 1 to 5000 ohms so that R may be varied in steps of 1 ohm from 1 to 11,110 ohms. In dial type arrangement shown in Fig. 23.17, the rheostat arm consists of one dial of ten 1-ohm coils, another of ten 10-ohms coils and a third of ten 100-ohms coils. Sometimes, a fourth dial of ten 1000-ohms coils is also provided. At the ends of each of the three arms P, Q and R of the bridge are provided binding screws for connecting the unknown resistance, the battery and the galvanometer as shown. The apparatus is also provided with two spring keys, one included permanently in the battery circuit and another in the galvanometer circuit. Sometimes, a special double key is provided by means of which, when the lever is depressed, the battery circuit is first closed and on a further movement of the lever the galvanometer circuit also.

To determine the value of an unknown resistance, the connections are made as shown in the Fig. 23.15. The galvanometer which should be highly sensitive is heavily shunted and the ratio arms are adjusted to 10 ohms each. the rheostat arm is set to 1 ohm. The battery key is pressed and then the galvanometer key and the direction of deflection in the galvanometer noted. Next, the rheostat arm is set to 10000 ohms and again the direction of deflection noted. If this direction of deflection is *opposite* to the previous one, the unknown resistance X has some value between 1 and 10000 ohms. If, however, the two deflections are in the same direction, there is either some faulty contact or wrong connection, or the value of X is either less than 1 ohm or greater than 10000 ohms. When the unknown resistance does not lie between 1 and 10000 ohms, P must be made smaller than Q if $X < 1$ ohm and greater than Q if $X > 10000$ ohms.

When X lies between 1 and 10000 ohms both P and Q are kept equal to 10 ohms and by varying the resistance in the rheostat arm *systematically*, two resistances in the rheostat arm differing by 1 ohm, say, n and $(n + 1)$ ohms, with which the deflections in the galvanometer are in *opposite* directions are found out by trial. Now, since $Q = P$, the unknown resistance X must lie between n and $(n + 1)$ ohms.

Next, to increase accuracy, Q is made equal to 100 ohms keeping P = 10 ohms and the value of galvanometer shunt resistance is increased. Since Q = 10 P, the resistance R of the rheostat arm now required for a balance will be 10 times the unknown resistance. Hence, the balance should now occur between 10 n and 10 $(n + 1)$ ohms. Intermediate resistances between these limits are then tried until the limits differ by 1 ohm, *i.e.*, two resistances differing by one ohm, say, n_1 and $(n_1 + 1)$ ohms are found by trial which give deflections in opposite directions. Then the unknown resistance X will be between $n_1/10$ and $(n_1 + 1)/10$ ohms. It should be noted that with the ratio $Q = 10\ P$, the above two resistances in the rheostat arm can be found only if X is less than $\Sigma/10$. where Σ stands for sum of all the resistances in the rheostat arm.

Now, to increase the accuracy still further, Q is made equal to 1000 ohms keeping P=10 ohms and the galvanometer shunt is also removed. Since $Q = 100\ P$, $R = 100\ X$ and hence the balance should now occur between $10n_1$ and $10\ (n_1+1)$ ohms. Intermediate resistance between these limits are tried and a resistance n_2 ohms in the rheostat arm is found out which produces no deflection in the galvanometer. Then the unknown resistance will be equal to $n_2/100$ ohms. If the rheostat arm cannot be adjusted to give the exact balance point, two resistances differing by one ohm in the rheostat arm are found by trial which give deflections in opposite directions and then the unknown resistance can be determined to the third place of decimals by the *method of proportional parts*. If, for example, 258 ohms give a deflection of 6 divisions in one direction and 259 ohms give a deflection of 8 divisions in opposite direction, a difference of 1 ohm in R causes a change of 14 divisions. Hence, a resistance of (258 + 6/14) ohms will produce a balance and therefore, the value of the unknown resistance is 258.4 ohms. But there is hardly any justification in going to third place to decimal for such a high degree of accuracy cannot be claimed from a post office box. It should be noted that with $Q = 100\ P$, the balance can be obtained only if X is less than $\Sigma/100$, where Σ stands for the sum of all resistances in the rheostat arm.

When the unknown resistance lies between 0.1 and 1 ohm, ratios Q = 100 ohms, P = 10 ohms and Q = 1000 ohms, P = 10 ohms should be tried while if X lies between 0.1 and 0.01, balance point can be obtained with Q = 1000 ohms only.

When the unknown resistance is greater than Σ but less than 10 Σ, ratios Q = 10, P = 100 and Q = 10, P = 1000 can be used and the value of X determined to the nearest 10 ohms. If the unknown resistance is greater than 10 Σ but less than 100 Σ, its value to the nearest 100 ohms can be determined by taking Q = 10 and P = 1000 ohms.

When X > 100 Σ or < 0.01 ohms, its value cannot be determined with the post office box and, hence, it is not suitable for the measurement of very low or high resistances.

Sometimes, it is found that when a change from a lower to a higher ratio is made, the resistance required in the rheostat arm for a balance does not lie between the limits expected from the results of the observations with the lower ratio. This may be due to contact resistance or to the inaccuracy in the value of some of the coils. In plug type of post office box contact resistances can be sufficiently reduced by tightening the plugs in the sockets but they cannot be avoided entirely. In the case of inconsistency of limits those obtained with *higher ratio* should be accepted.

23.15 Kelin's Method of determining Galvanometer Resistance

The galvanometer, the resistance G of which is to be determined, is connected as an unknown resistance in one of the four arms of the Wheat-stone bridge as shown in Fig. 23.18. The cell is connected as usual in the branch AC and a key is placed instead of the galvanometer in the branch BD. Let the current distribution in the bridge be as shown in the figure. Then applying Kirchhoff's second law to each of the meshes $CBAEC$, $CDAEC$, CBD and BDA in turn, we get :

$$(I - i_2)\,P + (I - i_g - i)\,Q + I\,r = E$$

$$i_g\,G + (i_g + i)\,R + I\,r = E$$

$$(I - i_g)\,P + i\,K - i_g\,G = 0$$

and $iK + (i_g + i)\,R - (I - i_g - i)\,Q = 0$

Rearranging we get :

$$I\,(P + Q + r) - i\,Q = E + i_g\,(P + Q) \qquad (23.18)$$

$$I\,r + iR = E - i_g\,(G + R) \qquad (23.19)$$

$$I\,P + iK = i_g\,(P + G) \qquad (23.20)$$

and

$$IQ + i\,(K + R + Q) = i_g\,(R + Q) \qquad (23.21)$$

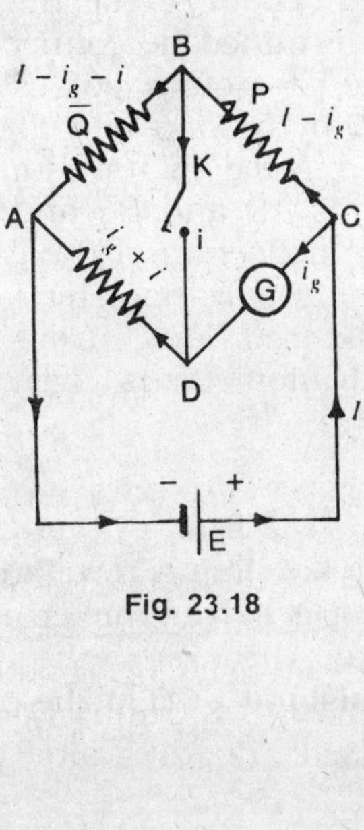

Fig. 23.18

Now let the current i_g in the galvanometer be *independent* of the resistance K in the branch BD, then it follows from Eqs. (23.18) and, (23.19) that I and i are also *constant*. When $K = \infty$, $i = 0$, and therefore, i is always equal to *zero*. Then putting $i = 0$ in Eqs. (23.20) and (23.21), we get :

$$IP = i_g \,(P + G)$$

and
$$IQ = i_g \,(R + Q)$$

whence,
$$\frac{P}{Q} = \frac{P + G}{R + Q}$$

or
$$\frac{P + G}{P} = \frac{R + Q}{Q}$$

or
$$\frac{G}{P} = \frac{R}{Q}$$

Thus, when the values of resistances P, Q and R are so adjusted as to satisfy the above relation, the current in the galvanometer is the same whether the key in the branch BD is closed or open.

Experiment 23.2

Object—To determine the resistance of a moving coil (suspended type) galvanometer with *post office box* by *Kelvin's method*.

Apparatus—A post office box, Leclanche cell, a moving coil galvanometer whose resistance is to be determined, a variable high resistance (a resistance box or a U-tube containing water), lamp and scale arrangement and connecting wires.

Theory—Let the galvanometer whose resistance is to be determined be connected in the 'unknown resistance arm' CD of the post office box, *short circuiting* the *usual* 'galvanometer arm' BD by a *short* and *thick* piece of copper wire. Then, if the cell be connected in its *usual* arm AC and the resistances of the ratio arms AB and BC and the 'rheostat arm' AD are so adjusted that the deflection of the galvanometer remains the *same* whether the 'usual galvanometer-key' K_2 is pressed or released, the bridge is *balanced* and from Section 23.15 the resistance of the galvanometer is given by

$$G = \frac{P}{Q} R \qquad \ldots \qquad (23.22)$$

Method —Level the base of the galvanometer by means of the levelling screws and release its coil. Throw light on the mirror of the galvanometer and get the spot of reflected light on the scale. Connect the galvanometer in the arm of 'unknown resistance' CD of the post office box. Join the *usual* galvanometer

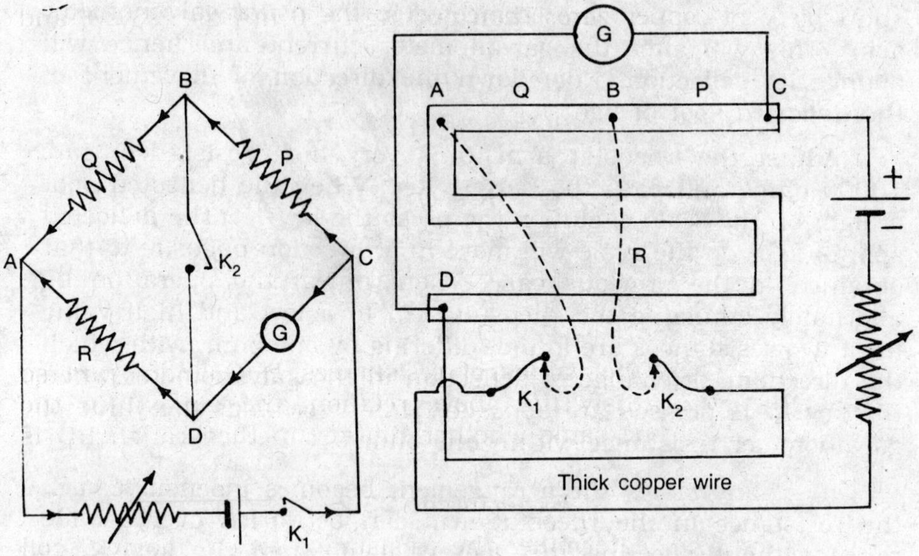

Fig. 23.19

terminals *B, D in the post office box* by a short piece of thick copper wire and make the rest of the connections as shown in Fig. 23.19 inserting *in the cell circuit a variable* high resistance, say, a resistance box adjustable upto 10,000 ohms or a U-tube containing water, *in series* with the cell. See that the ends of the connecting wires are *clean* and that all connections have been *firmly* made. If the post office box is of plug type, see also that the plugs and the sockets are clean and the plugs are not loose.

Adjust the ratio arms *P* and *Q* to 100 ohms each. Press the battery key. The coil of galvanometer will be deflected and as the galvanometer is very sensitive the spot of light will move *off* the scale. Adjust variable high resistance placed in series with the cell to a high value so as to send only a *small* current through the galvanometer and hence to produce a small deflection in it and thus bring the spot of light near the middle of the scale in its *deflected* state, *i.e.*, with the battery key K_1 *pressed.*

Adjust the 'rheostat arm' to a few ohms. Press the battery key, and when the deflected spot of light is *stationary* on the scale press the second key K_2 also. Since the bridge is out of balance, pressing the key K_2 will send a current through the short piece of copper wire connected in the *usual* 'galvanometer arm'. This will alter the galvanometer current and hence will change its deflection. Note down the direction of movement of the *deflected* spot of light.

Adjust the 'rheostat arm to a very high resistance, say, 10000 ohms and pass the battery key. When the deflected spot of light on the scale is stationary, press the key K_2. The deflected spot of light on the scale will move in a direction *opposite* to that obtained in the previous case. Then, repeat the operation by alternately adjusting the rheostat arm to a low and high value until two resistances are found, differing by one ohm, with which the directions of change of deflection in the galvanometer when the key K_2 is pressed are *opposite*. This determines the value of galvanometer resistance within one ohm.

If, at any stage, the arrangement becomes insensitive, *i.e.*, the resistance in the rheostat arm can be varied over a wide range without affecting the spot of light when the key K_2 is pressed, increase the current in the galvanometer by adjusting the value of high resistance placed in series with the cell. This will produce a *larger* deflection in the galvanometer and the arrangement will become more sensitive.

Now keeping P = 100 ohms, make Q = 1000 ohms. Since $Q = 10\ P$, the resistance in the rheostat arm which will produce balance will now be 10 times that of the galvanometer resistance. Adjust the 'rheostat arm' between 10 times the limits obtained for galvanometer resistances with the *former* ratio and find by trial as above a resistance in the 'rheostat arm' which, with K_1 *pressed*, will produce no shift of the deflected spot of light on the scale when the key K_2 is pressed. The bridge will be *balanced*. Note down the resistance R in the 'rheostat arm' and calculate the galvanometer resistance from Eq. (23.22). This gives the galvanometer resistance within one tenth of an ohm. If with P/Q = 100/1000, the bridge is not exactly balanced, the galvanometer resistance can be computed by the *method of proportional parts* (Section 23.14).

Sources of Error and Precautions

(1) The base of the galvanometer should be carefully levelled and the coil should be unclamped. This ensures the *free* movement of the coil in the space between the magnet and the soft iron piece.

(2) The post office box should be preferably of *dial pattern* for in it the variable contact resistances are sufficiently reduced. In the case of plug type P.O. box all sockets must be clean and the plugs made tight.

(3) When the bridge is *out of balance*, the degree of alteration in the *steady* deflection in the galvanometer by tapping the *key* K_2 will depend upon the current which flows along the branch *BD*. Hence, to ensure *sensitiveness* of the test of want of balance, the resistance of the branch *BD* should be made as *small* as possible. *B* and *D* should, therefore, be joined by a short piece of *thick* copper wire.

(4) To avoid unnecessary heating of the various coils and consequent alteration in their resistances, the battery key K_1 should be pressed only when observations are to be made.

(5) As the galvanometer is generally very sensitive, the deflection in it is very large and, hence, the spot of reflected light move *off* the scale when the battery key is closed. The deflection in the galvanometer can be decreased by sending in it a *small* current only. A variable high resistance, say, a resistance box adjustable up to 10000 ohms should, therefore, be inserted in the cell circuit *in series with the cell* and the value of the resistance adjusted to get the spot of light on the scale in its *deflected* position. When the current through the galvanometer is very *small*, the spot of light can also be brought on the scale by turning back the coil, by twisting the suspension strip by rotating the torsion head which carries it, but it is in general inadvisable to interfere with the suspension of the coil.

(6) In order that the bridge may be highly sensitive, the resistances of the four arms should be of the *same* order of magnitude. Hence, if the galvanometer resistance lies between 100 and 200 ohms, each of the ratio arms should be adjusted to 100 ohms.

(7) *The bridge is balanced when there is no change of steady deflection in the galvanometer whether the key K_2 is pressed or released*. Hence, the battery key K_1 should be pressed first in order to produce steady deflection in the galvanometer; and only when the spot of light on the scale is *stationary* with the key K_1 kept pressed that the key K_2 should be pressed to observe any change of direction and extent of the steady deflection.

Observations

S. No.	Ratio arms P Ohm	Ratio arms Q ohm	Rheostat arm R *ohm*	Direction of change of deflection when K_2 is pressed	Galvanometer resistance G
1	100	100	1	Left	Greater than 1 ohm
2	100	100	10000	Right	
3	100	100	156	Left	Lies between 156 and 157 ohm
4	100	100	157	Right	
5	100	1000	1563	Left	G = 156.4 ohm
6	100	1000	1564	No change	
7	100	1000	1565	Right	

Calculations

$$G = \frac{P}{Q} R$$

$$= \frac{100}{1000} \times 1564$$

$$= 156.4 \text{ ohm}$$

Result—The galvanometer resistance = ohm

Criticism of the method—When performed carefully, the method gives a fairly satisfactory value of the galvanometer resistance. The main difficulty in the method is that as the galvanometer is usually very sensitive, the steady current in it is large and as a result the spot of reflected light goes off the scale, and if the deflection in the galvanometer is reduced by placing a high resistance in series with the cell, the sensitiveness of the test of proper balance of the bridge decreases and the resistance in the rheostat arm can be varied over a wide range without changing the steady galvanometer deflection when the key K_2 is pressed. The result is also affected by thermo-electric and heating effects in the circuits.

The best method of determining the galvanometer resistance is to clamp its coil and connect it in the unknown resistance arm of the post office box and then to balance the bridge in the usual manner using *another* galvanometer as detector. Kelvin's method is not suitable for moving-magnet mirror galvanometer in which the deflection can be reduced without sacrificing the sensitivity by adjusting the position of the control magnet or if that is not sufficiently strong by an external bar magnet.

Exercise—*To determine the resistance of a Weston galvanometer metre bridge by Kelvin's method.*

Connect the Weston galvanometer whose resistance is to be determined into the left gap of the metre bridge by *short* pieces of *thick* copper wire and connect in the right gap a resistance box. Connect a Leclanche cell with a variable high resistance in series with it in the *usual* position of the battery, *i.e.*, between *A* and *C* including a plug key in the cell circuit as shown in (Fig. 23.20). Join *B* to the jockey by a *thick* copper wire. Adjust the resistance box in the right gap of the bridge to a suitable value, say, 53 ohms. Close the cell circuit and adjust the variable high resistance to produce a convenience deflection in the galvanometer. Press the jockey over the bridge wire; the galvanometer deflection will, in general, change. Shift the jockey and determine by trial the position of the balance point *D* on the bridge wire such that *when the jockey is pressed at that point there is no change of deflection in the galvanometer.* The balance point should be obtained as near the middle of the wire as possible by adjusting the resistance box in the right gap to a suitable value. Let l be the reading on the scale of the position of the balance point on the bridge wire and let R be the resistance in the right gap of the bridge. Then, the resistance G of the galvanometer from Section 23.15 is given by :

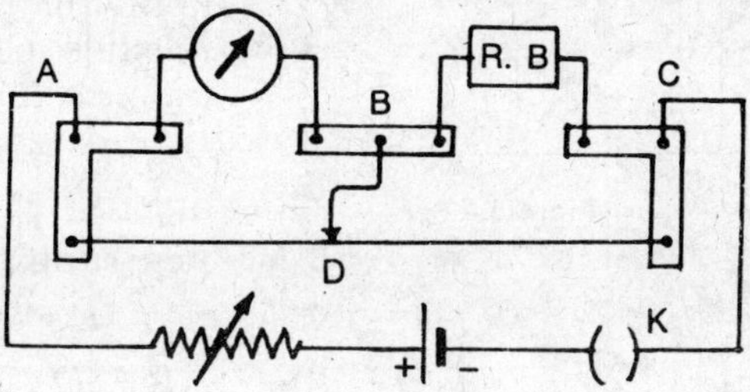

Fig. 23.20

$$G = \frac{l}{100 - l} R$$

In this method, the galvanometer remains *deflected* the *whole* of the time during the determination of the balance point and often becomes insensitive with the result that the point of contact of the jockey with the bridge wire can be shifted through an appreciable distance without making any noticeable change in the deflection. Consequently, the accuracy obtainable by this method in the value of galvanometer resistance is not very great.

23.16. Mance's Method of Determining Cell Resistance

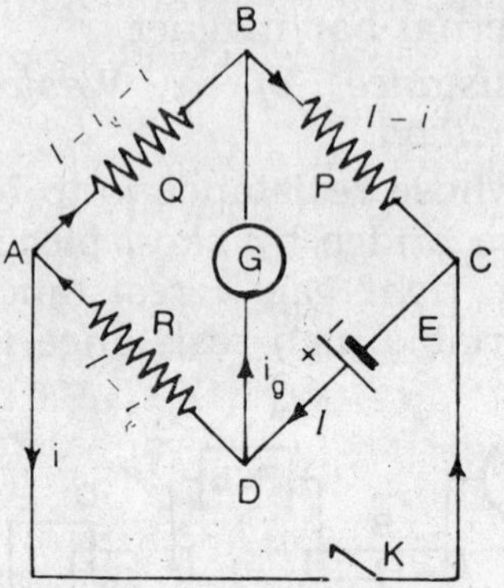

Fig. 23.21

The cell, the internal resistance of which is to be determined, forms one of the four arms of the Wheatstone bridge as shown in Fig. 23.21. The galvanometer is connected as usual in the branch *BD* and a key is placed *instead of the cell* in the branch *AC*. Let the current distribution in the bridge be as shown in the figure. Then applying *Kirchoff's second law* to the meshes *CBD* and *DAB* which do not include *K*, we have :

$$IX + i_g G + (I - i) P = E$$

and $(I - i_g) R + (I - i_g - i) Q - i_g G = 0$

Rearranging, we get :

$$I (P + X) + i_g G = E + i P \qquad (23.23)$$

and

$$I (Q + R) - i_g (R + Q + G) = iQ \qquad (23.24)$$

Multiplying Eq. (23.23) by $Q + R$ and Eq. (23.24) $(P + X)$ and then subtracting, we get :

$$i_g [G (Q + R) + (R + Q + G) (P + X)] = E (Q + R) + i [P (Q + R) - Q (P + X)]$$

Now, let the current i_g in the galvanometer be *independent* of the resistance *K* in the branch *AC*. Then, the above expression for i_g must be independent of the current, *i*, for which since *i* is *not equal to zero*, we must have :

$$[P (Q + R) - Q (P + X)] = 0$$

or

$$PR - QX = 0$$

whence,

$$\frac{X}{P} = \frac{R}{Q}$$

Thus, when the values of resistances *P*, *Q* and *R* are so adjusted as to satisfy the above relation, *the current in the galvanometer is the same whether the key in the branch AC is closed or open*. This also follows from the fact that when the above relation is satisfied *BD* and *AC* become *conjugate conductors* so that the current in *BD* must be *independent* of the resistance of the *conjugate* conductor *AC*.

Experiment 23.3

Object—To determine the internal resistance of a Leclanche cell with *post office box* by *Mance's method*.

Apparatus—A post office box, Leclanche cell, a moving coil (suspended type) galvanometer with lamp and scale arrangement, variable high resistance (a resistance box or a *U*-tube containing water) a plug key and connecting wires.

Theory—Referring to Fig. 23.22, let the cell whose internal resistance is to be determined be connected in the 'unknown resistance arm' *CD* of the *P.O.* box, *short-circuiting* the *usual* 'battery arm' *AC* by a *short* and *thick* piece of copper wire. Then, if the galvanometer be connected in its usual arm *BD* and the resistances of the ratio arms AB and BC and the 'rheostat arm' *AD* are so adjusted that the deflection of the galvanometer remains the *same* whether the usual battery key K_1 is pressed or released, the bridge is *balanced* and from Section 23.16, the internal resistance of the cell is given by :

$$X = \frac{P}{Q} R \tag{23.25}$$

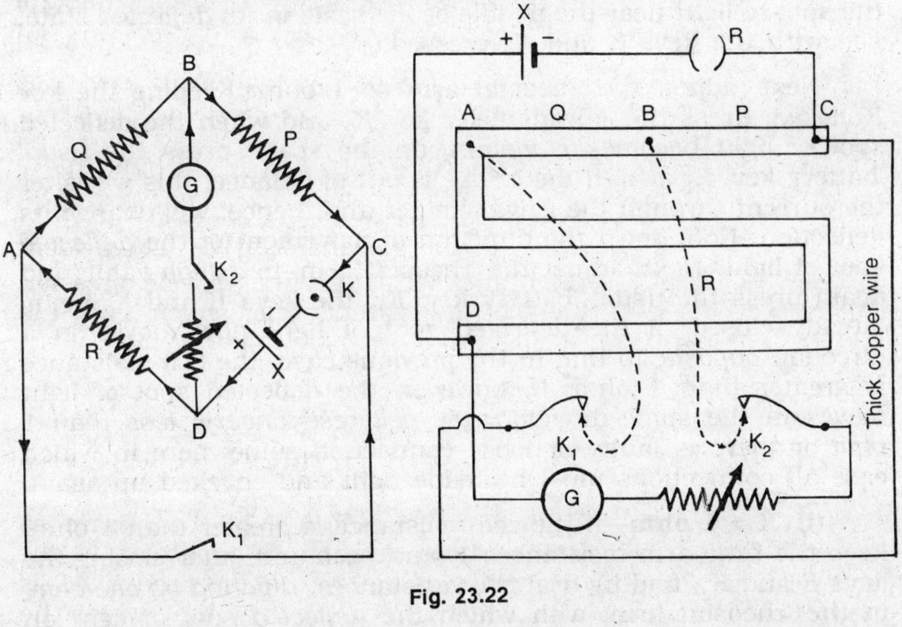

Fig. 23.22

Method—Level the base of the galvanometer by means of the levelling screws and unclamp its coil. Throw light on the mirror of the galvanometer and get a bright spot of reflected light on the scale. Connect the Leclanche cell whose internal resistance is to be determined in the 'unknown resistance arm' *CD* of the post office box by *short* pieces of *thick* copper wire including a plug key K in the circuit. Join the usual cell terminals A, C by a *short* piece of *thick* copper wire and make the rest of the connections as shown in Fig. 23.22 inserting in series with the *galvanometer*, a variable high resistance, say, a resistance box adjustable upto 10,000 ohms or a U-tube containing water. Take care that the ends of the connecting wires are *clean* and that all connections are *tight*. In the case of a plug type post office box see also that the sockets are clean and the plugs are tight.

Adjust the ratio arms P and Q to 1 ohm *each*, close the key K included in the arm CD and the galvanometer key K_2. A current will flow through the galvanometer and its coil will be deflected. As a consequence, the spot of reflected light on the scale will move. If the deflection in the galvanometer is *excessive*, the spot of light will move off the scale. In such a case, adjust the variable resistance connected in series with the galvanometer to a *high* value so as to send only a *small* current through the galvanometer and, hence, to produce a small deflection in it and thus bring the spot of light near the middle of the scale in its *deflected* state, *i.e.*, with the keys K and K_2 pressed.

Next, adjust the 'rheostat arm' to 1 ohm. Keeping the key K closed, press the galvanometer key K_2 and when the deflected spot of light becomes *stationary* on the scale, press the *usual* 'battery key' K_1 also. If the bridge is out of balance, this will alter the current through the galvanometer and, hence, will change its deflection. Note down the direction of movement of the *deflected* spot of light. Next, adjust the 'rheostat arm' to a *high* value and again press the usual 'battery key' K_1, the keys K and K_2 being already closed. If the *deflected* spot of light now moves in a direction *opposite* to that in the previous case, the cell resistance is greater than 1 ohm. If, however, the deflected spot of light moves in the *same* direction, *the cell resistance* is *less than* 1 *ohm* or there is faulty or loose connection somewhere in which case all connections must be made tight and checked up again.

(i) X > 1 ohm—If the cell resistance is greater than 1 ohm, keep the ratio arm resistances 1 ohm *each* and, after closing the keys K and K_2, find by trial two resistances, *differing by one ohm*, in the 'rheostat arm, with which the *deflected* spot of light on the scale moves in *opposite* directions when the usual 'battery key' K_1 is pressed. This determines the cell resistance within one ohm.

Now, keeping P = 1 ohm, make Q = 10 ohms. Since Q = 10 P, the resistance in the 'rheostat arm' which will produce balance will now be 10 times that of the cell resistance. Adjust the 'rheostat arm' between 10 times the limits obtained for cell resistance with the former ratio and find again by trial two resistances in the 'rheostat arm' *differing by one* ohm, which move the *deflected* spot of light in *opposite* directions on the scale when the key K_1 is pressed. This determines the cell resistance within 1/10 of an ohm.

Next, keeping P = 1 ohm, adjust Q = 100 ohms. Since Q = 100 P, the resistance in the 'rheostat arm' which will produce balance will now be 100 times that of the cell resistance. Thus, find by trial a resistance in 'rheostat arm', lying between 10 times the limits obtained for cell resistance with the ratio 1 : 10, with which there is *no shift* of the deflected spot of light on the scale when the usual 'battery key' K is pressed. The bridge will then be balanced. Note down this resistance R in the 'rheostat arm' and calculate the cell resistance from Eq. (23.25). This gives the cell resistance within 1/100 of an ohm.

(ii) X < 1 ohm—If the cell resistance is less than 1 ohm as is often the case with a *freshly* prepared cell, keep P = 1 ohm and adjust Q = 10 ohms so that P : Q = 1 : 10. Close the keys K and K_2 and by trying resistances from 1 to 10 ohm *only* in the 'rheostat arm' find two resistances, *differing by one ohm*, which move the deflected spot of light in *opposite* directions on the scale when the *usual* 'battery key' K_1 is pressed. This gives the cell resistance within 1/10 of an ohm.

Now, keeping P = 1 ohm, make Q = 100 ohm so that P : Q = 1 : 100. Then, keeping the keys K and K_2 closed, find by trial a resistance lying between 10 and 100 ohm in the 'rheostat arm' which produces *no shift* of the deflected spot of light on the scale when the usual 'battery key' K_1 is pressed. Note down this resistance R in the 'rheostat arm' and calculate the value of cell resistance from Eq. (23.25). This gives the cell resistance within 1/100 of an ohm.

Sources of Error and Precautions

(1) The base of the galvanometer should be carefully levelled and the coil should be unclamped. This ensures the *free* movement of the coil in the space between the magnet and the soft iron piece.

(2) The post office box should be preferably of *dial pattern* for in it the variable contact resistances are sufficiently reduced. In the case of plug type *P.O.* box all sockets must be clean and the plugs made tight.

(3) X represents the resistance of the cell and its connecting

wires in the 'unknown resistance arm' *CD* of the *P.O.* box. Hence, in order that it may be equal to the cell resistance, the cell should be connected by *short* pieces of *thick* copper wire.

(4) A plug key *K* should be included in the arm *CD* along with the cell and should be closed only when observations are being made. This would prevent polarisation in the cell as well as unnecessary heating of the various coils and consequent alteration in their resistances.

(5) During the course of the experiment the zinc rod of the Leclanche cell should *never* be taken out or the cell otherwise *disturbed* in any way, for this would alter the internal resistance of the cell. The starting or stopping of current from the cell should be accomplished by the use of the key *K* only.

(6) When the bridge is *out of balance*, the degree of alteration in the *steady* deflection of the galvanometer by tapping the key K_1 will depend upon the current which flows along the branch *AC*. Hence, to ensure *sensitiveness* of the test of want of balance, the resistance of the branch *AC* should be made as *small* as possible. *A* and *C* should, therefore, be joined by a *short* piece of *thick* copper wire.

(7) As the galvanometer is generally very sensitive, the deflection in it is very large and hence the spot of reflected light moves off the scale when the galvanometer key K_2 is closed. To overcome this difficulty a variable high resistance would be connected *in series with the galvanometer* and its value adjusted to get the spot of light on the scale with the key K_2 kept pressed.

(8) In order that the bridge may be highly sensitive, the resistance of the four arms should be of the *same* order of magnitude. Hence, each of the ratio arms should be adjusted to 1 ohm.

(9) *The bridge is balanced when there is no change of steady deflection in the galvanometer whether the usual battery key* K_1 *is pressed or released.* Hence, after closing the keys, K_1 *and* K_2, *it is only when the spot of light on the scale becomes stationary* that the key K_1 should be pressed to observe any change of direction and extent of the steady deflection.

(10) While testing the balancing of the bridge by pressing the key K_1, it is often observed in the case of a Leclanche cell that the spot of light first moves with a jerk in one direction and then after a moment begins to move *gradually* in the *opposite* direction on account of polarisation of the cell. In such a case, the direction in which the spot of light moves with a jerk *immediately after* the key K_1 is pressed in the correct direction.

Observations

If $X > 1$ ohm.

S. No.	Ratio arms		Rheostat arm R ohm	Direction of *change* of deflection when K_1 is pressed	Cel resistance X
	P ohm	Q ohm			
1	1	1	1	Left	Greater than 1 ohm
2	1	1	100	Right	
3	1	1	2	Left	Lies between 2 and 3 ohm
4	1	1	3	Right	
5	1	10	21	Left	Lies between 2.1 and 2.2 ohm
6	1	10	22	Right	
7	1	100	215	Left	
8	1	100	216	No change	X = 2.16 ohm
9	1	100	217	Right	

If $X < 1$ ohm.

S. No.	Ratio arms		Rheostat arm R ohm	Direction of *change* of deflection when K_1 is pressed	Cell resistance X
	P ohm	Q ohm			
1	1	1	1	Right	Less than 1 ohm
2	1	1	100	Right	
3	1	10	7	Left	Lies between 0.7 and 0.8 ohm
4	1	10	8	Right	
5	1	100	8	Left	
6	1	100	77	No change	X = 0.77 ohm
7	1	100	78	Right	

Calculations

$$X = \frac{P}{Q} R$$

$$= \quad \text{ohm}$$

Result—The internal resistance of the given Leclanche cell

= ohm.

Criticism of the Method.—The method is not very suitable for the determination of the internal resistance of a Leclanche cell owing to polarisation effects and the variation of internal resistance with the current. When the current in the galvanometer is reduced, by adjusting the variable high resistance in series with it, to get the deflected spot of light on the scale, the galvanometer often becomes insensitive. The difficulty can, however, be overcome by the use of a moving-magnet mirror galvanometer or by connecting a condenser in series with the galvanometer as suggested by Lodge.

23.17. Lodge's Modification of Mance's Method

Referring to Fig. 23.23, in this method a condenser of about 0.5 microfarad capacitance is placed in series with the galvanometer which should preferably have high resistance. The condenser prevents any steady current to flow through the galvanometer. If there is any change of *P. D.* between B and D due to flow of current in the branch AK_2C when the key K_2 is pressed the charge on the condenser alters and this produces a momentary kick or throw in the galvanometer. The resistance in the various arms of the bridges are so arranged that there is no kick in the galvanometer when the key K_2 is closed. Then, the bridge is balanced and the usual Wheatstone bridge relationship holds. By the use of the condenser the balance point can be accurately determined on account of the now increased sensitivity of the arrangement, and all chances of damaging the galvanometer by the passage of large currents have been avoided.

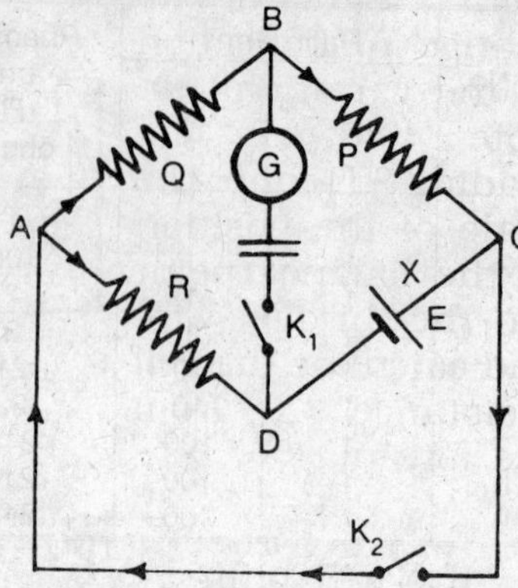

Fig. 23.23

Exercise—*To determine the internal resistance of a Leclanche cell with metre bridge by Mance's method.*

As shown in Fig. 23.24, connect the cell whose internal resistance is to be determined in the *left* gap of the metre bridge by means of *short* pieces of *thick* copper wire, including a plug key *K* also in the circuit, and in the right gap of the bridge, connect a decimal-ohm box. Connect a Weston galvanometer, with a *variable* high resistance in series with it, in the *usual* position of the *cell, i. e.*, between *A* and *C*, and join *B* to the jockey by a *thick* copper wire. See that the end of the connecting wires are *clean* and that all connections have been *firmly* made.

Adjust the decimal-ohm box to a suitable value, say, $R = 2$ ohm. Close the cell key *K*. *A* current will flow through the galvanometer and hence its needle will be deflected. If the deflection in the galvanometer is *excessive*, reduce it by adjusting the variable high resistance placed in series with the galvanometer. Then, press the jockey over the bridge wire. The current in the galvanometer will, in general, be altered and hence the deflection in it will change. Note down whether the deflection increases or decreases. Shift the jockey to another point on the bridge wire and find by trial the position of the balance point *D* on the bridge wire such that when the jockey is pressed at that point, there is no change of deflection in the galvanometer. Note down the reading l on the scale of the position of the balance point on the bridge wire and calculate the cell resistance X from the formula :

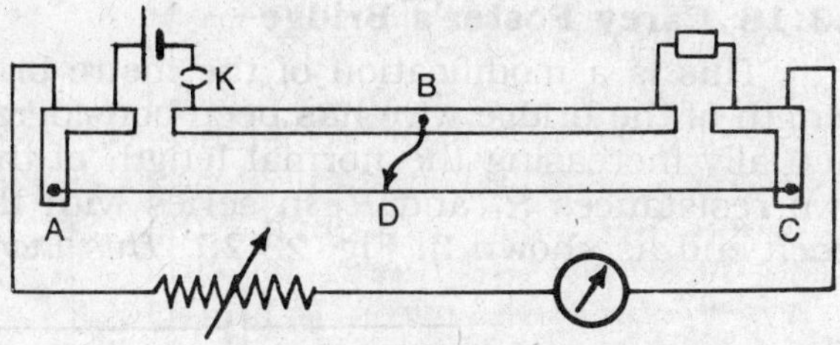

Fig. 23.24

$$X = \frac{l}{100 - l} R$$

Next, adjust the decimal ohm box to the above calculated value of the cell resistance. Shift the jockey to near the *middle* of the bridge wire and find as accurately as possible the position of the balance point on the wire. Note down the reading on the scale of the position of *this* balance point on the wire and calculate the value of the cell resistance. Then, slightly decrease or increase the value of the known resistance R and take at least two more sets of observations and calculate the value of cell resistance *separately* for *each* set of observation and take the *mean*.

If with a particular value of known resistance R in the right gap, no balance point is obtained over the entire bridge wire, either there is a faulty contact or wrong connection somewhere in the circuit or the balance point lies in the end strip. In the latter case, the known resistance R in the right gap is either very much smaller or greater than the cell resistance.

This method of determining cell resistance is not capable of giving very accurate result, firstly, because the internal resistance of the cell varies with the current drawn from it and, secondly, on account of polarisation effects, there is a small change of steady deflection of the galvanometer when the jockey is pressed even if the conjugate relation holds. Another difficulty is that the galvanometer often becomes insensitive. This difficulty can, however, be overcome by using a moving-magnet mirror galvanometer in which case the deflection in the galvanometer can be reduced without affecting its sensitivity by adjusting the position of the control magnet. The other sources of error which affect the result are the thermo-electric effects, heating effects, the non-uniformity of the bridge wire, the unknown resistances at the ends of the non-coincidence of the point of contact of jockey with the pointer on it.

23.18 Carey Foster's Bridge

This is a modification of the metre bridge in which *effective* length of the bridge wire has been considerably increased without actually increasing the normal length of the wire, by connecting two resistances R_1 and R_2 in series with the bridge wire, one at *each* end as shown in Fig. 23.25. *This increases the accuracy of*

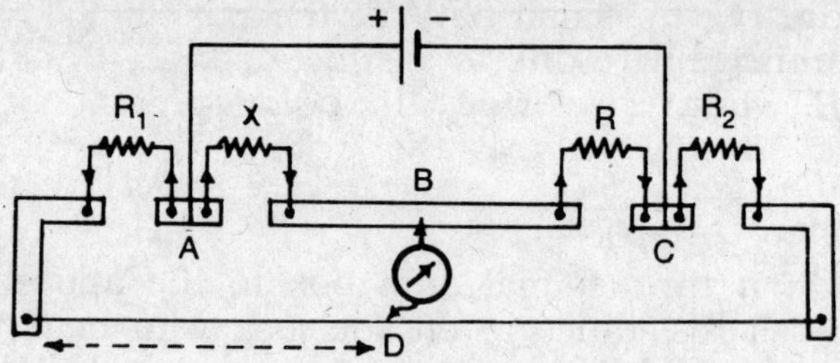

Fig. 23.25

the result due to any error in reading the position of the balance point on the scale as well as the sensitiveness of the bridge become less significant.

If dX be the small error produced in the value of the unknown resistance X due to a small error dl of reading l, the relative inaccuracy in the result is (prec. 6 expt 27.3) given by :

$$\frac{dX}{X} = \frac{Ldl}{l(L-l)} = \frac{dl}{(l-l/L)\,l} \qquad (23.26)$$

dX is minimum when $l = L/2$. Therefore, the *least inaccuracy* in the result will be given by :

$$\frac{dX}{X} = \frac{4}{L}dl = \frac{2}{5L}$$

since dl = 1 mm., the *smallest* division on the scale. For *metre bridge* L = 100 cm. and, hence,

$$\frac{dX}{X} = \frac{1}{250}$$

For *Carey Foster's bridge* the length is *virtually increased.* Let the resistances in the two outer gaps be equivalent to n times the length of the bridge wire. Then,

$$L = 100n + 100 = 100\,(n+1)$$

and

$$\frac{dX}{X} = \frac{2}{5L} = \frac{1}{250\,(n+1)}$$

For $n = 1$, *i.e.*, when L virtually increased to twice the length of the bridge wire :

$$\frac{dX}{X} = \frac{1}{500}$$

For = 2 $$\frac{dX}{X} = \frac{1}{750}$$

and so on.

Thus, the accuracy in the result increases as L increases, being directly proportional to L.

Now, if dl be the small shift in the position of the balance point on the bridge wire due to a *small* change dR of the *known* resistance R when the bridge is *balanced*, then dl/dR will measure the *sensitiveness* of the bridge. When the bridge is balanced

$$\frac{X}{R} = \frac{1}{L-l}$$

or $$\text{R} = \frac{L-l}{l} X = \left(\frac{L}{l} - 1\right) X$$

$\therefore$ $$dR = -\frac{Ldl}{l^2} X$$

whence, $$\frac{dl}{dR} = \frac{l^2}{LX}$$

Let $l = L/n$ where n is any number greater than unity. Then,

$$\frac{dl}{dR} = -\frac{L}{n^2 Y}$$

This shows that sensitiveness for the bridge varies directly as the length of the bridge wire, being greater with larger length of the wire. For balance point in the middle of the wire, $n = 2$ and then $dl/dR = -L/4\ X$. The –ve sign in the above equation shows that when the value of R is increased by dR the value of l will decrease by an amount dl.

23.19. Comparison of Two Nearly Equal Resistance with a Carey Foster's Bridge

Carey Foster's bridge is specially suited for the comparison of two *nearly equal* resistance *whose difference is less than the resistance of the bridge wire*. As shown in Fig. 23.26, two resistances X and Y to be compared are connected in the *outer* gaps of the bridge *in series* with bridge wire. these two resistances together with the bridge wire form the two arms of the Wheatstone bridge, one composed of X plus a length of the bridge wire upto the balance point and a second composed of Y plus the rest of the bridge wire. The remaining two arms are formed by two *early*

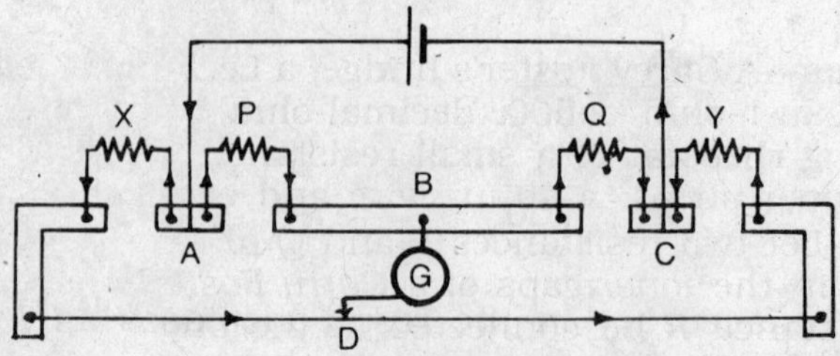

Fig. 23.26

equal resistances P and Q which are connected in the *inner* gaps of the bridge. If l_1 be the reading on the scale of the position of the null point, we have, from usual Wheatstone bridge formula :

$$\frac{P}{Q} = \frac{X + \sigma(l_1 + \alpha)}{Y + \sigma(100 - l_1 + \beta)}$$

or
$$\frac{P}{Q} + 1 = \frac{X + Y + \sigma(100 + \alpha + \beta)}{Y + \sigma(100 - l_1 + \beta)} \qquad (23.27)$$

where α and β, units of length of the bridge wire, are the *end corrections* at the left and right ends of the bridge wire respectively, and σ is the *resistance per unit length* of the bridge wire.

If, now, X and Y are *interchanged* and l_2 be the reading on the scale of the position of the *new* null point, we have :

$$\frac{P}{Q} = \frac{Y + \sigma(l_2 + \alpha)}{X + \sigma(100 - l_2 + \beta)}$$

$$\frac{P}{Q} + 1 = \frac{X + Y + \sigma(100 + \alpha + \beta)}{X + \sigma(100 - l_2 + \beta)} \qquad (23.28)$$

Comparing Eqs. (23.27) and (23.28) we see that the fractions on the right hand side are equal and since their numerators are identical, their denominators must also be equal. Hence, equating the denominators of the right hand sides of Eqs. (23.27) and (23.28), we have :

$$Y + \sigma(100 - l_1 + \beta) = X + \sigma(100 - l_2 + \beta)$$

whence,
$$X - Y = \sigma(l_2 - l_1) \qquad (23.29)$$

Thus, *the difference between the resistances X and Y can be obtained by determining the resistance of the bridge wire between the two null points.*

Experiment 23.4

Object—To determine the *resistance per unit length of a Carey Foster's bridge* wire and then to compare the resistance of a given one-ohm coil with a standard one-ohm resistance.

Apparatus—A Carey Foster's bridge, a Leclanche cell, Weston galvanometer, a 1-ohm coil, a decimal-ohm box preferably dial pattern, sliding rheostat of a small resistance, a single way plug key, thick copper strips, a shunt wire and connecting wires.

Theory—Let two resistances *P* and *Q* of *nearly equal* values be connected in the *inner* gaps of a *Carry Foster's bridge* and let a known resistance *R* be connected in the *outer* left gap of the bridge. Then, if a thick copper strip be connected in the *outer* right gap of the bridge and l_1 and l_2 be the readings on the scale of the positions of the null point on the bridge wire before and after interchanging the known resistance *R* and the thick copper strip in the outer gaps, we have from Eq. (23.29) by putting $X = R$, $Y = 0$:

$$R = \sigma (l_2 - l_1)$$

$$\sigma = \frac{R}{l_2 - l_1} \qquad (23.30)$$

Now, let the coil of *unknown* resistance *X* be connected in the *outer* left gap and a standard known resistance *Y* of *nearly the same* value in the *outer right* gap of the bridge. Then, if l'_1 and l'_2 be the readings on the scale of the positions of the null point before and after interchanging *X* and *Y*, we have, from equation (23.29) :

$$X - Y = \sigma (l'_2 - l'_1)$$

whence, $$X = \sigma (l'_2 - l'_1) - Y \qquad (23.31)$$

This equation can be used to calculate X, if σ is determined from Eq. (23.30).

Method (*a*) **Determination of** σ—Connect a decimal-ohm box in the *left* gap of a Carey Foster's bridge and a *thick* copper strip in its outer *right* gap as shown in Fig. 23.27. Next, connect the two fixed terminals at the base of a sliding rheostat *MN* to *A* and *C* and its *sliding contact B* through the Weston galvanometer to

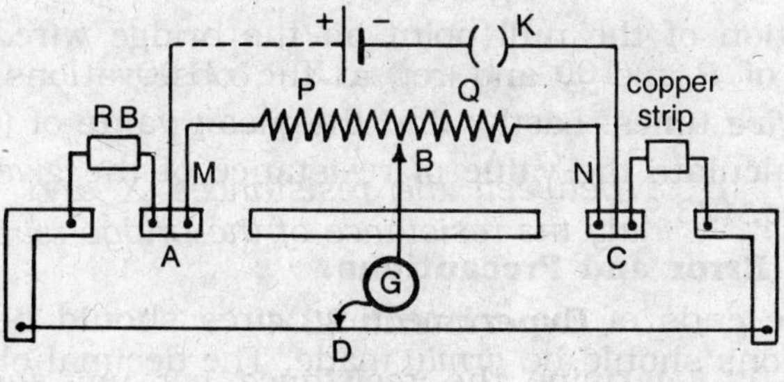

Fig. 23.27

the jockey. Finally, connect the Leclanche cell between *A* and *C* including a plug key *K* in the circuit.

Now, adjust the *sliding contact B* of the rheostat *MN* in the *middle* so that the resistances *P* and *Q* introduced with its help in the *inner* gaps of the bridge may be *nearly equal.* Next adjust the decimal ohm box to a value *R slightly* less than the resistance of the bridge wire, say, about 2 ohm. Then, shunt the galvanometer and close the key *K*. Shift the jockey towards the left end of the bridge wire and find by trial the null point on the bridge wire, the exact position of the null point being determined with *full* galvanometer sensitivity by removing the shunt. Note down the reading l_1 on the scale of the position of the null point of the bridge wire.

Then, interchange the two resistances in the *outer* gaps of the bridge and determine the value of l_2 as for l_1 above. This time, the balance point will lie near the right end of the bridge wire. Alter the values of *P* and *Q slightly* by moving the sliding contact *B* of the rheostat *MN* and repeat the observations and determine the mean value of $(l_2 - l_1)$. Next, alter the value of *R* and as above, take at least *five* sets of observations for $(l_2 - l_1)$ with different values of *R*, say, 1, 1.2, 1.4, 1.6, 1.8 ohm, Calculate the value of σ for *each* set of observations *separately* from Eq. (23.30) and then find the *mean* value of σ.

(*b*) **Comparison of resistance**—Next, connect in place of the copper strip, the one ohm coil whose resistance *X* is to be compared with a standard one ohm resistance. Adjust the decimal-ohm box to $Y = 1$ ohm. Shift the jockey to the middle of the bridge wire and find by trial the exact position of the null point l'_1 on the bridge wire, the final adjustment being done without any shunt across the galvanometer.

Next, interchange the resistance in the *outer* gaps of the bridge and shifting the jockey to the other side of the middle of the bridge wire determine as before the reading l'_2 on the scale of the position of the null point on the bridge wire. Then alter the values of *P* and *Q* and repeat the observations for l'_1 and l'_2 at least *five* times. Lastly, find the mean value of $(l'_2 - l'_1)$ and use it to calculate the value of resistance of the given one ohm coil from Eq. (23.31).

Sources of Error and Precautions

(1) The ends of the connecting wires should be *clean* and all connections should be *firmly* made. The decimal-ohm box and the given one-ohm resistance coil should be connected by *thick* copper strips.

(2) A rheostat should be used to introduce the resistances P and Q in the *inner* gaps of the bridge and the *sliding* contact should be adjusted to be approximately in the *middle*. It is not absolutely necessary that P and Q should be exactly equal except for high sensitiveness of the bridge, nor should their values be known. If $P = Q$, the positions of null point before and after interchanging the resistances in the outer gaps will be at equal distances from the middle point of the bridge wire, provided, of course, the wire is uniform. If P and Q differ very much it will not be possible to obtain the two positions of the null point on the bridge wire.

(3) The use of rheostat to introduce P and Q in the inner gaps possesses several advantages. Besides being cheap, it is flexible, for it can be used to obtain the null point in any part of the bridge wire and also enables us to take several sets of readings for $(l_2 - l_1)$ for the same values of X and Y. With fixed values for P and Q this could not have been possible.

(4) In order that the bridge may have high sensitiveness, the resistances of the four arms should be of the same order.

(5) In order to reduce the inaccuracy in the result due to a small error in reading the position of the null point to minimum, the null points while comparing X and Y should lie as near the middle of the bridge wire as possible.

(6) While determining the value of σ the value of of R should be comparable with the resistance of the bridge wire so that the two positions of the null point before and after interchanging the resistances in the outer gaps lie near the ends of the bridge wire. The value of $(l_2 - l_1)$ will then be almost equal to the entire length of the bridge wire and the error in the value of σ due to non-uniformity of the bridge wire will be reduced to minimum.

(7) A plug key should be included in the cell circuit and should be closed when observations are being made.

(8) The galvanometer should be shunted by a low resistance wire to avoid excessive deflection in it when the bridge is out of balance. The exact position of the null point should be determined with full galvanometer sensitivity by removing the shunt wire from it.

(9) The cell circuit should be closed before depressing the jockey over the bridge wire, but when breaking, reverse order should be followed.

(10) The jockey should always be pressed gently and the contact between the jockey and the bridge wire should not be made while the jockey is being moved along.

Observations

(*A*) **Determination of σ**

S. No.	R ohm	Position of balance point with copper strip in the		$(l_2 - l_1)$ cm.	Mean $(l_2 - l_1)$ cm.	σ $\Omega\ cm^{-1}$
		Right gap l_1 cm.	*Left* gap l_2 cm.			
1						
2						
					Mean	

(*B*) **Comparison of given one-ohm coil with standard one ohm**

S. No.	Position of balance point with the one-ohm coil *X* in the		$(l'_2 - l'_1)$ cm.	Resistance *X* of given one-ohm coil ohm
	Left gap l'_1 cm.	Right gap l'_2 cm.		
1				
2				
3				
			Mean	

Calculations

$$\sigma = \frac{R}{l_2 - l_1}$$

$$= \text{ohm/cm.}$$

$$X = \sigma\,(l'_2 - l'_1) \qquad Y = 1 \text{ ohm}$$

$$= \quad \text{ohm}$$

Result—The resistance of the given one-ohm coil = ohm.

Percentage error = %.

Criticism of the method—The method is very suitable for the comparison of two *nearly equal* resistances and yields quite accurate results. In the derivation of the expression X – Y = σ $(l_2 - l_1)$ the *end corrections* have cancelled out and hence such unknown factors have no influence on the result. The sensitiveness of the bridge and the accuracy in the result have been increased considerably by virtually increasing the length of bridge wire.

It is assumed that the bridge wire is uniform in cross-section. This may, however, not be the case in practice. Hence, in all exact work the wire should be first calibrated and the value of resistance of the bridge wire between the two positions of the null point before and after interchanging X and Y obtained from the calibration curve. The result is also affected by heating of resistors and the bridge wire by the current and thermo-electric effects.

From the Eq. (23.31) it is evident that the difference between the two resistances, to be compared is equal to the resistance of the bridge wire between the two positions of the null point. From this it follows that the method fails if this difference is greater than the total resistance of the bridge wire for then the two positions of the null point cannot be obtained. Note that in this method of comparison of resistances, the two resistances are never compared directly.

23.20. Calibration of Carey Foster's Bridge Wire

A length of manganin or constantan wire is taken and its resistance measured with a Carey Foster's bridge. Then, a length of it 2 cms. in excess of bridge wire is cut off. If a wire of the same material and cross-section as the bridge wire is available, a piece of it 7 cms. in length can be cut off without measuring its resistance. This is then soldered to two stout copper connecting strips as shown in Fig. 23.28, the two cm. excess of the wire being soldered to the strip. Then, the resistance X of the wire between the strips will be nearly equal

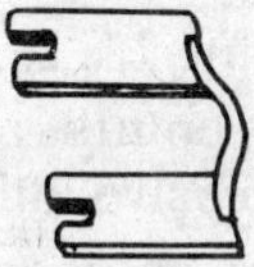

Fig. 23.28

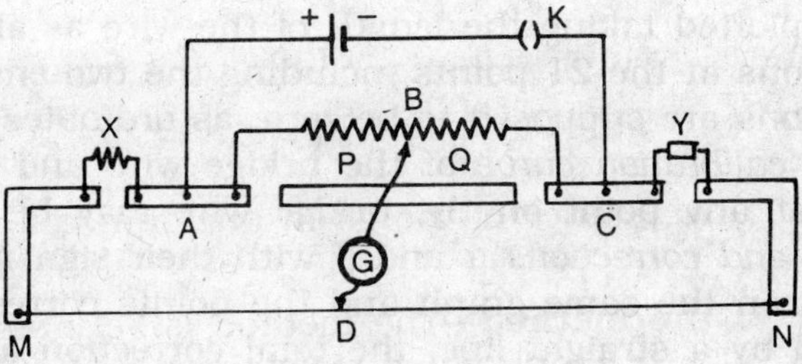

Fig. 23.29

to that of the bridge wire. This resistance X is connected in the outer left gap of the Carey Foster's bridge (Fig. 23.29) and a *thick* copper strip Y of practically zero resistance is connected in the outer right gap of the bridge. Then, the two fixed terminals at the base of a stable rheostat are connected to the points A and C of the bridge, its *sliding* contact B being contacted through a galvanometer to the jockey. The cell is connected in its usual place between A and C.

The cell circuit is closed and the rheostat adjusted so that the null point lies as close to the left end M of the bridge wire as possible and the reading l_1 on the scale of its position of the bridge wire noted. X and Y are then interchanged and the reading l_2 on the scale of the position of the new null point determined. Then, $(l_2 - l_1)$ is the first section of the wire whose resistance is equal to X. Next, keeping the jockey *undisturbed* X and Y are placed in their first position and the rheostat adjusted again until the null point lies at the point of contact of jockey on the bridge wire, *i.e.*, at l_2. Then X and Y are again interchanged and the process repeated until the right end N of the bridge wire is reached, noting down the readings on the scale of the positions of the successive null points on the bridge wire.

Then, successive differences x_1, x_2, x_3, ...x_{20} between the reading l_1 , l_2 , l_3 etc. on the scale of the position of the successive null points are found out. Let their *mean* value be x. Each successive difference x_1 , x_2, etc., is then subtracted from this mean value x which gives the error over each section. To find out the correction which is to be applied at a point, say, P distant l, from the left end of the bridge wire, these errors are added up algebraically from the left end of the bridge wire to the point P. Next a graph is plotted taking the length of the wire as abscissa and the corrections at the 21 points including the two ends at which the corrections are supposed to be zero, as ordinates. The graph will be the *calibration curve* of the bridge wire and from it the correction at any point on the bridge wire may be found out. Now, if the *end corrections* α and β with their sign *reversed* are also plotted on the same graph and the points corresponding to them joined by a straight line, the total correction at any point of the bridge wire can then be read off on the curve with reference to this line.

23.21. Effect of Temperature and other Factors on Resistance

The resistance of a conductor also depends upon the purity and density of its material, drawing and annealing of the material and its temperature. The presence of even slight trace of another metal has an enormous effect on the resistance, and hence copper used for electrical purposes must be exceptionally pure. The *resistivity of an alloy is much greater than that of any of its constitutent metals.* This is a characteristic property of alloys and is used in the preparation of wires of high resistivity. Annealing diminishes the resistance of metals but gives it permanence, and drawing increases the resistance.

The resistance of metals is also affected by magnetic fields but the effect in most cases is very small. In the case of bismuth the resistance can be doubled if a strong magnetic field is applied perpendicular to the direction of the current. This remarkable property of bismuth can be used to measure strengths of magnetic fields. The resistance of solid crystals, *e.g.*, of selenium, tellurium and carbon are affected by light or X-rays. On account of this property of selenium, its cells are used in talking films and telephony by means of light *beam.* In the case of the single crystals not belonging to the regular system, *e.g.*, tin (tetragonal) the resistivity changes with direction also.

The resistivity and hence the resistance varies with temperature. For pure metals the resistance increases with rise of temperature. If R be the resistance of a metallic conductor at t°C and R_0 that at 0°C, then $R = R_0\ (1 + \alpha t)$, where α is called the *temperature coefficient of resistance* for the material of the conductor and is numerically equal to the increase in unit resistance at 0°C for 1°C rise in temperature. Here, the original resistance has been taken at 0°C, but in the general case it can be at any temperature and then *the temperature coefficient is numerically equal to the fractional increase in resistance per °C rise in temperature.* The value of α for common metals (except mercury) is about 1/273 or 3.76×10^{-3} per °C, *i.e.*, the same as the volume or pressure co-efficient for gases, and hence *the resistivity for pure metals is almost directly proportional to their absolute temperature.* The value of α for iron is very high and increases suddenly at red heat to an abnormally high value. On account of this property of iron, it is used for automatic regulation of current in Nernst lamps. The presence of an impurity decreases the value of α but the product of temperature coefficient and the specific resistance at any temperature is the same for the pure and the impure metal.

The simple expression $R = R_0 (1 + \alpha t)$ for variation of resistance with temperature in the case of pure metals is true for *small* ranges of temperature only, for the graph between R and t is not really a straight line but a *parabolic* curve. For larger ranges of temperature or for an accurate work, the variation of R with t should, therefore, be represented by more exact expression $R = R_0 (1 + \alpha t + \beta t^2)$, where β is a very small additional constant depending upon the material of the conductor. The variation of resistance with temperature is used to measure temperature. The metal usually employed is platinum, *e.g.*, in platinum resistance thermometer and bolometer. For lower temperature lead and gold are also employed.

As the resistivity of pure metals is almost proportional to their absolute temperature, it was thought that the resistance of pure metals would vanish at absolute zero (–273°*C*) but the researches of Kamerlingh Onnes, Dewar and others have shown that the above rule changes rapidly near absolute zero for the resistance of many pure metals drops rather suddenly near absolute zero to a value too small to be measurable. *The state at which the metal offers no measurable resistance to the passage of current is called superconducting state.*

The resistance of alloys also increases with temperature but the temperature coefficient is much smaller in alloys than in pure metals, especially in alloys containing nickel and manganese in certain proportion, *e.g.*, constantan, manganin, etc. The temperature coefficient of manganin becomes *negative* temperature increases, the point at which the change occurs depend as upon the iron content. On account of *negligible temperature coefficient and high resistivity*, manganin and constantan are widely used in the construction of standard resistances.

The resistance of non-metallic conductors, *e.g.*, carbon, silicon, tellurium, etc., electrolytes, and insulators like glass, marble, slate, etc. which become partial conductors at high temperatures, *decreases* with rise of temperature, *i.e*, they have a *negative* temperature coefficient. The resistance of a carbon filament lamp at room temperature is about 1 to 2.4 times its resistance when heated under full voltage. The resistance of electrolytes decreases by about 2.5 per cent per degree near room temperature. The non-metallic element boron has an abnormally high negative temperature coefficient. Very pure graphite, however, has a *positive* temperature coefficient.

23.23. Standard Resistances, Resistance Boxes and Rheostats

For many practical purposes, *e.g.*, measurement or comparison of resistances, standard resistances are required. The mercury standard described in Section 23.2 is not convenient for general use in the laboratory and hence standard resistances consisting of wire of suitable materials are constructed. The most important properties of such standard resistance should be :

(a) permanence, *i.e.*, no variation of the resistance value of the finished standard with time.

(b) robust construction and convenient size,

(c) negligible variation with temperature,

(d) very small thermo-electric effect.

(e) very low inductance and self-capacitance, and

(f) capacity for carrying appreciable current without over-heating.

A consideration of these requisites of standard resistances shows that the material to be used in their construction would be of *high resistivity* so that a reasonably compact standard of any given value with a convenient size can be made up without using either an abnormally great length or dangerously thin section. The *temperature coefficient of resistance* for the material should be *practically nil* and it should have a very small thermo-electric effect with copper. In addition, it should not easily oxidise, should be unaffected by moisture, acids, etc. and should be easily worked and jointed.

These requirements are best satisfied by alloys manganin, therlo, constantan, eureka, etc. The alloy manganin of the composition 84% Cu, 12% Mn. 3.5% *Ni* and 0.5% Fe has an extremely low temperature coefficient 0.000004 per °C at 20 °C, a very small thermo-electric E.M.F. against copper 3 to 8 micro-volts per °C, and a high resistivity 50 micro ohms× cm. at 20°C and is, therefore, used in high-grade standard resistances.

Properties of Resistance Materials

Material	Resistivity ρ micro-ohms × cm.	Temp. coefficient α per °C	Thermo-electric E.M.F. against copper micro-volts	Remarks
METALS				
Copper ...	1.7	0.0040		

Iron ...	9—15	0.0045 to 0.006		
Mercury	95	0.00092		
Platinum (pure) ...	10.8	0.0039		
Steel	15—50	0.0052 to 0.006		
ALLOYS				
Brass	7–9	0.0015		
German Silver (60% Cu, 21% Ni, 19% Zn)	15–40	0.00022 to 0.0007	35	Zn produces unstable properties.
Nickelin (58% Cu, 41% Ni, 1% Mn)	42	0.00023		
Constantan (35-55 Ni, 65–45% Cu 0–20% Zn)	49–52	0.00001	40	Cheap, high thermo–electric E.M.F.
Eureka (60% Cu, 40%)	50	"	40	
Platinum silver (1 part platinum, 2 part silver)	31.6	0.0003	Small	High temp. coefficient.
with addition of about 1% tungsten	34–40	0.0003	20	High temp. coefficient. Tungsten improves permanence
Manganin (70% Cu, 30 % Mn)	42	0 to 0.00003	very low	High resistivity and very low temp. coefficient.
Therlo (71% Cu, 16.5% Al, 10.5%Mn, 2% F)	47	0.00005	Very low	Properties similar to manganin.
Nichrome ...	93	0.0004		Very high resistivity

Various forms of standard resistances have been designed. In one form a coil of manganin wire is wound *non-inductively* on a metal bobbin. In non-inductive winding the wire is first doubled on itself and then wound (Fig. 23.30) which gives the effect of two coils, side by side, carrying currents in *opposite* directions so that the magnetic flux in one coil neutralise the magnetic flux through the other. The *residual* flux is, therefore, *zero* or so small that the self inductance may be neglected. In this type of winding the *self-capacitance* is also *small*. This wire is always silk covered in order to save space and is insulated from the metal bobbin by a double layer of shellaced silk which is baked before the wire is wound on to remove moisture. The wire is laid in a single layer so that the heat produced by the

Fig. 23.30

current in it may be removed as efficiently as possible. After winding, the coil is shellaced and baked at 140°C. this dries the coil as well as anneals the wire. Annealing of the material is also done during its manufacture. Annealing removes bending strains in the wire and hence ensures greater permanence of the resistance of the coil. The bobbin and the coil are enclosed in a cylindrical case with an ebonite cover, the space between the coil and the wall of the cylinder being filled with paraffin wax to prevent the absorption of moisture. The leads from the coil are two copper rods which are silver-soldered to the coil and serve as the terminals. Their ends are amalgamated, and when electric connections are made, they are dipped in mercury cups to establish good contact.

In the construction of low resistances (Fig. 23.31) *strips* of suitable alloys are used instead of coils of wires, and since they are often required to carry large currents as is the case with those used for potentiometer work, adequate cooling must be provided which is done by immersing the resistance in first grade paraffin oil. This oil should be free from acid and water in order to avoid corrosion of the resistance alloy. The oil should be stirred by a motor-driven stirrer and cooled by some water-cooling arrangement. These low resistance standards used with large currents are provided with potential terminals as well as current terminals. The nominal value of the standard is measured between the potential terminals and the standard is connected to a circuit by the current terminals.

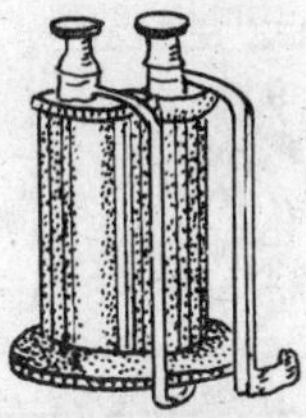

Fig. 23.31

For varying the resistance of a circuit in *steps* of known amount, resistance boxes are used. In the plug type arrangement resistance coils of various values are arranged in series like the weight in a weight box by means of which any multiple of the unit of the box up to the sum of all the coils can be obtained by removing the appropriate plugs whereby the corresponding coils will be put into the circuit. The values of coils are such as to give a particular resistance with the smallest number of coils. The series commonly adopted is 1, 2, 2, 5, 10, 20, 20, 50, up to 500 or 5,000. There is no coil

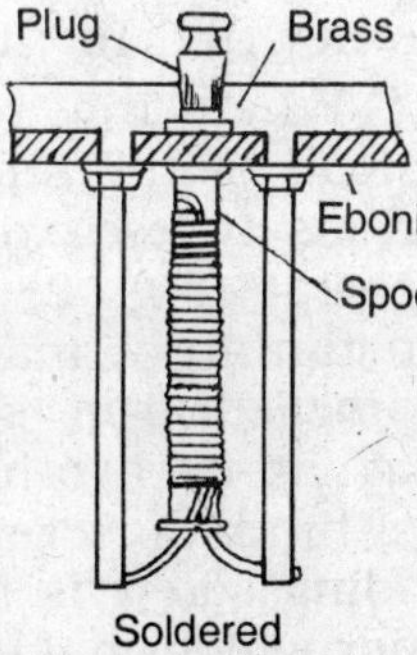

Fig. 23.32

between blocks separated by the gap marked 'infinity' so that by withdrawing the 'infinity plug' the circuit which includes the box is broken. Fig. 23.32 shows the method of mounting coils in such boxes. As the space available for each coil is the same, finer wires are used for large resistances. For accurate work it is necessary that the *contact resistances* in these boxes should be *small* and constant in value. This is achieved by making the plugs *tapering*, *i.e.*, conical in shape, and pressing them firmly in the sockets before any measurements are made. Further, they should always be inserted (or withdrawn) with a screwing motion which ensures good contact and prevents the plug from fastening tight in the socket.

In the dial type arrangement there is a group of ten 1-ohm coils in series, a group of ten 10-ohm coils in series and so on. The coils in a group are arranged in a circle (Fig. 23.33) and the junctions between the *consecutive* coils are connected to brass studs fixed in circle over the ebonite cover of the box. By rotating a laminated copper brush or arm round each dial and making contact with the appropriate stud, any required number of units, tens and hundreds can be obtained. This pattern of resistance boxes has almost *negligible contact resistances* and hence is preferable to the plug type arrangement in which the contact resistances are appreciable and depend upon the pressure with which the plugs are inserted in the sockets.

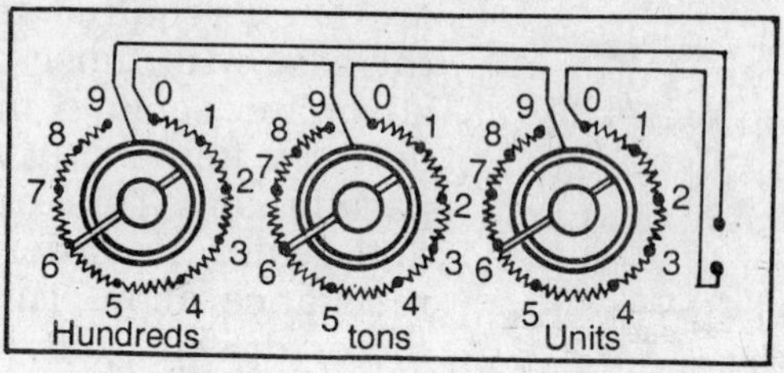

Fig. 23.33

When the resistance of a circuit is varied *continuously* instead of the steps, a rheostat is used. These rheostats are of various types, *e.g.*, rotatory rheostat, etc. In one form of sliding rheostat (Fig. 23.34) a wire of the suitable material such as constantan or manganin, etc., is wound on a cylinder of slate or enamelled iron so that its turns are close together but *not touching* each other. The two ends of the wire are connected to two binding screws at the two ends of the cylinder. A third binding screw is fixed to metal rod along which slides the metal slider which makes contact with the turns of wire. By connecting the binding screw fixed to the metal rod and one of the screws at the base and then sliding the metal slider, any desired part of the total resistance of the rheostat can be introduced in the

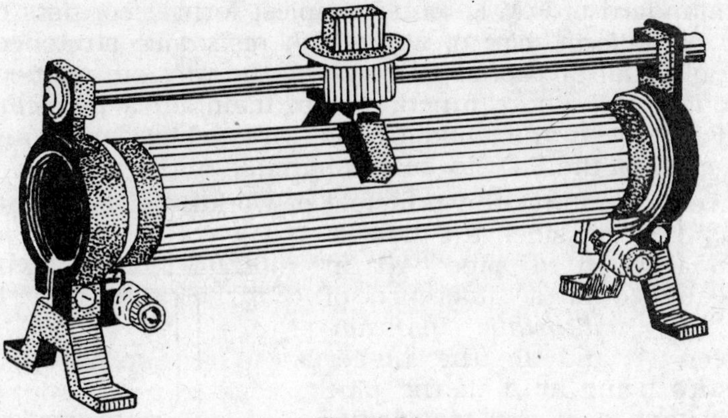

Fig. 23.34

circuit. For introducing the *whole* of the resistance, the two binding screws *at the base* may be used. The rheostat can also be used as a *potential divider*. For this the current is passed through the whole rheostat by using the screws at the base when there is a fall of potential from one end of the rheostat to the other, from which a portion may be tapped off by using one of the terminals at the base and the screw attached to the rod along which the metal slider moves.

In using the resistance boxes and the rheostats, care should be taken never to pass currents of values greater than the maximum which the coils in the boxes or rheostats can safely carry otherwise the coils or the wire in the rheostat will be over-heated and burnt.

Bulb or Lamp Resistances—When the current is taken directly from electric mains, the strength of the current in the circuit is regulated by lamp resistances. These lamp resistances consist of several lamps connected *in parallel* (Fig. 23.35) the combination being inserted in series with the circuit.

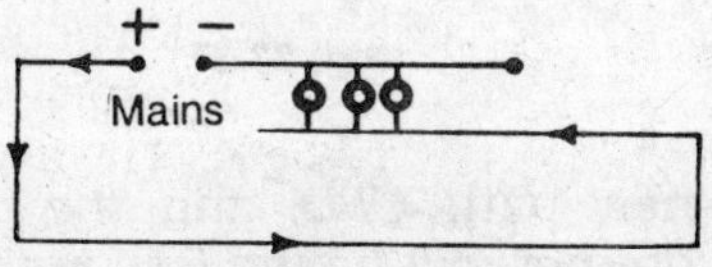

Fig. 23.35

The lamp resistances are very cheap and convenient.

23.24. Potentiometer

A potentiometer is a device for comparison or measurement of E.M.F. or P.D. by balancing the unknown against a variable potential difference the value of which may be known in terms of a standard of E.M.F. In its simplest form it consists of a long piece of *uniform wire* of fairly *high resistance* stretched over a scale of equal divisions. The ends of the wire are connected to a secondary cell whose function is to maintain a *perfectly steady* E.M.F. between the ends of the wire, which must always be greater than the P.D. to be measured.

To understand the principle of working of a potentiometer, let us first consider the flow of water through a pipe *AB* in consequence of a difference of level or *hydrostatic pressure* between *A* and *B*, the higher pressure being at *A*. If the pipe *AB* be tapped at any two points between *A* and *B*, say, at *C* and *D* (Fig. 23.36) and a *branch pipe CPD* attached, the current of water flowing through the pipe *AC* will now divide at *C* into two parts both flowing towards *D* one through the *main pipe CD* and the other through the *branch pipe CPD and the greater the distance between C and D greater will be the pressure difference tending to urge the water through the branch pipe.*

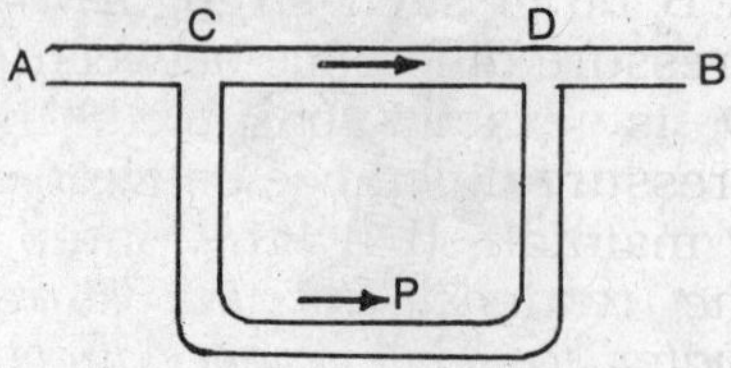

Fig. 23.36

Now, let us consider the flow of a current of electricity along a conductor *AB* in consequence of a *P.D.* between *A* and *B*, the potential at *A* being *higher*. Then, if at any two points *C* and *D* between the ends of the conductor, a branch conductor *CPD* is connected to the conductor *AB* (Fig. 23.37), the current flowing along *AC* will now divide at *C* into two portions both flowing towards *D*, one along *CD* and the other along the new path *CPD*, and *the greater the distance between C and D greater will be the P.D. tending to urge the current along the branch conductor.*

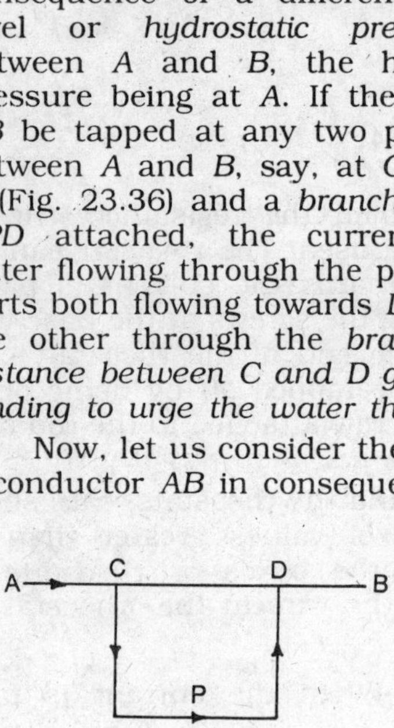

Fig. 23.37

Next referring again to Fig. 23.36, let us suppose that something of the nature of a rotatory or force pump be inserted in the branch *CPD* (Fig. 23.38) and let it be driven at constant speed by some power and thus maintain a steady pressure difference between N and *P*. The pump may evidently be driven in either direction and will assist or oppose the flow of water through the branch pipe in the direction *CPD* according as the pressure at *N* is maintained lower or higher than that at *P*. Now, let the pressure at *N* be higher than that at *P*. Then, as the pump now opposes the flow of water through the branch pipe in the direction *CPD*, the current of water in the branch pipe will flow in the direction *CPD* or in the opposite direction according as the pressure difference between *N* and *P* is smaller or greater than that between *C* and *D*. If the current of water through the pipe A B be of such strength that the pressure difference between *C* and *D* is exactly balanced by the pressure difference between *N* and *P* maintained by the pump, then *the tendency for the current of water to flow in one direction is neutralised by the tendency to flow in the opposite direction and hence there will then be no flow of water through the branch pipe*. This absence of current of water in the branch pipe can be shown by some sort of flow indicating instrument inserted in the branch *CPD*.

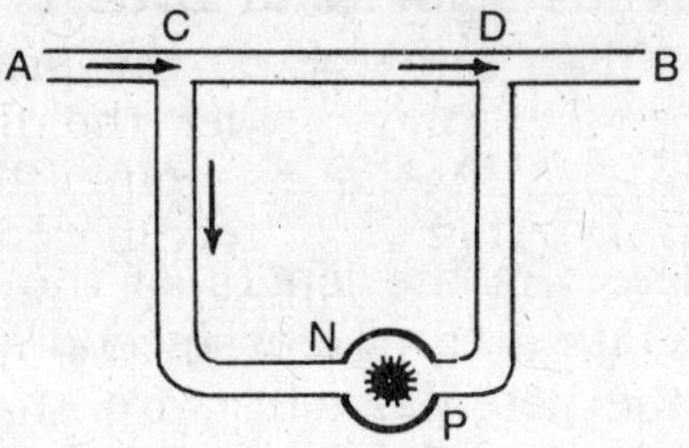

Fig. 23.38

Now, referring to Fig. 23.37, let us suppose that a cell of constant E.M.F. be inserted in the branch *CPD* (Fig. 23.39) and let *N* be the *positive* pole of the cell. Then, since the potential at *N* is higher than that at *P*, the current flowing through the branch conductor in consequence of the P.D. between *C* and *D* will be opposed by the P.D. between *N* and *P*, and evidently the current in the branch conductor will flow in the direction *CPD* or in the opposite direction according as the P.D. between *N* and *P* is smaller or greater than that between *C* and *D*. If there be no flow of current through the branch conductor *CPD*, *the tendency for the current to flow in the direction CPD is neutralised by the tendency to flow in the opposite direction* and hence in that case, from the analogy of flow of water through a pipe as given above, we conclude that the *potential difference between C and D tending to urge the current in direction CPD is exactly balanced by the potential difference between N and P tending to urge the current in the opposite direction*. The absence of current in the branch conductor can be shown by a galvanometer inserted in the branch *CPD*.

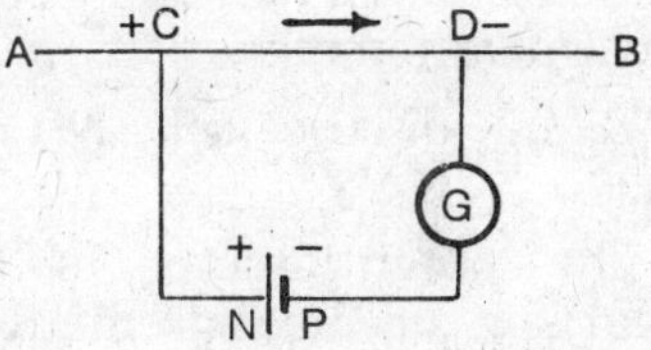

Fig. 23.39

Now, let r be the resistance of the portion of the potentiometer wire between C and D and let i be the *steady* current flowing through it. Then, the potential difference between C and D is equal to ir. When there is no deflection in the galvanometer, *the potential difference between C and D is exactly balanced by the potential difference between N and P. Hence*, P.D. between N and P is given by :

$$V = ir$$

But $$r = \rho \frac{l}{S}$$

where l is the length of the portion of the wire between C and D, S its area of cross-section and ρ the *resistivity* of the material of the potentiometer wire.

Thus, $$V = \frac{i \rho l}{S}.$$

If the wire be of *uniform* cross-section throughout and the current in the wire is of constant strength, $i \rho / S$ will be constant. Since ρ / S is the resistance of unit length of the wire, $i\rho/S$ will be the fall of potential per unit length of the wire, *i.e.*, *the potential gradient* of the potentiometer wire. Hence, if K be the *potential gradient* of the wire,

$$V = i\rho l/S = Kl$$

When there is no current in the branch CPD, the cell is on *open circuit* so that the potential difference between N and P equals the E.M.F. of the cell. Thus, when there is no deflection in the galvanometer.

E.M.F. of the cell $NP = Kl$ = potential gradient × length of CD.

The potentiometer is the *most accurate* device for the comparison or measurement of E.M.F. or P.D. The accuracy of the instrument depends upon, (i) the *constancy* of the difference of potentials between its ends and (*ii*) the *uniformity* of cross-section of the wire. The accuracy in obtaining balance evidently depends upon the *sensitiveness* of the galvanometer inserted in the branch CPD. For a given potential difference between the ends of the potentiometer wire, the *sensitiveness* of the potentiometer increases with the length of its wire, for the greater the length of the wire, smaller will be the potential gradient and hence greater will be the sensitiveness.

But it is very inconvenient to use a very long wire. Hence, instead of one wire several parallel wires, connected in series are often used, the first and second wires being connected by copper strips fixed at one end of the board and the second and third similarly connected at the other end, and so on. But the use of many wires involves two difficulties, (*a*) the apparatus becomes cumbersome and (*b*) it is difficult to get a very long wire of absolutely *uniform* cross-section throughout. These difficulties are overcome by using a short wire connected in series with a number of resistance coils, the resistance of *each* coil being *equal* to that of the wire. A potentiometer of this type is shown in Fig. 23.40.

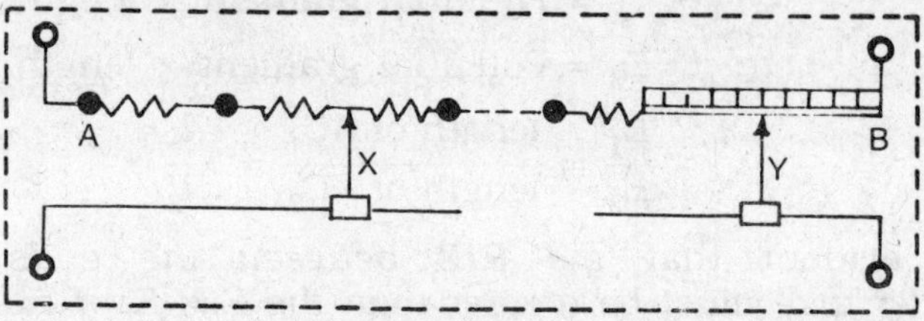

Fig. 23.40

It is very important that there should be no appreciable thermo-electric E.M.F. within the potentiometer itself. Accordingly *manganin which has a very low thermnio-electric E.M.F. with copper* is usually selected as the material for the construction of potentiometer coils and slide wire. *Manganin has also negligible temperature coefficient of resistance and high resistivity.* The contacts and joints in the potentiometer are often made of a special gold-silver alloy which is not affected by acids. to protect the contacts from atmosphere they are in some cases included within the case of the instrument. This also ensures a uniform temperature of all parts.

23.25. Comparison of E. M. Fs. of Two Cells

As shown in Fig. 23.41, the potentiometer wire is connected to a battery of constant E.M.F. in series with a suitable rheostat so that a *steady potential difference* is maintained between the ends of the wire. The + *ve* poles of the two cells whose E.M.Fs. E_1 and E_2 are to be compared are joined to the end A of the potentiometer wire which is at *higher* potential while their –*ve* poles are connected, through a two-way key, to a galvanometer the latter being connected to a jockey which can

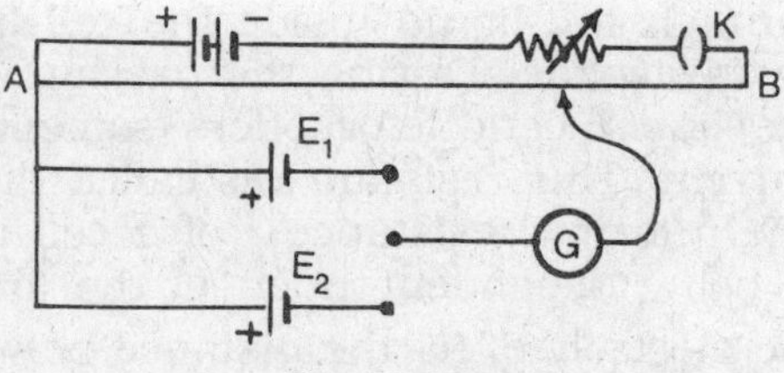

Fig. 23.41

be slid over the potentiometer wire. The two cells are connected in turn to the galvanometer by means of the two-way key and the point of contact of jockey with the potentiometer wire altered until the galvanometer shows no deflection. Let the points of contact be C_1 and C_2 for cells of E.M.Fs. E_1 and E_2 respectively when there is no deflection in the galvanometer. Then,

$$E_1 = \text{P.D. between } A \text{ and } C_1$$

$$= \text{Potential gradient} \times \text{length of } AC_1$$

Similarly, $$E_2 = \text{Potential gradient} \times \text{length of } AC_2$$

$$\therefore \quad \frac{E_1}{E_2} = \frac{\text{length of } AC_1}{\text{length of } AC_2} = \frac{l_1}{l_2}$$

It is evident that *the P.D. between the ends of the potentiometer wire must be greater than the E.M.F. of each of the two cells under examination.* The advantages of this method are :

(*a*) When there is no deflection in the galvanometer, the cells are on *open circuit* and hence we compare the true values for their E.M.Fs.

(*b*) Since the cells are not giving out current during the measurements except *very small* currents during adjustments, we get accurate results even with those cells which polarise readily, and

(*c*) It is a *null method, i.e.*, it depends upon reducing the deflection in the galvanometer to zero, a process which can be carried out with much greater exactness than the reading of a deflection.

23.26 Internal Resistance of a Cell

When a resistance is connected across a cell, the current which results circulates right round the circuit from the positive plate to the negative plate in the *external* circuit and back through the liquid inside the cell from the negative plate to the positive plate. During the passage of the current *inside* the cell the liquid conductor offers some resistance to the flow of the current. This resistance is called the *internal resistance* of a cell. The internal resistance r of a cell is made up of two parts, one r_1 being dependent upon (*i*) the area of the plates *immersed* in the electrolyte, (*ii*) the distance between them and (*iii*) the nature of the electrolyte and the other r_2 being dependent upon the strength of the current only. If S be the area of the plates immersed and l the distance between them, then $r_1 = l/kS$ where k is a constant which depends only on the nature of the electrolyte and is known as its *specific conductivity*; r_2 is associated with the passage of current out of the plates into the electrolyte and *varies directly as the strength of the current.* For *very small* current r_2 may be assumed to be *negligible.*

Thus, if R be resistance of the external circuit, the total resistance of the circuit is $(r + R)$ and hence the strength of the current in the circuit is given by :

$$i = \frac{E}{r + R}$$

where E is the E.M.F. of the cell.

Now, if V be the *terminal* P.D., *i.e.*, P.D. across the external resistance, R, we have from *Ohm's low:*

$$Vi = R$$

or
$$i = V/R$$

Equating the above two expressions for i, we have :

$$\frac{V}{R} = \frac{E}{r + R}$$

or
$$\frac{r + R}{R} = \frac{E}{V} \qquad (23.36)$$

whence,
$$r = \frac{E - V}{V} R \qquad (23.37)$$

This equation enables us to calculate *internal resistance* of the cell, if E and V are determined.

Experiment 23.6

Object—To determine the *internal resistance* of a *Leclanche cell by* means of a potentiometer.

Apparatus—A potentiometer, a storage battery, a rheostat, a resistance box, a high resistance, a Weston galvanometer, a plug key, a tapping key and connecting wires.

Theory—Let a *constant P.D.* be maintained between the two ends A and B of a potentiometer wire, the potential of A being *higher* than that of B. Let the +*ve* terminal of the cell whose internal resistance is to be determined be connected to A and the –*ve* terminal through a galvanometer to a jockey which slides over the potentiometer wire. Further, let l_1 be the length of the potentiometer wire for no deflection in the galvanometer. Then, since the cell is on *open circuit*, we have :

$$E = Kl_1 \qquad (23.38)$$

where E is the E.M.F. of the cell and K the *potential gradient* of the potentiometer wire.

Now, let the cell be *short-circuited* by a resistance R. Since the cell is now supplying current, the potential difference V between its terminals is less than the E.M.F. of the cell and hence the length of the potentiometer wire for balance will be *less* than l_1. Let l_2 be the length of the potentiometer wire in this case when the galvanometer shows no deflection. Then,

$$V = Kl_2 \qquad (23.39)$$

Now, if r be the internal resistance of the cell, we have, from Eq. (23.37) :

$$r = \left(\frac{E}{V} - 1\right) R$$

Substituting the values of E and V from Eqs. (23.38) and (23.39) in the above equation, we get :

$$r = \left(\frac{l_1}{l_2} - 1\right) R \qquad (23.40)$$

from which the internal resistance of the cell can be calculated if l_1 and l_2 are determined.

Method—Referring to Fig. 23.42, connect the two ends A and B of the potentiometer wire to a storage battery in series with a rheostat R_1 including a plug key K_1 in the circuit. Connect the $+ve$ pole of the Leclanche cell E whose internal resistance is to be determined to the *higher* potential end A of the potentiometer wire and connect its $-ve$ pole to the jockey of the potentiometer through a Weston galvanometer and a high resistance R_2 all in series. Finally, connect a resistance box through a tapping key across the poles of the Leclanche cell E.

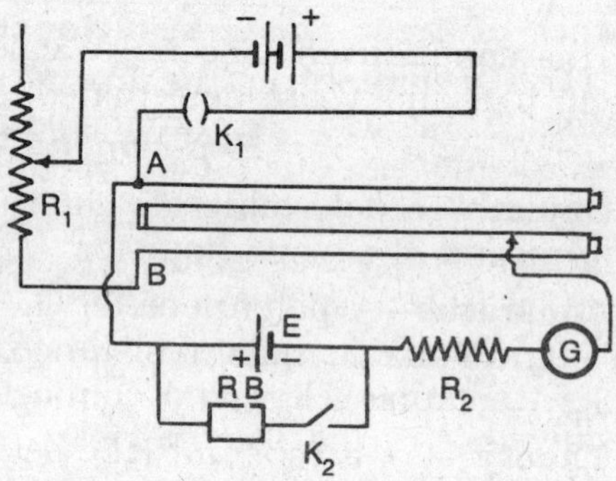

Fig. 23.42

Close the key K_1. Press the jockey over the first wire near the end A and then on the last wire near the end B and see that the deflections of the galvanometer are in *opposite* directions. If it is not so, then, if the connections are correct, the potential difference between the ends of the potentiometer wire is less than the E.M.F. of the cell E. In that case, reduce the resistance in the battery circuit by means of the rheostat R_1 and adjust its value until the null point is obtained roughly on the last wire. Now, remove the high resistance R_2 connected in series with the galvanometer and determine the *exact* position of the null point. Open K_1 and note down the length of the wire from the end A up to the position of the null point on the wire which gives the value of l_1.

Next, adjust the resistance box connected across the Leclanche cell E to a suitable resistance R, say, 1, 2, 5, 10, 20, 30, 50 or 100 ohms. Close K_1. First press K_2 and then the jockey over the potentiometer wire and determine as before the exact position of the new null point keeping the current in the battery circuit the same, open K_1 and note down the length of the wire from the end A to the position of the new null point on the wire which gives the value of l_2.

Repeat the experiment with different values of R *taking observations for open and closed circuits of the Leclanche cell alternately.* Then, slightly increase the resistance in the battery circuit and again repeat the experiment with the same values of R. Finally, calculate the value of internal resistance r of the Leclanche cell from Eq. (23.40) for each set of observations separately.

Sources of Error and Precautions

(1) The battery in the main circuit must be of practically *constant* E.M.F. and preferably of large capacity so that *the current in the potentiometer wire may remain constant throughout the test.*

(2) The magnitude of the P.D. between the ends of the potentiometer wire must be *greater* than the E.M.F. of the cell whose internal resistance is to be determined.

(3) A suitable rheostat should be connected in series with the battery in the main circuit *in order to increase the sensitiveness of the potentiometer* by decreasing the total fall of potential between the ends of the potentiometer wire, and should be adjusted so that the null point lies roughly on the last wire.

(4) A key should be included in the main circuit *which should be closed only when observations are to be made.*

(5) A high resistance, say, of 10000 ohms should be connected in series with the galvanometer and should be kept until the approximate position of the null point has been found. This will reduce the current in the galvanometer, thus preventing it from excessive deflections as well as *minimise polarisation in the cell* while the approximate position of the null point is being determined. The exact position of the null point should be

determined by finally removing this resistance which increases the sensitiveness of the test of null point by increasing the current in the galvanometer. Note that the presence of high resistance in the galvanometer circuit does not in any way alter the position of the null point.

(6) A tapping key should be connected in series with the resistance box connected across the cell E and should be pressed only when observations are to be made. This will avoid continuous flow of current from the cell into the box, thus minimising polarisation effects in the cell.

(7) The cell whose internal resistance is to be determined *should not be disturbed during the course of the experiment* as it may alter its internal resistance.

(8) The contact between the jockey and the potentiometer wire should be momentary and should not be made while the jockey is being moved along, otherwise the wire will be worn unevenly in various parts and no longer remain uniform in cross-section.

(9) If the balance point is obtained on the connecting strip between the wires, it should be shifted on to any of the two wires connected to it by slightly changing the resistance in the battery circuit. But note that the total resistance of the battery circuit for *same* set of observations of the P.D. between the terminals of the cell on open or closed circuit must remain the same.

Observations

S. No.	Length of the potentiometer wire with the cell on						R ohm	Internal resistance r ohm
	Open circuit			Closed circuit				
	No. of complete wires	Position of null point cm.	Total length of wire l_1 cm.	No. of complete wires	Position of null point cm.	Total length of wire l_2 cm.		
1.							1	
2.							2	
3.							5	
4.							10	
5.							20	
6.							30	
7.							50	
8.							100	

Calculations—Set I.

$$r = \left(\frac{l_1}{l_2} - 1\right)$$

$$= \quad \text{ohm.}$$

[Similarly, calculate r from other sets of observations]

Result—The resistance of the Leclanche cell *varies with the current drawn* from it and lies *between*...and...ohm.

Criticism of the method—When the Leclanche cell is short-circuited, it discharges at a considerable rate and rapidly becomes polarised. The result is that the internal resistance of the cell increases. The greater the current drawn from the cell, the greater is the polarisation in it and hence greater is the increase in its internal resistance. The value of internal resistance of the cell will, therefore, be different values of resistance short-circuiting the cell. If the time taken to determine the null point when the cell is short-circuited is large, *the null point will continuously shift on the slide wire due to its internal resistance changing on account of polarisation.*

The value of E.M.F. of the cell can, however, be determined quite accurately as the cell is then on open circuit and in this respect this method is superior to voltmeter method of measuring the E.M.F. of the cell. It is a null method and is, therefore, comparatively more accurate than the voltmeter method. But since the method possesses all the defects due to polarisation, it does not yield satisfactory results.

Experiment 23.7

Object—To calibrate a given voltmeter of 1-volt range by means of a potentiometer.

Apparatus—A potentiometer, two storage batteries, two rheostats, a standard cadmium cell, a Weston galvanometer, a voltmeter to be calibrated, a two-way key, two single-way keys connecting wires.

Theory—Referring to Fig. 23.43, let a *constant* P.D. be maintained between the two ends *A* and *B* of potentiometer wire, the potential of *A* being *higher* than that of *B*. Let the +*ve* pole of a cadmium cell *D* be connected to the end *A* and the −*ve* through a galvanometer to the jockey of the potentiometer. Then, if l_1 is the length of the potentiometer wire for no deflection in the galvanometer, we have :

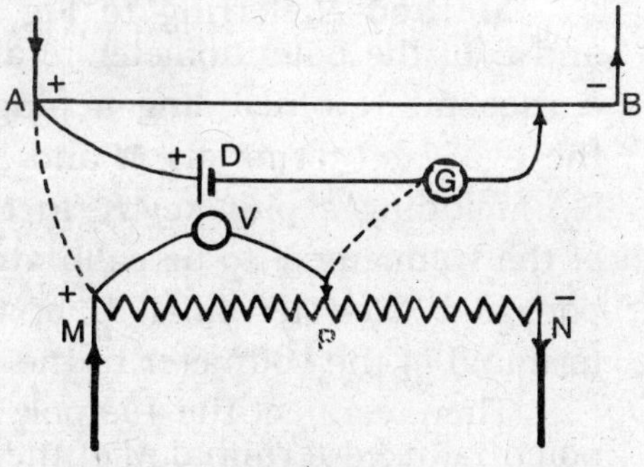

Fig. 23.43

$$E = Kl_1$$

where E is the E.M.F. of the cadmium cell and K is the *potential gradient* of the potentiometer. This gives the value of K.

Now, let the cadmium cell be replaced by, portion MP of a rheostat MN through whole of which a *steady* current is maintained, the *higher* potential point M of the rheostat being connected to the end A of the potentiometer. Then, if l_2 is the length of the potentiometer wire at the null point, the P.D. across the portion MP of the rheostat is given by :

$$V = Kl_2 \quad \ldots \quad (23.41)$$

If the P.D. across the *same* portion MP of the rheostat as measured by a voltmeter is V', the error in the reading V of the voltmeter, is $(V' - V)$.

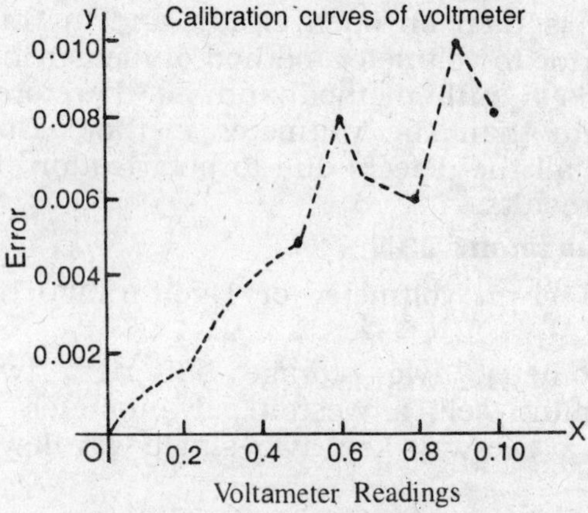

Fig. 23.44

Now if the P.D. between the points M and P of the rheostat is made *variable* by making one of the points, say, P variable, and the errors corresponding to the various readings of the voltmeter *thus* obtained be determined by the potentiometer, a *calibration curve* of the voltmeter may be plotted by taking V' as abscissae and $(V' - V)$ as ordinates.

Method—Referring to Fig. 23.45, connect the two ends A and B of the potentiometer to a storage battery S_1 in series with a rheostat R_1, including a plug key K_1 in the circuit. Connect the two *fixed* terminals M and N of a rheostat at R_2 to a battery S_2, including a plug key K_2 in the circuit. Connect one terminal of the voltmeter V to be calibrated through a key K_3 to the *higher* potential *fixed* terminal M of the rheostat R_2 and the second terminal of the voltmeter to the *sliding* terminal P of the rheostat.

Then, connect the *+ve* pole of a cadmium cell and the higher potential fixed terminal M of the rheostat R_2 to the sliding contact maker X of the potentiometer which is nearer to the higher potential end A of the potentiometer. Finally, connect the *–ve* pole of the cadmium cell and the *sliding* terminal P of the rheostat

R_2 through a two-way key and a galvanometer to the jockey Y which slides over the potentiometer wire.

Close the key K_1 and connect the *–ve* pole of a cadmium cell by means of the two-way key to the galvanometer. Place the slider X at the end of the 10th coil and the jockey at 18.3 div. on the slide-wire. Adjust the rheostat R_1 in the main *circuit* till there is no deflection in the galvanometer. *This gives a potential gradient of 1 millivolt/div. along the potentiometer slide wire.*

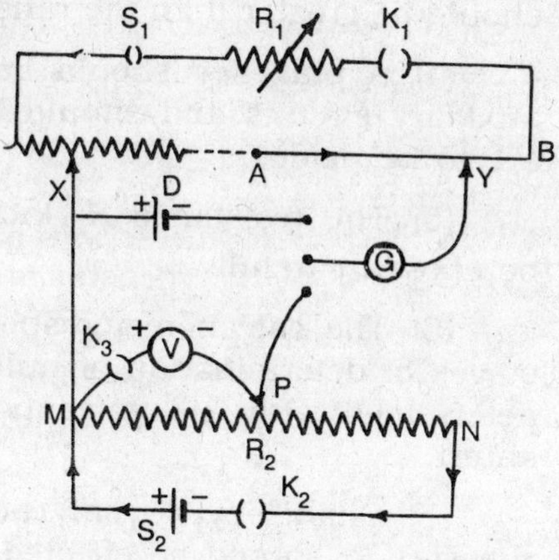

Fig. 23.45

Next, connect the sliding terminal P of the rheostat R_2 by means of the two-way key to the galvanometer and determine the *equivalent* length l_2 of the potentiometer wire corresponding to the P.D. across the portion M.P. of the rheostat R_2.

Note down the reading V' of the voltmeter, open the keys K_1 and K_2 and calculate the *true* value V of the P.D. corresponding to the reading V' of the voltmeter and then the error $(V' - V)$ in the voltmeter reading.

Next, alter the P.D. between M and P by moving the *sliding* terminal P of the rheostat R_2 and determine as before, the errors $(V' - V)$ corresponding to the various readings V' of the voltmeter thus obtained covering the *entire range* of the instrument. Finally, taking voltmeter readings as abscissae and corresponding errors as ordinates, plot the calibration curve of the voltmeter as depicted in Fig. 23.44.

Sources of Error and Precautions

(1) The main circuit battery S_1 must be of practically *constant* E.M.F. and preferably of large capacity so that *the current through the potentiometer coils and wire may remain constant throughout the expt.*

(2) The magnitude of the P.D. between A and B must be *greater* than the *maximum* P.D. to be measured by means of the potentiometer.

(3) The E.M.F. of the battery S_2 in the auxiliary circuit should be greater than the range of the voltmeter to be calibrated.

(4) A plug key should be included in both the main and auxiliary circuits and should be closed only when observations are to be made.

(5) The resistance R_1 included in the main circuit should be chosen carefully.

(6) The galvanometer should be shunted during the initial stages of determination of null point, the exact null point being determined with full galvanometer sensitivity by removing the shunt.

(7) Change-over from cadmium cell to the portion *MP* of rheostat R_2 should be effected by the use of a two-way key.

(8) The contact between the jockey and the potentiometer wire should be momentary and should not be made while the jockey is being moved along.

(9) A key should be connected in series with the voltmeter and should be kept open when P.D. across the portion *MP* of the rheostat is measured by the potentiometer. In such a case no current will be drawn by the voltmeter and the potentiometer will measure *accurately* the P.D. across *M.P.*

(10) The voltmeter should be calibrated over its *entire* range.

Observations

[A] *Standardisation of potentiometer wire.*

E.M.F. of standard cadmium cell = 1.0183 volts.

Length of potentiometer wire corresponding to E.M.F. of cadmium cell			
	No. of coils	Length of slide wire div.	*Equivalent* length l_1 div.
At start of expt.	10	18.3*	1018.3**
At end of expt.	10	18.3	1018.3**

*Length of slide wire = 100 divisions.

**These must be *equal.*

[B] *Calibration of voltmeter.*

Least count of voltmeter = volts.

S. No.	Length of potentiometer wire corresponding to P.D. across the portion MP of the rheostat R_2			Accurate value of P.D. V volts	Reading of voltmeter V' volts	*Error* in the reading of voltmeter $(V' - V)$ volts
	No. of coils	Length of slide wire div.	*Equivalent* length l_2 div.			
1.						
2.						
3.						
4.						
5.						
6.						
7.						
8.						
9.						
10.						

Calculations— $K = \dfrac{1.0183}{1018.3} = 10^{-3}$ Volts per div.

Reading No. 1 $\quad V = Kl_2 \quad =$ volts

(N.B. Make similar calculations for other readings).

Result—The graph so obtained by plotting the *errors* against the *voltmeter readings* is the *calibration curve* of the given voltmeter.

Criticism of the Method—The method gives fairly accurate results. The accuracy in the calibration of voltmeter depends upon the constancy of *potential gradient* of the potentiometer, accurate knowledge of E.M.F. of the cadmium cell and *uniformity* of the slide wire. With the use of a rheostat R_2 as described above, the current in the *auxiliary circuit* can be kept *very low* and still the *entire* range of the voltmeter can be calibrated. This avoids unnecessary heating of conductors in the auxiliary circuit as well as prevents waste of energy of the battery. But standardising the potentiometer to give a potential gradient of 1 milli-volt/div., the calculations are very much simplified.

23.27. Measurement of Current by Means of a Potentiometer

The potentiometer can be used to measure currents. In fact ammeters are accurately calibrated by means of a potentiometer. The current I to be measured passes through a standard resistance R usually of one ohm *already* included in the circuit in which the current I flows and the potential difference V across it is determined by means of a potentiometer accurately standardised in terms of the E.M.F. of a standard cell. Then, the current through the standard resistance is given by $I = V/R$.

Experiment 23.8

Object—To calibrate a given *ammeter* of 1 ampere range by means of a potentiometer.

Apparatus—A potentiometer, two storage batteries, two rheostats, a standard cadmium cell, a Weston galvanometer, standard one-ohm resistance, ammeter which is to be calibrated, a two-way key, two single way keys, shunt wire and connecting wires.

Theory—Let a steady P.D. be maintained between the two ends A and B of a potentiometer wire, the potential of A being *higher* than that of B. Let +ve pole of cadmium cell be connected to the end A and the –ve through a galvanometer to the jockey of the potentiometer. Then if l_1 be the length of the potentiometer wire for no deflection in the galvanometer, we have :

$$E = Kl_1$$

where E is the E.M.F. of cadmium cell supposed to be known and K is the *potential gradient* of the potentiometer wire. Thus, the value of K is known.

Now, let the cadmium cell be replaced by a *one-ohm standard* resistance through which a *steady* current I is flowing, its higher potential terminal being connected to the end A of the potentiometer wire, and let the current I *also* flow through the ammeter to be calibrated connected *in series* with the standard resistance. Then, if l_2 be the length of the potentiometer wire for no deflection in the galvanometer, the P.D. between the terminals of the standard one-ohm resistance is given by :

$$V = Kl_2$$

and the current through it by :

$$I = V/R = Kl_2 \qquad (23.42)$$

Let the reading of the ammeter be, I'. Then, the *error* in it is $(I' - I)$. If, now, the current through the standard resistance be varied and the values of $(I' - I)$ corresponding to different readings I' of the ammeter be determined, *calibration curve* of the ammeter may be drawn.

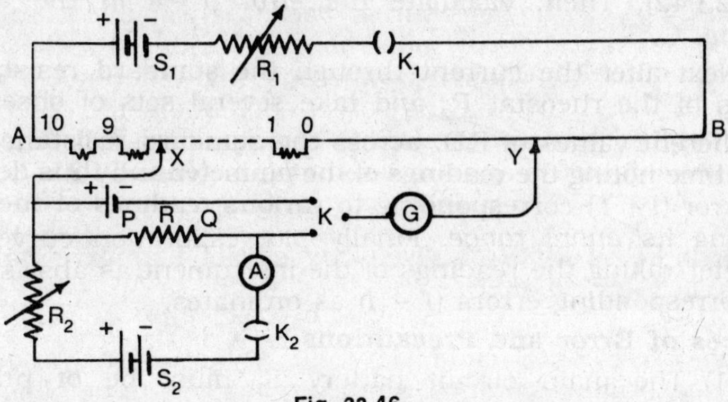

Fig. 23.46

Method—Referring to Fig. 23.46, connect the two ends *A* and *B* of the potentiometer to one of the storage batteries S_1 in series with a rheostat R_1, including a plug key K_1 in the circuit. Connect the *standard one-ohm resistance R* to the other storage battery S_2 in series with a rheostat R_2 and the ammeter which is to be calibrated, including a plug key K_2 in the circuit. Finally, connect the *+ve* pole of the cadmium cell and the *higher* potential terminal *P* of the standard one-ohm resistance together to the *sliding* contact maker *X* of the potentiometer which is nearer to the *higher* potential end *A* of the potentiometer and the *–ve* pole of cadmium cell and the lower potential terminal *Q* of the standard one-ohm through a two-way key *K* to one terminal of the galvanometer the second terminal of which is connected to the jockey *Y* which slides over the potentiometer wire.

First, *standardise* the potentiometer wire with the cadmium cell. For this connect *–ve* pole of cadmium cell by means of the two-way key *K* to the galvanometer, put the slider *X* at the end of the 10th coil and the jockey at 18.3 div. on the slide wire and adjust the rheostat R_1 in the *main circuit* till there is no deflection in the galvanometer. Now, the *potential gradient K* of the potentiometer equals 10^{-3} volts/div.

Next, close K_2 also. Connect the lower potential terminal *Q* of the standard one-ohm resistance by means of the two-way key *K* to the galvanometer and determine the equivalent length l_2 of the potentiometer wire corresponding to the potential difference *V* between the terminals of the standard resistance. Note down the reading of the ammeter. Open K_2 and K_1, and calculate the *true* value of current corresponding to reading I' of ammeter from Eq. (23.42). Then, calculate the *error* $(I' - I)$ in the ammeter reading.

Next alter the current through the standard resistance by means of the rheostat R_2 and take several sets of observations for different values of P.D. across the standard resistance at the same time noting the readings of the ammeter and thus determine the error $(I - I')$ corresponding to various readings of the ameter covering its *entire range*. Finally plot calibration curve of the ammeter taking the readings of the instrument as abscissae and the corresponding errors $(I' - I)$ as ordinates.

Sources of Error and Precautions

(1) The main circuit battery S_1 must be of practically *constant* E.M.F. and preferably of large capacity so that *the current in the potentiometer coils and wire may remain constant throughout the test.*

(2) The magnitude of the P.D. between A and B must be *greater* than the *maximum* P.D. to be measured by means of the potentiometer:

(3) The battery S_2 in the auxiliary circuit should be of constant E.M.F. and ample capacity. Its E.M.F. should be slightly *higher* than that necessary to produce *full-scale* deflection in the ammeter.

(4) The rheostat R included in the main circuit should be carefully chosen.

(5) A key should be included in both the main and the auxiliary circuits and should be closed only when observations are to be made.

(6) The galvanometer should be shunted. The shunt should be removed when the exact position of the null point is to be determined in order to increase the sensitiveness of the test of null point by increasing the current in the galvanometer.

(7) The calibration of the potentiometer wire with the cadmium cell and determination of different values of P.D. across the standard one-ohm resistance corresponding to various readings of the voltmeter should be made *in as short a time as possible* in order to minimise the chance of error due to variation of steady current in the potentiometer wire. Accordingly, it is convenient to use a two-way key so that the change from the cadmium cell to the standard resistance or *vice versa* may be effected quickly.

(8) The contact between the jockey Y and the potentiometer wire should be momentary and should not be made while the jockey is being moved along.

(9) the ammeter should be calibrated over its *entire* range.

Observations

[A] *Standardisation of potentiometer wire.*

E.M.F. of standard cadmium cell = 1.0183 Volts

Length of potentiometer wire corresponding to E.M.F. of the cadmium cell

	No. of coils	Length of slide wire div.	*Equivalent* length l_1 div.
At start of expt.	10	18.3*	1018.3**
At end of expt.	10	18.3	1018.3**

*Length of slide wire = 100 divisions.

**These must be equal

[B] *Calibration of ammeter*

Least count of ammeter = amp.

S. No.	Length of Potentiometer wire corresponding to P.D. (V) across the standard one-ohm resistance			*Accurate* value of		Reading of ammeter	*Error in* reading of ammeter
	No. of coils	Length of slide wire div.	*Equivalent* length l_2 div.	PD V (Volt)	Current I (Amp.)	I' amp.	$(I'-I)$ amp.
1 2 3 4 9 10							

Calculations—$K = \dfrac{1.0183}{1018.3} = 10^{-3}$ volts per div.

Reading No. 1. $V = Kl_2 =$ = volts.

and $I = V/R =$ = amp.

[*N.B.* Make similar calculations for other readings].

Result—The graph so obtained by plotting the *errors* in the *readings of ammeter* against the corresponding *ammeter readings* is the *calibration curve* of the given ammeter.

Criticism of the method—The accuracy in the calibration of ammeter depends upon the constancy of the *current* through the standard one-ohm resistance during the measurements, the constancy of the *potential gradient* of the potentiometer, the *uniformity* of the slide wire and the accurate knowledge of the value of standard resistance and the E.M.F. of the cadmium cell. The method yields sufficiently accurate results. For greater accuracy, a more sensitive potentiometer, *e.g., Crompton potentiometer* should be used and the ordinary one-ohm resistance should be replaced by a standard resistance provided with separate current and potential terminals.

The method can also be used to calibrate a *voltmeter*, if the P.D. across the one-ohm standard resistance is measured by both the voltmeter and the potentiometer.

23.28. Measurement of Resistance by a Potentiometer

The potentiometer is also used to determine the value of an unknown resistance. A *steady* current of *known* strength I is allowed to pass through the resistance R to be measured and the potential difference V across it is determined by means of an accurately calibrated potentiometer. then obviously $R = V/I$.

Experiment 23.9

Object—To compare two resistances by means of a potentiometer.

Apparatus—A potentiometer, two resistances to be compared, a *special key K* as shown in Fig. 23.48, a Weston galvanometer, a shunt wire, two storage batteries, two rheostats, two single-way plug keys and connecting wires.

Theory—Let a *steady* potential difference be maintained between the two ends A and B of a potentiometer wire, the potential of A being *higher* than that of B. Let a *steady* current of strength I be allowed to pass through the two resistances R_1 and R_2 to be compared connected *in series*. Further, let the higher potential terminal of one of the resistances, R_1, be connected to the end A of the potentiometer wire and the lower potential terminal through a galvanometer to the jockey of the potentiometer. Then, if l_1 be the length of the potentiometer wire for no deflection in the galvanometer, the potential difference between the terminals of R_1 is given by :

$$V_1 = IR_1 = Kl_1 \quad = \quad (23.43)$$

where K is the *potential gradient* of the potentiometer.

Now, let the experiment be repeated with the other resistance R_2, and if l_2 be the length of the potentiometer wire corresponding to the P.D. V_2 across its terminals when there is no deflection in the galvanometer. Then,

$$V_2 = IR_2 = Kl_2 \quad \ldots \quad (23.44)$$

Dividing Eq. (23.43) by Eq. (23.44), we get :

$$\frac{R_1}{R_2} = \frac{l_1}{l_2} \quad \ldots \quad (23.45)$$

from which the ratio of two resistances may be calculated. If a graph is plotted between l_1 and l_2, the *slope* of the straight line thus obtained will also give the ratio of the two resistances.

Method—Referring to fig. 23.47, connect the two ends, A and B, of the potentiometer to the storage battery B_1 in series with a rheostat R, including a key K_1 in the circuit. Then, connect the two resistances R_1 and R_2 to be compared *in series* and send through them a steady current from the storage battery B_2 including a rheostat R' and a key K_2 in the circuit. Next, connect the terminal e of the key K to *sliding* contact maker X which is nearer to the *higher* potential end A of the potentiometer and the

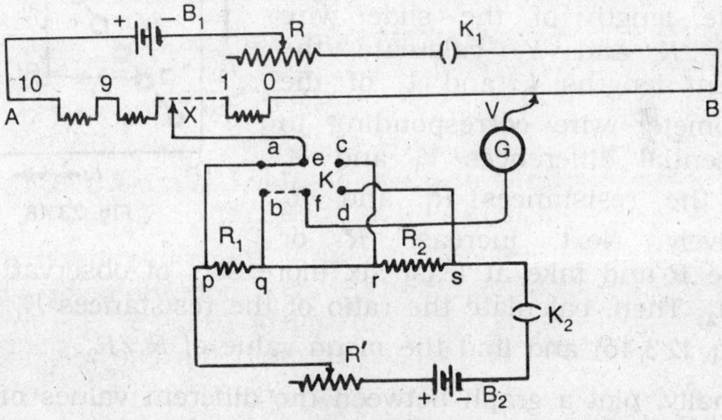

Fig. 23.47

terminal f through the galvanometer to the jockey Y of the potentiometer. Finally, connect the *higher* terminals p are r of the two resistances R_1 and R_2 to the terminals a and c of the key K respectively and the lower potential terminals q and s to the terminals b and d respectively.

Next, close K_2 and adjust the rheostat R' to produce small potential difference across the resistances R_1 and R_2. Now, shunt the galvanometer, close K_1 and connect *e* to *a* and *f* to *b*. Determine the approximate equivalent length of the potentiometer wire corresponding to the P.D. across R_1. Then, connect *e* to *c* and *f* to *d* and determine the approximate equivalent length of the potentiometer wire corresponding to the P.D. across R_2 and thus determine which of the resistances R_1 and R_2 has the *greater* value.

Next, connect the terminals *e* and *f* to the larger resistance and adjust the rheostat *R* in the *main circuit* so that when the contact maker *X* lies near the end *A* of the *last* coil of the potentiometer, an approximate balance point is obtained on the slide wire. *This adjusts the P.D. between A and B to be greater than the potential differences* V_1 *and* V_2 *across the resistances* R_1 *and* R_2, *at the same time making the potentiometer to have maximum sensitiveness for the given value of* V_1 and V_2.

Now, remove the shunt from the galvanometer and determine by connecting R_1 and R_2 *in turn* to *e* and *f*, the *exact* positions of the null point corresponding to the potential differences V_1 and V_2 and after noting down the number of coils and the length of the slide wire between *X* and *Y*, calculate the *equivalent* lengths l_1 and l_2 of the potentiometer wire corresponding to the potential differences V_1 and V_2 across the resistances R_1 and R_2 respectively. Next, increase R' or decrease *R* and take at least six more sets of observations for l_1 and l_2. Then, calculate the ratio of the resistances R_1 and R_2 from Eq. (23.45) and find the *mean* value of R_1/R_2.

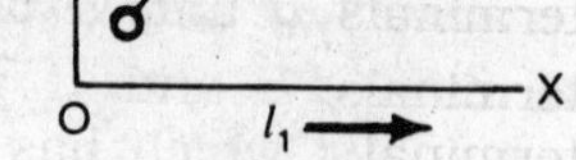

Fig. 23.48

Finally, plot a graph between the different values of l_1 and the values of l_2 corresponding to the same set. This will come out to be a straight line as shown in Fig. 23.48. Find the *slope* of this line which will also give the ratio of the two resistances.

Sources of Error and Precautions

(1) The supply batteries B_1 and B_2 should be of practically *constant* E.M.Fs. and ample capacities.

(2) The steady P.D. between the ends A and B of the potentiometer must always be *greater* than the potential differences V_1 and V_2 across the resistances R_1 and R_2.

(3) A suitable rheostat R should be included in the main circuit and should be adjusted to have the potentiometer *maximum sensitiveness* for given values of V_1 and V_2 ensuring at the same time that the P.D. between A and B is *greater* than V_1 and V_2.

(4) A key should be included in *each* of the main and auxiliary circuit and should be closed only when observations are to be made.

(5) The two resistances R_1 and R_2 to be compared should not be handled or disturbed during the experiment, and while determining the potential differences V_1 and V_2 across them, change over from one resistance to the other should be effected *in a very short time* with the help of a special key as described above, in order to diminish the effects of heating and any variations in the batteries.

(6) The *higher* potential terminals p and r of the resistances R_1 and R_2 should be connected to the terminals a and c respectively of the key K which are *collinear* with the terminal e connected to the sliding contact maker X near to the *higher* potential end A the potentiometer; and the lower potential terminals q and s of R_1 and R_2 should be connected to the terminals b and d respectively which are *collinear* with the terminal f which has been connected to the galvanometer.

(7) The galvanometer should be shunted with a low resistance wire. The exact position of the null point should be determined with the full galvanometer sensitivity without any shunt across it.

(8) While determining the values of l_1 and l_2 for a *particular* set of observations, both R and R′ should be kept *constant.*

(9) The contact between the jockey Y and the slide wire should be momentary and should not be made while the jockey is being moved along.

(10) The graph between l_1 and l_2 will come out to be a straight line and should be smoothly drawn.

Observations

S. No.	Length of potentiometer wire corresponding to the P.D. across the resistances.						
	R_1			R_2			
	No. of c oils	Length of slide wire div.	*Equivalent* length l_1 div.	No. of coils	Length of slide wire div.	*Equivalent* length l_2 div.	$\frac{l_1}{l_2}$
1							
2							
...							
5							
8							

Calculations *I* set : $\frac{R_1}{R_2} = \frac{l_1}{l_2}$

(*N.* B.—Make similar calculations for other sets).

$\therefore$ *Mean* value of $\frac{R_1}{R_2}$ =

From the graph (23.9)

PR = QR =

$\therefore$ $\frac{R_1}{R_2} = \frac{QR}{PR} =$

Result—The ratio of the values of the given resistances R_1 and R_2 as obtained by calculations =

and as obtained from graph =

Criticism of the method—This method of comparing resistances gives a very accurate result. The slight inaccuracy in the result may be due to the inconsistency of the E.M.Fs. of the batteries, the non-uniformity of the slide wire and the heating of the resistances. The method is especially suited to the comparison of two low resistances for which the Wheastone bridge method as ordinarily employed is not suitable. But the two low resistances to be compared should be of the *same order* of magnitude. To compare two resistances of different orders, say, one of the order of 1/100 ohm and the other of the order of 1 ohm, the method has to be slightly modified as described below. For very accurate results, the comparison should be made by means of *Crompton's potentiometer.*

Exercise—*To determine a low resistance by means of a potentiometer.*

Connect the two ends *A* and *B* of the potentiometer to a

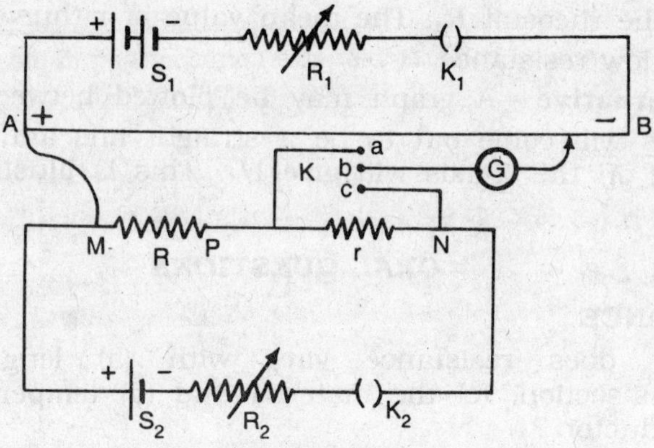

Fig. 23.49

storage battery S_1 in series with a rheostat R_1, including a key K_1 in the circuit. Next, connect the low resistance r to be measured *in series* with a known resistance R of about 1 to 5 ohms, and then insert the *combination in series* with a circuit containing a battery S_2, rheostat R_2 and a key K_2. Connect the *higher* potential terminal M of the resistance R to the *higher* potential end A of the potentiometer. Finally, connect the lower potential terminals P and N of the resistances R and r respectively through a two-way key and a galvanometer, to the jockey of the potentiometer, as shown in Fig. 23.49.

Next, connect a to b and determine the length l_1 of the potentiometer wire at the null point corresponding to the *P.D.*, V_1 across the resistance R. Then, connecting c to b, determine the length l_2 of the potentiometer wire corresponding to the *P.D.*, V_2 across the *series combination* of the two resistances R and r. If I is the *steady* current flowing through the resistances R and r, we have :

$$\frac{V_1}{V_2} = \frac{IR}{I(R+r)} = \frac{l_1}{l_2}$$

whence,
$$r = \left(\frac{l_2}{l_1} - 1\right) R$$

from which r may be calculated.

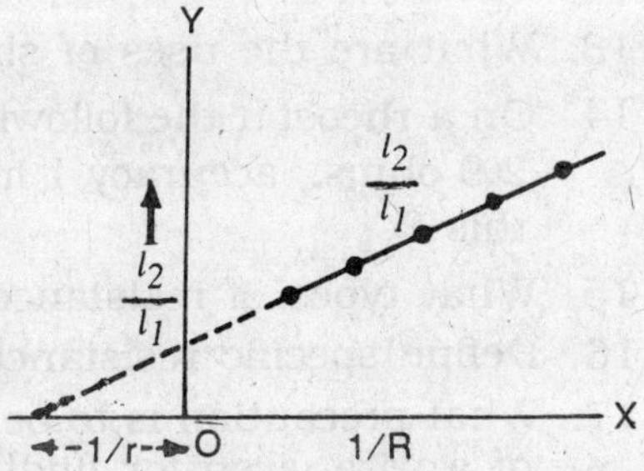

Fig. 23.50

Repeat the experiment with at least three values of R taking at least three sets of observations with each value of R by varying the *P.D.* across the resistances with the help of the rheostat R_2. The mean value of r thus obtained will give the low resistance.

Alternative —A graph may be plotted between l_2/l_1 and $1/R$. This will come out to be a straight line and its *negative* intercept on the X-axis will give $1/r$. This is illustrated in Fig. 23.50.

ORAL QUESTIONS

RESISTANCE

1. How does resistance vary with (a) length, (b) the cross-section, (c) the material and (d) temperature of the conductor ?
2. Do you know any case in which resistance decreases with temperature ?
3. What are the laws of resistances in series and in parallel ?
4. How is the resistance in the box made up ?
5. Why is the wire first doubled and then wound ?
6. What is the material of which the resistance coils are usually made ?
7. What is the composition of manganin and constantan ? Why do you select this particular material for the coil ?
8. Of what material are the coils of galvanometer made ? Why do you not use manganin or constantan there ?
9. What is meant by S.W.G. ? Of what materials are the plugs of the box made ?
10. Why are the studs put in ebonite board ?
11. Why are the plugs made tapering ?
12. What is the length of the wire connecting the blocks where infinity is written ?
13. What are the uses of slide wire rheostat ?
14. On a rheostat the following specifications are given 1.4 arms, 2.9 ohms : accuracy 1 in 100 at 160C : how do you interpret this ?
15. What types of resistance is ceiling fan regulator ?
16. Define specific resistance and state its units ?
17. What precaution is to be taken in the measurement of length of a wire given for finding out its specific resistance ?

METRE BRIDGE

1. What is a Wheatstone's bridge ?

2. What is a metre bridge and why is it so called ? Show the four arms in your apparatus. Point out the two points which are of the same potential.
3. What do you understand by conjugate conductors ?
4. What is the advantage in taking observation by interchanging the positions of the unknown resistance and the resistance box ?
5. Is it essential that the wire should be of uniform cross-section ?
6. If it is not, how will you modify your formula to get correct result ?
7. What condition should be satisfied to get a null point ?
8. Why does the deflection change direction in the galvanometer when the jockey passes over the null point ?
9. Why is it desirable to get the null point as near the centre as possible ?
10. What do you understand by end- corrections of a metre bridge ? What are they due to ?
11. Is the value of end-correction always positive ? If not in what case can it be negative ? Can it be ever zero ? If so, when ?
12. How can you determine the end corrections of a metre bridge ?
13. Why should the resistance boxes be preferably of the dial pattern ?
14. What is the advantage of using the interchanging commutator ? Explain its working. Why is it connected by thick wires to the metre-bridge ?
15. What should be the ratio of P and Q while determining the end-corrections and why ?
16. Sometimes, there is no null point obtained on the whole of the wire; what are the reasons ? How will you readjust your apparatus to get it ?
17. Trace the direction of current (i) at the null point, (ii) when the deflection in the galvanometer is to the right and (iii) when it is to the left.
18. Which key do you press first and why ?
19. How is null point affected by introducing a resistance R at null point in the cell circuit or in the galvanometer circuit ?
20. What is the use of the battery key ?

21. How can thermo-electric effects be got rid off in the metre bridge ?
22. Why do you shunt the galvanometer in the beginning and then cut it off when the approximate null point is obtained ?
23. Define sensitivity of a galvanometer.
24. What kind of galvanometer do you use? Explain its working.
25. How is it that the needle of a galvanometer comes to a dead stop immediately and does not oscillate ?
26. If your galvanometer is not dead beat how can you make it so ? Can you use a tangent galvanometer here ?
27. Can you replace your galvanometer by a telephone ? If so, is there any advantage in doing so ? why do you use Leclanche cell here ?
28. What maximum resistance can you measure with a metre-bridge ?
29. Can you find the resistance of a galvanometer or a cell with metre-bridge by using no extra galvanometer or a cell ? If another galvanometer were used what care will you take to avoid damage to the galvanometer ?
30. What do you understand by the sensitiveness of a Wheatstone's bridge ? On what factors does it depend ? When is it most sensitive ?
31. Are the positions of the battery and the galvanometer interchangeable ?
32. When the positions of the battery and the galvanometer are interchanged does the arrangement remain equally sensitive ? If not, which of the two arrangements is more sensitive and why ?

POST OFFICE BOX

1. Why is the instrument called P.O. box ?
2. Where are the four arms of the bridge ?
3. How are the resistances in P.O. box arranged ?
4. How can you find an unknown resistance with a P.O. box ?
5. What is the minimum and maximum resistance that can be measured with your P.O. box ?
6. How can you measure the resistance of a galvanometer with a P.O. box by Kelvin's method ?
7. Can you get a true null point in this case ?

8. When is the bridge said to be balanced ? Why is there no change of deflection in the galvanometer at the balance point when the key short-circuiting the usual galvanometer arm is pressed ?
9. Why should the galvanometer be properly levelled ?
10. Why does the spot of light shift on changing the resistance in the rheostat arm of the bridge ? Why should a thick copper wire be used to short-circuit the usual galvanometer arm ?
11. Which of the two keys would be pressed first and why ? If we take the ratio 10 : 10 in one case and 100 : 100 in the other, will it make any difference ? If at any stage the arrangement becomes insensitive, how can you make it more sensitive ?
12. If the bridge is not exactly balanced in the ratio 10 : 1000, how will you compute the resistance of the galvanometer ?
13. Why are there only three resistances in each of the ratio arms of a P.O. box ?
14. How can the resistance of the galvanometer be best determined ?
15. Why is the former method called Kelvin's method ?
16. Describe Kelvin's galvanometer.
17. What part does the control magnet play in the determination of resistance as such a galvanometer by Kelvin's method ?
18. Does the control magnet affect the deflection for given want of balance in the bridge ?
19. How can you determine the internal resistance of a cell with a P.O. box by Mance's method ?
20. What is the underlying principle of the method ?
21. When is bridge said to be balanced in this case ?
22. On pressing the key short-circuiting the usual battery arm, why does the deflection in the galvanometer not change when the bridge is balanced ?
23. What is the principal difference underlying Mance's and Kelvin's methods ?
24. Why is a thick copper wire used to short circuit the usual battery arm ?
25. What is the advantage of using a plug key in series with the cell ? Can you get a true null point in this case ?

26. How will you get exact value of resistance even if you do not get balance point with the ratio 100/100 ?
27. What is the range between which resistances of primary cells lie ?
28. What is the relation between internal resistance and efficiency ?
29. Which galvanometer will serve better in Mance's method : a moving coil or Kelvin's galvanometer ?
30. What part does the control magnet play in Mance's method when a Kelvin's galvanometer is used ?
31. How does it differ from that in Kelvin's method ?
32. What is Lodge's modification of Mance's method ? Explain the action of the condenser here.
33. What is the capacitance of the condenser used ?
34. What type of mirror is attached to the galvanometer coil ?
35. Which arrangement is better : a lens and plane mirror or a concave mirror alone ?
36. If you are using a concave mirror, how can you roughly estimate its radius of curvature ?
37. How is the image of the filament of the lamp formed on the scale ?
38. What type of image does a plane mirror form ?
39. How is it then that you get a real image on the scale with the help of the plane mirror ?

CAREY FOSTER'S BRIDGE

1. On what factors do the sensitiveness and accuracy of a metre bridge depend ?
2. How can these be increased ?
3. What is Carey Foster's bridge ?
4. In what respects is it an improvement over a metre bridge ?
5. How can you determine an unknown resistance with a Carey Foster's bridge ?
6. How will you determine σ, the resistance per unit length of the bridge wire ?
7. How does the accuracy in the determination of σ depend upon the difference between the known resistances in the outer gaps ?

8. What can be the maximum value of this difference which you can take ?
9. If the wire be not uniform, will this value of σ give the correct result ? If not, how will you modify your method of determining σ ?
10. Should the resistances in the inner gaps be fixed and known ?
11. Should these two resistances be exactly equal ? If they are unequal, what is the harm when their differences is (*a*) very small (*b*) very large ? In the latter case where may the balance points lie ?
12. What is the advantage in making the resistances in the inner gaps equal ?
13. Can you not use a rheostat for this purpose ? If so, what advantage does it possess over fixed known resistances in the inner gaps ?
14. What should be the order of the resistance in the inner gaps of the bridge and why ?
15. When is the bridge said to be most sensitive ?
16. What do you understand by calibration of the bridge wire and how is it done ?
17. What is the function of the rheostat introduced between the inner gaps of the bridge in this experiment ?
18. What should be the order of the resistance of the rheostat ?
19. What should be the order of the difference of resistances in outer gaps and why ?
20. What is the resistance of the bridge wire between two successive balance points ?
21. What is the resistance of the bridge wire ?
22. How will you test the uniformity of the bridge wire ?
23. How will you draw the calibration curve for the bridge wire ?
24. If you also know the values of end corrections, how can you get the value of the total correction at any point from the calibration curve ?

PLATINUM RESISTANCE THERMOMETER

1. Describe a platinum resistance thermometer.
2. What is the function of the compensating leads ?
3. Define temperature-coefficient of resistance.
4. Explain how it can be determined for platinum by Carey Foster's bridge.

5. What are the sources of error in your experiment and what precautions do you take to eliminate them ?
6. Why have the usual positions of the galvanometer and the battery been interchanged ?
7. Why should the four arms be non-inductive ?
8. Why should the null point be determined observing the immediate deflection in galvanometer on pressing the jockey ?
9. Discuss the accuracy in the result obtainable by this method.
10. How can you determine resistance of the platinum resistance thermometer graphically ?

POTENTIOMETER

1. What is a potentiometer ? What is it used for ?
2. What do you understand by e.m.f. of a source ?
3. In what units is it measured ?
4. What is the difference between P.D. and e.m.f. ?
5. What is the principle of a potentiometer ?
6. Is it used for the comparison of e.m. fs. or P.D.s. of two cells ?
7. What are the requirements of a potentiometer wire ?
8. Why must it be of high specific resistance ?
9. Why should the wire be of uniform diameter ?
10. What would the P.D. be equal to if the wire were not uniform throughout ?
11. What do you mean by potential gradient of a potentiometer and how will you determine it ?
12. What do you understand by sensitiveness of a potentiometer? How can you increase it ?
13. What do you understand by the accuracy of a potentiometer and upon what factors does it depend ?
14. Without sacrificing accuracy how can you increase its sensitiveness ?
15. How does the accuracy of observations differ from that of the instrument ?
16. What sort of galvanometer is most suitable in a potentiometer circuit ? Can you use a tangent galvanometer instead ? If it is not very sensitive, how does increasing the sensitiveness of the potentiometer affect the determination of the exact null point ?

17. What are the requirements of the cell used in the main circuit of the potentiometer and why ?
18. Why is it that a Leclanche cell can be used in Wheatstone bridge experiments and not with a potentiometer ?
19. What is the function of the key in the main circuit of a potentiometer ?
20. What is the use of the rheostat in the main circuit ?
21. It is sometimes found that the deflection is in the same direction all over the potentiometer wire, what is it due to ?
22. In case the P.D. across the potentiometer wire is less than the P.D. to be measured, how can you increase the former ?
23. Is it necessary that positive terminal of the cells should be connected to one and the same end of the potentiometer wire ?
24. Can all the negative terminals be connected at one end of the potentiometer wire instead of the positive ones ? If so, how ?
25. What is the effect on the null point if (*a*) the number of cells in the main circuit is increased, (*b*) the resistance in the main circuit is altered, (*c*) a resistance is introduced, in series with the galvanometer ?
26. Trace the current (*a*) when the null point has been obtained, (*b*) when the deflection is towards the left, (*c*) when the deflection is towards the right.
27. What are the uses of a potentiometer.
28. Why is a potentiometer used for comparison of E.M.Fs. of two cells ?
29. What other methods are there for the comparison of E.M.Fs. of cells ? Compare their relative merits. How can you measure the internal resistance of a cell with a potentiometer ?
30. Why should the cell, whose internal resistance is to be determined, be not disturbed during the course of the experiment ?
31. Why should a resistance of about a thousand ohms be kept connected in series with a galvanometer until the approximate position of the null point has been found ?
32. Does this resistance affect the position of the null point ?

33. Do you get a consistent value for the internal resistance of a cell by this method ? If not why ?
34. Compare this method of measuring internal resistance of a cell with Mance's method ?
35. Can you measure internal resistance of an accumulator by this method ?
36. How can potentiometer be used for the comparison of two resistances ?
37. Why should the supply batteries in the main and auxiliary circuits be of constant E.M.Fs. and ample capacities ?
38. Why would the P. D. between the ends of the potentiometer wire be always greater than the P.D. across each of the two resistances to be compared ?
39. Describe the working of the special key used for changing over from one resistance to the other in the auxiliary circuit.
40. Compare this method of measuring resistances with Wheatstone's bridge method.
41. How can you use a potentiometer for measurement of current ?
42. What do you understand by calibration of an ammeter and that of a voltmeter.
43. Why is it done at all ? How do you calibrate an ammeter and a voltmeter with a potentiometer ?
44. How do you find out the potential gradient of the potentiometer wire ?
45. Why should a standard cell be used for the purpose ?
46. Will a Daniell cell serve the purpose ?
47. What is the use of known standard resistance in the auxiliary circuit ?
48. Why should it be preferably of one ohm ?
49. What is the function of the keys in the auxiliary circuits ?
50. What is the range of the voltmeter or the ammeter you are calibrating ?
51. What should be the P.D. across the ends of the potentiometer wire in order to calibrate the entire range of the instrument and why ?
52. How will you adjust this to be so ?
53. Discuss the accuracy of the result obtainable by this method.

24
Heating Effects of Currents

24.1. Laws of Heating Effects of Currents

We have seen in Section 24.1 that when a current is allowed to pass through a conductor, work is done at the expenses of part of electrical energy of the circuit which appears as heat in the conductor and that, if the potential difference between the ends of the conductor be V volt, the work done or the energy transformed into heat when one unit current flows through it for one second is V joule. Hence, if the conductor carries a current of I ampere, the energy transformed into heat in one second will be VI joule and in t seconds VIt joule. If this energy is equivalent to H calories of heat, we have :

$$H = \frac{VIt}{J} \qquad (24.1)$$

where J is the *mechanical equivalent of heat*. Now, $J = 4.18$ joule per cal. and from *Ohm's law* $V = IR$, where R is the resistance of the conductor, hence, the heat produced in calories is given by the expressions :

$$H = \frac{VIt}{4.18} = \frac{I^2Rt}{4.18} = \frac{V^2t}{4.18R} \qquad (24.2)$$

or

$$H = 0.24\ VIt = 0.24\ I^2Rt = 0.24\frac{V^2t}{R} \qquad (24.3)$$

From these equations it follows that :

(1) The quantity of heat developed in a conductor of given resistance by the passage of a current of given strength is directly proportional to the time for which the current flows.

(2) The quantity of heat developed in a conductor of given resistance in a given time is directly proportional to the square of the current strength or to the square of the potential difference at the ends of the conductor and product of the current strength and the potential difference.

(3) For a given current strength, the quantity of heat produced in a given time is directly proportional to the resistance of the conductor.

(4) For a given potential difference, the quantity of heat

produced in a given time is inversely proportional to the resistance of the conductor.

These relationships are known as *laws of heating effects of currents*. The relationship H=0.24 I^2Rt was first experimentally established by Joule and is known as *Joule's law*. The other expressions for H in equation (24.3) are *alternative* forms of Joule's law.

Experiment 24.1

Object—To determine the value of *mechanical equivalent of heat* with *Joule's calorimeter* using a copper voltameter.

Apparatus—Joule's calorimeter, a copper voltameter, two jars, a battery of accumulators, a sliding rheostat of about 50 ohms, a voltmeter, a weight box, a stop-watch, a thermometer, a physical balance, a chemical balance and connecting wires.

Description of Apparatus—The Joule's calorimeter consists of a copper calorimeter of about half-a-litre capacity silvered on its outer surface, and fitted with an ebonite cover which supports the heating coil suspended inside the calorimeter. The heating coil is made of a high resistance wire, *e.g.*, eureka and is connected to two terminals fixed to the ebonite cover by thick copper leads. The cover is provided with three holes, the central one for the thermometer and the two outer ones for the stirrer. The calorimeter is placed in a copper container which is silvered on its inner surface, the calorimeter being thermally insulated from the container by a thick layer of felt.

Theory—Let a resistance coil be immersed in water contained in a calorimeter and let a current of I amperes be allowed to pass through the resistance coil. Then, if V volts be the potential difference across the coil the energy transformed into heat in t seconds will be equal to VIt joule. If this energy transformed be equivalent to H calories of heat,

$$H = \frac{VIt}{J} \tag{24.4}$$

where J is the mechanical equivalent of heat. Now, if the current I flowing through the resistance coil also passes through a copper voltameter placed in series with the coil and if it deposits m gm. of copper on the cathode plate in t seconds, we have :

$$m = ZIt \tag{24.5}$$

where Z is the electro-chemical equivalent of (cupric) copper. Eliminating I between Eqs. (24.4) and (24.5) we have :

$$H = \frac{Vm}{JZ}$$

Now, let M be the mass of water contained in the calorimeter and let W be the water equivalent of the calorimeter and the stirrer. Then, if θ be the rise in temperature of the calorimeter and its contents, the heat taken up by them is $(M+W)\,\theta$ calories. Assuming that all the heat produced in the heating coil is used up in raising the temperature of the calorimeter and its contents, we have, by equating the two expressions for H :

$$(M+W)\,\theta = \frac{Vm}{JZ}$$

whence,

$$J = \frac{Vm}{Z(M+W)\,\theta} \qquad (24.6)$$

This equation enables us to calculate J, if other quantities are determined.

Method—Weigh the empty calorimeter with stirrer in a physical balance. Fill it with water sufficient to cover the heating coil and again weigh it and thus determine the mass of the water contained in the calorimeter. Then, cover the calorimeter with its ebonite lid immersing the heating coil in water and place the calorimeter in its metal container lined with felt.

Now, clean the copper cathode plate of the copper voltameter thoroughly on its both sides with sand paper and a rug or emery cloth until its surface is quite bright and weigh it in an accurate chemical balance correct upto 1/10 of a milligram. Then, make connections as shown in Fig. 24.1, connecting the heating coil to a storage battery in series with a suitable sliding rheostat R and a copper voltameter C, and placing a voltmeter in parallel with the heating coil. Note down the temperature of water contained in the calorimeter correct upto 1/5 or preferably 1/10 of a $^{\circ}C$. Then, start the current and simultaneously with it the stop-watch. The resistance coil in the calorimeter will get heated and the temperature of the calorimeter and its contents will gradually rise. Stir the water constantly and efficiently. Note down the potential difference across the heating coil with the voltmeter and keep it constant by adjusting the rheostat, if necessary. When the temperature of the calorimeter

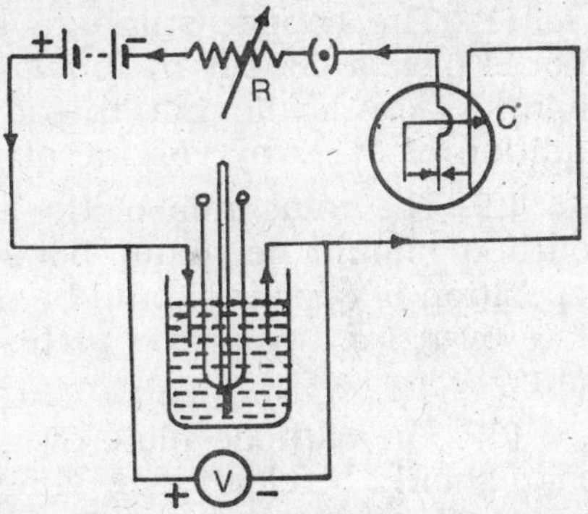

Fig. 24.1

and its contents has risen by about 10°C, switch off the current. Note down the exact time for which the current has been passed. Also note down the final temperature of water. Now, allow the water to cool for the *same* time for which it was heated and determine the fall in the final temperature during this interval. Add half the fall in temperature to the final temperature of the calorimeter and its contents which will give the final temperature corrected for heat losses by radiation.

Now, remove the cathode plate from the voltameter and immediately immerse it in a jar of tap water already placed near the voltameter. This will remove the copper sulphate solution left on the plate. Then, transfer the plate to another jar containing distilled water to which two or three drops of sulphuric acid per litre have been added. Next press the plate without rubbing between sheets of filter paper or clean blotting paper to remove the moisture as far as possible. Then dry the plate finally in warm air coming out of a hot air blower. Weigh the cool dry plate in the chemical balance to the nearest tenth of a milligram with the help of a rider and thus find out the mass m of the copper deposited. Finally, calculate the value of J from equation (28.6) above, taking $Z = 0.0003295$ gms per coulomb.

Sources of Error and Precautions

(1) The copper sulphate solution in the copper voltameter should have a density of about 1.18 gm/c.c., and should be made slightly more acidic than the aqueous solution of the salt by the addition of 0.1% by volume of concentrated sulphuric acid.

(2) The total area of the cathode surface immersed in the solution should be about 50 sq. cm. per ampere passing. The deposition of copper should be made on both sides of the cathode plate by using two anode plates one on each side of the cathode plate.

(3) The cathode plate over which deposit of copper is to be made should be clean and should never be touched with fingers. It should always be handled by interposing a double strip of paper and should always be lifted by the attached wire or the hook at the top.

(4) The current must leave the copper voltameter by the cathode plate, *i.e.*, the plate over which the deposit of copper is to be made should be connected to the negative of the battery or to a point at a potential lower than that of the point to which the anode plate is connected.

(5) The rise in temperature of the calorimeter and its contents should be about 10°C.

(6) The strength of the current should be between 1.5 and 2 amperes so that during the interval the temperature of the calorimeter and its contents rises by about 10°C, a weighable amount of copper must be deposited on the cathode plate.

(7) The potential difference across the heating coil should be small, say, about 6 volts. If the potential difference exceeds 8 volts the electrolysis of water which takes place when the current passes through the heating coil will affect the result seriously.

(8) The strength of current should be kept constant while deposition of copper is being made on the cathode plate. The voltmeter reading should be kept constant by adjusting the sliding rheostat, if necessary.

(9) The water in the calorimeter should be constantly and efficiently stirred during the heating of the coil in order to equalise the temperature throughout.

(10) The final temperature of the calorimeter and its contents should be corrected for loss of heat by radiation.

(11) When the cathode plate is taken out of the voltameter after the deposition of copper, it should be immediately immersed in a jar of tap water. The plate should then be transferred to a jar containing distilled water to which a few drops of sulphuric acid per litre have been added.

(12) While pressing the cathode plate between sheets of blotting or filter paper, it should never be rubbed.

Observations

Mass of the calorimeter, and stirrer	=	gm =....kg
Mass of the calorimeter, stirrer and water	=	gm =....kg
Least count of thermometer.	=	$^\circ$C
Initial Temperature of Water.	=	$^\circ$C
Final temperature of water after the current has passed for t seconds.	=	$^\circ$C
Fall in final temperature of water in t seconds.	=	$^\circ$C
Mass of cathode plate before deposition of copper.	=	gm =....kg
Mass of cathode plate afrer deposition of copper.	=	gm =....kg
Potential difference across the heating coil.	=	volt.

Calculations

Mass of water in the calorimeter.	=	gm =...kg
Water equivalent of calorimeter and stirrer.	=	gm =....kg
Final temperature of water corrected for loss of heat by radiation.	=	$^\circ$C
Rise in temperature of calorimeter and its contents.	=	$^\circ$C
Mass of copper deposited on the cathode plate	=	gm =....kg

$$J = \frac{Vm}{Z(M + W)\,\theta} \times$$

$$= \quad \times \text{ Joule/cal}$$

Result—The value of mechanical equivalent of heat

= × Joule/cal.

Standard value of J = 4.18 joule/Cal.

Percendage ϕ Error = ...%

Criticism of the Method—The value of mechanical equivalent of heat obtained by this method is always slightly higher than its correct value. The total amount of heat produced in the resistance coil cannot be accurately determined for (*i*) heat losses to the surroundings cannot be completely checked, (*ii*) the change in temperature of the calorimeter and its contents cannot be followed as accurately as we wish and (*iii*) some heat is also taken up by the thermometer and the heating coil and it is not easy to determine the true value of the water equivalent of the calorimeter, the stirrer, the heating coil and thermometer. Loss of heat to the surroundings can be eliminated if the calorimeter is placed in a double-walled vessel, the space between the walls being filled with water at a constant temperature equal to the mean of the initial and final temperatures of calorimeter, for then the calorimeter gains heat from the surroundings during the first half of the experiment and loses to them an equal amount during the second half. Loss of heat by evaporation can be avoided by using a liquid of low vapour pressure, *e.g.*, paraffin oil or aniline. These liquids have a small specific heat and hence with them the rise in temperature will also be quick and comparatively large. With these liquids, higher voltages can be applied across the heating coil as there is no fear of electrolysis.

Now, unless the heating coil is made of a material of negligible temperature coefficient, its resistance changes appreciably with temperature and it is difficult to keep both the current and the potential difference across the coil constant. A part of the main current flows through the voltmeter and unless its resistance is very high the current taken by it will not be negligible and in that case the current through the heating coil will be (I—V/R) where R is the resistance of the voltmeter. This difficulty can, however, be overcome by measuring the potential difference with a potentiometer with which the accuracy will also be increased.

Most of the above defects are removed in Callendar and Barne's continuous flow calorimeter in which the heat produced in the coil is removed by means of constant flow of water as rapidly as it is produced.

ORAL QUESTIONS

JOULE'S CALORIMETER

1. State Joule's law for heating effects of a current.
2. Why is heat at all produced when a current is allowed to pass through a conductor ?
3. How does the amount of heat developed in a conductor by the passage of the current in it vary with (*a*) the resistance of the conductor, (*b*) the strength of the current and (*c*) the time ?
4. Define mechanical equivalent of heat. How is work done in this experiment ?
5. Describe Joule's calorimeter. Of what material is the resistance coil made ? What is its approximate resistance ? Why is the calorimeter placed in a copper vessel silvered on its inner side and lined with felt?
6. Explain how you can determine mechanical equivalent of heat by means of a Joule's calorimeter.
7. What are the sources of error in this experiment ? What precautions do you take to eliminate them ?
8. Why should the P.D. across the resistance coil be small ?
9. How can you increase the rise in temperature without producing any electrolysis ?
10. Name any liquid of low specific heat.
11. How do you prevent the loss of heat to the surroundings ?
12. How can loss of heat by radiation be prevented ? What are the unavoidable sources of error in this experiment ?

25

Thermo-Electricity

25.1. Seebeck Effect

Seebeck, discovered in 1826, that if in a circuit formed by joining two dissimilar metals, the two junctions are kept at *different* temperatures, a current flows round the circuit. Such currents are termed *thermo-electric currents* and the E.M.F. producing them is called *thermo-electric force*. The energy required to maintain the current is derived from the surroundings and the direction of current in the circuit depends upon the metals constituting the thermo-couple. A thermo-couple constructed of antimony and bismuth is one of the most sensitive thermo-couples available and in this case the direction of the current is from antimony to bismuth through the cold junction. In the case of copper-iron couple the current flows from copper to iron through the hot junction. If the temperature of the hot junction be continually increased, thermo-electric current increases until maximum value is reached when it begins to decrease. The temperature of the hot junction at which maximum current flows is constant for a given couple and is known as the *neutral temperature* for that couple. On further raising the temperature of the hot junction, the current decreases to zero and is then reversed. The temperature at which the current is zero and its direction is reversed depends upon the temperature of the cold junction and is always as much above the neutral temperature as the cold junction below it. This is illustrated in Fig. 25.1.

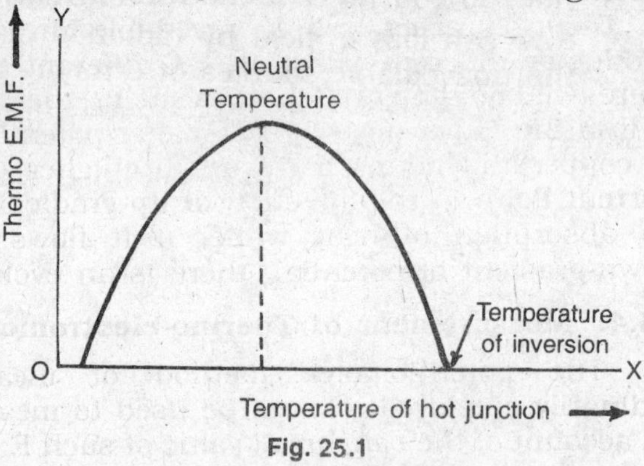

Fig. 25.1

25.2. Peltier Effect

In 1834, Peltier discovered that if a current be allowed to pass through a circuit consisting of two dissimilar metals, there is either an evolution or absorption of heat at the junctions, *i.e.*, the junctions are either heated or cooled, the condition whether a junction will be heated or cooled being determined by the direction of the current. This effect, known as *Peltier effect*, is quite distinct from Joule's heating effect which takes place at all points of the circuit and arises from its resistance. Joule's effect is independent of the direction of the current, *i.e.*, it is irreversible, while the Peltier effect is reversible for if the current be reversed, the junction which previously evolved heat now absorbs it and *vice versa*.

The Peltier effect arises on account of the existence of contact potential difference at the junction of two dissimilar metals. When the current traverses a junction where there is a slight up-gradient of potential due to contact, the current gains energy. Consequently, there is an absorption of heat at that junction as a result of which it is cooled. On the other hand, if the current traverses a junction in the direction of the down-gradient of potential due to contact, work is done and the current gives out some energy producing a heating effect at the junction.

25.3. Thomson Effect

In 1856, Thomson showed that whenever a current is allowed to pass through an unequally heated conductor, *i.e.*, whose different parts are at different temperatures, there is an evolution or absorption of heat in the conductor, the condition whether there will be an evolution or absorption of heat being determined by the direction of the current. This effect is known as *Thomson effect* and is reversible. In an unequally heated conductor different parts are at different potentials and hence there will be a potential gradient along the conductor. In the case of iron the colder parts are at higher potentials, while in the case of copper the hotter parts are at higher potentials. When the current flows in the direction of up-gradient of potential there is an absorption of heat while, if it flows in the direction of down-gradient of potential, there is an evolution of heat.

25.4. Measurement of Thermo-electromotive Force

The potentiometer method of measuring E.M.Fs, as ordinarily employed, cannot be used to measure thermo E.M.Fs. on account of the *very small* value of such E.M.Fs. The magnitude of the thermo E.M.Fs. is of the order of a few millivolts, *e.g.*, the E.M.F. of a copper-iron couple, when the junctions are kept at 0°C and 100°C, is only about 0.0013 volts. The thermo E.M.Fs. can be measured by a suitable modification of the ordinary potentiometer method so as to produce a P.D. of about a microvolt across each cm. length of the potentiometer wire which is small enough to admit of sufficiently accurate measurement.

Experiment 25.1

Object—To study the variation of thermo E.M.F. of a copper-iron circuit with temperature.

Apparatus—A potentiometer, an accumulator, a standard cadmium cell, a resistance box, a high resistance of 15,000 ohms, a rheostat, a sensitive galvanometer, two thin glass tubes, two small test tubes, iron wire, a two-way key, two single way plug keys, connecting wires and heating apparatus.

Theory—Let a high resistance *R* of about 1000 ohms be

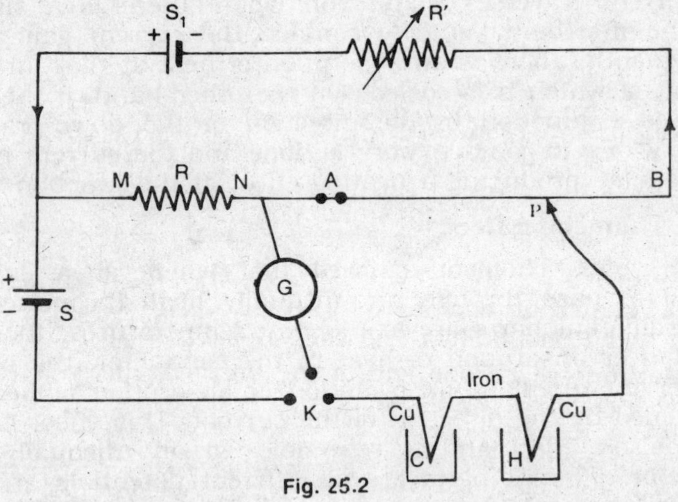

Fig. 25.2

connected in series with the potentiometer wire *AB* (Fig. 25.2) and let the combination be connected to an accumulator in series with a rheostat *R*′, the current entering the combination of *R* and the potentiometer wire at *M* and leaving at *B*. Further, let the +*ve* pole of a standard cell *S* be joined to the higher potential terminal *M* of the resistance *R*, and the hot junction *H* of the copper-iron thermo-couple to the jockey; and let the −*ve* pole of the standard cell and the cold junction C of the thermo-couple be connected through a two-way key to one terminal of the galvanometer, the other terminal of which is joined to the lower potential terminal *N* of the resistance *R*. Now, if the standard cell circuit be closed by means of the two-way key and the rheostat *R*′ so adjusted that there is no deflection in the galvanometer, the P.D. across the resistance *R* will be balanced by the E.M.F.

of the cell. Hence, if E be the E.M.F. of the standard cell, and I the steady current through the resistance R or the potentiometer wire, we have,

$$E = IR \tag{25.1}$$

Now, let the standard cell circuit be broken and the cold junction of the thermo-couple connected to the galvanometer. Then, if P be the position of the null point on the potentiometer wire, the thermo E.M.F. of the copper-iron couple is equal to the P.D. between A and P. Hence, if e be the thermo-electric E.M.F., and r the resistance of the portion of the wire between A and P:

$$e = Ir \tag{25.2}$$

Dividing Eq. (25.2), by Eq. (25.1), we get :

$$\frac{e}{E} = \frac{r}{R} \tag{25.3}$$

Let ρ be the resistance per unit length of the potentiometer wire and let l be the length of the wire between A and P. Then, as the wire is uniform $r = \rho l$. Putting this value of r in the above equation we get :

$$e = \rho \frac{E.l}{R} = K\,l \tag{25.4}$$

where $K = \rho\ E/R$

This equation enables us to calculate the thermo E.M.Fs. e if E, ρ and R are known and the value of l determined.

Method —Pass one end of the iron wire down a thin glass tube G (Fig. 25.3) and join to it at J a copper wire AJ. This ensures that the wires are in contact at the junction only. Then, to ensure good contact at J, place the combination in a test tube containing a small amount of mercury and dip the junction J into the mercury. Next, pass the other end of the iron wire down another thin glass tube G and prepare in a similar manner another junction J' of iron and copper. Place one of the juncitons in air and the other in an oil bath which may be kept at different temperatures.

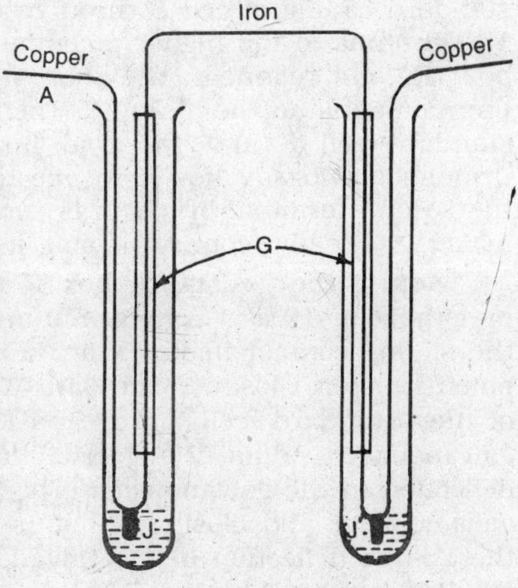

Fig. 25.3

Now, connect a resistance box R in series with the potentiometer at the end A (Fig. 25.4). Then, connect the combination to an accumulator S' in series with a rheostat R' of high value, including a plug key K_1 in the circuit, the +*ve* pole

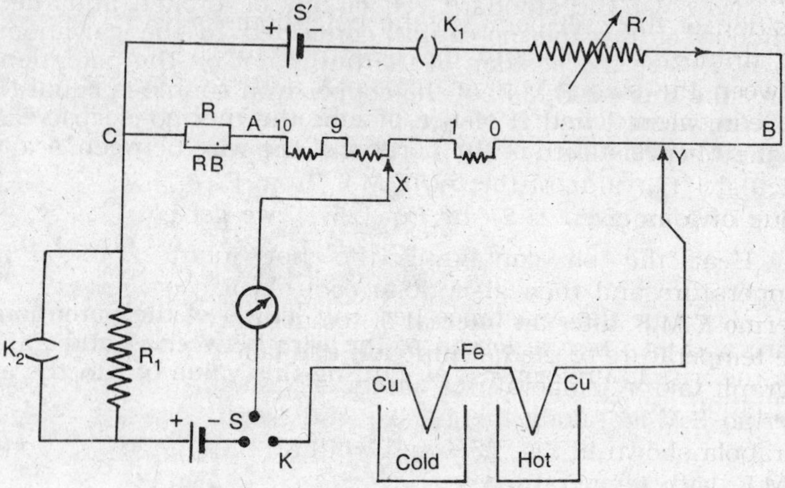

Fig. 25.4

of the accumulator being joined to the end C of the combination of R and the potentiometer. Next, connect the +*ve* pole of the standard cadmium cell S through a high resistance R_1 of about 15000 ohms to the higher potential terminal C of the resistance box R and connect the hot junction of the copper-iron thermo-couple to the jockey Y. Then, connect the −*ve* pole of the standard cell S and the cold junction of the thermo-couple through a two-way key K to one terminal of the galvanometer, the second terminal of which is connected to the sliding contact maker X. Finally, connect a plug key K_2 across R_1.

Adjust the resistance box R to about a thousand ohms, preferably to 1018.3 ohms. Shunt the galvanometer and place the sliding contact maker X at the end A of the last coil of the potentiometer. Close K_1 and with K_2 open, connect the −*ve* pole of the standard cell by means of the two-way key to the galvanometer. Adjust the rheostat R' until there is practically no deflection in the galvanometer. Then, remove the shunt from the galvanometer and closing K_2 adjust the rheostat R' finally until there is no deflection in the galvanometer. This exactly balances the P.D. across R by the E.M.F. of the standard cell.

Next, open K_2, shunt the galvanometer again and connect the cold junciton of the thermo-couple by means of the two-way key K to the galvanometer. Then, find by trial by placing the sliding contact maker X on various studs corresponding to the different coils and pressing the jockey over the stretched wire, an approximate position of the null point on the wire. Now, remove the shunt from the galvanometer and determine the exact position of the null point on the potentiometer wire. Note down the number of coils and the length of the potentiometer wire between the sliding contact maker X and the jokey Y and find the equivalent length l of the potentiometer wire corresponding to the thermo-electric E.M.F. of the copper-iron couple. Then calculate the value of thermo E.M.F. from Eq. (25.4), taking the value of ρ as given.

Heat the oil containing the hot junction to a high temperature and then allow it to cool. As it cools, measure the thermo E.M.F. after an interval of temperature of about 5 °C until the temperature of the hot junction has fallen to about 30 °C. Plot a graph taking temperatures along X–axis and the corresponding thermo E.M.Fs. along the Y–axis. The graph will form part of a parabola shown in Fig. 25.1 and will give the variation of thermo E.M.F. with temperature.

Sources of Error and Precautions

(1) The ends of the connecting wires should be clean and the connections should be firmly made. As the thermo E.M.F. to be measured is small, bad contacts will lead to troublesome difficulties.

(2) The accumulator should be fully charged and its E.M.F. should remain constant while observations are being made.

(3) A key should be included in the main circuit of the potentiometer which should be closed only when observations are to be made.

(4) The P.D. between the ends of the stretched wire of the potentiometer should not be greater than a millivolt otherwise the length l of the potentiometer wire corresponding to the thermo E.M.F. of the copper-iron couple will be very small and the value of l cannot be measured sufficiently accurate. Accordingly, the high resistance R and the rheostat R' should each be about a thousand ohms. If the resistance of stretched wire of the potentiometer is of exactly one ohm and there are 1000 divisions

on the measuring scale, then with R=1018.3 ohms, the potential difference across each division will be exactly equal to 1 microvolt and the calculations will then be also simplified.

(5) To protect the standard cell, a high resistance of about 15000 ohms should be connected in series with it. This will prevent large currents from being taken from the cell. This resistance should be removed from the cell circuit when its E.M.F. has been nearly balanced, so that the final adjustment of the rheostat *R'* may be made more accurately. Note that this resistance does not affect in any way the position of balance.

(6) While balancing the P.D. across the resistance *R* by the E.M.F. of the standard cell, the sliding contact maker *X* must lie at the end *A* of the *last* coil of the potentiometer.

(7) In the thermo-couple, the wires of the two metals should be in contact at the junctions only and there should be good contact at the junctions. To ensure this, the junctions should be made especially in the manner described under the method.

(8) The copper wires connecting the cold and hot junctions of the thermo-couple to the two-way key and the jockey should be long enough to ensure that their junctions with the key and hot jockey do not differ appreciably in temperature.

(9) The galvanometer should be a very sensitive one. A shunt should be connected across it to avoid damage to the instrument by excessive deflections in it during determination of the approximate position of the null point. The exact position of the null point should be determined with full galvanometer sensitivity by removing the shunt from it.

(10) The contact between the jockey and the wire should always be momentary. The contact between the jockey and the wire should not be made while the former is being moved along.

Observations

(*i*)	Room temp.	=	°C
(*ii*)	E.M.F. of the standard cell	=	volt
(*iii*)	High resistance *R*	=	ohm.
(*iv*)	Resistance per unit length ρ of the potentiometer wire	=	ohm

S.No.	Temperature of hot junction °C	Length of potentiometer wire corresponding to the thermo E.M.F.			Thermo E.M.F.
		No. of coils	Length of stretched wire div.	Equivalent length l div.	microvolt
1.					
2.					
3.					
..					
...					

Calculations

$$K = \rho E/R$$
$$=$$
$$= \text{ microvolt/div.}$$

I reading e $= Kl_1$

$= \text{ microvolt}$

[Make similar calculations for other readings].

Result—The graph between thermo E.M.F. and temperature for copper-iron couple is a parabola as shown in Fig 25

rom graph Neutral temperature of Cu-Fe thermo Couple = ...

Temp of inversion = ... °C

Criticism of the method—The method yields sufficiently accurate result. The accuracy of the result depends upon the constancy of the E.M.F. of the accumulator maintaining a steady current through the potentiometer wire, the uniformity of the wire and the accuracy of knowledge about the value of E.M.F. of the standard cell at the time of experiment and the resistance per unit length of the wire. For very accurate results, the thermo E.M.F. should be measured with a Crompton potentiometer.

Exercise—*To determine the melting point of paraffin wax by measuring thermo E.M.Fs. of copper-iron circuit.*

Place one junction of copper-iron couple in melting ice and the other in boiling water. Determine the thermo E.M.F. of the circuit by means of the potentiometer and note down the temperature of boiling water. Allow the water to cool and determine the thermo E.M.Fs. after an interval of temperature of about 5°C until the temperature of the junction has fallen to about 30°C noting down the temperature of the hot junction at which the thermo E.M.F. is measured. Then, place the hot junction in melting paraffin wax and when the wax solidifies, measure the thermo E.M.F. of the circuit. Next, plot a graph taking the temperatures of the hot junction as abscissae and the corresponding thermo E.M.Fs. expressed in microvolts as ordinates. The thermo-electric curve so obtained will be a small portion of a parabola (Fig. 25.1). From the curve find out the temperature corresponding to the thermo E.M.F. when the hot junction was placed in melting wax. This will be the melting point of wax..

ORAL QUESTIONS

1. What is a thermo-electric effect ? Who discovered it ?
2. What is Seebeck effect ?
3. Where from does the energy come to maintain thermo-electric current ?
4. On what does the direction of thermo-electric current depend ? What is the direction of current in the case of *(a)* antimony-bismuth couple and *(b)* a copper, iron couple ?
5. What happens to the current when the temperature of the hot junction is continuously increased?
6. What is neutral temperature ?
7. What is thermo-electric inversion ?
8. What is Peltier effect ? How does it differ from Joule's heating effect ?
9. What determines whether there would be the evolution of heat or its absorption at a junction?
10. What is contact potential ?
11. How do you explain Peltier effect ?
12. What is Thomson effect ? How does it differ from Peltier effect ?
13. How do you explain Thomson effect by the existence of a potential gradient along an unequally heated conductor ?
14. What is thermo E.M.F. ? What is the order of magnitude of a thermo E.M.F.? How do you measure it?
15. Why does the potentiometer method of measuring E.M.Fs ordinarily employed cannot be used to measure thermo E.M.F.?
16. Describe the method of measuring thermo E.M.F. with a potentiometer.
17. What is the function of the high resistance connected in series with the potentiometer wire? What should be its

approximate value and why ? Why do you require a potential gradient of about a micro-volt per div. in this case ?

18. What precaution do you take in this experiment ?
19. Why has a high resistance been connected in series with the standard cell ? Why should it be removed from the strandard cell circuit when the E.M.F. of the cell nearly balances the P.D. across the high resistance ?
20. How do you prepare the thermo-couple ? How do you ensure good contact at the junctions only ?
21. Which of the two junctions, hot or cold, you will connect to the jockey and why ?
22. Discuss the accuracy of result obtainable by this method.

26
Measurement of Capacitance

26.1 Capacitance of a Conductor and Its Unit

When an insulated conductor is charged with a certain quantity of electricity, the potential of the conductor is raised, but the extent to which the potential is raised depends not only upon the magnitude of the charge but also upon the size of the conductor and the nature of its surroundings. If these conditions remain unchanged the potential acquired by the conductor is directly proportional to the quantity of charge given to it. Similarly, the charge given to a conductor of given size is proportional to the potential acquired by it. Thus, we may write :

$$Q = CV$$

where Q is the charge given to the conductor, V its potential and C is a constant for the conductor under the specified conditions.

From the above expression it is evident that the greater the value of C, greater will be the charge required to produce a given potential. For this reason, the factor C is called electrical capacity or capacitance of the conductor. *The capacitance of a conductor is measured by the ratio of the charge given to the conductor to the potential which that charge produces,* or alternately *by the quantity of charge necessary to change the potential of the codnuctor by one volt.*

The capacitance of a conductor depends upon its size, being greater, the greater the size. Further, if a conductor carrying a charge Q is immersed in a medium of dielectric constant K, its potential will be $1/K$ of its value in air and hence to bring it to the same potential, the charge on the conductor must be made $K\,Q$. This shows that the capacitance of the conductor has now become K times its value in air. Thus, the *capacitance of a conductor depends on the dielectric medium, being greater, the greater the vlaue of K.* The capacitance of a conductor is also affected by neighbouring conductors as discussed in the next section.

Now, when $Q = 1$ and $V = 1$, C is also equal to unity. This gives us the definition of unit of capacitance. A conductor has a capacitance of one farad if one coulomb of charge raises its potential by one volt. The practical unit of capacitance is called the *farad* (named after Faraday) and *it is the capacitance of a body whose potential is raised one volt by a charge of one coulomb.* For practical purposes, farad is a very large unit, and hence a submultiple of it, called the micro farad μF which is one-millionth of the farad is often employed. A still smaller unit called the micro-micro farad $\mu\mu f$ which is one-millionth of a micro farad, is also used.

One micro farad $\mu F = 10^{-6}$ Farads

One micro-micro farad $\mu\mu F = 10^{-6}\ \mu f = 10^{-2}\ F = 1$ pico Farad

26.2. Principle of a Condenser

Let a metal plate A be connected to positive pole of a battery [Fig. 1.1 (a)] and let it acquire the potential of the pole of the

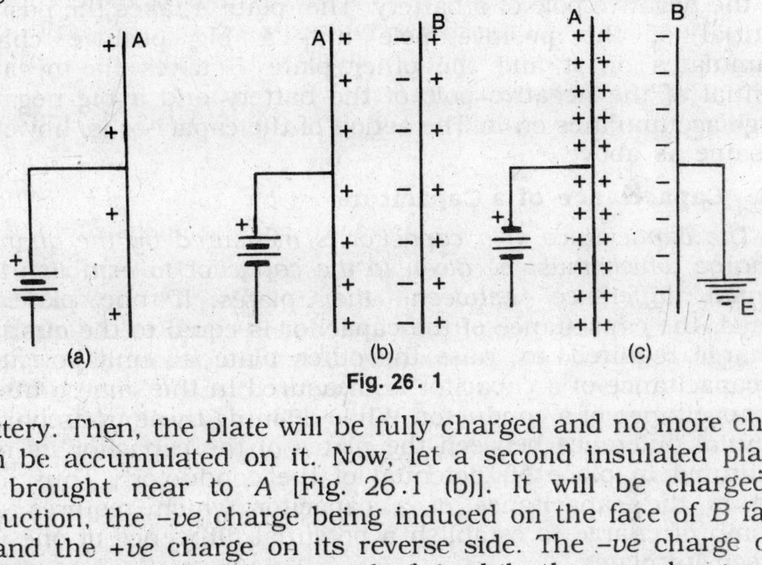

Fig. 26. 1

battery. Then, the plate will be fully charged and no more charge can be accumulated on it. Now, let a second insulated plate B be brought near to A [Fig. 26.1 (b)]. It will be charged by induction, the *–ve* charge being induced on the face of B facing A and the *+ve* charge on its reverse side. The *–ve* charge on B will tend to lower the potential of A while the *+ve* charge on B will tend to raise the potential A, and since the *–ve* charge is nearer to A its effect on the potential of A will be greater than that of *+ve* charge even though they are equal in amount. The result is that the potential of A is lowered a little and more charge flows from the battery to A in order to raise its potential to its original value, *i.e.*, the capacitance of A increases.

Now, let the plate B be earthed [Fig. 26.1 (c)]. Its *+ve* charge will go to earth and the remaining *–ve* charge will slightly increase as the nearby *+ve* charge which tendeed to decrease it has been removed. This increased *–ve* charge on B lowers the potential of A considerably so that still more charge can be accumulated on it in order to bring its potential to its original value, *i.e.*, the capacitance of *A* has further increased. Thus, the capacitance of a charged conductor increases when an insulated conductor is brought near it and more so when the latter is earthed. Such an arrangement of increasing the capacitance of a conductor is called a condenser or capacitor. It consists of two conductors separated by an insulated substance. The conductors are called the plates and the insulating medium between them the dielectric of the condenser.

In laboratories, the condenser is usually charged by connecting one plate A to the postive pole and the other plate *B* to the negative pole of a battery. The plate *A* takes the positive potential of the positive pole and a big positive charge accumulates on it and the other plate *B* takes the negative potential of the negative pole of the battery and a big negative charge accumulates on it. The action of the capacitor is, however, the same as above.

26.3. Capacitance of a Capacitor

The capacitance of a capacitor is measured by the quantity of charge which must be given to the capacitor to establish unit potential difference between the plates. If one plate be earthed, the capacitance of the capacitor is equal to the quantity of charge required to raise the other plate to unit potential. The capacitance of a capacitor is measured in the same units as the capacitance of a conductor. While defining these units phrase "potential difference between the plates of the capacitor" is now substituted in place of "potential of the conductor". Thus, one farad is the capacitance of a capacitor which requires one coulomb of charge to establish a potential difference of one volt between its plates.

The capacitance of a capacitor depends upon three factors :

(*a*) The size of the plates. The capacitance of a capacitor is directly proportional to the size of plates—the greater the size, the greater the capacitance.

(*b*) The distance between the plates. The capacitance of a capacitor is inversely proportional to the distance between the plates—the greater the distance, the less the capacitance.

(*c*) The dielectric medium between the plates. The capacitance of a capacitor is directly proportional to the dielectric constant or permitivity *K* of the dielectric medium between its plates—the greater the dielectric constant *K*, the greater the capacitance.

Below are given the values of capacitance of a few types of condenser :

(i) Parallel plate capacitor. Area of each plate A and distance between the plates d. $C = \frac{KA}{d}$

(ii) Spherical capacitor. Inner sphere of radius a insulated and outer sphere of radius b earthed, $C = 4 \pi K \frac{ab}{b-a}$

(iii) Spherical capacitor. Inner sphere of radius a earthed and outer sphere of radius b insulated, $C = 4 \pi K \frac{b^2}{b-a}$

(iv) Cylindrical capacitor. Length l, radii of inner and outer surfaces a and b respectively, $C = \frac{4 \pi Kl}{2.3026 \times 2 \log_{10} (b/a)}$

Experiment 26.1

Object—To compare the capacitances of two capacitors by means of a ballistic galvanometer.

Apparatus—Two capacitors whose capacitances are to be compared, a ballistic galvanometer, an accumulator, a rheostat, a Morse key, a two-way key, a tapping key and connecting wires.

Theory—Let the two capacitors of capacitances C_1 and C_2 respectively be charged to the same potential difference V between the plates, then, if Q_1 and Q_2 be the respective charges on the plates of the two capacitors, we have :

$$Q_1 = VC_1 \text{ and } Q_2 = VC_2$$

or

$$Q_1 / Q_2 = C_1 / C_2 \qquad (26.1)$$

Now, let the capacitors be discharged separately through a ballistic galvanometer and let θ_1 and θ_2 be the values of the first throw of the galvanometer in the two cases corresponding to movement of the spot of light through d_1 and d_2 cm. Then, we have :

$$Q_1 = K \theta_1 (1 + \lambda/2) \text{ and } Q_2 = K \theta_2 (1 + \lambda/2)$$

or

$$Q_1 / Q_2 = \theta_1 / \theta_2$$

where λ is the logarithmic decrement.

But

$$\frac{\tan 2\theta_1}{\tan 2\theta_2} = \frac{d_1}{d_2}$$

or if throws are small :

$$\theta_1/\theta_2 = d_1/d_2$$

and hence, $Q_1/Q_2 = \theta_1/\theta_2 = d_1/d_2$ (26.2)

Comparing with Eq. (14.1), we get :

$$C_1/C_2 = d_1/d_2 \qquad (26.3)$$

Alternatively, a graph may be plotted taking d_2 along *X*-axis and d_1 along *Y*-axis. This will come out to be a straight line as shown in (Fig. 26.2) and its slope PR/QR will give the ratio C_1/C_2 of the two capacitances.

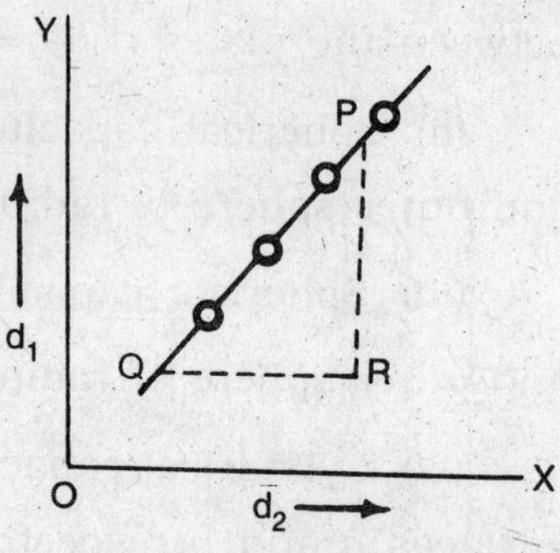

Fig. 26.2

Morse key—Morse key makes a good charge and discharge key and at the same time does not permit the capacitor to be insulated from the circuit. It consists of a stiff metal lever *AK*, one end *A* of which is held by a spring in contact with a metal stud connected with '*a*'. When the other end *K* of the lever is depressed, the contact of *A* with the stud below it is broken

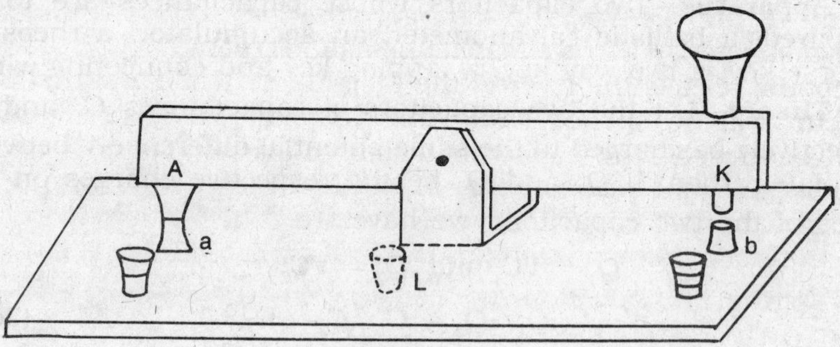

Fig. 26.3

and the contact is made between *K* and the other stud connected with *b*. When the lever is released, the spring breaks contact at *b* and restores at *a*. If the capacitor is connecred to the fulcrum *L* and the galvanometer and the battery at *a* and *b* respectively, then, when *K* is depressed the capacitor-battery circuit is completed and the capacitor is charged, and when *K* is released the capacitor-battery circuit is broken while the condenser-galvanometer circuit is made and the condenser is discharged through the galvanometer.

Method—Level the base of the galvanometer by means of the levelling screws and release its coil. Throw light on the mirror of the galvanometer and get the spot of reflected light on the scale. Next, make connections as shown in Fig. 26.4. Adjust the sliding contact *O* of the rheostat *XY* to get a suitable voltage between *XO*. Adjust the position of spot of light on the scale at zero. Connect *P* to *M*. This connects the capacitor C_1 to the galvanometer. Depress the end *K* of the lever of the Morse key the capacitor-battery circuit is completed and the capacitor is charged by the battery. Note that while the capacitor is being charged, one terminal of the galvanometer is kept insulated from it and hence the charge does not pass through the galvanometer. When the capacitor has been charged for about 20 seconds, release the lever of the Morse key. The capacitor-battery circuit will be opened while the capacitor galvanometer circuit will be closed. The capacitor will be discharged and a quantity of charge equal to the charge of one of its plate will pass through the galvanometer. Note down the movement of the spot of light on the scale corresponding to the first throw of the galvanometer.

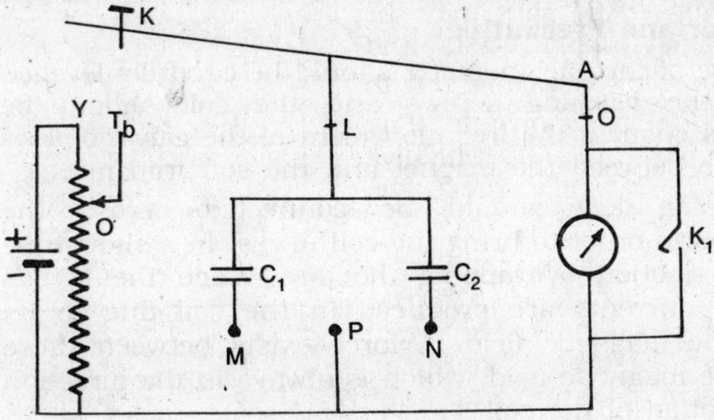

Fig. 26.4

Bring the galvanometer coil to rest by tapping the key K_1. When the spot of light on the scale is stationary substitute the second capacitor for the first by connecting *P* to *N* and determine the movement of spot of light corresponding to the first throw of the galvanometer as above by first charging the capacitor for about 20 seconds and then discharging it through the galvanometer. The ratio of the two throws gives the ratio of the capacitance of the two capacitors. Repeat the experiment at least six times by applying different voltages to the capacitors with the help of the rheostat *XY* and determine the mean ratio of the capacitances of the two capacitors.

Finally, plot a graph taking d_2 along *X*-axis and the

corresponding values of d_1 along Y-axis. This will come out to be a straight line as shown in Fig. 26.2. Measure its slope PR/QR which will give C_1/C_2.

Sources of Error and Precautions

(1) The base of the galvanometer should be carefully levelled by means of the levelling screws and the coil should be unclamped. This ensures the free movement of the galvanometer coil in the space between the magnet and the soft iron piece.

(2) A tapping key should be connected across the galvanometer coil in order to bring the coil to rest in a short time by damping its motion by tapping the key. When the key is tapped induced currents are produced in the coil due to its movement in the magnetic field. A force exists between these currents and the magnetic field, which is always in the direction opposing the motion of the coil.

(3) A morse key should be used in charging and discharging the capacitor in order to avoid the risk of connecting the battery and the galvanometer directly. The use of two separate tapping keys is dangerous for they are liable to be pressed down simultaneously and thus connect the battery to the galvanometer.

(4) The e.m.f. of the battery used for charging the two condensers should be constant. An accumulator may be used for the purpose for the e.m.f. remains sensibly constant. If with the accumulator the throws of the galvanometer are very large, a fraction of E may be used by potentiometer arrangement.

(5) In absorptive capacitors the charging current dies away gradually over a long period of time after the capacitor has acquired the applied potential difference, and the charge continues to flow into the capacitor for some time after the charge necessary to produce the required potential difference has been taken up. In such capacitors, therefore, on account of charge absorption the capacitance is rather indefinite and depends upon the time of charging. The shorter the charging time, the nearer the measured capacity approaches its true value, provided of course the time is long enough to charge the capacitor to the potential difference applied. Thus, the capacitors should be charged for short intervals only, especially if the dielectric be oil or paraffined paper, and their charging time should be nearly equal.

(6) The graph between d_1 and d_2 will come out to be a straight line and it should be smoothly drawn.

Observations

Rest position of spot of light = zero

So. No.	Deflection d_1 corresponding to throw θ_1 mm.	Deflection d_2 corresponding to throw θ_2 mm.	d_1/d_2
1.			
2.			
3.			
...			
...			
...			
		Mean	

I set. $\frac{C_1}{C_2} = \frac{d_1}{d_2}$

(make similar calculations for other sets).

$\therefore$ Mean value of C_1/C_2 =

From Graph (26.2) PR = QR =

$$\frac{C_1}{C_2} = \frac{PR}{QR} =$$

Result—The capacitances of the two capacitors are in the ratio of.........

Criticism of the method—Since the ratio of capacitances is equal to the ratio of first throws of the galvanometer, quite accurate results can be obtained by this method by using a sensitive galvanometer. When the deflections, however, are large the approximate Eq. (26.3) will not hold good and then the exact equation $C_1/C_2 = \theta_1/\theta_2$ should be used for calculating the ratio of the capacitances. The values of θ_1 and θ_2 can be easily obtained from the relationships $\tan 2\theta_1 = d_1/L$ and $\tan 2\theta_2 = d_2/L$ where L is the distance of the scale from the mirror. For more accurate results, methods which depend upon adjusting for no deflection of the galvanometer should be used.

Exercise 1—*To compare the e.m.fs. of two cells by means of a ballistic galvanometer.*

Take a standard capacitor. First charge it fully with the cell of lower e.m.f., and then discharge it through the ballistic galvanometer noting down the first throw θ_1 of the galvanometer (Fig. 26.5). Next charge the capacitor fully with the cell of higher e.m.f. E_2 and then discharge it through the galvanomter noting down the first throw θ_2 of the galvanometer. Then, the ratio of the e.m.fs. of the two cells is given by $E_1/E_2 = \theta_1/\theta_2$

It is advisable to obtain the throw with the cell of lower e.m.f. first otherwise if the capacitor is not dielectric loss-free, the residual charge left over after experimenting with the cell of higher e.m.f. may appreciably affect the throw with the cell of lower e.m.f. This method of comparing e.m.fs. is absolutely free from errors due to polarisation. Its only disadvantage is that it is not a null method and hence its possible accuracy is less than that obtainable with a potentiometer.

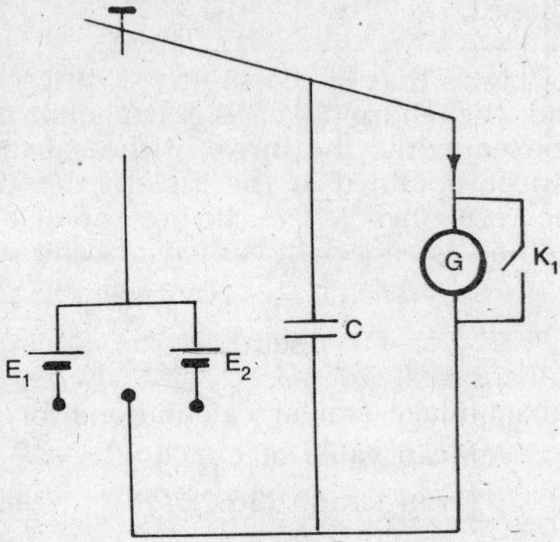

Fig. 26.5

Exercise 2—*To determine the internal resistance of a cell by means of a ballistic galvanometer.*

Connect a standard capacitor through a tapping key to the cell of e.m.f. E in series with a ballistic galvanometer as shown in Fig. 26.6. Place a resistance box across the cell. Remove the infinity plug of the resistance box. Tap the key and observe the first throw θ_1 of the galvanometer due to the charging of the capacitor. Then, $E \propto \theta_1$. Next insert the infinity plug in the resistance box and remove plug of resistance R equal to 2 or 3 ohm. Tap the key and again note the first throw θ_2 of the galvanometer due to charging of the capacitor to the p.d. V between the poles of the cell. Then $V \propto \theta_2$. Hence,

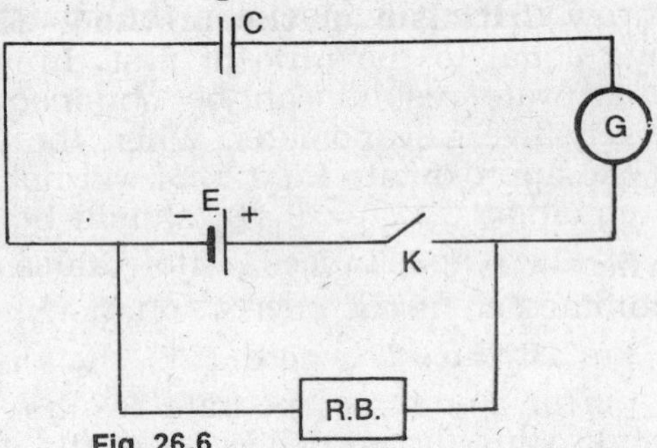

Fig. 26.6

$$\frac{E}{V} = \frac{\theta_1}{\theta_2}.$$

But from Eq. (26.36)

$$V = ER/(R + r)$$

Thus, $\frac{E}{V} = \frac{R+r}{R} = \frac{\theta_1}{\theta_2}$

whence, $r = \frac{\theta_1 - \theta_2}{\theta_2} R$

Note that by closing the key *K*, the closing of the cell circuit and the charging of the capacitor is affected simultaneously. Consequently, the throw of the galvanometer corresponding to *V* can be obtained at the instant the circuit is completed through the resistance *R*, *i.e.*, before the cell gets time to polarise. This method is especially suited to cells which polarise quickly.

Experiment 26.2

Object—To determine the ballistic constant *K* of moving coil ballistic galvanometer with a standard capacitor of known capacitance.

Apparatus—The moving coil ballistic galvanometer of suspended type, a capacity box, an accumulator, a voltameter, a Morse key and a tapping key.

Theory—Let a standard capacitor of known capacitance *C* be charged fully to a known potential difference *E* and let it be then discharged through the ballistic galvanometer whose constant K is to be determined. Then, if θ_1 be the first observed throw of the galvanometer, the charge passing through it is given by :

$$Q = CE = K\theta_1 (1 + \lambda/2)$$

whence, $$K = \frac{CE}{\theta_1 (1 + \lambda/2)} \quad (26.4)$$

where λ is the logarithmic decrement of the circuit and can be obtained from the expression :

$$\lambda = \frac{1}{10} \times 2.3026 \log_{10} \frac{\theta_1}{\theta_n}$$

where θ_n stands for the *n*th throw of the galvanometer.

Method—Level the base of the galvanometer by means of the levelling screws and release its coil. Set up a scale in front of the galvanometer distant one metre from it. Throw light on the mirror of the galvanometer and get a bright spot of reflected light on the scale. Adjust the spot of light on the scale at zero. Connect one terminal of the capacity box through a Morse key to one terminal of an accumulator and one terminal of the galvanometer. Connect the second terminal of the capacity box directly to the second terminal of the accumulator and the galvanometer as shown in Fig. 26.7. Connect a tapping key across the galvanometer.

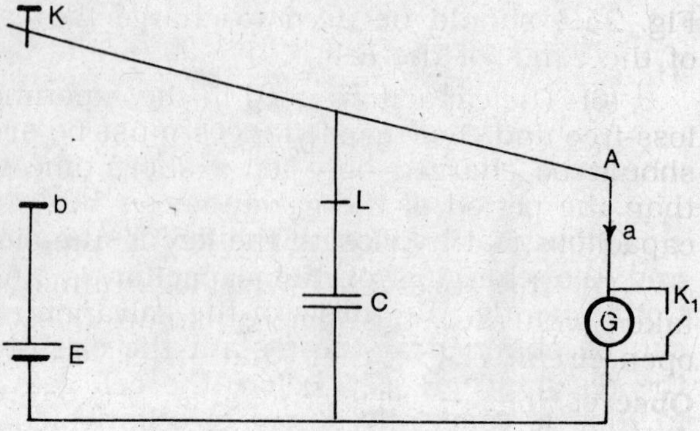

Fig. 26.7

Adjust the capacity box to a *suitable* capacitance. Press the knob *K*. The cell circuit will be closed and the capacitor will be charged. During the charging of the capacitor one terminal of the galvanometer remains insulated from the capacitor and hence the charge does not pass through it. When the capacitor has been charged for about 20 sec., release *K* : the cell circuit will be broken while that of the galvanometer closed. The capacitor will discharge through the galvanometer. Note down the deflection of spot of light on the scale corresponding to first and the eleventh throws of the galvanometer. Measure the e.m.f. of the accumulator with a voltameter. Calculate the constant *K* of the galvanometer from Eq. (26.4). If even with the least capacitance available in the capacity box, the spot of light goes off the scale, the capacitor shall be charged with a fraction of the cell by the potential divider arrangement as depicted in Fig. 26.4. Repeat the experiment several times with different values of *C* and find the mean value of *K*.

Sources of Error and Precautions

(1) The base of the galvanometer should be carefully levelled by means of the levelling screws and its coil unclamped before taking any observations with it.

(2) A tapping key should be connected across the galvanometer.

(3) A Morse key should be used in charging and discharging the condenser.

(4) The e.m.f. of the cell used to charge the capacitor should be constant. An accumulator may be used for the purpose.

(5) The capacitance of the capacitors chosen should be of such a magnitude as to give suitable throw of light spot on the scale. In case this is not possible and the spot of light moves off the scale the potential divider arrangement as depicted in

Fig. 25.4 should be used to charge the capacitor by a fraction of the e.m.f. of the cell.

(6) The capacitors used in the experiment must be dielectric loss-free and their capacitances must be accurately known. They should be charged only for a short time which should be less than the period of the galvanometer but sufficient to charge the capacitors to the potential difference applied.

(7) The observations for logarithmic decrement should be taken with the galvanometer circuit closed, for the damping on open circuit is less than that on closed circuit.

Observations— Determination of θ_1 and K

Rest position of spot of light = zero

S. No.	Capacitance C μF	P.D. across the capacitor E in volts	Deflection of spot of light corresponding to throws of galvanometer		λ	K Coulomb per mm
			θ_1	θ_2		
1.						
2.						
3.						
					Mean	

Calculations—Set I : $\lambda = \frac{1}{10} \times 2.3026 \log_{10} \times \frac{\theta_1}{\theta_{11}}$

$$K = \frac{CE}{\theta_1 (1 + \lambda/2)}$$

$$=$$

(N.B. Make similar calculations for other sets).

Result—The constant K of the ballistic galvanometer

= coulombs/mm.

Criticism of the method—The method gives a fairly satisfactory value of K. If the deflection d_1 of the spot of reflected light on the scale is large, then instead of taking θ_1 equal to d_1, its value should be calculated from the expression $\tan 2\theta_1 = d/L$, where L is the distance of the scale from the mirror. The accuracy of result is affected by the dielectric loss and the inaccuracy in the knowledge of true value of the capacitance of the capacitors. The value of K depends upon the nature of the working conditions and upon the distance of the scale from the mirror if θ_1 is taken to be equal to d_1. Hence, while expriment with a ballistic galvanometer either K is eliminated from the formula used to determine the unknown quantity or it is evaluated in the course of the work.

Experiment 26.3

Object—To compare the capacitances of two capacitors by *de Sauty's method.*

Apparatus—Two capacitors, two high-resistance boxes, a battery of accumulators of e.m.f. 6 to 9 volts, Morse key, a high-resistance galvanometer and connecting wires.

Theory—As depicted in Fig. 26.8, the two capacitors whose capacitances C and C' are to be compared form the two *adjacent* arms of Wheatstone bridge. The remaining two arms XY and YZ of the bridge consist of *non-inductive* high resistances R and R' respectively. A high-resistance galvanometer G is inserted between X and Z and a battery of rather high e.m.f. E is connected between Y and W through a Morse key K.

When the knob a of the key K is pressed, it makes contact at b and the capacitors are charged. When the knob a is released, contact is made at d and the capacitors get discharged. The value of the resistances R and R' are so adjusted that the galvanometer coil remains *undeflected* whether the key K is pressed or released. Then obviously the points X and Z remain at the same potential during charging as well as discharging of the capacitors.

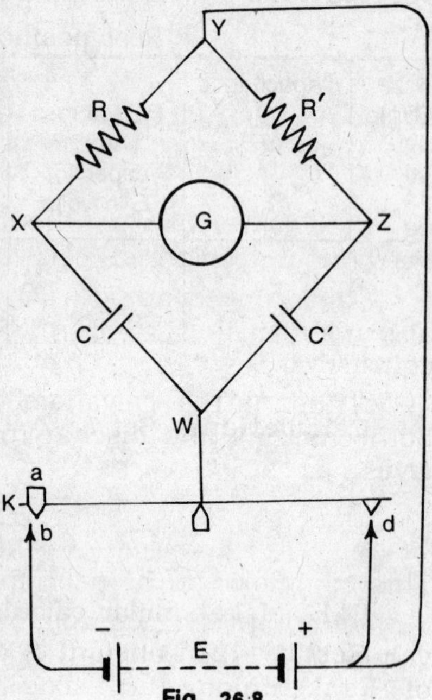

Fig. 26.8

Let at any instant t, reckoned from the instant the key K is depressed, the charge on the condenser C is q and that on C' is q'. Then, from the theory of charging of a condenser with a resistance in series, we have :

$$q = q_0 \, e^{-t}/RC$$

and
$$q' = q'_0 \, e^{-t} \, t/R'C'$$

where q_0 and q'_0 are charges on the two capacitors respectively when they are *fully* charged :

Now, let V and V' be the p.d. across the capacitors C and C' respectively at the instant t. Then, obviously,

$$V = \frac{q}{C} = \frac{q_0}{C} e^{-t/RC}$$

and

$$V = \frac{q'}{C'} = \frac{q'_0}{C'} e^{-t/R'C'}$$

But if X and Z are to be at the *same* potential at any instant t during the charging of the capacitors,

$$V = V$$

or

$$\frac{q_0}{C} e^{-t/RC} = \frac{q'_0}{C'} e^{-t/R'C'}$$

Now $q_0/C = q'_0 \; C' = E$, the e.m.f. of the battery. Hence, the above equation gives :

$$\frac{t}{CR} = \frac{t}{C'R'} \tag{26.5}$$

i.e., the time-constant for the two capacitor-circuits are equal when the bridge is balanced.

The same condition for the balancing of the bridge would also result if the process of discharging of the two capacitors is considered.

Thus, if the galvanometer-coil remains stationary during both charging and discharging of the two capacitors, Eq. (26.5) gives :

$$\frac{C'}{C} = \frac{R}{R'} \tag{26.6}$$

This equation can be used to compare the two capacitances.

Alternatively, a graph may be plotted between the various values of R taken along the X-axis and the corresponding values of R' taken along the Y-axis. Then, the slope of the straight line thus obtained will give the ratio C/C' of the two capacitances.

Method—Level the base of the high resistance galvanometer by means of the levelling screws and unclamp its coil. Throw light on the mirror of the galvanometer and get a bright spot of reflected light on the scale. Form a Wheatstone bridge as shown in Fig. 26.8 by connecting the two given capacitors C and C' between W and X, and W and Z respectively and two high-resistance boxes, one between X and Y and another between Y and Z. Next, connect the galvanometer between X and Z and a battery of 6-9 volts between Y and W through a Morse key K as shown.

Adjust the resistance box in XY to a suitable value R, say,

10,000 ohm and the resistance box in *YZ* to the *same* order *R′*. Depress the knob *a* of the key *K*, when the capacitors will be charged and, in general, a throw wil be produced in the galvanometer and hence the spot of light will move on the scale. Release the key *K* when the capacitors will get discharged and there will again be a movement of the spot of light on the scale.

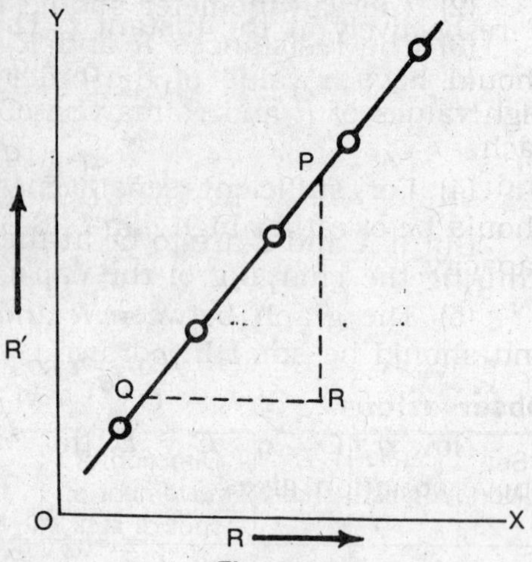

Fig. 26.9

Now, keeping *R* fixed, adjust the value of *R′* such that there is no throw in the galvanometer, *i.e.*, the spot of light remains *stationary* on the scale, whether the key *K* is pressed or released, *i.e.*, when the capacitors are charged or discharged. To effect this adjustment of *R′* first give *R′* some value and observe the direction of movement of the spot of light on the scale during charging of the two capacitors. Then, give *R′* the maximum available in the resistance box and again observe the direction of movement of the spot of light on the scale during charging of the capacitors, when it will be found to be *opposite* to that in the previous case. Gradually, narrow down the gap between these two values of *R′* until a value of *R′* is found out with which the bridge is balanced. In the initial stages it may be found convenient to alter *R′* in steps of 1000 ohm, then 100 ohm and finally, if necessary, in steps of 10 ohm.

Note down the values of *R* and *R′* when the bridge is balanced and calculate the ratio $\frac{C}{C'}$ from Eq. (26.6). Repeat the above steps by altering the value of *R* and find the mean value of $\frac{C}{C'}$ after calculating its value *separately* from *each* set of observations. Finally, plot a graph between *R* and *R′*, taking *R* along the *X*-axis and *R′* along the *Y*-axis. The graph will come out to be a *straight line* as shown in Fig. 26.9. Find the *slope* *PR/QR* of *this* line which will give $\frac{C}{C'}$.

Sources of Error and Precautions

(1) The base of the galvanometer should be carefully levelled and its coil should be unclamped before use.

(2) The galvanometer should be of *high* resistance.

(3) The resistances R and R' should be non-inductive and should have a value of the order of 10,000 ohms. Sufficiently high values of R and R' may be obtained by using a P.O. box of each.

(4) For sufficient sensitiveness of the bridge, the battery should be of rather high e.m.f., 6-9 volts but need not be of large capacity.

(5) The graph between R and R' should be a straight line and should be smoothly drawn.

Observations

Set No.	R ohms	R' ohms	Direction of movement of spot of light	R' for galv. balance	Point on	$\frac{R'}{R}$
1	10,000	9000 9100 9200	Left No. deflection Right	9100	A	
2					B	
...					...	
...					...	
					Mean	

Calculations

Set I. $\frac{C}{C'} = \frac{R'}{R} \qquad =$

(N.B.—Make similar calculations for orther sets).

$PR \qquad = \qquad QR =$

$$\frac{C}{C'} = \frac{PR}{QR} =$$

Result—The ratio of the capacitances of the given two capacitors :

(*i*) from calculations : $\frac{C}{C'} =$ .....

(*ii*) from graph : $\frac{C}{C'} =$ ......

Criticism of the Method—This is a simple method of comparing two capacitances. It is a *null* method and gives the best results when the resistances R and R' are considerable and the capacitors have large capacitances and are dielectric loss-free. But the sensitiveness is not great for the quantity of charge which flows through the galvanometer, is only a *small* part of the difference of charges on the capacitors which themselves are *small*. Consequently, the deflection of the galvanometer is very small and the adjustment of the resistances R and R' can be varied usually over a *wide* range, without causing any appreciable change in the deflection of the galvanometer.

The experiment can also be performed by using an a.c. source such as an audio frequency oscillator instead of the battery and some a.c. detector, *e.g.*, head-phones, in place of the galvanometer. In such a case the resistances R and R' are so adjusted that there is perfect silence in the head-phones.

26.4. Combination of Capacitors

(1) *Law of capacitances in parallel*—When the capacitors are so arranged that their similarly charged plates are connected together (Fig. 26.10), they are said to have been connected in parallel ; and the total capacitance of the system of capacitors arranged in parallel is equal to the sum of the individual capacitances.

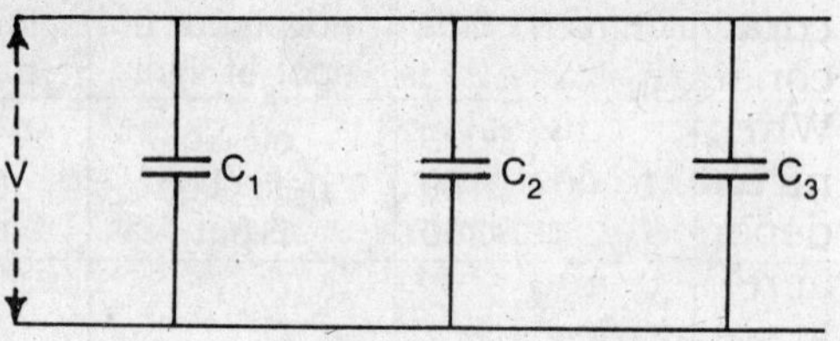

Fig. 26.10

$$C_t = C_1 + C_2 \ldots\ldots\ldots\ldots C_n$$

or $$C_t = \Sigma\, C_n$$

It is evident from above that the capacitance increases where capacitors are connected in parallel.

(2) *Law of capacitances in series*—When the capacitors are so arranged that one plate of one capacitor is connected to the oppositely charged plates of the other and so on (Fig. 26.11), they are said to have been connected in series and the reciprocal of the total capacitance of the system of capacitors arranged in series is equal to the sum of the reciprocals of the individual capacitances. Thus, mathematically,

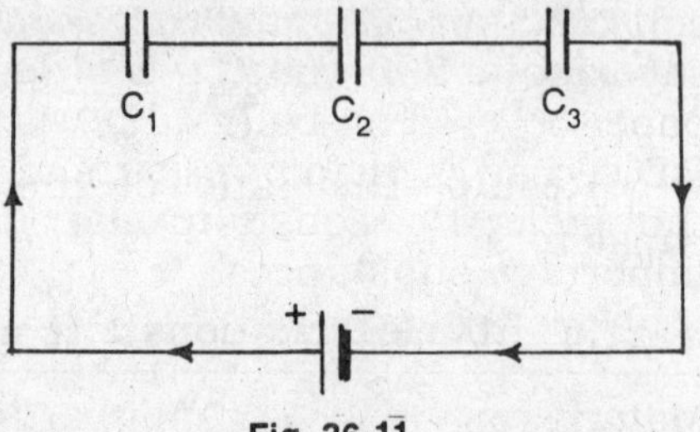

Fig. 26.11

$$\frac{1}{C_t} = \frac{1}{C_1} + \frac{1}{C_2} + \frac{1}{C_3} \ldots\ldots\ldots \frac{1}{C_n}$$

or $$\frac{1}{C_t} = \Sigma \frac{1}{C_n}$$

It is evident from above that the capacitance is decreased by connecting capacitors in series. In a series arrangement note also that the total capacitance is *less* than the least of the individual capacitances.

26.5. The Dielectric Constant or Permitivity

We have stated in 26.3 that the capacitance of a capacitor increases K times when the space between its plates is filled with a medium of dielectric constant K instead of air. Hence, *the dielectric constant K of any substance can be measured by the ratio of the capacitance of a capacitor with that substance filling the space between its plates to the capacitance of the same capacitor with air in the space between the plates.* The dielectric constant for air is equal to unity for most practical purposes. When a dielectric is traversed by lines of force, there is a mechanical strain in it. When this strain exceeds a certain limit depending upon the nature of the dielectric, a spark passes across it and the dielectric is punctured. Hence, while selecting a substance for the dielectric of a capacitor, we must consider not only its cost and value of dielectric but also its dielectric strength, *i.e.*, its ability to stand the application of high potential differences without being punctured. Thus, the break down potential difference is approximately proportional to the thickness of the dielectric, the dielectric strength is usually measured by the P.D. necessary to puncture one millimetre thickness and is expressed in volts or kilovolts per mm. Mica is the best dielectric for it has high dielectric constant and offers great resistance to rupture by electric discharge and is, therefore, used in the construction of standard condensers. For relatively small P.Ds., say, about 200 volts, the dielectric used is generally paraffined paper covered on both sides with tin or aluminium foil. For higher P.D.s, use is made of gas and oil. The following table contains the dielectric constant and the dielectric strengths of some important substances.

Dielectric constant and dielectric strength

Material	Approx, Dielectric Strength in kilo-volts per mm.	Approx. Dielectric Constant K
Asbestos	3-4.5	...
Bakelite	20-25	5-6
Cotton	3-4	...

Ebonite	10-40	2-3
Glass	5-12	3-8
Gutapercha	10-20	3-5
India rubber	10-25	2-3
Mica	40-150	3-8
Paper	4-10	2
Paraffin wax	8	2
Porcelain	9-20	4-7
Shellac	5-20	2.5-3.5
Slate	3	7
Oil	25-30	2-3.1
Water	...	40-90
Air	3	1.000585

ORAL QUESTIONS

COMPARISON OF CAPACITANCES

1. What do you understand by capacitance of a conductor and upon what factors does it depend ? What is its unit ? Define it.
2. How does the capacitance of conductors vary with the nature of the neighbouring conditions ?
3. What is a capacitor ? Explain how the capacitance of a conductor can be increased by placing near it an earthed conductor ?
4. Upon what factors does the capacitance of a capacitor depend ?
5. How can you increase the capacitance of a capacitor ?
6. Is the capacitance of a capacitor defined in the same way as the capacitance of a conductor ?
7. How can you determine or compare the capacitances of capacitors.
8. How do you charge a capacitor ? What precautions do you observe in your experiment ?
9. What is the use of Morse key ? Explain the charging and discharging of a capacitor with its help ?
10. Why don't you use a Leclanche cell in this experiment ?
11. Why do you charge the capacitors for a short time only ? Is the result in any way affected by the time of discharge also ?
12. Why should the throws be small ? If the throws are large, how can you reduce them ?
13. How will the formula for comparing the capacitances be modified if the throws are large ?

14. If the throw obtained with the capacitor of large capacitance is beyond the scale, how will you modify your method to get it on the scale keeping the throw with the other capacitor the same ?
15. How can you verify the laws of combination of capacitors by a ballistic galvanometer ?
16. How are standard capacitors constructed ? Why is mica generally used as dielectric in standard capacitors ?
17. What should be the qualities of a good dielectric for a standard capacitor ?
18. How do you define dielectric constant and dielectric strength of a material ?

CONSTANT OF A BALLISTIC GALVANOMETER

19. What is a ballistic galvanometer ? What are its chief features ? To what use is it generally put ?
20. What do you understand by the constant of a ballistic galvanometer ? How can you determine it with a standard capacitor of known capacitance ? What precautions do you take in this experiment ?
21. Do you know any other method of calibrating the ballistic galvanometer ? Compare them.
22. Does the value of *K* depend upon the working conditions and the distance between the scale and the mirror ?
23. What do you understand by logarithmic decrement ? Why and how do you determine it ?
24. Do you take observations for logarithmic decrement on closed or open circuit of the galvanometer and why ?
25. On what factors does the damping depend ? How can you increase or decrease the damping ?
26. What is the use of the tapping key connected across the galvanometer ? Explain how by tapping this key the galvanometer coil can be brought to rest in a short time ?
27. How can you compare the e.m.fs. of two cells by a ballistic galvanometer ?
28. Compare this method with the potentiometer method of comparing the e.m.fs. of the cells.
29. Can you determine the internal resistance of a cell by a ballistic galvanometer ? If so, how ?
30. Compare this method with other methods of determining the internal resistance of a cell.

27

Electromagnetic Induction and Varying Current

27.1. Laws of Electromagnetic Induction

In 1831, Faraday made the most important discovery that when the number of lines of magnetic induction passing through a conducting circuit is changed, a current is produced in the circuit. This phenomenon is called *electromagnetic induction* and the current produced by induction is termed the *induced current.* Faraday summed up the result of his experiments in the following laws of electromagnetic induction :

(1) *Whenever the magnetic flux threading or linked with a conductor is changed an induced e.m.f. acts round the conductor.* If the conductor be a closed circuit, an induced current flows round the circuit. The induced e.m.f. or the current lasts only while the change is taking place.

(2) *The magnitude of the induced e.m.f. is equal to the rate of change of the flux linked with the conductor,* that is,

$$e = -\frac{d\phi}{dt} \tag{27.1}$$

Note that the magnitude of the induced e.m.f. is absolutely independent of (*i*) the shape or size of the conductor, (*ii*) the nature of its material, or even (*iii*) the existence of a closed circuit. If the conductor forms a part of a closed circuit, the magnitude of induced current will be equal to the induced e.m.f. divided by the resistance of the circuit. The magnitude of induced current, therefore, depends upon the shape, size and the material of the conductor as well.

The change in magnetic flux linked with a conductor may be due to motion, relative to the conductor, of magnets or of conductors carrying currents, or, without relative motion, to the starting or stopping or variation of the currents in the neighbouring conductors or in the conductor itself. When the induced e.m.f. in a circuit has been produced on account of relative motion of a neighbouring circuit or on account of change of current in it, the circuit in which the e.m.f. is induced is called the secondary and the neighbouring circuit, the primary.

The negative sign in Eq. (27.1) shows that *the direction of induced e.m.f. (and current) is such that it tends to oppose the very cause which produces it, i.e.,* the induced current establishes a magnetic flux which tends to neutralize the change in flux which causes it. This statement is known as *Lenz's law* or third law of electromagnetic induction and is very helpful in determining the direction of the induced current. From this law it is evident that when the current in the primary circuit is increased or the primary moved nearer to the secondary, the current induced in the secondary will be in a direction opposite to the primary current, *i.e., inverse*, and when the current in the primary is decreased or the primary moved away from the secondary the current induced in the secondary will be in the same direction as the primary current, *i.e., direct.* In the case of a single conductor moving in a magnetic field at right angles to itself and the field, direction of induced e.m.f. (and current) can be conveniently determined by Fleming's Right Hand Rule which runs as follows—*Hold the thumb and the first two fingers of the right hand mutually at right angles. Then if the forefinger points in the direction of the magnetic field and the thumb in the direction of motion of the conductor, the finger will point in the direction of the induced e.m.f. (and current).*

27.2. The Self-Inductance (L) of a Circuit

When the current in a closed circuit is started, stopped or varied, the magnetic flux linked with the circuit changes and hence an e.m.f. (and current) is induced in the circuit. This effect is known as *self-induction*, when the permeability of the medium is constant, the value of the effective magnetic flux linked with a circuit is proportional to the primary current in the circuit. Thus, if ϕ be the effective magnetic flux linked with a circuit when a current I is flowing through it.

$$\phi \propto I$$

or

$$\phi = LI \tag{27.2}$$

where L is constant for the circuit and is known as *coefficient of self induction* of the circuit. If the circuit consists of a coil of wire with *n turns*, each of the magnetic lines of force threads through all of the turns. Hence, the effective flux in the case is equal to n times the flux through each turn. If $I = 1$, $\phi = L$. Hence *coefficient of self-induction or self-inductance of a circuit is equal to the effective magnetic flux or linkages of the circuit when unit current is flowing through it.*

When the magnetic flux linked with a circuit is changing, owing to the current changing, then the e.m.f. induced in the circuit is :

$$e = -\frac{d\phi}{dt}$$

where $\frac{d\phi}{dt}$ is the rate of change of flux through the citcuit.

Putting $\phi = LI$, the above equation is modified to :

$$e = -\frac{d}{dt}(LI)$$

$$= -L\frac{dI}{dt} \qquad (27.3)$$

where $\frac{dI}{dt}$ is the rate of change of current in the circuit.

If $\frac{dI}{dt} = I$, $e = L$. Thus, coefficient of self-induction *or inductance of a circuit is numerically equal to the induced e.m.f. in the circuit when the current is changing at the rate of one unit per second.* The negative sign in Eq. (27.3) shows that the e.m.f. is inverse or opposing when the rate of change of current is positive, *i.e.*, current is increasing. Eq. (27.3) can be used to define unit of self-inductance. *The self–inductance of a coil is one henry if a current changing at a rate of one Ampere Sec^{-1} causes an induced e.m.f. of one volt in the coil.* The self-inductance of an iron-free coil depends only upon the geometry of the coil and the number of turns in it.

The self-inductance is analogous to inertia for when the current is started in a circuit, the induced e.m.f. tends to retard the growth of current. The result is that the current does not reach its steady maximum value instantaneously but takes some time which may, in some cases, be several minutes. For instance, when a d.c. motor is started it takes a fairly considerable time, on account of large self-inductance of field coils, before the field current attains its full strength. Similarly, when the current is stopped, the induced e.m.f. on account of self-induction tends to make it keep on, and if the self-inductance is large and the break very quick, the magnitude of the induced e.m.f. is very great. For example, when a d.c. motor is shut off, the e.m.f. induced is very great so that an arc is formed at the switch which, some time, burns the contacts away.

A loop of wire and a short straight wire have negligible self-inductance and are said to be practically non-inductive. In a loop of wire the induction effect of every small portion of it is neutralised by the part diametrically opposite to it. It should be noted that no circuit is absolutely non-inductive, for even with a straight wire, some field must exist around it. If a coil of wire consists of many turns closed together, the self-inductance is very large, *i.e.*, it is highly inductive. If a coil is wound on an iron core, the self-inductance will be considerably increased, *e.g.*, in field coils of d.c. motors. In coils of resistance boxes the coils are wound almost completely non-inductively by first doubling the wire along its length and then winding it on the bobbin. This bifilar winding gives the effect of two wires side by side but carrying currents in opposite directions ; the result is that the magnetic field due to current in one is neutralised by the magnetic field due to opposite current in the other and the coil has a negligible self-inductance.

Experiment 27.1

Object—To determine the self-inductance of a given coil by Rayleigh's method.

Apparatus—The coil whose self-inductance is to be determined, a Post Office box, a moving coil ballistic galvanometer of suspended type, a decimal ohm box containing resistances of the order of 0.1 and 0.01 ohm, a platinoid wire and a binding screw, an accumulator, a special double key as shown in Fig. 27.2, a tapping key and connecting wires.

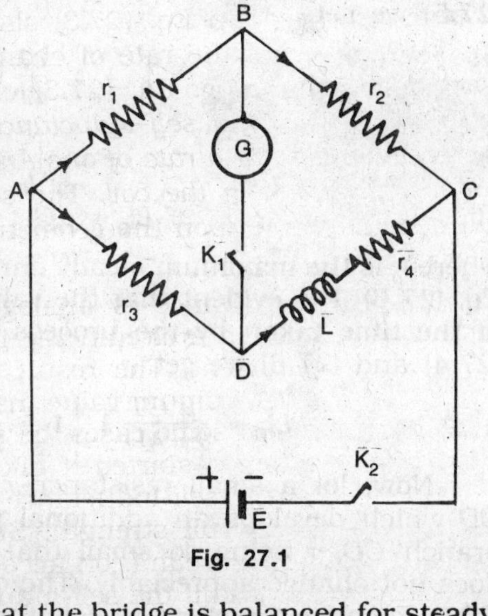

Fig. 27.1

Theory—Let the coil whose self-inductance L is to be determined be placed in one arm of a Wheatstone brige as shown in Fig. 27.1, the resistance in the other arms being non-inductive, and let the resistances of the various arms be so adjusted that the bridge is balanced for steady currents, the galvanometer used being ballistic. Then, if the battery circuit is opened while the galvanometer circuit remains closed, during the decay of current an e.m.f. $-L\frac{dI}{dt}$ will be induced in the coil in the branch CD, where I is the current in CD at any instant. This will send an instantaneous current in the galvanometer, the result being a momentary throw in the galvanometer. Let θ_1 be the first observed throw of the galvanometer, then the quantity of charge passing through the galvanometer due to the above cause will be given by

$$Q = \frac{C\,T}{2\pi\,AH}\,\theta_1\left(1+\frac{\lambda}{2}\right) \tag{27.4}$$

where C is the torsional reaction per unit radian twist, H the field of the permanent magnet, λ the logarithmic decrement, T the period of the galvanometer and A the effective face area of the coil of the galvanometer.

The instantaneous current i in the galvanometer due to induced e.m.f. in L will be proportional to $-L\,\frac{dI}{dt}$ that is,

$$-L\,\frac{dI}{dt} = \mu\,i \tag{27.5}$$

where μ is a constant for the circuit depending upon the relative values of resistances in the various branches. Integrating Eq. (27.5) we get :

$$\int_0^t -L\,\frac{dI}{dt}\,dt = \int_0^t \mu i\;dt$$

or

$$-L\int_{I_0}^{0} dI = \mu\int_0^t i dt$$

or

$$LI_0 = \mu Q \tag{27.6}$$

where I_o is the maximum steady current flowing through CD. From Eq. (27.6) it is evident that the value of charge Q is independent of the time taken by the process of induction. Combining Eqs. (27.4) and (27.6) we get :

$$LI_0 = \frac{\mu\,CT}{2\pi\,AH}\,1\left(1+\frac{\lambda}{2}\right) \tag{27.7}$$

Now, let a small resistance r be introduced in the branch CD which develops an additional potential difference rI_0 in the branch CD, r being so small that the current I_o in the branch does not change appreciably. The effect of this additional P.D. in CD is to cause a current to flow through the galvanometer, which is given by :

$$i' = \frac{rI_0}{\mu}$$

If the steady deflection produced in the galvanometer by this current i' be ϕ we have :

$$AHi' = C\phi$$

or
$$AH \frac{rI_0}{\mu} = C\phi$$

whence
$$\mu = \frac{AHrI_0}{C\phi}$$

Substituting the value of μ in Eq. (27.7), we get :

$$L = \frac{T}{2\pi} .r . \frac{\theta_1}{\phi} \left(1 + \frac{\lambda}{2}\right) \tag{27.8}$$

The value of logarithmic decrement can be obtained from the expression :

$$\lambda = 2.3026 \log \frac{\theta_1 + \theta_2 + \theta_2 \ldots \theta_{n-1}}{\theta_2 + \theta_4 + \ldots \theta_n} \tag{27.9}$$

where θ_n is the *n*th observed throw of the galvanometer.

Description of Double Key—The double key shown in Fig. 27.2 is very helpful in breaking the galvanometer circuit the moment the discharge has passed through it. It consists of three strips *PP′, QQ′ and RR′ and a block SS′* of brass mounted horizontally on ebonite block and connected to separate terminals T_1, T_2, T_3 and T_4 as shown. The battery is connected between T_2 and T_4 and the galvanometer between T_1 and T_3. When *M* is depressed *R* and *S* are connected and the battery circuit is closed on depressing *P* the contact is made between *P* and *Q* and the galvanometer circuit is also closed. The strips *Q* and *R* remain insulated by the ebonite strips in between them. Thus, *M* and *P*

Fig. 27.2

are simply battery and galvanometer key respectively and with their help the bridge can be balanced for steady currents. Now, if keeping *P* depressed *M* is released this will not break any of the two circuits but if now *P* is also released the battery circuit will be broken first and then after some time the galvanometer circuit, thus opening the galvanometer circuit immediately after the discharge has passed through it and the impulse given to it. In this case, the throw of the galvanometer will almost be independent of electromagnetic damping.

Method—Level the base of the galvanometer by means of the levelling screws and release its coil. Throw light on the mirror of the galvanometer and get a bright spot of reflected light on the scale. Adjust the spot of light on the scale at zero. Connect the coil *L* whose self-inductance is to be determined in the unknown resistance arm of the Post Office box inserting in series with it a decimal ohm box containing resistances of the order of 0.1 and 0.01 ohms. In series with the rheostat arm of the P.O. box put a platinoid wire the length of which can be varied by slipping it through a binding screw. Connect the galvanometer and the battery directly to their places in the P.O. box through the special double key shown in Fig. 27.2, thus avoiding the use of the P.O. box keys. Insert a rheostat of high value in the battery circuit and adjust it to a suitable value. Connect a tapping key across the galvanometer.

Adjust the decimal-ohm box in series with the coil *L* to zero ohm. Adjust the ratio arms of the P.O. box to 10 : 10 or to 1 : 10 if available according as whichever order is nearer the resistance of the coil *L*. Close the battery circuit by depressing *M* and then the galvanometer circuit by depressing P. Adjust the rheostat in the battery circuit if necessary until a convenient deflection is obtained and the spot of light appears on the scale. Now, adjust the rheostat arm of the P.O. box until by changing the resistance in it by one ohm the direction of deflection in the galvanometer is reversed. Then, adjust the length of the platinoid wire in the circuit until a perfect balance of the bridge for steady currents is obtained, the final adjustment being done by passing the current only for a short time and thus avoiding the heating of the coil *L*.

Having obtained perfect balance for steady currents, release *M* keeping *P* depressed. Note that this will not break either the battery circuit or the galvanometer circuit. Next, release *P*, *R* and *S* will separate first and then *P* and *Q*. This will first break the battery circuit and then immediately after it the galvanometer circuit and a throw will be produced in the galvanometer. Note down the deflections of the spot of light on the scale corresponding to successive throws of the galvanometer. Then, determine the period of the galvanometer ; and after it bring it to rest by tapping the key connected across it.

Now, test the bridge again for balance for steady currents. If the balance has not remained perfect, alter the length of platinoid wire until the balance of the bridge is again perfect. Next, introduce in the arm containing the coil *L* a suitable small resistance r of the order of 0.01 ohm with the help of decimal-ohm box connected in series with the coil L. Depress the key M and then key P. A steady deflection will be produced in the galvanometer. Note down the deflection of spot light on the scale corresponding to this steady deflection.

Now, calculate the logarithmic decrement λ from Eq. (27.9) and then the value of self-inductance L of the given coil from Eq. (27.8). Repeat the experiment with different values of r and take at least three sets of observations. Find the mean value of L.

Sources of Error and Precautions

(1) The connecting wires should be short and straight. No coiled wire should be used for making connections.

(2) The base of the galvanometer should be carefully levelled by means of the levelling screws and the coil should be unclamped.

(3) A tapping key should be connected across the galvanometer in order to bring the galvanometer coil to rest in a short time by tapping the key, when so desired. The motion of the coil in the magnetic field induces in it an opposing current which brings the coil to rest.

(4) The e.m.f. of the cell used should be constant while the induced e.m.f. $-L\dfrac{dI}{dt}$ is being developed in the coil L in CD. Accumulator may be used for its e.m.f. remains sensibly constant.

(5) To get a suitable deflection a high adjustable resistance should be connected in series with the cell.

(6) Since the resistances r_1 and r_3 or r_2 and r_4 are in a way shunting the galvanometer, throw obtained in it is small, specially when the resistance of coil L is small. A measurable and reliable throw can be obtained by opening the galvanometer circuit immediately after the discharge has passed through it with the help of a special double key.

(7) In order to have high sensitiveness of the bridge, it is advisable to insert a suitable small non-inductive resistance in series with the coil L in the same arm CD.

(8) The current should be allowed to pass through the bridge only for as short a time as possible, otherwise the balance point will continually change due to the change of the resistance of the coil L in CD. The coil generally consists of copper wire

which has a large temperature coefficient for resistance. The key, therefore, should be pressed only when observations are to be made.

(9) The balancing of the bridge for steady currents must be perfect.

(10) The resistance r introduced in the arm CD in series with the coil L to produce steady deflection ϕ should be very small so that it may not affect the value of the steady current I_o in that branch appreciably. It may be preferably of such a magnitude as to produce ϕ equal to θ at the same time not affecting the value of I_o.

(11) For the determination of logarithmic decrement λ, as many successive throws of the galvanometer as it is possible to observe should be noted after the discharge has passed through it.

(12) While determining the period T of the galvanometer, the galvanometer circuit must be kept *open*.

Observations—(A) Determination of θ_1, ϕ, and λ

Rest position of spot of light = zero

S.No.	Scale deflection corresponding to successive throws of galvanometer						r ohms	ϕ in scale divisions	λ
	θ_1	θ_2	θ_3	θ_4	θ_5	θ_6			
1.									
2.									
3.									

(B) Determination of period of the galvanometer.

S. No.	No. of oscillations	Time taken		Period T Sec.
		Min.	Sec.	
1.				
2.				
3.				
			Mean	

Calculations—Set 1 : $\lambda = 2.3026 \log \dfrac{\theta_1 + \theta_2\, \theta_8 + \ldots \theta_{n-1}}{\theta_2 + \theta_3 + \theta_4 + \ldots + \theta_n}$

$$L = \frac{T}{2\pi}\ r\ \frac{\theta_1}{\phi}\left(1 + \frac{\lambda}{2}\right)$$

$$= \quad \text{henry}$$

Result—The self-inductance of the given coil = henry.

Criticism of the method—The method gives accurate results of the self-inductance of a coil specially when the double key is used to break the galvanometer circuit immediately after the discharge has passed through it. For more accurate results null methods, *e.g.*, Anderson's method in which the throw due to self-induction is neutralised by an equal and opposite effect due to a known capacitance, should be used. Rayleigh's method is suitable for the determination of self-inductance of coils not containing any magnetic material and having a fairly large value of self-inductance. When the coil has an iron core the *make and break* methods must not be used. With iron core the magnetic flux threading the circuit is not proportional to the current.

27.3. The Mutual Inductance (M) of Two Circuits

When two closed circuits are in close proximity, then, whenever, the current in any one of them is started, stopped or varied, there is a change in the magnetic flux threading the second circuit due to current in the first circuit and an e.m.f. (and current) is induced in the second circuit. This effect is known as *mutual induction.* Clearly, any one of the two coils can be used as the primary and the other as the secondary. If the permeability of the medium be constant the magnetic flux ϕ linked with secondary due to the current in the primary is directly proportional to the magnitude of the current in the primary. Thus,

$$\phi \propto I$$

or

$$\phi = MI$$

where M is a constant of the two circuits and for iron free circuits depends only upon the geometrical shapes and the relative positions of the two circuits. It is called the coefficient of mutual induction or simply the mutual inductance of the two circuits. If $I = 1$, $\phi = M$. Thus, *the* coefficient of mutual induction *or mutual inductance of two circuits is equal to the effective magnetic flux linked with one circuit due to unit current flowing in the other.*

When the magnetic flux threading the secondary circuit is changing owing to the current in the primary changing, then the e.m.f. induced in the secondary is given by :

$$e = -\frac{d\phi}{dt}$$

where $\frac{d\phi}{dt}$ is the rate of change of flux through the secondary owing to the current in the primary changing. Putting $\phi = MI$ in the above equation, we have :

$$e = -\frac{d}{dt}(MI)$$

$$= -M\frac{dI}{dt}$$

where $\frac{dI}{dt}$ is the rate of change of the current in the primary circuit.

When $\frac{dI}{dt} = 1$, $e = -M$. Thus, the coefficient of mutual induction *or mutual inductance of two circuits is numerically equal to the e.m.f. induced in one circuit when the current strength in the other varies at the rate of one unit per second.* The mutual inductance is also measured by the mutual potential energy of the two circuits when unit current is flowing in each. There is only one value of mutual inductance between two coils for any one of them can be used as the primary circuit.

The S.I. unit of mutual inductance is henry (H). The mutual inductance or coefficient of mutual induction of two coils is said to be one henry, when a current change of one ampere per second in one coil induces an e.m.f. of one volt in the other.

The mutual inductance of two circuits can be increased by increasing the number of turns in the coils as well as by winding them on iron cores. When iron core is used, the iron gets magnetised when the current passes through the primary. The change in magnetic flux is, therefore, much greater and hence the induction effects are considerably increased. The magnitude of the induced e.m.f. in the secondary varies directly as the number of turns in the secondary for the e.m.f. produced in different turns are added up. Hence, the e.m.f. induced in the secondary can be made as large or as small as we please by using manyor few turns. This accounts for the large number of turns used in the secondary of a transformer or an induction coil. Further, the greater the rate of change of current in the primary, greater will be the e.m.f. induced in the secondary. Hence, if the current in the primary is stopped suddently, large e.m.f. can be developed in the secondary. This explains the production of very high e.m.f. in the secondary of an induction coil at break which is very sudden.

Experiment 27.2

Object—To determine the mutual inductance of two coils by means of a ballistic galvanometer.

Apparatus—The two coils whose mutual inductance is to be determined, a moving coil ballistic galvanometer of suspended type, rheostat, a resistance box, an accumulator, a plug-type reversing commutator and two tapping keys.

Theory—Let the secondary winding of the mutul inductance be connected to a ballistic galvanometer and let a current *I* be allowed to pass through the primary winding. Then, as the current grows from 0 to *I* in the primary circuit, an e.m.f. equal to $-M \frac{di_1}{dt}$ is induced in the secondary, where *M* is the mutual inductance of the two coils and i_1 the current in the primary at any instant. This induced e.m.f. produces an instantaneous current in the secondary circuit which is given by :

$$i_2 = \frac{e}{R} = -\frac{M\, di_1}{R\,.\,dt}$$

where *R* is the resistance of the secondary circuit including that of the galvanometer. The charge passing through the galvanometer during the growth of current *I* in the primary is given by :

$$Q = \int_0^t i_2\,.\,dt = -\frac{M}{R}\int_0^t \frac{di_1}{dt.}\,dt$$

or

$$Q = -\frac{M}{R}\int_0^I di_1 = \frac{MI}{R} \qquad (27.10)$$

If $\theta_1, \theta_2 \ldots \theta$ be the consequent successive throws of the galvanometer, the charge passing through it is also given by :

$$Q = K\theta_1\,(1 + \lambda/2) \quad \ldots \qquad (27.11)$$

where *K* is the ballistic constant of the galvanometer and λ the logarithmic decrement.

Equating the two Eqs. (27.10) and (27.11) for Q, and simplifying, we get :

$$M = \frac{R}{I} K\theta_1\,(1 + \lambda/2)$$

If *T* is the period of oscillation of the galvanometer and *k* its figure of merit,

$$K = \frac{T}{2\pi.}\,k$$

$$\therefore \quad M = \frac{R}{T.}\,k\,\frac{T}{2\pi.}\,\theta_1\,(1 + \lambda/2) \quad \ldots \qquad (27.12)$$

Now, let the secondary winding in series with the ballistic galvanometer be connected across a small resistance *r*. This causes an e.m.f. *rI* to act in the secondary circuit which sends round current *I'* producing a steady deflection ϕ in the galvanometer so that,

$$I' = k\phi$$

or
$$k = \frac{I'}{\phi} = \frac{rI}{R} \cdot \phi \tag{27.13}$$

Putting this value of k in Eq. (27.12), we have :

$$M = \frac{T}{2\pi} \cdot \frac{r}{\phi} \cdot \theta_1 \;(1 + \lambda/2) \tag{27.13a}$$

from which M may be calculated. The value λ can be obtained from the expression :

$$\lambda = 2.3026 \log \frac{\theta_1 + \theta_2 + \theta_3 + \ldots\theta_{n-1}}{\theta_2 + \theta_3 + \theta_4 + \ldots + \theta_n}$$

Method—Level the base of the galvanometer by means of the levelling screws and unclamp its coil. Throw light on the

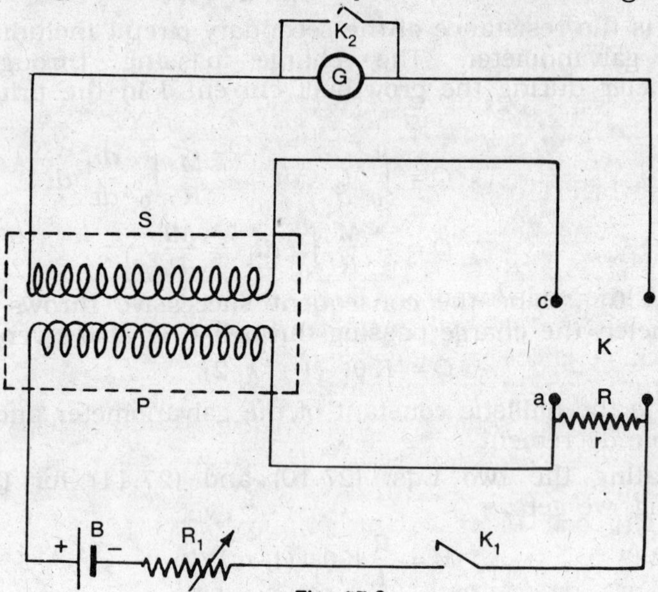

Fig. 27.3

mirror of the galvanometer and get a bright spot of light on the scale. Adjust the spot of light on the scale at zero. Connect the primary winding P (Fig. 27.3) to an accumulator B in series with a rheostat R_1, a resistance box R and a tapping key K_1. Next, connect the terminals of the resistance box R to the two collinear terminals a and b of a plug type commutator K, to the remaining terminals c and d to which connect the secondary winding in series with a ballistic galvanometer. Connect a tapping key K_2 across the galvanometer.

Adjust the resistance box R to a small value of resistance, say, r ohm. Close the primary circuit and adjust the current to a suitable value. Connect c to d and when the galvanometer coil is at rest, release the key K_1, and note down the deflections of the spot of reflected light on the scale corresponding to successive throws of the galvanometer. Bring the galvanometer coil to rest by tapping the key K_2.

Next, connect c to a and d to b and observe the steady deflection produced in the galvanometer.

Now, alter the value of r and repeat the experiment taking at least three sets of observations with different values of r. Determine the period of the galvanometer. Calculate the value of M from Eq. (27.13) for each set of observation separately and then take the mean value of M.

Sources of Error and Precautions

(1) The connecting wires should be short and straight. No coiled wire should be used for making connections.

(2) The base of the galvanometer should be carefully levelled by means of the levelling screws and the coil should be unclamped before passing current through it.

(3) A tapping key should be connected across the galvanometer.

(4) If the coils used are large, they should, if possible, be placed as far from the galvanometer as possible, so that there is no appreciable effect on the galvanometer coil due to direct action of maximum current in the primary circuit.

(5) A rheostat should be included in the primary circuit to adjust the value of the current in it to a suitable value, say, about one ampere.

(6) The cell used in the primary circuit should be an accumulator.

(7) For determining the successive throws of the galvanometer, its circuit should be kept closed for the damping on open circuit is not the same as that on closed circuit. Further, as many throws of the galvanometer as it is possible to observe should be noted.

(8) The resistance r should be very small compared to the resistance of the secondary circuit.

(9) While determining the period of the galvanometer its circuit should be kept open.

Observations—(A) Determination of θ_1 , ϕ and λ

Rest position of spot of light = zero

S.No.	Scale deflection corresponding to successive throws of galvanometer						r ohm	ϕ in scale divisions	λ
	θ_1	θ_2	θ_3	θ_4	θ_5	θ_6			
1.									
2.									
3.									

(B) Determination of period of the galvanometer.

S. No.	No. of oscillations	Time taken		Period T Sec.
		Min.	Sec.	
1.				
2.				
3.				
			Mean	

Calculations—Set I : $\lambda = 2.3026 \log \dfrac{\theta_1 + \theta_2 + \theta_3 + \ldots \theta_{n-1}}{\theta_2 + \theta_3 + \theta_4 + \ldots + \theta_n}$

$$=$$

$$M = \frac{T}{2\pi} \, r \, \frac{\theta_1}{\phi} \left(1 + \frac{\lambda}{2}\right)$$

$$= \quad \text{Henry}$$

Result—The mutual inductence of the given coil = Henry

Criticism of the Method—The method gives a fairly satisfactory value of mutual inductance of two coils. The main source of error is the effect of self-inductance in the primary. The method is not applicable to coils which have an iron core, for then the magnetic flux linked with secondary due to current in the primary will not be proportional to the current in the primary. For more accurate results, alternating current bridges employing vibration galvanometer as detector should be used.

Experiment 27.3

Object—To determine a high resistance by the method of leakage of a capacitor.

Apparatus— capacitor, high resistance, accumulator, ballistic galvanometer, Morse key, a two-way key and two one-way keys.

Theory—If a charged capacitor of capacitance C is connected across a resistance R, the charge leaks through the resistance. If the charge on the condenser at any time t is Q, the

P.D. between its plates is Q/C and that across the resistance is $R\dfrac{dQ}{dt}$. Hence, as there is no applied voltage in the circuit, the potential equation of the circuit is :

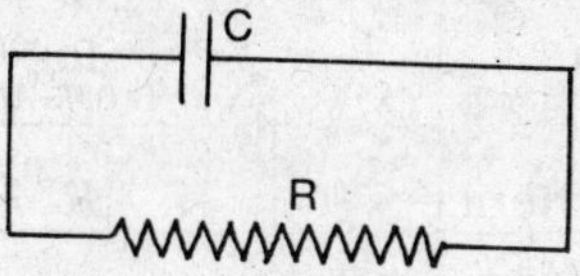

Fig. 27.4

$$\frac{Q}{C} + R\frac{dQ}{dt} = 0$$

or

$$\frac{dQ}{Q} = -\frac{dt}{CR}$$

Integrating, this gives :

$$\log Q = -\frac{t}{CR} + k \qquad (27.14)$$

If *initially* the charge on the capacitor is Q_0, we have $Q = Q_0$, when $t = 0$, and therefore,

$$k = \log Q_0$$

Substituting this in Eq. (27.14) and simplifying, we have :

$$Q = Q_0\, e^{-t/CR} \qquad (27.15)$$

Now, let a capacitor of capacitance C be given a charge Q_0 and then immediately discharged through a ballistic galvanometer. If θ_1 is the first throw of the galvanometer and λ the logarithmic decrement, we have :

$$Q_0 = K\,\theta_1\,(1 + \lambda/2) \qquad (27.16)$$

Next, let the same capacitor be again given the same charge Q_0 and then after having allowed the charge to leak through a resistance R for a known time t, let the capacitor be discharged through the same ballistic galvanometer. If α_1 is the first throw of the galvanometer in this case, we have :

$$Q = K\alpha_1\,(1 + \lambda/2) \qquad (27.17)$$

Dividing Eq. (27.17) by Eq. (27.16), we have :

$$\frac{Q}{Q_0} = \frac{\alpha_1}{\theta_1}$$

But from Eq. (27.15), we have :

$$\frac{Q}{Q_0} = e^{-t/CR}$$

$$\frac{\alpha_1}{\theta_1} = e^{-t/CR}$$

or $$\log \frac{\theta_1}{\alpha_1} = \frac{t}{CR}$$

whence, $$R = \frac{t}{C \log (\theta_1 / \alpha_1)}$$

or $$R = \frac{t}{2.3026\ C\ \log_{10} (\theta_1 / \alpha_1)} \qquad (27.18)$$

This equation can be used to determine a high resistance.

Rearranging Eq. (27.18), we have :

$$\log_{10} \frac{\theta_1}{\alpha_1} = \frac{t}{2.3026\ RC}$$

This equation shows that if the charge is allowed to leak through the resistance for different intervals of time and then a graph plotted taking the various values of t as abscissae and the corresponding values of $\log_{10} \dfrac{\theta_1}{\alpha_1}$ as ordinates, the slope of the graph, as illustrated in Fig. 27.5 will be given by :

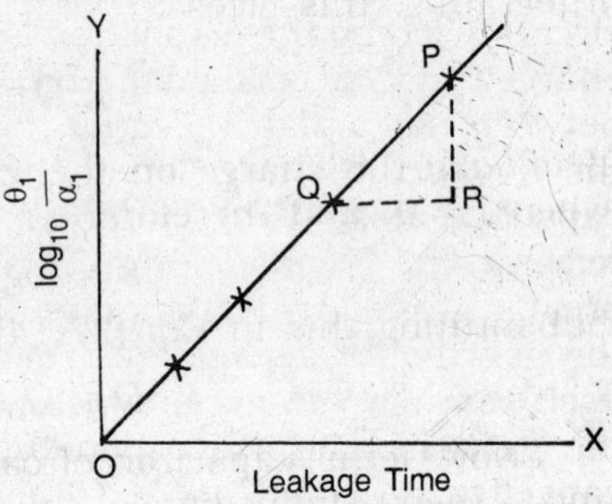

Fig. 27.5

$$\tan \phi = \frac{1}{2.3026\ RC}$$

where $$R = \frac{1}{2.3026\ C\ \tan \phi}$$

If a capacitor has a self-leakage resistance, the experiment should be performed in two parts :

(1) By allowing the charge to leak *without* the external resistance R connected across the capacitor. This will determine the self-leakage resistance r of the capacitor.

(2) By allowing the charge to leak *with* the external resistance R connected across the capacitor. This will determine the equivalent resistance R' of the parallel combination of r and R.

The external resistance R can then be calculated from the formula :

$$\frac{1}{R'} = \frac{1}{r} + \frac{1}{R} \tag{27.19}$$

Method—Level the base of the galvanometer by means of the levelling screws and release its coil. Throw light on the mirror of the galvanometer and get a bright spot of light on the scale. Adjust the spot of light on the scale at zero. Make electrical connections as shown in Fig. 27.6. Join a to b and then press the Morse key at P. This completes the cell-capacitor circuit whereby the capacitor is charged to a constant P.D. between its plates equal to the e.m.f. of the cell. When the capacitor is fully charged, release the Morse key. This breaks the cell-capacitor circuit but at the same time completes the galvanometer-capacitor circuit. The capacitor discharges itself through the galvanometer. Note down the first throw, θ_1 of the galvanometer corresponding to the charge, say, Q_o on the capacitor.

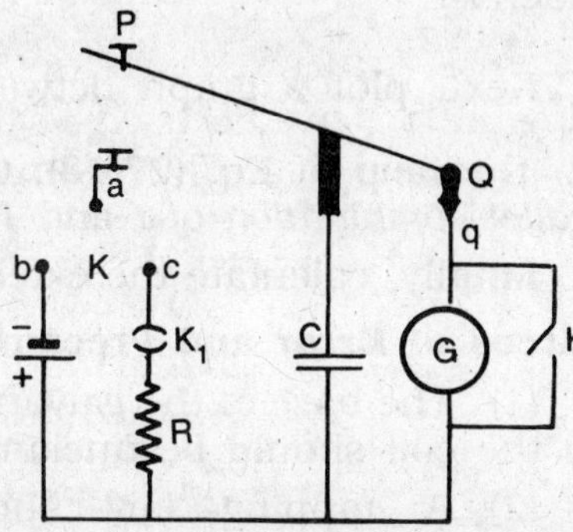

Fig. 27.6

Again, charge the capacitor to the same voltage by pressing the Morse key at P and then keeping the knob P depressed and the plug key K_1 open, turn the key K from the position ab to the position ac, and immediately start a stop-watch. When the charge on the capacitor has leaked, if at all, for a known time t, say, 5 seconds, release the Morse key and observe the first throw α_1 of the galvanometer. Repeat this self-leakage experiment several times for different values of leakage time t, noting down the value of a_1 for each value of t.

Next, close the key K_1, turn the key K into the position ab and press the Morse key at P. When the capacitor is again fully charged, turn key K from the position ab to the position ac keeping the Morse key depressed. When the charge on the capacitor has leaked for a known time t, say, 5 seconds, release the Morse key and note down the first throw α_1 of the galvanometer. Repeat this leakage experiment with the resistance R connected across the capacitor, several times with different values of t observing the value of α_1 for each value of t.

Plot a graph taking the various values of t along the X-axis and the corresponding values of $\log_{10} \dfrac{\theta_1}{\alpha_1}$ along the *Y*-axis. Find out the slope of the graph and then with the help of Eq. (27.18), calculate the value of *r* the self-leakage resistance of the condenser.

Next, plot a graph between *t* and $\log_{10} \dfrac{\theta_1}{\alpha_1}$ and determine with the help of Eq. (27.18), the equivalent resistance *R′ of the parallel combination of r* and *R*.

Finally, calculate the external resistance *R* from Eq. (27.19).

Sources of Error and Precautions

(1) The base of the galvanometer should be carefully levelled and the coil should be unclamped before its use.

(2) A tapping key should be connected across the galvanometer.

(3) The cell used should be an accumulator.

(4) The capacity of the capacitor should be such as to ensure a satisfactory rate of leak.

(5) There should be no leakage of charge through the keys *K* and K_1.

(6) The graphs should be smoothly drawn.

Observations

(A) (*i*) Capacitance of the capacitor *C* =...F

(*ii*) Rest position of spot of light = zero

(*iii*) First throw of the galv. corresponding to the full charge Q_0 on it θ_1 = cm.

(B) Measurement of *t*, α'_1 and α

S.No.	Leakage time sec.	First throw of galv. after the charge has leaked through		$\log_{10} \dfrac{\theta_1}{\alpha'_1}$	$\log_{10} \dfrac{\theta_1}{\alpha_1}$
		Self leakage resistance of the capacitor, *i.e.* with K_1 open α'_1 cm.	Parallel combination of self leakage resistance and external resistance, *i.e.* with K_1 closed α_1 cm.		
1.	5				
2.	10				
3.	...				

Calculations—From the graph between t and $\log_{10} (\theta_1/\alpha'_1)$,

$$\tan \phi' = \frac{PR}{QR} = \qquad =$$

$$\therefore \quad r = \frac{1}{2.3026\, C \tan \phi} \qquad = \quad \text{ohm}$$

From the graph between t and $\log_{10} (\theta_1/a_1)$, we have :

$$\tan \psi = \frac{PR}{QR} \qquad =$$

$$\therefore \quad R' = \frac{1}{2.3026\, C \tan \psi} \quad = \quad = \quad \text{ohm}$$

From Eq. (27.19), the external resistance R is given by :

$$\frac{1}{R} = \frac{1}{R'} - \frac{1}{r}$$

$$=$$

$$= \qquad \text{ohm}$$

Result—The value of the given resistance as determined by the method of leakage of a capacitor = ohm.

Criticism of the Method—The experiment, if performed carefully, gives accurate results. The method is, however, suitable for determinaiton of *high* resistances only, of the order of 20 megohms. The method can be used to measure the resistance and capacity of a capacitor if a suitable known high resistance is available.

Experiment 27.4

Object—To study the current flow during charging and discharging of a capacitor.

Apparatus—A 9-volt battery, a microammeter (0-100 microamperes), resistances and capacitors on board.

Theory—During the process of charging of a capacitor, the current through the R-C circuit is given by :

$$i = i_0\, e^{-t/CR} \qquad (27.20)$$

While during discharge

$$i = -\, i_0\, e^{-t/CR} \qquad (27.20a)$$

The direction of the current is opposite in the two cases.

Method 1—Make up the connections as shown in Fig. 27.7.

Either a zero centre microammeter be used or else use may be made of a commutator to let the current through the microammeter always pass in the same direction. Thus, keeping the connections at no. 1 in the commutator, the microammeter may be used to give instantaneous value of the current during charging while with the connections at no. 2 in the commutator,

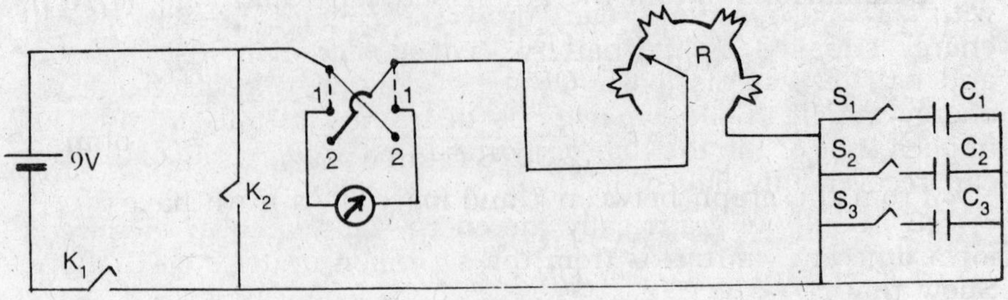

Fig. 27.7

the microammeter may be used to record the instantaneous value of current during the discharging of the capacitor.

2. Choose suitable values of R and C to have a large time constant, say, C = 500 microfarad, R = 100 kilo-ohm, giving a value of CR = 50 sec.

3. Switch on the key K_1 and observe the current at regular intervals of 10 to 15 sec. during the process of charging the capacitor.

4. When the capacitor has been charged, switch off K_1. Alter the connections in the commutator to no. 2 and close switch K_2 when the capacitor starts discharging through the same resistance R. Record the discharge current at regular intervals of 10 to 15 seconds in the initial stages and at intervals of 20 to 30 seconds at later stages.

5. Keep C constant, alter R, say, to a value 150 kilo-ohm and repeat your observations.

6. Repeat your observations for another set of values of C and R such that CR is the same as in the first case. Thus, we could choose C = 250 microfarad and R = 200 Kilo-ohm.

7. Plot a graph between t along X-axis and i along Y-axis, both for charge and discharge of the capacitor. Find the value of time constant τ from the graph. Also calculate it from the relation :

$$\tau = 0.69\,RC =$$

and compare the two.

8. Plot a graph between i_2 and t. Compute the area under the curve which gives $i_2 \cdot dt$. Multiply this by R, the resistance of the circuit. This gives the energy dissipated in the circuit.

Repeat this for the same value of C but an altered value of R. It will be noticed that the energy dissipated is independent of the resistance R. Compare the energy dissipated in the circuit with $\frac{1}{2}CV_0^2$. Thus, one can draw the conclusion that of the energy supplied by the battery, half is stored in the capacitor and half of it is dissipated in the circuit. During discharge the energy stored in the capacitor is dissipated in the circuit. This implies that of the total energy used in charging a capacitor, one can recover only half of it.

9. Compute graphically the energy dissipated in the circuit for a different value of C from the set taken under step 6 above. Show that :

$$\frac{E_1}{C_1} = \frac{E_2}{C_2} \tag{27.21}$$

where E_1 and E_2 are the energies dissipated for charging capacitors C_1 and C_2 at constant V_0.

Observations—Set I—C = 500 microfarad, R = 100 Kilo ohm

Value of *i* in micro-amps.

Time (Sec.)	During Charge	During Discharge	i^2 (amp.)^2
0	90		81×10^{-10}
10	74		54.7
20	64		40.9
30	49		24.0
40	40		16.0
50	33		10.9
60	27		7.3
80	18		3.24
100	12		1.44
120	8		0.64
150	4		0.16

SET II —C = 500 microfarad R = 150 Kilo-ohm

Value of *i* in microamps

Time (Sec.)	During Charge	During Discharge	Value of i^2 (amp.)^2
0		60	36×10^{-10}
15		49	24
30		40	16
45		33	10.89
60		27	7.39
75		22	4.84

190	18	3.24
120	12	1.44
150	8	0.64
280	5	0.25
10	3	0.09

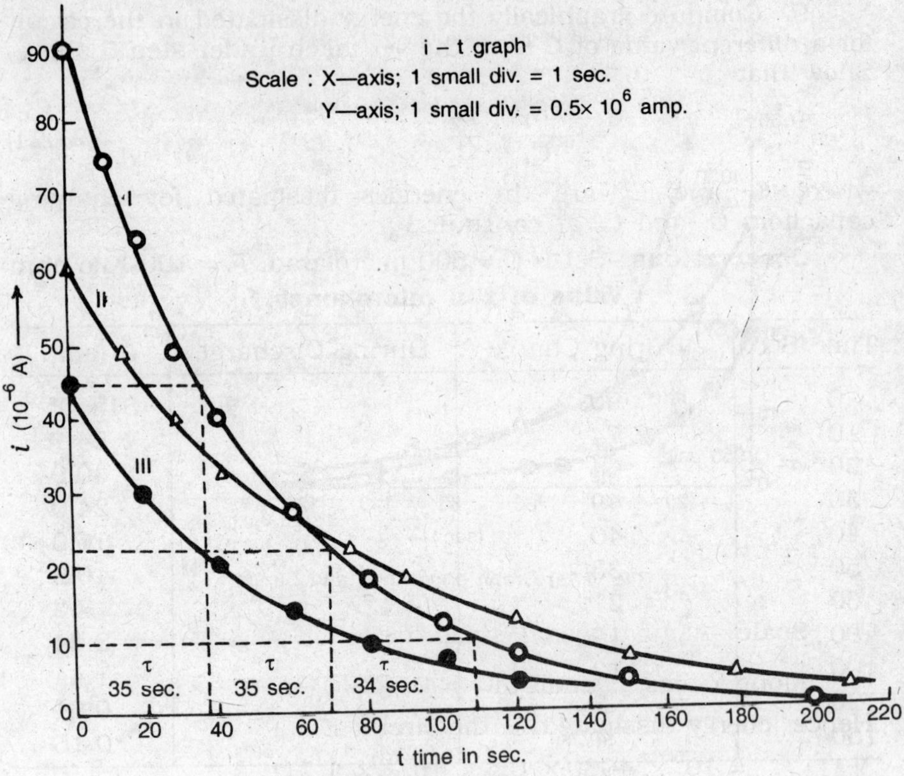

Fig. 27.8 Graph between *i* and *t*

Calculations—I. From the graph between i and t for set I,

$$\tau =$$

Calculated value of $\tau = 0.69$ CR

$$= 0.69 \times 500 \times 10^{-6} \times 100 \times 10^{3} \text{ sec.}$$

$$= 34.5 \text{ sec.}$$

Similar calculations of τ for set II and III.

II. From the i^2 and t graph for set I

Area under the curve = 4030 small divisions

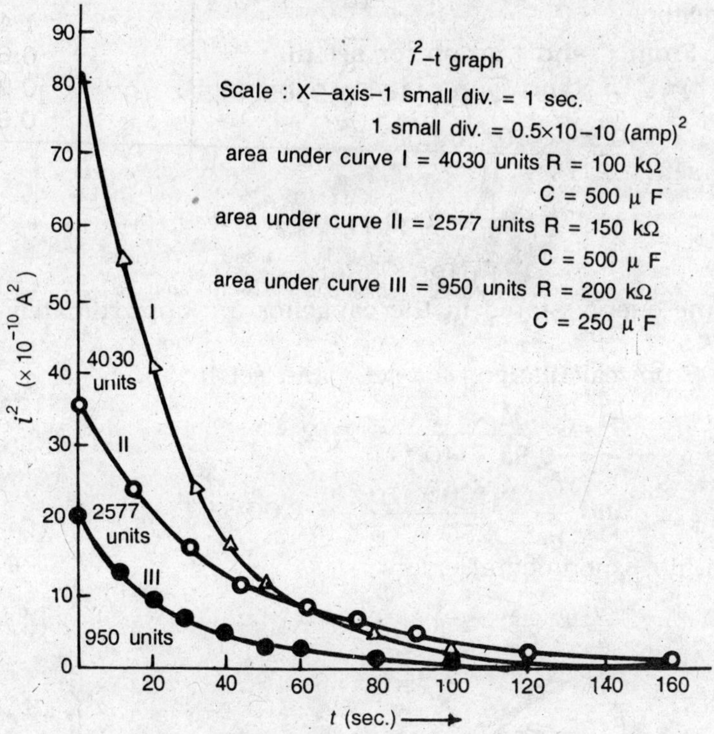

Fig. 27.9 Graph between i^2 and t

Scale—Along x-axis 1 small div. = sec.

Along Y-axis 1 small div. = 0.5×10^{-10} $(amp.)^2$

Hence, energy dissipated in the circuit E = R. i^2 dt.

$$= 10^5 \times 4030 \times 0.5 \times 10^{-10} \times 1$$

$$= 2.015 \times 10^{-2} \text{ Joule.}$$

From set II calculate the energy dissipated in the circuit. Thus,

$$E_1 = R\, i^2 \,.\, dt = 150 \times 10^3 \times 2577 \times 0.5 \times 10^{-10} \times 1$$

$$= 1.94 \times 10^{-2} \text{ Joule.}$$

Also the energy stored in the capacitor is

$$E = \frac{1}{2} CV_0 = \frac{1}{2} \times 500 \times 10^{-6} \times 9 \times 9$$

$$2.02 \times 10^{-2} \text{ Joule.}$$

Thus, we find that $E \approx E_1 \approx E_0$; the energy dissipated in the circuit is independent of R and equals the energy stored in the capacitor.

III. From i^2 and t graph for set III

$$E_2 = R\,i^2\,dt = 200 \times 10^3 \times 950 \times 0.5 \times 10^{-10} \times 1$$

$$= 0.95 \times 10^{-2} \text{ Joule}$$

For this set,

$$\frac{1}{2}\,CV_0^2 = \frac{1}{2} \times 250 \times 10^{-6} \times 9 \times 9$$

$$= 1.0125^{-2} \text{ Joule}$$

E_2 and the energy stored in the capacitor are approximately the same.

IV. From calculations for set I and set II :

$$\frac{E_1}{E_2} = \frac{1.94 \times 10^{-2}}{0.95 \times 10^{-2}} = 2.12$$

$$\text{and } \frac{C_1}{C_2} = \frac{500 \times 10^{-6}}{250 \times 10^{-6}} = 2.00$$

Thus, within experimental errors :

$$\frac{E_1}{E_2} = \frac{C_1}{C_2}$$

or
$$\frac{C_1}{E_1} = \frac{E_2}{C_2}$$

for a fixed charging potential V_0

Result—(*i*) $\tau = 0.69\,RC$ has been verified.

(*ii*) In charging or discharging a capacitor, an energy equal to $\frac{1}{2}\,CV_0^2$ is dissipated in the circuit and is independent of the resistance of the circuit.

(*iii*) $\frac{E_1}{E_2} = \frac{C_1}{C_2}$ for fixed value of charging potential V_0

Experiment 27.5

Object—To study charge and discharge of a capacitor through a resistance using a neon bulb.

Apparatus—A power supply with variable d.c. output upto 300 volts, a set of resistors in series, a set of capacitors which could be connected in parallel, a neon lamp (as the one used in line testers) and stopwatch.

Theory—Consider the circuit depicted in Fig. 27.10. It consists of a power supply with a variable output (0-300 volts

D.C.), a high resistance R and a cpacitor C, having a neon flashing lamp in parallel with it.

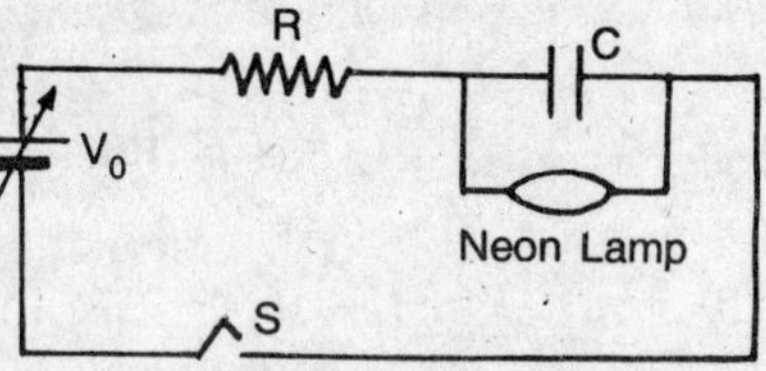

Fig. 27.10

If the switch S is closed the p.d. across the capacitor builds up and when it has reached a value V_s, the striking or flashing potential of neon, a discharge passes through it. Let us calculate the time required for the first flash to be seen. Evidently, this time t_1 is given by the relation :

$$V_s = V_0 \int (1 - e^{-t_1/CR})\, dt$$

where CR is the time constant of the circuit, we then have, on rearranging,

$$t_1 = CR \log \frac{V_0}{V_0 - V_s} \qquad (27.22)$$

Let us now make some simplifying assumptions :

(*i*) the discharge takes place almost instantaneously, the neon bulb circuit offering almost no resistance (the resistance of the lamp may be of the order of 2000 ohm and thus this discharge time is very short compared with the charging time).

(*ii*) it ceases when the p.d. across the capacitor has fallen to V_e, the extinction potential for neon, when we can suppose that the neon bulb offers infinite resistance.

Thus, after the first flash, the p.d. falls to V_e and then again builds upto V_s when a flash is seen again, and this process goes on and on. Thus, the flashes are seen at regular intervals of time.

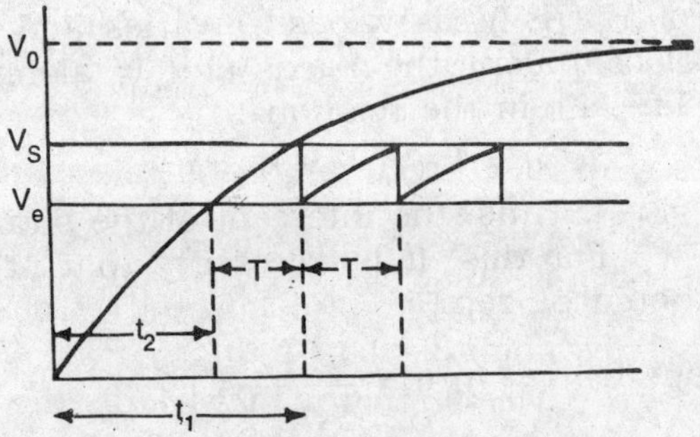

Fig. 27.11

We can calculate the time period T of these flashes. Evidently, T is the time difference between the time t_1 for the p.d. to rise to V_s and the time t_2 for the p.d. to rise to V_e . We then have :

$$t_1 = CR \log \frac{V_0}{V_0 - V_s} \tag{27.23}$$

$$t_2 = CR \log \frac{V_0}{V_0 - V_e} \tag{27.24}$$

Thus, $$T = t_1 - t_2 = CR \log \frac{(V_0 - V_e)}{(V_0 - V_s)} \tag{27.25}$$

It is evident that (*i*) for C and V_0 fixed, T varies as R and (*ii*) for R and V_0 fixed, T varies as C.

Further, from Eq. (27.25) we have :

$$e^{T/RC} = \frac{V_0 - V_e}{V_0 - T_e} = 1 + \frac{V_s - V_e}{V_0 - V_s}$$

whence,

$$e^{T/RC} - 1 = \frac{V_s - V_e}{V_0 - V_s}$$

or $$\frac{1}{e^{T/RC} - 1} = \frac{V_0 - V_s}{V_s - V_e}$$

or $$y = k\,(V_0 - V_s)$$

where $y = \frac{1}{e^{T/RC} - 1}$ and $k = \frac{1}{V_s - V_e}$, a constant because V_0 and V_e are constants.

Hence, for a fixed known value of RC, T may be measured by noting the number of flashes and the corresponding time for a known value of V_0. As V_0 is varied, T varies and so also the value of y. A plot of y along the Y-axis and V_0 along the X-axis gives a straight line. From the relation :

$$y = k\,(V_o - V_s)$$

if $y = 0, V_O = V_S$. Thus, the intercept of the straight line on the x-axis gives V_s. For this, it is necessary that at the origin, the value of y should be zero.

Also the slope being $\tan\theta = \frac{1}{V_s - V_e}$

we can calculate V_e from a knowledge of V_s.

Description—The apparatus is depicted below with the power supply of output variable in the range 0-300 volts d.c. S is the selector switch to introduce resistance in the circuit in

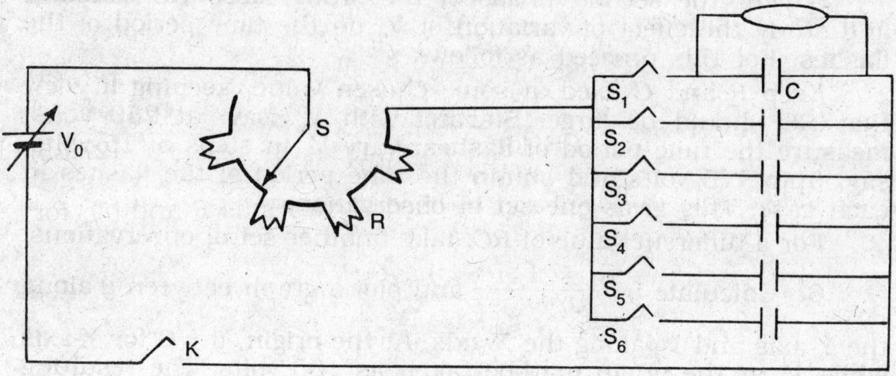

Fig. 27.12

steps of 1 megaohm. Switches S_1 through S_6 are for introducing capacitance in the circuit in steps of 0.5 microfarad.

Method—1. Set the power supply to give a suitable output voltage in the vicinity of 150 volts and record its value V_o. Fix the value of the capacitance in the circuit by closing S_1 and, if necessary, more of the switches S_2 to S_6. With the help of the band switch S introduce a suitable resistance in the circuit.

2. Close switch k and see the flashes of the neon bulb. With the help of a sensitive stop watch record the time taken for a sufficiently large number of flashes (not less than 50) and compute T.

3. Alter the value of R in steps of 1 megaohm with the help of the switch S and repeat observations for finding the time period of the flashes of the neon bulb each time, thus, yielding one complete set of observations.

Keeping the capacitance in the circuit fixed, take two more sets by altering V_o.

Plot a graph between T and R for one value of V_o. On the same sheet, plot the graphs between T and R for other values of V_o.

4. Another approach to the study of the same phenomenon of the effect of varying RC on the charging and discharging of a capacitor is to keep R constant at some fixed arbitrary value and to vary C, recording each time the period of flash of the neon bulb. Repeat the above set of observations for two other values of V_o.

Plot a graph between T and C for each set of observations. This completes the first phase of study of RC Circuit.

5. For the second phase of the study, keep RC constant and study the effect of variation of V_0 on the time period of the flashes. For this proceed as follows :

Keep R and C fixed at some chosen value, keeping in view that RC should be large. Starting with V_0, say, at 250 volts, measure the time period of flashes. Vary V_0 in steps of 15 volts, say, upto 175 volts and obtain the time period of the flashes in each case. This gives one set of observations.

For a different value of RC, take another set of observations.

6. Calculate $y = \frac{1}{e^{T/RC} - 1}$ and plot a graph between y along the Y-axis and V_0 along the X-axis. At the origin, $y = 0$ for Y-axis while V_0 at the origin may be taken as 100 volts. The resulting graph is a straight line.

(*i*) Produce the straight line back to cut the X- axis. This value of V_0 on the X-axis gives the value of V_s because for $y = 0$, $V_0 = V_s$.

(*ii*) Measure the slope of the straight line. Equate this to $\frac{1}{V_s - V_e}$ and compute V_e, the extinction voltage.

Sources of Error and Precautions

(1) It is desirable to use a stopwatch reading upto 0.1 sec. to determine the period of flashes accurately, and the number of flashes to be timed should not be less than 50.

(2) Switch S should be closed after waiting for the power supply to warm up for about a minute or so.

Observations—Table 1. C = 0.5 microfarad.

V_0	R (in megaohm)	Flashes		Periodic time T (sec.)
		Number	Time (sec.)	
165	1			
	2			
	3			
	4			
180				
	1			
	2			
	3			
	4			
	...			

Table 2

R = 2 megaohm

V_0	C (in microfarad)	Flashes Number	Flashes Time (sec.)	Periodic time T (sec.)
165	0.5			
	1.0			
	1.5			
	2.0			
180	0.5			
	1.0			
	1.5			
	2.0			
...	...			

Table 3

Comparison table. For fixed value of *Vo*

RC (sec.)	T from Table 1	T from table 2
1. sec.		
2. sec.		

Table 4

R = 3 megaohm; C = 1.0 microfarad

RC = 3.1 sec.

V_0	Flashes Number	Flashes Time (Sec.)	Time Period T (Sec.)	T/CR	$e^{T/RC}$	$y = \frac{1}{e^{T/RC} - 1}$
250	40	32.7	0.818	0.2727	1.31	3.226
235	50	47.0	0.94	0.3133	1.36	2.777
220	40	43.4	1.08	0.36	1.43	2.308
205	40	51.9	1.3	0.4333	1.54	1.861
190	40	63.8	1.59	0.53	1.71	1.408

Calculations—From the graph between y and V_o, the intercept on X-axis = volts.

Hence, V_s (= V_o for y = 0) = volts

Now, the slope of the straight line = tan θ =

Equating this to $\frac{1}{V_s - V_e}$ we get :

$$\therefore \quad \frac{1}{V_s - V_e} =$$

Since V_s = volts,

$$V_e = \quad \text{Volts}$$

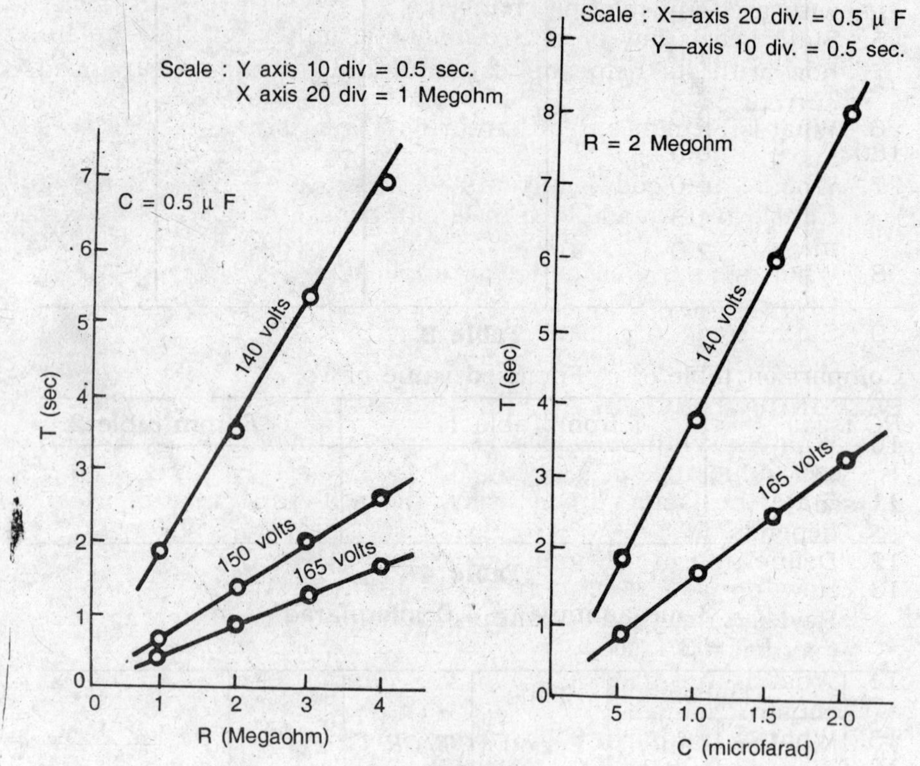

Fig. 27.13

Result

(*i*) From the comparison Table 3, it is evident that the p.d. builds up across the capacitor depends upon the time constant *CR*.

(*ii*) Sparking voltage of neon = volts

(*iii*) Extinction voltage of neon = volts.

ORAL QUESTIONS

ELECTROMAGNETIC INDUCTION

1. What is electro-magnetic induction ? State the laws of electro-magnetic induction.

2. Upon what factors does the magnitude of the induced e.m.f. depend ? If the conducting circuit is not closed will there be any induced e.m.f. produced in it when the magnetic flux threading the circuit varies ?
3. Upon what factors does the strength of the induced current depend ?
4. What are the various ways in which the magnetic flux linked with a circuit may be changed ?
5. State Lenz's law of electro-magnetic induction and explain how with its help you can find the direction of induced current.
6. What is Fleming's right hand rule ? Is it different from Lenz's law ?
7. Where are eddy currents ? Why are iron cores of transformers made of thin laminations instead of one soild mass ?
8. What are the chief practical applications of eddy current ? Give examples.
9. State some applications of the phenomenon of electro-magnetic induction.

SELF-INDUCTANCE BY RAYLEIGS'S METHOD

10. What is self-induction ? How do you define co-efficient of self-inductance of a circuit ?
11. State the factors upon which the self-inductance of a coil depends.
12. Define unit of self inductance ?
13. How do you determine the self-inductance of a coil by Rayleigh's method ? What precaution do you observe in your experiment ?
14. What is the special use of the double key ? Describe its construction.
15. What is the harm if the connecting wires are coiled ?
16. Why don't you use a Leclanche cell in this experiment as you do in other experiments with P.O. box ?
17. Why should the ratio arms be adjusted to 10 : 0 for preferably to 1 : 1 if possible ?
18. If the resistance of the coil is small how can you adjust the four arms to the same order ?
19. Why should the balancing of bridge for steady current be perfect ?
20. What is the use of platinoid wire connected in the rheostat arm?
21. If the battery key is kept depressed for a long time, why does the balance point alter ?

22. How do you prevent the heating of the coil while the bridge is being balanced ?
23. Why should the resistance introduced in the arm of the coil to produce steady deflection be small ? What should be its approximate value ?
24. How and why do you determine logarithmic decrement ?
25. While determining the period of the galvanometer why should the galvanometer circuit be kept open ?
26. Do you know other method by which more accurate results can be obtained ?
27. Is Rayleigh's method suitable for determining the self-inductance of a coil having an iron core ? if not, why ?
28. State some practical applications of self-inductance'.
29. Compare self-inductance with inertia.
30. Why does the current take some time to attain its steady maximum value when allowed to pass through a circuit containing self-inductance?
31. Why does an arc form at the switch when a motor is switched off ?
32. How are induction effects reduced to minimum in bifilar winding of resistance coils ?

DETERMINATION OF MUTUAL INDUCTANCE

33. What is mutual induction ? Define coefficient of mutual induction or mutual inductance of two coils.
34. Upon what factors does it depend ? Is it the same for both the coils when any one of them acts as the primary and the other as the secondary ?
35. What is the unit of mutual inductance ? Define it.
36. Describe how you find the mutual inductance of two coils. What precautions do you take in the experiment ?
37. Why should the coils be placed away from the galvanometer ?
38. Why should the connecting wires be straight ?
39. What is logarithmic decrement ? Do you determine it on closed circuit or open circuit of the galvanometer and why ?

40. What are the main sources of error in this experiment ? Is the method also applicable to coils having iron cores ? If not, why ?

41. How can you determine the mutual inductance of two coils accurately ?

42. State some practical applications of a mutual induction.

43. Why are there a large number of turns in the secondary of a transformer or an induction coil ?

44. Why is the e.m.f. at break in the induction coil very high compared to that at make ?

28

Galvanometers, Ammeters and Voltmeters

28.1. Galvanometers

Galvanometers are instruments which are intended primarily to indicate the existence of a current and which may, under certain circumstances, be capable of measuring it. The construction of galvanometers is based on the interaction between coils carrying currents and magnets. They may be, in general, divided into two classes as given below according as the moving part is the magnet or the coil.

(A) *Moving-magnet Galvanometer*—In this type the current is passed through a fixed coil which produces a magnetic field under which the magnet moves.

(B) *Moving coil Galvanometer*—In this type the current is passed through a movable coil placed between the poles of a powerful magnet. The coil moves on account of the interaction between its magnetic field and that of the permanent magnet.

Galvanometers are also classified as pointer galvanometers and reflecting or fixed to the moving part to note down its movement.

28.2.

(*a*) *Tangent Galvanometer*—As described in Section 26.5, it consists essentially of a small magnetic needle pivoted or suspended at the centre of a circular coil and provided with a long light aluminium pointer which moves over a circular scale of degrees. To use the instrument the plane of the coil is set in the magnetic meridian and the current to be measured passed through it. If then θ be the deflection of the needle from the magnetic meridian, the current is from Eq. (28.8) given by

$$i = B_H \frac{2r}{\mu_0 n} \tan\theta$$

where r is the radius of the coil, n the number of turns in the coil B_H the horizontal component of the earth's magnetic field. If the values of r and n are accurately known, the tangent galvanometer can be used to measure current absolutely, for the value of B_H can be evaluated at any place by a quite independent method. The instrument possesses several disadvantages.

(1) The instrument can be used only in a particular orientation, *viz.*, with the plane of its coil in the magnetic meridian.

(2) The deflection is not proportional to the current. It uses the tangent relation and hence becomes inaccurate at angles greater than 70°.

(3) The reduction factor $\left(B_H \dfrac{2r}{\mu_0 n}\right)$ depends upon the value of B_H which is affected by the presence of the magnetic substances and current bearing conductors. The evaluation of B_H requires a separate and highly accurate experiment.

(4) The magnetic needle has a fine size and its poles move out of the coil as θ increases. As a result, the deflecting field is not proportional to the current.

(5) Unless the coil is exactly circular and has only a single layer, the value of *r* is also uncertain.

(6) The instrument is bulky and the system of supporting the needle is not robust.

The tangent galvanometer is not a sensitive instrument and hence is not suitable for use in null methods, *i.e.*, in Wheatstone bridge and potentiometer.

(*b*) *Sine Galvanometer*—This instrument closely resembles tangent galvanometer and consists essentially of a small magnetic needle pivoted or suspended at the centre of a coil of large radius. The coil in this case is capable of rotation about a central vertical axis, the rotation being read off on a horizontal circular scale at the base. When the current is passed through the coil, the needle is deflected. The coil is then rotated about the vertical axis until the coil overtakes the deflected needle and both the needle and the coil lie in one common plane. The deflecting field *B* is then perpendicular to the magnetic needle and in the equilibrium position :

$$B_H = \frac{\mu_0 n i}{2r} \sin\theta$$

whence,
$$i = \frac{2rB_H}{\mu_0 n} \sin\theta$$

or
$$i = K \sin\theta \qquad \textbf{(28.1)}$$

where $K = \dfrac{2rB_H}{\mu_0 n} \sin\theta$ is the reduction factor of the sine galvanometer.

Since the maximum value of sin θ is 1, it is clear from Eq. (28.1) that the maximum current which may be measured by means of this instrument is equal to *K*. For currents greater than *K*, the coil cannot catch up the needle when the former is rotated and hence it is then impossible to make the needle lie in the plane of the coil. The sensitiveness of sine galvanometer increases as θ increases and is maximum when $\theta = 90^o$. The instrument possesses the advantage that deflections up to 90° may be used.

(c) *Helmoholtz or Double-coil Tangent Galvanometer*—As described in Section 26.6, it is a modification of the ordinary tangent galvanometer in which the deflecting field has been made uniform by placing the magnetic needle midway between equal circular coils, mounted vertically at a distance apart equal to the radius of either and connected in series.

In the case of above galvanometers, the deflections are generally measured with the help of a pointer. The accuracy in reading deflections in such a case can be increased by increasing the length of the pointer. For great accuracy, however, the pointer-over-scale system of measuring deflection is dispensed with and deflections are measured with the help of a mirror and lamp-and-scale arrangement. In a mirror galvanometer, if the deflections are not large, the current is proportional to the displacement of the index produced by it. This is a very great advantage in practical work.

Method for increasing sensitiveness—There are three methods of increasing the deflection due to a current of given strength in the case of moving-magnet galvanometer:

(1) *By increasing the deflecting field*—This can be done by increasing the number of turns *n* in the coil and by decreasing its radius *r*. But number of turns cannot be increased beyond a certain limit, first, because this increases the resistance of the coil and secondly, because the value of *r* then becomes uncertain as the turns are not coincident. Further, it is not desirable to make *r* very small otherwise the magnetic needle will not lie in uniform field and the tangent relation will not hold good.

(2) *By decreasing the control field* B_H—This is done with the help of a control magnet so placed that its field at the centre of the coil is in the opposite direction to that of the earth. By adjusting the position of the control magnet with respect to the coil, the controlling field B_H can be reduced as much as you please. When the field due to the control magnet is made equal and opposite to the earth's field, the resultant controlling field is zero. In such a case the torsion of suspension provides the necessary restoring couple.

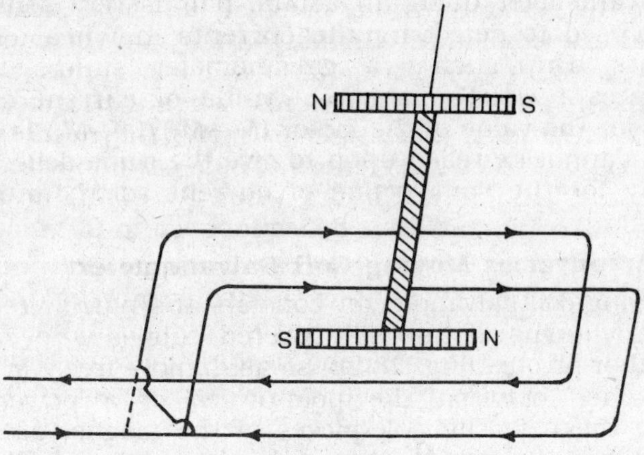

Fig. 28.1

(3) *By the use of an astatic pair of needles*—An astatic pair consists of two light magnetic needles (Fig. 28.1) fitted in a light frame with their axes parallel to each other and with their dissimilar poles pointing in the same direction. The needles are magnetised nearly equally and are suspended one inside and the other outside a flat coil of a large number of turns, the needle lying parallel to the plane of the coil. Flattening of the coil is equivalent to a reduction of the radius and thus increases the sensitiveness of the instrument.

When a current passes through the coil, the two parts of the coil field, one around the needle outside the coil and the other around the needle inside the coil, are in opposite directions and since dissimilar poles of the needles are also pointing in the same direction the needles are acted upon by couples tending to deflect them in the same sense. If M_1 and M_2 be the magnetic moments of the two needles, the deflecting couple exerted on the astatic pair by the coil field is proportional to $(M_1 + M_2)$ roughly. Further, since the controlling field is practically uniform and similar poles of the needles are oppositely directed, the controlling couple is proportional to B_H (M_1-M_2). Thus, by using an astatic pair, the sensitiveness can be increased by $(M_1+M_2)/(M_1-M_2)$ times roughly. If $M_1 = M_2$, the restoring couple due to the controlling field is zero; in such a case a torsion suspension must be used to provide the necessary restoring couple, otherwise the system will become unstable and all current weak or strong would produce a deflection of $90°$ and there would be no return of the needle to zero. Sometimes, the needles are placed in separate coils one above the other wound oppositely so as to produce deflecting couple on each of the needle in the same sense.

A galvanometer using an astatic pair is very sensitive and be chiefly used to detect minute currents in Wheastone bridge experiments. But such a galvanometer suffers from two disadvantages : (1) The absolute value of current cannot be measured for the value of the factor $(M_1+M_2)/(M_1-M_2)$ is unknown and (2) it cannot be relied upon to give the same deflection from day to day for the same value of current as M_1 and M_2 may change.

28.3. D'Arsonval or Moving Coil Galvanometer

A moving-coil galvanometer consists of a narrow rectangular coil of many turns of fine silk-covered copper wire suspended (Fig. 28.2a) or pivoted (Fig. 28.2b) so as to move freely in a narrow annular space between the pole-pieces of a powerful steel horse-shoe magnet. The pole-pieces of the magnet are made of soft iron and are hollowed out to make them cylindrically concave in shape. This makes the field *radial* (Fig. 28.2c). Within the coil is fixed a cylinder of soft iron which serves to concentrate the lines of force in the gap thus making the field in the gap strong and practically uniform. The coil is wound on a light frame-work which is either of metal or of non-conducting material, according as a dead-beat or a ballistic action is required.

In the suspended coil type, the suspension is a fine rectangular strip of phosphore-bronze which also serves as one of the current leads to the coil. These rectangular strips have very small torsional rigidity and hence the twist for a given couple is much greater and the stress on the material much less than with a wire of circular cross section. They possess an additional advantage of offering a comparatively large surface area for the dissipation of heat produced by the current. Conneciton to the other end of the coil is made by means of a very light metal spring attached to the bottom of the coil by a wire. The spring is made of a very

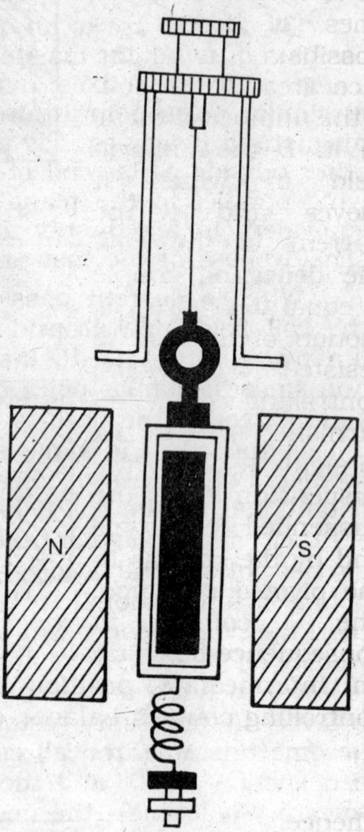

Fig. 28.2 (a)

fine strip and consists of relatively large turns so that it shall exert only a small controlling couple on the coil.

In the pivoted coil type, the coil rests vertically on two jewelled pivots and working loosely within two brass pillars. The motion of the coil is controlled by two light metallic springs attached to the bottom and top of the coil and coiled in opposite directions to neutralise the effect of rotation of the coil due to variation of temperature in the springs. The current leaves and enters the coil through these two springs.

When a current flows through the coil, forces act in opposite directions on the opposite sides of the coil, *i.e.*, a couple acts on the coil which tends to make it set at right angles to the field so as to enclose as many lines of forces as possible. If A be the face area of the coil, N the number of turns in it, B the uniform field in which it moves and i the current in the coil, the deflecting couple is equal to $B\,ANi$. The motion of the coil is resisted by the controlling couple $C\theta$ furnished by the torsion of the suspension in the suspended coil type and by the springs in the pivoted coil type. The coil in consequence takes up an intermediate position in which the deflecting and the controlling couples balance each other, *i.e.*, where,

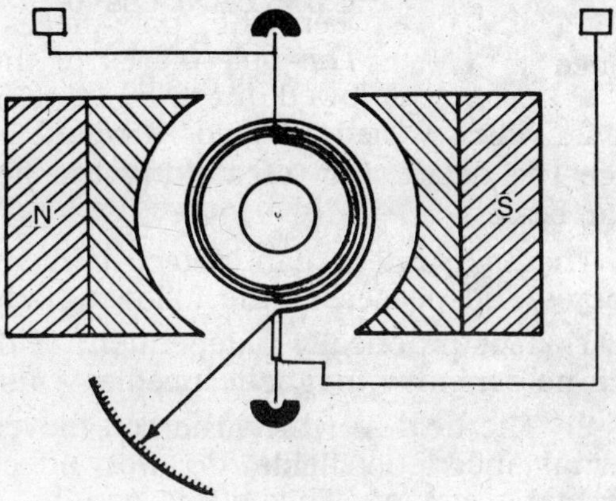

Fig. 28.2(b)

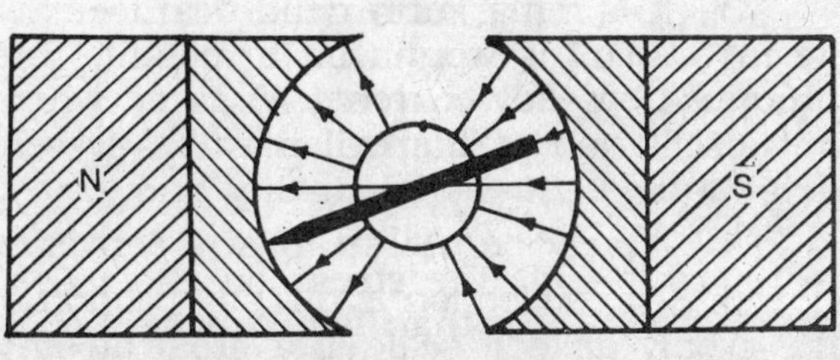

Fig. 28.2(c)

$$BANi = C\,\theta$$

whence, $$i = \frac{C}{BAN}\,\theta$$

or $$i = K\,\theta$$

where $K = \frac{C}{BAN}$

Thus, for radial field the deflection is directly proportional to the strength of the current. The deflections are observed by means of a pointer or a beam of light reflected from a small mirror attached to the rigid support of the coil.

If the field is not radial, then as soon as the coil rotates from its rest position in which it lies with its face parallel to the field, the couple is reduced to $BANi \cos \theta$, where θ is the inclination of the plane of the coil to the field, and

$$C\theta = BANi \cos \theta$$

whence

$$i = \frac{C\theta}{ABN \cos \theta}$$

A scale which varied as $\theta/\cos \theta$ would be very inconvenient and hence the desirability of making the field radial which gives a linear scale.

The instrument can detect currents of the order of 10^{-8} amperes and possesses the following advantages :

1. It is practically independent of the earth's magnetic field since no sensitive magnetic needle is used.

2. The field in which the coil moves is so strong that other external magnetic fields do not affect the readings to any appreciable extent. This effect can be still further reduced by magnetic screening.

3. It is remarkably dead-beat (see Section 28.5) in its action when the coil is wound on a metallic frame, the damping being produced by eddy currents. Further, when the current is stopped any oscillation of the coil may be prevented by short-circuiting it for a moment.

4. It may be set up in any convenient position.

5. The deflection is directly proportional to the current, and the evenly divided scale is a great advantage in practice.

6. The instrument can be made very sensitive by increasing the number of turns in the coil and decreasing the value of C, the controlling couple per unit twist.

28.4. Sensitiveness of Figure or Merit of a Galvanometer

In the case of pointer or non-reflecting galvanometers, *the sensitiveness is defined as the deflection in degrees produced by a current of one micro-ampere*. Alternatively, the figure of merit is defined as *the current in amperes necessary to produce a deflection of one degree*. In the case of reflecting galvanometer, *the sensitiveness is usually defined as the number of mms. deflection on a scale one metre away by one micro-ampere* ; and the figure of merit *as the number of amperes required to produce one mm. deflection on a scale one metre from the galvanometer mirror.*

Exercise—*To determine the figure of merit of a reflecting galvanometer.*

(I) Method—Shunt the reflecting galvanometer with a low resistance wire and then connect it through a commutator to an accumulator of known e.m.f., *E*. connected in series with a variable high resistance *R* (Fig. 28.3), usually of the order of megohm. Adjust the resistance *R* so that the spot of light remains on the scale when the current is started. Note down the deflection in mm. on the scale with the current direct and reversed and find its mean value. If *S* be the resistance of the shunt and *G* that of the galvanometer, the main current is given by :

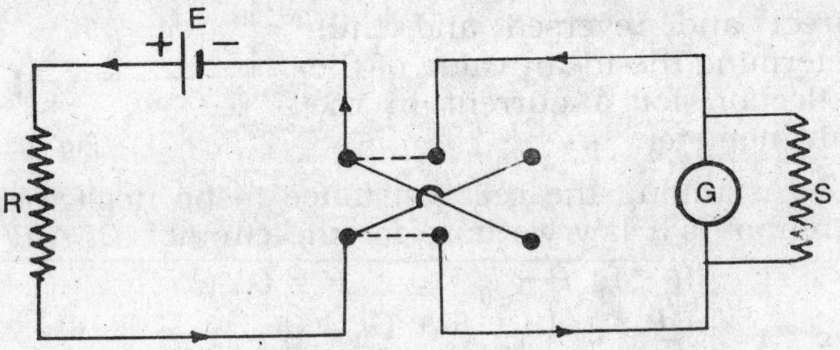

Fig. 28.3

$$i = \frac{E}{R + \dfrac{SG}{S+G}}$$

the resistance of the accumulator, being very small when compared with *R*, has been neglected. Since the current divides through the shunt and the galvanometer in the inverse ratio of their resistances, the galvanometer current is given by :

$$ig = \frac{E}{R + \dfrac{SG}{S+G}} \times \frac{S}{S+G} \text{ Ampere}$$

$$= \frac{ES}{R(S+G) + SG} \text{ Ampere}$$

If the deflection on the scale be δ mm., the figure of merit of the galvanometer is given by :

$$\eta = i/\delta \text{ amp. per mm.}$$

(II) Method—Take a P.O. box and connect an accumulator

of known e.m.f., E in the usual 'battery arm' AC through a tapping key K_1. Connect the 'galvanometer in the usual galvanometer arm BD through a commutator K_2. Adjust P = 10 ohm, Q = 1000 ohm and R = 5000 to 10000 ohm, leaving the usual 'unknown resistance arm' disconnected. Press the key K_1 and note down the deflection on the scale in mm. with current direct and reversed and thus determine the mean value of the deflection for a current in the galvanometer.

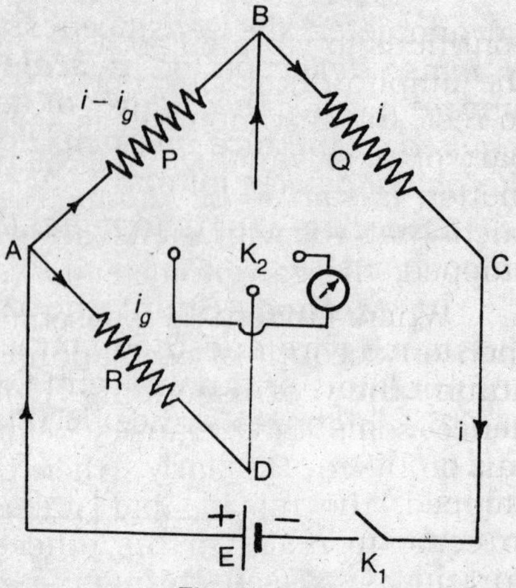

Fig. 28.3 (a)

Assuming the cell resistance to be negligible and applying Kirchhoff's II law we have for the circuit $ABDA$ (Fig. 28.3a) :

$$(i - i_g)\ P - i_g\ G - i_g\ R = 0$$

or

$$i\ P - i_g\ (P + R + G) = 0 \tag{28.2}$$

and for the circuit $ABCEA$

$$(i - i_g)\ P + i\ Q = E$$

$$i\ (P - Q) - i_g\ P = E \tag{28.3}$$

Multiplying Eq. (28.2) by $(P + Q)$ and Eq. (28.3) by P and then subtracting Eq. (28.2) from Eq. (28.3), we have :

$$i_g\ [(P + Q)\ (P + R + G) - P^2] = EP$$

or

$$i_g\ [(P + Q)\ (R + G) + PQ] = EP$$

whence

$$i_g = \frac{EP}{(P + Q)\ (R + G) + PQ}$$

Since $P << R$, we can neglect PQ in the denominator and hence

$$i_g = \frac{EP}{(P + Q)\ (R + G)} \tag{28.4}$$

If δ mm. be the mean deflection, the figure of merit of the galvanometer is given by :

$$\eta = i_g/\delta \text{ amp. per mm.}$$

The value of i_g can be known from Eq. (28.4).

28.5. Meaning of the Term 'Dead-beat' and Methods of Damping

When the moving part of a galvanometer, if disturbed,

continues to vibrate about its equilibrium position for a long time, the amplitude of vibration gradually diminishing until it comes to rest, the galvanometer is said to be *periodic*. But, if the moving part of a galvanometer, if disturbed, rapidly comes to rest, its motion is said to be 'damped' and if the damping is critical, *i.e.*, such that the vibrations are almost instantaneously completely stopped, the galvanometer is known as '*dead-beat*' or 'aperiodic'.

When current is passed through a periodic galvanometer, then on account of oscillation, the moving part does not at once attain the corresponding deflection, and hence some time elapses before a reading can be taken. Similarly, when the current is stopped, the moving part takes some time to come to rest. On the other hand, when current is passed through a dead-beat or aperiodic galvanometer the moving part almost *instantaneously* reaches its steady position without overshooting it, and when the current is stopped, it returns to zero with the same rapidity. Consequently, dead-beat galvanometers are very suitable for the measurement of steady deflections and are, therefore, especially used in all null methods. It should be clearly understood that in dead-beat galvanometers, the damping does not alter the final position of the moving part, but simply enables it to attain that position in a shorter time.

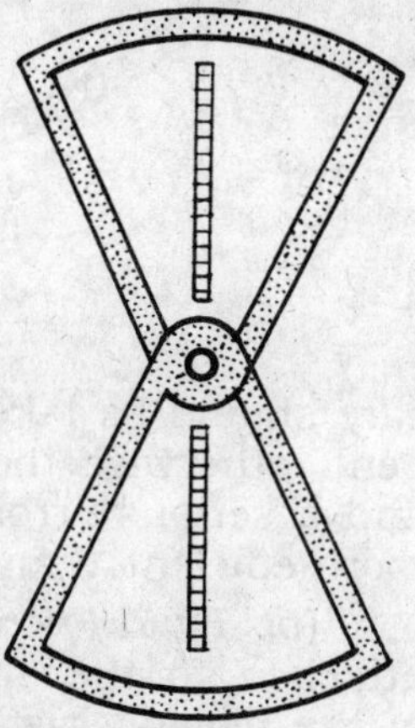
Fig. 28.4

There are three systems of damping in general use :

(*a*) *Air Friction Damping*—There are two methods of applying air friction damping. In the first method two vanes (Fig. 28.4) made of thin aluminium sheet are mounted on the spindle of the moving system and move in a closed, sector-shaped box.

In the second method a light aluminium platon *P* (Fig. 28.5) is attached to the spindle *S* of the moving system and moves in a rectangular or circular air chamber *C* closed at one end, the clearance between the piston and the walls of the chamber being about a few thousandths of an inch. When the piston moves rapidly into the chamber, the air in the closed space is compressed as a result of which it exerts pressure and opposes the motion of the piston and hence that of the whole moving system. When the piston moves out of the chamber rapidly, the pressure in the closed space falls, and as the pressure on the open side of the piston is then greater than that on the other side, it opposes the motion of the piston. In this damping system,

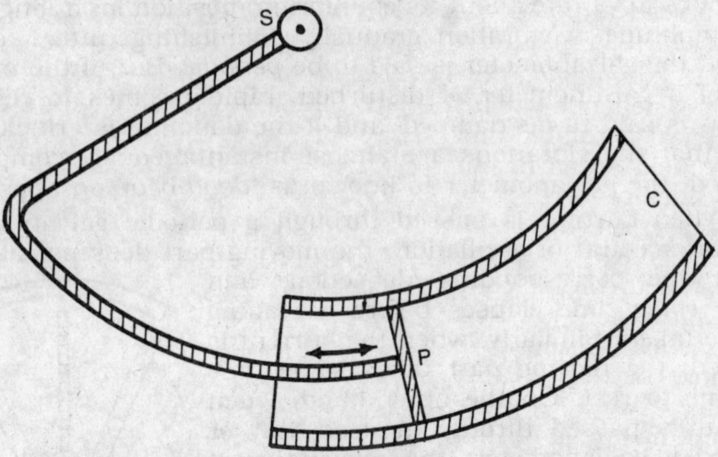

Fig. 28.5

care should be taken to ensure that the arm of the piston is not bent, otherwise the piston will touch the walls of the chamber and a serious error will result in the deflection. This method is more efficient than the first one.

(*b*) *Fluid Friction Dumping*—In this method a disc (Fig. 28.6) attached to the spindle of the moving system is dipped completely into a pot of damping oil. As the disc rotates, its motion is opposed by the viscous drag which increases with the speed of the rotation of the disc. In order to avoid surface tension effects, the suspending stem of the disc should be cylindrical and of small diameter where it enters the oil surface.

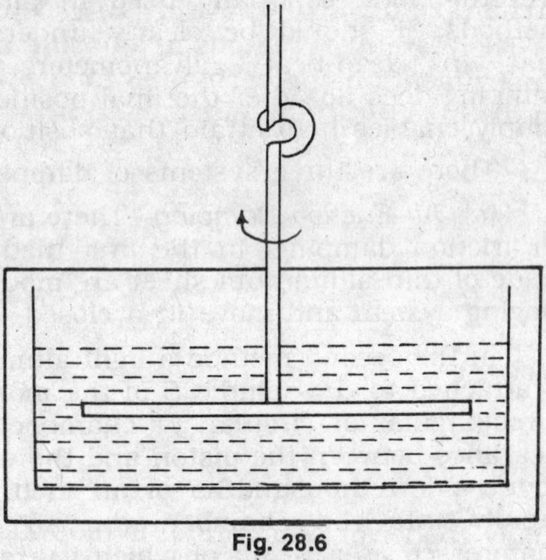

Fig. 28.6

Increased fluid friction damping can be obtained by the use of vanes in vertical planes, carried on a spindle and immersed in oil as shown in Fig. 28.7.

The method of fluid friction damping suffers from two disadvantages: first, it can be used only in instruments which are used in a vertical position and secondly, owing to creeping of the damping oil, it is difficult to keep the instrument clean.

(c) *Electromagnetic or Eddy Current Damping*—This method of damping depends upon the fact that when a sheet of conducting material is rotated in a magnetic field so as to cut through lines of force, eddy currents are set up in it, their magnitude being proportional to the velocity of the movement of the conductor. On account of interaction between these currents and the permanent magnetic field, a force acts on the conductor in a direction opposing its motion. The magnitude of the force is proportional to the magnetic field and to the currents, and hence to the velocity of movement of the conductor. The damping force is, therefore, zero when there is no movement of the conductor.

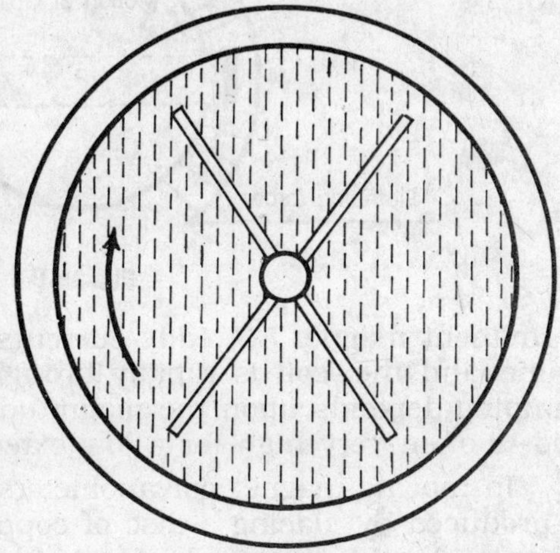

Fig. 28.7

There are two methods of applying electro-magnetic damping in instruments using permanent magnets. In the first method, a thin disc *D* (Fig. 28.8) of conducting but non-magnetic material,

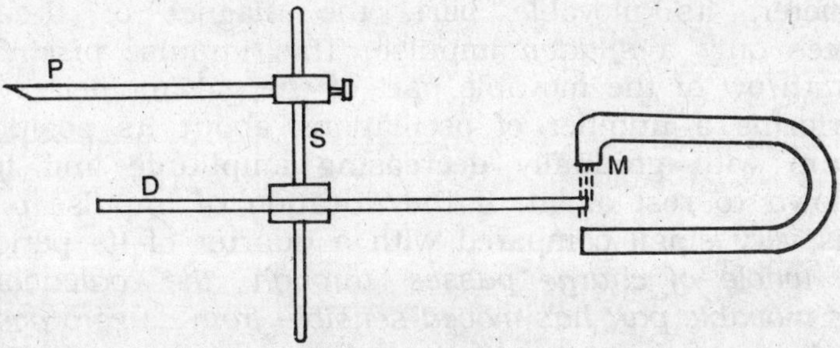

Fig. 28.8

usually copper or aluminium, is mounted on the spindle *S* which carries the pointer *P* of the instrument, the edge of the disc being in the gap of the permanent magnet *M*. When the spindle rotates, the edge of the disc cuts through the lines of force in the gap of the magnet. This produces eddy currents in the disc which damp the motion of the disc and hence that of the spindle.

In the second method, the coil is wound on a light metal frame *P* (Fig. 28.9). As the coil rotates in the field of the

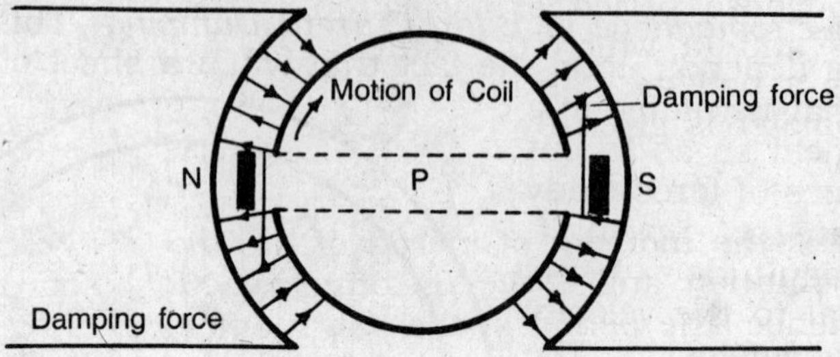

Fig. 28.9

permanent magnet *NS*, eddy currents are produced in the metal frame and the coil is rapidly brought to rest. In this case the damping depends upon the magnitude of the external resistance and is often very high for a low external resistance.

In moving-magnet galvanometers, electro-magnetic damping is produced by placing a disc of copper or aluminium under the magnet. As the magnet moves, eddy currents are produced in the disc which oppose the motion of the magnet. The damping can be increased by enclosing the magnet in a metal case.

28.6. Ballistic galvanometer

When a transient current, *i.e.*, a current of very short duration, *e.g.*, the discharge of a capacitor, passes through a galvanometer, its movable part, the magnet or the coil, experiences only a *sudden* impulse. This impulse produces a swing or throw of the movable part of the galvanometer which then performs a number of oscillations about its position of equilibrium with gradually decreasing amplitude and finally settles down to rest again. If the duration of impulse on the system is very small compared with a quarter of its period so that *the whole of charge passes through the galvanometer before its movable part has moved sensibly from its zero position*, the magnitude of the swing of the galvanometer is independent of the duration of impulse, and further if *the motion of the movable part is entirely undamped* so that no energy is lost during the swing, the magnitude of the *first* swing or throw, for a given galvanometer, is dependent only upon the quantity of charge passing through it. A galvanometer which satisfies the above two conditions is known as a *ballistic galvanometer* and is used to measure a quantity of charge passed through it.

It is evident from above that a ballistic galvanometer is not a galvanometer of any special type or construction, but any pattern whose period of oscillaiton is large, usually 10 to 15 seconds and in which the damping is negligibly small can be used as a ballistic galvanometer. The period of oscillation of a galvanometer is given by :

$$f = 2\pi \sqrt{\frac{I}{C}}$$

where I is the moment of inertia of the movable part about the axis of rotation and C the restoring reactional couple per unit twist. Hence, the period of the galvanometer can be made large by increasing the moment of inertia of the movable part and by reducing the controlling forces. The moment of inertia is increased by loading the movable part while the controlling forces are reduced by the use of quartz fibres or phospho-bronze strips as suspensions which have small torsional rigidity. In moving coil ballistic galvanometers which use non-conducting materials for suspension, the current is led into the coil by delicate spirals of very thin copper strip.

The damping of the motion of the movable part of the galvanometer consists of air damping and eddy current or electro-magnetic damping. The air damping is usually very small and the electro-magnetic damping is minimised by winding the coil on a non-conducting frame, *e.g.*, of wood or ebonite and by using a non-metallic case. In the moving coil galvanometer electro-magnetic damping is dependent upon the resistance of the external circuit and is sufficiently high when the external resistance is low.

As indicated above, both the moving-magnet and moving-coil galvanometers are used for ballistic purposes. In the moving-magnetic type, the magnet is set initially at right angles to the axis of the coil so that when the current passes, the deflecting and the controlling fields may be mutually at right angles. If then θ be the magnitude of the first swing or throw of the magnet when a quantity of charge Q is passed through the coil, it can be shown that

$$Q = \frac{B}{G} \cdot \frac{T}{\pi} \sin \frac{\theta}{2}$$

where B is the intensity of the restoring field, G the galvanometer constant, *i.e.*, field due to unit current in the coil and T the period of oscillation of the magnet. Thus, the quantity of electricity passing through a moving-magnet ballistic galvanometer is proportional to the sine of half the angle of the first throw.

For a moving-coil galvanometer, the relation between the

quantity of charge passed through the galvanometer and the consequent magnitude of the first throw is given by :

$$Q = \frac{C}{NAB} \cdot \frac{T}{2\pi}\ \theta$$

where C is the torsional rigidity of the suspension, B the field of the permanent magnet, N the number of turns in the coil, A the face area of the coil and T the period of the galvanometer. Thus, the quantity of electricity passing through a moving-coil ballistic galvanometer is proportional to the magnitude of the first throw.

The above relations between Q and θ are derived on the assumption that the motion of the movable part of the galvanometer is entirely undamped and that no energy is lost during the swing. But this is not the case in practice for there is always some air damping present. The result is that the observed throw is less than what would be if there were absolutely no damping. Hence, to get the value of undamped throw of the galvanometer, the observed value of θ should be multiplied by 'damping factor'. It can be shown that the value of this correcting factor is $(1 + \lambda/2)$, where λ is the logarithmic decrement. Hence, the complete expressions for the quantity of charge passing through the two types of galvanometer are :

$$Q = \frac{B}{G} \cdot \frac{T}{\pi}\ \sin. \left[\frac{\theta}{2}\left(1 + \frac{\lambda}{2}\right)\right] \tag{28.5}$$

for a moving-magnet ballistic galvanometer ; and

$$Q = \frac{C}{NAB} \cdot \frac{T}{2\pi}\ \theta\left(1 + \frac{\lambda}{2}\right) \tag{28.6}$$

for a moving-coil ballistic galvanometer. The value of logarithmic decrement λ may be calculated from the expression :

$$\lambda = 2.3026 \log_{10} \frac{\theta_1 + \theta_2 + \theta_3 + \ldots + \theta_{n-1}}{\theta_2 + \theta_3 + \theta_4 + \ldots + \theta_n}$$

where θ_n is the nth throw of the galvanometer. While determining the value of λ, the galvanometer should be in the circuit of the experiment.

In actual practice, ballistic galvanometers are reflecting instruments and the value of θ under ordinary circumstances does not exceed 6 or 7 degrees. Hence, if d be the movement of the spot of reflected light on the scale corresponding to θ, we have, since the angular motion of the spot of light is twice that of the mirror :

$$\text{Sin}\ \theta/2 \cong \theta/2 \text{ and } 2\theta = d/L$$

where L is the distance of the scale from the mirror. Thus, for small angles :

$$\theta = d/2L \text{ and } \sin \theta/2 \cong \theta/2 = d/4L$$

Putting these values of θ and $\sin \theta/2$ in Eqs. (28.5) and (28.6) we have, for a moving-magnet galvanometer,

$$Q = \frac{B}{G} \cdot \frac{T}{\pi} \frac{d}{4L} (1 + \lambda/2) = K_1 d (1 + \lambda/2)$$

where $K_1 = \frac{B}{G} \cdot \frac{T}{\pi} \frac{1}{4L}$ is the ballistic constant of the moving-magnet galvanometer ; and for a moving-coil galvanometer,

$$Q = \frac{C}{nAB} \cdot \frac{T}{2\pi} \cdot \frac{d}{2L} (1 + \lambda/2) = K_2 d (1 + \lambda/2)$$

where $K_2 = \frac{C}{NAH} \cdot \frac{T}{2\pi} \cdot \frac{1}{2L}$ is the ballistic constant of the moving-coil galvanometer.

In fact, the equation which holds generally for all types of galvanometers, provided that the deflections are sufficiently small, can be written as :

$$Q = Kd (1 + \lambda/2)$$

where K is the ballistic constant of the type of instrument used.

While noting down the deflection of the spot of light on the scale, the scale is usually kept at a distance of one metre from the galvanometer mirror. Hence, using the above equation we may define the ballistic constant *of a galvanometer as the charge required to produce a deflection of one millimetre of the spot of light on a scale placed at a distance of one metre from the galvanometer mirror, in the absence of damping.*

Experiment 28.1

Object—To determine the *ballistic constant* of a ballistic galvanometer by the *steady deflection method.*

Apparatus—A ballistic galvanometer whose ballistic constant is to be determined, a lamp and scale arrangement, a Post Office box, an accumulator, a voltmeter, a commutator and a stop-watch.

Theory—We know that the ballistic constant K of a ballistic galvanometer is connected to its figure of merit η by the relation:

$$K = \eta \frac{T}{2\pi} \qquad (28.7)$$

where T is the period of the galvanometer. Thus, if the figure of merit η of the ballistic galvanometer is determined and its period

T observed, the ballistic constant *K* of the galvanometer can be calculated from the above equation.

To determine the figure of merit η of the galvanometer, it is connected in the usual 'galvanometer arm' *BD* of a post office box as shown in Fig. 28.10. Then, keeping the usual 'unknown

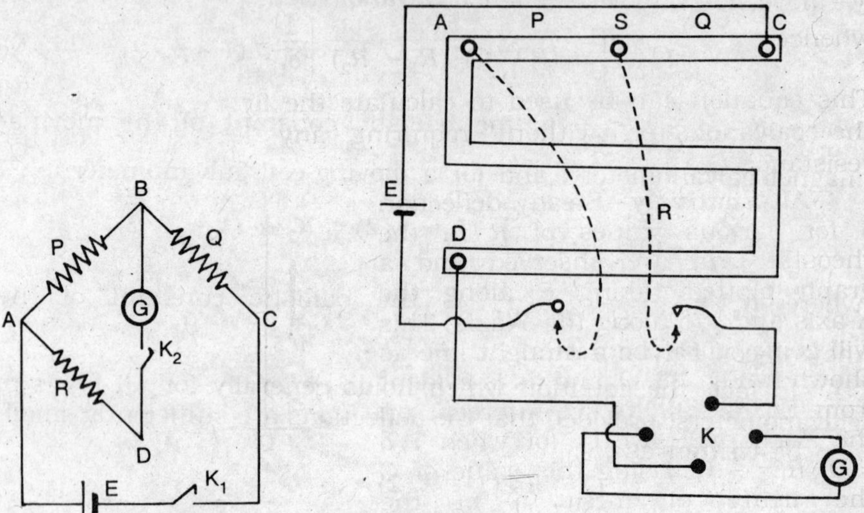

Fig. 28.10

resistance arm' *CD* of the post office box *disconnected,* if an accumulator of known e.m.f., E connected in the usual 'battery arm' *AC*, produces a steady deflection δ on the galvanometer-scale, the figure of merit of the galvanometer is from Eq. (28.4) given by :

$$\eta = \frac{EP}{(P+Q)(R+G)}\frac{1}{\delta} \quad (28.8)$$

where *P* and *Q* are the resistances in the ratio arms, *R* the resistance in the rheostat arm of the post office box and *G* the resistance of the galvanometer.

The above equation can also be written as :

$$R + G = \frac{EP}{(P+Q)\eta} \cdot \frac{1}{\delta} \quad (28.9)$$

so that, if δ_1 and δ_2 are the steady deflections on the galvanometer-scale for two values R_1 and R_2 of the resistance in the rheostat arm, we have :

$$R_1 + G = \frac{EP}{(P+Q)\eta} \cdot \frac{1}{\delta_1}$$

and $$R_2 + G = \frac{EP}{(P+Q)\,\eta} \cdot \frac{1}{\delta_2}$$

which on subtraction gives :

$$R_1 - R_2 = \frac{EP}{(P+Q)\,\eta}\left(\frac{1}{\delta_1} - \frac{1}{\delta_2}\right)$$

whence $$\eta = \frac{EP}{(P+Q)\,(R_1 - R_2)}\left(\frac{1}{\delta_1} - \frac{1}{\delta_2}\right) \qquad (28.10)$$

This equation can be used to calculate the figure of merit η of the galvanometer, without requiring any knowledge of its resistance G.

Alternatively—Steady deflection δ for various values of R in the rheostat arm are observed and a graph plotted taking R along the X-axis and $1/\delta$ along the Y-axis. This will come out to be a straight line as shown in Fig. 28.11 and as is evident from Eq. (28.9), its –ve intercept on the X-axis will give G, for when $1/\delta = 0$, $R = -G$. Using this value of G the figure of merit η of the galvanometer can then be calculated from Eq. (28.8).

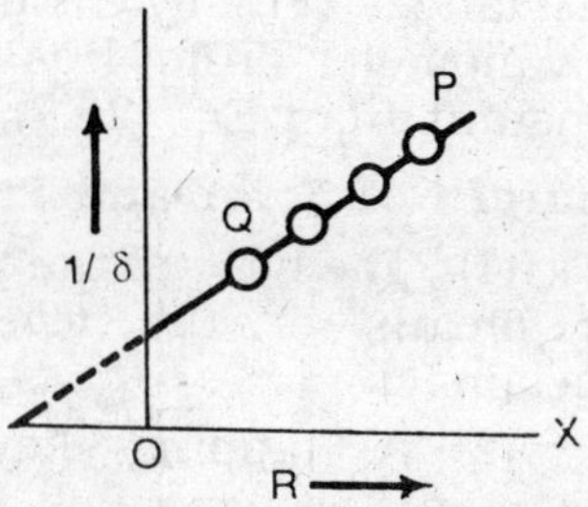

Fig. 28.11

Method—Level the base of the ballistic galvanometer by means of the levelling serews and release its coil. Throw light on the mirror of the galvanometer and get the bright spot of reflected light on the scale set up at a distance of *one metre* from it. Adjust the spot of light on the scale at zero. Connect the galvanometer in the usual 'galvanometer arm' BD of the post office box through a commutator K and an accumulator in the usual 'battery arm' AC, leaving the arm of 'unknown resistance' CD disconnected as shown in Fig. 28.10.

Adjust the ratio arm P to 10 ohm, the ratio arm Q to 100 ohm and the rheostat arm R to a high resistance, say, 5000 ohm. Press the battery key and note down the *steady* deflection of the spot of light produced on the scale. Then, reverse the current through the galvanometer with the help of the commutator K and again note down the steady deflection of the spot of light on the scale. Find the *mean* value of the galvanometer deflection δ for the resistance R = 5000 ohm in the rheostat arm.

Now, keeping the resistances P and Q in the ratio arms the *same*, increase the resistance R in the rheostat arm in steps to 1000 ohm and determine the mean value of δ as above for at least six values of R.

Next, measure the e.m.f. of the accumulator with a voltmeter and noting down the time of oscillation of the galvanometer coil for a known number of osciliations, also determine the period of the galvanometer.

Plot a graph taking the values of resistance in the rheostat arm of the post office box along the X-axis and the corresponding values of *reciprocal* of deflection δ along the Y-axis. This will come out to be a straight line as shown in Fig. 28.11. Measure its *negative* intercept on the X-axis which gives G, the galvanometer resistance. Finally, calculate the figure of merit of the galvanometer from Eqs. (28.8) and (28.10), and the ballistic constant from Eq. (28.7).

Sources of Error and Precautions

(1) The base of the galvanometer should be carefully levelled by means of the levelling screws and the coil should be unclamped.

(2) A tapping key should be connected across the galvanometer terminals in order to bring the galvanometer coil to rest in a *short* time by tapping the key, when so desired.

(3) In the case of plug type of post office box all sockets must be *clean* and the plugs made *light*.

(4) The accumulator connected in the usual battery arm of the post office box should be fully charged so that it may have a *constant* e.m.f., thus, producing *steady* deflection in the galvanometer.

(5) The scale should be set up at a distance of *one metre* from the galvanometer mirror.

(6) The resistance R in the rheostat arm should be *very large* when compared to the resistance P in the ratio arm so that the latter may be neglected in comparison to the former.

(7) The graph between $\frac{1}{\delta}$ and R should be *smoothly* drawn with (0,0) at the origin.

(8) While determining the period T of the galvanometer, the galvanometer circuit must be kept *open*.

Observations—(A) *Determination of figure of merit of galvanometer.*

(*i*) Measurement of δ for different values of R.

Rest position of spot of light = zero.

S. No.	Ratio arms		Rheostat arm R ohms	Deflection δ in galvanometer in mm.	1/δ in $(mm)^{-1}$	$\frac{1}{\delta_1} - \frac{1}{\delta_2}$ for $R_1 - R_2$ 3,000 ohms
	P ohm	Q ohm				
1.	10	1000	5,000		*a*	
2.	10	1000	6,000		*b*	
3.	10	1000	7,000		*c*	
4.	10	1000	8,000		*d*	
5.	10	1000	9,000		*e*	
6.	10	1000	10,000		*f*	
					Mean	$x =$

(*ii*) e.m.f. of the accumulator = volts

(B) *Determination of period of the galvanometer.*

S. No.	No. of oscillations	Time		Period T Sec.
		Min	Sec.	
			Mean	T =

Calculations—From Eq. (28.10)

$$\eta = \frac{EP}{(P+Q)(R_1+R_2)}\left(\frac{1}{\delta_1} - \frac{1}{\delta_2}\right)$$

$$= \frac{EP}{(P+Q)(R_1+R_2)} \cdot x$$

$$= \quad \text{Amp./mm.}$$

From Graph (28.11)

Galvanometer resistance G = intercept OR = ohms

[Using the value of G prepare the following table.]

S. No.	R ohm	$(R + G)$ ohm	δ mm.	$(R + G)\,\delta$ ohm × mm
1.	1,000			
2.	6,000			
3.	7,000			
4.	8,000			
5.	9,000			
6.	10,000			
			Mean	

$$\eta = \frac{EP}{(P+Q)(R+G)\delta}$$

$$= \frac{EP}{(P+Q) \cdot y}$$

$$= \quad \text{Amp./mm.}$$

Mean value of η = Amp./mm.

Hence $$K = \eta \cdot \frac{T}{2\pi}$$

$$= \quad \text{Coulombs/mm.}$$

Result—The ballistic constant of the ballistic galvanometer = Coulombs/mm.

28.7. Ammeters

Amperemeters or ammeters, as they are usually called, are instruments used for the measurement of current directly in amperes. To measure a current, an ammeter is inserted in the circuit so that the whole current to be measured passes through it. This introduction of the instrument into the circuit must have least possible influence on the circuit, *i.e.*, the instrument must be able to record the value of the current without changing it. In addition, the 'voltage drop' IR across the instrument and the power loss I^2R and consequent heat produced in the instrument must, in spite of large currents, be negligible. This requires that an ammeter should be a *low* resistance instrument. Consequently, the coil or the wire used in an ammeter is always of low resistance. When the resistance of the coil is not small, it is shunted with a wire of very low resistance. The shunt makes the total resistance low as well as allows most of the current to pass through it so that only a safe part goes through the coil.

There are three types of ammeters in common use : (*a*) Moving iron, (*b*) Hot wire and (*c*) Moving coil.

(*a*) **Moving iron ammeter**—(*i*) *Repulsion type*— It consists of two rods A and B (Fig. 28.12) of soft iron lying inside and parallel to the axis of a long coil C. One of the rods A is fixed and the other B is pivoted and attached to a spindle carrying a pointer. When the current to be measured is passed through the coil, the iron rods are similarly

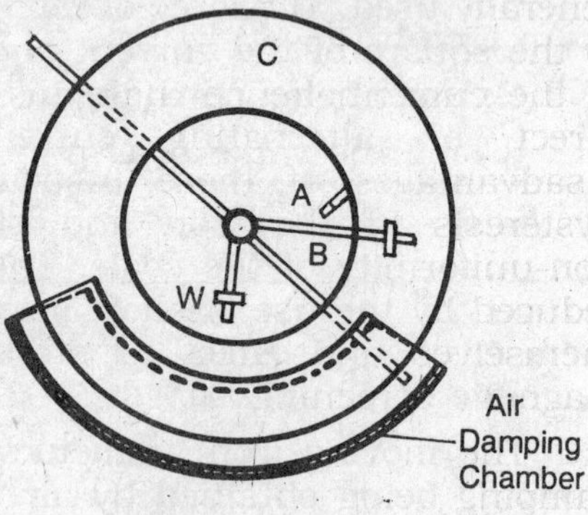

Fig. 28.12

magnetized and the repulsion of the movable rod from the fixed one ensues. With the movement of the movable rod, the pointer is deflected on the scale until equilibrium is established by controlling forces provided either by a balance weight *W* or a spring. The scale is graduated in amperes so that by noting the position of the point on it, the current can be read off directly.

(*ii*) *Attraction type*—It consists of a coil *A* and a soft iron core *C* mounted eccentrically (Fig. 28.13). The iron core is made of several thin discs of soft iron. When the current to be measured is passed through the coil, a magnetic field is set up and the iron core moves from weaker magnetic field outside the coil into the stronger field inside it. With the iron core moves the pointer attached to it until equilibrium is established by controlling forces which may be gravitational or due to a spring, and the current is read off directly on the scale. The shape of the discs in the iron core is such that a suitably divided scale is obtained.

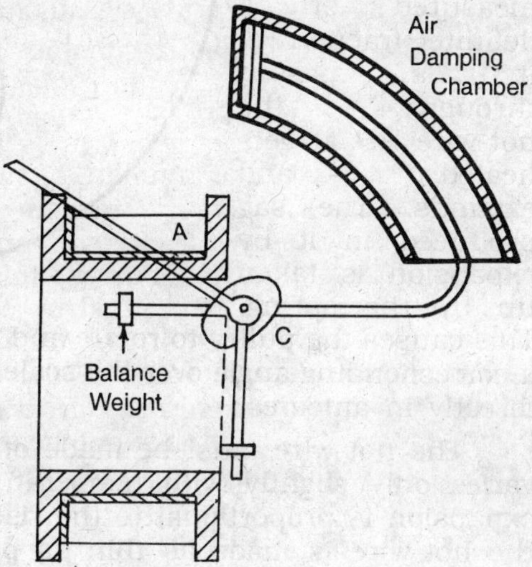

Fig. 28.13

The moving iron ammeters are very cheap and can stand much rough usage without damage and are, therefore, most generally used. The force of repulsion or attraction is proportional to the square of the current and is independent of the direction of the current; hence these instruments can be used for either direct or alternating current measurements. The main disadvantages of these instruments are: (*i*) errors due to hysteresis in the iron and stray magnetic fields and (*ii*) the non-uniformity of the scale. The error due to hysteresis can be reduced by the use of short iron pieces so that they demagnetize themselves and effect of strong fields can be minimised by magnetic screening.

The moving iron ammeters are dead-beat instruments, the damping being obtained by air friction device.

(*b*) Hot wire ammeter—The commonest form of double- sag hot wire ammeter consists of a wire *AB* (Fig. 28.14) of a platinum-iridium, at the middle of which is attached another wire *W* of phosphor bronze, the latter being attached to a fine silk thread which passing round a smail pulley *P* is fastened to a spring *S* which keeps the whole system taut. A light pointer is carried by the spindle upon which the pulley is mounted. When the current to be measured or a definite fraction of it, is passed through the hot-wire *AB*, it gets heated and expands. The sag produced in it by expansion is taken up by the spring.

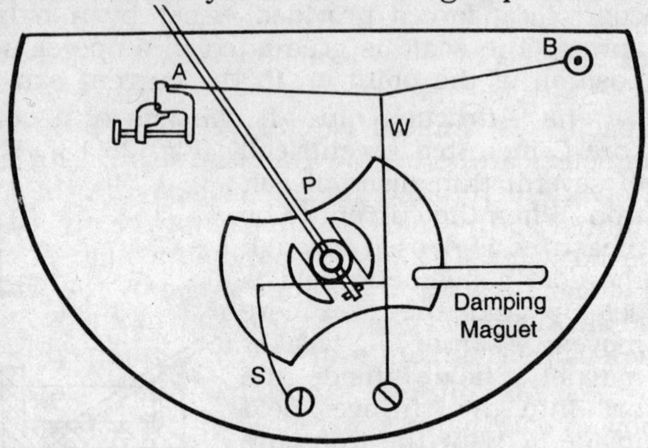

Fig. 28.14

This causes the pulley to rotate and the pointer to deflect through a corresponding angle over the scale calibrated to give the current directly in amperes.

The hot-wire must be made of a material whose resistance varies only slightly with temperature and whose coefficient of expansion is proportional to the rise of temperature. Further, as the hot-wire is made as thin as possible in order that it may attain a steady temperature quickly, its material must have sufficient mechanical strength to withstand the stresses placed upon it by the tension of the spring system. The box of the instrument must be made of materials whose coefficient of expansion is equal to that of the hot-wire so that changes in temperature other than those caused by the current through the hot-wire may not affect the deflection.

The instrument is dead-beat. The damping is electro-magnetic and is caused by eddy currents produced in a thin aluminium disc carried by the spindle of the pulley, the edge of the disc being situated in the air-gap of a permanent magnet. This damping is very essential to stop movements due to vibration or to sudden changes in the current, otherwise the hot-wire will be subjected to excessive stresses.

The hot-wire ammeter is very simple and cheap. Its sensitivity can be greatly increased by placing the hot-wire in vacuum. Since the heat produced is indepedent of the direction of the current the instrument can be used for either direct or alternating-current measurements. Moreover, since the heating effect by an alternating current is independent of its wave-form or frequency, the calibration of the ammeter is the same for both direct and alternating current. The working of the instrument is not affected by stray magnetic fields.

Since the hot-wire is always very fine, its current-carrying capacity is limited. Consequently, the double-sag hot-wire ammeter can be used without any shunt for current of the order of one ampere only. For higher currents the instrument must be shunted so that only a safe part of the current may flow through the hot-wire. The instrument has a high power consumption. It is sluggish owing to the time taken by the wire to heat up. Since the heating effect is proportional to the square of the current, the scale is not uniform, it being cramped at the bottom end.

(*c*) **Moving-coil ammeter**—It is simply a pivoted type moving-coil galvanometer (Fig. 28.15) whose coil has been shunted with a wire *R* of low resistance. When a definite fraction of the current to be measured is allowed to pass through the coil, it is deflected on account of interaction between the permanent magnet field and the magnetic field of the coil. The motion of the coil is controlled by one or two phosphor-bronze spring according as the instrument is a uni-pivot or bi-pivot one. The field of the permanent magnet is kept radial by the use of concave pole-pieces and hence the current is proportional to the deflection of the coil which it produces. The instrument is dead-beat, damping being produced by eddy currents induced in the aluminium formerd upon which the coil is wound.

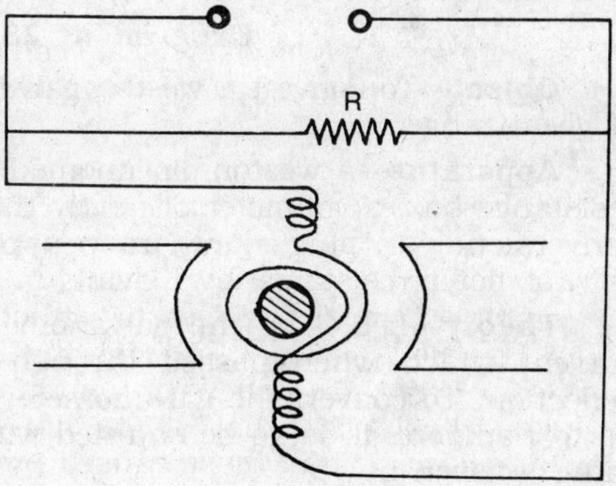

Fig. 28.15

In uni-pivot instrument, the friction torque at the pivot is reduced by the elimination of one point. This makes the instrument highly accurate. But the instrument requires careful handling and must be used only in a horizontal position.

The moving coil ammeter is sufficiently accurate. Its power consumption is low. The scale is remarkably uniform. The same instrument can be used to measure a large range of a current by the use of shunts. The instrument is free from errors due to hysteresis and from errors due to stray magnetic fields.

The instrument suffers from friction and heating errors which are, of course, present in other types of instruments also. Unless the permanent magnet is carefully aged during maunfacture, its weakening with time may introduce a considerable error. As the direction of deflection depends upon the direction of current the instrument can be used only for direct-current measurements.

The range of an ammeter can be increased by shunting the instrument with a suitable low resistance wire. If r is the resistance of an ammeter which gives a full-scale deflection with a current of i amperes, then the shunt resistance required to increase the range of the instrument from i amperes to I amperes is given by $s = \frac{i}{I - i} r$. These shunts used to increase the range of ammeters are known as range multipliers.

Experiment 28.2

Object—To convert a weston galvanometer into an ammeter of given range.

Apparatus—A weston galvanometer, an accumulator, a high resistance box, a voltmeter of nearly the same range as given for conversion, a plug key and apparatus for determining galvanometer resistance by Kelvin's method.

Theory—Let G be the galvanometer resistance and I_g the current which when passed through it produces a full-scale deflection. To convert the galvanometer into an ammeter reading up to I ampere, it is to be shunted with a low resistance whose value is given by :

$$S = \frac{I_g}{I - I_g} \cdot G \qquad (28.11)$$

The length of the shunt wire having a resistance S is given by :

$$l = \frac{\pi r^2 S}{\rho} \qquad (28.12)$$

where r is its radius and ρ the resistivity of the material.

Method—Determine the resistance of the Weston galvanometer by Kelvin's method.

Next, connect (Fig. 28.16) the Weston galvanometer to an accumulator in series with a high resistance box including a plug key K in the circuit. Adjust the resistance box to a high value, say, 5000 ohm. Note down the zero reading to the galvanometer. Close the key K and again adjust the resistance box so that almost full-scale deflection is obtained in the galvanometer. Note down the resistance R introduced into the circuit by the resistance box and also the deflection n produced in the galvanometer taking into account the zero error of the Instrument. Measure the e.m.f., of the accumulator with an accurate voltmeter and calculate the figure of merit of the galvanometer from the formula :

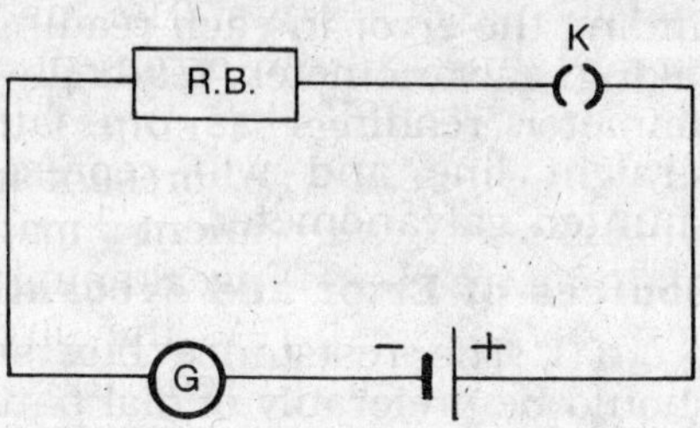

Fig. 28.16

$$K = \frac{e}{n\,(R + G)}$$

Repeat the experiment with different values of R and thus find the mean value of K. Multiply this by the total number of divisions N on one side of zero of the galvanometer scale which gives I_g the current required to produce full-scale deflection in the galvanometer.

Now, calculate the shunt resistance S from Eq. (28.11) and then selecting a proper gauge, say, S.W.G. 22, calculate from Eq. (28.12) the length l of a manganin or constantan wire which will have a resistance equal to S, the value of r and ρ being known from tables. Then cut off a length of the wire about 4 cm more than the calculated value and mark two points on it at the calculated distance apart, leaving some spare parts on each side. Connect the wire across the galvanometer such that the marked points are just outside the screws. Graduate the scale divisions in amperes taking N divisions equal to the range I ampere. This converts the Weston galvanometer into an ammeter of the given range.

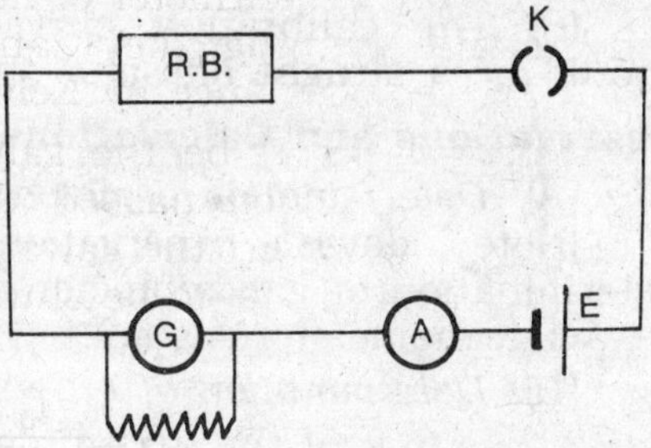

Fig. 28.17

Next, connect (Fig. 28.17) the shunted galvanometer in series with an ammeter of nearly the same range, a high resistance box, and an accumulator including a key K in the circuit. Adjust the resistance box to a high value. Close the key K and note down the reading of the galvanometer and the ammeter. Convert the galvanometer reading into ampere and find the difference between the readings of the two instruments, if any, which gives the error in the galvanometer reading. Alter the resistance of the circuit and in this way take at least six sets of observations, finding the error in each reading of the galvanometer. Plot a graph taking galvanometer readings as abscissae and corresponding ammeter readings as ordinates. The graph will be nearly a straight line and will represent the calibration curve of the shunted galvanometer.

Sources of Error and Precautions

(1) The resistance box should be of high resistance and should be preferably of dial pattern. At no stage of the experiment the resistance in the resistance box should be reduced to zero for then a heavy current will flow through the circuit damaging the galvanometer or ammeter or both.

(2) The accumulator used should be freshly charged so that its e.m.f. may remain constant throughout the determination of I_g.

(3) The ammeter used for calibration of the shunted galvanometer should be of nearly the same range and its readings should be corrected for zero error of the instrument.

(4) The +ve marked terminal of the ammeter should be connected to the positive pole of the cell or the high potential point of the circuit.

(5) While connecting the shunt across the galvanometer, care should be taken to see that only the exact required length of the wire lies between its terminals.

(6) The calibration curve of the shunted galvanometer should be a straght line and should be smoothly drawn.

Observations and Calculations

[A] *Determination of galvanometer resistance G.*

[Note down observations for Kelvin's method for determination of galvanometer resistance G]

Galvanometer resistance G = ohms.

[B] *Determination of I_g*

e.m.f. of the accumulator = volts

S. No.	Resistance introduced by the resistance box R ohm	Deflection in galvanometer n scaled div.	Figure of merit of galvanometer K Amp./div.
		Mean	

Total No. of divisions N on one side of zero of this galvanometer scale = ...

∴ The current I_g reqd, to produce full-scale deflection = Amp.

[C] *Calculation of length of the wire :*

$$S = \frac{I_g}{I - I_g} \cdot G$$

$$= \quad \text{ohm}$$

Gauge selected S.W.G. No...

From tables resistivity ρ of the material (.....) = ohm × cm

radius r of the wire = cm.

∴ $l = \pi r^2 \, S/\rho$ = = cm.

[D] *Calibration of the shunted galvanometer.*

S. No.	Galvanometer reading in		Ammeter reading I' Amp.	Error $I\text{-}I'$ Amp.
	Scale divisions	Amperes I		

Results—(*i*) The length of the shunt wire of.....S.W.G....required to convert the Weston galvanometer into an ammeter of the given range.

(*ii*) The graph shown in Fig.... is the calibration curve of the shunted galvanometer and is a straight line.

Criticism of the method—If performed carefully, the experiment yields quite satisfactory results. The accuracy of the result depends upon the constancy of e.m.f. of the accumulator and the accuracy of the measuring instruments, namely, the voltmeter and the ammeter. For higher accuracy the shunted galvanometer should be calibrated by means of a potentiometer.

28.8. Voltmeters

Voltmeters are instruments used for the measurement of p.d. directly in volts. To measure the p.d. betwen two points of a circuit, a voltmeter is connected across the two points whose p.d. is required. In order that the instrument should record the actual p.d. between two points, it is necessary that in connecting the instrument across the points, the p.d. between them should not be altered. To fulfil this condition the instrument must take only an infinitesimal current. In addition, the power consumption V^2/R and constant heat produced in the instrument should be negligible. Hence, it is clear that a voltmeter must have a high resistance whose value must always be very high in comparison with the resistance between the points whose p.d. is to be measured. In fact, the *resistance of an ideal voltmeter should be infinite.*

Voltmeters can be divided into four classes : (*a*) Electro-static (*b*) Moving iron, (*c*) Hot wire and (*d*) Moving coil.

The electro-static voltmeters are in general of two types : (*i*) The quadrant type which is used for voltages up to 10 or 20 kilovolts and (*ii*) The attracted disc type which is used for voltages above 10 or 20 kilovolts. The construction and working of these voltmeters is similar to those of Lord Kelvin's quadrant and attracted disc electrometers.

The electro-static voltmeters are ideally perfect in having infinite resistance. The power consumption is extremely small. They work equally well on both direct and alternating currents. Their working is free from errors due to hysteresis and are not affected by stray magnetic fields and eddy currents. They are, however, not sufficiently sensitive to read low voltages, their useful range being from about 500 volts upwards to several hundred kilovolts.

The moving iron voltmeters and the hot-wire voltmeters are constructed from moving iron ammeters and hot-wire ammeters respectively by connecting a high non-inductive resistance in series with the solenoid and the hot-wire respectively, and then calibrating the scale to read the p.d. directly in volts. These instruments are suitable for measuring either direct or alternating voltages.

The moving-coil voltmeters are constructed from moving-coil galvanometer or ammeter by connecting a high non-inductive resistance in series with the moving coil (Fig. 28.18). The higher the resistance connected in series with the coil, the more accurately will it indicate the p.d. to be measured. Cheap voltmeters usually have fairly low resistance and consequently they do not give reliable results. Moving-coil voltmeters can be used for direct voltages only.

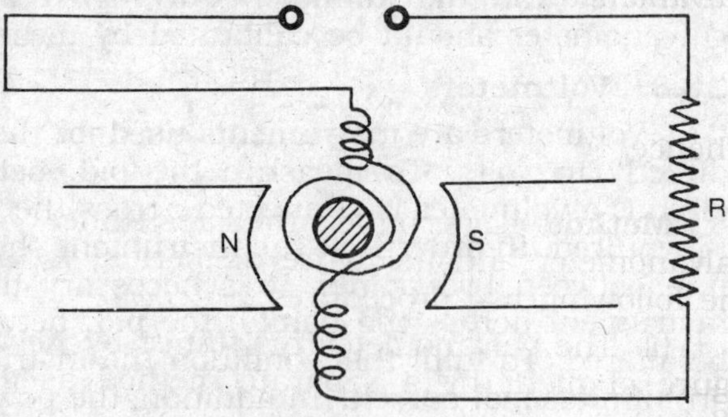

Fig. 28.18

The range of moving-iron, hot-wire or moving-coil voltmeters may be increased by using a suitable high resistance in series with it. Thus, if a voltmeter gives a full-scale deflection when a current passes through it and if its resistance is r, then the series resistance R required to increase the range of the instrument from υ volts to V volts is given by :

$$\frac{V}{R+r} = i = \frac{\upsilon}{r}$$

whence,
$$R = \frac{V-\upsilon}{\upsilon} \cdot r$$

The resistance connected in series with a voltmeter to increase its range is called 'multiplying resistance' or 'multiplier'. The value of these multipliers must be constant and hence they must have a low temperature coefficient and since they absorb appreciable power, ample provision for cooling must be made.

An ammeter can be converted into a voltmeter by placing a high resistance in series with it, and a voltmeter can be used as an ammeter if the former is shunted with a low resistance.

Experiment 28.3

Object—To convert a Weston galvanometer into a voltmeter of given range.

Apparatus—A weston galvanometer, an accumulator, a voltmeter of nearly the same range as specified for conversion, a plug key, a rheostat and apparatus for determining galvanometer resistance and its sensitivity.

Theory—To convert a Weston galvanometer into a voltmeter reading upto V volts, a *high* resistance is to be connected *in series* with it. If R be the magnitude of this series resistance, I_g the current required to produce full-scale deflection in the galvanometer and G the resistance of the galvanometer :

$$I_g = \frac{V}{R+G}$$

whence,
$$R = \frac{V}{I_g} - G \qquad (28.13)$$

Method—Determine the resistance G of the Weston galvanometer and its figure of merit η by adopting any one of the following two procedures :

(*i*) The galvanometer resistance by Kelvin's method and its figure of merit by a separate method as described in Section 28.4.

(*ii*) Both the galvanometer resistance and its figure of merit by a *single* method using P.O. box where galvanometer resistance may be obtained graphically and its figure of merit by way of calculations from Eq. 28.10.

Next, note down the total number of divisions N on one side of zero of the galvanometer scale and, multiplying it by the figure of merit of the galvanometer, calculate the current I_g required to produce full-scale deflection in the galvanometer. Then, calculate the resistance R required to convert the galvanometer into a voltmeter of given range V from Eq. (28.13). Procure a resistance of this value and connect it in series with the galvanometer. Graduate the galvanometer scale in volts taking full-scale deflection equivalent to V volts. This converts the galvanometer into the voltmeter reading upto V volts.

To calibrate the instrument, connect (Fig. 28.19) the whole of a rheostat S, using its fixed terminals at the base, to an accumulator including a plug key K in the circuit, connect the galvanometer together with its series resistance R, in parallel with a part of the rheostat using one of the fixed terminals at the base and the screw joined to the sliding contact maker. Connect between the same points a voltmeter of nearly the same range.

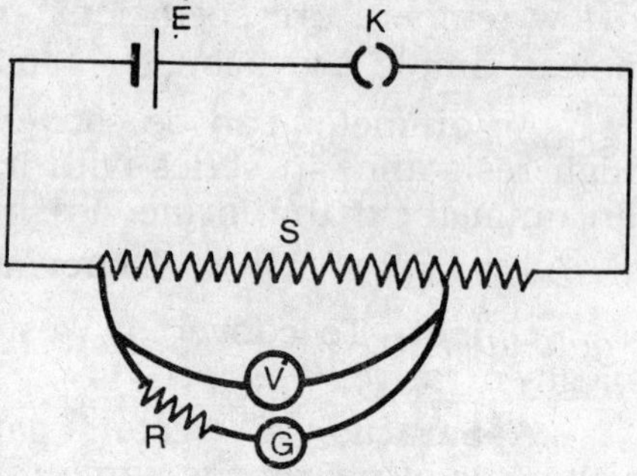

Fig. 28.19

Close the key K. Adjust the rheostat to get full-scale deflection in the galvanometer. Note down the reading of the instruments and find the difference between them which gives the error in the galvanometer reading. Decrease the deflection in the galvanometer with the help of the sliding contact maker of the rheostat and in this way take at least six sets of readings of the two instruments, finding error corresponding to each reading of the galvanometer. Plot the calibration curve of the galvanometer taking the galvanometer readings as abscissae and the corresponding readings of the voltmeter as ordinates. The graph will be a straight line.

Sources of Error and Precautions

Besides the precautions observed in the determination of the galvanometer resistance G and its figure of merit η, the following precautions should also be taken :

1. The voltmeter used for calibration of the galvanometer

after conversion into a voltmeter, should be of nearly the *same* range and its readings should be corrected for zero error of the instrument.

2. The +ve marked terminal of the voltmeter should be connected to the higher potential terminal of the rheostat.

3. The calibration curve of the galvanometer after its conversion into a voltmeter should be a straight line and should be smoothly drawn.

Observations and Calculations

Determination of G and η.

(Record observations for determining galvanometer resistance G and its figure of merit η)

Galvanometer resistance G = ohm.

Its figure of merit η = Amp./div.

Total no of divisions N on one side of the zero of the galv. scale =

∴ The current I_g required to produce full-scale deflection = Amp.

[B] *Calculation of resistance R required for conversion of galvanometer into voltmeter.*

$$R = \frac{V}{I_g} - G$$

=

= ohm

[C] *Calibration of galv-scale to read upto V volts.*

S. No.	Galvanometer reading in		Voltmeter reading v' in volts	Error $v - v'$
	Scale divisions	Volts v		

Results—(*i*) The series resistance required to convert the Weston galvanometer into a volmeter of the given range = ohm

(*ii*) The graph shown in Fig..... is the calibration curve of the galvanometer converted into a voltmeter and is a straight line.

ORAL QUESTIONS

GALVANOMETERS

1. What is a galvanometer ?
2. How many classes of galvanometers do you know of and what are the main principles on which they are constructed ? Give examples of each.
3. Describe the construction of a moving magnét galvanometer. How is the needle deflected in it ?
4. Describe a tangent galvanometer. What is it used for ?
5. What advantages does Helmholtz double coil tangent galvanometer possess over an ordinary tangent galvanometer ?
6. How can you increase the sensitiveness of a moving-magnet galvanometer ?
7. Why can you not use a large number of turns in the coil ? What is the harm if the radius of the coil is very small ?
8. What is control, magnet ? Explain how it is used to decrease the controlling field.
9. In case the controlling field is reduced to zero, how can the necessary restoring couple be provided ?
10. What is an astatic pair ? How does it increase the sensitiveness of the galvanometer ?
11. Which is the most sensitive moving-magnet galvanometer ?
12. What is the principle of construction of moving-coil galvanometer ? How many types are there ? How is the coil deflected ?
13. What provides the controlling couple ?
14. Why is a phosphor-bronze strip used for suspension ?
15. Why are springs in a pivoted type moving-coil galvanometer made thin ? Why have they large number of turns and why are they coiled in opposite directions ?
16. What is the core of the coil made of and what purpose does it serve ?
17. Why are concave pole-pieces used ? What is the harm if the field is not radial ?
18. What factors determine the sensitiveness of a moving-coil galvanometer ?

19. What are the advantages of a moving coil galvanometer over a moving-magnet galvanometer ?
20. What is the advantage of a mirror galvanometer ? What advantages does the lamp and scale method of measuring deflections possess over the use of mechanical pointer ?
21. What is the relative sensitiveness of the optical and mechanical methods of measuring deflections ?
22. What is meant by the sensitivity of a galvanometer ? Which is the most sensitive galvanometer ? Is it possible to have a galvanometer which may be regarded very sensitive from one point of view and very insensitive from the other ?
23. How can you determine the sensitivity of a sensitive moving coil galvanometer ?
24. Distinguish between periodic and aperiodic galvanometer.
25. What is a dead-beat galvanometer ? How can you make a galvanometer dead-beat ?
26. How many methods of obtaining damping do you know ? Describe them and mention the instruments in which they are used.
27. Which is the most efficient method of damping ?
28. What are eddy currents and how are they produced ?
29. What is a ballistic galvanometer ? What are its special features ? What is it used for ?
30. What is logarithmic decrement ? How and why do you determine it ?
31. How can you use a galvanometer (*i*) as an ammeter (*ii*) as a voltmeter ?
32. Can you use all types of galvanometers as ammeters and voltmeters ? If so, how?

AMMETERS AND VOLTMETERS

33. What are ammeters and voltmeters ? How do they differ from galvanometers ?
34. How many types of ammeters do you know and which is the best ?
35. Why should an ammeter have low resistance and a voltmeter a high resistance ? What is the harm if the resistance of the voltmeter is not so ?

36. Do you know of any voltmeter whose resistance is infinity ?
37. Why is an ammeter connected in series and a voltmeter in parallel ?
38. Why can you not connect a voltmeter in series with a circuit ?
39. Describe a hot-wire ammeter. Of what material is the hot wire made and why ? How is the damping produced in the instrument ?
40. Explain the construction and working of the moving-coil ammeter and voltmeter.
41. What advantages does an ammeter possess over a tangent galvanometer as a current measuring instrument ?
42. Why are voltmeters and ammeters marked + on or near the terminals ? Are all voltmeters and ammeters so marked ? If not, why ?
43. Why does the needle of these instruments move to some extent to give a deflection and then stop there ?
44. How can you test the accuracy of an ammeter or a voltmeter ?
45. How can you increase the range of an ammeter or a voltmeter ?
46. Can a voltmeter be used as an ammeter and *vice versa*, if so, how ?

29

A.C. Measurements

29.1. Circuit Containing Capacitance, Inductance and Resistance

When a harmonically varying e.m.f., $E_0 \sin \omega t$ is applied to a circuit containing a resistance R, an inductance L and a

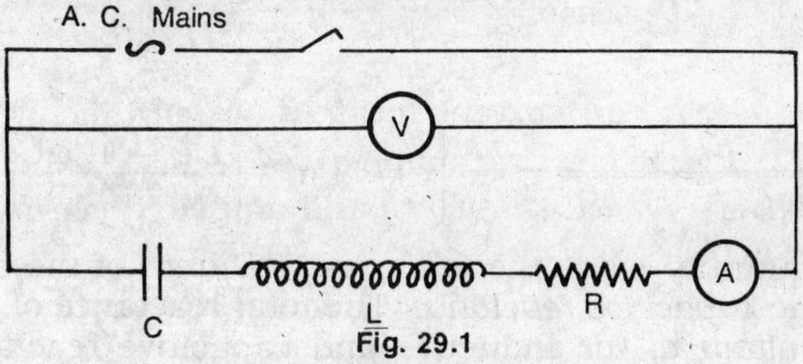

Fig. 29.1

capacitance C in series (Fig. 29.1), we have :

$$Ri - L\frac{di}{dt} + V = E_0 \sin \omega t$$

where i is the current and V the potential to which the capacitor is charged. Let q be the charge on the capacitor at the instant when its potential is V, then $V = q\,C$ and we get from the above equation :

$$Ri - L\frac{di}{dt} + \frac{q}{C} = E_0 \sin \omega t \qquad (29.1)$$

Let the solution be :

$$i = A \sin (\omega t - \alpha),$$

where A and α are to be determined.

$$\therefore \qquad \frac{di}{dt} = A\,\omega \cos (\omega t - \alpha)$$

$$\therefore \qquad dq = i\,dt = A \sin (\omega t - \alpha)\,dt$$

$$\therefore \qquad q = \int i\,dt = \frac{A}{\omega} \cos (\omega t - \alpha)$$

Substituting the values in Eq. (29.1), we get :

$$RA \cdot \sin (\omega t - \alpha) - [L\omega - 1/(\omega C)]\,A \cos (\omega t - \alpha) = E_0 \sin \omega t$$

Comparing the coefficients of $\sin \omega t$ and $\cos \omega t$ on the two sides of the above equation, we get :

$$R\,A\cos\alpha + [L\omega - 1/(\omega C)]\,A\sin\alpha = E_0 \qquad (29.2)$$

$$-R\,A\sin\alpha + [L\omega - 1/(\omega C)]\,A\cos\alpha = 0 \qquad (29.3)$$

Squaring and adding Eqs. (29.2) and (29.3), we have :

$$A^2\,[R^2 + \{L\omega - 1/(\omega C)\}^2] = E_0^2$$

$$\therefore \qquad A = \frac{E_0}{\sqrt{R^2 + [L\omega - 1/(\omega C)]^2}}$$

Also from Eq. (29.3)

$$\tan\alpha = \frac{L\omega - 1/(\omega C)}{R}$$

Hence,

$$i = \frac{E_0}{\sqrt{R^2 + [L\omega - 1/(\omega C)]^2}}\sin\left[\omega t - \tan^{-1}\left(\frac{L\omega - 1/\omega C}{R}\right)\right] \qquad (29.4)$$

The quantity $L\omega$ is the *inductive reactance* of the circuit and $1/(C\omega)$ the *capacitive reactance*. The total reactance of the circuit *is the resultant of* the inductive and capacitive reactances. The quantity $\sqrt{(L\omega - 1/\omega C)^2 + R^2}$ is called the *impedance* of the circuit and is denoted by Z. The total reactance of the circuit is denoted by *X*, the inductive reactance by X_L and the capacitive reactance by X_c. It is clear from the expression for the impedance of the circuit that,

$$Z^2 = X^2 + R^2$$

From Eq. (29.4) it is evident that the maximum value of the current is given by :

$$i_0 = \frac{E_0}{\sqrt{[(L\omega - 1/(\omega C)]^2 + R^2}} \qquad (29.5)$$

Thus, the less the impedance of the circuit the greater is the value of the maximum current in it. The unit of impedance is ohm. After all, impedance is the effective resistance of an A.C. circuit. For a circuit in which $L\omega = 1/(\omega C)$ is zero the current attains a value $i_0 = E_0/R$. Under such conditions the true ohmic resistance alone determines the current in the circuit.

The value of the angle α depends on the relative magnitudes of the inductive and the capacitive reactances in the circuit, ohmic resistance being constant. When $L\omega > 1/(\omega C)$ the current lags behind the applied e.m.f., when $L\omega < 1/(\omega C)$ the current leads the applied e.m.f. However, when $L\omega = 1/(\omega C)$ (the condition of resonance), the current is in phase with the applied e.m.f.

The vector diagram representing the p.ds. across the various components may be conveniently drawn. There will be three components to be laid off in the vector diagram.

(i) Ri the active component of the applied e.m.f. along OA.

(ii) $L\omega i$ drawn 90 ahead of Ri, and

(iii) $\frac{i}{C\omega}$ drawn 90° behind Ri.

This is shown in Fig. 29.2. These three vectors can be combined by obtaining the resultant of any two and then combining this resultant with the third one. In the figure $L\omega i$ and $i/C\omega$ have been first combined giving the resultant.

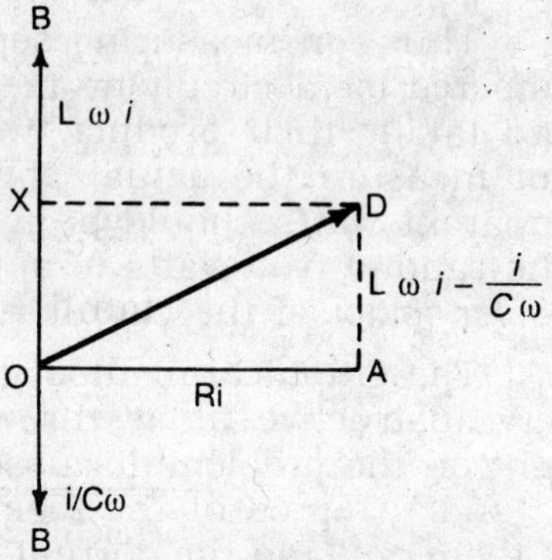

Fig. 29.2

$$OX = \left(L\omega - \frac{1}{C\omega}\right) i$$

which when combined with OA gives the resultant OD. Evidently,

$$OD = E = Zi = i\sqrt{R^2 + \left(L\omega - \frac{1}{C\omega}\right)^2}$$

In the case illustrated, the resultant e.m.f. is ahead of the current by the angle :

$$AOD = \alpha = \tan^{-1}\left(\frac{L\omega - \frac{1}{C\omega}}{R}\right)$$

and the circuit behaves like a slightly inductive circuit. The current may be ahead or lag behind the impressed e.m.f. E according as $1/C\omega$ or $L\omega$ has the larger value.

29.2. Power in an A.C. Circuit

In an *Inductive* circuit,

$$E = E_o \sin\omega\ t$$

$$i = i_o \sin\ (\omega\ t-\alpha)$$

where $\alpha = \tan^{-1}\ (L\omega/R.)$

Therefore, the rate of working at any instant is :

$$W = E_0\ i_0 \sin_0\ (\omega t - \alpha)$$

$$= E_0\ i_0 \cos\alpha \sin^2\omega t - \frac{1}{2} E_0\ i_0 \cos\alpha \sin 2\ \omega t$$

The mean value of $\sin^2\ \omega t$ for a cycle is $1/2$ and of $\sin 2\ \omega t$ is zero :

Mean rate of working $= 1/2\ /E_o i_o \cos\alpha$

$$= \frac{E_0}{\sqrt{2}} \frac{i_0}{\sqrt{2}} \cos\alpha$$

$$= \text{(virtual volts)} \times \text{(virtual amps.)} \times \cos\alpha$$

Thus, on measuring separately the virtual volts and virtual amperes for a circuit by means of a voltmeter and an ammeter and taking their product, we get the apparent watts. This does not measure the actual power absorbed in the circuit, for the apparent watts should be multiplied by cos α to get true watts. The ratio of true watts to apparent watts, *i.e.*, cos α is called the *power factor* of the circuit.

The graphical method given below is very instructive as it throws additional light on the problem discussed above. Let E_o and i_o represent the maximum values of the e.m.f. and the current in the circuit (Fig. 29.3). Resolving i_o parallel and perpendicular to E_0, we have the components $i_o \cos\alpha$ and $i_o \sin\alpha$ respectively. The component parallel to E_o gives the mean rate of working $1/2\ E_o i_o \cos\alpha$, whereas the latter component $i_o \sin\alpha$ as it lags 90° in phase behind the e.m.f. gives the mean rate of working.

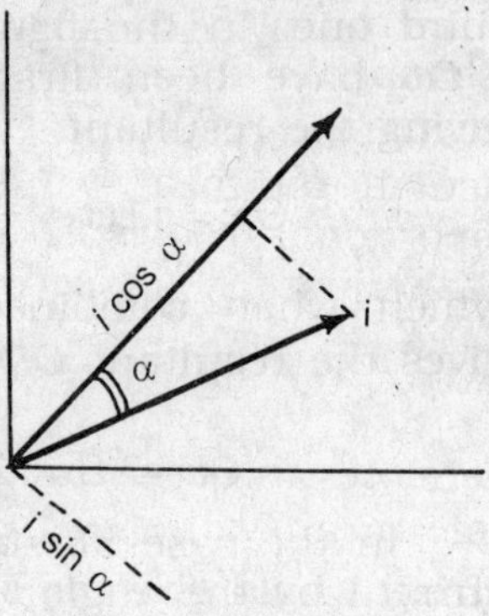

Fig. 29.3

$$\frac{E_0 i_0 \int_0^{2\pi} \sin\omega t \sin(\omega t - 90)\, d(\omega t)}{\int_0^{2\pi} d(\omega t)}$$

$$= -\frac{E_0 i_0}{2\pi} \int_0^{2\pi} [\frac{1}{2} \sin 2\omega t\, d(\omega t)]$$

$$= \frac{E_0 i_0}{8\pi} [\cos 2\omega t]_0^{2\pi} = 0$$

Thus, for the component $i_o \sin\alpha$ perpendicular to E_o, the mean rate of working is zero. This component is termed the *Idle* or *Wattless current*. The name is self-explanatory ; it does not contribute to the rate of working in the circuit. The current is entirely wattless when $\alpha = 90^\circ$ which is the limiting case when the inductance in the circuit is very large and R is negligible.

29.4. A.C. Measuring Instruments

The mean values of alternating e.m.f. and current for a complete cycle are each zero. Hence, an ordinary electromagnetic voltmeter or ammeter whose moving system is comparatively massive will indicate this mean value. Thus, such instruments cannot be employed for the measurement of alternating current and voltage. In the first half-cycle, the average current is $i_0\, 2/\pi$ (i_0 being the maximum value of a.c.) and in the latter half $i_0\, 2/\pi$ hence effects being directly proportional to the current, the suspended system receives equal and opposite impulses during a complete cycle and it is in such a quick succession that the needle fails to respond to it. Hence, a.c. measuring instruments should be independent of the direction of flow of the current. This is possible when the deflection is proportional to the square of the current, *e.g.*, hot wire instruments. Hence, only such instruments are employed for the measurement of a.c.

To interpret the readings of these instruments, let us calculate the mean value of $i_o^2 \sin^2 \omega t$ for it is this value that is proportional to the deflection of the instrument. The mean value of this quantity for a complete cycle

$$= \int_o^{2\pi} i_o^2 \sin^2 (\omega t)\, d\, (\omega t) \Big/ \int_o^{2\pi} d\,(\omega t) = i_o^2/2.$$

Therefore, the continuous current whose square would have the same mean value as that of the a.c. is $i_0/\sqrt{2}$; this continuous current would give the same reading on the hot wire instrument. This is termed the *virtual current* and is equivalent to a.c. of the maximum value i_o. Thus, the a.c. is measured by the strength of the steady current which would produce the same heating effect. Hence, it is also termed *effective current*.

If the steady value of the current be I, then,

$$I = i_0/\sqrt{2}$$

$$\therefore \quad I^2 = i_o^2/2 \text{ Mean value of the square of the current.}$$

Thus, $I = \sqrt{(i_0^2/2)}$ root of the mean of the square of the current. This value is, therefore, termed the *Root Mean Square value* (R.M.S. value) *of the current*. In the same manner $E = E_o/\sqrt{2}$.

From the above treatment, it is evident that a *virtual ampere* is one which will produce the same heating effect in a resistance as a steady current of 1 ampere will produce at the same time. A *virtual volt* is one which when, applied to the ends of a resistance, will produce the same heating effect as a steady p.d. of 1 volt applied across it for the same time.

Experiment 29.1

Object—To determine the impedance of a given a.c. circuit.

Apparatus—A capacitor, an inductance, resistance, an a.c. ammeter, an a.c. voltmeter, a step-down transformer having a number of tapping in the secondary and flexible cord for connections.

Theory—Let a capacitor of capacitance C, an inductance L and a resistance R be connected in series; and let an alternating e.m.f. E be applied across the combination as shown in Fig. 29.5. Then, if i virtual amperes is the current in the circuit, Eq. (29.5) gives :

$$\text{Impedance} = \sqrt{\left([L\omega - \frac{1}{C\omega]^2} + R^2\right)} = \frac{E_0}{i_0} = \frac{E}{I}$$

$$\text{or Impedance } Z = \frac{\text{e.m.f. as measured by an a.c. voltmeter}}{\text{Current as measured by an a.c. ammeter}}$$

Alternatively, the current i in the circuit may be measured by applying various known e.m.fs., E in the circuit and a graph plotted taking E along the X-axis and the corresponding values of i along the Y-axis. This will come out to be a straight line as shown in Fig. 29.4 and its slope QR/PR with the i-axis will give impedance of the circuit.

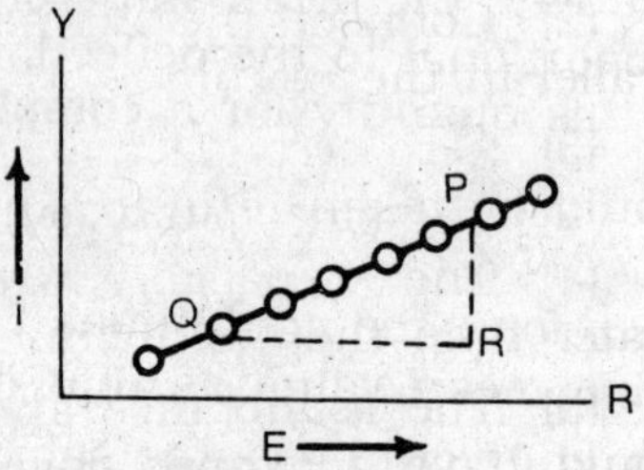

Fig. 29.4

Method—Connect the primary of a step-down transformer to the a.c. mains and connect to one of the tappings of the secondary a suitable capacitor, an inductance and a resistance, all *in series*, including an a.c. ammeter in the circuit, as shown in Fig. 29.5. Connect a.c. voltmeter across the tapping of the secondary of the transformer to measure the virtual e.m.f. applied in the circuit. Having properly made the connections, switch on the current and note down the reading of the voltmeter and the ammeter which give E and i respectively. Next, switching off the current, change the tapping of the secondary of the transformer and in this way take at least

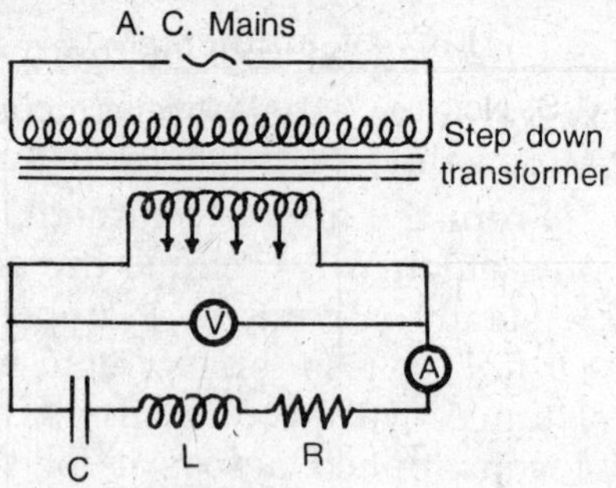

Fig. 29.5

six sets of observations for i and E. Calculate the impedance of the circuit from each set of the observations separately and then take the mean.

Next, plot a graph taking applied e.m.f. E along the X-axis and the corresponding values of i along the Y-axis. This will come out to be a straight line. Measure its slope QR/PR with the i-axis which will give impedance of the circuit.

Also calculate impedance Z from the formula :

$$Z = \sqrt{L\omega - 1/\omega C^2 + R^2}$$

if the values of L, C and R are known.

Sources of Error and Precautions

(1) Both the ammeter and the voltmeter must be a.c. measuring instruments.

(2) Connect the ammeter in series and the voltmeter in parallel in the circuit.

(3) Zero error, if any, in the above measuring instruments should be taken into account.

(4) The capacitor used should have a working voltage greater than the a.c. mains voltage.

(5) The inductance and the resistance used in the circuit should have a proper current rating so that they do not get over-heated and thereby get short-circuited and burnt.

(6) Before changing the tapping of the transformer see that the current has been switched off.

(7) The graph between i and E should be a straight line and should be smoothly drawn.

Observations

L.C. of voltmeter = Volts

L.C. of ammeter = Amp.

S. No.	Voltmeter reading E volts	Ammeter reading i amp.	Impedance ohm
1.			
2.			
3.			
...			
...			
...			
		Mean	

Calculations—I Set

$$Z = \frac{E}{i} = \qquad = \quad \text{ohm}$$

[Make similar calculations for other sets.]

∴ Mean value of $Z =$ ohm

From Graph (29.4)

$QR =$ $PR =$

∴ Impedance $Z = \frac{QR}{PR} =$ ohm

Now, $L =$ Henry, $C =$ Farad

$R =$ ohm and $\omega = 2\pi\nu$

$$\therefore \quad Z = \sqrt{\{L\omega - (1/\omega C)\}^2 + R^2}$$

$=$

$=$ ohm.

Result—The impedance of the circuit = ohm

Experiment 29.2

Object—To use an electrical vibrator to determine (*i*) the frequency of a.c. mains and (*ii*) the capacitance of a capacitor.

Apparatus—Electrical vibrator, table clamp pulley, pan, wt. box, fishing cord, 6-10 volt battery, a microammeter, a voltmeter (0-15*V*) and a capacitor.

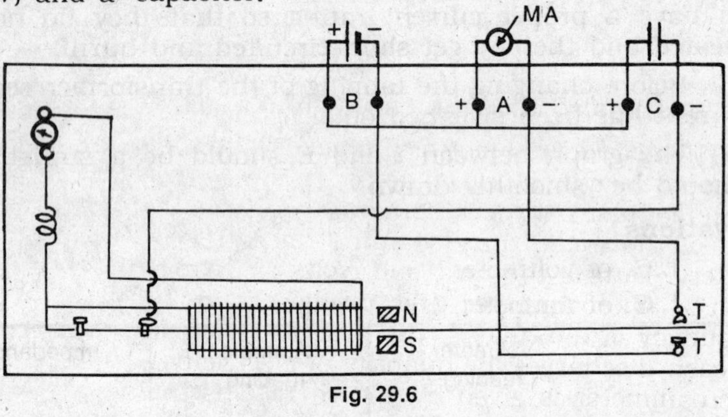

Fig. 29.6

Description of Apparatus—*The Electrical Vibrator.* The electrical vibrator (Fig. 29.6) consists of a solenoid through which passes a thin rod of steel which can be clamped at one end. At its free end the steel rod carries a hook to which a string under tension can be connected. The solenoid is connected in series with a 25-watt lamp and is used directly with a.c. mains. A permanent horse-shoe magnet is also mounted on the baseboard, the steel wire passing through the pole-pieces. When a.c. is fed through the solenoid, the rod is magnetised longitudinally with the polarity reversing with the change of sign of the current. Owing to the interaction of this with the field due to the permanent magnet the rod vibrates with the frequency of a.c. The length of the steel rod can be adjusted so as to get resonance indicated by a large amplitude of vibration of its free end. A continuous vibration is then maintained.

While using the vibrator for the determination of the capacitance of a capacitor, a battery, a microammeter and a capacitor can be connected to the terminals marked *B, A* and *C* on it. The steel rod carries a small iron piece with flat ends which makes contact alternately with the two flat steel discs T.T. ; this during one half cycle charges the capacitor and during the latter half discharges it through the microammeter.

Theory—(*i*) *Frequency of A.C. Mains.*—When a cord of mass per unit length *m*, is connected to the vibrating rod of the vibrator and stretched with a tension *T*, the cord vibrates in segments as in Melde's experiment. If the length of the cord is then adjusted until the nodes are clearly marked, the frequency of the stretched string is the same as of the vibrating rod which is vibrating with the frequency of a.c. mains. Then, if l be the length of one loop of the vibrating string, its frequency of vibration is given by :

$$\nu = \frac{1}{2l}\sqrt{\frac{T}{m}} \tag{29.7}$$

and this is also the frequency *f*, of a.c. mains.

(*ii*) *Capacitance of a Capacitor* —As depicted in the figure, if the connections are made as at *B, A* and *C*, the vibrating rod make contact between the battery and the capacitor during one half of the cycle, thus charging the latter to a voltage *V* of the battery. During the next half cycle the capacitor makes contact with a microammeter and discharges through it. If *f* be the frequency of a.c. mains, the process is repeated *f* times per sec. At each discharge, the quantity of electricity passing through the microammeter is given by *CV* where *C* is the capacitance of the capacitor.

∴ Quantity of charge flowing per sec. = $C.V.\,f$

But $q = i\ t$ and for *f* times charging per sec., $t = 1$ sec.

∴ $q = i$

or $q = i = CVf$

$$\therefore \quad C = \frac{i}{Vf} = \frac{\text{Current}}{\text{Voltage} \times \text{frequency}} \tag{29.8}$$

Thus, recording i and V and taking the value of f from the first experiment, and substituting these in Eq. (29.8), the value of C can be calculated.

Method—(*i*) *A.C. Mains Frequency*—Switch on the current and see that the rod of the electrical vibrator begins to vibrate. Adjust the length of rod till it is found that its free end attains the maximum amplitude. Now, switch off the current. Tie a fishing cord to the hook at the free end of the rod. Pass the cord over a table-clamp pulley and tie it to a light pan. Place some suitable weights on the pan, say, a gm. or two. Switch on the current when the string will be found to vibrate in segments as in Melde's experiment. Alter the length of the vibrating cord by shifting the vibrator backward till the nodes are sharply defined. Mark the positions of the extreme nodes leaving out the first and the last loop. Measure the length between them and divide it by the number of intervening loops to get the value of l. For same tension take two more sets by altering the length of the cord vibrating in resonance with the steel rod. Calculate the mean value of l for this fixed tension. Weigh the pan in the balance and then compute the total tension applied to the cord. Also weigh in a chemical balance, say, 200 cm. of the fishing cord used in the experiment and calculate m, its mass per unit length.

Repeat the experiment with different tensions. Finally, calculate the frequency from each set of observations and get the mean value of f.

(*ii*) *Capacitance of a* capacitor—Connect a 6 volt battery across the terminals marked B, a microammeter across the terminals marked *A* and the capacitor across *C*. Untie the cord for this part of the experiment. Adjust the distance of the flat steel discs *T*, *T* so that the small iron piece with flat ends (attached to the steel rod) makes contact with them when the vibrator is working. When this has been properly adjusted switch on the current. The capacitor is first charged to a voltage V (= e.m.f. of the battery) and then discharges through the microammeter and this process takes place f times per sec. This means a steady current in the microammeter. Take the reading of the microammeter. Also record the e.m.f. of the battery with a suitable voltmeter.

Repeat the experiment by altering the p.d. used to charge the condenser to, say, 8 and 10 volts. For each set calculate the value i/V and then with its mean value, calculate the value of the capacitance of the capacitor using Eq. (29.8).

Sources of Error and Precautions

(1) The length of the steel rod must be initially adjusted so that it vibrates in resonance with a.c. frequency. This is attained when the free end of the rod vibrates with maximum amplitude.

(2) There should be no friction in the pulley for then we cannot be sure of the tension applied to the cord.

(3) The nodes and antinodes on the cord should be sharply defined. The extreme loops should be left out while making measurement for *l*, as in these cases we cannot be certain of the extreme position of the nodes.

(4) The fishing cord should preferably be employed in this experiment as it possesses a fairly constant mass per unit length.

(5) Zero error of the microammeter, if any, should be taken into account or else the pointer should be initially adjusted on the zero mark.

(6) The contact of the vibrating rod with the flat steel disc *T,T.* should be properly adjusted so that the microammeter shows a steady deflection when the vibrator is working.

Observations—(A) Determination of A.C. Mains frequency.

S. No.	No. of loops	Length of cord	*l*	Mass of pan	Mass placed on pan	T (kg-wt.)
1.						
2.						
3.						

Mass of.....cm of the cord = gm. m = ... kg

(B) Determination of Capacitance of capacitor.

S. No.	Reading of the microammeter *i* amp.	Reading of the voltmeter *V* volt	*i/V*
1.			
2.			
3.			
		Mean	

Calculations—(A) A.C. mains frequency.

Set I. $$f = n = \frac{1}{2l}\sqrt{\frac{T}{m}}$$

$$=$$

$$= \quad \text{H}_z.$$

(B) Capacitance of a capacitor

$$C = \frac{i}{Vf}$$

$=$

$=$ microfarad.

Result—(*i*) The frequency of A. C. Mains = H_Z.

(*ii*) The capacitance of the given condenser = microfarad.

Experiment 29.3

Object—To study LR circuit with alternating current and thus to determine :

(*i*) the power factor, cos ϕ of the inductive load.

(*ii*) the equivalent power loss resistance of the inductor,

(*iii*) the inductance of the inductor,

(*iv*) the phase difference between the applied voltage V_o and that across *R*, V_R.

Apparatus—An inductance coil, a known resistance (non-inductive), an ammeter, a variac, three voltmeters and connection wires.

Theory—Refer to Fig. 29.7. A non-inductive resistance *R* is connected in series with an inductance and an ammeter and a.c. supply. The three voltages V_o, the supply voltage, V_R that across *R* and V_L and that across *L* are measured with the three voltmeters, all of the same specifications.

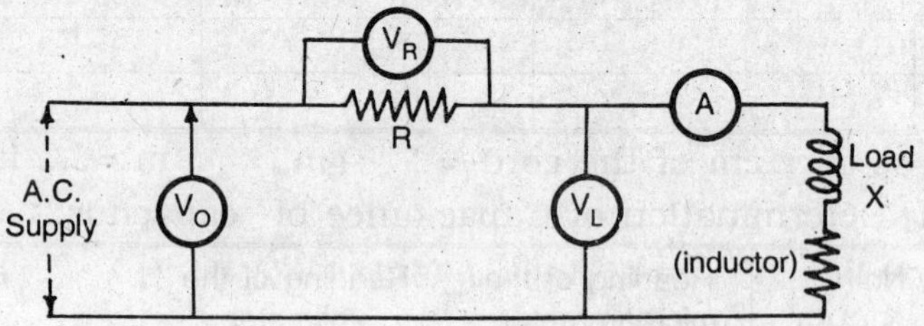

Fig. 29.7

If we have a pure inductor, *L*, then if the current in the circuit is i_o cos ωt, the e.m.f. across the inductor is given by :

$$V_L = \omega L i_0 \cos(\omega t + \pi/2)$$

meaning thereby that the voltage across it leads *i* and hence the voltage V_R across the resistor by $\pi/2$. But, in actual practice, every inductor has some resistance *r* associated with it; this is called its d.c., resistance. Apart from this there are iron losses and copper losses which play a significant role in a.c. These losses could be effectively represented by a series resistance, sometimes referred to as the a.c. resistance. Usually the series resistance r is to take into account the d.c. resistance as well as the a.c. resistance of the inductor. Thus, the vector diagram for the circuit is given in Fig. 29.8.

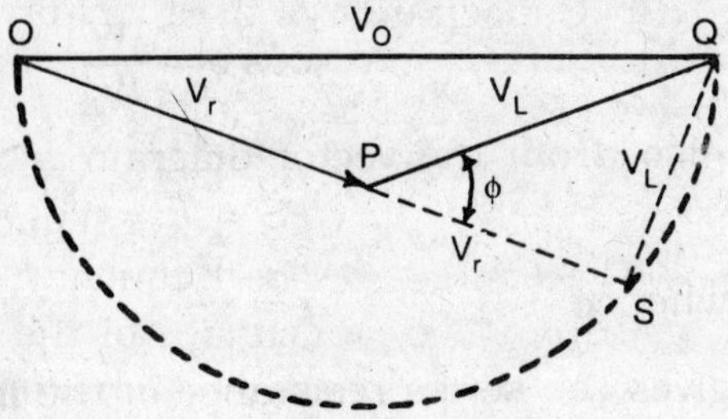

Fig. 29.8

The voltage V_L does not lead the current by $\pi/2$. The voltage across the inductor can be seen to lead the voltage V_R across the resistor by an angle ϕ less than $\pi/2$; $\cos \phi$ is the power factor of the inductive load.

If a semi-circle is drawn on V_o as diameter, OP produced cuts this at S. SQ then represents the p.d. across the inductive part (equal to $L\omega i$) while PS represents the p.d. across the resistive part of the inductance V_r (equal to $r.i$) where r is the series resistance of the inductance (the sum of the d.c. and a.c. resistance).

From the vector diagram, we have :

$$V_0^2 = V_R^2 + V_L^2 + 2V_R V_L \cos \phi$$

Neglecting the currents taken by the voltmeters across R and L the current in R is the same as the load current i. Hence, $V_R = R.i$. We then have :

$$V_0^2 = V_R^2 + V_L^2 + 2\ (V_L \cos \phi\ Ri)$$

$$= V_R^2 + V_L^2 + 2\ WR$$

where $W = i.V_L \cos \phi$ the power consumption of the inductive load, hence

$$W = i\ V_L \cos \phi = \frac{V_0^2 - V_R^2 - V_L^2}{2R}$$

Also,
$$\cos \phi = \frac{V_0^2 - V_R^2 - V_L^2}{2\ V_R V_L}$$

Thus, from the measurements of the three voltages and the resistance R we can calculate W and $\cos \phi$. Further, if the load current i is measured with an ammeter, the apparent rate of power consumption is $i\ V_L$, which can then be calculated. It will be found that this differs from the actual rate of power consumption W, $\cos \phi$ may also be computed from the relation :

$$\cos \phi = \frac{W}{iV_L} \quad (29.9)$$

Also, from the vector diagram :

$$PS = ir = V_r$$

whence,

$$r = \frac{V_r}{i} \quad (29.10)$$

gives the series resistance of the inductor. Compare this with the d.c. resistance of the inductor and account for the difference.

Further, $QS = L\omega i = V_L$ can enable us to calculate L, for

$$L = \frac{V_L}{2\pi fi} \quad (29.11)$$

Method—1. Measure the non-inductive resistance R with a multimeter and also the d.c. resistance of the inductor.

2. Make the connections as shown in Fig. 29.7. Adjust the variac, say, at 8.5 volts and switch on the power. Take the readings of the volmeters giving V_o, V_R and V_L. Measure the current i with the ammeter A. Care should be exercised in reading the voltmeters, where the pointer lies in between two graduations of a voltmeter scale, approximation must be made to read the fraction. This has to be done because the voltages occur in the second power in the expressions for W and cos ϕ.

3. Repeat the experiment by altering the input voltage V_o with the variac in steps of 20 volts.

4. For each set of observation, calculate W, cos ϕ, i, V_L and cos ϕ.

5. Draw vector diagrams for each set of observation in the manner depicted in Fig. 29.8. Measure angle ϕ and calculate cos ϕ from the cosine tables.

Also, calculate r and L by using the Eqs. (29.10) and (29.11) where V_r and V_L are determined from the vector diagram.

From each of the vector diagrams measure the angle ϕ' = angle QOP giving the phase difference between V_o and V_R. Calculate the mean value and show that,

$$\tan \phi' = \frac{L\omega}{R + Lr}$$

where L and r have been determined from the Vector diagrams. Cos ϕ is the power factor for the entire circuit.

Sources of Error and Precautions

(1) Since the squares of voltages are to be used, it is necessary to use accurate instruments. Due approximation must be made where the pointer falls between two graduations of the scale.

The possibility of error will certainly be reduced if only one voltmeter is used, a transfer switch connecting it to the points required as given in Fig. (29.9). Two equivalent resistances *U* and *W*, each equal to the resistance of the voltmeter, occupy the places of the missing voltmeters. With both switches *S* and *T* thrown to the right, the voltmeter *V* is put across *AC* and measures the supply voltage *V*. When both the switches are thrown to the left, the voltmeter is across *BC* and measures the p.d. V_L across the load *X*. When the switch *S* is thrown to the left and the switch *T* to the right, the voltmeter is across *AB* and measures the voltage V_R across the non-inductive resistance *R* only. The resistance *U* and *W* are simultaneously transferred to the positions not occupied by the voltmeter.

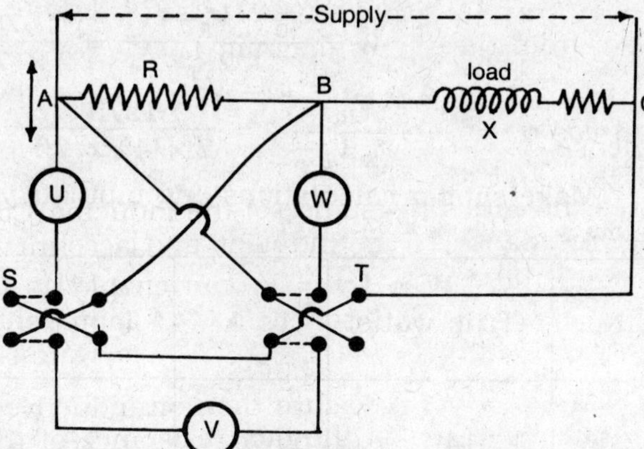

Fig. 29.9

(3) It is desirable that V_R and V_L should be as nearly equal as possible, hence, *R* must be reasonably chosen to satisfy this condition.

Observations—sample set—D.C. Resistance of the inductor = 7.6 ohms

Resistance *R* of the non-inductive coil = 70 ohms.

P.D. in volts across

S. No.	Ammeter reading *i* amp.	*L* and *R* combined V_o	*R* V_R	*L* V_L	$(V_o^2 - V_R^2 - V_L^2)$ (volts)^2
1.	0.6	65	45	25	1575
2.	0.8	85	60	35	2400
3.	1.0	103	70	45	3684
4.	1.3	120	80	50	5500
5.	1.45	135	90	60	6525
6.	1.75	160	110	70	8600

Calculations—From the first set :

$$W = \frac{V_0^2 - V_R^2 - V_L^2}{2R} = \frac{1575}{2 \times 70} = 11.25$$

and $\cos\phi = \frac{V_0^2 - V_R^2 - V_L^2}{V_R V_L} = \frac{1575}{2 \times 45 \times 25} = 0.70$

Make similar calculations and tabulate your results as given below :

S. No.	W (True watts)	Cos ϕ	$i \cdot V_L$ (apparent watts)	$\cos\phi = W/iV_L$
1.				
2.	17.14	0.57	28	0.61
3.	26.31	9.58	45	0.58
4.				
5.				
6.				

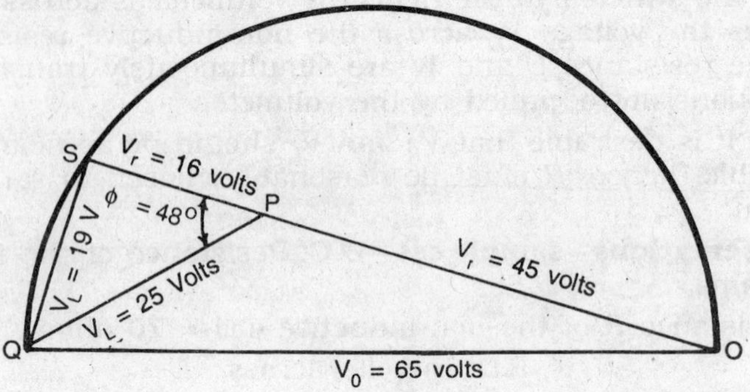

Fig. 29.10

From the vector diagram I, $\phi = 48°$, $\cos\phi = 0.67$

$\angle\, QOP = \phi' = 18°$

$V_r = SP = 16$ volts $= r.i$

$r = V_r/i = 16/0.6 = 26.6$ ohm

Also $SQ = V_L = 19$ volts $= L\omega i$

whence, $L = \frac{V_L}{\omega i} = \frac{19}{2\pi \times 50 \times 0.6} = 0.1008$ henry

Make similar calculations from other diagrams and tabulate your results as given below :

i (amp.)	r (ohm)	L (henry)	ϕ (degree)	cos ϕ	ϕ' (degree)
0.6	26.6	0.1008	48	067	18
0.8	27.5	0.1075	53	0.6018	19
1.0	26.0	0.1178	56	0.5592	21
1.3	27.5	0.0808	43	0.7314	16
1.45	25.5	0.0988	53	0.6018	21
1.75	22.8	0.1054	55	0.5736	21
Mean	25.98	0.1018		0.6229	19.33

tan 19.33 = 0.3502

Also, $\tan \phi' = \frac{L\omega}{R+r} = \frac{0.1018 \times 2\pi \times 50}{67.7 + 25.98} = 0.335$

Results I (*a*) $\cos \phi = \frac{V_0^{\,2} - V_R^2 - V_L^2}{2V_R\, V_L} =$

(*b*) $\cos \phi = W/iV_L =$

(*c*) cos ϕ (from vector diagrams) =

II. The series resistance r of the inductor = ohm.

III. The inductance of the inductor = henry.

Exercise—*To determine the power factor by three ammeter methods.*

Make the connections as shown in Fig. 29.10 (*a*).

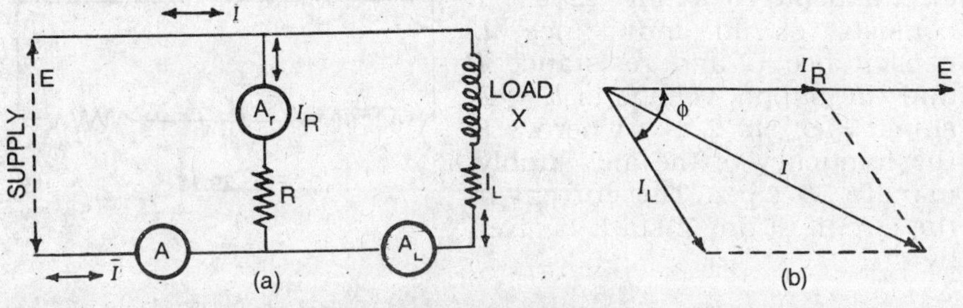

Fig. 29.10 (a) and (b)

Figure 33.10 (*b*) represents the current vectors I, I_R and I_L and the supply voltage E vector. The current measured by the ammeter A is the vector sum of the load current I_L and I_R that taken by the non-inductive resistance R, this being in phase with the applied voltage E, From the vector diagram,

$$I^2 = I_R^2 + I_L^2 + 2I_R \cdot I_L \cdot \cos \phi \qquad (29.13)$$

But $I_R = \frac{E}{R}$

$$\therefore \quad I^2 = I_R^2 + I_L^2 + 2.\frac{E}{R} I_L \cdot \cos\phi$$

$$= I_R^2 + I_L^2 + 2.\frac{W}{R}$$

where $W = E.\, I_L \cos\phi = \left[\frac{I^2 - (I_R^2 + I_L^2)}{2}\right] R$ (29.14)

Also, from Eq. (29.13),

$$\cos\phi = \frac{I^2 - (I_R^2 + I_L^2)}{2\, I_R\, I_L} \quad (29.15)$$

whence from Eqs. (29.14) and (29.15) W and $\cos\phi$ can be calculated from the observations of I, I_R, I_L and the known value of R.

Experiment 29.4

Object—To study a series resonant a.c. circuit.

Apparatus—A variac to give 110 volts a.c. or less, an inductance coil of an inductance a few henry, capacity box with a working voltage of 500 volts, one kilo-ohm resistor, a voltmeter and connection wires.

Theory—Consider the circuit depicted in Fig. 29.11. It consists of an inductance L, capacitance C and resistance R and the supply voltage is $E = E_o \sin \omega t = E_o \sin 2\pi \nu t$ where ν is the frequency of the a.c. supply, usually 50 c.p.s. The current in the circuit at any instant is given by :

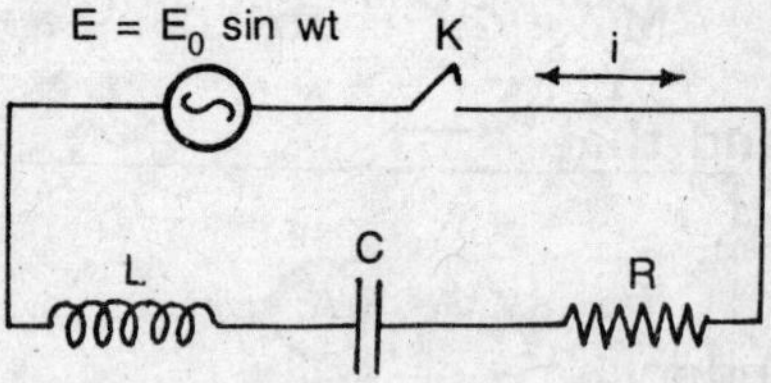

Fig. 29.11

$$i = \frac{E_0}{\sqrt{R^2 + \left(L\omega - \frac{1}{C\omega}\right)^2}} \sin\left\{\omega t - \tan^{-1}\left(\frac{L\omega - \frac{1}{C\omega}}{R}\right)\right\}$$

The peak value of the current is given by

$$i = \frac{E_0}{\sqrt{R^2 + \left(L\omega - \frac{1}{C\omega}\right)^2}}$$

and hence the r.m.s. value E and I are related by the equation :

$$I = \frac{E}{\sqrt{R^2 + \left(L\omega - \frac{1}{C\omega}\right)^2}}$$

and the angle of lag is given by :

$$\frac{\left(L\omega - \frac{1}{C\omega}\right)}{R} = \frac{\text{Reactance}}{\text{Resistance}}$$

For the circuit to be resonant,

$$L\omega - \frac{1}{C\omega} = 0$$

or $$\omega = 2\pi\nu = \frac{1}{\sqrt{LC}}$$

For the circuit to be resonant for an inductance value of, say, ten henry, at 50 c.p.s., the capacitance C should be of the order of 1 μ F (micro farad).

If I_r, be the current in the circuit at resonance,

P.D. across inductance $(V_l) = L\,\omega\, I_r$

P.D. across capacitor $(V_c) = \dfrac{I_r}{C\omega}$

and that $$L\,\omega I_r = \frac{I_r}{C\omega}$$

i.e., $$V_L = V_c$$

Hence, to study the condition of resonance, we can take an inductance of a fixed value L and a variable capacitance C. Observations of V_L and V_c are made by varying C, keeping R and E constant. For value of C less than that for resonance condition,

$$\frac{I_r}{C\omega} > L\,\omega I_r$$

or $$V_c > V_L$$

As the capacitance is increased in the circuit V_c goes on diminishing while V_L goes on increasing. At resonance

$$V_c = V_L$$

After the condition of resonance has been reached, if the capacitance is further increased in the circuit :

$$V_c < V_L$$

Hence, by plotting graphs between capacitance values along X-axis and the corresponding values of V_c and V_L along Y-axis, we can see how V_c and V_L vary with the value of C. The point of intersection of the two curves is the point where $V_c = V_L$ and hence the condition of resonance. The value of C corresponding to this point of intersection of the two graphs (Fig. 29.13) gives the capacitance at which the circuit becomes resonant.

Further, at resonance, the p.d., across the resistance,

$$V_R = RI_r$$

should be a maximum. Hence, the measure of p.d. across R may be simultaneously made. It will be found to increase with the increasing value of C in the initial stages, then reach a maximum, after which further increase of C decreases the value of V_R in the circuit. Hence, a plot of a graph between p.d. across resistance V_R against the capacitance values C can also be used to study for what value of C the given circuit of L and R is resonant. From the graph the value of C corresponding to the maximum value of V_R can be determined.

Further, at resonance, the p.d. across both L and C combined V_{CL} should be theoretically zero. This should be so if the choke coil is of negligible resistance and there are no other losses. Practically, V_{CL} is never zero at resonance and its value is a measure of the effective resistance of the choke. The term effective is used to indicate that the resistance is not just the d.c. resistance of the winding but also includes power losses in the iron due to hysteresis.

Hence, if V_{CL} is also measured simultaneously for varying values of C, V_{CL} will be found to vary in a manner opposite to that of V_R. At resonance V_R is a maximum while V_{CL} is a minimum. From the minimum value of V_{CL} on the (V_{CL}, C) graph, the value C can be computed that makes the given circuit, with constant values of L and R, resonant at the supply frequency.

It may be worthwhile plotting the set of graphs $\left(V_R, \frac{1}{\sqrt{C}}\right)$ and $\left(V_{CL}, \frac{1}{\sqrt{C}}\right)$ on the same sheet and the set of graphs $\left(V_C, \frac{1}{\sqrt{C}} \text{ and } V_L, \frac{1}{\sqrt{C}}\right)$ on another sheet. The point corresponding to the maximum value of V_R and the minimum value of V_{CL} gives the values of $\frac{1}{\sqrt{C}}$ for resonance. The point of intersection between the curves for V_C and for V_L, each plotted against $\frac{1}{\sqrt{C}}$ on the same scale, again gives the value of $\frac{1}{\sqrt{C}}$ for resonance.

Vector diagram may be drawn by taking the input voltage as the base *AB*. With centre *A* and radius equal to V_C+*R* draw an arc of a circle, with centre *B* and *d* radius equal to V_L+*R* draw an arc of a circle, to cut that for V_C in the point *C*. The process may be repeated for the whole set of observations with the fixed value of *R*. It will be found that the vertex *C* of the vector triangle lies on the segment of a circle such that angle *ACB* is constant (Fig. 29.15).

Method—Figure 29.12 depicts the circuit diagram wherein the output of an auto-transformer is fed on to the circuit

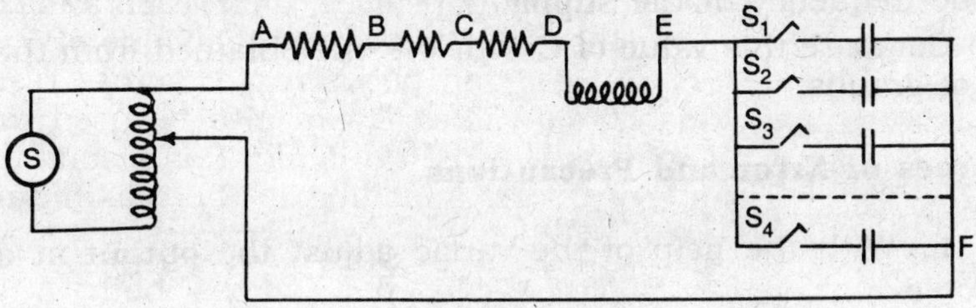

Fig. 29.12

containing a fixed value of *L* and variable values of *R* and *C*. *R* can be varied from 1 kilo ohm to 3 kilo ohm by shorting *AC* or *AB* respectively with a copper wire. Switches S_1, S_2... enable us to vary the capacity in the circuit.

Use may be made of a multimeter to measure the voltages V_R, V_L, V_C, V_{CL}, V_{LR} *i.e.*, the voltages across, *AD*, *DE*, *EF*, *DF* and *AE* respectively.

To begin with, introduce a capacitance of, say, 0.13 microfarad in the circuit, shortcircuit *A* and *C* so that *R* is one kilo-ohm, Switch on the power supply and measure the voltages V_R, V_L, V_C, V_{LC}, V_{LR} across points mentioned above. Repeat the observations by altering the value of *C* in the circuit in regular steps.

Plotting of Graphs.—I. On the same sheet of graph and with the same scale, draw graphs $\left(V_C, \frac{1}{\sqrt{C}}\right)$ and $\left(V_L, \frac{1}{\sqrt{C}}\right)$. Find the coordinates of the point of intersection of the two graphs. The value of *C* for this point is the one for which the circuit with the fixed value of *L* and *R* will be in resonance at the frequency of the supply (50 Hz). Graph has been drawn on the basis of the data given in the observation table.

II. Plot graphs with V_R and V_{LC} along *Y*-axis and $\frac{1}{\sqrt{C}}$ along the axis of X. Find the value of $\frac{1}{\sqrt{C}}$ corresponding to the maximum value of V_R from the $\left(V_R, \frac{1}{\sqrt{C}}\right)$ graph.

Also find the value of $\frac{1}{\sqrt{C}}$ corresponding to the minimum value of V_{LC} from the $\left(V_{LC}, \frac{1}{\sqrt{C}}\right)$ graph. These values of $\frac{1}{\sqrt{C}}$ determined from the two graphs will be very nearly the same. From this the value of C can be calculated which makes the circuit resonant at the frequency of the supply.

Compare this value of C with the one obtained from the first set of graphs.

Sources of Error and Precautions

1. With the help of the variac adjust the output at about 25 volts,

2. The graphs must be drawn smoothly, choosing a convenient scale for the values of p.d. and the value of $\frac{1}{\sqrt{C}}$.

Observations—(Here we reproduce one set of observations)
Input = 25.0 Volts. R = 1 Kilo-ohm.

C Microfarad	$\frac{1}{\sqrt{C}}(\times 10^3)$ volt	V_R volt	V_L volt	V_C volt	V_{LC} volt	V_{L+R} volt
0.12	2.88	1.2	9.0	33.0	24.0	9.0
0.14	2.67	1.5	11.5	36.0	24.0	12.0
0.16	2.5	1.7	13.0	37.0	24.0	14.0
0.18	2.36	2.2	18.0	40.0	24.0	18.0
0.20	2.24	2.5	21.0	43.0	23.0	22.0
0.22	2.13	3.2	27.0	48.0	23.0	28.0
0.26	1.96	4.4	37.0	54.0	22.0	37.5
0.30	1.82	6.2	50.0	60.0	20.0	52.0
0.32	1.77	7.2	55.0	67.0	10.0	58.0
0.38	1.62	9.4	65.0	74.0	16.0	72.0
0.48	1.44	8.4	63.0	52.0	18.0	65.0
0.52	1.39	7.0	55.0	42.0	20.0	58.0
0.56	1.34	6.4	53.0	36.0	20.5	55.0
0.58	1.31	6.2	51.0	33.0	21.0	52.0

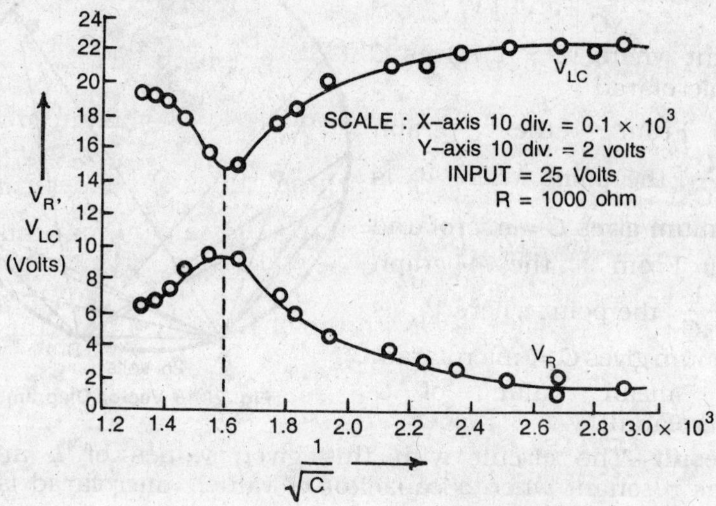

Fig. 29.13 Graphs–V_R, V_{LC} plotted against $\frac{1}{\sqrt{C}}$

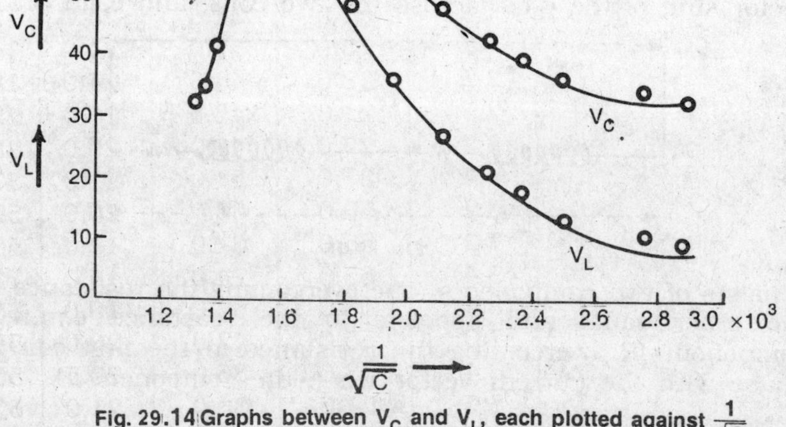

Fig. 29.14 Graphs between V_C and V_L, each plotted against $\frac{1}{\sqrt{C}}$

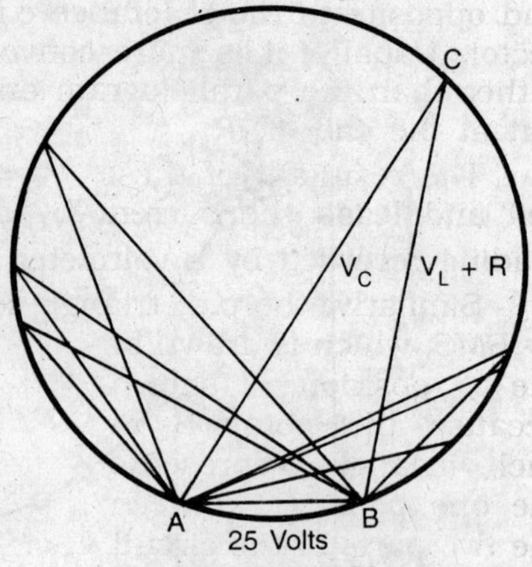

Fig. 29.15 Vector Diagram

Calculations—(*i*) From the graphs $\left(V_C, \frac{1}{\sqrt{C}}\right)$ and $\left(V_L, \frac{1}{\sqrt{C}}\right)$ the point where $V_C = V_L$ gives C = microfarad

(*ii*) From the graph $\left(V_R, \frac{1}{\sqrt{C}}\right)$ the point where V_R is a maximum gives C = microfarad

(*iii*) From the graph $\left(V_{LC}, \frac{1}{\sqrt{C}}\right)$ the point where V_{LC} is a minimum gives C = microfarad.

Hence, mean value of C = microfarad.

Result—The circuit with this given values of L and R becomes resonant when a capacitor of value....microfarad is put in series.

At resonance, (*i*) $V_C = V_L$

(*ii*) V_R is a maximum

and (*iii*) V_{CL} is a minimum.

29.4. Inductive Circuits in Series

When two inductive circuits are joined in series, the same current must flow through them both. But, in general, the e.m.f. over one will not be in phase with that over the other and as such the total e.m.f. required to maintain the current in the circuit will be less than the sum of the two e.m.fs.

The circuit is shown in Fig. 29.16. The supply p.d. is the vector sum of the p.ds. across the two coils, and each of these

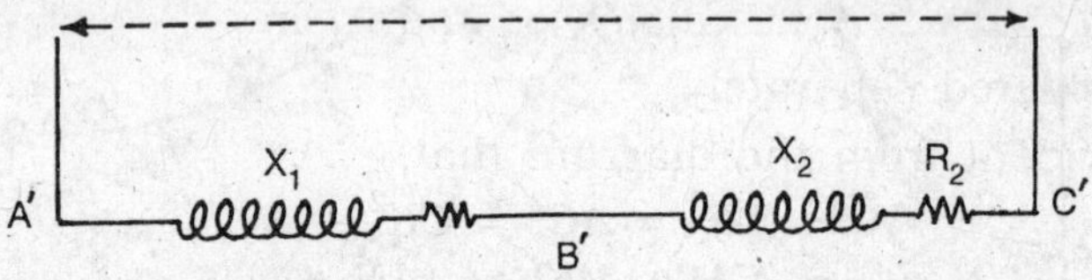

Fig. 29.16

consists of two components, one overcoming the resistance, and the other equal and opposite to the reactance e.m.f. The component IR, overcoming the resistance in the first coil is in phase with the current vector while the component IX_1, equal and opposite to the reactance e.m.f. is 90° ahead of the current vector. Usually, it is more convenient to use the vector triangle rather than the parallelogram and hence IX_1 is set up not at A, but at the end of IR_1.

The resultant of these $V_1 = IZ_1$ is the p.d. across the first coil and leads the current by an angle ϕ_1 V_1 is evidently the reading recorded by a voltmeter connected across $A'B'$.

Similarly, the p.d. triangle for the second coil $B'C'$ is shown by BMC, which is drawn in the position shown because the point B in each triangle represents the one point B' between the two parts of the circuit and as such it should occupy only one position in the diagram. At B, IR_2 has been set off horizontally and at its extremity is IX_2 is drawn vertically, the resultant being $BC = V_2 = IZ_2$, the p.d. across the second coil as measured by a voltmeter. This p.d. leads the current by a larger angle ϕ_2 The p.d. across the two coils in series, *i.e.*, the supply voltage is :

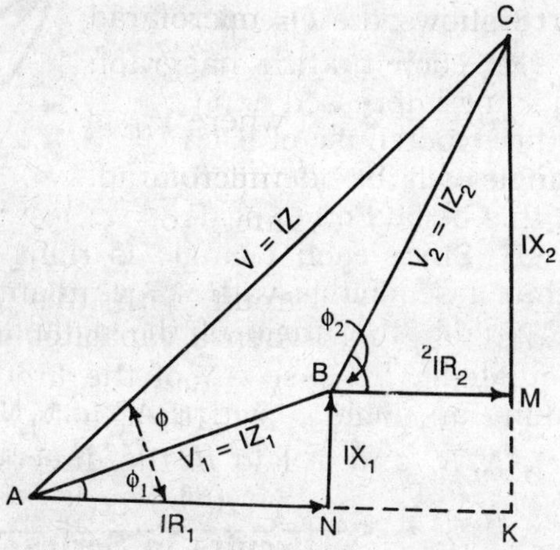

Fig. 29.17

$$V = IZ = \overrightarrow{AC},$$

the vector sum of V_1 and V_2 leading the current by an angle intermediate between the angles ϕ_1 and ϕ_2 in each part of the circuit considered separately.

It is evident from the diagram that,

$$V^2 = (IZ)^2 = AC^2 = CK^2 + AK^2$$

$$= (CM + MK)^2 + (AN + NK)^2$$

$$= (IX_2 + IX_1)^2 + (IR_1 + IR_2)^2$$

or $$Z^2 = (X_1 + X_2)^2 + (R_1 + R_2)^2$$

and $$\tan\phi = \frac{CK}{AK} = \frac{X_1 + X_2}{R_1 + R_2}$$

The total resistance in the circuit is the sum of the resistances of each part, and the total inductance is the sum of the separate inductances. Of course, the two parts are supposed to be far enough apart to avoid mutual induction between them.

29.5. Inductive Circuits in Parallel

The problem of a divided circuit, having two inductances in parallel, can be more conveniently solved graphically. Fig. 29.18 shows the circuit.

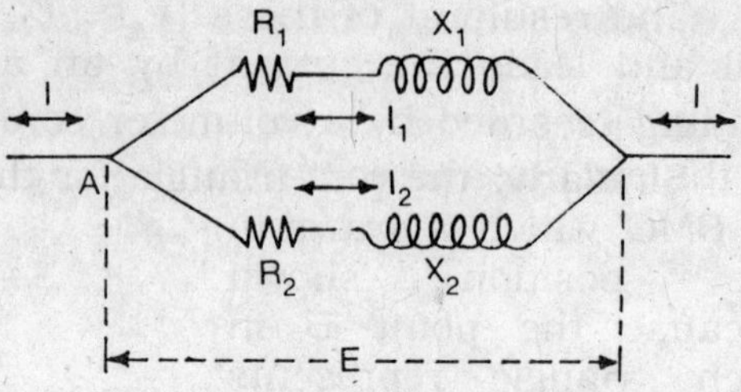

Fig. 29.18

As each branch has the same impressed e.m.f. E, the hypotenuse of each triangle will be identical. Let this be laid off to scale as AB. Since each triangle is right angled, it can be inscribed within a semicircle with AB as diameter. From the constants of the circuit, the angle of the ϕ_1 and ϕ_2 in each branch can be calculated. The base AN of the first triangle can be laid off by making an angle ϕ_1 with BA. Join NB. The value of the current $I_2 = AN/R_1$ and is laid in the direction AN.

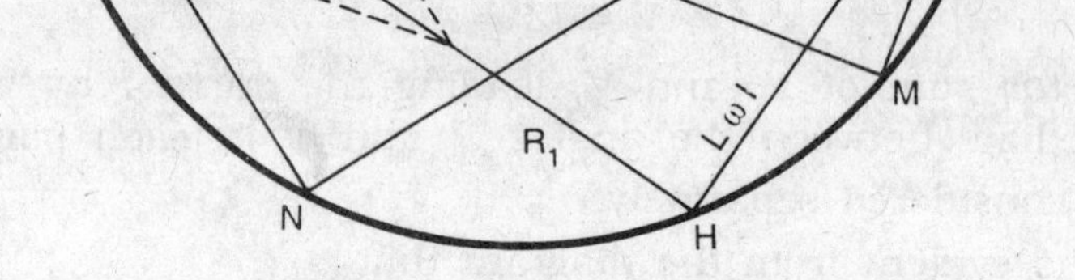

Fig. 29.19

In the same manner the triangle AMB is laid off for the second circuit R_2 L_2. The value of the current is determined by the relation $I_2 = AM/R_2$ and laid along AM. The current I in the circuit is the vector sum of I_1 and I_2 and is given by the vector AK. Produce AK to cut the semi-circle at H. Then, if R and L be the equivalent resistance and inductance of the divided circuit, these can be computed from the relation :

$$RI = AH$$

and $$L\omega I = BH,$$

triangle AHB representing the equivalent circuit.

29.6. Condenser Circuits in Series

Fig. 29.20 represents two capacitance circuits in series. The supply p.d. is the vector sum of the p.ds. across the two parts of the circuit, and each of these consists of two components, one

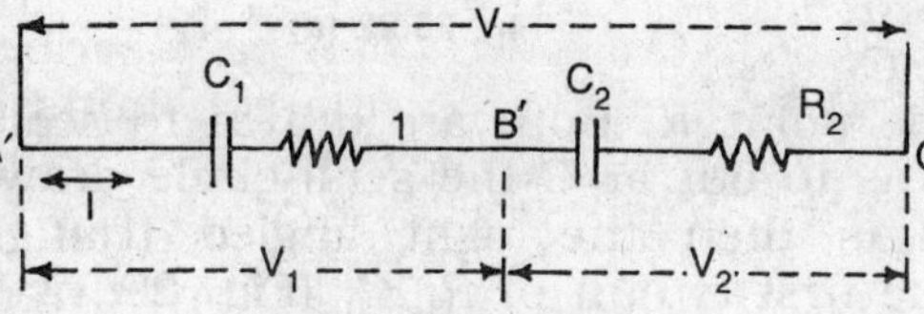

Fig. 29.20

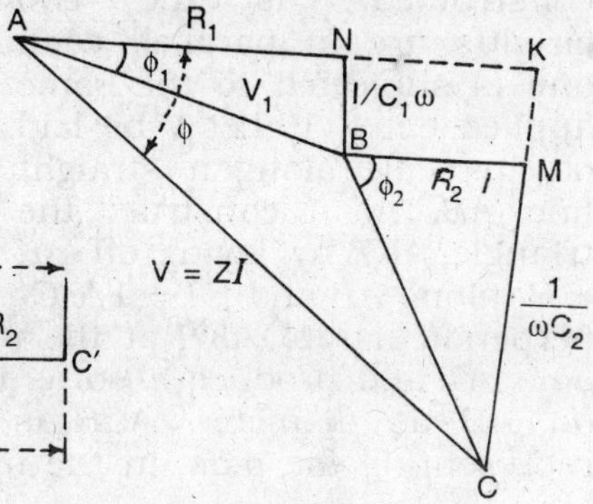

Fig. 29.21

overcoming the resistance, and the other equal and opposite to the reactance e.m.f. The component IR_1, overcoming the resistance in the first part of the circuit, is in phase with the current vector while the component IX_1 (= $I/C_1\omega$) equal and opposite to the reactance e.m.f. is 90° behind the current vector. This gives the vector triangle ABN.

Similarly, for the second part of the circuit we have the vector triangle BMC.

Evidently,

$$V = ZI = AK^2 + CK^2$$

$$= (R_1 I + R_2 I)^2 + \left(\frac{I}{\omega C_1} + \frac{I}{\omega C_2}\right)^2$$

or $$Z^2 = (R_1 + R_2)^2 + \left(\frac{I}{\omega C_1} + \frac{I}{\omega C_2}\right)^2$$

and $$\tan \phi = \frac{\frac{I}{\omega C_1} + \frac{I}{\omega C_2}}{R_1 + R_2}$$

Hence, it is evident that the total resistance of the circuit is the sum of the resistances of each part and that the total capacitance C is given by :

$$\frac{I}{C} = \frac{I}{C_1} + \frac{I}{C_2}$$

29.7. Condenser Circuits in Parallel

Figure 29.22a represents two capacitance circuits in parallel. The diagramatical solution for such a circuit is easily constructed. As the two circuits are in parallel, each one is subjected to the same applied e.m.f. V. Let V be laid off to scale along a straight line AB. At A construct the triangle AKT by laying off $AK = R_1$ along AB and $KT = 1/\omega C_1$ perpendicular to AKB at the point K. Both are drawn to scale. Join AT and produce it so as to cut at C the semi-circle drawn on AB as diameter. ACB is then the right angled triangle representing the p.ds. in the first circuit of R_1 C_1 (Fig. 29.22a)

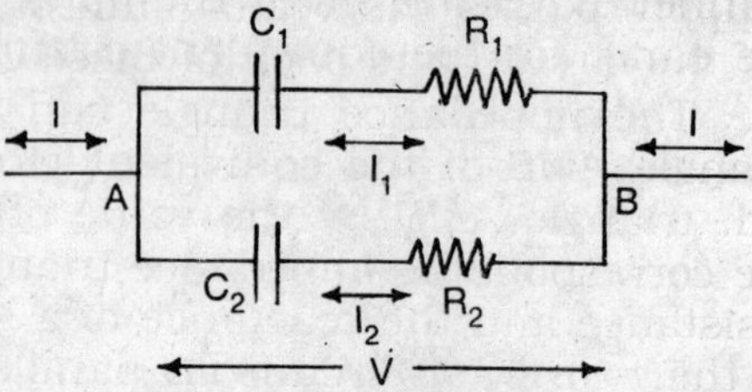

Fig. 29.22 (a)

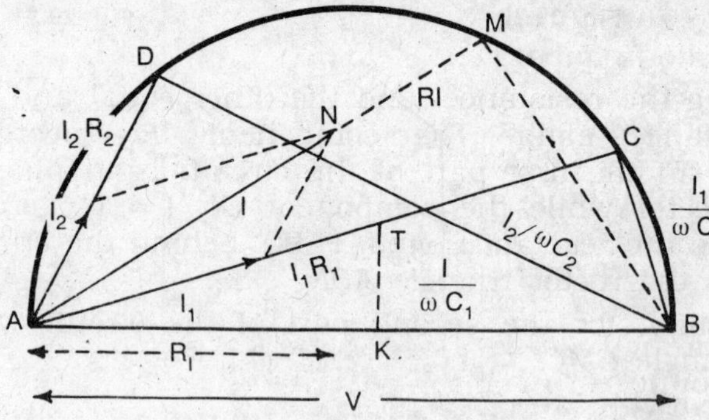

Fig. 29.22 (b)

The current I_1 in this branch will be in phase with AC, the active component, $V_{R1} = AC = R_1 I_1$, of the applied voltage V. This current can be shown on this diagram as I_1 laid off to an appropriate scale along AC.

In the same manner the p.d. triangle for the other circuit C_2R_2 can be drawn within the same circle with the same scale. This is shown by ADB. Here, the current in this branch is shown by I_2.

The equivalent circuit for this combination can be easily

determined. The current through such an equivalent circuit is I given by the vector sum of I_1 and I_2 and this is represented by the diagonal of the parallelogram by AN. Produce AN to meet the semi-circle at M. AM gives the active component, $V_R = AM$, of the applied voltage V. This value of V_R is obtained by measuring AM using the same scale as used in laying off AB. In the same manner, $V_x = MB$ gives the value of the reactive component of the e.m.f. for the equivalent circuit.

The impedance triangle can be constructed from the p.d. triangle AMB of the equivalent circuit. Dividing the sides of the p.d. triangle AMB by the value of the resultant current I_1 gives the corresponding impedance triangle, the sides of which give the resistance and the reactance of a single circuit that is equivalent to the two given circuits in parallel.

29.8. Comparison of Capacities—A.C. Bridge Method

The arrangement is depicted in Fig. 29.23 where R_1 and R_2 are non-inductive resistances and C_1 and C_2 are two capacitances, one of which is known and the other is to be determined. The output of the audio-frequency oscillator set at a frequency of 1000 c.p.s. is fed across AE. Between B and D is a headphone. Silence in the head-phone indicates that the current through it is zero. A sound in the headphone does not indicate whether R_2 is too large or too small for a balance, but R_2 can be varied in the direction that gives less sound until a minimum sound is heard. By varying R_2 back and forth from a value that is slightly too small to a value on the other side of the balance point that is just large enough to give the same loudness of sound as the smaller value, and decreasing this range in R_2 until the sounds are just audible, the balance point can be located.

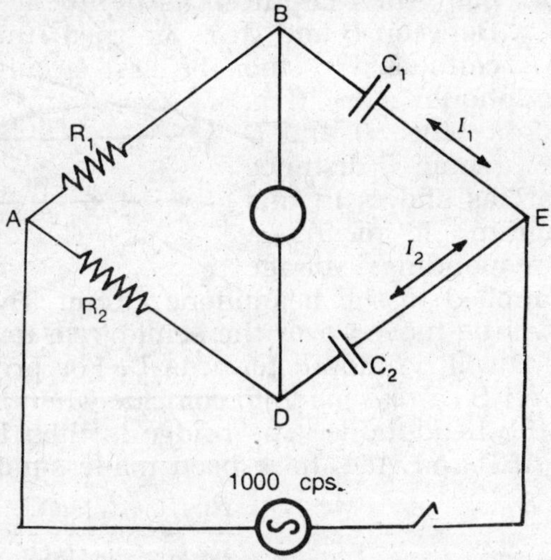

Fig. 29.23

The condition for balance is evidently.

$$\frac{R_1}{R_2} = \frac{-\dfrac{j}{\omega C_1}}{-\dfrac{j}{\omega C_2}}$$

or
$$\frac{R_1}{R_2} = \frac{C_2}{C_1}$$

whence,
$$C_1 = C_2 \frac{R_2}{R_1} \qquad (29.16)$$

This relationship can be conveniently established by a careful study of the corresponding vector diagram. Using the method for drawing the vector diagram for such a circuit as discussed in Art 29.7, we find that the vector Δ *ABE* for the circuit R_1C_1 and the triangle *ADE* for the circuit R_2C_2. Between *B* and *D* is connected the headphone *T*. (Fig. 29.24). Since *B* and *D* are some distance apart as shown in this diagram, it means a corresponding voltage is applied to the headphone circuit. By increasing R_2 the point *D* will be moved over the semi-circle towards *B* or by decreasing R_1, *B* will be moved towards *D*. For proper values of R_1 and R_2, *D* and *B* can be made to coincide when there would be no current in the headphone. The bridge is then balanced. In such a case Δ *ABE* and *ADE* have been made similar and hence,

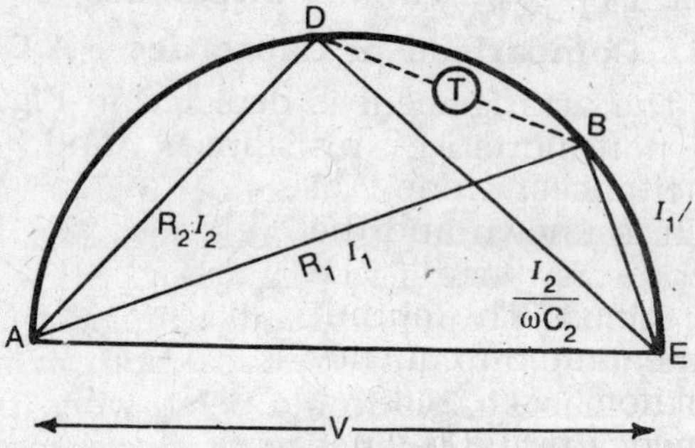

Fig. 29.24

$$\frac{R_1 I_1}{R_2 I_2} = \frac{I_1/\omega C_1}{I_2/\omega C_2}$$

or
$$C_1 = C_2 \frac{R_2}{R_1}$$

From the figure it is apparent that a slight change in R_2 will change the position of *D* by a greater amount when *AD* and *DE* are equal than when they are widely different. Hence, for greatest sensitiveness R_2 should be nearly equal to $1/\omega C_2$.

In this discussion it has been assumed that the capacitor arms of the bridge contain capacitance only. If either capacitor shows residual charge it will not be possible to obtain complete silence in the headphone. The nearest balance will be a minimum of sound. For a better balance it is necessary to use a modification of this—*The Wien Capacitance Bridge*.

29.9. Power Factor of a Capacitor

In practice, every capacitor dissipates some power when an alternating p.d. is applied across it. To account for this power loss, a capacitor may be regarded as made up of a capacitance associated with a resistance—either a small resistance in series

Fig. 29.25

or a large resistance in parallel. Fig. 29.25 shows these two modes of representation of a capacitor, the essential condition being that for the same applied voltage the current through the combination must be identical both in magnitude and phase in both the cases. This would be so when the vector impedances of the two alternative modes of representation are identical. We, thus, have :

$$r_s + \frac{I}{j\omega C_s} = \frac{r_p/j\omega C_p}{r_p + j\omega C_p}$$

which gives

$$\frac{r_s j\omega C_s + 1}{j\omega C_s} = \frac{r_p}{r_p j\omega C_p + 1}$$

$$\therefore \quad (-\, r_s r_p \omega^2 C_s C_p + 1) + j\,(r_p \omega C_p + r_s \omega C_s - r_p \omega C_s) = 0$$

Whence, we have :

$$r_s r_p \omega^2 C_s C_p = 1 \qquad (29.17)$$

and $$r_p \omega C_p + r_s \omega C_s - r_p \omega C_s) = 0$$

or $$r_p (C_s - C_p) = r_s C_s \qquad (29.18)$$

Dividing (29.18) by (29.17)

$$\frac{C_s - C_p}{r_s \omega^2 C_s C_p} = r_s C_s$$

$$\therefore \quad C_s - C_p = (r_s \omega C_s)^2 \cdot C_p$$

$$C_p = \frac{C_s}{1 + (r_s \omega C_s)^2} \tag{29.19}$$

This expresses C_p in terms of C_s, r_s and ω.

Again from Eq. (29.17) we have

$$r_p = \frac{\{1 + (r_s \omega C_s)^2\}}{r_s \omega^2 C_s \times C_s}$$

$$= \frac{r_s [1 + (r_s \omega C_s)^2]}{(r_2 \omega C_s)^2}$$

$$= r_s \left\{\frac{1}{r_s \omega C_s}\right\} \tag{29.20}$$

This equation expresses r_p in terms of r_s, C_s and ω.

The factor $r_s C_s \omega$ is called the *power factor of the capacitor.*

In the case of mica capacitor, the power factor is extremely small, say, 0.0001, hence C_s and C_p differ by a negligible amount whereas r_p is approximately $\frac{1}{r_s \omega^2 C_s^2}$ These values of C_s, C_p, r_s and r_p depend upon the frequency. In the case of ordinary capacitor using paraffin wax as dielectric, the power factor is of the order of 0.03 but even then the difference between C_s and C_p though sensible is not large.

Experiment 29.5

Object—To determine the capacitance of a capacitor with Wien's series resistance bridge for capacity measurement.

Apparatus—Two four-dial resistance boxes with the decimal ohm dial, one decimal ohm dial box having units and tenths of an ohm, audio-oscillator, headphone, a capacitor of known capacity, the experimental capacitor, a key and connection wires.

Theory—*Wien's Series resistance Bridge*—The bridge connections are shown in Fig. 29.26. R_1 and R_2 are pure ohmic resistances. C_2 is a standard mica capacitor in series with a variable standard resistance r_2. In the arm *BE* is a paraffin paper capacitor which will be regarded as equivalent to a perfect capacitor of capacitance C_1 with a resistance r_1 in series with it. Headphone is placed across *BD*. The equivalent series resistance of C_2 will be considered negligible. The output of an audio-frequency oscillator of frequency 1000 c.p.s. is connected across *AE* through a key or a switch *K*.

When the bridge is balanced *i.e.*, there is silence in the

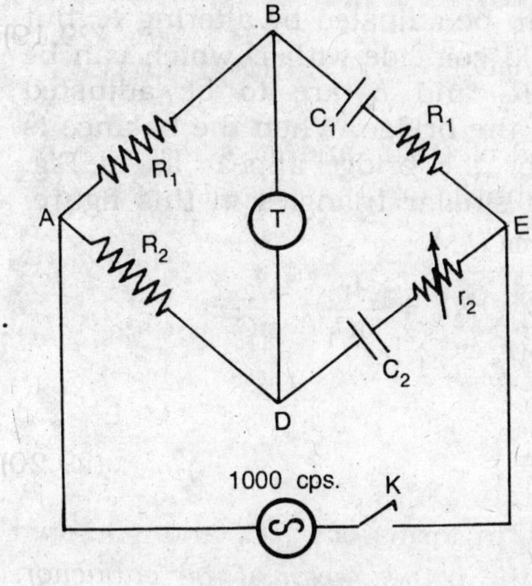

Fig. 29.26

head-phones, the condition for the balance is :

$$\frac{R_1}{R_2} = \frac{r_1 + \dfrac{1}{j\omega C_1}}{\dfrac{1}{r_2 + j\omega C_2}}$$

or $$\frac{R_1}{R_2} = \frac{(r_1\, j\omega C_1 + 1)\; j\omega C_2}{(r_2\; j\omega C_2 + 1)\; j\omega C_1}$$

$$= \frac{r_1\; j\omega\; C_1 C_2 + C_2}{r_2\; j\omega\; C_1\; C_2 + C_1}$$

whence, $$\frac{R_1}{R_2} = \frac{r_1}{r_2} \quad (29.21)$$

and $$\frac{R_1}{R_2} = \frac{C_2}{C_1} \quad (29.22)$$

Eqs. (29.21) and (29.22) represent the two conditions of balance. These may be satisfied by varying R_2 and r_2 alternately. Thus, when the bridge is balanced, we have from Eq. (29.22) :

$$C_1 = \frac{R_2}{R_1} \times C_2 \quad (29.23)$$

whence, C_1 can be calculated.

The vector diagram for the bridge may now be studied to throw additional light on the adjustments to be made for balancing the bridge. The vector diagram has been drawn in accordance with the method dealt with in the text earlier.

The vector diagram for the capacitor circuits $C_1\, r_1$ and $C_2 r_2$ are the triangles NBE and MDE and the telephone is connected

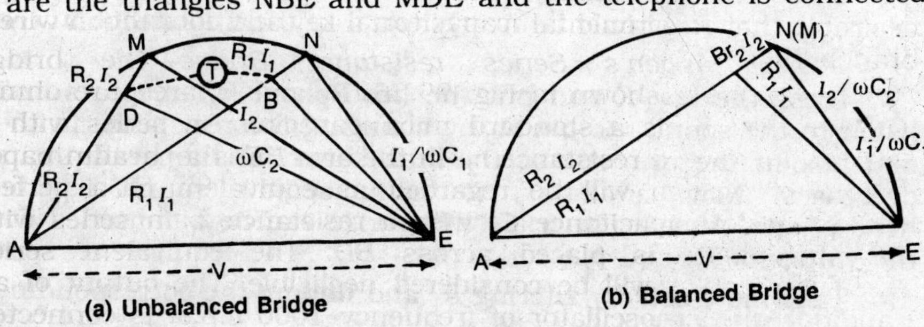

Fig. 29.27

across BD. For balancing the bridge it is not sufficient that M should coincide with N which can be adjusted by altering R_2 but it is also necessary that B should coincide with D which can be achieved by r_2. Hence, both R_2 and r_2 are to be adjusted alternately by trial for balancing the bridge. When the balance is achieved the vector diagrams for the bridge appear as in Fig. 29.27 (*b*). There are two sets of similar triangles in this figure, from which we have :

$$R_1 I_1 = R_2 I_2 = r_1 I_1 : r_2 I_2 = \frac{I_1}{\omega C_1} : \frac{I_2}{\omega C_2}$$

giving as before, $$\frac{R_1}{R_2} = \frac{r_1}{r_2}$$

and $$\frac{C_1}{C_2} = \frac{R_2}{R_1}.$$

whence, $$C_1 = \frac{R_2}{R_1} \cdot C_2$$

evidently, r_2 must be set equal to $r_1 \dfrac{C_1}{C_2}$ in order to find the value of R_2 that gives silence in the headphone.

Method—Make the connections as shown in Fig. 29.26. For R_1 and R_2 use two four-dial resistance boxes giving any resistance from 0.1 ohm to 1000 ohm. For r_2 use a decimal ohm box having two dials—units and tenths of an ohm. C_2 is a standard mica capacitor whose equivalent series resistance can be considered negligible. The experimental capacitor is an ordinary capacitor using paraffin wax as dielectric and is connected in the arm *BE*. The output of an audio-frequency oscillator is fed across *AE* and the frequency is set on its dial at 1000 c.p.s.

Adjust the resistance R_1 in the arm *AB* to 100 ohm. It is desirable that R_1 should be nearly equal to the impedance in the branch *BE*.

Press the key *K* and put the head-phone on the ears and listen to the sound. Adjust R_2 by changing the resistance in the dial box in the arm *AD* so that the sound in the head-phone decreases. Now, adjust r_2 to minimise the sound in the head-phone. After a few trials the bridge can be balanced. Note the values of R_2 and r_2.

Take more sets by altering R_1 and finding the corresponding values of R_2 and r_2.

Tabulate your observations as shown in the observation

table. Calculate C_1 from each set and find its mean value. Also from each set calculate r_1 which equals $\frac{C_1}{C_2}$ and find its mean value. From a knowledge of ω, C_1 and r_1, calculate the power factor $\omega C_1 r_1$.

Sources of Error and Precautions

(1) For the determination of C_1 it is not necessary that R_2 and r_2 should be specifically non-inductive ; but for the determination of the power factor it is essential to use non-inductive resistances.

(2) It is important to record the frequency of the oscillator, usually 1000 c.p.s., at which measurements are made since both the capacity C_1 and the resistance r_1 of the condenser are dependent on frequency.

(3) C_2, the standard capacitor, must be chosen of a capacity nearly equal to that of C_1 and R_1 should be set nearly equal to the impedance of C_1 branch of the bridge. The reason for this is that the bridge is most sensitive when the impedances of all the arms are equal.

(4) The headphone employed should preferably have an impedance of the same order as that of the other branches of the bridge.

(5) In obtaining the balance position of the bridge for a minimum of sound in the head-phone, a number of separate attempts should be made. These will vary from one another by slight amounts and then the arithmetic mean of the results so obtained should be taken.

Observations—Capacity C_2 of the standard capacitor = μF

Frequency at which measurements are made = c.p.s.

Set No.	R_1	R_2	Mean R_2	r_2	Mean r_2	Calculated Capacity C_1	Calculated resistance r_1
1		(1) (2) (3)		(1) (2) (3)			
2		(1) (2) (3)		(1) (2) (3)			
3		(1) (2) (3)		(1) (2) (3)			
					Mean		

Calculations—From set 1,

$$C_1 = \frac{R_2}{R_1} \times C_2 = \qquad = \mu F$$

Similar calculations from other sets,

$\therefore$ Mean value of $C_1 = \quad \mu F$

From set 1, $r_1 = r_2 \dfrac{C_2}{C_1} = \quad$ ohm.

....

$\therefore$ Mean value of $r_1 = \quad$ ohm.

Also $f = 1000$ c.p.s.

$\omega = 2\pi f \quad =$

$\therefore$ Power factor of the capacitor $C_1 = r_1 \, \omega C_1$

$=$

$=$

Result—The capacity of the unknown capacitor $= \mu F$

Power factor of the condenser $=$

Experiment 29.6

Object—To determine the self-inductance of a given coil by Maxwell's inductance Bridge.

Apparatus—Two four-dial resistance boxes (units to thousands), one three dial resistance box (units, tenths and hundredths), unknown inductance, a known standard self-inductance preferably having a value as close to the unknown as possible, audio-frequency oscillator, head-phone, a key or switch and connection wires.

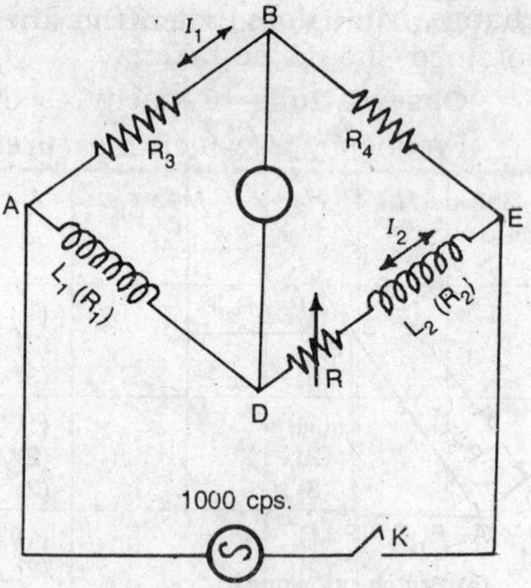

Fig. 29.28

Theory—Refer to Fig. 29.28. R_3 and R_4 are the resistances in the arms *AB* and *BE* which are provided with two four dial resistance boxes with non-inductive windings. In the arm *AD* lies the unknown inductance L_1 of resistance R_1 while in the arm *DE* is the known standard inductance L_2

having resistance R_2 along with a non-inductive variable resistance R. Across BD is the head-phone while across AE is fed the output of the audio frequency oscillator set usually at a frequency of 1000 c.p.s.

The condition for the balance of the bridge is :

$$\frac{R_3}{R_1 + jL_1\omega} = \frac{R_4}{(R + R_2) + jL_2\omega}$$

or $$R_3 (R + R_2) + jR_3 L_2\omega = R_1 R_4 + jL_1 R_4\omega$$

whence, $$R_3 (R + R_2) = R_1 R_4$$

or $$\frac{R_3}{R_4} = \frac{R_1}{(R + R_2)} \qquad (29.24)$$

Also $$R_3 L_2\omega = R_4 L_1\omega$$

or $$\frac{L_1}{L_2} = \frac{R_3}{R_4} \qquad (29.25)$$

or $$L_1 = L_2 \cdot \frac{R_3}{R_4} \qquad (29.25)$$

from which L_1 can be calculated from a knowledge of L_2, R_3, and R_4. When the conditions embodied in Eqs. (29.24) and (29.25) are satisfied, the sound in the headphone will be a minimum. Balance may be approached by alternate adjustment of R_4 and R_3. The two conditions of balance are then semi-independent for while a change of R_4 affects condition embodied in Eq. (29.24), a change in R does not affect the condition embodied in Eq.

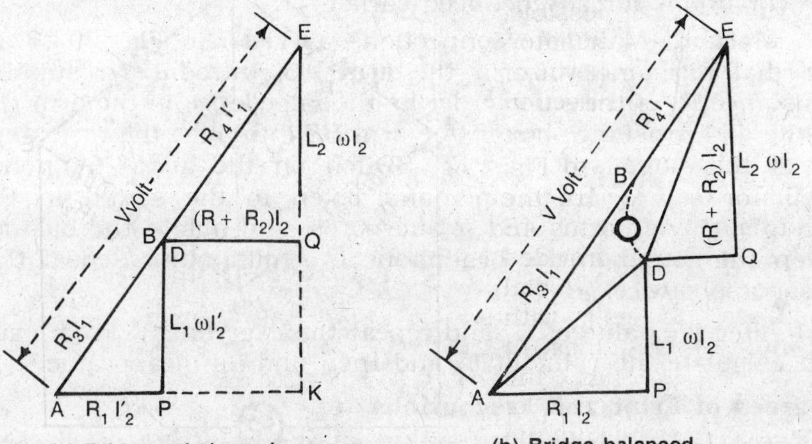

(a) Bridge unbalanced (b) Bridge balanced

Fig. 29.29

(29.25). If for L_2 we have an inductometer so that L_2 is a known variable inductance, with constant value of R_2, the two adjustments are quite independent and the exact balance is conveniently obtained by alternate adjustment of R_4 and L_2.

The vector diagram for the unbalanced bridge is of great help in understanding the adjustments to be made for obtaining the balance. The graphical representation of the inductive branch ADE is shown by two e.m.f. triangles APD and DQE (Fig. 29.29). The corresponding diagram for the non-inductive circuit ABE is a straight line commencing at *A* and ending at *E*, since this circuit is also subjected to the same impressed voltage *V*. The headphone is connected across *BD*. For balancing the bridge there should be no current in the headphone. This would happen if *B* and *D* coincide (Fig. 29.29b). This could be achieved by (*i*) altering R_4 so that *B* lies opposite *D* and (*ii*) then altering *R* so that *D* approaches *B*. With fixed inductances L_1 and L_2 the only possible adjustments for a known value of R_2 are the values of R_4 and *R*. With an inductometer, adjustments of R_4 and L_2 would have attained the balance conveniently. It is also apparent that if *B* and *D* are to coincide at the middle point of *AE*, the four arms of the bridge must have equal impedances.

From the figure for the balanced bridge, we have from similar triangles :

$$\frac{L_1}{L_2} = \frac{R_1}{R + R_2} = \frac{R_3}{R_4}$$

the conditions already deduced earlier.

Method—Make the connections as given in Fig. 29.28. In the dial resistance box in the arm *AB*, introduce a suitable resistance R_3 of the same order as the impedance in the arm *DE*. In the dial resistance box in the arm *BE* introduce the resistance R_4 of the same order as R_3. Switch on the audio frequency oscillator set at 1000 c.p.s and listen to the sound in the headphone. Adjust R_4 and *R* alternately and obtain the balance when the sound in the headphone is a minimum. Repeat this observation twice.

Alter the value of R_3 and repeat the experiment. From each set, calculate the value of L_1 and then find its mean value.

Sources of Error and Precautions

1. The resistances R_3 and R_4 should be of the same order of magnitude and equal to the order of magnitude of the impedance in either inductive arm of the bridge. This is to ensure

that the bridge is in the sensitive condition and as such can be accurately balanced.

2. In obtaining the balance position of the bridge for a medium of sound in the headphone, a number of separate attempts should be made. These will vary from one another by slight amount and then the arithmetic mean of the results so obtained should be taken.

3. Referring to Eq. (29.24), we have :

$$R_1 = \frac{R_3 (R + R_2)}{R_4}$$

$$= \frac{R_3}{R_4} R_2 + \frac{R_3 R}{R_4}$$

If R_1 is less than $\frac{R_3}{R_4} R_2$, R will have to be negative. When such is the case R will have to be placed in the same arm as L_1 and not in the arm in which we have L_2.

To avoid this difficulty it is desirable to have a decimal ohm box in each of the arms *AD* and *DE* and use both or one of them whichever is necessary.

(4) For different sets of observation, variation in R_3 should not be large.

Observations and Calculation

Known standard inductance L_2 = mH

Set No.	R_3 ohm	R_4 ohm	Mean value of R_4 ohm	R ohm	Mean R' ohm	$L_1 = L_2 \frac{R_3}{R_4}$
1		i ii iii	i ii iii			
2						
					Mean	

Result—The inductance of the given coil = mH.

29.10. Maxwell's Self-inductance Bridge—Coils in Parallel

When the two inductances, the unknown L_1 and the known L_2, are placed in adjacent and parallel arms of the bridge, as shown in Fig. 29.30, there is a non-inductive resistance in series

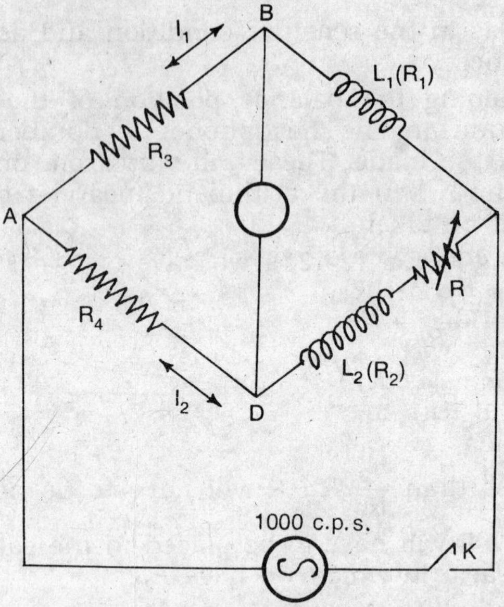

Fig. 29.30

with each coil. The condition for the balanced bridge is evidently,

$$\frac{R_3}{R_4} = \frac{R_1 + jL_1\omega}{(R_2 + R) + jL_2\,\omega}$$

or $$R_3\,(R_2 + R) + jR_3\,L_2\omega = R_1R_4 + jL_1\,R_3\omega$$

whence, equating real on each side and likewise unreal,

$$R_3\,(R + R_2) = R_1R_4$$

or $$\frac{R_3}{R_4} = \frac{R_1}{R + R_2} \qquad (29.27)$$

and $$L_2R_3 = L_1R_4$$

or $$\frac{L_1}{L_2} = \frac{R_3}{R_4} \qquad (29.28)$$

giving

$$L_1 = L_2 \times \frac{R_3}{R_4} \qquad (29.29)$$

Hence, for no sound in the headphone, the two conditions embodied in Eqs. (29.27) and (29.28) have to be satisfied. Balance may be approached by alternate adjustment of R_3 and R for a fixed value of R_4. Then, L_1 can be calculated in terms of L_2, R_3 and R_4.

The vector diagram of the bridge is given in Fig. 29.31. The p.d. triangle for the branch consisting of R_3, R_1 and L_1 is *ABE* while for the other parallel circuit is the triangle *ADE*. The headphone is connected across *BD*. For the balanced bridge *B* should coincide with *D*. This can be achieved by altering R_3 and *R* to the desired values for a fixed value of R_4. Fig. 29.32 depicts the vector diagram for the balanced bridge, and we then have :

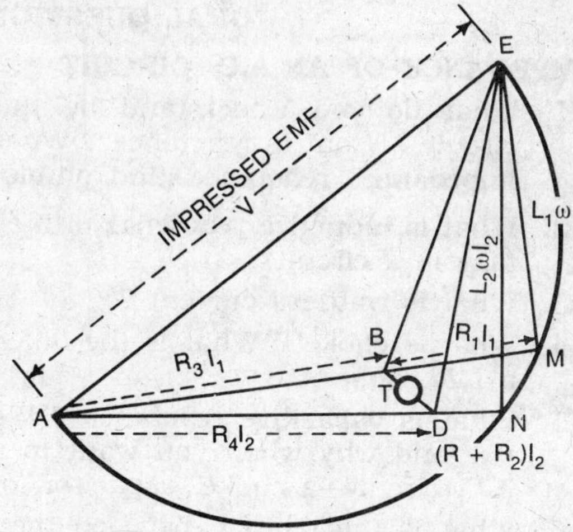

Fig. 29.31. Unbalanced Bridge

$$\frac{R_3}{R_4} = \frac{R_1}{R + R_2} = \frac{L_1}{L_2}$$

the same conditions as deduced earlier.

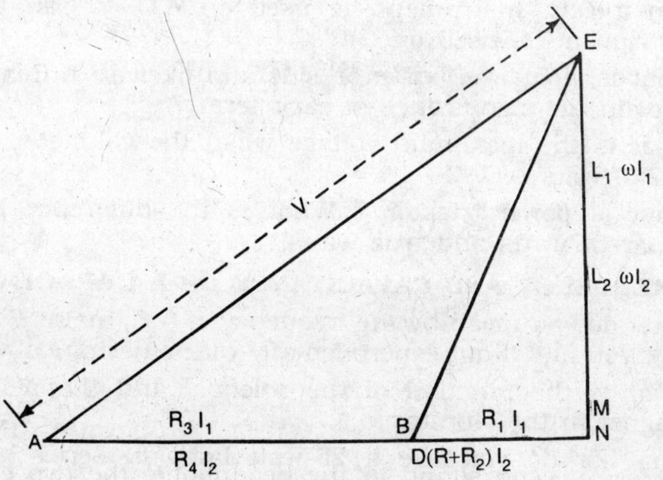

Fig. 29.32 Balanced Bridge

ORAL QUESTIONS

IMPEDANCE OF AN A.C. CIRCUIT

1. What do you understand by the impedance of an A.C. circuit ? What is reactance ? What is the relation between impedance, reactance and ohmic resistance ?
2. What is inductive reactance of a circuit ? What part does it play in a circuit ?
3. What is wattless current ?
4. What is choke ? What is the difference between a *H.F.* and a *L.F.* choke ?
5. What is capacitive reactance ? Which would you prefer to use and why when you want to reduce the current in an A.C. circuit—a choke, capacitor, of a resistance ?
6. What is the relation between the total applied *E.M.F.* in a circuit and *P.Ds.* across the resistance, inductance and capacitance in that circuit ?
7. Compare an A.C. circuit with a D.C. circuit. What is the difference in the behaviour of a capacitor when used on D.C. and when used on A.C. ?
8. What is the difference in the measuring instruments used for A.C., and D.C. ?
9. Can a D.C. instrument be used on A.C. ? Can an A.C. instrument be used on D.C. ?
10. What is virtual, effective or *R.M.S.* current in A.C. circuit ? Explain the significance of each term.
11. What is the maximum voltage when the voltmeter reading is 220 volts A.C. ?
12. What is power factor ? What is the difference between apparent watts and true watts.

A.C. FREQUENCY AND CAPACITANCE OF A CAPACITOR

13. What do you mean by the frequency of A.C. mains ? Explain how you find it out experimentally with an electrical vibrator.
14. What is the function of the solenoid and the permanent magnet in the vibrator ?
15. What for do you use a 25-watt lamp in series with the solenoid ?
16. Is it necessary to adjust the length of the steel rod before starting the experiment ? If so, why ?

17. What are the sources of error in the experiment ? What precautions do you observe and why ?
18. Explain how the capacitor is first charged and then discharged through the micrometer. Why do you get a continuous current as indicated by the microammeter ?
19. What is the function of the steel discs *T.T.* ? Can you use it to compare the capacitances of two capacitor ? If so, how ? Is this method of comparing capacitances superior to the ballistic galvanometer method ? Point out the relative merits and demerits of the two methods.

POWER FACTOR OF AN A.C. CIRCUIT

20. What is the expression for the power consumed in an ohmic resistance ? What is the expression for the power consumed in an inductive circuit ?
21. What do you mean by power factor ? What is the relation between apparent watts and true watts consumed in a circuit ? What is the power factor of a pure inductive circuit ?
22. What is the order of the ohmic resistance of the inductance that you are using ?
23. Draw the p.d. Vector diagram for the circuit. Can you calculate from this Vector diagram (*i*) the resistance or d.c. resistance of the inductive or load circuit and (*ii*) the inductance of the coil ? Will the calculated value of the resistance of the load circuit be the same as the d.c. resistance ?
24. If there is a difference, how do you account for this difference ? Which is greater—the calculated resistance from Vector diagram or the measured d.c. resistance ?
25. What precautions do you take in the experiment ?
26. Why should the values of p.d. across various points be measured very accurately ?
27. Should you make an approximation while taking the voltmeter reading when the pointer lies between two scale divisions.
28. Is there any other simple method of measuring the power factor ?

STUDY OF SERIES RESONANCE

29. What do you mean by resonance ? When does resonance occur in a series a.c. circuit ? When resonance occurs what is the relationship between p.d. across inductance and the

p.d. across capacitance ? Can these values of p.d. be greater than the supply e.m.f. at resonance ?

30. What is theoretically the value of p.d. across both *L* and *C*, and why ? Do you achieve this condition practically ? How do you account for this difference ?
31. At resonance, what is the p.d. across the resistance theoretically ? do you get this result practically ? How do you account for this difference ?
32. Why do you use an inductance coil of about two henry in the circuit ? How is such a coil made ?
33. How does the inductive reactance vary with frequency ?
34. How does the capacitive reactance vary with frequency ? Draw graphs to illustrate your above answers.
35. From the graphs, draw the graph of the net reactance with frequency. What conclusion can you draw from the net reactance graph ?
36. Explain the nature of graphs (V_C, C), (V_L, C) that you have drawn in this experiment. At what value of *C* does resonance occur ?
37. Explain the nature of graphs (V_R, C). (V_{CL}, C) that you have drawn. At what value of *C* does resonance occur ?
38. What precautions do you take in the experiment ?

WIEN'S SERIES RESISTANCE BRIDGE FOR CAPACITY

39. For what purpose do you use this bridge ?
40. Draw the Vector diagram for an unbalanced bridge. From the diagram, point out what adjustments would be necessary to balance the bridge.
41. What are the conditions for the bridge to be balanced ? How do you test this balance ?
42. What is the impedance of the headphone ? Does it vary with frequency ?
43. Why do you use a mica condenser of known capacity, why not a paper capacitor ?
44. Why do you consider a capacitor as equivalent to a capacitance with a resistance in series with it ?
45. Which of the resistances in the bridge should be specifically non-inductive for the determination of the power factor of the capacitor ?

46. Why is it important to record the frequency of the oscillator, usually 1000 c.p.s. ?
47. When is the bridge most sensitive ?
48. What precautions do you observe in the experiment.

MAXWELL'S INDUCTANCE BRIDGE (Coils in Series)

49. For what purpose do you use this bridge ?
50. Draw the Vector diagram for an unbalanced bridge. From the diagram point out what adjustments would be necessary to balance the bridge (*a*) if the standard inductance is fixed, (*b*) if this is variable in a known manner.
51. What are the conditions for the bridge to be balanced ?
52. What should be the order of the resistance R_3 and R_4 ? What is the purpose of choosing R_3 and R_4 of this order
53. Why do you use a decimal ohm box in series with both L_1 and L_2? Would one decimal ohm box in series with L_2 not serve the purpose ? What difficulty may you experience by using only one decimal ohm box in series with L_2 ?
54. What precautions do you take in the experiment ?
55. What is an inductometer ? Why is it more useful for this experiment than a fixed inductance ?

30
Semiconductor

30.1 Introduction

A substance having forbidden gap energy (E_g) of about 1 *eV*, is called a semiconductor. This value of band gap energy is much lower than that of an insulator ($\approx 6eV$). The two most important and widely used semiconductors are germanium and silicon, having the values of E_g equal to 0.785 *eV* and 1.21 *eV* (at 0K) respectively. For some specialized (*e.g.*, optoelectronic devices) uses GaAs, SiC have also been found to be useful semiconductors.

In pure semiconductor, at absolute zero (0K) the Fermi level lies in the middle of forbidden energy gap and the conduction band is totally empty. At low temperatures, electrons cannot acquire sufficient energy under an applied electric field to cross over into the conduction band. Hence, at low temperature, these materials are insulators. However, at room temperature, some valence electrons acquire enough thermal energy ($\geq E_g$) and cross over to the conduction band.

The fraction of electrons with greater than E_g at temp. T is

$$n = N \exp\left(-\frac{E_g}{2kT}\right) \tag{30.1}$$

where N is the total number of electrons in valence band.

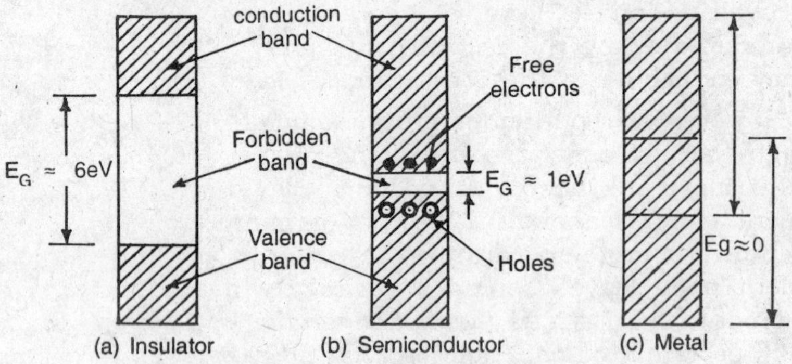

Fig. 30.1 Energy–band Structure

These electrons, often called conduction electrons, move under the influence of applied electric field and the semiconductor becomes slightly conducting.

An electron which cross over from valence band to conduction band, leaves a vacancy in the valence band. The vacancy created due to transfer of an electron from valence band to the conduction band is known as *hole* (Fig. 30.1). Thus, a hole is an empty energy level in otherwise filled valence band. As discussed in the following sections, the holes behave as positively charged current carriers.

30.2 Current Carriers (Electrons and Holes) in Semiconductor

The crystal structure of germanium, one of the most widely used semiconductor in electronic devices, is shown in Fig. 30.2. The crystal structure of all the elements of group IV of periodic table, like *silicon* are similar. The atoms in the structure are strongly held by covalent bonds. Each atom has four valence electrons, *i.e.*, the atom is tetravalent. Each of the valence electrons of a germanium atom is shared by one of its four nearest neighbours. That is, covalent bond exists between the neighbouring atoms and the atoms are strongly held together by the binding force resulting due to the covalent bonds. Because the valence electrons help to bind the atoms, they are themselves held tightly in the bond and they cannot contribute to the conduction at low temperatures.

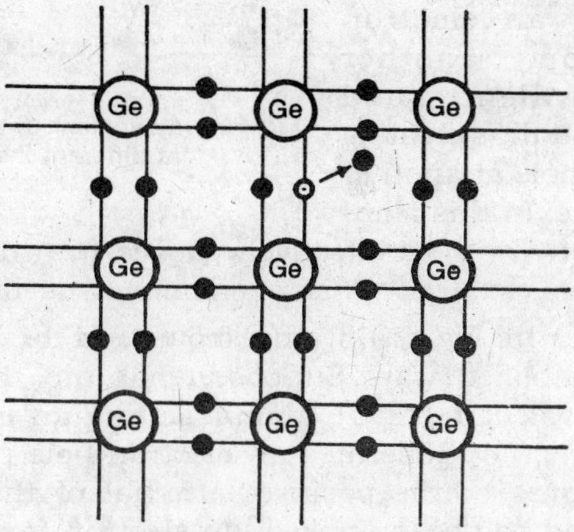

Fig. 30.2 Covalent bonds in Ge crystal

However, at room temperature, some of the electrons acquire enough thermal energy to break the covalent bond and become free to move in the crystal. The energy required to break a covalent bond is about 0.72 *eV* for germanium and 1.1 *eV* for silicon at room temperature. The electron moving out of the covalent bond leaves behind a vacancy which is called a hole. The process is known as *thermal generation of electron-hole pairs.* The importance of the hole is that it serves as a carrier of electricity almost as effectively as the free electrons. The mechanism by which a hole contributes to the conductivity may be understood from the following discussion :

An electron from a neighbouring atom can break away and fill the existing vacancy a—hole creating a hole at another site. This hole may now be filled by an electron from another covalent bond and thus creating a hole at another site. In this manner, the hole effectively moves in a direction opposite to that of the electron. This phenomenon is illustrated in Fig. 30.3.

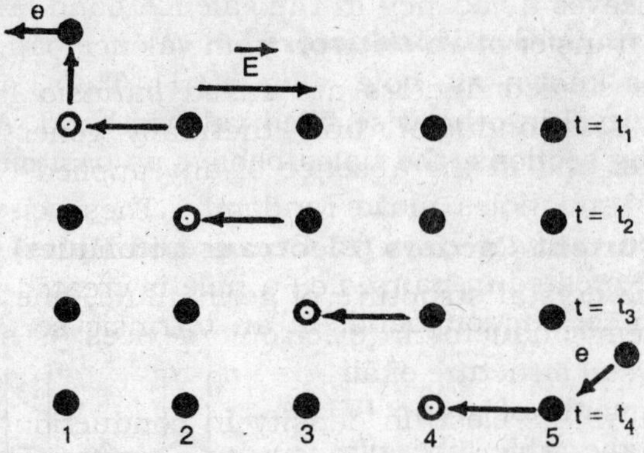

Fig. 30.3 **Movement of a hole under an electric field at different instants of time in a semiconductor**

In Fig. 30.3, the movement of hole is depicted in a very simplified way. Suppose, that the thermal excitation breaks a covalent bond at atom-1 setting an electron free and creating a hole, *i.e.*, generates an electron-hole pair. The electron generated moves to the positive terminal of the battery when an electric field has been applied, an electron from atom-2 may go to atom-1 and fill the vacancy in atom-1 but in the process creates a hole at atom-2. Similarly, an electron from atom-3 moves to atom-2 and the hole moves to atom-3. Continuing in this way, the hole moves to atom-5 which takes an electron from the negative terminal of the battery and the covalent bond is completed.

The above discussion indicates that the motion of the hole in one direction actually involves the transport of a negative charge *an equal distance* in the opposite direction. A *hole behaves as a positive charge* equal in magnitude to that of an electron as far as flow of electric current is concerned. The theoretical justification for a hole behaving like a free positive charge carriers have been given quantum mechanically. A hole is also shown quantum mechanically to possess an effective mass.

In a semiconductor, both electrons and holes contribute to

the conduction process. If I_e is the component of the current carried by electrons and *In* that by holes, then the total current is given by :

$$I = I_e + I_h \quad (30.2)$$

30.3 Intrinsic Semiconductors

Pure semiconductors are called *intrinsic semiconductors.* In intrinsic semiconductor, only thermally generated electron-hole pairs exist and in the absence of any applied electric field these electrons and holes move randomly. These electrons and holes are called *intrinsic carriers.* Because (conduction) electrons and holes are created in pairs, *i.e.*, a hole is created when an electron moves to conduction band, in an intrinsic semiconductor :

$$n_e = n_h = n_i \quad (30.3)$$

where, n_e is the electron density in conduction band, n_h is hole the density in valence band and n_i is the number density of intrinsic carriers called the *intrinsic concentration.*

Thermal agitation continues to produce new electron-hole pairs but at the same time recombination of holes and electrons also takes place. In equilibrium, the two processes take place at the same rate and the intrinsic concentration remains constant.

30.4 Doping of Semiconductor : Extrinsic Semiconductor

The intrinsic semiconductors at room temperature contain thermally generated free electrons and holes but their numbers are very small and conductivity is too small to be of any practical use. To increase their conductivity suitable impurities are added to intrinsic semiconductor. This deliberate addition of desirable impurity is called *doping* and the impurity atoms are called *dopants.* Depending upon the kind of impurities added, there are *two types of extrinsic semiconductor : n-type and p-type.* The amount of doping is normally small, *e.g.*, 1 part in 10^8 in Si or 1 impurity atom for every 10^8 atom of Si.

N-Type Semiconductor—If a tetravalent semiconductor having four valence electrons is doped with controlled amount of *pentavalent atoms* like arsenic (As) [or antimony (Sb) or phosphorus (P)] which have five valence electrons, the added atoms substitute the silicon atoms (Fig. 30.4). Out of the five valence electrons of an arsenic atom, four enter in covalent bonds with the four silicon atoms surrounding it while the fifth electron not being part of any bond is comparatively free. The energy required to detach this fifth electron from the atom is of the order of only 0.01 *eV*. This energy, at room temp. being of the order of thermal energy, all these electrons move into the conduction band becoming free. Obviously, the number of such free electrons will be equal to the number of atoms added. Thus, each pentavalent atom added releases one free electron to the host crystal and it is for this reason the pentavalent atoms are called the donor atoms. Because of these free electrons in the doped crystal, the doped crystal is called *n-type* crystal. The electrons set free are called *extrinsic carriers* and the semiconductor is called *n-type extrinsic semiconductor*.

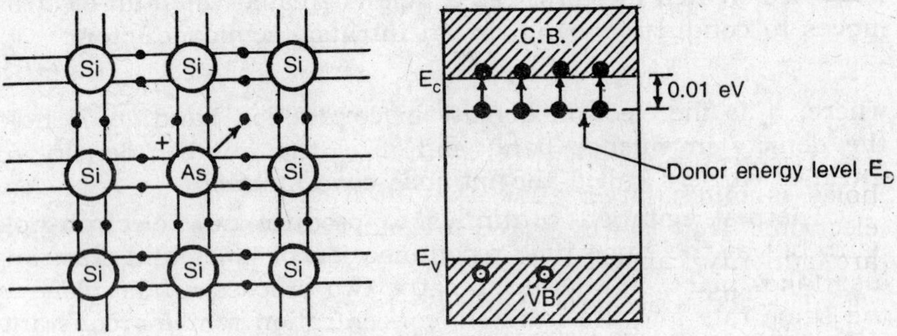

Fig. 30.4 N-type semiconductor

Like pure semiconductors, thermally generated electron hole pairs exist in doped semiconductors also. But due to excess (free) electron in the *n*-type semiconductors, the number of free electrons is much higher than that of holes. For this reason, in *n*-type semiconductors, electrons are called *majority carriers*. In an *n*-type semiconductor, the free electron concentration n_e (number per unit volume) is nearly equal to the density of donor atoms and is very large compared to concentration of hole n_h. That is,

$$n_e \cong N_d >> n_h \tag{30.4}$$

P-Type Semiconductor—If an intrinsic semiconductor, say, silicon, is doped with a controlled amount of trivalent atoms, *e.g.*, indium (In) [or boron (B) or aluminium (Al)] which has three valence electrons, the added atoms substitute Si atoms [Fig. 30.5 (a)]. Because the dopant atom has only three valence electrons, there is one incomplete bond with a neighbouring Si atom, due to deficiency of an electron. Thus, a hole is created. Such impurities make available positive carriers because they create holes and can accept electrons. These impurities are, therefore, known as *acceptor* or *p-type impurities* and the resulting material is called *p-type (extrinsic) semiconductor*.

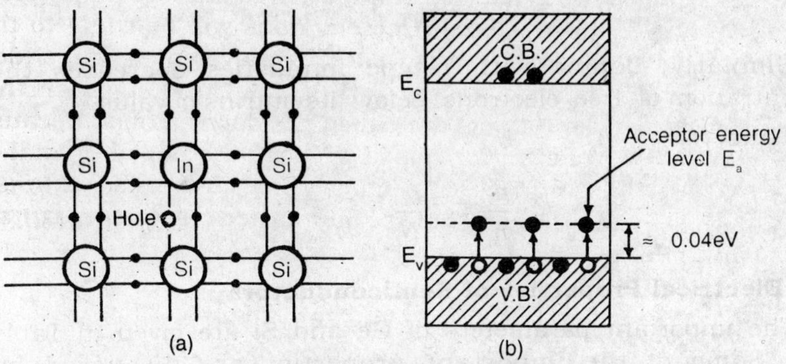

Fig. 30.5. p-type semiconductor

Because each dopant atom creates one hole, the number of holes is much larger than that of (thermally generated) free electrons. That is why, *holes are majority carriers* and electrons are minority carriers.

The added acceptor impurities creates an energy just above the valence band, as shown in Fig. 30.5 (b). Since a very small amount of energy (≈ 0.04 *eV*)is required for an electron to leave the valence band and occupy the acceptor energy level, the electrons move to the acceptor energy level creating large number of holes in the valence band.

In *p*-type semiconductors, the number density of holes n_h in valence band is approximately equal to the number density of added acceptor atoms N_a and is very large compared to the number density of (free) electrons n_e, that is,

$$n_h \cong N_a \gg n_e \qquad (30.5)$$

The Mass Action Law—According to this law under the equilibrium the product of the concentration of free negative and positive charge carriers is a constant independent of the amount of donor and acceptor doping and is equal to the square of the concentration of intrinsic carriers *(n_i)* that is,

$$n_e n_h = n_i^2 \qquad (30.6)$$

However, the intrinsic concentration n_i is itself a function of temperature. When n-type impurities are added, not only the concentration of free electrons increases but the number of holes decreases below its intrinsic value. Thus,

$$\because \quad n_e \cong N_d$$

$$\therefore \quad n_h = \frac{n_i^2}{n_e} = \frac{n_i^2}{N_a} \qquad (30.7)$$

Similarly, doping with *p*-type impurities decreases the concentration of free electrons below its intrinsic value.

Thus, $n_h \cong N_a$

$$\therefore \quad n_e \cong \frac{n_i^2}{n_h} = \frac{n_i^2}{N_a} \qquad (30.8)$$

30.5 Electrical Properties of Semiconductors

The important parameters of Ge and Si are given in Table 30.1. Some of the important properties are discussed by *Conductivity*. A semiconductor is *bipolar* as it contains charge carriers of both the polarities, *i.e.*, electrons and holes, unlike metals which are unipolar containing only electrons.

Table 30.1

Properties of Germanium and Silicon

Property	Ge	Si
Atomic number	32	14
Atomic weight	72.6	28.1
Atoms/cm^3	4.4×10^{22}	5.0×10^{22}
E_{go} (in eV at 0K)	0.785	1.21
E_g (in eV at 300 K)	0.72	1.1
n_i (cm^{-3}, at 300 K)	2.5×10^{13}	1.5×10^{10}
Intrinsic resistivity (Ω-cm, at 300 K)	45	2,30,000
μ_e (cm^2/V-s, at 300 K)	3,800	1,300
μ_h (cm^2/V-s, at 300 K)	1,800	500
Dielectric constant (relative)	16	12

Let the *mobility* ($\mu = \frac{v}{E}$ = drift velocity per unit applied electric field) of an electron be μ_e and that of a hole μ_h. Electrons and holes move in opposite directions in applied electric field E, but as they are oppositely charged, the current due to both of them is in the same direction. Hence, current density $\left(= \frac{I}{A}\right.$, I – current, A – area of cross section) is given by :

$$J = (n_e \mu_e + n_h \mu_h) eE = \sigma E \quad (30.9)$$

where, e = magnitude of charge on an electron = magnitude of charge on a hole

n_e = number density of electrons

n_h = number density of holes

σ = conductivity of the material

$$\therefore \text{ conductivity } \sigma = (n_e\mu_e + n_h \mu_h) e \quad (30.10)$$

For an intrinsic semiconductor $n_e = n_h = n_i$

$$\therefore \quad \sigma = (\mu_e + \mu_h) n_i e \quad (30.11)$$

Intrinsic Concentration—When temperature of a semiconducting material is increased, the number of thermally generated electron-hole pairs increases. This raises the concentration of electron n_e and holes n_h which in turn increases the conductivity. The variation of intrinsic concentration n_i with temperature T is found to be given by :

$$n_i^2 = A_o T^3 e^{-Eg/kT} \quad (30.12)$$

where E_g is the energy band gap at 0K (in eV), k is the Boltzman constant (eV/K) and A_o is a constant independent of temperature. As can be seen in Table 30.1, Ge has around 10^{22} atoms/cm^3, while at room temperature (300 K) $n_i \approx 10^{13}$/cm^3. Thus, *only 1 atom in about 10^9 atoms contributes a free electron-hole pair to the crystal.* For silicon, this ratio is 1 in 10^{12}.

Energy Gap—The forbidden energy gap Eg in a semiconductor is dependent on the temperature. It has been found experimentally that for :

Silicon : $E_g(T) = 1.21 - 3.60 \times 10^{-4} T$ (30.13)

At $T = 300\ K \quad E_g = 1.1\ eV$

Germanium $E_g(T) = 0.785 - 2.23 \times 10^{-4}\text{T}$ (30.14)

$T = 300\text{ K}, \quad E_g = 0.72\ eV$

Experiment 30.1

Aim—To determine the forbidden band gap energy of a semiconductor.

Apparatus required—Four-probe set-up, constant current source, millivoltmeter, semiconducting crystal chip with non-conducting botton surface, thermometer (0 – 200°C), etc.

Theory—The resistivity of a semiconductor is given by :

$$\frac{1}{\rho} = C \, exp\left(-\frac{E_g}{2kT}\right) \quad (30.15)$$

where constant C is dependent on temperature ($T^{3/2}$), mobilities of carriers and density of states function. Taking log of both sides and changing sign of the Eq. (30.15), we get :

$$\log \rho = \frac{E_g}{2kT} - \log C \quad (30.16)$$

From the above equations, it is implied that the resistivity is the function of the absolute temperature. To measure the resistivity and E_g, four-probe method is employed as it is more accurate than the conventional two-probe method.

Four-probe Method—The method is illustrated in Fig. 30.6. The four probes are placed collinearly with equal spacing between them on the sample. The current is passed through the two outer probes and the poetntial is measured between the two inner

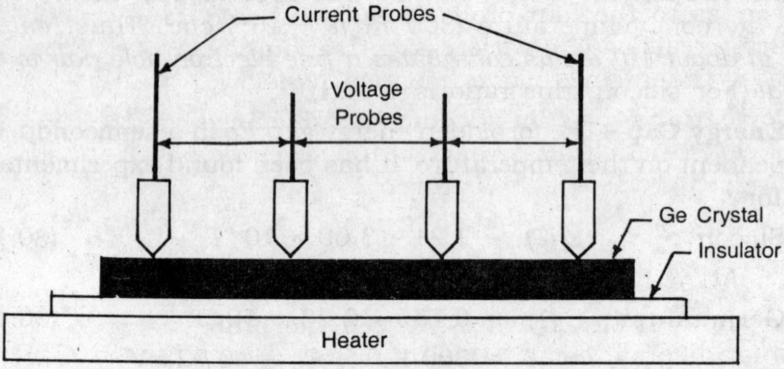

Fig. 30.6. Four-probe arrangement

probes. The sample used should be thin, less than half of the probe spacing and the resistivity of the material should be uniform. Also, the specimen should be large enough so that it may be considered to be of semi-infinite volume.

Resistivity measurement of a large sample—Current I is passed through the outer probes and the floating potential is measured between the two inner probes. The floating potential V_f at a distance r from an electrode carrying a current I in a material of resistivity ρ is given by :

$$V_f = \frac{\rho I}{2\pi r} \quad (30.17)$$

If the potential at probe 2 is V_{f2} and at probe 3 is V_{f3}

$$V = V_{f2} - V_{f3} \quad (30.18)$$

and the resistivity, if the distance between the probes is equal to s, is :

$$\rho = \frac{V}{I} \cdot 2\pi s \quad (30.19)$$

If the probe spacing is 0.159 cm, then $2\pi s = 1$, therefore,

$$\rho = \frac{V}{I} \quad (30.20)$$

For better results, current passed should be in microamperes (or at most in milliamperes) so that if the resistivity is in ohm-cm, the measured voltage is in microvolt (or in millivolt).

Very often, used samples are not large enough to consider them infinite and, therefore, Eq. (30.19) is not directly applicable. However, since the four-probe is the most accurate method for measuring resistivity, a variety of correction have been developed. The most general method is mentioned below.

Correction for a thin-slice non-conducting bottom surface— In this case, the bottom surface of the slices non-conducting and the corrected resistivity ρ' is given by :

$$\rho' = \frac{\rho}{F(w/s)} \quad (30.21)$$

where w/s is the correction factor, w is the width of the sample and s is the spacing between the probes. The values of $F(w/s)$ for different combination of w/s is given in Figs. 30.7 and 30.8 from where it can be taken.

From Eq. (30.16), we have :

$$\log_e \rho = \frac{E_g}{2kT} - \log_e C$$

A graph is plotted for $\log_{10}\rho$ V $\frac{1}{T(K)} \times 10^3$. The slope of the curve is given by :

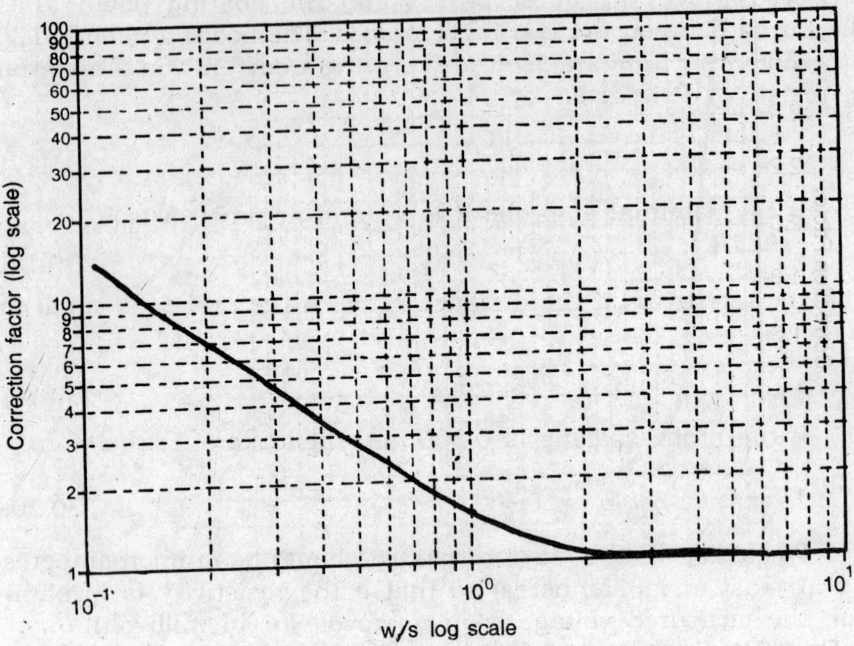

Fig. 30.7 Correction factor on log scale

$$\frac{\log_e \rho}{\frac{1}{T}} = \frac{E_g}{2k}$$

Hence, $$E_g = \frac{2k \log_e \rho}{\frac{1}{T}} \tag{30.22}$$

where, $k = 8.6 \times 10^{-s}{}_e V/\,deg$, is the boltzmann constant and $\log_e \rho = 2.3026 \log_{10} \rho$

Procedure

1. Arrange the base plate of the four-probe arrangement such that the non-conducting surface of the semiconducting sample is on the plate side and the probes are in the middle of crystal. Ensure that the contacts between the probes and the crystal are proper.

2. Place the arrangement in the oven properly and fix the thermometer in the hole provided in the oven.

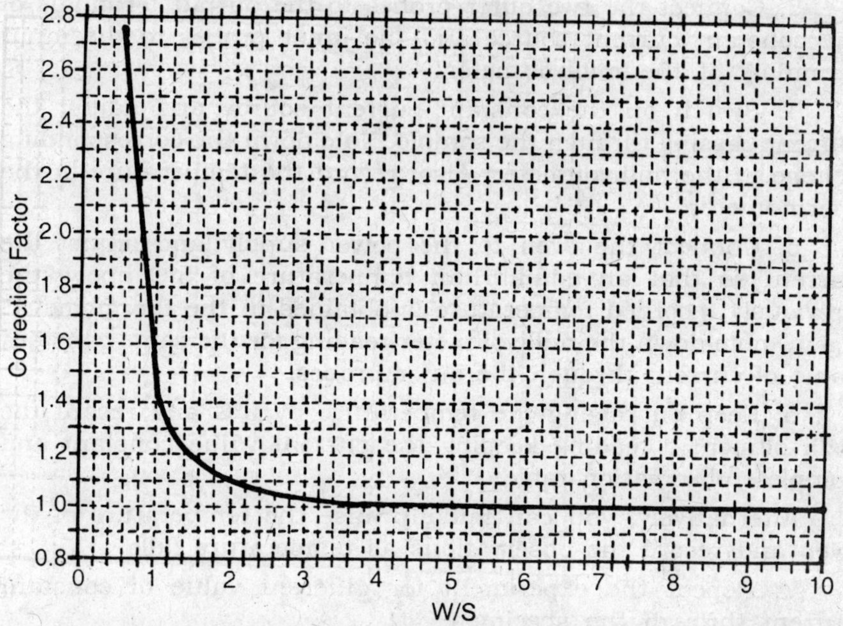

Fig. 30.8 (a) Correction factor in linear scale. Use this graph to find F(w/s), if it is likely to fall in the range of 0.8. Note, the scale is expanded for clarity.

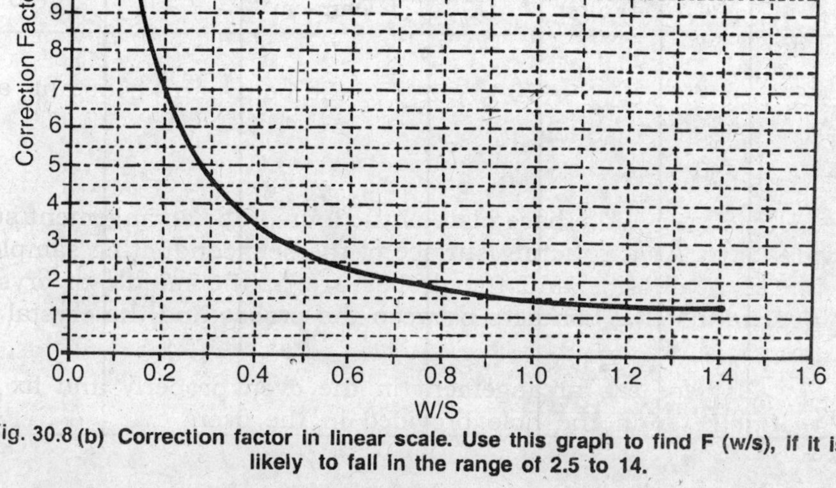

Fig. 30.8 (b) Correction factor in linear scale. Use this graph to find F (w/s), if it is likely to fall in the range of 2.5 to 14.

3. Connect the two outer probes to the output terminals of the constant current source and the inner probes to the input terminals of the millivoltmeter.

4. Switch on the constant current source and adjust the current passing through the sample. Note down the corresponding voltage in the millivoltmeter. Also, record the temperature in the thermometer.

5. Connect the oven to the power supply and adjust the current so that an equilibrium temperature of 5°C above the previously recorded temperature is obtained in the thermometer. Again, note down the millivoltmeter reading for the same constant value of current through the outer probes.

6. Keep on raising the temperature by 5°C and record the millivoltmenter reading keeping the current value constant and complete observation table.

Alternatively, you can initially raise the temperature of the oven and record the observations as temperature falls.

7. Repeat the experiment for different value of constant current through the specimen.

Observations

Current I = ... mA (Constant)

Distance between probes s = ... cm

Thickness of the crystal w = ... cm

Sr. No.	Temp. (°C)	Voltage (Volts)	Temp. (K)	Resistivity ρ Ohm-cm	$\text{Log}_{10}\rho$	$\frac{10^3}{T(K)}$
...	...	...	...	...	...	...
2.	25					
3.	30					
4.	35					
5.	40					
.						
.						
.						
.						
.						
.						
.						
17	150					

Calculations—Plot $\text{Log}_{10}\ \rho$ *Vs* $\frac{1}{T(K)} \times 10^3$

Find slope of this curve. Slope $= \frac{\log_e \rho}{\frac{1}{T}} = \frac{E_g}{2K}$

$$E_g = \frac{2k \log_e \rho}{T} = \text{........}\ eV$$

Actual value of E_g = ... eV

% Error = %.

Sources of Errors and Precautions

1. The contacts between the probes and crystal should be tight. Ensure proper contacts and connections.

2. It should be ensured that proper terminals of the probes are connected to the current source and voltmeter.

3. Do not pass very high current. It may spoil the semiconducting crystal.

4. The observations should be noted for each temperature when thermal equilibrium is attained and a stable temperature is achieved.

30.6 Hall Effect

American Physicist E. H. Hall observed that *when a current-carrying conductor (semiconductor or metal) is placed in a magnetic field perpendicular to the direction of current, a voltage is developed across the conductor in a direction perpendicular to both the current and magnetic field.* This effect is known as *Hall effect.* This effect is very useful in determining :

(i) The nature of charge carriers, *e.g.*, whether semiconductor is of *n*-type or *p*-type.

(ii) Carrier concentration or the number density of charge carriers.

(iii) Mobility of charge carriers.

Consider a conductor in the form of a flat strip (Fig. 30.9). Let a current I is made to flow through the strip along *x*-axis. *P* and *P'* are the two points on the opposite faces *a b c d* and *a' b' c' d'* respectively. If a galvanometer is connected between points *P* and *P'*, the galvanometer does not show any deflection indicating there is no potential difference between the two points. But, when a magnetic field is applied along *y*-axis, *i.e.*, perpendicular to the direction of current, a deflection is produced in the galvanometer indicating a potential difference between *P* and *P′*. This e.m.f. is known as *Hall e.m.f.* or *Hall voltage* or *Hall potential* V_H.

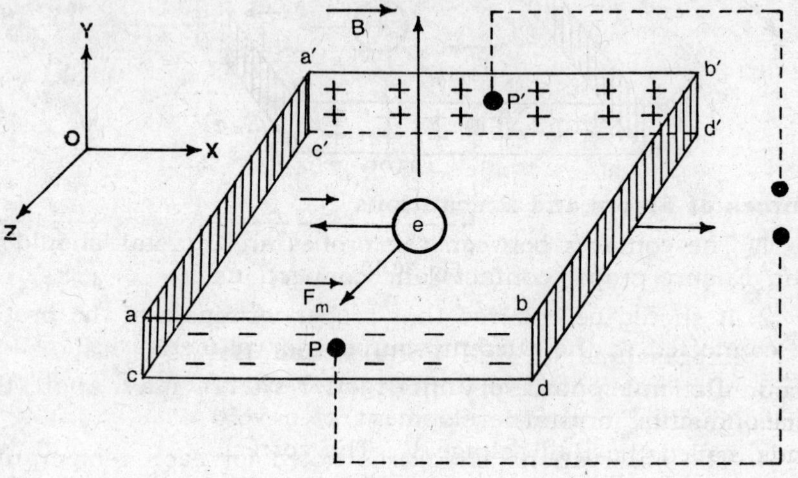

Fig. 30.9

As shown in Fig. 30.9, if the current is passed along *x*-axis, then the electrons move along negative direction of *x*-axis. The force on an electron due to the magnetic field *B* is :

$$\vec{F} = e\,(\vec{v} \times \vec{B})$$

or $$F = evB \sin 90° = evB \tag{30.23}$$

where, v and e are the drift velocity and charge of the electron. Using the Fleming left hand rule it is seen that the force on the electrons will be directed towards the face *a b c d, i.e.*, along positive *z*-axis, thereby making the face *a b c d* negative and *a'b'c'd'* positive.

If the current is carried by positively charged carriers, i.e., holes, the carriers move in the same direction that of the current. The magnetic force cause the positive charge carriers to move towards the face a b c d, thereby making the face a b c d positive and a'b'c'd' negative (Fig. 30.10)

Thus, by determining the polarities of the surface of the strip, we can determine the sign of the charge carriers. In fact, this is one of the most useful applications of the Hall effect.

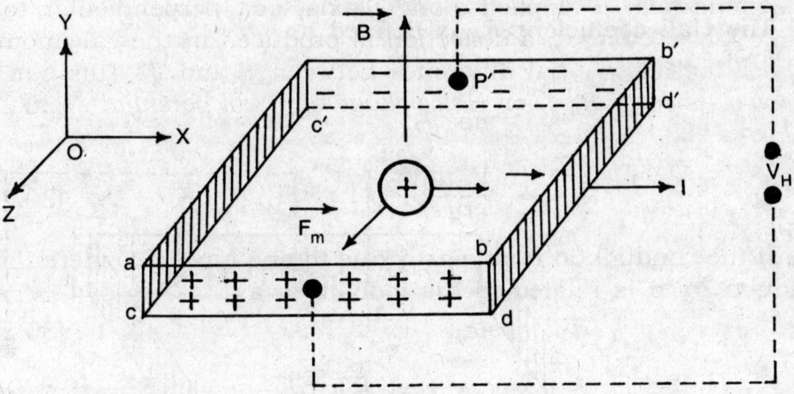

Fig. 30.10

Expressions of Hall Voltage and Hall Coefficient—The accumulation of positive and negative charges on the opposite faces, results in the development of a voltage across the two ends, called the Hall voltage V_H. The corresponding electric field is denoted by E_H.

The force due to the electric field on a current carrier of chargee is eE_H. While the force due to the transverse magnetic field B is evB ($\pm\, evB \sin 90^\circ$), where v is the drift velocity of the current carrier. The two forces being oppositely directed, for equilibrium, *i.e.*, when the accumulation of charges stops, we have :

$$eE_H = evB$$

or $$E_H = vB \tag{30.24}$$

If w is the width and d the thickness of the specimen, its cross-sectional area $A = wd$. Therefore,

$$\text{current density } J = \frac{I}{wd} = env \quad (I = envA) \tag{30.25}$$

where n is the number of carriers per unit volume :

From Eqs. (30.24) and (30.25), we have :

$$V_H = E_H d = vBd = \frac{BI}{enw} \tag{30.26}$$

Thus, knowing V_H, B, I, e and w the carrier concentration n

can be calculated.

The Hall coefficient R_H is defined as :

$$R_H = \frac{1}{ne} \tag{30.27}$$

or $$R_H = \frac{V_H w}{BI} \tag{30.28}$$

If the conduction is primarily due to one type of carriers then conductivity σ is related to the mobility μ as :

$$\sigma = ne\mu \tag{30.29}$$

Thus, knowing conductivity $\left(\sigma = \frac{J}{E}\right)$ the mobility of the charge carriers can be determined as :

$$\mu = \frac{\sigma}{ne}$$

or $$\mu = \sigma R_H \tag{30.30}$$

Experiment 30.2

Object—To study the Hall effect and to calculate : (i) the Hall coefficient R_H and (ii) the concentration of charge carrier *n*.

Appaturtus—Semiconducting specimen in the form of a strip mounted on a stand, microampere current source, *dc* millivoltmeter, electromagnet with power supply, magnetic flux density meter.

Theory—See Section 30.6

Hall Co-efficient $$R_H = \frac{V_H w}{BI}$$

Carrier concentration $$n = \frac{e}{R_H}$$

Procedure

1. Connect one pair of contacts of the specimen on the opposite faces to the current source and the other pair to the millivoltmeter (Fig. 30.11)

2. Switch on the power supply of electromagnet and measure

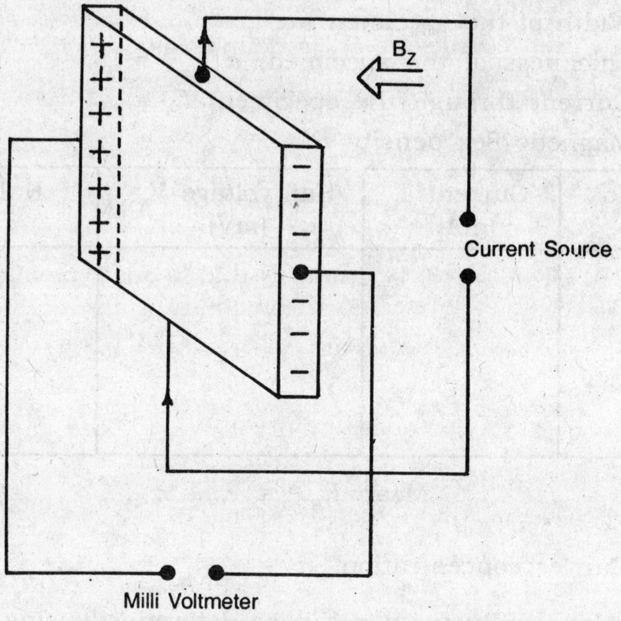

Fig. 30.11

the magnetic flux density at the centre between the pole faces by placing the tip of the flux meter there. Now, do not change the value of current in the electromagnet, *i.e.*, keep the magnetic field constant.

3. Place the specimen at the centre between the pole faces such that the magnetic field is perpendicular to the strip.

4. Pass a current (in mA) from the current source through the specimen and measure the resulting Hall voltage in the millivoltmeter. Enter these values in observation table.

5. Increase the current through the specimen gradually and measure the corresponding Hall voltages. Complete the observation table.

6. The entire process can be repeated for different values of the magnetic flux density.

7. If the power supply of the electromagnet and current source have reverse switches, the direction of the magnetic field and the current can be reversed and the process can be repeated. Find the mean of different R_H and H.

Observations

Width of the specimen w =

Thickness of the specimen, d =

Current through the specimen, I =

Magnetic flex density B =

Sr. No.	Current I (mA)	Hall voltage V_H (mV)	Hall coefficient R_H
1			
2			
3			
4			
5			

Mean R_H = ...

Carrier concentration, $n = \frac{1}{eR_H}$ = ...

[Also, conductivity σ can be determined using the relation.

$$\sigma = \frac{J_{(x)}}{E_{(x)}}$$

where, $J_x = \frac{I_x}{A} = \frac{I_x}{wd}$ and E_x = electric field along x-axis.

[The subscript x denotes that J and E should be in the same direction. Also, conductivity $\mu = \frac{\sigma}{en}$.]

Precautions

1. Ensure that the specimen is located at the centre between the pole faces and is exactly perpendicular to the magnetic field.

2. The contacts for passing the current and measuring Hall voltage should be proper.

3. To measure the magnetic flux density, place the tip of the fluxmeter at the centre between the pole faces and perpendicular to the field direction.

ORAL QUESTIONS

SEMICONDUCTOR

1. Why are semiconductors so called ?
2. What is forbidden energy gap ? How does it change with temperature in semiconductors ?
3. What is the value of E_g in *Ge* and *Si* at (i) *OK* and at (ii) 300 *K* ?
4. What is a hole ? Why does it behave as a positively charged particle ?
5. Does a hole possess a mass ? Explain.
6. What is thermal generation of electron-hole pair ?
7. What is an intrinsic semiconductor ? Why is the number of two-types of current carrier equal ?
8. What is doping ? Why is it done ?
9. What is an extrinsic semiconductor ? Name the two types of semi-conductors.
10. Name any two dopants used to produce *n*-type semiconductor. Why are such dopants calld *donor impurities* ?
11. Why are *n*-type semiconductors so called ?
12. Name any two dopants used to produce *p*-type semiconductors. Why are such dopants called *acceptor impurities* ?
13. Why are *p*-type semiconductors so called ?
14. How does doping increase the conductivity in semiconductor ? Explain with help of energy level diagram in both *n*- and *p*-types semiconductors.
15. State and explain mass-action law.
16. Does a *p*-type semiconductor contain electrons ? Explain.
17. What is the value of intrinsic carrier concentration in (i) *Ge* and (ii) *Si* ?
18. Define 'mobility'. Whose mobility is high—electron or hole ?
19. Why is four-probe method better than two-probe method for resistance measurement ?
20. How does the resistance of semi-conductor change with temperature ? Explain.

HALL EFFECT

1. What is Hall effect ? Explain its origin.

2. Name the information that can be obtained by studying the Hall effect.
3. Define 'Hall voltage'.
4. What is Hall Coefficient ?
5. Explain how carrier concentration can be determined in a specimen.
6. What is the unit of Hall coefficient ?

31

P-N Junction Diode

31.1 P-N Junction Diode

A *p*-type or *n*-type silicon (or germanium) can be grown by adding suitable and controlled amount of impurity in (the melt of) pure silicon (or germanium). The crystal grown are cut into thin slices called *wafer*.

If on a wafer of *n*-type silicon, a thin film of trivalent (or acceptor impurity), say, aluminium, is placed and heated to a high temperature (≈575°C), aluminium diffuses into silicon. Consequently, a *p*-type semiconductor is formed on an *n*-type semiconductor. The surface of contact or region of transition between *p*-type and *n*-type regions is called the *p-n junction.* The wafer becomes a *p-n junction diode* or simply *junction diode.* Electrical contacts are taken out from *p* and *n* regions.

Alternately, a *p-n* junction diode can also be grown by diffusing a pentavalent (or doner) impurity, say, phosphorus, into a *p*-type semiconductor.

Fig. 31.1 (a) shows a *p-n* junction diode while its symbolic

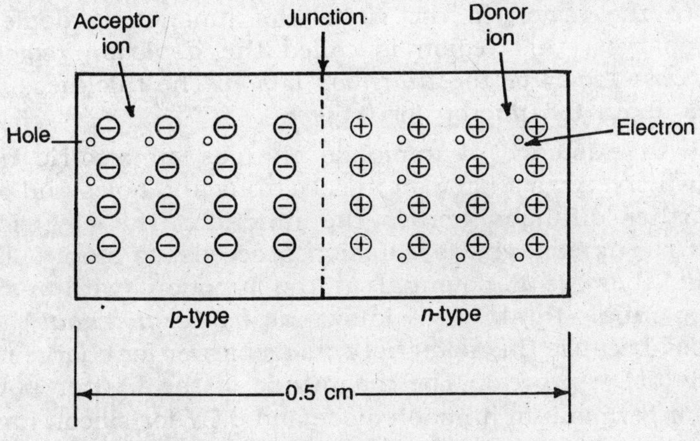

Fig. 31.1 (a)

representation is shown in Fig. 31.1(b). The *p*-side is also known as the *anode* and the *n*-side the *cathode*. The donor ion is represented by a plus sign because this impurity atom donates an electron and becomes positively charged. The acceptor ion is shown by a negative sign because this impurity atom accepts an electron and becomes negatively charged.

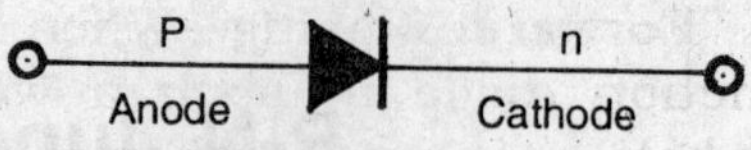

Fig. 31.1(b)

Depletion (or space charge) Region and Barrier Potential

In a *p-n* junction diode, the *p*-region has high concentration of holes and the *n*-region that of electrons. Holes from the *p*-region start diffusing across the junction into the *n*-region while the electrons from the *n*-region diffuse across the junction into the *p*-region. When an electron diffuses into *p*-region, it falls into a vacancy; in other words, it recombines with a hole. This process is called *recombination*. This recombination results in the production of an ionised acceptor atom, as it has accepted an electron. The ionised acceptor atom becomes negatively charged as is immobile. In the *p*-region, in the immediate vicinity of the junction, *immobile excess negative charges* are formed.

Similarly, holes from the *p*-region recombine with electrons in *n*-side on crossing the junction. This results in the formation *immobile excess positive charges* in the vicinity of the junction on the *n*-side. There are also some unneutralized ions in the neighbourhood of the junction called the uncovered charges.

Since the region in the vicinity of junction is depleted of mobile charges, this region is called the *depletion region*, the *space-charge region* or the *transition region*. The thickness of this region is about 0.5 micron (or 0.5μm).

Due to existence of immobile charges an electric field is created which exerts a force on the electrons and holes and oppose their further diffusion across the junction. The electric field becomes strong enough, as diffusion proceeds, to stop it. Due to this field, there is a potential at the junction (with *n*-side at higher potential—think why?) known as *barrier potential* V_B. This is so called because this potential acts as barrier for charge motion in the diffusion process. The magnitude of the barrier potential is 0.3V for germanium junction diode and 0.7V for silicon junction diode. Besides the nature of the crystal, the amount of doping and temperature of the crystal are other factors which determine the magnitude of V_B. Due to this a very large field (of the order of 10^6 Vm^{-1}) exists across the depletion region.

Biasing of P-N Junction Diode

Forward Biasing—When a battery is connected to a *p-n* junction diode in such a way that the positive terminal of the battery is connected to the *p*-side and the negative terminal to the *n*-side, the junction diode is said to be forward biased. A forward biased junction diode is shown in Fig. 31.2

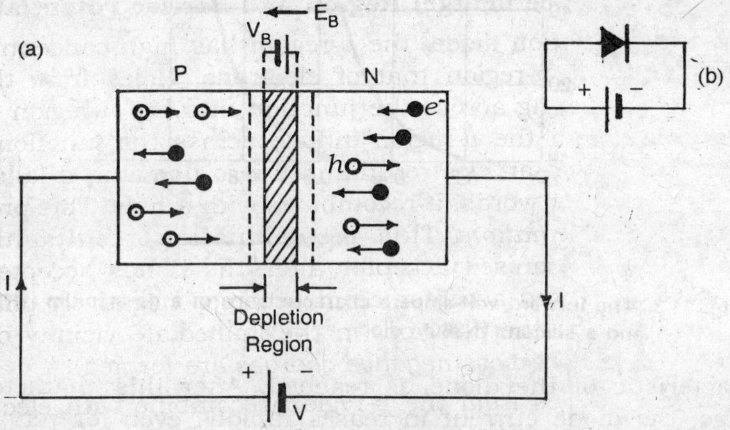

Fig. 31.2. A forward biased P-N junction diode. The dotted lines indicate original (unbiased) depletion layer. V_B–Barrier potential, V-applied potential

When the diode is forward biased, the externally applied voltage creats an electric field across the junction in a direction opposite to the field due to barrier potential. Therefore, the electric field and potential across the junction decreases, *i.e.*, potential barrier is lowered and the diffusion of electrons and holes across the junction thereby increases. Due to increased movement of majority carrier towards the junction, the thickness of the depletion region decreases. A reduced depletion region offers a low resistance to the flow of current. When the forward bias (the applied potential) exceeds the barrier potential, electrons and holes diffuse easily resulting in a sharp increase in current through the junction diode.

Forward Bias Characteristic—The variation of current *I* with applied forward bias (or voltage) *V*, called the forward bias characteristic is shown in Fig. 31.3 for a silicon diode. The current increases very slowly across the diode till a certain voltage, a

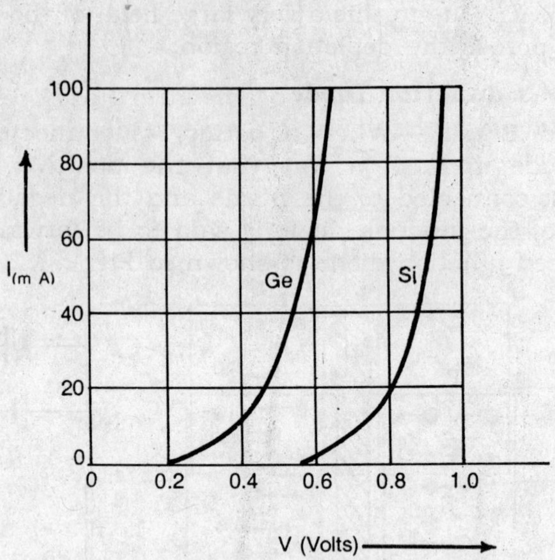

Fig. 31.3. The forward volt-ampere characteristics of a germanium (1N270) and a silicon (1N3605) diode at 25°C

characteristic of the diode, is reached. After this characteristic voltage, the diode current increases rapidly, even for very small increase in the diode bias voltage. This voltage is called the thereshold (or *cut-in* or *knee*) voltage V_γ. The value of V_γ is 0.2V for a Ge diode and 0.6 V for a Si diode.

The variation of current I with voltage V for a diode is given by :

$$I = I_o \left(e^{\frac{V}{\eta V_T}} - 1 \right) \tag{31.1}$$

A positive value of *I* means that current flows from the *p*-side to the *n*-side. For a forward biased diode, *V* is positive indicating that the *p*-side of the junction is positive with respect to n-side. The value of $\eta = 1$ for Ge and $\eta = 2$ for Si. The voltage equivalent of Temperature V_T is equal to

$$V_T = \frac{T}{11600} \tag{31.2}$$

At room temperature (T = 300K), V_T = 0.026V = 26 *mV*. When diode is forward biased, *i.e.*, V is positive and several times V_T, the unity in Eq. (31.2) may be neglected. Therefore, except for very small V (near the origin), the current increases exponentially with voltage.

Reverse Biasing—*When a battery is connected to p-n junction diode in such a way that the positive terminal of the battery is connected to n-side and the negative terminal to the p-side, the junction diode is said to be reverse biased.* A reverse biased junction diode is shown in Fig. 31.4.

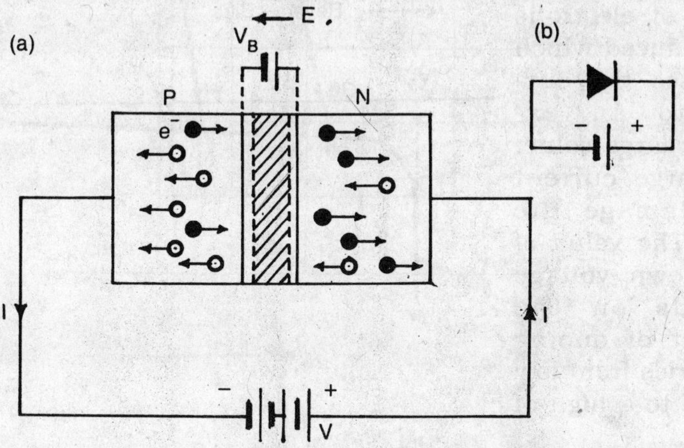

Fig. 31.4. A reverse biased PN junction diode. The dotted lines indicate original depletion region

When the diode is reverse biased, the externally applied potential and the barrier potential are in the same direction. Due to this, *thickness of depletion region increases or resultant barrier potential increases* making the diffusion of majority carriers across the junction more difficult. The resistance offered by the depletion region becomes very high and consequently current drops to a negligibly small value. However, a very small amount of current flows due to the motion of (thermally generated) minority carriers, *i.e.*, holes in the *n*-region and electrons in the *p*-region. This small current in the diode is called reverse saturation current (≈ μ *A* for a *Ge* diode and ≈ *nA* for *Si* diode, at room temperature). This current, involving falling down the potential of charge carriers across the junction, is uninfluenced by magnitude of barrier potential.

Reverse Bias Characteristics—The variation of current with applied reverse bias (or voltage), called the reverse bias characteristic is shown in Fig. 31.5. When the diode is reverse biased, very small current, called *reverse saturation (or leakage) current* flows through it. However, when the applied reverse voltage reaches a characteristic value called the *breakdown voltage* (also called *peak-inverse voltage*), the current suddenly increases to a large value. At the breakdown voltage, covalent bonds of the semiconducting material gets ruptured and an *avalanche* of electrons are produced which increases the current to an extremely large value. This large current may damage the diode. The value of breakdown voltage depends on the amount of doping and varies from low of ≈ 5V to a high of ≈ 200V.

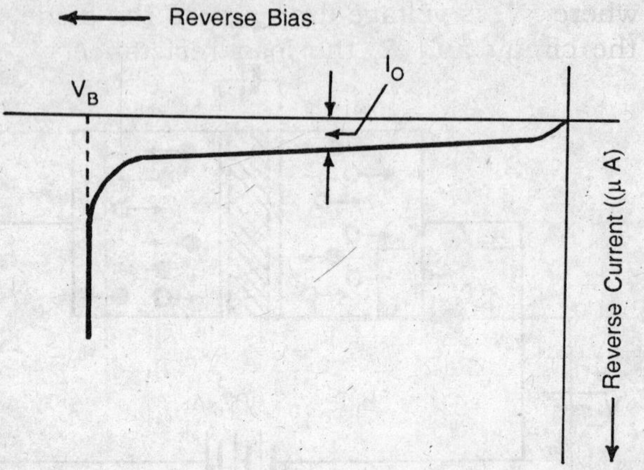

Fig. 31.5.

In Eq. (31.1), for reverse bias V is negative and it is several times V_T, therefore, $I = -I_o$. The reverse current is, therefore, constant and independent of the applied voltage. Consequently, I_o is called the reverse *saturation* current.

The complete characteristic curve of a *p-n* junction diode is shown in Fig. 31.6.

Fig. 31.6. (a) The volt-ampere characteristic of an ideal p-n diode (b) The volt ampere characteristic for a germanium diode redrawn to show the order of magnitude of currents. Note the expanded scale for reverse currents. The dashed portion indicates breakdown at V_Z.

Load-Line of a Diode Circuit—Consider a forward biased diode circuit diagram shown in Fig. 31.7 (a). If the resistance of source is negligibly small so that it can be taken as zero, then

$$V = V_D + IR_L \quad (31.3)$$

where, V_D is voltage drop across the diode, I the current through the circuit and R_L the load resistance.

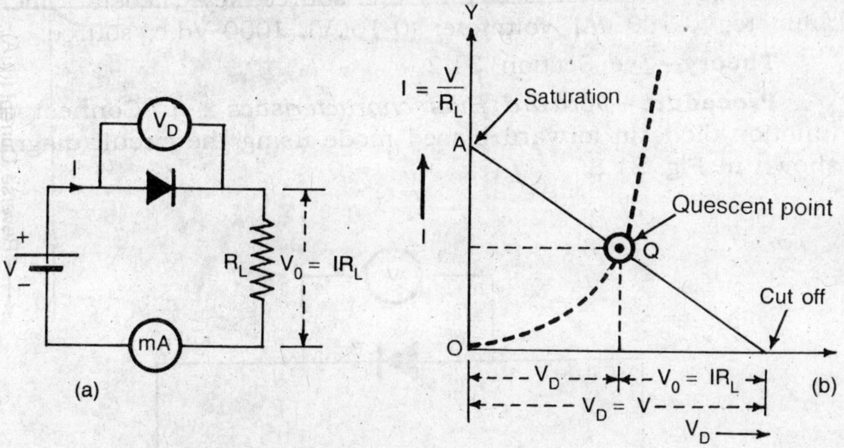

Fig. 31.7

From Eq. (31.3) we have :

$$I = -\frac{1}{R_L}V_D + \frac{V}{R_L} \quad (31.4)$$

This is of the form $y = mx + c$, the equation of a straight line. Hence, a graph plotted between I (along Y-axis) and V_D (along X-axis) is a straight line [(Fig. 31.7 (b)].

If $V_D = 0$, then $I = \frac{V}{R_L}$. This gives the point A on the I (or Y) axis and represents the maximum possible value of current, called saturation current, in the circuit. Thus, $I_{sat} = \frac{V}{R_L}$.

If $I = 0$, then $V_D = V$. This gives the point B on the V_D (or X-) axis and represents the maximum possible value of V_D corresponding to zero current called the *cut-off voltage*.

The straight line joining AB is called the *load-line*. The point where the forward bias characteristic curve intersects the load line gives the *operating* (or *quiescent*) point *Q*.

Experiment 31.1

Aim—To draw the *V-I* characteristics of a forward and reverse biased *p-n* junction diode and to draw the load line.

Apparatus—A junction diode (1N270 or 1N3605), milliammeter (0-500 *mA*), voltmeter (0-3V), 5V d.c. source, key, rheostat, micro-ammeter (0-100 μA), voltmeter (0-100V), 1000 Vd.c. source.

Theory—See Section 31.2.

Procedure—*Forward Bias characteristics* : 1. Connect the junction diode in forward biased mode using the circuit diagram shown in Fig. 31.8.

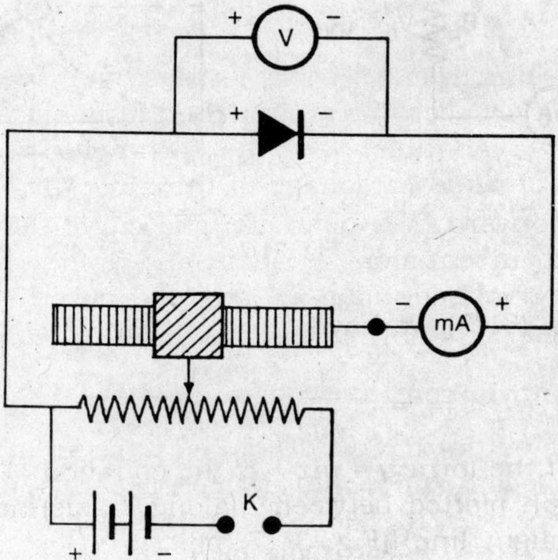

Fig. 31.8

2. Adjust the position of the variable contact of the rheostat so that initially the voltmeter shows zero. Now, increase the voltage in steps of 0.1V and record the voltmeter and milliammeter readings in the observation table.

Reverse Bias characteristics :

3. Connect the junction diode in reverse bias mode using the circuit diagram shown in Fig. 31.9.

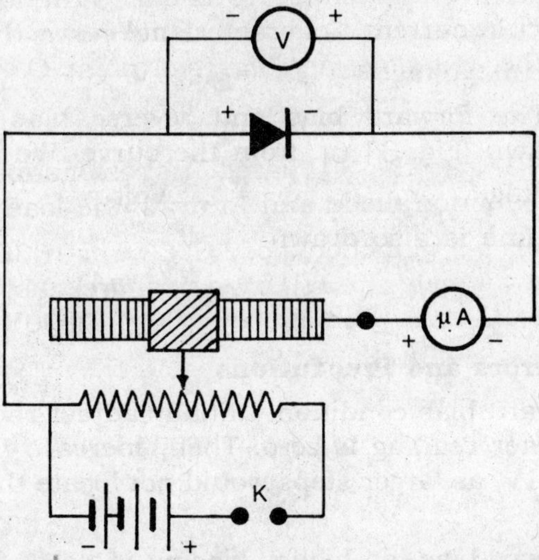

Fig. 31.9

4. Starting from zero volt, increase the reverse bias potential in steps of 2V. Record the voltmeter and micro-ammeter readings in the observation table. Normally, the applied reverse bias should not exceed breakdown voltage of the diode.

Draw forward bias characteristic curve taking voltage along +*X*-axis and current along + *Y*-axis, and reverse bias characteristic curve taking voltage along –*X*-axis and current along -*Y*-axis.

Load line : 5. Short circuit the diode (by connecting two ends of the diode by a connecting wire) so that $I_{sat} = \frac{V}{R_L}$ and measure the current in the circuit. This gives the print A, as discussed earlier. To locate the point B, measure the voltage across the source in open circuit. Join the points A and B on the forward bias characteristic curve. This gives the load line and the point of interaction gives its Q-point.

Observations

Sr No.	Forward Bias		Reverse Bias	
	Voltage (*V*)	Current (*mA*)	Voltage (*V*)	Current (μA)
1.				
2.				
.				
.				
.				
.				

Short circuit current I_{sat} = ...*mA*

Open circuit voltage V_{oc} = ...V

Result—The forward bias and reverse bias characteristic curves are drawn (Fig. 31.6). from the curve. We find :

cut-in voltage = ...Volt

The load line is also drawn.

At *Q*-point : I = ...*mA*

V = ...volt

Sources of Errors and Precautions

1. In forward bias condition, initially adjust rheostat position so that voltmeter reading is zero. Then, increase the voltages in the steps of 0.1V, as larger steps would not locate the graph in the knee region.

2. In reverse bias, as higher biasing is used, initially adjust the voltage at 0.5 volt then increase the voltage in steps of 2V.

3. Unless permitted by your teacher, do not increase the reverse bias so much that breakdown occurs as it may damage the diode.

Exercise—Use the forward bias characteristic curve to estimate the dynamic resistance R_{ac} of the junction diode (in the linear region of the curve), using the relation :

$$R_{ac} = \frac{\Delta V}{\Delta I}$$

Also, calculate d.c. resistance at the *Q*-point using the relation

$$R_{dc} = \frac{V}{I}$$

31.2 Band Gap by Reverse Saturation Method

The current I through a p-n junction diode, when a potential V is applied across the junction, is given by

$$I = I_0 \left[\exp\left(\frac{eV}{kT} - 1\right) \right] \quad \ldots \quad (31.5)$$

where e is the electronic charge, k the Boltzmann constant and T is the temperature. V is positive when the applied voltage biases the junction in the forward direction and V is negative when biasing is reverse.

When forward bias voltage is applied such that $V > \frac{kT}{e}$,

$$\exp\left(\frac{eV}{kT}\right) >> 1 \text{ and } I \approx I_o \exp\left(\frac{eV}{kT}\right)$$

When reverse bias voltage is applied such that

$$V > \frac{kT}{e}, \left(-\frac{eV}{kT}\right) << 1,$$

$$I = -I_o \qquad \ldots \qquad (31.6)$$

This current is known as *reverse saturation current* which remains constant at a small value until the junction breaks down. This current flows due to thermally generated minority carriers.

It can be shown that in terms of parameters of semiconductors,

$$ln\, I_o = ln\left[-AeN_nN_p\left(\frac{v_n}{p_p} + \frac{v_p}{n_n}\right)\right] + \left(\frac{-Eg}{kT}\right) \qquad \ldots \qquad (31.7)$$

where, A = area of the junction, v_n = Velocity of electron

v_p = velocity of hole, P_p = hole density in the p-region

n_n = electron density in the n-region, Eg = band gap energy over a small temperature range, I_o depends only on the term containing T and the first term in Eq. (31.7) can be treated as constant. Therefore, Eq. (31.7) can be expressed as

$$2.303 \log_{10} I_o = \text{Constant} - \frac{Eg}{kT}$$

If I_o is in microamperes and Eg in eV, then

$$2.303 \log_{10} (I_o \times 10^{-6}) = \text{Cost} - \frac{Eg \times 1.6 \times 10^{-19}}{kT}$$

Taking $k = 1.38 \times 10^{-23}$ J/T , we have :

$$\log^{10} I_o = \text{Const.} \frac{Eg \times 1.6 \times 10^{-19}}{2.303 \times 1.38 \times 10^{-23}\, T}$$

$$= \text{Const.} - 5.036 \times E_g \times \frac{10^3}{T} \qquad \ldots \qquad (31.8)$$

If a graph is plotted between log I_o and $\frac{10^3}{T}$, its slope $\left(= \frac{\log_{10} I_o}{10^3 / T}\right)$ is given by

$$\text{slope} = -5.306 \times Eg$$

$$\therefore \text{ Band gap of the semiconductor } Eg = \frac{\text{Slope of the line}}{5.036}$$

Experiment 31.2

Aim : To study variations of reverse current through PN junction and estimate the band gap of a semiconductor.

Apparatus : A junction diode (e.g. DR25-germanium), DC power supply, a small controlled heat oven with temperature indicator (or thermometer), nanometer, connecting wires, etc.

Theory : See Section 31.2.

Procedure : 1. Connect the junction diode in reverse bias mode (Fig. 31.10) to a DC source of 5V. Sometimes, adjustment in the value of bias voltage may be required so that at room temperature reverse current is 10 μ A.

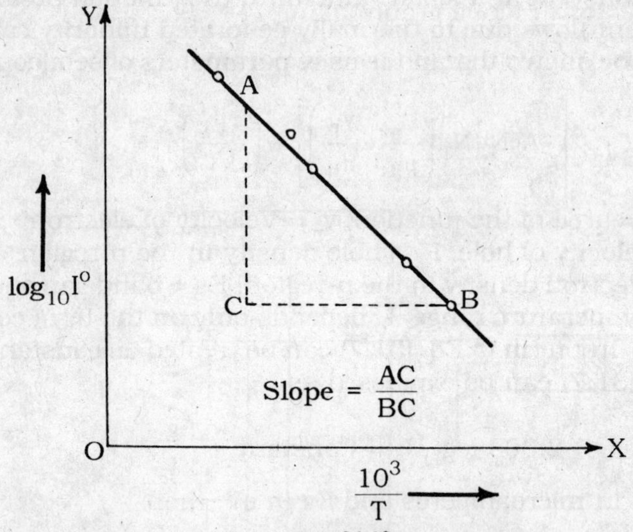

Fig. 31.10

2. Keep the ammeter in the 100 microampere range.

3. Switch on the heat oven and adjust it so that the temperature increases slowly. Slower is the rise of temperature better will be your observations as if heated rapidly the temperature of the sample may not be the same as that in the indicator due to thermal inertia.

Do not exceed the temperature above 70°C. For this purpose, if your oven is automated set the temperature at 65°C in cool/fixed position.

4. Note down the temperature and corresponding current at an interval of 5°C. If needed, the range of ammeter should be adjusted.

You may repeat your observation while the oven is cooling. The cooling process should also be slow by adjusting cooling control or switching off the oven and allowing the oven to cool on its own.

Observations

Sr. No.	Temp. (^oC)	Temp. T (K)	Current (I_o)	$\frac{10^3}{T}$	$Log_{10} I_o$

Calculations : Plot a graph taking $\frac{10^3}{T}$ along x-axis and $\log_{10} I_o$ along y-axis. A typical graph is shown in Fig. 31.11. Find the slope of the graph.

$$\text{Band Gap Eg} = \frac{\text{Slope}}{5.036} \quad \ldots \quad \text{eV}$$

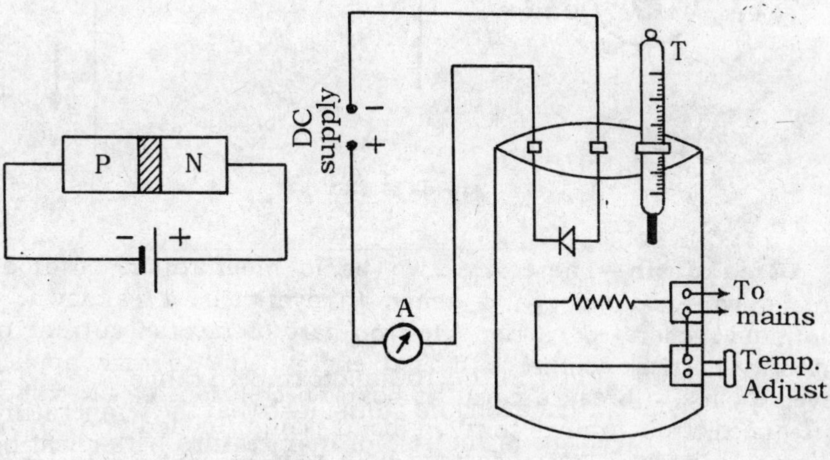

Fig. 31.11

Result : The energy band gap of Ge Crystal is ... eV.
Standard value of Eg = ... eV
Percentage Error = ... %

Sources of Errors and Precautions

1. It should be ensured that the junction diode is reverse biased.
2. Both heating and Cooling processes should be slow.
3. The ammeter should be kept in the proper range.
4. The value of temperature and current should be recorded simultaneously.

31.3 Zener Diode

If the reverse bias voltage across a *p-n* junction diode is increased, at a particular voltage called *breakdown voltage*, the (reverse) current suddenly increases to a large value which may even damage the diode. The *zener diode is specially made junction diode for use in reverse breakdown region which can be used continuously without being damaged.* The breakdown voltage in a zener diode is called the *zener voltage*. A zener diode is represented by the symbol shown in Fig. 31.12 (a) and its characteristic curve is shown in Fig. 31.12 (b).

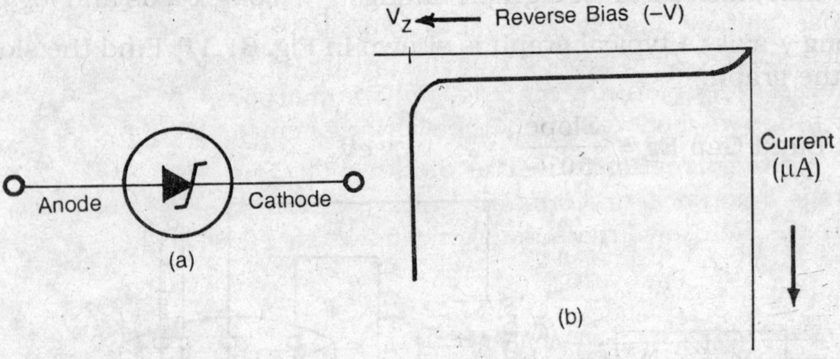

Fig. 31.12

Breakdown—There are two basic mechanism of diode breakdown—*avalanche* and *zener*. In avalanche breakdown, a thermally generated carrier (electron-part of reverse current or minority carrier) acquire sufficient energy, due to high applied reverse bias, to break a covalent bond on collision, *i.e.*, to knock out an electron from a covalent bond. Thus, in addition to the original carrier, a new electron-hole pair is generated. These carriers also acquire sufficient energy due to applied field and cause further generation of electron-hole pairs due to collisions. If process goes on, avalanche multiplication or breakdown occurs.

In zener diode both p- and n-sides are heavily doped. Due to high dopant densities, in these diodes, the depletion junction width is small and therefore the junction electric field is extremely high. Due to strong force on the bound electrons (in covalent bonds) because of strong electric field, electrons break out from the covalent bonds, increasing the current enormously. The zener breakdown occurs at a field of about 2×10^7 V/m. This value of field is reached at a voltage below 6V for heavily doped diodes . For lightly doped diodes, the breakdown voltage is higher. Zener breakdown voltage decreases with increasing temperature because an increase in temperature increases the energy of the valence electrons and hence makes it easier for these electrons to escape from the covalent bonds.

At the zener breakdown, there is large increase in current and the voltage the junction diode remains constant over a fairly large variation of current. Due to this reason, zener diode is used as *voltage regulator* to provide a constant voltage from a source whose voltage may vary over a sufficient range. The value of zener voltage depends on the (i) nature of the semiconducting material, (ii) amount of doping, (iii) the method of fabricating the *p-n* junction diode and (iv) the working temperature.

Voltage regulation—The ability of a zener diode to act as voltage reference or constant voltage device can be understood with the help of circuit diagram shown in Fig. 31.11.

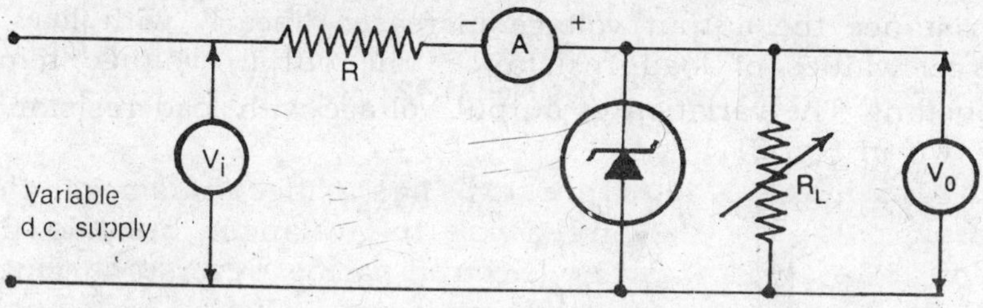

Fig. 31.13

When the input voltage is less than the zener breakdown voltage, almost no current (only reverse saturation current) flows through the zener diode and the (approximately) entire input voltage appears across the load resistance. Thus, till the breakdown occurs, the output voltage increases linearly with the input voltage. However, the value of R is so chosen that the diode operates in the breakdown region. When the diode operates in breakdown, $V_z = V_o$ *i.e.*, the output voltage across the load is equal to the zener voltage. After zener breakdown any change in the input voltage appears across the resistance R and the voltage across the diode and hence the load resistance remains practically equal to the zener voltage V_z. The voltage regulation curve of zener diode is shown in Fig. 31.14.

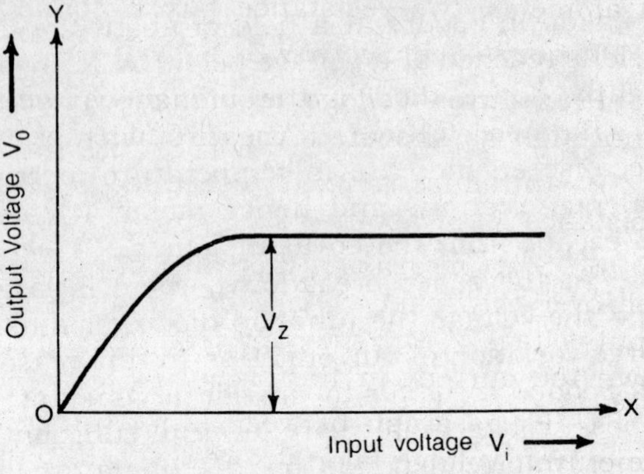

Fig. 31.14

To study the voltage regulation property of zener diode with load resistance R_L, the input source is maintained at a voltage greater than the zener voltage of the diode. On slowly increasing the load resistance it is found that for small values of load resistance the output voltage increases linearly with load. For larger values of load resistance, the output voltage remains constant. The variation of output voltage with load resistance is shown in Fig. 31.15.

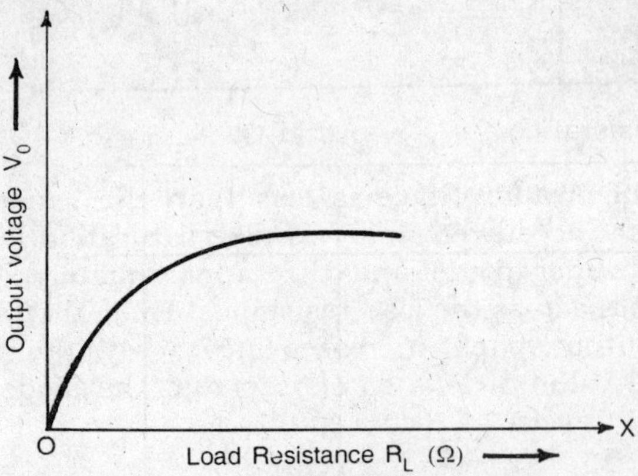

Fig. 31.15

Experiment 31.2

Aim—To study the voltage regulation characteristics of a zener diode.

Apparatus—A variable d.c. supply, a zener diode, a voltmeter, an ammeter, two resistance boxes, connecting wires, etc.

Theory—See Section 31.3.

Procedure—*Voltage regulation characteristic* : 1. Make the connection as shown in the circuit diagram in Fig. 31.13.

2. Introduce a fixed resistance R (100 to 500Ω) and load resistance R_L of about 1 kΩ.

3. Start increasing the value of input voltage in small steps, well over the zener voltage of the given diode. Note down each input voltage V_i and the corresponding output voltage V_o. Also note down the current in the circuit.

4. Plot a graph between input voltage V_i (along *X*-axis) and the output voltage (along *Y*-axis) (Fig. 31.14).

5. Determine zener voltage V_z from the drawn curve.

Voltage regulation with changing load : 6. Keep the input voltage over the zener voltage and the value of *R* same as earlier. Starting from a small value of about 50 Ω increase the value of R_L in steps of 50 Ω and note down the value of R_L and the corresponding value of the output voltage.

7. Draw a graph between R_L (along +*X*-axis) and V_o (along +*Y*-axis) (Fig. 31.15).

8. Determine minimum load resistance for voltage regulation, i.e., minimum load resistance for which output voltage becomes constant.

Observation

Resistance = ...Ohm

Load resistance = ...Ohm

Sr. No.	Input voltage V_i (Volts)	Current (*mA*)	Output voltage V_0 (Volts)
1	0.0V		
2	0.5V		
3	1.0V		
.	.		
.	.		
.	.		

Variation of output voltage with load resistance

Resistance = ...Ohm

Input voltage (fixed) = ...Volt

Sr. No.	Load resistance R_L (Ohm)	Output voltage V_o (Volt)
1	50V	
2	100V	
3	150V	
.	.	
.	.	
.	.	
.	.	

Result—1. The drawn regulation curves are shown in Figs.31.14 and 31.15. It is evident from the graph between load resistance R_L and the output voltage V_o that the output voltage increases linearly with load but for large values of R_L, the output voltage remains constant.

2. The zener diode of the given diode, V_z = ...Volt.

3. The minimum load resistance for voltage regulation = ...Ohm.

Sources of Errors and Precautions

1. The zener diode must be connected in reverse bias.

2. A constant finite value of resistance R must be there in the circuit to limit the value of current at breakdown.

3. The input voltage should be increased from zero in small steps well above the zener breakdown voltage.

4. To study the variation of output voltage with resistance, the input voltage must be fixed above the zener voltage.

31.4 Photovoltaic Effect : Solar cell

The barrier potential at the junction of a *p-n* junction diode causes the energy level of the *p*-side to be displaced upward and those of *n*-side downward. Fig. 31.16 shows the energy band diagram of the diode under the equilibrium in the absence of any externally applied electric voltage.

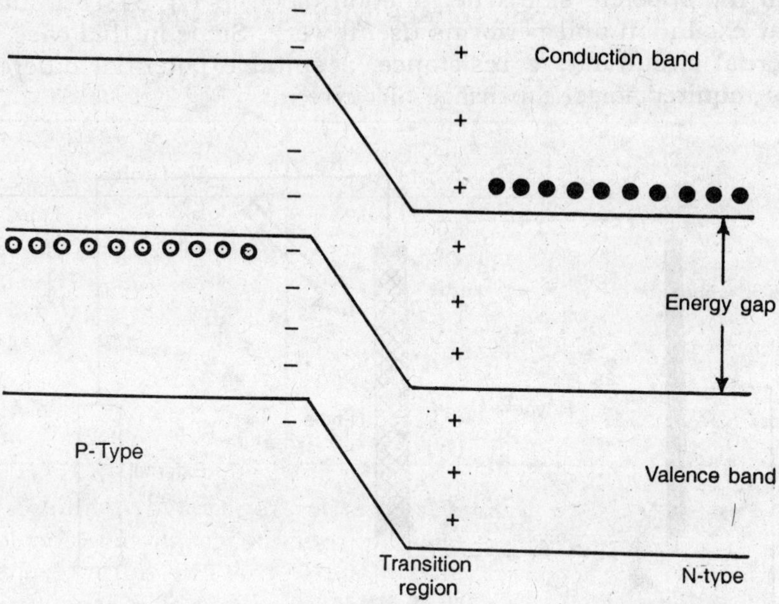

Fig. 31.16

When photons of solar radiation with energy greater than the forbidden energy gap E_g strikes a *p-n* junction, electrons in the valence band can acquire enough energy to jump to the conduction band, thus producing electron-hole pairs. The action of the junction is to separate the pairs, electrons pass down the potential hill into *n*-side, whereas the holes go up the hill to the p-side resulting in the reduction of the height of the potential hill or barrier. A new equilibrium condition would be finally reached across the junction with the appearance of a potential difference. This is called the open circuit voltage V_{oc}, which is the function of intensity of the incident radiation.

If the junction is connected by an external wire with negligible resistance (*i.e.*, it is shorted), a definite (non-zero) current is obtained without any applied voltage. This is due to separation of solar-generated electron-hole pairs, going across the potential hill. This current called *short-circuit current* is a function of the intensity of the incident light.

To make use of the solar generated electron-hole pairs, one has to lead the electrons from the *n*-side, take them through an external circuit to do useful work, and then return them to do the *p*-side, where they can meet and recombine with the holes coming from the opposite side. The junction diode in Fig 31.15 is under solar excitation and performs useful work. Since in this case, the external circuit has a resistance, a suitable potential difference V is required for the passage of current.

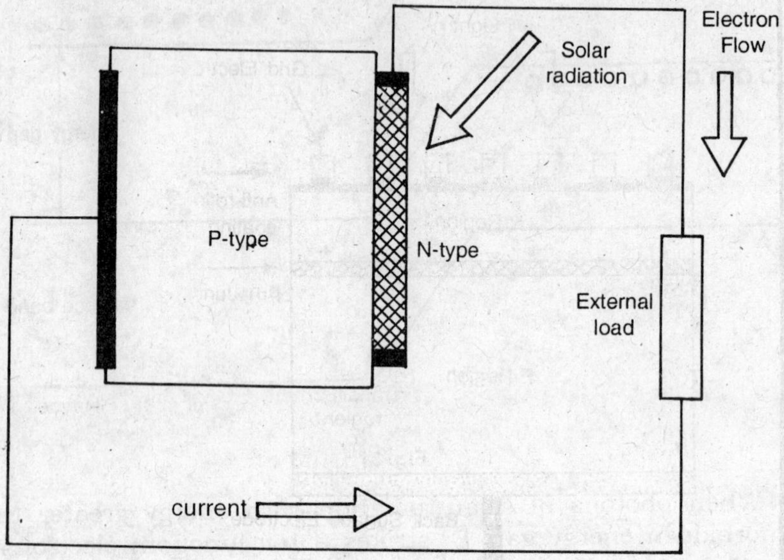

Fig. 31.17

Solar Cell—*A solar cell is a photovoltaic device used to convert sunlight into electrical energy.* In diodes to be used as solar cells, one of the regions (p or n) is made very thin so that light can reach junction without being appreciably absorbed. Incident light produces charge carriers, electron-hole pairs, which eventually decrease the potential barrier, as discussed above. This decrease, does not affect the minority current (since these particles fall down the barrier), but when the potential hill is sufficiently low, some majority carriers cross the junction. These carriers correspond to a forward current and hence such a flow will reduce the net (reverse) current. When the majority current equals the minority current, the total current is reduced to zero. The voltage at which zero resultant current is obtained is called the *photovoltaic potential.* This potential is about 0.5V for silicon diode and 0.1V for germanium.

Performance of Solar Cell—A solar cell is a *p-n* junction diode in the configuration shown in Fig. 31.18. Silicon is the most widely utilized basic material because of its suitability and

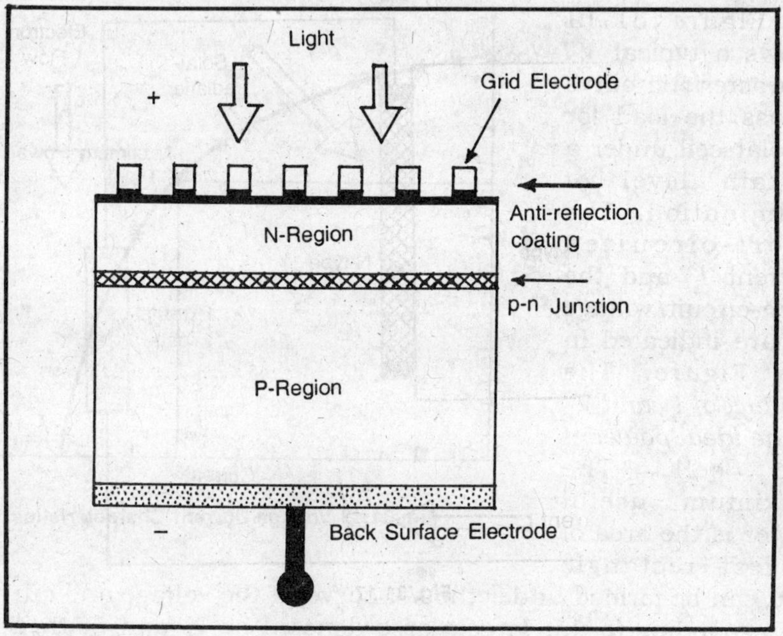

Fig. 31.18 Physical configuration of the solar cell

its abundance on the earth. In the melt of basic material, boron is added in an induction furnace to convert it into a *p*-type crystal. The *p*-type crystal is then placed in a diffusion furnace containing a gaseous *n*-type dopant such as phosphorus, allowing the *n*-dopant to diffuse into the surface and thus forming a *p-n* junction diode.

The conversion (from light energy to electrical energy) efficiency of silicon solar cell ranges between 10 to 20%. This small efficiency is mainly due to the fact that the cell respond to only a portion of the wavelengths available in the solar spectrum. The photon with wavelengths greater than 1.11 μm do not have sufficient energy to create electron-hole pairs in a silicon cell. Also the excess energy (over E_g) of a photon does not contribute to the pair (or current) production. Moreover, the incident light energy not used in pair production heat the cell. The increased thermal energy of carriers enable them to cross the potential barrier and pass the junction in the wrong way, thereby reducing the efficiency. Therefore, it is important to keep the cell temperature down by reflecting away the unwanted portion of the incoming spectrum and by quickly transferring the waste heat out.

Figure 31.19 shows a typical *V-I* characteristic curve across the load for a solar cell under a certain level of illumination. The short-circuited current I_{sc} and the open-circuit voltage V_{oc} are indicated in the Figure. The *product of* I_{sc} *and* V_{oc} *is the ideal power* of the cell. The maximum useful power is the area of largest rectangle that can be formed under the V-I curve. If the voltage and current corresponding to this largest area denoted by V_m and I_m, then the maximum useful power is $V_m I_m$. The ratio of useful power to the ideal power is called the *fill factor*. Thus,

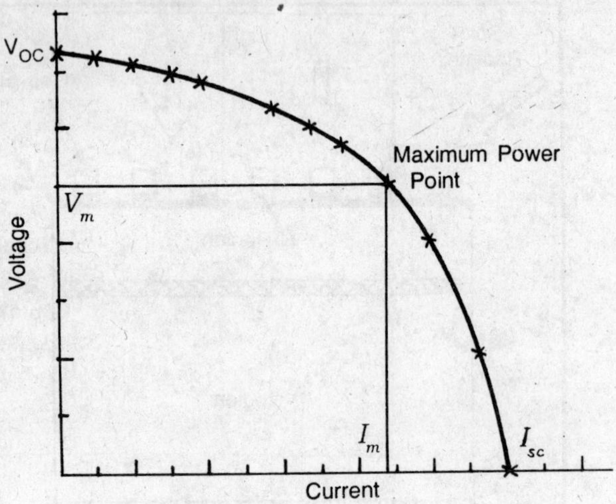

Fig. 31.19 Voltage-Current Characteristics

$$\text{Fill factor} = \frac{\text{Maximum useful power}}{\text{Ideal power}} = \frac{V_m I_m}{V_{oc} I_{sc}} \qquad (31.9)$$

Typical values for a silicon based solar cell :

V_{oc} — 450 to 600 mV

I_{sc} — 30 to 40 mA/cm^2.

Fill factor — 0.65 to 0.80.

Experiment 31.3

Aim—To study the characteristics of a solar cell.

Apparatus—Solar cell enclosure, solar simulator current source (0 to 5A), d.c. microammeter, d.c. voltmeter, a resistance box, connecting wires, etc.

Theory—See Section 31.4.

Procedure—1. Make the connection as shown in the circuit diagram of Fig. 31.20.

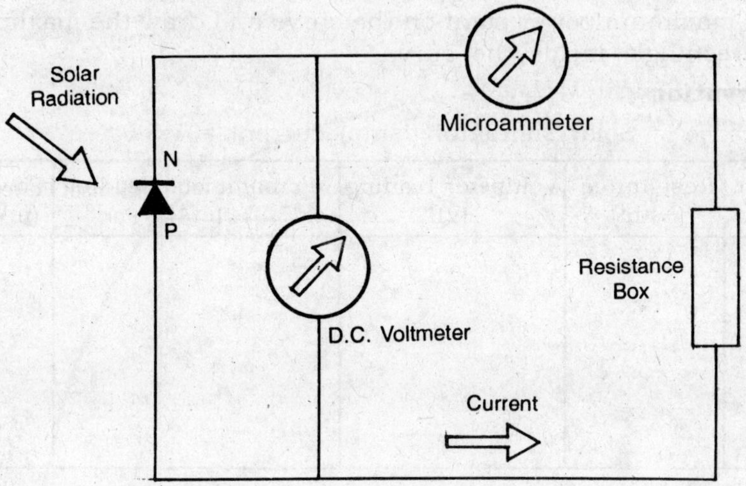

Fig. 31.20

2. Keep the d.c. microammeter in 100 mA range and d.c. voltmeter in 2V range (later in 20V range if the voltage exceeds 2V).

3. Initially set zero resistance in the resistance box.

4. Switch on the d.c. micrommeter, d.c. millivoltmeter and the lamp driver, *i.e.*, solar simulator current source.

5. Adjust the solar simulator current at 1.5A.

6. Read the current in the microammeter and the corresponding value of the voltage in the millivoltmeter. The voltage under the zero resistance condition should be zero. However, resistance may not be zero, due to the presence of various bulk and contact resistance.

7. Record short circuit current, *i.e.*, current under zero resistance condition.

8. Increase the resistance in such steps that there is gradual decrease in the value of the current.

9. Finally, record the current for infinite resistance, *i.e.*, under open circuit condition. Also, record the open circuit voltage which is the maximum voltage.

10. Repeat the experiment for other values of the lamp currents—2A, 2.5A and 3A.

11. Draw I-V characteristics for different values of lamp current, taking current along +*X*-axis and voltage along +*Y*-axis. Make maximum power point on the curve and draw the maximum area rectangle inside the curve.

Observation

Solar simulator (lamp) current = 1.5A

Sr. No.	Resistance (Ohm)	Voltmeter reading (V)	Microammeter reading (mA)	Power (mW)
1.				
2.				
.				
.				
.				
.				

Calculations

I_{sc} = V_{oc} =

I_m = V_m =

$$\text{Fill factor} = \frac{V_m I_m}{V_{oc} I_{sc}}$$

Find fill factor for different simulation current and find the mean.

Precautions

1. Correction for zero error in voltmeter and ammeter, if any, should be taken.

2. Near the short circuit current and open circuit voltage, change the value of the resistance in such a way that there is gradual change in the voltage or current, so that more points are obtained for plotting the curve.

3. Normally, the simulation current should not exceed 4A.

4. The light should fall normally on the cell.

Exercises

1. Draw a graph to show the variation of power with resistance taking R along +*X*-axis and *P* along +*Y*-axis.

2. Draw a graph showing the variation of resistance with voltage taking *R* along +*X*-axis and *V* along +*Y*-axis.

ORAL QUESTIONS

1. What is a *p-n* junction diode?
2. How is a junction diode (i) forward and (ii) reverse biased?
3. What is depletion region? What is depleted in this region?
4. What is potential barrier? What is its value in a silicon diode and germanium diode?
5. What is the order of magnitude of the electric field in the junction region?
6. What is knee voltage? Write its value for a Ge diode and Si diode?
7. Write an expression for the variation of current I with voltage V for a diode.
8. Show the forward bias and reverse bias characteristics of a *p-n* junction diode.
9. What is reverse saturation current? Explain.
10. Explain the term breakdown voltage.
11. What is load line? Define the terms saturation current and cut-off voltage.
12. Differentiate between avalanche and zener breakdowns.
13. Name the factors which determine the value of zener voltage.
14. Explain how zener diode can be used to regulate voltage.
15. What is photovoltaic effect?
16. What is a solar cell? How does it differ from a usual junction diode.
17. Why is the efficiency of a solar cell low?
18. Define fill factor. What is desirable—a smaller fill factor or a larger fill factor? Why?
19. Define the terms short circuit current and open circuit voltage for a solar cell.
20. How does the value of solar simulation current affect the experimental condition?

32

Transistors

32.1 The Junction Transistor

In 1948, John Bardeen, William Shockley and Waltair Brattain discovered a new semiconductor device called *transistor* which revolutionized the field of electronics. They were jointly awarded Nobel Prize in Physics for their discovery. A transistor is also called *Bipolar Junction Transistor* (BJT) and semiconductor triode. It is called bipolar since current is carried by two types of charge carriers, *i.e.*, electrons and holes.

The word transistor is contraction of words transfer and resistor, *i.e.*,

Transistor = Transfer + Resistor

and is indicative of its basic operation, as explained in the following discussions. The transistor is capable of producing *amplification* and this is the basis of its large number of applications.

The BJTs have almost completely replaced vacuum triodes due to the following advantages : (i) small size, (ii) no warm up time required, (iii) no power required for heating filament as in tubes, (iv) mechanically rugged, (v) light weight, (vi) shock proof, (vii) longer life, etc.

Construction—A junction transistor is fabricated by sandwiching a thin layer of a *p*-type semiconductor between two layers of *n*-type semiconductor for an *n-p-n* transistor and by sandwiching a thin layer of a *n*-type semiconductor between two layers of *p*-type semiconductors for a *p-n-p* transistor. The semiconductor sandwitch is extremely small and is hermatically sealed against moisture inside a plastic or metallic case.

In Fig. 32.1 (a), an *n-p-n* transistor is shown while *p-n-p* transistor is shown in Fig. 32.1 (b). The middle layer called *base*, is *very thin* and is *lightly doped* as compared to the other two layers. One of the outer layers is *heavily doped* and is called *emitter*. The other outer layer is moderately doped and is called *collector*. Usually, the emitter-base contact area is smaller than the collector base contact area. Terminals are taken out from the emitter, base and collector for external connections. Thus, a transistor is a three-terminal device. In the symbolic representation, the direction of arrow represents the direction of flow of conventional current which is opposite to the direction of motion of electrons. It may be noted that the direction of the arrow in a p-n-p transistor is opposite to that of an *n-p-n* transistor. In an *n-p-n* transistor, electrons are majority carriers and current is largely carried by electrons while in a *p-n-p* transistor, holes are majority carriers.

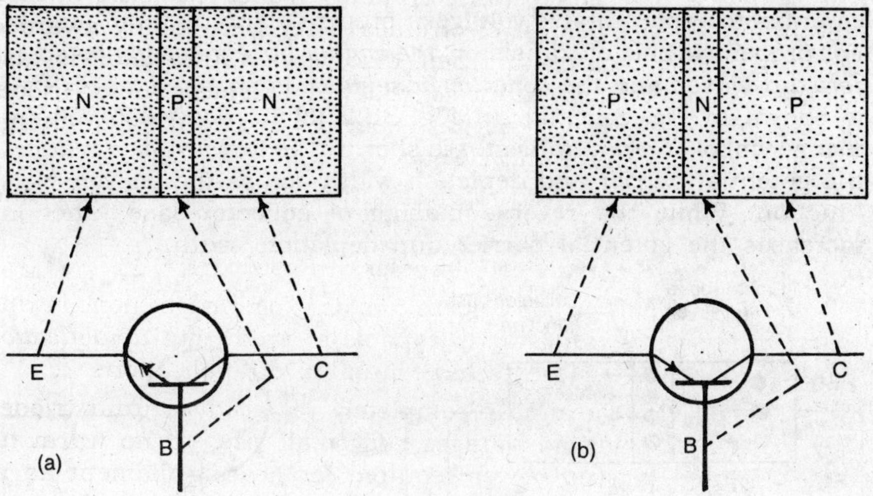

Fig. 32.1

The function of the emitter is to emit or eject the majority carriers into the base. To achieve this, *the emitter is heavily doped compared to the collector and the base.* The base is very thin, of the order of 10^{-3} cm, and very lightly doped. So, most of the charge carriers injected into it by the emitter are passed on to the collector. The function of the collector is to collect the charge carriers from the base. The doping level of this region is intermediate between that of the emitter and the base. *Apart from the differences in the level of doping, the emitter and collector have different sizes. The collector is thicker than the emitter so as to be able to collect the charge carriers and dissipate heat. Therefore, the emitter and the collector cannot be interchanged.*

A transistor has two *p-n* junctions, one between the emitter and the base, the emitter-base junction, and the other between the collector and the base, the collector-base junction. At the depletion layer, the contact potential is about 0.3 V in a Ge transistor and 0.7 V in a *Si* transistor at 25°C. Silicon transistors are most widely used because they have higher voltage and current ratings and are less temperature sensitive, *i.e.*, less temperature induced noises.

32.2 Working of Transistor

N-P-N Transistor—For the working of a transistor, it is essential to apply suitable voltage or proper polarity to its terminals. For proper working of transistor, *the emitter base junction should be forward biased* and the *collector-base junction reverse biased.* This is shown in Fig. 32.2 (a). In Fig. 32.2 (b) its equivalent circuit using symbol of *n-p-n* transistor is shown. The forward bias reduces the potential barrier and depletion width across the emitter-base junction. While the reverse biasing of collector-base junction increases the potential barrier and depletion width.

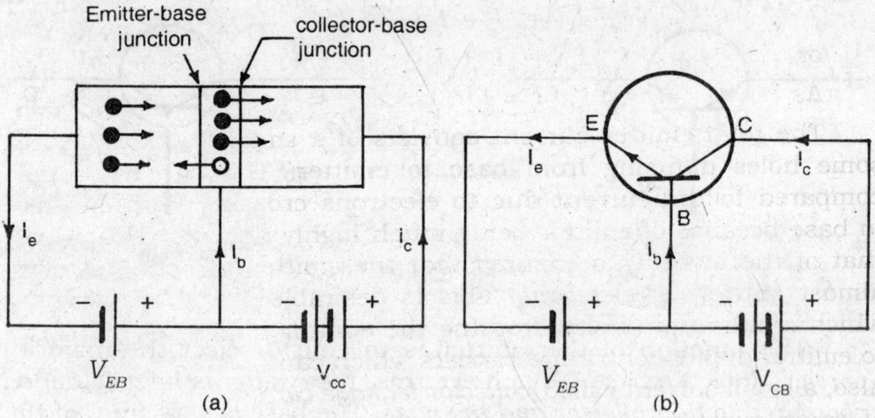

Fig. 32.2

Electrons in the emitter are repelled by the negative bias V_{EB} and the forward bias of the emitter-base junction helps the electrons in injecting into the *p*-type base region.. Since the base is thin and lightly doped, only a small fraction (upto 5%) of the electrons recombine with the holes in this region and the remaining electrons diffuse into the collector. The positive bias of the collector pulls these electrons into the collector. The electrons going from the battery V_{EB} to the emitter constitute the emitter current I_e, in the opposite direction. Some of the electrons entering the base from the emitter, recombine with the holes in the base and get neutralised. As the base region is very thin and lightly doped, the number of electrons recombining in this region is small. Moreover, due to a high positive potential at the collector, the electrons rush towards the collector. The electrons collected by the collector flow towards the *+ve* terminal of battery V_{CB}. The electrons going from the collector to the battery V_{CB} constitute the collector current I_e.

The deficiency of the electrons inside the emitter is made up by the electrons flowing from *-ve* terminal of the battery V_{EB}. Thus, the current is carried by the electrons both inside the transistor as well as in the external circuit. In the emitter-base external circuit the flow of electrons is from *P* to *N*. Hence, the conventional current is from *N* to *P*. In the collector-base circuit the flow of electrons in the external circuit is from *N* to *P*, resulting in conventional current I_c from P to N. The difference between I_e and I_c is the base current I_b which is in the same direction as that of I_c. Thus,

$$I_b = I_e - I_c$$

or $$I_e = I_b + I_c$$

As $$I_b << I_c, \quad I_e \approx I_c$$

The total emitter current consists of a small fraction due to some holes diffusing from base to emitter. This is very small compared to the current due to electrons crossing from emitter to base because of emitter being much highly doped compared to that of the base. This ensures that the emitter current consists almost entirely of electrons. This is desirable since the current which results from holes crossing the emitter junction from base to emitter do not contribute carriers which can reach the collector. Also, a small current called *collector leakage current* due to diffusion of some minority carriers (electrons) from the collector into the base.

P-N-P Transistor—In this transistor also the emitter-base junction is forward biased while the collector-base junction is reverse biased, as shown in Fig. 37.3. In this case, majority carriers are holes and they are repelled towards the base due to positive potential applied on the emitter and the forward bias make the movement of holes across the junction easier. The base is thin and lightly doped so it has comparatively low concentration (or number density) of electrons. Therefore, only a fraction of holes (upto 5%) coming from emitter recombine with electrons and rest of the holes cross-over to the collector region.

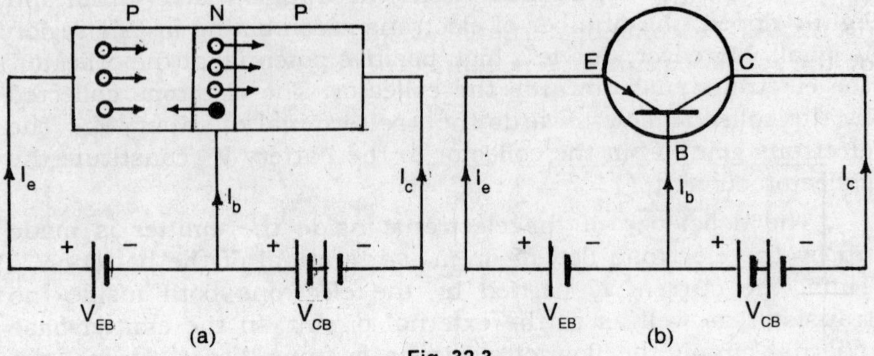

Fig. 32.3

The reverse bias across the collector junction helps these holes in crossing over into the collector region. As one hole reaches the collector, an electron comes from the negative terminal of the battery V_{CB} and combines with it. At the same time, an electron from some covalent bond is released, which moves to the positive terminal of the battery V_{EE} and a hole is created in the emitter. Thus, inside a *p-n-p* transistor current is carried by hole while electrons carry current in the external circuit.

In the emitter base external circuit, the flow of hole is from P (emitter) to *N* (base), the conventional current I_e flows from *N* to *P*, *i.e.*, into the emitter. At the collector, electrons enter into it from outside and, therefore, the conventional current I_c flows out of the collector. The difference between I_e and I_c is the base current I_b, thus,

$$I_b = I_e - I_c$$

or $$I_e = I_b + I_c$$

As $$I_b << I_c, \quad I_e \approx I_c$$

In this case also, the total emitter current consists of a small fraction due to some of the electrons diffusing from base to emitter. Also, a small current called *collector leakage current* due to diffusion of some holes from the collector into the base.

32.3 Transistor Configurations

In any electrical and electronic circuit, input and output require two terminals each. A transistor has only three terminals. Hence, one terminal of the transistor is made common to the input as well as the output side. The resulting configurations are : (i) common-emitter (CE), (ii) common-base (CB) and (iii) common-collector (CC). The three configurations are shown in Fig. 32.4. The common terminal is taken as the reference point for the whole circuit and is connected to ground.

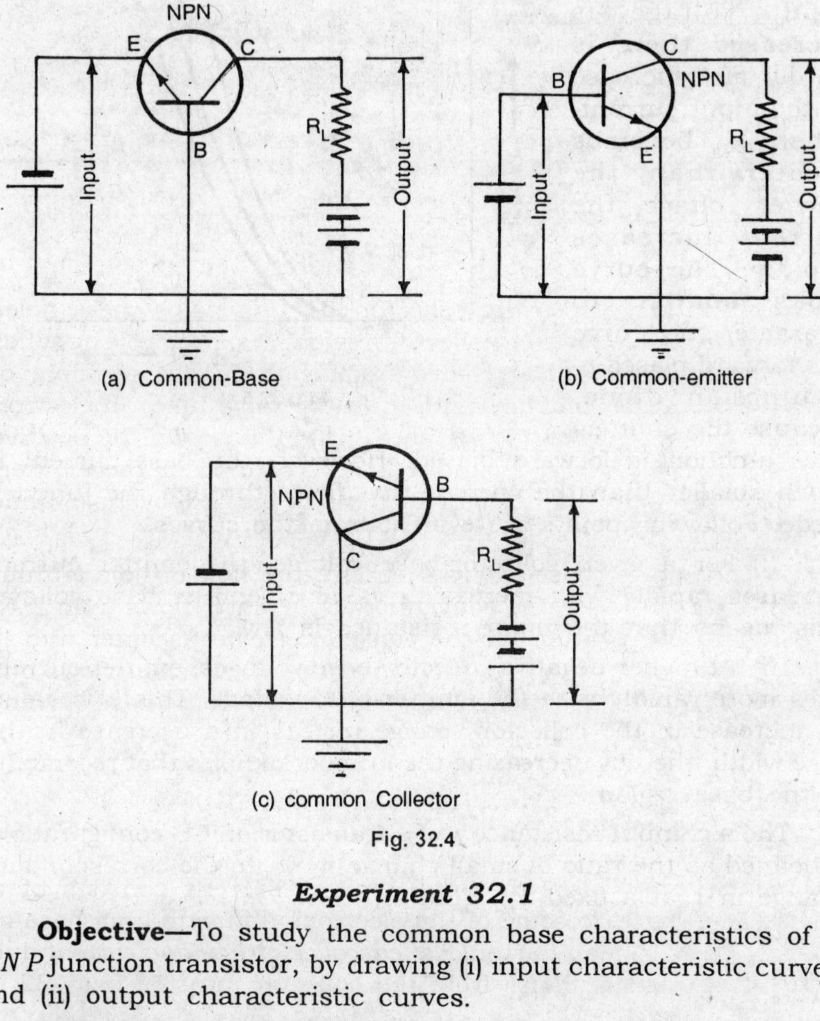

Fig. 32.4

Experiment 32.1

Objective—To study the common base characteristics of a *PNP* junction transistor, by drawing (i) input characteristic curves and (ii) output characteristic curves.

Apparatus—A P-N-P transistor, a voltmeter of range 0-3 V, a voltmeter of range 0-10 V, two ammeters of range 0-25 mA, a potentio meter of total resistance of the order of one mega ohm, a potentiometer of total resistance of the order of 25 kΩ, a 3V and a 9V battery, keys, connecting wires, etc.

Theory—*Input characteristic—These are curves showing the* *variation of input voltage or emitter-base voltage and input current or emitter current at different fixed values of collector-to emitter V_{cb} voltages.* A few typical input characteristic curves are shown in Fig. 32.5. Initially as input voltage is increased there is hardly any increase in the input current. When V_{eb} becomes greater than the barrier voltage, the current increases rapidly. The curve looks similar to characteristic curve of a forward biased *p-n* junction diode, because the emitter-base junction is forward biased. However, the base current is much smaller than the current that flows through the junction diode. Following points are evident from the curves :

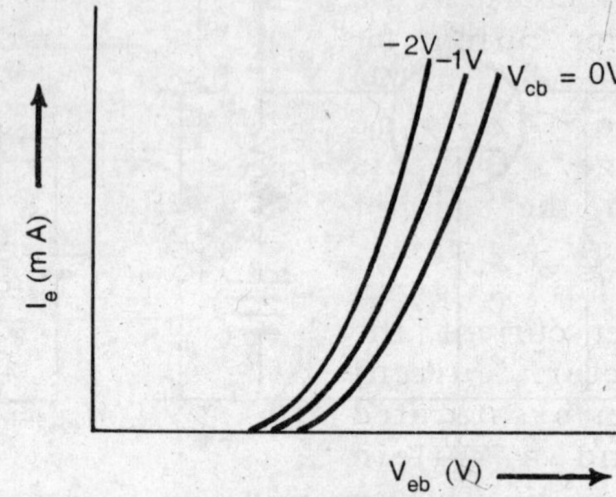

Fig. 32.5

(i) For a given collector-base voltage, the emitter current increases rapidly with increasing value of emitter-base voltage. This means that the input resistance is low.

(ii) At higher negative collector-base voltages, emitter current rises more rapidly with the emitter-base voltage. This is because an increase in the collector voltage results in a decrease in the base width, thereby decreasing the number of holes that recombine in the base region.

The a.c. input resistance r_i of a transistor, in CB configuration, is defined as the ratio of small change in V_{eb} to the corresponding change in I_e at a fixed V_{cb}. That is,

$$r_i = \left(\frac{\Delta V_{eb}}{\Delta I_e}\right)_{V_{cb}} \qquad (32.1)$$

Output characteristics—*These are curves representing the variation of output (or collector) current I_c with output (or collector-base) voltage V_{cb} at different fixed values of I_e.* A few typical output characteristic curves are shown in Fig. 32.6. The following points can be noted from the curves : (i) For a given value of emitter current, the collector current is not zero when collector base voltage is zero. (ii) For a given emitter current, there is rapid increase in the collector current for an increase in low collector-base voltage. In this region, the collector has low resistance. (iii) For a given emitter current, the collector current becomes saturated beyond a certain value of V_{cb}. This occurs when the collector voltage is sufficient high so that the collector collects almost all the majority carriers coming from the emitter-base region.

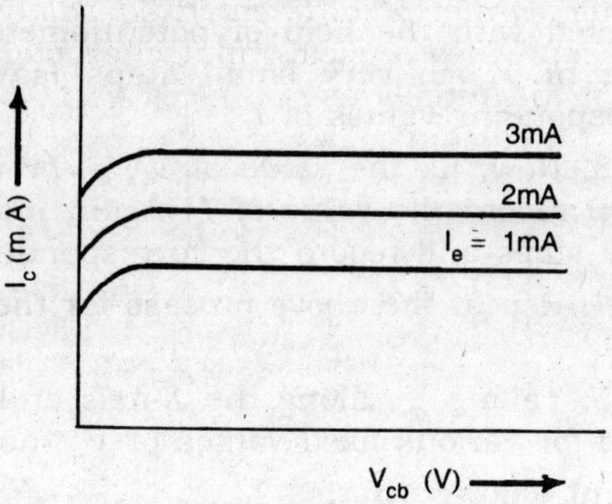

Fig. 32.6

The a.c. output resistance r_o is defined as the ratio of small change in V_{cb} to the corresponding change in I_c at a fixed value of I_e. That is,

$$r_o = \left(\frac{\Delta V_{cb}}{\Delta I_c}\right)_{I_e}$$

Procedure—1. As shown in the circuit diagram in Fig. 32.7, complete the circuit diagram.

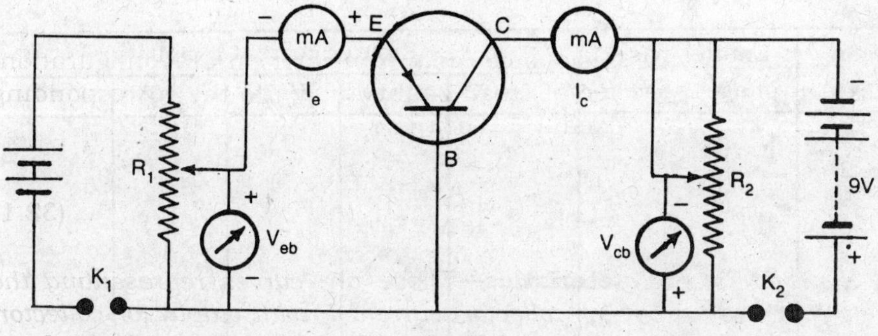

Fig. 32.7

Input characteristics

2. Adjust the value of V_{cb} = 0 with the help of potentiometer

R_2. Note down the value of I_e at V_{eb} = 0. The value of V_{eb} can be adjusted with the help of potentiometer R_1. Now, increase the value of V_{eb} in very small steps (say, 0.*1V*) and record the corresponding values of I_e.

3. Now, fix the value of V_{cb} = *1V* and keeping the value of V_{eb} = 0 record the value of I_e. Again increase the value of V_{eb} in small steps and record the corresponding values of I_e.

4. Repeat the above process for the fixed values of V_{cb} = *2V*, 3V,....

5. Taking V_{eb} along the *X*-axis and I_e along *Y*-axis, plot the curve for various fixed values of V_{cb}, as shown in Fig. 32.5.

Output characteristics

6. Adjust the value of I_e = 1 mA with the help of potentiometer R_1. Keeping the value of V_{cb} = 0 note down the value of I_c. Now, keeping I_e fixed, increase the value of V_{cb} in small steps (say, 0.1V) and note down the corresponding values of I_c.

7. Next, adjust the value of I_e = 2mA and keeping the value of V_{cb} = 0 note down the value of I_c. Now, keeping the value of I_e constant increase the value of V_{cb} in small steps and record the corresponding values of I_c.

8. Repeat the above process for fixed values of I_e = 3 *mA*, *4 mA*, ...

9. Taking V_{cb} along the *X*-axis and I_c along the *Y*-axis, plot the curves for various fixed values of I_e, as shown in Fig. 32.6

Observations—(i) *Input characteristics* : I_e Vs. V_{eb} at constant V_{cb}.

Sr. No.	Emitter-base voltage V_{eb} (volts)	Emitter current I_e (*mA*)				
		V_{cb}=0V	V_{cb}=1V	V_{cb}=2V	V_{cb}=3V	V_{cb}=4V
1						
2						
.						
.						
.						
.						
.						
.						
15						

(ii) *Output characteristics* : I_c Vs V_{cb} at constant I_e

Sr. No.	Collector-base voltage V_{cb}	Collector current I_c				
		I_e=1 mA	I_e=2mA	t_e = 3 mA	I_e = 4 mA	I_e=5 mA
1						
2						
.						
.						
.						
.						
.						
15						

Calculations—From the input and output characteristic curves calculate :

Input resistance r_i =

output resistance r_o =

Calculate output resistance separately in the two regions, i.e., linear region and saturated region.

Sources of Errors and Precautions

1. Wrong polarity connection may damage the transistor and waste your time. Ensure proper connection of all transistor terminals.

2. Apply correct biasing potential in each transistor.

3. Excessive collector-base voltage may cause transistor breakdown, do no exceed the safe voltage limit.

4. Do not pass a large emitter current, it can burn the transistor.

Exercise—*To study the common base characteristics of an N P N junction transistor.*

The biasing polarities of emitter-base junction and collector-base junction should be reversed with respect to that used in P-N-P transistor, so as to still keep the emitter-base junction forward biased and collector base junction reverse biased. Rest of the procedure remains identical.

Experiment 32.2

Objective—To study the common emitter characteristics of an *N P N* junction transistor by drawing (i) input and (ii) output characteristic curves.

Apparatus—Same as in Expt. 32.1 except that in this case *NPN* transistor and a microammeter (0-100 μA) for I_b measurement is to be used.

Theory— *Input characteristics* : These are the curves showing variation of input (or base-emitter) voltage V_{be} and input (or base) current I_b at different fixed values of collector-emitter V_{ce} voltages. A few typical input characteristic curves are shown in Fig. 32.8. From the curves, we can observe that as long as V_{be} is less than the barrier voltage input current is low and when V_{be} exceeds the barrier voltage, base current increases first slowly and then rapidly as in the case of a forward biased diode.

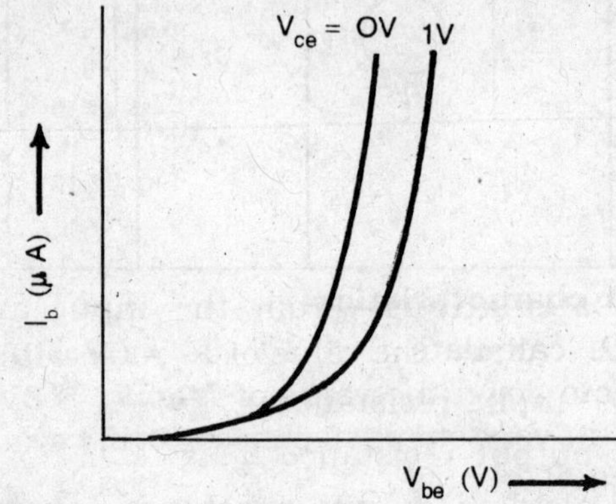

Fig. 32.8

Output characteristics : These are the curves showing the variation of output (or collector-emitter) voltage V_{ce} and output (or collector) current I_c at different fixed values of base current I_b. A few typical characteristic curves are shown in Fig. 32.9. From the curves, we can observe that the collector current increases very rapidly with increasing V_{ce} initially but soon it gets saturated and becomes almost independent of collector-emitter voltage. Further, for a given value of V_{ce}, I_c is large for larger value of I_b.

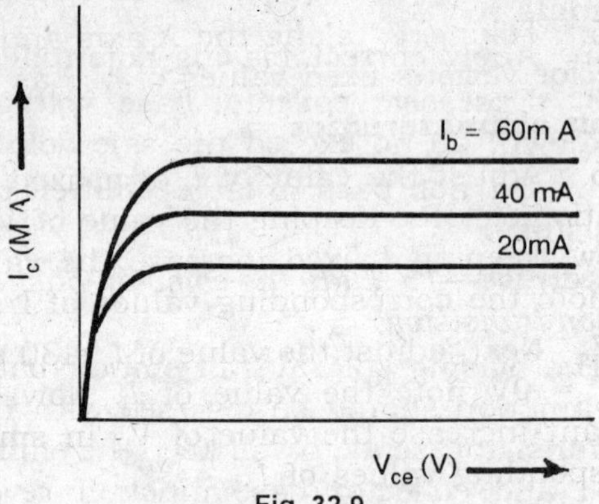

Fig. 32.9

Procedure—1. As shown in the circuit diagram in Fig. 32.10, complete the circuit.

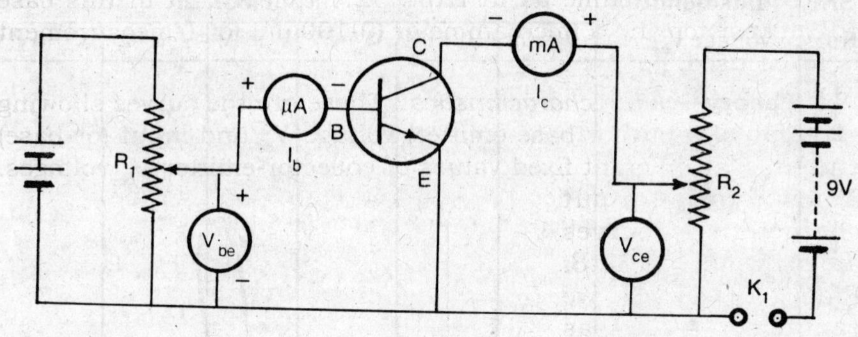

Fig. 32.10

Input characteristics

2. Adjust the value of V_{ce} = 0V with the help of potentiometer R_2. Note down the value of I_b at V_{be} = 0V. Now, increase the value of V_{be} in very small steps (say, 0.1V) and record the corresponding values of I_b.

3. Now, fix the value of V_{ce} = 1V and keeping the value of V_{be} = 0V record the value of I_b. Increase the value of V_{eb} in small steps and record the corresponding values of I_b.

4. Repeat the above process for the fixed values of V_{ce} = 2V, 3V,......

5. Taking V_{be} along the X-axis and I_b along Y-axis, plot the curve for various fixed values of V_{ce}, as shown in Fig. 32.8.

Output characteristics

6. Adjust the value of I_b at a fixed value, say, 20 μA, using potentiometer R_1. Keeping the value of V_{ce} = 0V note the value of I_c. Now, keeping I_B fixed increase the value of V_{ce} in small steps and note the corresponding values of I_c.

7. Next, adjust the value of I_b = 30 μA and keeping the value of V_{cb} = 0V note the value of I_c. Now, keeping the value of I_b constant increase the value of V_{ce} in small steps and record the corresponding values of I_c.

8. Repeat the above process for fixed values of I_b = 40 μA, 50 μA,...

9. Taking V_{ce} along the X-axis and I_c along the Y-axis, plot the curves for various fixed values of I_e, as shown in Fig. 32.9.

Observations— *Input characteristics* : I_b Vs V_{be} at constant V_{ce}

Sr. No.	Base-emitter voltage V_{be}	Input current I_b (μA)					
		V_{ce}=0v	V_{ce}=1V	V_{ce}=2V	V_{ce}=3V	V_{ce}=4V	V_{ce}=5V
1							
2							
3							
.							
.							
.							
.							
15							

Output Characteristics : I_c *Vs* V_{ce} at constant I_b

Sr No.	Collector-emitter voltage V_{ce} (*V*)	Output current I_c (mA)				
		I_b=2μA	I_b=4μA	I_b=6μA	I_b=8μA	I_b=10μA
1						
2						
3						
.						
.						
.						
.						
.						
15						

Calculations—From the input and output characteristic curves obtain :

input resistance r_i =

Output resistance r_o =

Sources of Errors and Precautions

Same as in Expt. 32.1.

Exercise—*To study the common emitter characteristics of a P N P junction transistor by drawing (i) input and (ii) output characteristic curves.*

The biasing polarities of the two junctions should be reverse compared to that used for *NPN* transistor. The remaining procedure is identical.

ORAL QUESTIONS

1. Why is a transistor so called?
2. Mention the advantages of a transistor over vacuum triode.
3. Why is emitter most heavily doped?
4. Are collector and emitter regions interchangable? Explain.
5. Which région is the most wide in a transistor? Why?
6. Why is base region thin and lightly doped?
7. In the symbolic representation of a transistor, what does the arrow indicate?
8. Why is the magnitude of base current much smaller than that of emitter and collector currents?
9. For normal operation of a transistor the emitter-base junction should be forward biased while the base-collector junction should be reverse biased. Why?
10. What is leakage current? Explain.
11. Define a.c. input and output resistance.
12. Explain the observed input and output characteristic curves of a transistor in (i) common base mode and (ii) common-emitter mode.

33
Measurement of Electronic Charge and *e/m*

The charge on an electron *e* is a fundamental quantity of physical science. Its value is 1.60 × $10^{-19}C$. The charge to mass ratio of an electron or *e/m* is also a basic physical quantity. Its value for an electron is 1.76 × 10^{11}C/kg or 1.76 × 10^7*e.m.u.*/gm. or 5.23 × 10^7 e.s.u./gm. In this chapter we will learn a few experiments for determining these quantities.

Experiment 33.1

Object—To determine the electronic charge *e* by Millikan's oil-drop method.

Apparatus—Millikan's oil-drop apparatus, atomiser, oil, power-supply, voltmeter.

Description of apparatus—*Millikan's oil-drop Apparatus.* The laboratory form of the apparatus is depicted in Fig. 33.1. *A* and *B*

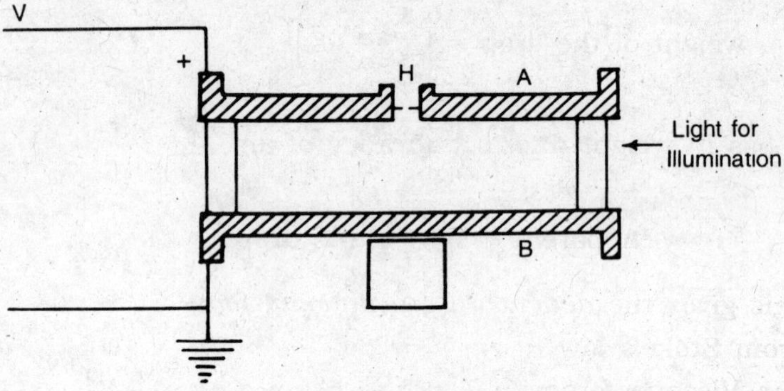

Fig. 33.1

are two optically worked metal plates which form the plates of a parallel plate copacitor. These are separated by three optically plane parallel plates of glass about 1.5 cm. thick. A large constant electrostatic field may be maintained between *A* and *B* by supplying the necessary p.d. across *A* and *B* from a power supply unit. Drops of oil from an atomiser are sprayed above *A* and fall through the small holes at H into the capacitor. Some of the drops are ionized carrying one or more electrons. The space between the plates is illuminated and the motion of the charged drops is observed through a telescope and timed through a given distance with the help of the graduated glass graticule placed in the position of the crosswires of the eyepiece of the telescope. The speed of the charged drop is measured (a) under gravity alone, and (b) under gravity and a known electric field E between the plates A and B.

Theory—(*i*) *Motion under gravity alone.* When no electric field is applied, the charged drop between the capacitor plates A and B moves down under the action of gravity alone. This motion in the beginning is accelerated and is opposed by the viscous forces of the air and as such the drop acquires a steady downward velocity υ_1 when the upward viscous force equals the downward gravitational force.

Let, the radius of the drop be a.

the density of the oil be ρ.

the density of air be σ.

The coefficient of velocity of air be η.

The steady velocity under gravity alone be υ_1.

Hence,

$$\text{weight of the drop} = \left(\frac{4}{3}\pi a^3 \rho g\right)$$

$$\text{Loss in weight due to buoyancy of air} = \left(\frac{4}{3}\pi a^3 \rho g\right)$$

$$\therefore \quad \text{Apparent weight of the drop} = \frac{4}{3}\pi a^3 (\rho - \sigma) g.$$

This gives the *downward gravitational force* on the drop.

From Stoke's law,

$$\text{The Viscous force} = 6\,\pi\,\eta\,a\,\upsilon_1$$

$$\text{Hence,} \quad \frac{4}{3\pi} a^3 (\rho - \sigma) g. = 6\,\pi\,\eta\,a\,\upsilon_1 \tag{33.1}$$

$$\text{which gives} \quad a = \left[\frac{9\eta \upsilon_1}{2(\rho - \sigma) g}\right]^{\frac{1}{2}} \tag{33.2}$$

(*ii*) *Motion under electric field and Gravity.* Let an electric field E of sufficient intensity be applied between the capacitor plates A and B so that it is sufficiently high to overcome the apparent weight of the drop. Let the charge on the drop be ne where e is the electronic charge. Let υ_2 be the constant speed of the same drop in the upward direction. Then,

the net upward force acting on the drop = $ne\,E - 4/3\,\pi a^3(\rho - \sigma)g$

while the viscous force opposing this = $6\,\pi\,\eta\,a\,\upsilon_2$

Hence,

$$ne\,E - 4/3\,\pi a^3\,(\rho - \sigma)g = 6\pi\eta\,a\,\upsilon_2 \qquad (33.3)$$

Adding Eqs. (37.1) and (37.3), we have

$$ne\,E = 6\pi\,\eta\,a\,(\upsilon_1 + \upsilon_2)$$

$$\text{or} \quad ne = \frac{6\pi\,\eta\,a(\upsilon_1 + \upsilon_2)}{E}$$

Substituting for a from Eq. (33.2) in the above,

$$ne = \frac{6\pi\eta}{E}(\upsilon_1 + \upsilon_2)\left[\frac{9\eta\upsilon_1}{2(\rho - \sigma)g}\right]^{\frac{1}{2}}$$

$$\text{or} \quad ne = \frac{9\sqrt{2.\pi\eta^{3/2}}}{(\rho - \sigma)}, \frac{(\upsilon_1 + \upsilon_2)\sqrt{\upsilon_1}}{E} \qquad (33.4)$$

If V be the potential difference between the capacitor plates A and B and d the distance between them,

$$E = \frac{V}{d} \text{ units.}$$

Hence,

$$q = ne = \frac{9\sqrt{2\pi\eta^{3/2}} \times d}{\sqrt{g(\rho - \sigma)}} \times \frac{(\upsilon_1 + \upsilon_2)\sqrt{\upsilon_1}}{V} \qquad (33.5)$$

$$= Kx\,\frac{(\upsilon_1 + \upsilon_2)\sqrt{\upsilon_1}}{V} \qquad (33.6)$$

where K is a constant equal to

$$\frac{9\sqrt{2.\pi\eta^{3/2}} \times d}{\sqrt{g(\rho - \sigma)}}$$

Method—Focus the light from a lamp by means of a lens L into a region between the plates *A* and *B*. Insert a thin pin in one of the holes in the middle of the upper plate *A*, rotate the telescope and focus it so that a well defined image of the pin is formed in a dark background. This evidently implies that the telescope is not in line with the source of light. The darkness of the background should not be such that readings of the glass graticule in the telescope eyepiece may be difficult. When the adjustment has been made, take out the pin.

With the help of the atomiser, spray drops of oil in *H*. Some of these drops due to friction get charged and enter the region between the plates through the holes in *A* at *H*.

Observe some of these drops through the telescope. To ascertain which of these are charged and which uncharged, switch on the electric field by applying a.p.d. of, say, 450 volts across *AB*, with *A* at a positive potential. Drops which start moving up are charged. Examine one such drop. When it approaches the upper plate *A*, *switch off* the power supply, the drop then starts descending down under the action of gravity alone. When the drop approaches the plate *B*, *switch on* the power supply; it then starts moving up. This alternate process of switching on and off of the power supply can keep the drop under observation for any length of time.

Having thus chosen a drop, examine its motion with the telescope. When it is moving down under the action of gravity alone, find the time it takes to move, say, through 40 divisions of the glass graticule in the telescope eyepiece. Knowing the value of one scale division of the glass graticule, calculate v_1 from its time of descent for 40 such divisions.

When the drop is reaching *B*, switch on the power supply and again find the time it takes to move up, say, 40 divisions of the scale. From this timing calculate v_2, the velocity under the action of both the electrostatic field and gravity.

Repeat these observations a number of times and compute the values of v_1 and v_2 for the particular field E between the plates. To evaluate E, measure the p.d. V across the plates with a voltmeter.

Repeat the experiment with different values of V and choosing different drops.

For each set of observation, calculate *ne*. It will be found to be different in each case for the simple reason that *n* may be different for drops examined. By rounding up the values find the higher common factor for these calculated values of *ne*, then assign the value to *n* in each case and find *é* for each set of observation. Finally, compute the mean value of *e*. Compare this with the standard value and compute the percentage error.

Sources of error and precautions

(1) The telescope for observing the drop should be focused in the manner described. Care should be taken to see that the dark background in which observations are to be made is not so dark as to make it difficult to read the glass graticule.

When the telescope has been focused, the pin used in focusing should be taken out.

(2) A drop once chosen for observation should be kept moving up (by switching on the power supply) and down (by switching off the power supply) so that at least five sets are taken for υ_1 and υ_2, potential *V* being constant for the whole set.

(3) The timing of the drop for calculations of υ_1 and υ_2 should be done with an accurate stop watch reading to 1/10 of a second.

(4) Many drops should be examined and under different electric fields *E*.

(5) If a lateral displacement of the drop occurs, it is due to collision and this drop should be then rejected. Whenever the speed of the drop suddenly changes, it may have acquired more charge or mass. When this happens, the drop cannot be used to repeat its earlier observations but may be used as a fresh drop for a fresh set of observations.

Observations

I. *Constants*

Viscosity of air η = 18.1×10^{-5} poise

Density of air, σ = 0.00124 gm. per c.c.

Density of oil, ρ = (0.874) gm. per c.c.

Value of *g* at – = cm./sec.^{2}

Room temperature = °C

Value of one scale division of glass graticule = 0.01 cm

Distance between the plates, *d* = cm.

Least count of the stop watch = 0.1 sec.

II. *Table for v_1 and v_2*

Set No.	P.D. across AB	Time for Travelling 40 Divisions				Calculated value of $q = ne$ coulomb
		Under Gravity Alone		Under Gravitational & Electric Fields		
		Time	Mean	Time	Mean	
I	450 volts	6.8 sec.	6.6 sec.	3.0 sec.	3.2 sec.	
II						
... ... VII						

Calculations

I. Common to all sets :

$(\rho - \sigma) =$

$$\therefore \quad \text{Constant } K = \frac{9\sqrt{2\pi\eta^{3/2}}d}{\sqrt{g(\rho-\sigma)}}$$

II. For each set separately :

$v_1 =$

$v_2 =$

$V =$

$$\therefore \quad \frac{(v_1+v_2)\sqrt{v_1}}{V} =$$

$$=$$

$$\therefore \quad ne = q = K_1 \frac{(v_1+v_2)\sqrt{v_1}}{V}$$

$$=$$

$=$ coulomb

In this manner, calculate *ne* for each set and then tabulate your result. You will find that the charge *q* on drops are multiples of a quantity of charge (*e*) which is the electronic charge. Find the mean value of *e* by taking large number of observations.

Result. The electronic charge as determined experimentally by Millikan's oil drop method = ... coulomb.

Standard value = 1.6023×10^{-19}C

$\therefore$ Percentage Error =

Experiment 33.2

Object—To determine the value of e/m of an electron by Thomson's method.

Apparatus—Cathode ray tube fitted on a wooden stand provided with two arms perpendicular to the axis of the tube and having scales with zero at the point where the central line of the arms cuts the axis of the tube, a power supply unit capable of giving the accelerating d.c. voltage for the CRT tube and the deflecting d.c. voltage for the electric field through a reversing key with suitable voltmeters, one pair of bar magnets, compass box, etc.

Description of Apparatus—Thomson's cathode ray tube, shown in Fig. 33.2 consists of the following three essential parts:

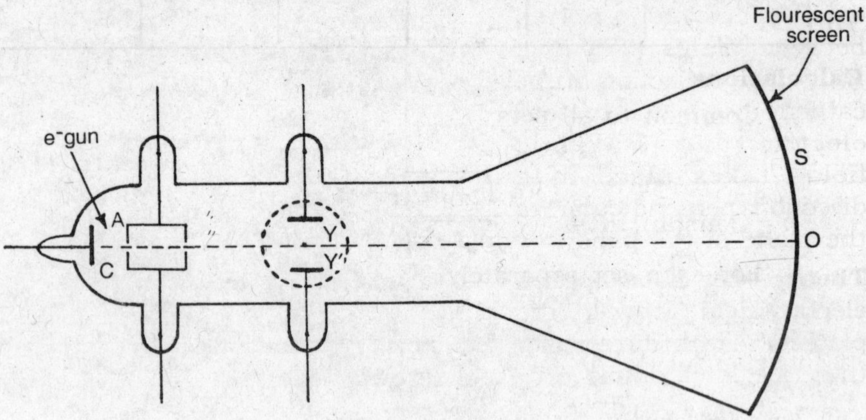

Fig. 33.2

(i) *The electron gun.* This consists of the cathode C and the anode A. It produces, accelerates and focuses the electrons into a fine beam so as to produce a small bright spot on the screen S. Narrow slits are cut in opposite plates at A so that the cathode rays passing through are limited to a narrow beam.

(ii) *The Screen.* The screen S has a coating of some fluorescent materials like zinc sulphide, etc. so that a greenish blue spot of light is observed where the electron beam impinges upon the screen. The screen has a graduated X-Y (mm) scale on its surface to note the exact position of the fluorescent spot.

(iii) *The electric and magnetic deflecting system.* Two plates YY' are fitted in the tube symmetrically on either side of the electron beam so that an electric field perpendicular to the plane of the paper can be applied between them. The cathode beam may also be *magnetically* deflected by placing a magnet on the wooden stand. When the bar magnets are placed with their lengths parallel to the arms with their opposite poles towards CRT, these produce a uniform magnetic field at the same place where electric field is produced when potential difference is applied between plates (YY') (Fig. 33.3). The deflection of the cathode beam due to both electric and magnetic fields takes place in a direction perpendicular to the plane of the paper.

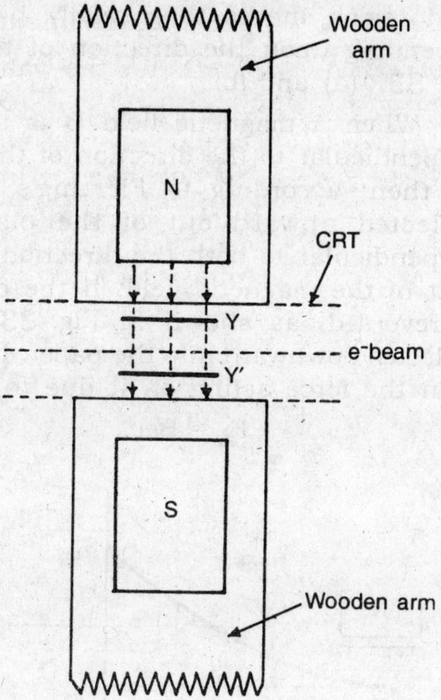

Fig. 33.3

Theory—Let E be applied electric field between the plates YY' in a direction perpendicular to the plane of the paper then the force acting on an electron of charge *e*.

$$F_E = eE$$

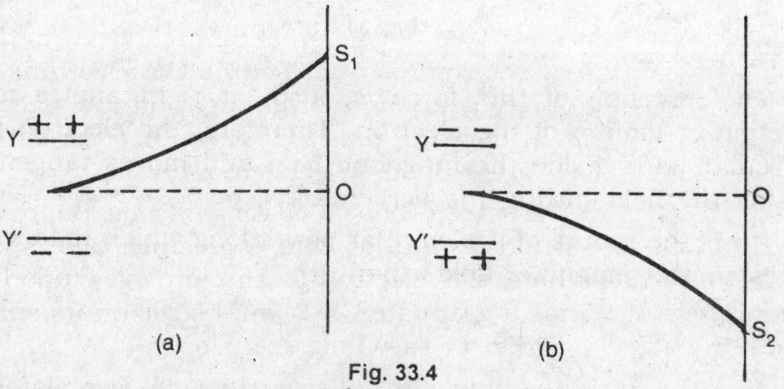

Fig. 33.4

The direction of this electrostatic force is at right angles to the direction of motion of the electron. The electron, therefore, takes a circular path within the electric field and on leaving the field region moves tangentially hitting the screen at S_1 or S_2 depending upon the direction of the electric field, as shown in Fig. 33.4(a) and (b).

When a magnetic field B is applied in the plane of paper perpendicular to the direction of the beam as shown in Fig. 33.5 (a) then, according to Fleming's left hand rule, the beam is deflected upward out of the plane of paper in a direction perpendicular to both the direction of motion of the electron and that of the magnetic field. If the direction of the magnetic field is reversed, as shown in Fig. 33.5 (b), the electron beam is deflected downward into the paper. If v is the velocity of an electron, then the force acting on it due to the magnetic field :

$$F_B = Bev.$$

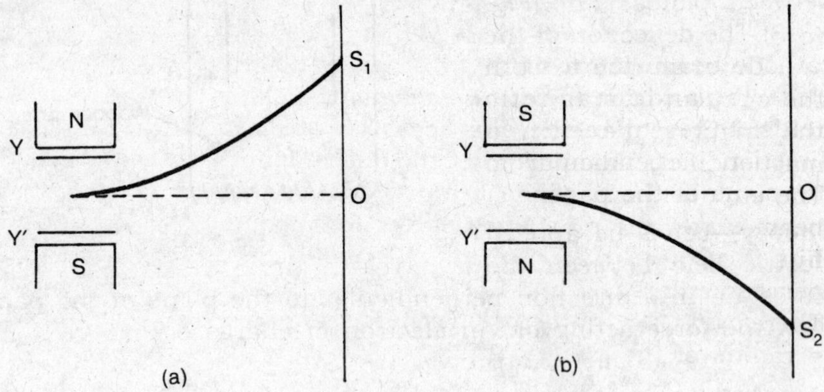

Fig. 33.5

The direction of this force is, also, at right angles to the direction of motion of the electron. Therefore, the electron takes a circular path inside the magnetic field and moves tangentially outside the field hitting the screen at S_1 or S_2.

If r is the radius of the circular path along which the electron moves in the magnetic field, then,

$$Bev = \frac{mv^2}{r}$$

or $$\frac{e}{m} = \frac{v}{rB} \qquad (33.7)$$

Determination of *v*—In order to determine v, one of the fields, say, the magnetic field is first applied so that the spot of light moves to S_1. Then the electric field is also applied simultaneously so as to produce deflection of the electron beam in the opposite direction. The value of the electric field is now so adjusted that the spot of light comes back to its initial position. In such a case, the forces acting on the electron due to the electric and magnetic fields are equal and opposite.

$$\therefore \quad Bev = eE$$

or $$v = \frac{E}{B}$$

Substituting for v in Eq. (39.7), we get :

$$\frac{e}{m} = \frac{E}{rB^2} \qquad (33.8)$$

Determination of r—In order to determine the radius r of the circular path inside the magnetic field, note the position of the bright spot in the absence of any field and again when the magnetic field is applied. Let the deflection of the spot be $y = OS_2$. The circular path of the beam OR is as shown in Fig. 33.6. The beam leaves the magnetic field at R tangentially to the path and hits the screen at S_2 so that S_2R produces meets O'O at T. The perpendiculars O'C and RC to O'O and RS_2 respectively meet at C. Therefore, C is the centre of the circular arc O'R and O'C = CR = radius of the circle = r.

Now $$\angle O'CR = \angle OTS_2 = \theta$$

$$\therefore \quad \tan\theta = \frac{OS_2}{OT} = \frac{O'X}{O'C}$$

$$\therefore \quad r = O'C = \frac{OT \times O'X}{OS_2}$$

Now, O'X is very nearly equal to the length of the region in which the magnetic field is applied. This is also equal to the length of the region in which the electric field is applied, *i.e.*, the length of the plates Y and Y'. Let it be l, *i.e.*, O'X = l; this is normally given by the manufacturer in the catalogue. OT is the

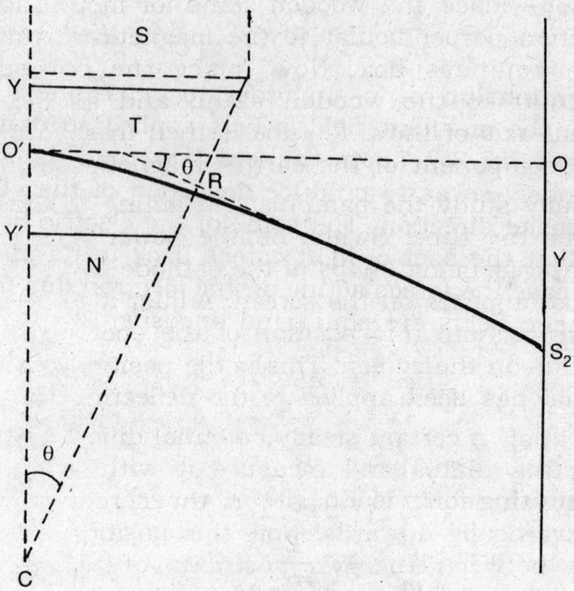

Fig. 33.6

distance of the centre of the screen from the centre of the region of the magnetic field (or the electric field). Let it be L, *i.e.*, OT = L,

$$\therefore \qquad r = \frac{Ll}{y}$$

Substituting this in Eq. (33.8), we get :

$$\frac{e}{m} = \frac{yE}{LlB^2} \tag{33.9}$$

The electric field E = $\frac{V}{d}$, where V is the applied potential difference between Y and Y' and d the separation between the plates (given by the manufacturer). Substituting for E, we have :

$$\frac{e}{m} = \frac{yV}{LldB^2} \tag{33.10}$$

Thus, knowing V (Volts), B (Tesla), d, l, L and y (all in metres), e/m (coulombs/kg) can be calculated using Eq.(33.10).

Method—Place the wooden stand for mounting the magnet in a direction perpendicular to the magnetic meridian with the help of a compass box. Now, place the cathode ray tube perpendicular to the wooden stand and at its centre. The longitudinal axis of the C.R. tube is then in the direction of the horizontal component of the earth's magnetic field.

Carefully study the pamphlet pertaining to the operation of the cathode ray tube. Switch on the power supply. Adjust the intensity and focusing knobs of the cathode ray tube until a well defined spot appears on the screen. Adjust it to lie in the centre of the screen. Note this position of the spot on the graduated vertical scale on the screen. This is the position of the spot when no potential has been applied to the deflecting plates.

Now, apply a certain steady potential difference V volt across the deflecting plates and measure it with an accurate high resistance voltmeter. The spot on the screen is found to be displaced vertically upwards. Note this position of the spot. The difference between the two positions of the spot gives the displacement y, for the applied *p.d.* of V volt.

Place two equal large bar magnets in the F W. direction, one on each arm of the wooden stand along the graduated scale with their proper poles pointing towards the cathode ray tube so as to reduce the electrostatic deflection. The magnets should lie symmetrically from the axis of the C.R. tube and should be displaced along the scales symmetrically so as to reduce the displacement of the spot to zero. Note the positions of the poles of the magnets nearest to the C.R. tube on the scale fixed to the wooden frame. Let these be r_1 and r_2 respectively. This adjustment implies that the field H produced by the magnets in their respective positions produces an equal and opposite displacement of the electron beam.

Remove the magnets. Reverse the applied voltage when the spot on the screen will be deflected vertically downwards. Note the position of the deflected spot and find the displacement y_2 for the same applied voltage V volts. Calculate the mean value of y_1 and y_2; this gives the mean displacement y for the applied potential V volts.

Place the magnets again on the wooden stand in the direction required to reduce the deflection and as before, arrange the poles of these magnets symmetrically and at such a distance apart that the field is of sufficient strength to reduce the deflection to zero. Again note the positions of the poles of the magnet nearest to the C.R. tube, on the scale of the wooden stand. Let these be r_1 and r_2. Switch off the power supply.

The value of B to balance the electrostatic deflection may be determined in the following manner :

Remove the cathode ray tube from the wooden stand which carries the magnets and place a compass box such that its centre lies on the common axis of the magnets. Adjust the pointer at zero-zero mark. Place the bar magnets with the same opposite poles towards the C.R. tube at the same distances r_1 and r_2 respectively as in the first part of the experiment when the deflection was reduced to zero. Observe the deflection of the pointer., reading both ends. Find the mean deflection θ_1. Reverse the magnets pole and adjust their distances at r_1' and r_2', the distances of the poles when the deflection was reduced to zero. Again read both ends of the pointer and measure the mean deflection θ_2. Take the mean of θ_1 and θ_2. Let this be θ.

If B_H be the horizontal component of the earth's magnetic field, the field B due to magnets is given by :

$$B = B_H \tan\theta$$

This gives B to balance the effect of the electrostatic field E.

Repeat the experiment by altering V and adjusting and measuring B to balance.

If the constants of the apparatus l, L and d are known and B_H is known at the place of observation, using Eq. (ua.10) we can calculate e/m.

Sources of Error and precautions. (1) The cathode ray tube should be placed with its longitudinal axis in the direction of the horizontal component of the earth's magnetic field.

(2) The spot on the screen must be adjusted to be sharp and intense. It should not be allowed to remain on the screen at a given position for a long time.

(3) When the potential difference V is applied to the deflecting plates, its value must be so adjusted as to give the displacement Y in exact mms.

(4) As the plates are not too far apart we can take the electric field E so produced as a uniform field and of an extent coinciding with the limit of the plates, *i.e.*, we can neglect the end corrections. We assume the field B_H as uniform in strength which is essentially not correct.

Observations

I. *Known Constants*

Length of the deflecting plates, l = cm. = m

Distance of the screen from the end of the plates, L = cm. = m

Distance between the plates, d = cm. = m

Value of B_H = oersted. = Tesla

II. *Determination of y and V.*

Initial position of the spot = cm.

S.No.	Applied Potential Difference *V* volt	Deflected position of spod when *V* is				*Mean* deflection	When *y* is restored to zero	
		Direct	y_1	Reversed	y_2	$\frac{y_1+y_2}{2}$	Position of magnet on east arm Distance of the nearest pole to C. T. Tube	Position of the other magnet on west arm. Distance of the nearest pole to C.
		cm	cm	cm	cm	m	cm	cm
1	Direct Reverse						N. pole r_1 S. pole r_2	*S.* pole r_1' *N.* pole r_2'
2								
3								
4								
5								
6								

III. *Determination of B*

N =	For y restored to Zero		Deflection of the Compass needle		Mean	Mean deflection $\theta = (\theta_1 + \theta_2)/2$	$B = B_H \tan\theta$
	Position of the magnet on *E* arm.	Position of the magnet on *W* arm.					
	Distance of the pole nearest to C.R. tube.		one end deg.	other end deg.			
1	N. Pole r_1 S. Pole r_2	S. Pole r_1' N. Pole r_2'			θ_1= θ_2=		
2							
3							
4							
5							
6							
7							

Calculations

(i) $e/m = \dfrac{y_1 V_1}{L l d B_1^2}$ = ... C/kg

(ii) $e/m = \dfrac{y_2 V_2}{L l d B_2^2}$ = ... C/kg

Mean e/m = ... C/kg

Result—The value of e/m of an electron = C/kg

Standard value = 1.76×10^{11} C/kg (= 1.76×10^{7} e.m.u./gm)

Percentage error =

Experiment 33.3

Object—To determine the value of e/m of an electron by helical (long solenoid) method.

Apparatus—The helical set up containing a cathode ray tube, a long closely wound solenoid, suitable higher resistance, dc and ac voltmeters and ammeters with control knobs, connecting cable.

Theory—If an electron of mass m is accelerated through a potential V_0 acquiring a velocity v then,

$$\frac{1}{2}mv^2 = eV_0$$

or

$$v = \sqrt{\frac{2eV_0}{m}}$$

Let l be the distance between the *centre of deflecting plates* and the *screen*. So the time t taken by the electron to traverse this distance is :

$$t = \frac{l}{v} = \frac{l}{\sqrt{\dfrac{2eV_0}{m}}} \qquad (33.11)$$

When an alternating potential difference is applied either across the X or Y deflecting plates, then due to the alternating electric field, a force acts on the electron beam at right angle to the direction of the beam. Under the action of two perpendicular forces, one due to the accelerating voltage and the other due to the alternating potential, the electron beam has a *parabolic path*. The electrons travel along the parabolic path with increased transverse velocity as they pass through the deflecting plates after which they travel along straight lines so that the beam on reaching the screen sweeps back and forth producing a straight line on the screen. If the accelerating potential is applied across the *X* plates the line produced is *horizontal* and it is *vertical* if the accelerating potential is applied across the *Y* plates. The length of the line depends upon the magnitude of the alternating potential difference.

A magnetic field is applied, by switching on the solenoid, in such a way that the lines of magnetic force are parallel to the axis of the *CRT*. Therefore, the applied magnetic field does not affect the forward motion of the electrons, *i.e.*, their longitudinal velocity, but acts on the transverse motion of the electrons and the electron takes a *circular path* due to this *transverse velocity*. The resultant path of the electrons is *helical* due to combination of uniform circular and uniform linear motion.

Let B be the strength of the applied magnetic field and v the transverse velocity of the electron, then,

$$Bev' = \frac{mv'^2}{r}$$

where r is the radius of the resulting circular path of the electron.

$$\therefore \quad \frac{v'}{r} = \omega = B\frac{e}{m}$$

The angular velocity $\omega = \frac{v'}{r}$ depends only upon B and is a constant for a given value of B.

$$\therefore \quad \text{Time period } T = \frac{2\pi}{\omega} = \frac{2\pi m}{Be} \qquad (33.12)$$

The value of B is adjusted to make the time T equal to the time t to travel to the screen from the centre of deflecting plates, given by Eq. (33.11). As this condition is independent of the transverse velocity, it holds for all the electrons and therefore all the electrons again collect on the axis and focussed at a sharp point on the screen, *i.e.*, the beam reduces to a point focus.

Such a condition may be obtained again for higher values of B when the time period $T = 1/2$ or $1/3$, etc. of the time t. Thus, we may obtain a number of foci by varying B. Let B_0 be the minimum value of B for which a point focus is obtained, then,

$$\frac{2\pi m}{B_0 e} = \frac{l}{\sqrt{\frac{2eV_0}{m}}}$$

$$\therefore \quad \frac{e}{m} = \frac{8\pi^2 V_0}{B_0^2 l^2} \qquad (33.13)$$

For a solenoid of length L having N turns when a current I is needed to be passed, for the magnetic field B_0, then,

$$B_0 = \frac{\mu_0}{2}\frac{NI}{L}(\cos\theta_1 - \cos\theta_2)$$

where θ_1 and θ_2 are the angles (Fig. 33.7) at the centre by the end points of the solenoid and $\mu_0 = 4\pi \times 10^{-7}$ Weber/Ampere-metre.

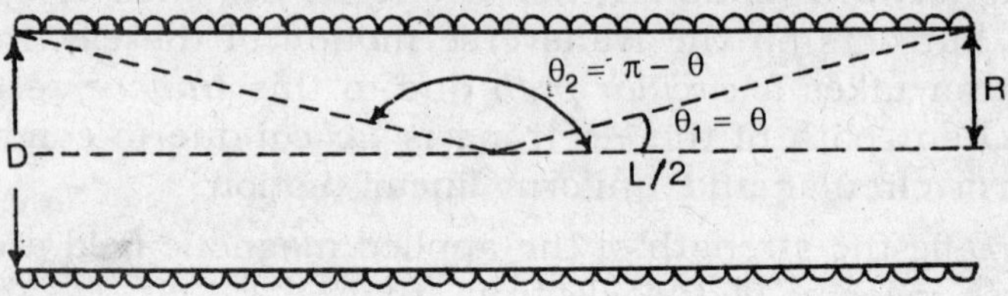

Fig. 33.7

When the centre of the solenoid coincides with the centre of the deflecting plates,

$$\theta_1 = \theta \text{ and } \quad \theta_2 = \pi - \theta$$

$$\therefore \quad B_0 = \frac{\mu_0 NI}{2L} 2\cos\theta = \frac{\mu_0 NI}{L}\cos\theta \qquad (33.14)$$

If R is the radius of the solenoid and D its diameter, then,

$$\text{Cos }\theta = \frac{L/2}{\sqrt{R^2 + (L/2)^2}} = \frac{L/2}{\sqrt{(D/2)^2 + (L/2)^2}} = \frac{L}{\sqrt{D^2 + L^2}}$$

Substituting the value of B_0 from Eq.(33.13) in Eq.(33.14) we get :

$$\frac{e}{m} = \frac{V_0}{I^2}\cdot\frac{1}{2}\left[\frac{4\pi}{\mu_0}\frac{L}{Nl\cos\theta}\right]^2$$

We know $\frac{\mu_0}{4\pi} = 10^{-7}$

$$\therefore \quad \frac{e}{m} = \frac{V_0}{I^2}\left[\frac{L}{Nl\cos\theta}\right]^2 \times 5 \times 10^{13}$$

$$= K\frac{V_0}{I^2} \qquad (33.15)$$

where $K = 5 \times 10^{13}\left(\frac{L}{Nl\cos\theta}\right)^2$.

If l_x is the distance from the centre of *X*-deflection plates to the screen and l_y the corresponding value for *Y*-plates, then for

X-plates : $k_x = 5 \times 10^{13}\left(\frac{L}{Nl_x\cos\theta}\right)$

Y-plates : $k_y = 5 \times 10^{13}\left(\frac{L}{Nl_y\cos\theta}\right)$

Description of apparatus—This set-up consists of three units : (*i*) a cathode ray tube, (*ii*) a power supply unit and (*iii*) a long closely wound solenoid in which the cathode ray tube can be placed so that the axis of the cathode ray tube and that of the solenoid coincides.

The front panel of the power supply unit is shown in Fig. 33.8. The cathode ray tube is connected to the power supply unit through a cable. The cable encloses the leads originating from the filament, grid, accelerating anodes and deflecting plates. The other end of the leads are connected internally to suitable points in the power supply unit. When the power supply is switched on, the filament of the cathode ray tube gets heated up and starts emitting electrons. Accelerating potential is applied by the accelerating voltage control. Separate knobs are provided for coarse and fine adjustment. The value of accelerating voltage can be read in the accelerating voltage d.c. voltmeter. The knobs for *focus control* and *intensity control* are used for focusing and to vary intensity respectively.

When a sharp point is obtained on the screen, the deflecting voltage switch is turned to the X-plates and with the help of X-voltage control regulator the a.c. voltage applied to the X-plates is adjusted to obtain a line of about 2 cm. Similarly, the Y-deflection switch and control can be used. The value of a.c. voltage can be read in the a.c. voltmeter.

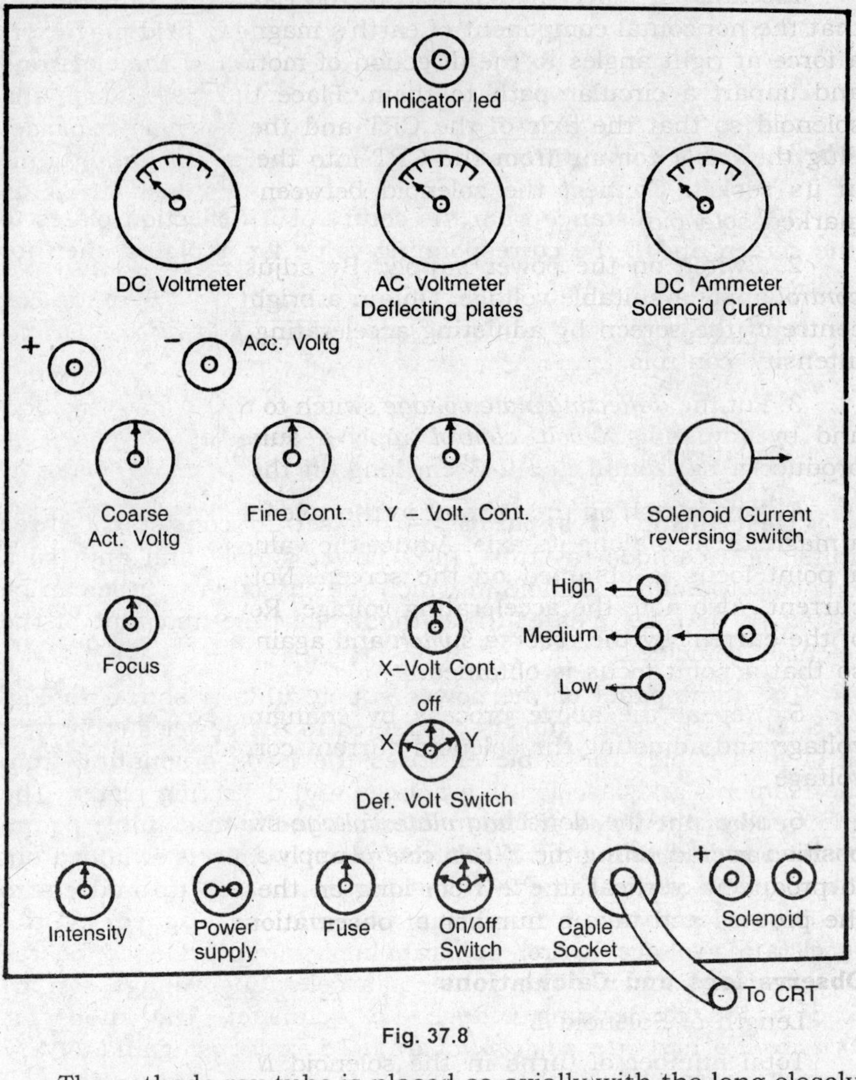

Fig. 37.8

The cathode ray tube is placed co-axially with the long closely wound solenoid suitably in order to apply a uniform magnetic field. The solenoid is connected to the power supply unit between the two terminals marked *solenoid*. The magnitude of *solenoid* current can be set by a regulating knob. The current can be reversed with a switch marked *reverse.* The value of current is read in a d.c. ammeter.

Method—1. Place the solenoid in the East-West direction so that the horizontal component of earth's magnetic field may exert a force at right angles to the direction of motion of the electrons and impart a circular path to them. Place the CRT inside the solenoid so that the axis of the CRT and the solenoid coincide. Plug the cable coming from the CRT into the power supply unit in its socket. Connect the solenoid between the two terminals marked *solenoid.*

2. Switch on the power supply. By adjusting the *acce. volt. control* apply a suitable voltage. Obtain a bright point spot on the centre of the screen by adjusting accelerating voltage, focus and intensity controls.

3. Put the *deflecting plate voltage* switch to the X-plate position and by adjusting *X-volt. control* apply a suitable a.c-voltage to produce a horizontal line 2-3 cm long on the screen.

4. Now switch on the current in the solenoid so as to produce a magnetic field along its axis. Adjust the value of the current till a point focus is obtained on the screen. Note the value of the current. Also note the accelerating voltage. Reverse the direction of the current by the *reverse, switch* and again adjust the current so that a point focus is obtained.

5. Repeat the above process by changing the accelerating voltage and adjusting the solenoid current corresponding to each voltage

6. Now put the *deflecting plate voltage* switch to the Y-plate position and adjusting the *Y-volt. control* apply a suitable a.c.voltage to produce a vertical line 2-3 cm long on the screen and repeat the process and take a number of observations.

Observations and Calculations

Length of Solenoid L =

Total number of turns in the solenoid N =

Outer radius of the solenoid tube R_1 =

Inner radius of the solenoid tube R_2 =

Mean effective radius of the solenoid R = $\frac{R_1 + R_2}{2}$ =

Mean effective diameter of the solenoid $D = 2R$ =

Mean distance of the X-plates from the screen l_x =

Mean distance of the Y-plates from the screen l_y =

$$\therefore \quad \text{Cos}\theta = \frac{L/2}{\sqrt{R^2 + \left(\frac{L}{2}\right)^2}} = \frac{L}{\sqrt{D^2 + L^2}} =$$

$$K_x = 5 \times 10^{13} \left(\frac{L}{Nl_x \text{ Cos}\theta}\right) =$$

$$K_y = 5 \times 10^{13} \left(\frac{L}{Nl_y \text{ Cos}\theta}\right) =$$

S.No.	Accelerating voltage V_0 (volts)	Horizontal X-deflection Current I (Amp.)				$\frac{V_0}{I}$	$\frac{e}{m} = K_x \frac{V_0}{I^2}$
		Forward	Reverse	Mean I	I^2		
1							
2.							
.							
.							
.							
.							
8.							

S.No.	Accelerating voltage V_0 (volts)	Vertical Y-deflection Current I (Amp.)				$\frac{V_0}{I}$	$\frac{e}{m} = K_y \frac{V_0}{I_0^2}$
		Forward	Reverse	Mean I	I^2		
1.							
2.							
.							
.							
.							
.							
8							

Mean $\frac{e}{m}$ for horizontal (X-) deflection = ... C/kg

Mean $\frac{e}{m}$ for vertical (Y-) deflection = ... C/kg

Mean e/m = ... C/kg

Graphical Method—Plot a graph between the *accelerating voltage* V_0 along X-axis and square of the *current through the solenoid* I^2 along Y-axis, as shown in Fig. 33.9. The graph obtained is a

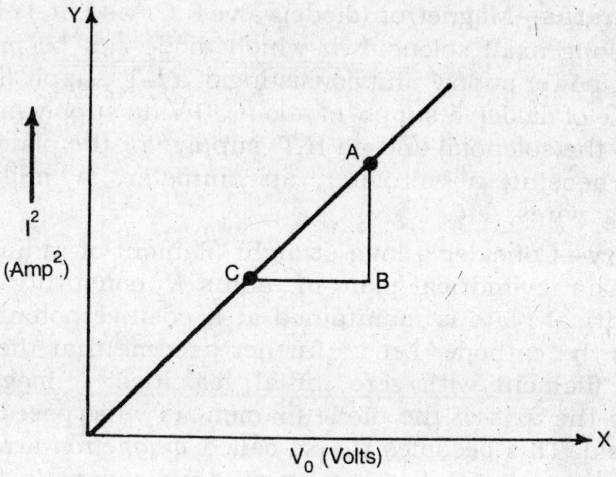

Fig. 33.9

straight line. Mark two points A and C on the graph and draw a right angle triangle ABC, then,

$$\text{Slope of the graph} = \frac{AB}{BC} = \frac{I^2}{V_0}$$

$$\therefore \quad \frac{BC}{AB} = \frac{V_0}{I^2}$$

$$\frac{e}{m} \text{ for horizontal (X-) deflection} = k = \frac{V_0}{I^2} = \text{............C/kg}$$

Repeat the above process for vertical (Y-) deflection and calculate $\frac{e}{m}$ for vertical (Y-) deflection = ...C/kg

Find mean e/m = ...C/kg

Result—The value of $\frac{e}{m}$ of an electron = ...C/kg

Standard value of e/m of an electron = 1.76×10^{11} C/kg

Percentage error = ...%

Experiment 33.4

Object—To determine the value of e/m of an electron by magnetron method.

Apparatus—Magnetron (diode) valve R.C.A. 41 or 42 or *E* 1189, standard long axial solenoid in which diode can be mounted at the centre, power supply unit consisting of a L.T. supply (for heating the cathode of diode), a supply of about 30 V (to supply magnetising current to the solenoid) and an H.T. supply (for the plate of diode), suitable rheostat, a voltmeter, an ammeter, a milliammeter, connecting wires, etc.

Theory—Consider a long straight filament of radius r ($r << R_a$) and a coaxial cylindrical plate of radius R_a constituting a diode. The cylindrical plate is maintained at a positive potential V with respect to the cathode. Let us further assume that the electrons leave the filament with zero initial velocities. A magnetic field parallel to the axis of the diode filament is superposed upon the electric field. This becomes the so-called *magnetron* arrangement, first introduced by Hull. This longitudinal magnetic field along the axis of the filament may be obtained by placing the diode inside a long straight solenoid such that its axis is coincident with the filament; the strength of the magnetic field may be varied by controlling the current through the solenoid.

Let us now investigate the electronic paths. Suppose, to begin with, the magnetic field is zero. The electrons, each of mass m and charge e, are accelerated towards the plate by the electric field. The velocity υ acquired by the electrons under a potential difference υ is given by the relation.

$$\frac{1}{2}m\upsilon^2 = Ve$$

$$\upsilon = \sqrt{\frac{2Ve}{m}} \qquad (33.16)$$

If a weak magnetic field is present, then due to Lorentz force the electron path will be curved, although the electron will strike the plate. This is illustrated in Fig.(33.10). As the field is further increased, the path becomes more and more curved until a critical value of the field B_c is reached when the path of the electron is tangential to the anode. For fields greater than B_c the electron does not strike the plate at all but returns to the filament. Thus, for a fixed anode potential V, if the anode current is studied as a

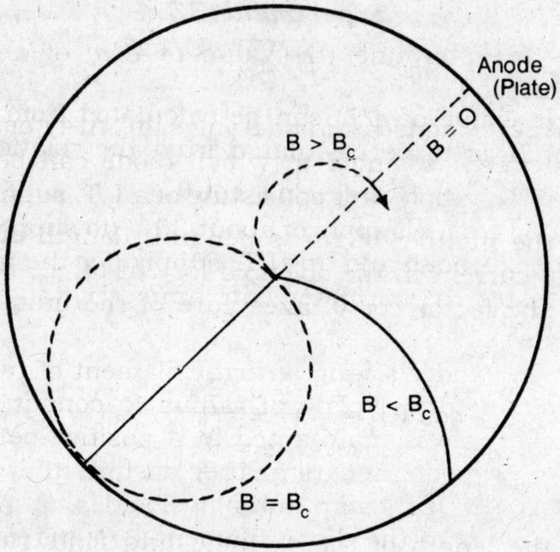

Fig. 33.10

function of B, the applied magnetic field, the plate current should be unaffected for values B less than the critical value B_c, whereas for values of B greater than the critical value B_c, the anode current should reduce to zero. Thus, B_c can be determined graphically by noting the cut-off. The equilibrium equation for the critical field is given by :

$$B_c . ev = \frac{mv^2}{p}$$

where p is the radius of the circular path, which for B_c equals half the radius of the cylindrical anode, *i.e.*, $p = \frac{r_a}{2}$

Thus, $$B_c . ev = \frac{2mv^2}{r_a}$$

or $$v = \frac{B_c . e . r_a}{2m} \qquad (33.17)$$

Equating (33.16) and (33.17), we have :

$$\frac{B_c . e . r_a}{2m} = \sqrt{\frac{2\,Ve}{m}}$$

$$\frac{e}{m} = \frac{8V}{B_c^2 . r_a^2} \quad C/kg \qquad (33.18)$$

Thus, the value of e/m can be calculated from this relation. The value of B_c is to be calculated from the relation :

$$B_c = \mu_0 n I_c . \cos\theta \qquad (33.19)$$

where n is the number of turns per unit, length of the solenoid, I_c the critical current in the solenoid for cut-off $\mu_0 = 4\pi \times 10^{-7}$ Wb/A – m and the factor cos θ takes care of the finite length of the solenoid.

$$\cos\theta = \frac{L/2}{\sqrt{R^2 + \left(\frac{L}{2}\right)^2}} \qquad (33.20)$$

where L is the length of the solenoid and R its radius.

Method –1. The mounted diode tube is placed well within the solenoid symmetrically which should be in E-W direction.

2. The anode is given a constant potential and the anode current is adjusted to give the maximum deflection in the milliammeter for zero current in the solenoid.

3. Make the connections to the solenoid and adjust the current through it with the help of the rheostat. Read the anode current.

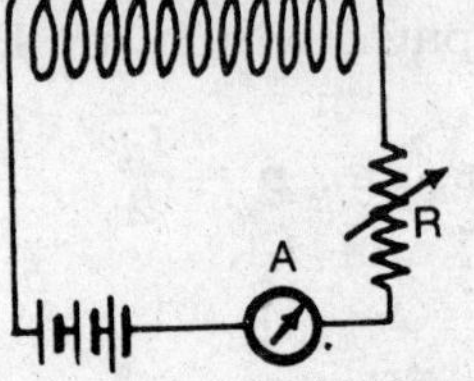

Fig. 33.11 Solenoidal Connection

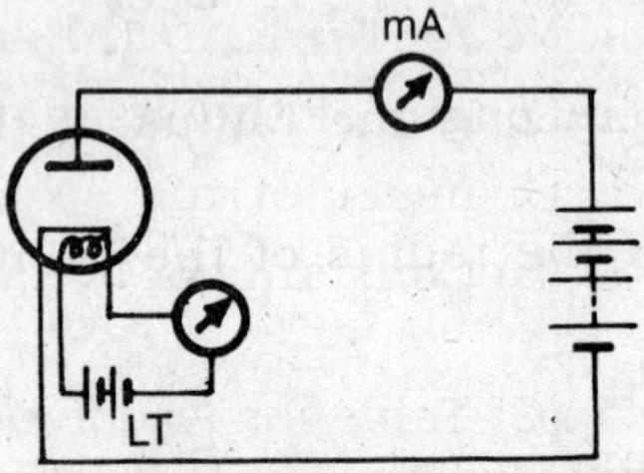

Fig. 33.12 Diode Connection

4. Increase the current in the solenoid gradually and for each setting of solenoidal current, note the anode current. The process is to be carried on until the anode current is considerably reduced.

5. Repeat the experiment with different anode potentials.

6. For each anode potential value, plot a graph between the plate current I_a and the solenoidal current I_s. The intersection of the tangent to the sloping curve with the steady I_a value for small values of B, gives the critical solenoidal current I_c corresponding to B_c. Calculate B_c from the relation.

$$B_c = \mu_0 n I_c . \cos\theta$$

7. Calculate e/m from the relation :

$$e/m = \frac{8V}{B_c^2 . r_a^2} \text{ C/kg}$$

for each anode potential value and finally calculate the mean value.

Sources of Error and Precautions—1. The diode should be placed well within the solenoid symmetrically.

2. The correction factor $\cos\theta$ while calculating B_c should be applied to take into account the finite size of the solenoid.

3. The plate voltage should be kept constant during one set of observations.

4. I_c should be very carefully determined from the graph.

5. For exact range of solenoid current and anode potential refer to catalogue provided by the manufacturer.

Observations

(A) Constants of the solenoid :

No. of turns in the solenoid = 1400

Length of the solenoid = 35 cm. = 0.35 m

Diameter of the solenoid = 7 cm = 0.07 m

(B) Constants of the Magnetron : Radius of the anode = 3 mm = 3×10^{-3} m.

(C) Table for Anode current variation with solenoidal current

Anode potential	5 volts	Anode potential	8 volts
Solenoidal current (Amp)	Anode current (mA)	Solenoidal current (Amp.)	Anode current (mA)
0.5	0.24		
0.6	0.24		
0.7	0.24		

0.8	0.24
1.0	0.22
1.2	0.20
1.3	0.18
1.4	0.16
1.5	0.14
1.6	0.12
1.7	0.10
1.8	0.08
1.9	0.07
2.0	0.06
2.2	0.04
2.4	0.02
2.5	0.01

Graph : Anode Potential = 5 volts

Calculations

$$\cos\theta = \frac{L/2}{\sqrt{(L/2)^2 + R^2}} =$$

$$\mu_0 \, n \cos\theta =$$

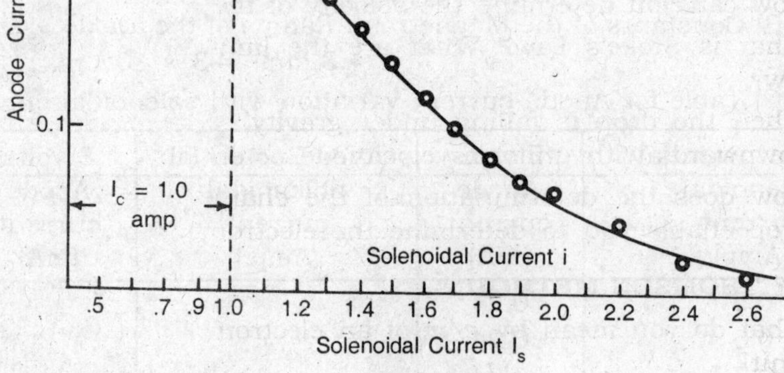

Fig. 33.13

From the graph (Fig. 33.13), for constant anode potential 5 *V* volts,

$$I_c = 1.0 \text{ amp.}$$

$$\therefore \quad B_c = (\mu_0 n.\cos\theta).I_c = ...$$

$$e/m = \frac{8V}{B_c^2\, ra^2} = ... \text{ C/kg}$$

Make similar calculations for other sets.

Mean e/m = ...C/kg

Result—The value of e/m of an electron = ...C/kg

Standard value of e/m of an electron = 1.76×10^{11} C/kg

% Error = ...

ORAL QUESTIONS

'*e*' BY MILLIKAN'S OIL DROP METHOD

1. Describe the construction of Millikan's Oil Drop apparatus.
2. How do you introduce liquid drops in between the two plates? How do these drops get charged?
3. What type of liquid are you using and why?
4. Why are the plates made of metal?
5. How do you select a particular drop for working?
6. When you establish an electric field, some of the drops move against gravity, why is it so?
7. What is the condition that a drop may remain stationary? How can you determine the velocity of the drop?
8. What is Stoke's Law? What are the limitations of Stoke's law?
9. When the drop is falling under gravity, why does it move downwards with uniform velocity?
10. How does the determination of the charge on the various drop enable you to determine the electronic charge?

e/m BY THOMSON METHOD

11. What do you mean by e/m of an electron? What is its S.I. Unit?

12. What is a cathode ray tube? Give the construction of the tube.
13. Is the ratio of the electronic charge to mass dependent on the material of the cathode?
14. How do you produce cathode rays in your experiment?
15. What is the principle of the experiment?
16. How is the electron beam deflected from its path when the magnetic field is applied?
17. Does the application of the magnetic field alter the kinetic energy of the moving electron?
18. How do you produce a magnetic field and how do you measure it?
19. How do you apply the electrostatic field and how do you calculate it?
20. How can you proceed to get a well defined spot on the screen?
21. What is meant by accelerating voltage?
22. What value of accelerating voltage are you using here? How do you get such a large voltage?
23. Why are the deflecting plates short and curved at the end towards the screen?
24. Why do you have the plates so close to each other?
25. Is the value of e/m of an electron constant?
26. Why is the helical method so called? Explain.
27. Instead of solenoid, can any other source of magnetic field be used in this experiment?
28. Why CRT must be placed along the axis of the solenoid?
29. Write the expression for the magnetic field along the axis of a long closely wound solenoid.

e/m BY MAGNETRON

30. What is a solenoid? What is its role in this experiment? What is the direction of the magnetic field?
31. What is a magnetron valve?
32. What is the critical magnetic field? Explain the variation of anode current with the applied magnetic field?

34

PHOTO-ELECTRIC EFFECT

34.1 Photo-electric Effect

In 1888, Hallwachs shone ultraviolet light on to an electrically charged metal plate, standing on an electroscope. He found that the plate gradually lost its charge, if it had been negatively charged at the start, but not if it had been positively charged. Many years later, other experiments showed that ultraviolet light caused metal plate to emit negatively charged particles, which had the same charge to mass ratio as the cathode rays, *i.e.*, they were electrons. Hence, the emission is called *photo-electric emission* and the effect is known as *photoelectric effect*.

Monochromatic light falling on a metal plate A will liberate photoelectrons. These can be detected as a current if they are attracted to metal electrode B by means of a potential difference V applied between A and B such that B is at positive potential. A galvanometer could serve to measure this photo-electric current.

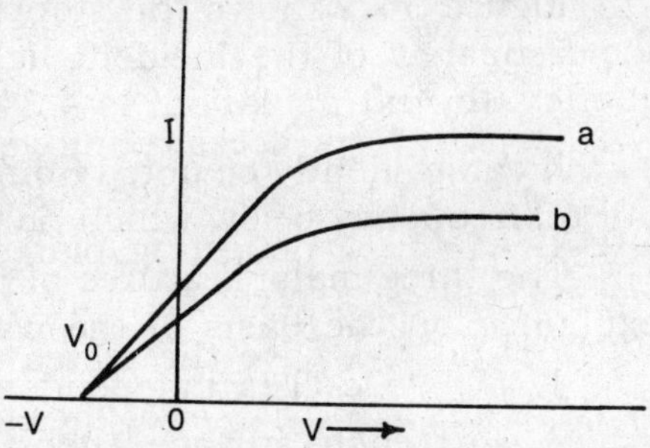

Fig. 34.1

Fig. 34.1 is a graphical representation of the data between the photo-electric current I and the corresponding potential difference V. If V is made large enough, the photo-electric current reaches a saturation value at which all photo-electrons emitted by the plate A are collected by the anode B. If V were reversed in sign, the photo-electric current does not immediately drop to zero; this proves that photo-electrons are emitted with a finite velocity. Some do reach the anode B despite the fact that the electric field opposes them. However, when this reverse potential is made large enough at a value V_0, the *stopping potential*, is reached at which the photo-electric current is zero. Evidently, eV_0 where e is the electronic charge measures the kinetic energy K_{max} for the fastest ejected photo-electrons, *i.e.*,

$$K_{max} = eV_0$$

From the graphs *a* and *b* corresponding to different intensities of the incident beam of light, it can be clearly inferred that K_{max} is independent of the intensity of light.

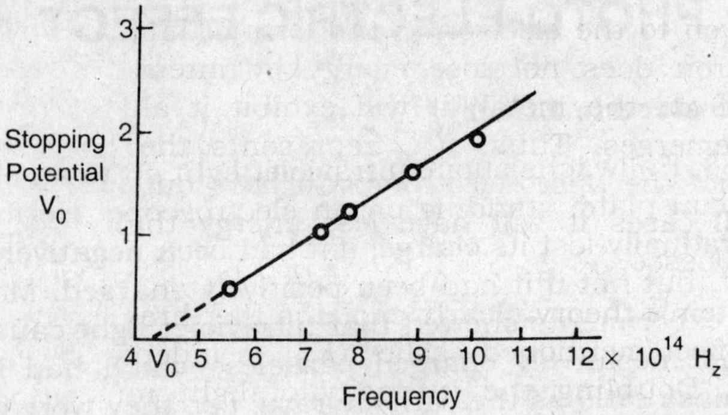

Fig. 34.2

Figure 34.2 shows the stopping potential V_o as a function of the frequency of the incident light for sodium. The cut off frequency (by extrapolation) is 4.39×10^{14} Hz.

It can evidently be noted from the graph that there is a definite cut off frequency below which no photo-electric effect occurs.

The three major features of photo-electric effect could not be explained on the basis of the wave theory.

1. K_{max} ($= eV_0$) is independent of the intensity of light.

2. For each surface there exists a characteristic cut off frequency ν_0. For frequencies less than this, the photo-electric effect disappears, no matter how intense the illumination.

3. No detectable time lag has never been measured between the falling of light and the ejection of the photo-electrons.

It was Einstein who successfully explained the photo-electric effect by making an assumption that energy in a light beam travels in concentrated bundles called PHOTONS. The energy E of a single photon is given by :

$$E = h\nu$$

where ν is the frequency and h, the Planck's constant.

Applying the photon concept to the photo-electric effect, Einstein wrote :

$$h\nu = W_0 + K_{max} \tag{34.1}$$

where $h\nu$ is the energy of the incident photon. The equation above expresses the fact that the photon carries the energy $h\nu$ into the surface. Part of this energy W_0 is used in causing the electrons to escape the metal surface. The excess energy ($h\nu - W_0$) is given to the electron in the form of kinetic energy; in case the electron does not lose energy by internal collisions, as it escapes from the metal, it will exhibit it all as kinetic energy after it emerges. Thus, K_{max} represents the maximum kinetic energy that the photo-electrons can have outside the surface, in nearly all cases it will have less energy than this because of internal losses.

Einstein's theory clearly explains the three features of photo-electric effect mentioned earlier. K_{max} is independent of the light intensity. Doubling the intensity of light merely doubles the number of photons and thus doubles the photo-electric current; it does not change the energy $h\nu$ of the individual photons or the nature of the individual photo-electric processes described by Eq. (40.1) above.

The existence of a characteristic frequency is explained in this manner. If K_{max} equals zero,

$$h\nu_0 = W_0$$

which asserts that the photon has just enough energy W_0 to eject the photo electrons but none extra to appear as kinetic energy. W_0 is called the work function of the substance. If ν is reduced below ν_0, the individual photons, no matter how many of them there are, (*i.e.*, no matter how intense the beam of light) will not have enough energy to eject photo-electrons.

Further, there is no question of a time lag between the falling of light and the ejection of a photo-electron because the requisite energy is supplied in a concentrated bundle.

34.2 Photo-electric Cells

There are three main types of photo-electric cells :

(i) Photo-conductive.
(ii) Photo-voltaic
(iii) Photo-emissive.

The oldest type is the photo-conductive type which suffers a change of resistance when illuminated. This is now of limited industrial importance. The photo-voltaic type is a generating cell and pass a current when illuminated. The photo-emissive type emits electrons from an illuminated cathode and requires a positive collecting anode like a thermionic valve.

34.3 Photo-emissive Cells

Figure 34.3 represents a common arrangement of electrodes. The cathode, in the form of a semi-cylindrical plate, has an emissive coating such as sodium, potassium or calcium, on the side facing the anode. For the use with ultraviolet light cadmium is used. The anode A is in the form of loop of wire of platinum (generally). To increase the number of photo-electrons per photon of incident light composite materials such as caesium on silver oxide (Ag. O–Cs) antimony caesium alloy, etc., are used as photo-cathode material. The electrodes are supported in an evacuated glass bulb fitted with a base like a thermionic valve; the anode and cathode connections being brought out at opposite ends of the envelope to reduce the possibility of leakage.

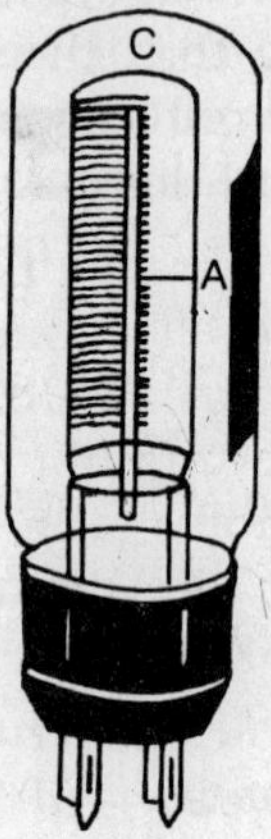

Fig. 34.3

The magnitude of the current in a vacuum type emissive cell is small. In order to increase it, the cell is filled with inert gas like neon or argon at a pressure of a few millimetre of mercury. Such a cell is known as a *gas-filled cell.*

34.4 Spectral Response

The emission varies with wavelength of the incident light and the alkali metals show a selective effect, with a maximum in or near the visible part of the spectrum, as shown in Fig. 34.4. The emission is practically independent of temperature, suggesting that the effect is in reality more like

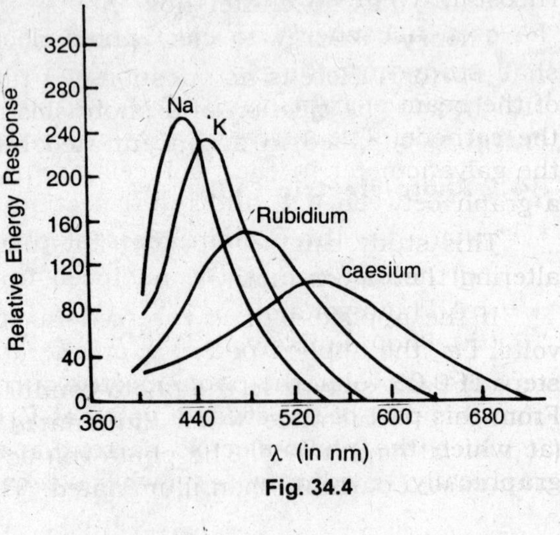

Fig. 34.4

ionisation than like thermionic emission, and that the electrons concerned are really those within the atom and not the free conductivity electrons.

Curves of these types are obtained experimentally in the following ways. Light from an incandescent lamp is passed through a prism of a monochromator for dispersion, a narrow band of wavelengths being selected by means of an approximately placed system. The current given by the photo-electric surface when exposed to the light passing through the system of slits is noted. The current given by a blackened thermo-pile when exposed to the same light is also noted. The ratio of these two readings is plotted vs. the mean wavelength of the incident light. Blackened thermopiles are used because they absorb all radiation incident upon them equally, regardless of the wavelength. This permits the measure of the energy contained in any part of the spectrum to be made. An automatic spectral sensitivity curve tracer has been designed for obtaining these curves quickly with the aid of a cathode ray tube.

Experiment 34.1

Object—To draw the characteristic curves of a photocell and to find the maximum velocity of the emitted electrons.

Apparatus—Photo-emissive cell mounted inside a blackened wooden box with a wide slit, a sensitive suspended coil galvanometer lamp and scale arrangement, an optical bench, power supply for d.c. anode potential or a dry battery 0-45 volts, a rheostat, a plug-key, a resistance box and connection wires.

Theory—To draw the characteristic curves of a photocell, we shall study how the photo-electric current, for a given intensity of the beam of light, varies with the p.d. between the anode and the cathode. The current being proportional to the deflection of the galvanometer in the photocell circuit, it would suffice to plot a graph between p.d. and the deflection.

This study can be repeated for different light intensities by altering the distance of the photocell from the lamp.

If the applied p.d. is reversed in small steps of, say, 0.05 volts, *i.e.,* the applied potential on the anode become negative in steps of 0.05 volts, the photo-electric current gradually decreases. From this part of the study a value of V_0 of the stopping potential (at which the photo-electric current is zero), can be computed graphically. Evidently, eV_0 where e is the electronic charge measures the kinetic energy K_{max} for the fastest ejected photo electrons, whence,

$$K_{max} = \frac{1}{2} m v_{max}^2 = eV_0$$

or $$v_{max} = \sqrt{\frac{2eV_0}{m}}$$

If e/m is taken as 1.76×10^{11} C/kg and V_0 in volts, then,

$$v_{max} = \sqrt{2 \times 1.76 \times 10^{11} \times V_0} \text{ m./sec.} \qquad (34.2)$$

Method—Make the connections as depicted in Fig. 34.5. Mount the photocell at one end of an optical bench. Arrange the source of light from the cell, say, at a distance of 60 cm. Place a suitable filter in the path of the rays coming from the lamp. Allow this light to fall on the cathode of the photocell, its anode being at a positive potential adjusted with the help of the potential divider R_1. Adjust the p.d., as recorded with the voltmeter to be, say, 20 volts. When light falls on the cathode, note the deflection of the galvanometer. If the deflection is large, control the same with the variable resistance R, usually a dial resistance box, placed in series with the galvanometer. The apparatus is now ready for use; other voltages used will have to be less than 20 volts.

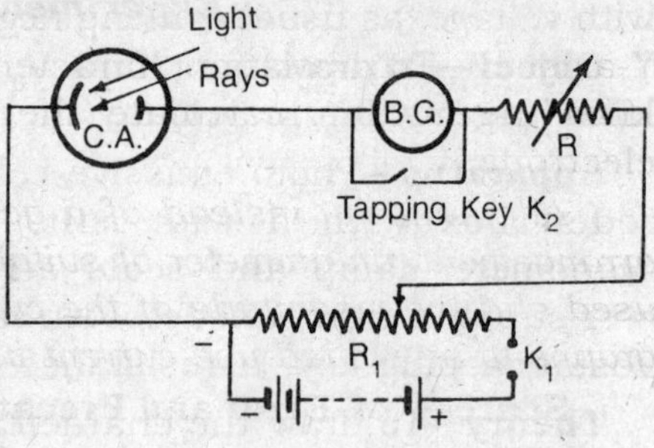

Fig. 34.5

(a) Anode voltage positive—Keeping the intensity constant we wish to study the relationship between anode voltage and anode current. Hence, in the first set of observations, keep the lamp at a distance of, say, 60 cm.

Allow the light to fall on the cathode, and note the deflection of the galvanometer. This corresponds to the anode voltage of 20 volts as recorded by the voltmeter. Repeat this by reducing the anode voltage first in steps of 2 volts and later in steps of one volt. Plot a graph between the anode potential and the corresponding deflection of the galvanometer.

This completes one set of observation.

Keep *R* fixed; increase the distance of the photocell from the source to, say, 80 cm, and repeat the whole set of observations and plot the graph.

Keeping *R* fixed, repeat the whole set of observations for distances between the source and the cell, say, 90 cm., 100 cm.....and draw corresponding graphs.

(b) Anode voltage negative—For a distance of 60 cm. from the source to the cell, repeat the experiment by applying a negative anode potential in steps of 0.05 volt. Make the necessary changes in the connections for this and use a low range voltmeter for reading the voltage. With the increase in negative potential the current and hence the deflection goes on rapidly falling. Take readings of the deflection for such negative potentials upto the stage when the deflection is reduced to 0.1 cm.

Repeat the experiment for a distance of, say, 100 cm, between the source and the cell.

Plot both these graphs on a different sheet of graph paper with voltage, as usually, along negative X-axis and deflection along Y-axis. By extrapolation, find V_0 the stopping potential. From a knowledge of V_0, calculate the velocity of the fastest ejected electrons.

Alternatively, instead of a galvanometer with lamp and scale arrangement, an ammeter of suitable range (10^{-3} to 10^{-12} A) can be used and the magnitude of the current can be read directly. In the graphs to be plotted use current in place of deflection.

Sources of Error and Precautions—(1) There should be no stray light in the room. It is desirable to perform the experiment in a dark room.

(2) It is often found that different portions of the same emitting surface may possess different sensitivities. It is advisable, therefore, to illuminate a large part of the cathode uniformly rather than to focus the light source on only a portion of the photo-emissive surface.

(3) For positive anode potential observations, readings of deflection should be taken by altering the anode potential from a maximum to lower values first in steps of 2 volts and later in steps of one volt.

(4) Negative anode potential should be varied in steps of 0.05 volt and the corresponding graph should be plotted on a separate graph sheet.

Observations

Distance between source and photocell = 60 cm.

Initial position of the spot of the gal. on the scale = 0.0 cm.

Table 1. Anode Potential Positive

Anode Potential volts	Deflection of Galvanometer cm.	Anode Potential volts	Deflection of Galvanometer cm.
18	24.8	6	20.4
16		5	19.7
14	23.8	4	18.7
12	23.3	3	17.4
10	22.0	2	15.1
8	21.5	1	5.3

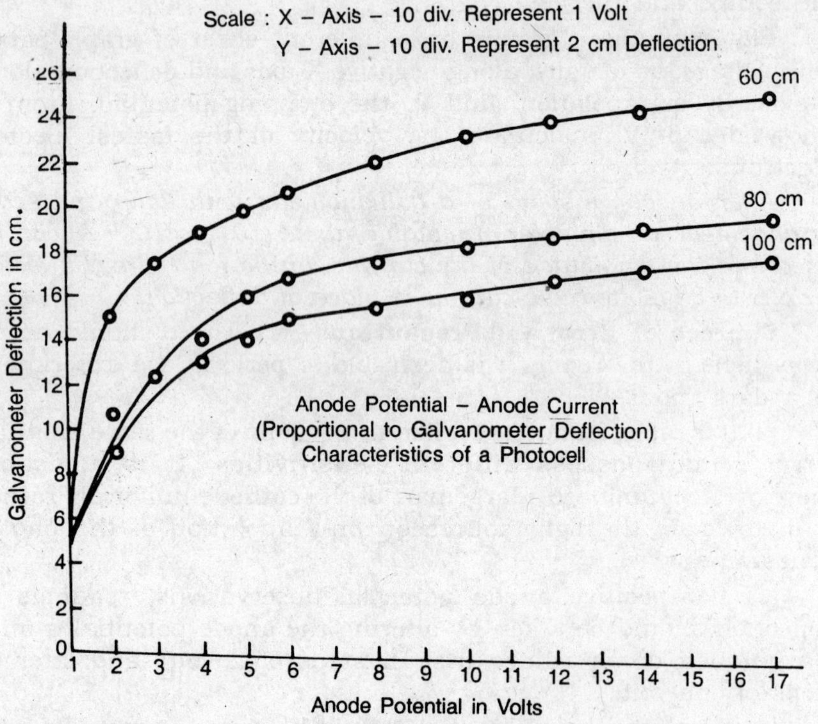

Fig. 34.6

Table 2. Anode Potential Negative

Anode Potential Volts	*Deflection of Galvanometer* cm.
–0.05	...
–0.10	...
–0.15	...
–0.20	...
–0.25	...

Similar Tables for distances of 80, 90.....cms.

Calculations—Plot graph between anode potential in volts and the corresponding deflection of the galvanometer in cms. of different distances between the source and the cell. This is shown in Fig. 34.6.

From Fig. 34.7, the negative stopping potential :

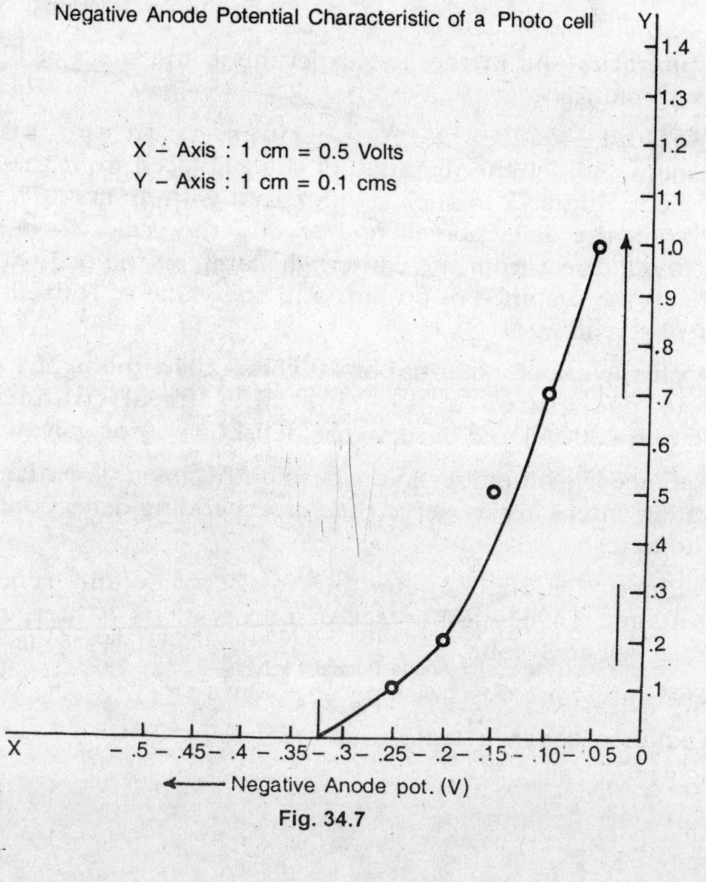

Fig. 34.7

$$V_0 = \quad \text{Volt}$$

$$\therefore \quad \upsilon_{max} = \sqrt{\frac{2eV_0}{m}}$$

$$= \sqrt{2 \times 1.76 \times 10^{11} \times V_0}$$

This gives the velocity of the fastest ejected electrons.

Result

1. Characteristic curves (anode voltage vs. galvanometer deflection—proportional to anode current) are shown in Fig. .....

2. The stopping potential at which the photo-electric current is zero is V_0 = volt.

3. The velocity υ_{max} of the fastest ejected electrons is given by υ_{max} = m/sec.

Exercise—To study the frequency response of a photo-electric cell.

Apparatus—As in the last experiment and a set of filters of different colours.

Method—Make the connections as in the previous experiment. Adjust the distance of the photocell from the source, say, 60 cm. Adjust R to such a value that with an anode potential of 20 volts the deflection is not beyond the scale when light is made to fall direct from the white light lamp on the cathode of the cell. Keep the distance of 60 cm. and the value of R fixed for the set of observations.

With the anode potential of 20 volts, allow the light from the lamp to pass through the blue filter (mean transmission wavelength 4300Å) and observe the deflection of the galvanometer.

Replace the blue filter in succession by green, greenish yellow and yellow filters and observe the corresponding deflections of the galvanometer.

Reduce the anode potential to 12 volts and repeat the observations. Take another set of observations with the anode potential set at 8 volts.

Tabulate your observations as follows :

Distance between lamp and cell = 60 cm.

Set No.	Anode Potential volts	Galvanometer Deflection in cms. with			
		Blue filter $\lambda =$ A^o	Green filter $\lambda =$ A^o	Greenish yellow filter $\lambda =$ A^o	Yellow filter $\lambda =$ A^o
1					
2					
3					

Plot a graph between mean wavelength transmitted by the filter along the *X*-axis and the corresponding galvanometer deflection along the *Y*-axis for a fixed anode potential. On the same sheet draw the similar graphs for other fixed anode potentials. Draw the conclusions and write them as results.

Experiment 34.2

Object—To determine the value of Planck's constant and work function of the material of the cathode of a photo-electric cell.

Apparatus—A vacuum type photo-electric cell (R.C.A. 935 or its equivalent) enclosed in a light-tight box containing a high intensity mercury arc light source, pico-ammeter, d.c. voltage source 6V, light filters (red, yellow, green, blue, violet), key, connecting wires.

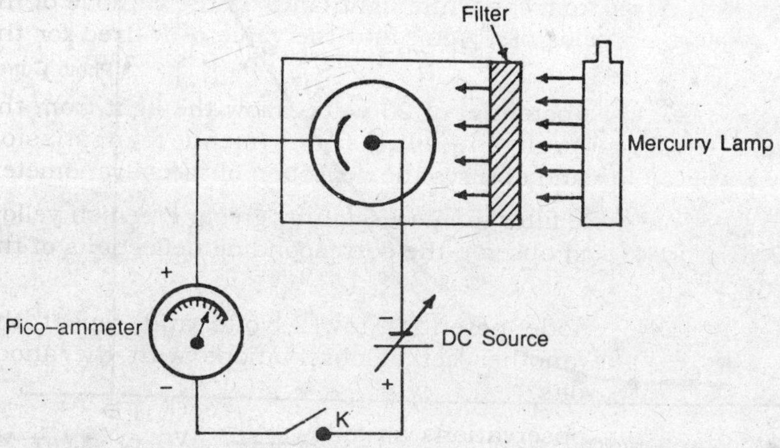

Fig. 34.8

Theory—The schematic circuit diagram of this experiment is shown in Fig. 38.8. When the light is incident through the filter onto the cathode, photoelectric emission begins and the current is measured in the pico-ammeter. The retarding potential, *i.e.*, negative potential on the anode is slowly increased till the stopping (or cut-off) potential is reached reducing the current to zero. In fact, due to some stray emissions, some time, the current may not actually reduce to zero but will acquire a small constant value (Fig. 34.9). Filters of different colours are used successively so that light of different frequencies are incident on the cathode. When the stopping potential is found for each frequency, a graph is plotted between *stopping potential vs. frequency* : (Fig. 34.10).

From Einstein's photoelectric equation, we have :

$$K_{max} = h\nu - W_0$$

work function $W_0 = h\nu_0$, Where ν_0 is the *threshold frequency* of the emitting surface. Also, $K_{max} = eV_0$, where V_0 is the stopping potential.

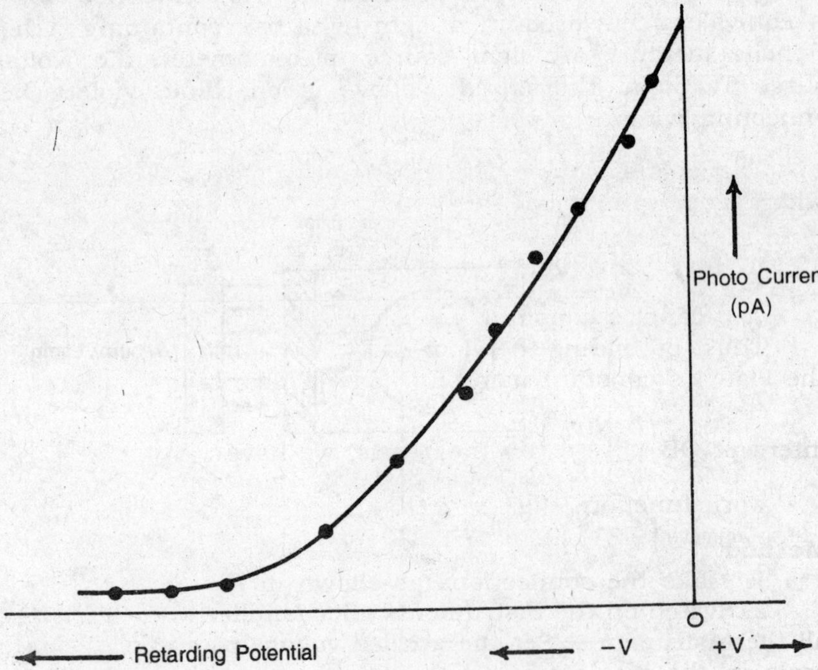

Fig. 34.9

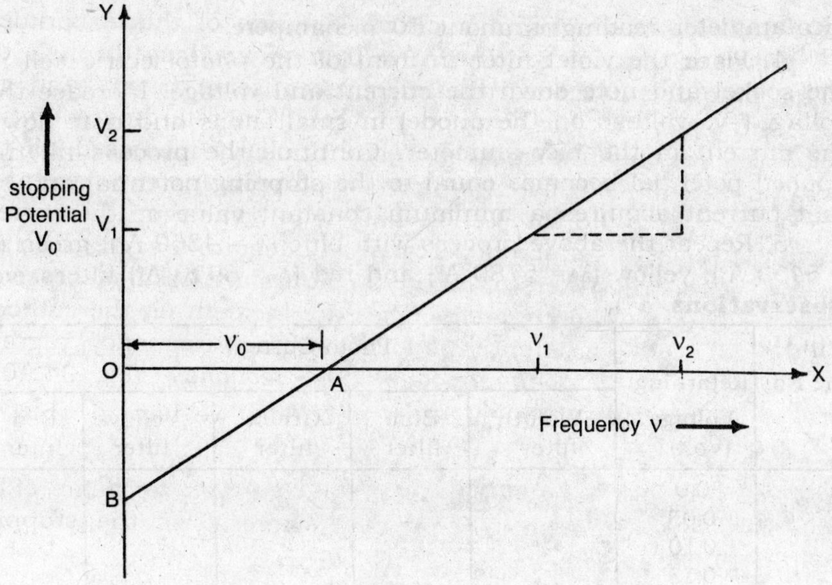

Fig. 34.10

$$\therefore \quad eV_0 = h\nu - h\nu_0$$

or
$$V_0 = \frac{h}{e}\nu - \left(\frac{h}{e}\right)\nu_0 \qquad (34.3)$$

This is analogous to the equation of a straight line $y = mx + c$, where slope m is equal to :

$$\text{slope} = \frac{h}{e}$$

$\therefore$ Planck's constant = e. slope (34.4)

Thus, by finding the slope of V_o vs. ν curve we can calculate the Planck's constant using Eq. (34.4). Also, by noting down the intercept OB $\left(= \frac{h\nu_0}{e}\right)$ on the y-axis, we have :

$$\text{work function} \quad W_0 = e.OB \qquad (34.5)$$

Method

1. Make the connections as shown in Fig. 34.8.
2. Switch on the instruments after familiarizing yourself with all the instruments. Set the applied voltage zero and select 1μ A scale in the pico-ammeter.
3. Adjust the mercury light source and the photo electric ce'' (if adjustable) so that light falls on the photocell window and th

pico-ammeter reading is about 50 picoampere.

4. Place the violet filter in front of the photoelectric cell in the socket and note down the current and voltage. Increase the voltage (–ve voltage on the anode) in small steps and note down the current in the pico-ammeter. Continue the process till the applied potential becomes equal to the stopping potential and so that current acquires a minimum constant value.

5. Repeat the above process with blue (λ = 4360 A°), green (λ = 5460 A°), yellow (λ = 5780 A°) and red (λ = 6910 A°) filters.

Observations

(μA) Sr. No.	Retarding Voltage (Volt) V	Photo-Current				
		Violet filter	Blue filter	Green filter	Yellow filter	Red filter
1.	0.0					
2.	0.05					
3.	0.10					
4.	0.15					
.	.					
.	.					
.	.					

Plot the graph between photocurrent and retarding potential *i.e.*, the potential when the photocurrent becomes nearly constant and minimum for all the filters. Then prepare the following table :

Filter	Violet	Blue	Green	Yellow	Red
Wave length λ	4050 A°	4360 A°	5460 A°	5780 A°	6910 A°
Frequency $\upsilon = \frac{c}{\lambda} Hz$					
Stopping potential V_o					

Plot a graph between frequencies and stopping potentials and calculate its slope.

$$\text{Shape} = \frac{V_2 - V_1}{\upsilon_2 - \upsilon_1} = \ldots\ldots \text{ volt/Hz} = \ldots\ldots\ldots \text{ volt sec.}$$

$\therefore$ Planck's constant h = e. slope

= 1.6×10^{-19} C × ...... Volt sec.

= .... Joule sec.

Work function W_0 = $e.\dfrac{h\nu_0}{e}$

= e.OB Joule

= .... electron-Volt

Result

1. Planck's constant h = ... Joule sec.
   Standard value = 6.63×10^{-34} Joule sec.
   % Error = %
2. Work function W_0 = ... eV
   Standard value = ... eV
   % Error = ... %

Precautions

1. It should be ensured that when no light is falling on the cathode the reading in the pico-ammeter is zero for the applied retarding potential.

2. When the value of retarding potential for which the photo-current becomes zero or a minimum, at least three more readings must be taken by further increasing the negative potential in very small steps.

3. If not fixed, it should be ensured that distances between light source, filter and photo-cell remain constant during the experimentation.

ORAL QUESTIONS

1. What is photo-electric effect? Who discovered this effect?
2. What are the laws of photo-electric emission?
3. What are the factors on which photo-electric current depends?
4. Is the photo-electric emission possible for all wavelengths of light?
5. What is photo-electric work function?
6. What is the difference between photo-electric emission and thermionic emission.

7. Explain Einstein's photo-electric equation.
8. What is threshold frequency?
9. What is a photo-cell? How many types of photo cell do you know?
10. What is the difference between gas-filled and vacuum photo cell?
11. Of what material is the cathode of a photo-voltaic cell made?
12. What type of cell are you using? What is the shape of the cathode and why? What is the shape of the anode and why?
13. What do you mean by spectral response of a photo cell?
14. What do you mean by characteristics of a photo cell and how do you study them?
15. How do you calculate the velocity of the fastest ejected electron?
16. Why is the cut off voltage negative? Is the cut off voltage dependent on the intensity of light?
17. To what uses is a photo cell put?
18. What is Planck's constant? What are its units?
19. Does a photon of blue light have more energy or red light?
20. What is the order of photocurrent in a photo electric cell?

35
Radioactivity – Geiger Muller Counter

35.1 Geiger Muller Counter

A schematic diagram of a G.M. counter is shown in Fig. 39.1. It consists of a glass cylinder *G* with a metal coating on the inside and a co-axial wire anode *A* (normally of tungsten) placed along the axis of the cylinder. The cylinder is filled with a gas such as argon mixed with halogen vapour at a low pressure of about 100 mm of mercury. The cylinder has a thin mica window through which the incident particles enter the tube. The central wire is maintained at a high positive potential (≈ 1 k V) w.r.t. the outer cylinder. The applied potential should be greater than certain *threshold* value but less than the voltage which cause a continuous discharge of electricity through the enclosed gas. Because the central wire is very thin, the electric field in its vicinity becomes very high.

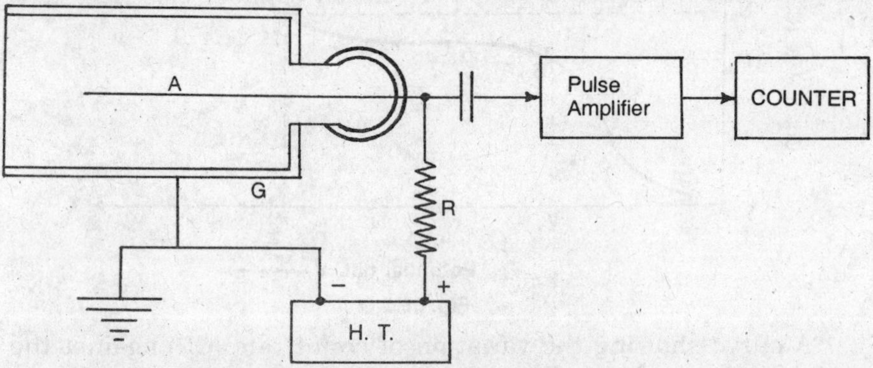

Fig. 35.1

When a radioactive source is placed near the window, inonising particles like β-particles or γ-ray photons enter the cylinder and cause *ionization* of gas molecules. Suppose, an incident ionizing particle produces one single electron-ion pair, the resulting electron will be accelerated towards the central positive wire and will attain a relatively high speed. Due to repeated collisions with other gas molecules it produces a large number of additional ion pairs. These new electrons are also accelerated and produce many more ion pairs. As a result, an *avalanche* occurs and a very large number of electrons reach the anode. In this way, the initial formation of a single ion pair results in a very *large pulse of current* through the resistance *R*. The resulting drop in potential across *R* decreases the potential across the tube below the threshold value and the discharge in the tube is extinguished and the current decays rapidly owing to small time constant of the circuit. It takes some time to recover to a new particle. This period is called *recovery* or *dead time* of the counter as during this period it becomes insensitive to the incident particle. The dead time sets an upper limit to the *counting rate* (CR) of the tube. The successive current pulses generated in the resistance corresponding to the ionizing particles into the tubes are detected and counted by a suitable arrangement.

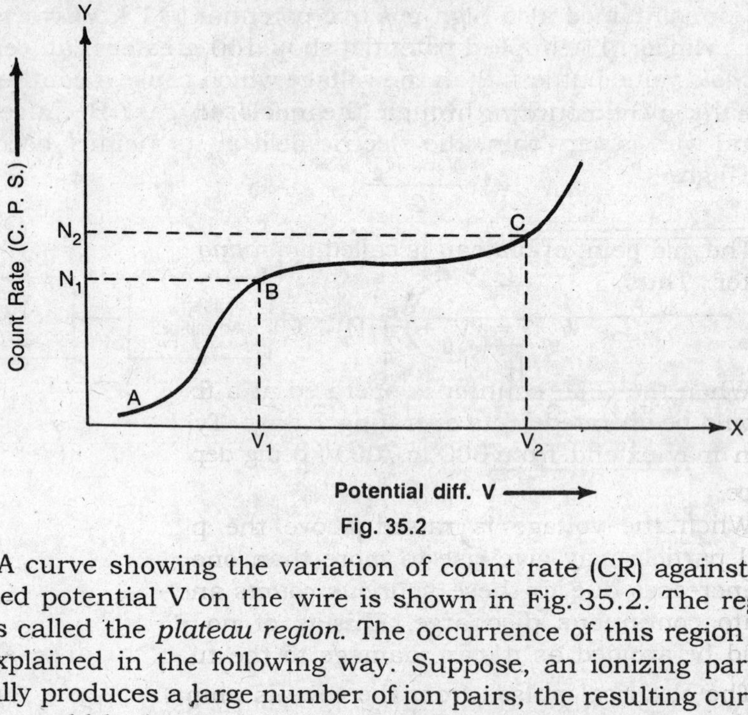

Fig. 35.2

A curve showing the variation of count rate (CR) against the applied potential V on the wire is shown in Fig. 35.2. The region BC is called the *plateau region*. The occurrence of this region can be explained in the following way. Suppose, an ionizing particle initially produces a large number of ion pairs, the resulting current pulse would be larger than that originating from a single ion pair. For moderate values of the potential difference between anode and cathode this is found to be the case. However, as the potential difference is increased, this variation in the size of the current pulse gradually disappears. Ultimately, it is found that the size of current pulse becomes effectively *independent* of the initial density of ionization for quite a range of potential difference. The counter then gives a full avalanche of nearly constant size for any number particle counted. In this region, the counter is said to be in the *plateau* or *Geiger region.*

In Fig. 35.2, the voltage corresponding to point A at which pulses begin to appear is known as *threshold voltage* or *Geiger Muller threshold potential.* Beyond this potential, the counting rate increases linearly upto the point B and then the counting rate remains largely constant upto the point C. This region BC is the plateau region and the voltage from B to C, $(V_2 - V_1)$ is the *length of plateau.* The slope of plateau is the average percentage change in counting rate per unit increment of applied voltage and is known as *relative plateau slope* (R.P.S.). Thus,

$$\text{R.P.S.} = \frac{(N_2 - N_1)/N}{V_2 - V_1} \times 100 \qquad (35.1)$$

where N_1 is the counting rate at V_1 and N_2 that at V_2 while

$$N = \frac{N_1 + N_2}{2}$$

The mid point of plateau is called *operating voltage* of the G.M. counter. Thus,

$$V_{op} = V_1 + \frac{1}{2}(V_2 - V_1) \qquad (35.2)$$

When the G.M. counter is operated at a fixed voltage, ideally it should be operated at its operating voltage. Typically, the plateau region may extend from 300 to 700 V, being dependent on the type of tube.

When the voltage is raised above the plateau region one initial particle may give rise to more than one count. The count rate increases due to these *spurious counts* and the counter can go into continuous discharge. This is of no practical use and should be avoided as it can damage to the tube.

The electrical pulses from the tube is amplified before being fed to a suitable counter. The front panel of a microprocessor controlled G.M. counter is shown in Fig. 35.3.

Experiment 35.1

Object—To plot the characteristics of a G.M. counter and determine its plateau region and the operating voltage.

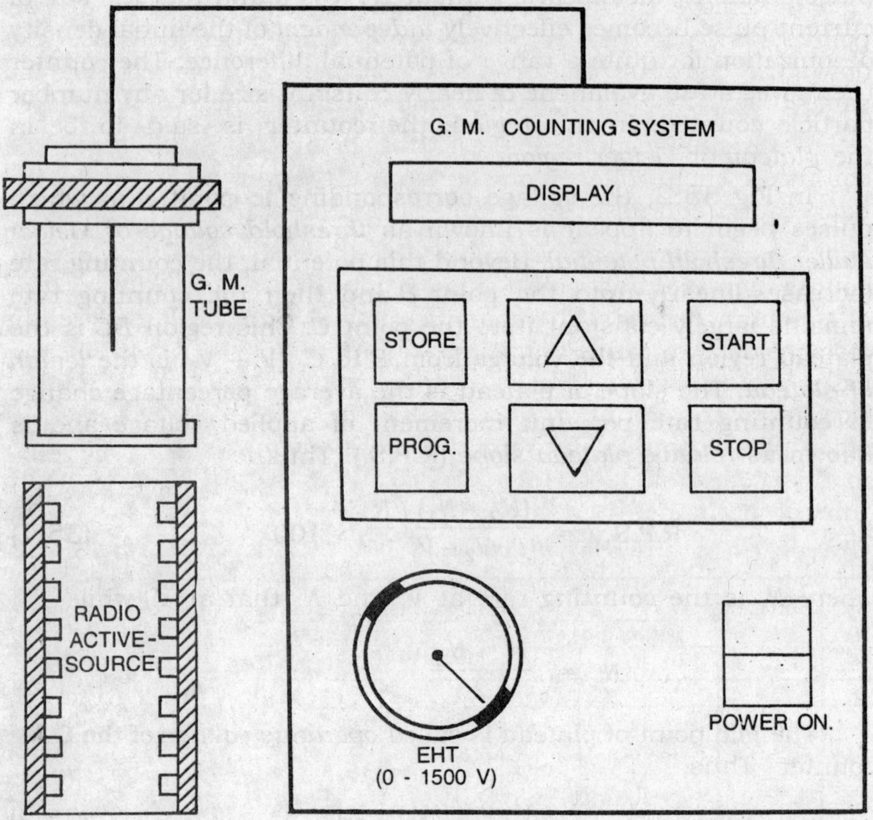

Fig. 35.3

Apparatus—G.M. counter with high voltage d.c. supply radioactive source (e.g., Sr^{90}), time measurement device (inbuilt or stopwatch), a source holder.

Theory—As discussed in Section 35.1.

Method

1. Switch on all the components after making proper connections. Place the radioactive source near the counter tube.

2. If your system is microprocessor based, select a desired time (say, 30 sec.) and press START button to start the counting. Gradually increase the HT to around 300V–350V to find the counting as displayed on the displayer. This point is called STARTING VOLTAGE of the G.M. tube. For other types of GM counter with the help of stopwatch do the above process for a fixed time.

3. Remove the radioactive source to a considerable distant point so that particles from the source do not enter the tube. Note down the counting on the display at the same voltage for the same time. This is the background count at that voltage.

4. Gradually increase the HT by 10 or 15 V and record the counts at each voltage for the same duration. Also record the background counts at each voltage for the same duration.

5. Continue the above process. It will be observed that initially counts increase drastically. This portion is called the *knee portion* of the characteristics. Thereafter, it remains more or less constant for considerable range of voltage (say, upto 750 V). This region is called *plateau region*. Thereafter there will be steep rise in counts. Now, stop the counting and reduce the voltage. This region of steep rise in counts is called *discharge region* and the operation in this region should be avoided as it can reduce the life of the tube or permanently damage it.

Observations and Calculations

Sr. No.	Voltage	Counts for t (=30 sec)					Counting rate $N = \frac{N'-N''}{t}$ (Sec^{-1})
		with radioactive source				without source	
		(i)	(ii)	(iii)	Mean N'	N"	

Plot a graph between N Vs t and determine the following parameters :

(i) Threshold voltage = voltage at A = ... Volt.

(ii) Voltage at which plateau region starts = voltage at B = V_1 = ... Volt.

(iii) Voltage at which plateau region ends = voltage at C = V_2 = ... Volt.

Plateau region = $V_2 - V_1$ = ... Volt

$$\text{Operating voltage} = V_1 + \frac{1}{2}(V_2 - V_1)$$

$$\text{Plateau slope} = \frac{(N_2 - N_1)/N}{V_2 - V_1} = \ldots\ldots = \ldots\ldots \times 100\%$$

Sources of Errors and Precautions

1. Keep the HT voltage low before switching off the system.
2. The radioactive source should be handled carefully by a pair of tongs or forceps.
3. The counting switch should always be kept ON so that counting rate do not exceed 10% of the upper limit of the plateau region as it can damage the tube.
4. The radioactive source should be placed in the source holder just below the G.M. Tube.

35.2 Dead Time of a G.M. Counter

The time interval during which a G.M. counter is completely insensitive to incident particles after recording a count due to ionisation inside the tube is called its dead time.

Suppose, a radioactive source gives n counts per second in a counter with zero dead time and the same source gives N counts per second in a counter with dead time T sec., then we can write :

$$T = \frac{1}{N} - \frac{1}{n}$$

or $$n - N = nNT \tag{35.3}$$

That is, the counter does not count $n - N$ counts owing to its being dead for the period T.

In order to obtain an expression for T, suppose, we have two radioactive sources of β particles (say, Sr^{90} and Tl^{204}). The two sources give N_1 and N_2 counts per second separately and N_3 counts per second when placed together. Therefore,

$$n_1 - N_1 = n_1 N_1 T \tag{35.4}$$

$$n_2 - N_2 = n_2 N_2 T \tag{35.5}$$

$$(n_1 + n_2) - N_3 = (n_1 + n_2) N_3 T \tag{35.6}$$

where n_1, n_2 are counts per second recorded by a counter with zero dead time for the two sources separately and $(n_1 + n_2)$ counts per second recorded when placed together.

From Eq. (35.4), we have :

$$n_1 (1 - N_1 T) = N_1 \Rightarrow n_1 = \frac{N_1}{1 - N_1 T} \quad (35.7)$$

Similarly, from Eq. (35.5) we have : $n_2 = \dfrac{N_2}{1 - N_2 T}$ (35.8)

Substituting Eqs. (35.7) and (35.8) in Eq. (35.6) we have :

$$\left(\frac{N_1}{1 - N_1 T}\right) + \frac{N_2}{(1 - N_2 T)} - N_3 = \left(\frac{N_1}{1 - N_1 T} + \frac{N_2}{1 - N_2 T}\right) N_3 T$$

or $$N_1 + N_2 - N_3 - 2 N_1 N_1 T + N_2 N_3 T + N_1 N_3 T - N_1 N_2 N_3 T^2 = N_1 N_3 T - N_1 N_2 N_3 T^2 + N_2 N_3 T - N_1 N_2 N_3 T^2$$

Neglecting the term containing T^2, we get,

$$N_1 + N_2 - N_3 - 2 N_1 N_2 T = 0$$

or $$T = \frac{N_1 + N_2 - N_3}{2 N_1 N_2} \quad (35.9)$$

Experiment 35.2

Object—To find the dead time of a G.M. counter.

Apparatus—A G.M. counter with high voltage d.c. supply, time measurement device (inbuilt or stopwatch), a double source holder, two radioactive source (e.g., Sr^{90} and Tl^{204})

Theory—As discussed in Article 35.2.

Method

1. As discussed in Expt.35.1, find out the operating voltage of the given G.M. tube. Adjust the power supply at the operating voltage.

2. Find out the background counting rate N_b' at the operating voltage and record it. Do not change this voltage (i.e., operating votage) throughout the experiment.

3. Place the first source S_1 on one of the source holders (Fig. 35.4) and determine the counting rate N_1.

4. Place the second source S_2 on the other source holder and determine the counting rate N_3 of the two sources together.

5. Remove the first source S_1 without disturbing the source S_2 and determine counting rate N_2 for the second source.

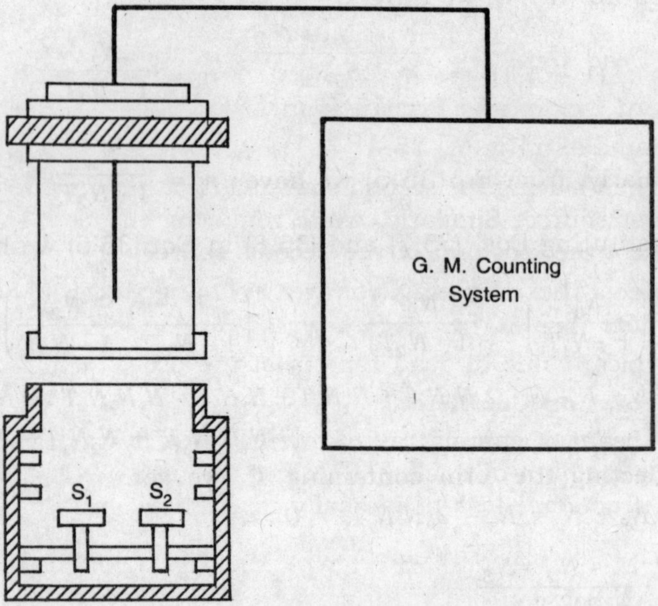

Fig. 35.4

6. Remove the second source now and again determine the background counting rate N_b''.

7. Repeat the above process three times for the same operative voltage.

Observations and Calculations

Operative voltage of the G.M. tube = ... Volt

Background counting rate at the beginning N'_b = ... Sec^{-1}

Background counting rate at the end N''_b = ... Sec^{-1}

Average background counting rate $N_b = \dfrac{N'_b + N''_b}{2}$ = .. Sec^{-1}

Sr.No.	Source	Counts in t (t = 30 sec.)				Observed Counting rate (sec^{-1})	Corrected Counting rate (sec^{-1})
		(i)	(ii)	(iii)	Mean		
1	S_1						N_1
2	$S_1 + S_2$						N_3
3	S_2						N_2

$$\text{Dead time T} = \frac{N_1 + N_2 - N_3}{2N_1N_2} = \text{Sec.}$$

Sources of Errors and Precautions

1. Same as in Expt. 35.1

2. When placing the second source do not disturb the position of the first source. Similarly, while removing the first source do not disturb the position of the second source.

3. Keep the operating voltage unchanged throughout the experiment.

Note : Typical value of dead time is of the order of 10^{-3} sec.

35.3 Absorption Coefficient

If a beam of gamma ray of intensity I is passed through an absorbing medium of thickness dx and its intensity is reduced by dI, then fractional loss of intensity is given by :

$$-\frac{dI}{I} \propto dx$$

$$\text{or} \quad -\frac{dI}{I} = adx$$

where a is *absorption coefficient* or *li·ear attenuation coefficient.* Integrating the above expression taking $I = I_0$ at $x = 0$ and $I = I$ at $x = 0$, we have :

$$\int_{I_0}^{I} \frac{dI}{I} = -a \int_0^x dx$$

$$\text{or} \quad \left|\log_e I\right|_{I_0}^{I} = -a\left|x\right|_0^x$$

$$\text{or} \quad \log_e I - \log_e I_0 = -ax \qquad (35.10)$$

$$\text{or} \quad \log_e \frac{I}{I_0} = -ax$$

$$\text{or} \quad \frac{I}{I_0} = e^{-ax}$$

$$\text{or} \quad I = I_0 e^{-ax} \qquad (35.11)$$

That is, the intensity falls exponentially with thickness.

Eq. (35.10) can also be expressed as :

$$\log_e I = \log_e I_o - ax$$

This equation is identical with the equation of a straight line $y = mx + c$. Thus, a graph between I and x is a straight line with its slope equal to the absorption coefficient a of the material with a negative sign.

The β-particles being charged particles interact with the absorbing material. However, its penetrating power is large for material like aluminium and its intensity decreases exponentially with thickness for small thickness. Hence, the above relation :

$$I = I_o e^{-ax}$$

or $$\log_e I = \log_e I_o - ax$$

also applies to absorption of β-particles by thin foil of aluminium. The value of I_o and that of I for different values of thickness x is determined with the help of a *G.M.* counter. The value of I is proportional to the number of β-particles in the beam which in turn is proportional to the number of counts recorded in a G.M. counter. If the counts N_0 and N are proportional to I_0 and I, we have :

$$N = N_o e^{-ax}$$

or $$\log_e N = \log_e N_o - ax$$

Thus, a graph between N and x is a straight line with its slope equal to the absorption coefficient with a negative sign.

$$\therefore \quad \text{Slope} = a = \frac{\log_e N}{x}$$

or $$a = \frac{2.3026 \log_{10} N}{x} \qquad (35.12)$$

If x is taken in metre, then, a is in m^{-1}. However, more commonly x is measured in gm/cm^2 or mg/cm^2.

Experiment 35.3

Object—To find the absorption coefficient of β-particles in aluminium using a G.M. counter.

Apparatus—In addition to those mentioned in Expt. 39.1, a set of (containing at least five) aluminium foils having different mass per unit area.

Theory—As discussed in Article 35.3.

Method

1. Determine the mass per unit area (mg/cm^2) of the set of aluminium square or rectangular foil, if not marked on it. Arrange the foils in increasing order of mass per unit area.

2. Determine the operating voltage of the G.M. tube and adjust the power supply at the operating voltage. Using the forceps, place

the β-source on the source holder below the G.M. tube and determine the counting rate without placing any aluminium foil piece, i.e., with zero thickness of the absorbing material.

3. Now, place the aluminium foil having minimum mass per unit area in between the source and G.M. tube and determine the counting rate.

4. Determine the counting rate with all other aluminium foil in succession.

5. Note the operating voltage and determine the background counting rate at the end.

Observations and Calculations

Operative voltage of the G.M. tube = Volts

Initial background rate $N'_b = \frac{\text{counts}}{\text{time}} = \ldots\ldots\ \text{Sec}^{-1}$

Initial background rate $N''_b = \frac{\text{counts}}{\text{time}} = \ldots\ldots\ \text{Sec}^{-1}$

Average background counting rate $N_b = \frac{N'_b + N''_b}{2} = \ldots\ \text{Sec}^{-1}$

Sr.No.	Thickness of foil (mg/cm²)x	Counting rate $\frac{\text{counts}}{\text{time}}$ (Sec⁻¹)	corrected counting rate $N = N_o - N_b$	$\log_{10}N$
1				
2				
.				
.				
.				
.				
.				
5				

Plot a graph between thickness (along x-axis) and $\log_{10}N$ (Fig. (35.5). If corresponding to x_1 and x_2 the values are $\log_{10}N_1$ and $\log_{10}N_2$, then,

$$\text{slope of the graph } m = \frac{\log_{10} N_1 - \log_{10} N_2}{x_1 - x_2}$$

$\therefore$ Absorption coefficient $a = 2.3026\,(-m) = \ldots\ldots\ldots\ \text{cm}^2/\text{mg}$.

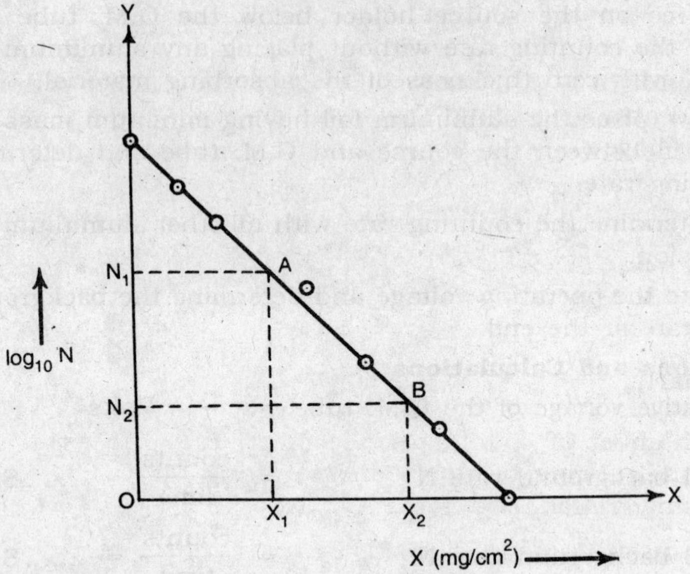

Fig. 35.5

Precautions

1. Same as in Expt. 35.1

2. The location of all the foil w.r.t. source and G.M. tube should be exactly identical.

35.4 Determination of Half-life Period of a Radioactive Source

A radioactive source decays according to the relation :

$$N = N_0 e^{-\lambda t} \tag{35.13}$$

where N_0 is the number of radioactive atoms at t = 0, N is the number of atoms at the time t and λ is a constant called *disintegration* or *decay constant.* The *half-life period* $T_{1/2}$ of a radioactive element is the time in which half of the original atoms get decayed, i.e., at $t = T_{1/2}$, $N = \dfrac{N_0}{2}$

$$\therefore \quad \frac{N_0}{2} = N_0 e^{-\lambda T_{1/2}}$$

$$\text{or} \quad \frac{1}{2} = e^{-\lambda T_{1/2}} \quad \Rightarrow \quad 2 = e^{\lambda T_{1/2}}$$

$$\therefore \quad \lambda T_{1/2} = \log_e 2$$

or $$T_{1/2} = \frac{\log_e 2}{\lambda} = \frac{0.6931}{\lambda} \tag{35.14}$$

$T_{1/2}$ using a G.M. counter—Let n_1 be the number of counts per second at a time t_1 and that n_2 at the time $t = t_2$. Using Eq. (35.13) we have :

$$N_1 = N_0 e^{-\lambda t_1}$$

$$\therefore \quad \frac{dN_1}{dt} = -\lambda N_0 e^{-\lambda t_1} = -\lambda N_1 \tag{35.15}$$

Similarly, $$\frac{dN_2}{dt} = -\lambda N_0 e^{-\lambda t_2} = -\lambda N_2 \tag{35.16}$$

The number of counts per second in the G.M. counter is equal to the number of atoms disintegrating per second or the rate of disintegration. Therefore, from Eqs. (35.15) and (35.16), we have :

$$n_1 = -\lambda N_0 e^{-\lambda t_1} \tag{35.17}$$

$$n_2 = -\lambda N_0 e^{-\lambda t_2} \tag{35.18}$$

Dividing Eq. (35.17) by (35.18), we have :

$$\frac{n_1}{n_2} = e^{\lambda(t_2 - t_1)}$$

or $$\log_e\left(\frac{n_1}{n_2}\right) = \lambda\,(t_2 - t_1)$$

$$\therefore \quad \lambda = \frac{\log_e\left(\frac{n_1}{n_2}\right)}{t_2 - t_1} = \frac{2.303\log_{10}\left(\frac{n_1}{n_2}\right)}{t_2 - t_1}$$

$$\therefore \quad \text{Half-life } T_{1/2} = \frac{0.6931}{\lambda} = \frac{0.6931}{0.2303}\,\frac{t_2 - t_1}{\log_{10}\left(\frac{n_1}{n_2}\right)}$$

or $$T_{1/2} = \frac{0.30096\,(t_2 - t_1)}{\log_{10}\left(\frac{n_1}{n_2}\right)} \tag{35.19}$$

Experiment 35.4

Object—To determine the half-life period of given radioactive source using a G.M. counter.

Apparatus—Same as in Expt. (35.1) with a suitable radioactive source.

Theory—As discussed in Article 35.4.

Method

1. Set the power supply voltage of the G.M. tube at its operating voltage.

2. Determine the background counting rate.

3. Place the source using forcep just below the G.M. tube and switch on the counter and timer. Exactly after one minute, stop the counter but let the time recording continue. Note down the counts in initial one minute.

4. Be careful and as soon the timer records 5 minutes, restart the counter and again measure the counts in the next one minute, i.e., as soon as the time completes 6th minute, switch off the counter and record it while the timer is again to be kept on.

5. Similarly, measure the counts between 10th and 11th minute. Keep on the process for one hour.

6. Remove the source and again measure the background counting rate.

Observations and Calculations

Operative voltage of the G.M. tube = ... Volts

Background counting rate at the beginning n'_b = ... min^{-1}

Background counting rate at the end n''_b = ... min^{-1}

Average background counting rate $n_b = \dfrac{n'_a + n''_b}{2}$ = ... min^{-1}

Sr.No.	Time from the start t in minutes	counting rate per minute	
		Observed n_0 (min^{-1})	Corrected $n = n_0 - n_b$ (min^{-1})
1	0		
2	5		
3	10		
.	.		
.	.		
.	.		
12	60		

Plot a graph between t (along x-axis) and the corrected counting rate (along y-axis). By taking two suitable points in the middle portion of the graph determine the counting rate per minute (Fig. 35.6)

$$\text{Half-Life period } T_{1/2} = \frac{0.30096\,(t_2 - t_1)}{\log_{10}\left(\dfrac{n_1}{n_2}\right)}$$

$$= \ldots\ldots\ldots\ldots \text{ min.}$$

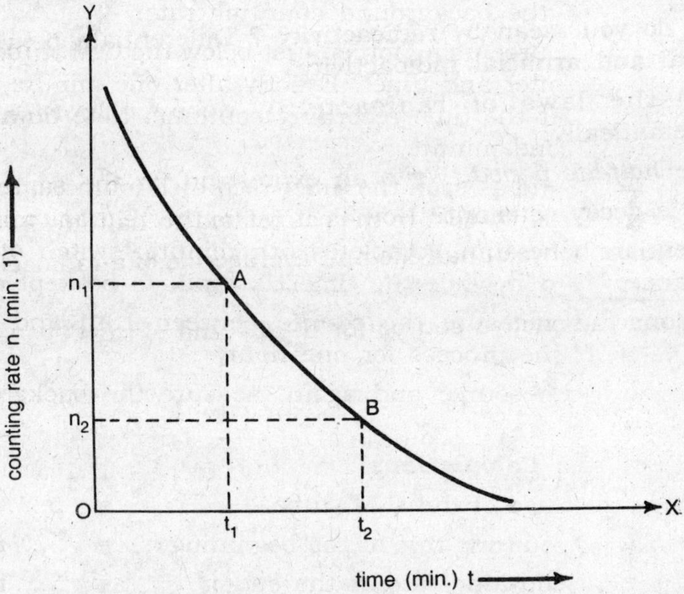

Fig. 35.6

Precautions

1. Same as in Expt. 35.1

2. The radioactive source selected should have a half-life period longer than the time taken for noting radioactive decay.

3. Care should be taken to start the timer and counter simultaneously.

ORAL QUESTIONS

G.M. COUNTER

1. What is a G.M. counter ? Explain its working.

4. What is G.M. threshold potential ?
5. Define relative slope of plateau for a G.M. counter.
6. Define *dead time* of a G.M. tube. Explain its origin.
7. Mention the advantages and disadvantages of a G.M. counter.
8. What is *gas multimplication* ?
9. What do you mean by avalanche ? How does it occur ?
10. What is discharge in G.M. Tube ? How is it overcome ?
11. Define absorption coefficient of a material for the incident nuclear radiations. What is its unit ?

RADIOACTIVITY

12. What do you mean by radioactivity ? Differentiate between natural and artificial radioactivity.
13. State the laws of radioactivity decay. Express it mathematically.
14. Define *half-life period.* Write an expression for the same.
15. What is decay constant? How is it related to half-life ?
16. Differentiate between half-life and average-life of a radioactive substance.
17. Name one radioactive source for α–, β– and γ, rays.

36

Cathode Ray Oscilloscope

36.1 Cathode Ray Oscilloscope

For the study of any transient phenomena, *i.e.*, either a potential or a current that varies with time, it is necessary to have an instrument that draws out the potential or current on, say, a vertical axis while the elapsed time is run out along the corresponding horizontal axis. This was initially achieved by mechanical devices. But with the discovery of the fact that beams of electrons can be deflected by transverse electric and magnetic fields and that electrons have very little inertia, it has been possible to so use the electron beam in oscillographs. This led to the development of the cathode ray oscilloscope.

The cathode of an oscilloscope tube is usually an indirectly heated nickel cylinder *C* whose flat end is oxide coated. The cylindrical control electrode *G* lies coaxial with *C* and encloses a diaphragm through which the electrons pass on their way towards

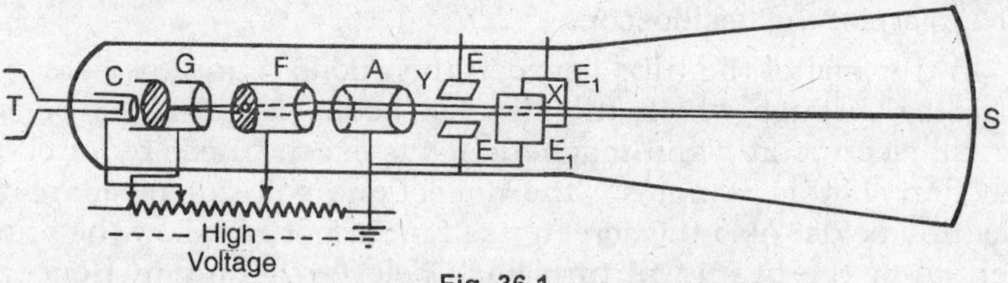

Fig. 36.1

the anode *A*. This control electrode is often called the grid and it has a negative potential with respect to the cathode. By varying the grid potential, the intensity of the electron beam and the brilliance of the spot on the screen can be controlled. As the electrons travel between the cathode and the anode *A*, they are accelerated by the positive anode potential, which may be a thousand volts or more with respect to the cathode. The anode *A* is an open cylinder and the electrons because their high speed pass through it to the screen *S*. The focusing may be achieved by magnetic means or by an electrostatic field if it has a radial component which causes the electrons to be deflected towards the axis. This component is obtained by an additional cylindrical electrode *F* called the *focusing electrode* at a potential lower than that of *A*. *F* and *A* are sometimes referred to as the first and the second anode respectively and because of their difference in potential, there is an electrostatic field inside each cylinder, near the gaps between them. The non-uniformity of this field gives it the radial component necessary for focusing.

Between the anode *A* and the screen *S* are a system of *deflecting plates*. These consist of two small pairs of metal plates *EE* and E_1E_1. In the figure they are shown separated along the beam axis, though in many oscilloscopes they are coterminous. *EE* are horizontal and represent a parallel plate capacitor system with a field *Y* determined by the plate configuration and the potential V_y applied to it. This potential V_y causes the vertical deflection and comes from the amplifier connected to the transient that is desired to be studied. The vertical plates E_1E_1 give an electric field *X* between them when a potential V_x is provided across them and cause a horizontal deflection of the electron beam. One plate of each pair is connected to the anode. If this is not done there would be an electrostatic field between them and the electron gun. Further, since it would be dangerous to have the deflector plates at a high potential, the anode is earthed, and the cathode and the other electrodes are given negative potential with respect to the earth. The free plate in each deflector system is usually referred to the *X*- and *Y*-plate of the oscilloscope.

In the part of the tube between the anode *A* and the screen *S*, the space through which the electron beam passes must be free from all electrostatic and magnetic fields except those required for deflection. If this is not so, the deflections will not measure the deflecting fields. Also the screen itself must not be below the anode potential or the beam will turn back before reaching it. Hence, to make this space equipotential, the inner walls of the tube are coated with graphite which is connected to the second anode *A*. To exclude stray magnetic fields the tube is surrounded by a shield of high-permeability alloy. The screen is coated with a fluorescent material such as zinc orthosilicate and when the electron beam is focused, a brilliant spot is visible on the screen due to fluorescence.

Time Base for C.R.O. Suppose, an alternating p.d. is applied to the *Y*-plates *EE*. The spot on the screen traces out a vertical line proportional to the applied voltage. In order to convert this line into a wave, some other deflecting force, whose direction is perpendicular to that of the alternating one, is required to provide a "*time base*". If a steady voltage is applied to the *X*-plates E_1E_1 (which are not connected to the alternating supply), the effect will be merely to displace the line traced out by the spot from its initial position by a certain definite amount. However, if this voltage can be made to increase in magnitude at a constant rate, the displacement of the spot in the direction perpendicular to that of the deflection due to the alternating field, will be at a uniform rate and a wave with a linear time base will be the result. In this manner the wave form of the alternating p.d. across *Y*-plates can be obtained on the screen of the oscilloscope.

One time base circuit is described below. Refer to Fig. 36.2. The capacitor *C* is charged through the high resistance R_1 at a nearly constant rate until the p.d. across *C* is sufficient to fire the thyratron *V*.

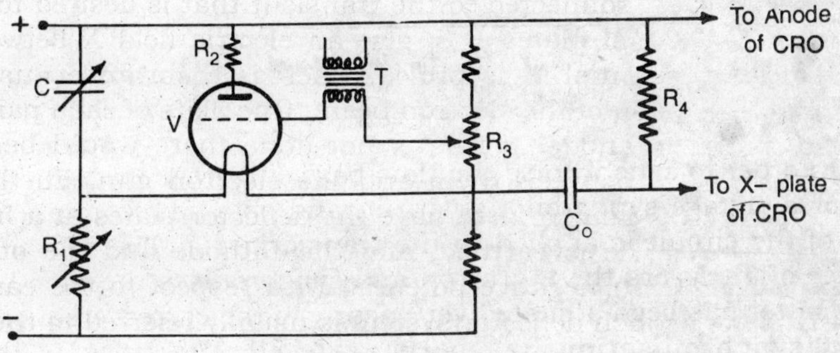

Fig. 36.2 Time Base with Thyratron Discharge Value

The capacitor *C* then discharges rapidly through *V*, the current being limited to a safe value by R_2. When the capacitor voltage is no longer high enough to maintain the discharge, the discharge is stopped and the capacitor commences to recharge. The action then repeats, the resulting potential variation across *C* having the form as shown in Fig.36.3. The voltage is applied to *X*-plate via C_0.

During the charging period the spot is swept across the screen at a nearly uniform rate and suddenly returns during the discharge period. The periodic time depends on the product CR_1 and the magnitude of V_2-V_1. It is usually adjusted in coarse steps by changing *C*. For fine setting R_1 is adjusted. The length of the sweep is proportional to V_2-V_1 and the deflection sensitivity of the cathode ray tube. It is controlled by means of R_3 which alters the grid potential of the thyratron and hence its firing voltage V_2. The capacitor charging curve is of course exponential but a sufficiently small section of it, which is used here, can be regarded as almost linear. Greater linearity can be obtained if R_1 is replaced by a pentode valve.

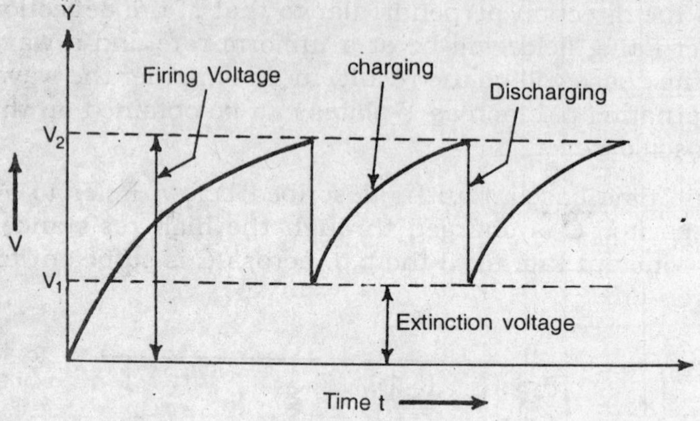

Fig. 36.3

In order to synchronise the time base either the voltage under test or a voltage synchronous with it, and obtained from another part of the circuit is applied to the transformer *T*. The secondary voltage of *T* triggers the thyratron and synchronises the time base if the latter has been adjusted very nearly to the correct frequency. When synchronisation has been achieved, the trace is then repeated again and again in the same position and appears stationary on the screen.

To study the wave-forms of different frequencies, the sweep frequency can be suitably varied with the help of two controls—a *range switch* giving different range of frequencies in coarse step by changing the capacitance *C* and marked *frequency selector* and a *frequency control switch* marked *frequency vernier* which by gradually varying the resistance R permits fine adjustment of sweep frequency within the limits of setting of any one of the range switch.

The CRO unit also contains in-built amplifiers to amplify weak signals, to increase the deflection in the vertical and horizontal directions without distortion of incoming signals, the *X*- and *Y*-plates producing the horizontal and vertical deflections are connected to the amplifiers. Regulating knobs provided on the panel are respectively labelled *horizontal gain* and *vertical gain.* The size of a given signal can be varied on the screen with the help of these gains. The front panel of a typical CRO is shown in Fig. 36.4.

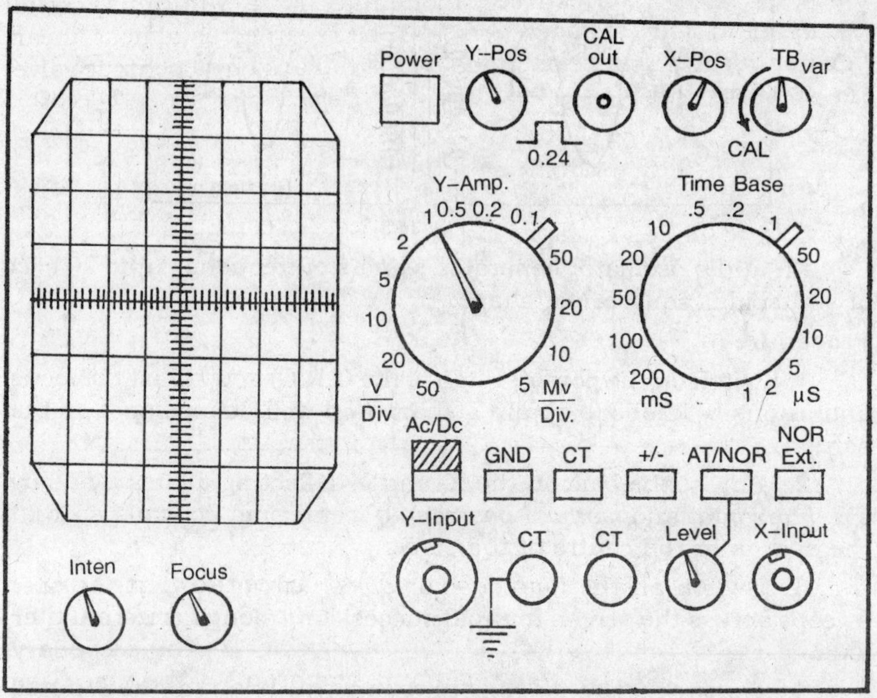

Fig. 36.4

36.2 Applications of Cathode Ray Oscilloscope

Periodic and transient phenomena can be observed with this instrument, provided, in the latter case, that the phenomenon exists for a period which is long enough to make an impression on the retina of the observer's eye.

Some applications to the study of periodic phenomena are given below.

Experiment 36.1

Object—To observe the wave-form of (i) a.c. mains supply and (ii) an oscillator using cathode ray oscilloscope.

Apparatus—Cathode ray oscilloscope, step-down transformer audio oscillator.

Theory— The a.c. mains supply is 220V-50 Hz sinusoidal signal The strength of supply can be reduced to a suitable value (say, 220 V to 5 V) using a suitable step-down transformer (Fig. 36.5).

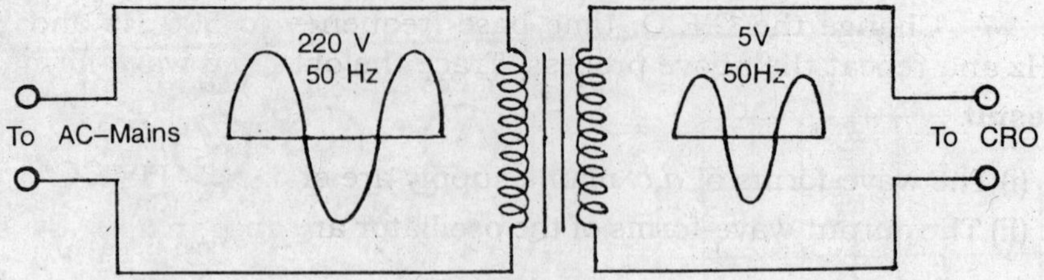

Fig. 36.5

An audio oscillator generates signals of frequency upto 20 kHz of sinusoidal, square, etc. shape.

Procedure

1. Switch on the power switch of the C.R.O. and adjust the focus and intensity knobs to obtain a sharp and well-defined spot on the screen.

2. Adjust the bias on the *X*- and *Y*-deflecting plates by using the *horizontal* and *vertical positioning* (centering) controls so that the spot is in the centre of the screen.

3. Switch on the *time base* or *sweep circuit* so that the spot sweeps across the screen in rapid successions along horizontal time axis.

4. Connect the secondary terminals of a step-down transformer to the *vertical input* terminals of the C.R.O. The primary terminals of the transformer is connected to the a.c. mains. Adjust the *coarse* frequency control of the sweep circuit the C.R.O. with the help of *frequency selector* control to 50 Hz and adjust the *fine frequency* control carefully to make the obtained wave-form *stationary*. Adjust the vertical and horizontal gains to obtain a proper amplitude of the wave-form on the screen. If the wave-form does not remain stationary, use *synch. amp* control to freeze the wave-form. Trace the wave-form.

5. Change the C.R.O. time base frequency to 100 Hz and 25 Hz and repeat the above process. The contracted and expanded wave-forms will be obtained, trace them.

6. For the second part, *i.e.*, to obtain wave-form from an audio oscillator, connect the oscillator to the supply and switch it on. Adjust its frequency of your choice, say, 1 kHz, and connect it to the C.R.O. Set the *coarse frequency* control of the sweep circuit of the C.R.O. to 1 kHz and do fine frequency adjustments to make the obtained wave-form stationary. Trace the wave-form.

7. Change the C.R.O. time base frequency to 500 Hz and 2 kHz and repeat the above process. Trace the obtained wave-forms.

Result

(i) The wave-forms of *a.c.* mains supply are shown in Fig. 36.5.

(ii) The output wave-forms of the oscillator are shown in Fig. 36.5.

Precautions

1. Do not allow the spot of light on a particular point on the screen for a long time, it may damage the screen.

2. Use of horizontal and vertical gains of C.R.O. should be made to obtain proper size wave-form on the screen.

Experiment 36.2

Object—To observe the action of *p-n* function diode as (i) half wave rectifier and (ii) full-wave rectifier.

Apparatus—A sine wave signal generator or oscillator, a half wave rectifier using *p-n* junction diode, a full wave rectifier using function diodes, a resistance box, butter paper for tracing the wave-form.

Theory—As discussed earlier, a rectifier is a device which converts an a.c. signal into a d.c. signal. A half wave rectifier contains only one junction diode which conducts during the positive half cycle of the input wave-form and, therefore, an output is available

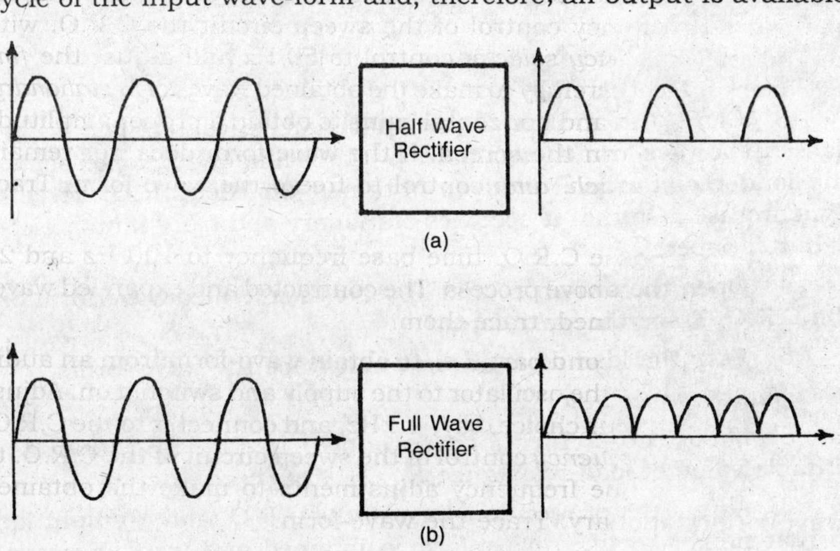

Fig. 36.6

only corresponding to the positive half cycle as shown in Fig. 36.6(*a*). The full wave rectifier consists of two junction diodes with each conducting during alternate half cycle. The output of a full wave rectifier is shown in Fig. 36.6 (*b*). For the complete description of the diode action, refer the relevant topic discussed earlier in this book.

Procedure— *Half wave rectifier*

1. Make the connection as shown in Fig. 36.7.

2. Adjust the proper time base of the C.R.O. so that the wave form is obtained on the screen. Also adjust the *Y*-amplitude of the C.R.O. so that proper waveform is obtained.

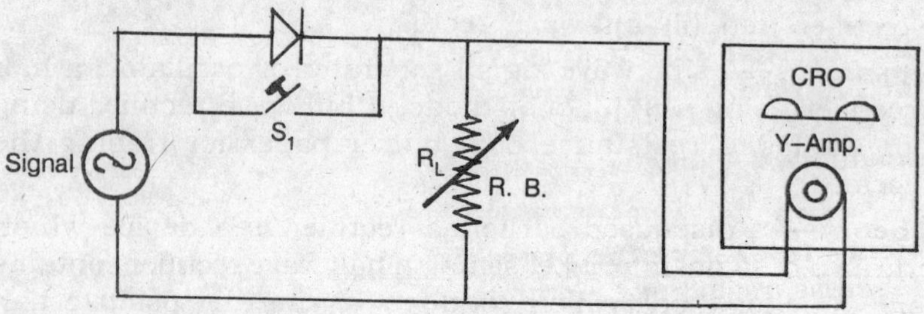

Fig. 36.7

3. Close the switch S_1 so that half wave rectifier is by-passed. The wave-form as produced by the oscillator developed across the load resistance R_L will be obtained on the C.R.O. screen. Synchronise it so that it becomes stationary and trace it on a piece of butter paper.

4. Open the switch S_1 and observe the rectified wave-form on the C.R.O. Trace the wave-form.

5. Vary the load resistance R_L and trace the wave-forms.

Full wave rectifier

6. Make the connections as shown in Fig. 36.8 taking a suitable value of load resistance R_L.

7. Adjust the proper time base of the C.R.O. and *Y*-amplitude so that proper rectified wave-form is obtained and trace the wave-form.

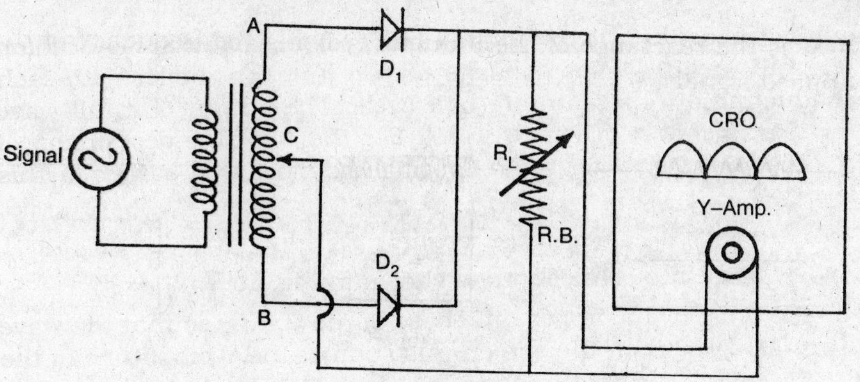

Fig. 36.8

8. Vary the load resistance R_L and trace the wave-forms.

Result—The half-wave rectified waveforms are shown in Fig.36.8 while the full wave rectified wave-forms are shown in Fig. 36.8

Precautions—Same as in Expt. 36.1

Experiment 36.3

Object—To observe the phase difference between the signals applied across resistance, inductance and capacitance and to measure their values.

Apparatus—A resistor R (1 $k\Omega$), an inductor L (0.5 H), a capacitor C (1μF), a variable resistance (0-1 $k\Omega$) , a signal (sine wave) generator, *C.R.O.*, keys, connecting wires.

Theory—We have discussed earlier that in an a.c. circuit containing only a resistance the current I and *e.m.f.* across it V_R are *in phase* as shown in Fig. 36.9(a). In an a.c. circuit containing an inductor only the current *lags behind* the *e.m.f.* V_L by $\frac{\pi}{2}$ radian as shown in Fig. 36.9(b). In the case of a pure capacitor current *leads* the voltage V_c by 90° as shown in Fig. 36.9(c).

In an a.c. circuit containing R and L in series, the voltage V_L across L leads the voltage V_R across R (or the current I) by $\frac{\pi}{2}$ radian so that the resultant voltage V_{LR} across LR *leads* the voltage V_R by a phase angle ϕ [Fig. 36.10(a)] given by $\tan\phi = \frac{\omega L}{R}$ where $\omega L = 2\pi f L$

$= X_L$ is the reactance of the inductor, f being the frequency of the applied signal.

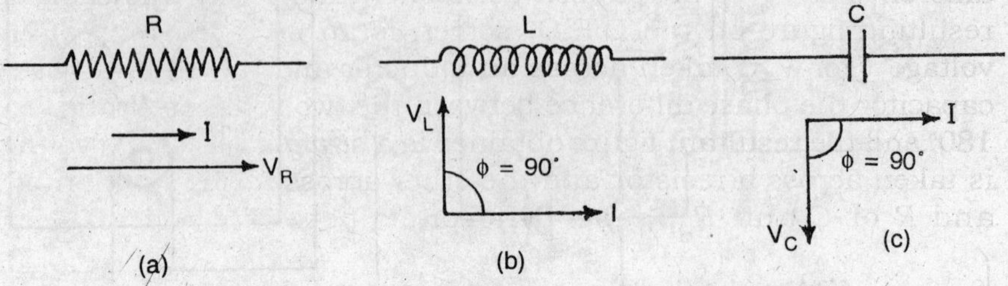

Fig. 36.9

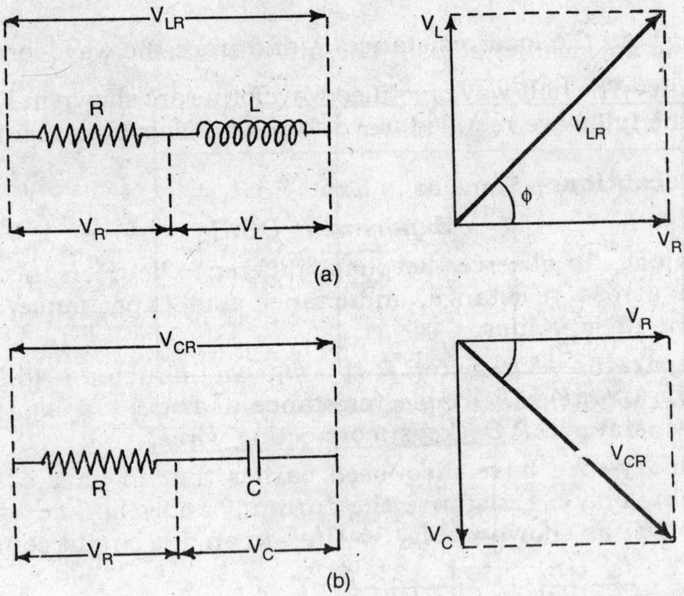

Fig. 36.10

As shown in Fig. 36.10(b), in a series *RC* circuit the resultant voltage V_{CR} *lags* behind the voltage V_R across *R* (or the current *I*) by a phase angle ϕ given by $\phi = \tan^{-1}\left(\frac{1}{\omega CR}\right)$ where $\frac{1}{\omega C} = \frac{1}{2\pi fC} = X_C$ is the reactance of the capacitor.

When an a.c. signal is applied to the *X-X* plates (V_x) and *Y-Y* plates (V_y) of a C.R.O. through a resistor, the phase difference between V_x and V_y is zero and the resultant figure obtained on the

C.R.O. screen is a *straight line.* If the signal V_y on *Y-Y* plates (or *X-X* plates) is applied either through an inductor or a capacitor a phase difference of 90° is introduced between V_x and V_y. In this case the resulting figure on the C.R.O. screen is an *ellipse.* If one of the voltage V_x or V_y is taken across an inductor and the other across a capacitor the phase difference between the two voltages V_x and V_y is 180° and the resultant figure obtained is a *straight line.* If one voltage is taken across a resistor and the other across a combination of *L* and *R* or *C* and *R* phase difference between V_x and V_y is ϕ $\left(\phi = \tan^{-1}\dfrac{\omega L - \dfrac{1}{\omega C}}{R}\right)$ not equal to $\dfrac{\pi}{2}$. In this case the resultant figure is an *inclined ellipse.* The above discussion is summarized below in a tabular form :

S.No.	X-X plates	Y-Y plates	Phase diff.	Resulting figure
1	V_R	V_R	0	Straight line
2	V_R	V_L	90°	ellipse
3	V_R	V_C	90°	ellipse
4.	V_L	V_C	180°	straight line
5.	V_R	V_{LR}	$\phi \neq 90°$	inclined ellipse
6.	V_R	V_{CR}	$\phi \neq 90°$	inclined ellipse

Method—1. Make the connections as shown in Fig. 36.11.

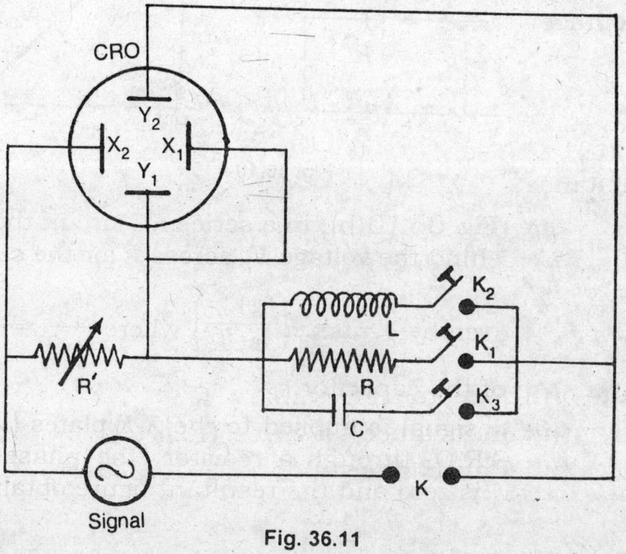

Fig. 36.11

2. Close the key K and press the key K_1. Adjust the value of variable resistance R' so that a straight line is obtained on the screen of C.R.O. The straight line is inclined 45° to the horizontal indicating that phase difference between V_x and V_y is 0° and $R = R'$ as $V_R = V_R'$. Note down the value of R' and let it be denoted by R'_R.

3. Now by pressing the key K_2 bring the inductor L in the circuit (key K_1 is off so R is not in the circuit). The applied voltage across R' and L differ in phase by 90° and an ellipse is obtained on the screen. Adjust R' so that ellipes becomes a circle. This happens when $X_L = R'$. Note down the value of R' let it be denoted by R'_L.

4. Remove the key K_2 and press the key K_3 bringing the capacitor in the circuit. Now, again the phase difference between potentials across X-X and Y-Y plates is 90° and so the figure obtained on the screen is an ellipse. Adjust the variable resistance R' so that ellipse again becomes a circle. This happens when $X_c = R'$. Note down the value of R' and denote it by R'_C.

5. Repeat the above processes at least three times.

Observations

Sr. No.	$R'_R(\Omega)$	$R'_L(\Omega)$	$R'_C(\Omega)$
1.			
2.			
3.			
Mean			

Calculations— Resistance $R = R'_R = \ldots\ldots\ldots\ \Omega$

Inductance $X_L = R'_L$

$\therefore \quad 2\pi f L = R'_L$

or $$L = \frac{R'_L}{2\pi f} = \ldots\ldots\ldots\ \text{H}$$

Capacitance $X_C = R'_C$

$$\therefore \quad \frac{1}{2\pi f C} = R'_C$$

or $$C = \frac{1}{2\pi f R'_C} = \ldots\ldots\ldots\ \text{F}$$

Result— (i) Resistance $= \ldots\ldots\ldots\ \Omega$

(ii) Inductance $= \ldots\ldots\ldots$ H

(iii) Capacitance $= \ldots\ldots\ldots$ F

The phase differences between the signals applied across R, L and C are observed.

Precautions—Same as in Expt. 36.1

Exercise — *To measure the frequency of a.c. mains using a C.R.O.*

Use an inductor L or a capacitance C of known value and adjust the value of variable resistance R' so that a circle is obtained on the C.R.O. screen. Employ the circuit shown in Fig. 36.11. In this situation :

(i) When inductor is used $\quad X_L = \omega L = 2\pi fL = R'$

or $$f = \frac{R'}{2\pi L}$$

(ii) When capacitor is used $\quad X_C = \frac{1}{\omega C} = \frac{1}{2\pi fC} = R'$

or $$f = \frac{1}{2\pi CR'}$$

Experiment 36.4

Object—To measure the phase difference between current and voltage for (i) R-C circuit and (ii) R-L circuit.

Apparatus—A variable step-down transformer, a variable resistor (resistance box), a capacitor, an inductor, a C.R.O., keys, etc.

Theory—As discussed earlier, the phase difference between current and voltage :

for a R-L circuit is $$\phi = \tan^{-1} \frac{\omega_L}{R}$$

for a R-C circuit is $$\phi = \tan^{-1} \frac{1}{\omega CR}$$

When the voltage applied to the X-X or Y-Y plates of a C.R.O. has a resistor and an inductor or a resistor and a capacitor as shown in Fig. 40.12, the figure obtained on the C.R.O. is an inclined ellipse. It can be shown that the phase difference is given by :

$$\phi = \text{Sin}^{-1} \frac{b}{a}$$

where a and b are the intercepts shown in Fig. 36.13.

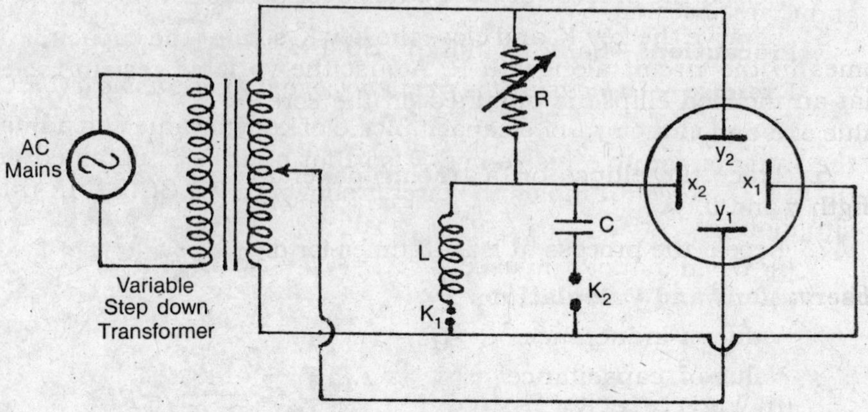

Fig. 36.12

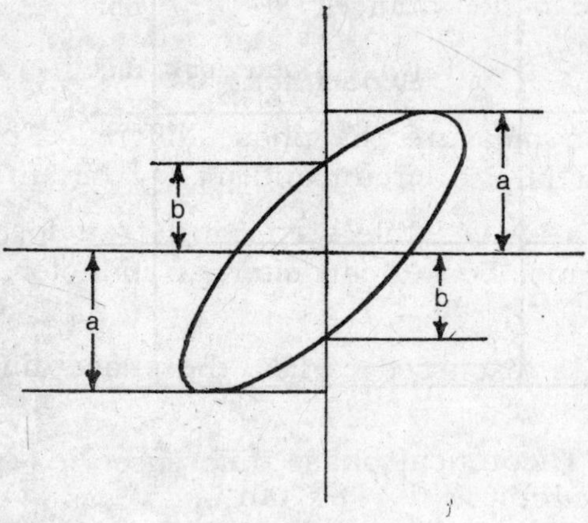

Fig. 36.13

Method

1. Make the connections as shown in Fig.36.12.

2. Close the key K_1 bringing the inductor in the circuit. The a.c. signal applied to the *X-X* and *Y-Y* plates is of the same frequency but differ in phase due to current and voltage differing in phase in an inductor. Adjust the variable resistor so that an inclined ellipse is obtained on the screen. Note down the value of *R* and *L*.

3. Trace the ellipse on a tracing paper and determine the length *a* and *b*.

4. Repeat the process at least 3 times for different values of R.

5. Remove the key K_1 and close the key K_2 so that the capacitor comes in the circuit alongwith R. Adjust the variable resistor so that an inclined ellipse is obtained on the screen. Note down the value of R and C.

6. Trace the ellipse on a tracing paper and determine the length a and b.

7. Repeat the process at least 3 times for different values of R.

Observations and Calculations

Value of inductance L = ... H

Value of capacitance C = ... F

Frequency of a.c. voltage f = ... Hz

$\therefore$ Angular frequency $\omega = 2\pi f$ = ... Sec^{-1}

Sr. No.	Component used	Resistance R (Ω)	a (cm)	b (cm)	observed phase diff. $\phi = \sin^{-1}\frac{b}{a}$	Mean ϕ
1.		...	...	...	...	
2.	Inductor L	...	...	...	...	...
3.		...	...	...	...	
1.		...	...	...	...	
2.	Capacitor C	...	...	...	...	...
3.		...	...	...	...	

R-L circuit : Theoretical phase difference $\phi = \tan^{-1}\frac{\omega L}{R}$ = ...

Observed mean phase diff. ϕ = ...

Error = ...

R-C circuit = Theoretical phase difference $\phi = \tan^{-1}\frac{1}{\omega CR}$ = ...

Observed phase difference ϕ = ...

Error = ...

Precautions—Same as in Expt.36.1.

36.3 Lissajous Figures

When two simple harmonic oscillations are superimposed the resulting configuration is called a *Lissajous Figure.* Simple harmonic motion plotted against time gives sinusoidal configurations. Two sinusoidal signal inputs to an oscilloscope give lissajous patterns on the screen. The shape of pattern depends upon the frequency, amplitude and phase relationship of the two inputs.

(i) *S.H.Ms of equal frequencies*—Let the two simple harmonic motion, having the same frequency or period take place along x- and y-axes respectively. They can be represented as :.

$$x = a \sin \omega t \qquad (36.1)$$

$$y = b \sin (\omega t + \phi) \qquad (36.2)$$

where a and b are the respective amplitudes. The S.H.M. along y-axis is ahead by a phase angle ϕ *w.r.t.* that along x-axis.

If $\phi = 0$, then $x = a \sin \omega t$ and $y = b \sin \omega t$

$$\therefore \quad \frac{y}{x} = \frac{b}{a} \quad \text{or} \quad y = \frac{b}{a} x$$

This is identical to the equation of a straight line ($y = mx + c$) passing through origin ($c = 0$) and is represented in Fig. 36.14(a). The slope of this straight line is given by :

$$m = \tan \theta = \frac{b}{a}$$

If $\phi = \frac{\pi}{2}$, then $x = a \sin \omega t$ and $y = b \sin (\omega t + \frac{\pi}{2}) = b \cos \omega t$.

$$\therefore \quad \frac{x^2}{a^2} + \frac{y^2}{b^2} = 1 \qquad (36.3)$$

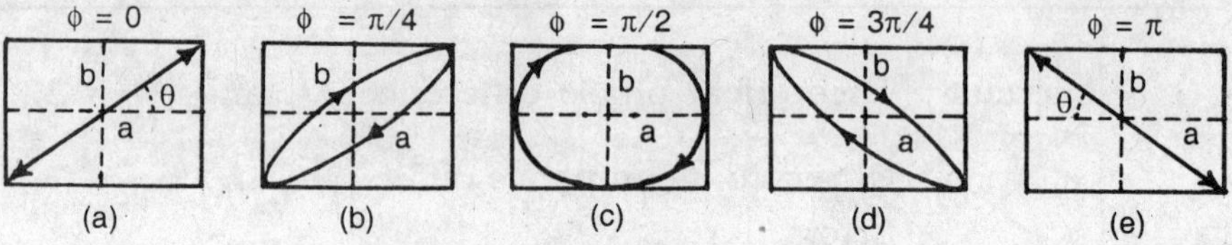

Fig. 36.14

This is the equation of an ellipse or a symmetrical ellipse whose major and minor axes coincide with x- and y-axes respectively as shown in Fig. 36.14(c). If $a = b$, the above equation changes into an equation of a circle.

$$x^2 + y^2 = a^2 \qquad (36.4)$$

and, therefore, figure obtained will be a circle. (Fig. 36.14(d).

If $\phi = \pi$, then $x = a \sin \omega t$ and $y = b \sin (\omega t + \pi) = -b \sin \omega t$.

$$\therefore \quad \frac{y}{x} = -\frac{b}{a} \quad \Rightarrow \quad y = -\frac{b}{a}\,x \qquad (36.5)$$

This again represents a straight line with slope $= -\frac{b}{a}$, as shown in Fig. 36.14(*e*).

Similarly, you can work out to see that for $\phi = \frac{\pi}{4}$ and $\frac{3\pi}{4}$ the resulting figures are oblique ellipse represented in Fig. 36.14(b) and 36.14(d) respectively.

For $\pi \leq \phi < 2\pi$ the whole cycle is repeated in the reverse order.

(ii) *S.H.Ms. with frequencies in the ratio 1 : 2*— When the frequency of the S.H.M. along *y*-axis (or vertical axis) is double that of the S.H.M. along *x*-axis (or horizontal axis) and the two S.H.Ms. have no phase difference ($\phi = 0$), the resulting figure obtained is shown in Fig. 36.15(b). In the reverse situation, *i.e.*, the frequency along the *x*-axis being double to that along the *y*-axis, the resulting figure is shown in Fig. 36.15(c).

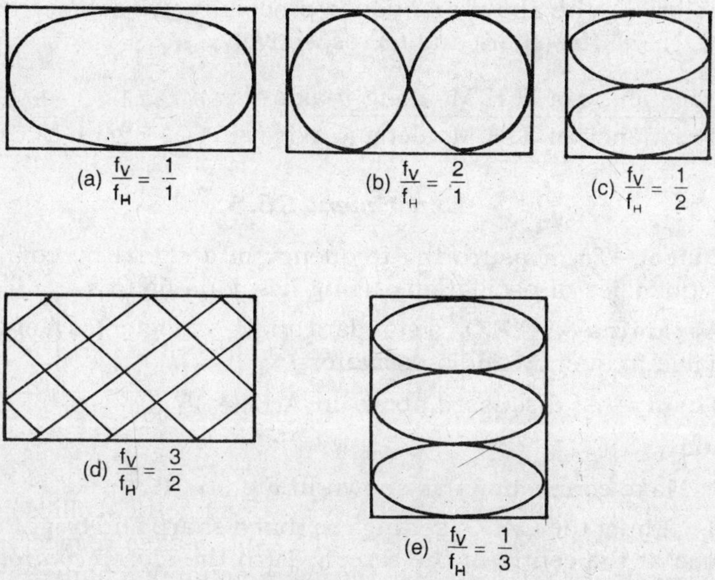

Fig. 36.15

In general, the frequency ratio of the inputs may be determined from an analysis of the Lissajous figure produced. *If a Lissajous figure is enclosed in a rectangle whose sides are parallel to the formation axes of the figure, the frequency ratio of the two inputs may be determined by counting the number of tangency points of the rectangle (x-axis and y-axis) enclosing the pattern. The ratio of the number of tangency points is in the inverse ratio to the two input frequencies.* That is,

$$\frac{\text{Number of tengency points along } x\text{-axis}}{\text{Number of tangency points along } y\text{-axis}} = \frac{\text{Frequency of S.H.M. along } y\text{-axis}}{\text{Frequency of S.H.M. along } x\text{-axis}} \quad (36.6)$$

Alternatively, count the number of points at which the vertical and horizontal lines cross the figure, that is,

$$\frac{\text{Frequency of S.H.M. along } y\text{-axis}}{\text{Frequency of S.H.M. along } x\text{-axis}} = \frac{\text{Number of crossing along } x\text{-axis}}{\text{Number crossing along y-axis}} \quad (36.7)$$

Following the above mentioned procedure, we can see that for Fig. 36.15 (*a*), (*b*), (*c*), (*d*) and (*e*) respectively,

$$\frac{\text{Frequency of S.H.M. along } y\text{-axis}}{\text{Frequency of S.H.M. along } x\text{-axis}} = \frac{1}{1}, \frac{2}{1}, \frac{1}{2}, \frac{3}{2}, \frac{1}{3}.$$

Experiment 36.5

Object—To measure the frequency of a signal by comparing the frequencies of oscillations using lissajous figures.

Apparatus—A C.R.O., a standard audio-oscillator (say, of 1 kHz), a variable frequency audio-oscillator.

Theory—As discussed above in Article 36.3.

Method

1. Make connections as shown in Fig. 36.16.

2. Adjust the C.R.O. settings so that a sharp and bright spot is obtained at the centre of the screen. Turn the *sweep control* to *off* position by setting it to *ext. input* position. Set the frequency of the audio-oscillator (variable) to the same frequency that of the standard audio oscillator, *i.e.*, 1 kHz.

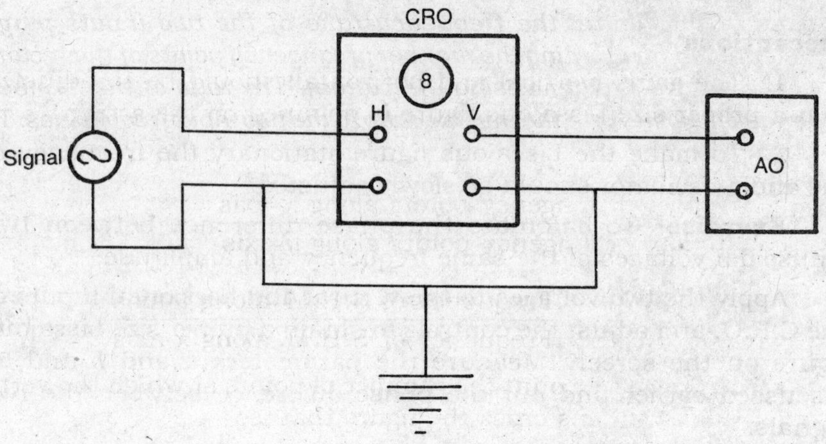

Fig. 36.16

3. Switch on both the oscillators and adjust the controls of oscilloscope to get a good Lissajous figure, *i.e.*, an ellipse (1 : 1) on the screen. Make fine adjustments to make the ellipse stationary. Now, the frequency of the two oscillators are equal, *i.e.*, 1 kHz. Note down the dial reading of the variable oscillator. It may be slightly different, say, 990 Hz.

By switching off the audio-oscillator allow the ellipse to change phase and note the various shapes that appear on the screen. Adjust the amplitude of the signal and obtain *circular* trace on the screen.

4. Set the frequency of audio-oscillator to 500 Hz and do fine adjustments so that 1 : 2 Lissajous figure (Fig. 36.15(c)) is obtained on the screen. Make adjustments to make the figure stationary. Note down the dial reading.

5. By changing the frequency of oscillator obtain Lissajous figures with ratio 1 : 3, 2 : 3....

Observations

Vertical input standard frequency = 1 kHz

Sr. No.	Hor. input (dial) freq.	Figure shape	Number of tangency points on hor. (x-axis)	on vert. (y-axis)	$\frac{fv}{fh}$	Actual hor. freq. (Hz)
1	990		1	1	1	1000/1 = 1000
2	510		2	1	2/1	1000/2 = 500
.						
.						

Precautions

1. The gains (vertical and horizontal) should be so adjusted that a proper size Lissajous figure is obtained on the screen.

2. To make the Lissajous figure stationary the frequency of the audio-oscillator should be slowly adjusted.

Exercise—To calculate the phase difference between two sinusoidal voltages of the same frequency and amplitude.

Apply the two voltages to the vertical and horizontal inputs of the C.R.O. and adjust the controls to obtain a proper size Lissajous figure on the screen. Measure the parameters a and b and as discussed earlier find out the phase difference between the two signals.

ORAL QUESTIONS

CATHODE RAY OSCILLOSCOPE

1. What for C.R.O. stands ? Why is it called so ?
2. What are cathode rays ? How are they produced ?
3. Name the various parts of C.R.T. and briefly explain their role.
4. With what material is the screen in C.R.T. coated ? Name a few of them.
5. What do you mean by horizontal and vertical gain ?
6. What is the function of time base circuit ?
7. What is a thyratron ?
8. Why should the spot of light not be allowed to stay on a particular point on the screen of C.R.O. for a long time?
9. What are Lissajous figure ? Explain how they can be used to compare the two frequencies.
10. Draw Lissajous figures for frequency ratio 1 : 2, 1 : 3 and 3 : 2 for two S.H.M. with a phase difference of $\pi/2$.
11. State the conditions when Lissajous figure is (i) a straight line, (ii) a circle and (iii) an ellipse.

Appendices

APPENDIX I

CONVERSION FACTORS

To convert		Multiply by	Logarithm
Length :			
In. into cm.	...	2.540	0.4048
Ft. into metres	...	0.3048	1.4840
Yds. into metres	...	0.9144	1.9611
Miles into km.	...	2.6093	0.2066
Area :			
Sq. in. into sq. cm.	...	6.451	0.8096
Sq. ft. into sq. metres	...	0.0929	0.9680
Sq. yds. into sq. metres	...	0.8361	0.9223
Acres into sq. metres	...	4047	3.6071
Volume :			
Cub. in. into c.c.	...	16.39	1.2143
Cub. in. into litres	...	0.01639	2.2145
Cub. ft. into litres	...	28.32	1.4521
Gallons into litres	...	4.546	0.6677
Gallons into cub. ft.	...	0.1606	1.2057
Mass :			
Grain into m. gm.	...	64.8	1,8116
Oz. (*av.*) into gm.	...	28.35	1.4525
lb. into gm.	...	453.6	2.6567
Ton into kgm.	...	1016	3.0069
Force, work, *etc.* :			
Gm.-wt. into dynes	...	981	1.9917
lb.-wt. into poundals	...	32.2	1.5079
Poundals into dynes	...	1.3825×10^4	4.4070
Cm.-gm. into ergs	....	981	2.9917

CONVERSION FACTORS

To convert		Multiply by	Logarithm :
Ft. lbs. into Joules	...	1.356	0.1324
Horse Power into ft. lbs./sec.	...	550	2.7404
Horse Power into Watts	...	745.7	2.8726
K. W.H. into Joules	...	3.600×10^6	6.5563
Miscellaneous :			
Radians into degree	...	57.296	1.7582
$\log_{10} N$ into $\log_e N$	...	2.303	.3623
Volume of water at 62°F into c.c.	...	454.6	2.6577
Volume of water at 62°F into gallons	...	.100	1.0000
atmospheres into dynes per sq. cm.	...	1.014×10^6	6.0059
atmospheres into lbs. per sq. in.	...	14.70	1.6730
P. h into cm. per sec.	...	44.70	1.5030

USEFUL DATA AND FORMULAE

π = 3.14159

1 dyne = force which, acting on a mass of 1 gm. produces an acceleration of 1 cm. per sec.

1 Bar = 10^6 dynes per sq. cm

1 Atmosphere = 760 mm. of Hg.
= 1.0136 bar.

1 erg = work done by a force of 1 dyne acting through 1 cm.

1 Joule = 10^7 ergs

1 Watt = 1 Joule per sec.

R, the gas constant = 8.305 Joules per gm. molecule
= 1.988 cal. per gm molecule.

1 gallon of water Weighs 10 lbs approx.

1 cub. ft. of water weighs 62.43 lbs.

Area of a circle, radius r...= πr^2

Area of an ellipse of semi-axes a and b ... ... = $\pi a b$

Area of the curved surface of a cylinder of length l, radius $r = 2\pi rl$.

Area of the curved surface of a cone of height $b = \pi r.\sqrt{r^2 + b^2}$

Area of the curved surface of a sphere of radius r ... = $4 \pi r^2$

Volume of a cylinder ... = $\pi r^2 l$

Volume of a cone = $\frac{1}{3} \pi r^2 l$

Volume of a sphere ... = $\frac{4}{3} \pi r^3$

Volume of an ellipsoid-axes $2a$, $2b$, $2c$ ... ... = $\frac{4}{3} \pi abc$

APPENDIX II

PHYSICAL CONSTANTS

BOILING POINT OF WATER UNDER VARIOUS BAROMETRIC PRESSURES

+ mm. →	0	1	2	3	4	5	6	7	8	9
mm.										
700	97.71	.75	.79	.83	.87	.91	.95	.99	98.03	.07
710	98.11	.14	.18	.22	.26	.30	.34	.38	.42	.45
720	98.49	.53	.57	.61	.65	.69	.72	.76	.80	.48
730	98.88	.91	.95	.99	99.03	.07	.10	.14	.18	.22
740	99.25	.29	.33	.37	.41	.44	.48	.52	.56	.59
750	99.63	.67	.70	.74	.78	.81	.85	.89	.93	.96
760	100.00	.03	.07	.11	.15	.18	.22	.26	.29	.33

ELECTRO-CHEMICAL-EQUIVALENT

Element		E.C.E. (gm./coulomb)	Chemical equivalent
Silver	...	.0011180	107.88
Copper	...	.0003295	31.88
Lead	...	.0010731	103.61
Hydrogen	...	.0000105	1.0081
Oxygen	...	.9000829	8.0000

SPECIFIC RESISTANCE AND TEMPERATURE COEFFICIENT

Element	Specific Resistance ohm-m ($\times 10^{-8}$)	Temp. Coefficient α ($\times 10^{-4}$)
Metals		
1. Aluminium	2.8	49
2. Copper	1.78	42.8
3. Iron	9.8	62
4. Lead	22.1	41
5. Zinc	5.9	35
6. Silver	1.65	40
7. Platinum	11	37.
8. Gold	2.42	40
9. Nickel	11.8	27
10. Mercury	95.6	8.9
11. Tungsten	66.1	—
Alloys		
1. Brass	6.6	10
2. German silver	26.6	2.2-6.0
3. Phosphor Bronze	5.0-10.2	35
4. Constantan (Cu - 58.80% Ni-40.00% Mn-1.2%)	49.1	–0.4 to 0.1
5. Manganin (Cu-85% Mn-12% Ni-3%)	44.5	0.02-0.5
6. Nichrome (Ni-67.5% Cr-15% Fe-16% Mn-1.5%)	110	1.7-3.5

UNIVERSAL CONSTANTS

1. Planck's Constant $h = 6.624 \times 10^{-34}$ Joule-sec
2. Avogrado's Number $N = 6.023 \times 10^{-23}$ Per mole
3. Charge on an electron $e = 1.6008 \times 10^{-19}$ Coulomb
4. Boltzman's Constant $k = 1.38 \times 10^{-23}$
5. Mass of an electron $m_e = 9.1 \times 10^{-31}$ kg
6. Specific charge of electron $e/m = 1.76 \times 10^{11}$ C/kg
7. Mass of a proton $m_p = 1.6726 \times 10^{-27}$ kg
8. Permittivity of free space $E_o = 8.85 \times 10^{-12}\ C^2\ N^{-1}\ m^{-2}$.

RESISTANCE OF WIRES OF VARIOUS GAUGES AND MATERIALS

S.W.G. No.	Diameter in mm.	Area of Gross-section Sq. cm.	Resistance per metre (ohm/metre)			
			Copper	Eureka	Manganin	German Silver
10	3.25	.08302	.0021	.057	.051	.049
12	2.64	.05480	.0032	.086	.077	.041
14	2.03	.03243	.0054	.146	,131	.070
16	1.63	.02076	.0083	.228	.204	.109
18	1.22	.01168	.0148	.405	.361	.193
20	.914	.006567	.0260	.722	.645	.345
22	.711	.003973	.0435	.120	1.07	.57
24	.559	.002452	.070	1.93	1.73	.92
26	.457	.001642	.105	2.89	2.58	1.38
28	.374	.001110	.155	4.27	3.82	2.02
30	.315	.000779	.222	6.08	5.45	2.90
32	.274	.000591	.293	8.02	7.18	3.83
34	.234	.000429	.404	11.1	9.90	5.27
36	.193	.000293	.590	16.2	14.50	7.74
38	.152	.0001824	.950	26.0	23.2	12.4
40	.122	.0001170	1.48	40.6	36.3	19.4

MAGNETIC ELEMENTS AND-*g*

Station	Declination	Angle of Dip	Horizontal component H	Vertical component V	*g*
Agra	0° 10′ E	40° 40′	.3484	.2985	979.06
Ajmer	1° 10′ E	39° 20′	.3483	.2861	978.98
Aligarh	0° 20′ E	41° 50′	.3455	.3091	978.08
Allahabad	0° 20′ W	7° 10′	.3629	.2758	978.94
Bareilly	0° 20′ E	42° 20′	.3436	.3136	...
Varanasi	0° 30′ W	37° 10′	.3635	.2764	...
Mumbai	0° 20′ W	25° 30′	.3761	.1792	978.63
Kolkata	0° 00′ W	31° 30′	.3819	.2342	978.78
Kanpur	0° 00′	38° 39′	.3628	.2901	978.97
Chandausi	0° 30′ E	42° 40′	.3429	.3167	...
Dehradun	0° 50′ E	45° 50′	.3315	.3405	979.06
Delhi	0° 40′ E	42° 52′	.3453	.3204	979.15
Gorakhpur	0° 20′ W	43° 40′	.3576	.2972	978.94
Gwalior	0° 20′ E	39° 00′	.3531	.2857	978.96
Jaipur	0° 30′ E	40° 30′	.3470	.2961	978.52
Jodhpur	0° 00′	39° 10′	.3482	.2835	978.92
Khurja	0° 30′ E	42° 10′	.3426	.3109	978.08
Lucknow	0° 10′ W	40° 00′	.3536	.2965	979.00
Meerut	0° 40′ E	43° 30′	.3389	.3212	979.15
Udaipur	0° 00	35° 50′	.3626	.2613	978.9
Equator					978.3
Poles					983.2

MATHEMATICAL TABLES

LOGARITHMS

	0	1	2	3	4	5	6	7	8	9	Mean Differences								
											1	2	3	4	5	6	7	8	9
10	0000	0043	0086	0128	0170	0212	0253	0294	0334	0374	4	8	12	17	21	25	29	33	37
11	0414	0453	0492	0531	0569	0607	0645	0682	0719	0755	4	8	11	15	19	23	26	30	34
12	0792	0828	0864	0899	0934	0969	1004	1038	1072	1106	3	7	10	14	17	21	24	28	31
13	1139	1173	1206	1239	1271	1303	1335	1367	1399	1430	3	6	10	13	16	19	23	26	29
14	1461	1492	1523	1553	1584	1614	1644	1673	1703	1732	3	6	9	12	15	18	21	24	27
15	1761	1790	1818	1847	1875	1903	1931	1959	1987	2014	3	6	8	11	14	17	20	22	25
16	2041	2068	2095	2122	2148	2175	2201	2227	2253	2279	3	5	8	11	13	16	18	21	24
17	2304	2330	2355	2380	2405	2430	2455	2480	2504	2529	2	5	7	10	12	15	17	20	22
18	2553	2577	2601	2625	2648	2672	2695	2718	2742	2765	2	5	7	9	12	14	16	19	21
19	2788	2810	2833	2856	2878	2900	2923	2945	2967	2989	2	4	7	9	11	13	16	18	20
20	3010	3032	3054	3075	3096	3118	3139	3160	3181	3201	2	4	6	8	11	13	15	17	19
21	3222	3243	3263	3284	3304	3324	3345	3365	3385	3404	2	4	6	8	10	12	14	16	18
22	3424	3444	3464	3483	3502	3522	3541	3560	3579	3598	2	4	6	8	10	12	14	15	17
23	3617	3636	3655	3674	3692	3711	3729	3747	3766	3784	2	4	6	7	9	11	13	15	17
24	3802	3820	3838	3856	3874	3892	3909	3927	3945	3962	2	4	5	7	9	11	12	14	16
25	3979	3997	4014	4031	4048	4065	4082	4099	4116	4133	2	3	5	7	9	10	12	14	15
26	4150	4166	4183	4200	4216	4232	4249	4265	4281	4298	2	3	5	7	8	10	11	13	15
27	4314	4330	4346	4362	4378	4303	4409	4425	4440	4456	2	3	5	6	8	9	11	13	14
28	4472	4487	4502	4518	4533	4548	4564	4579	4594	4609	2	3	5	6	8	9	11	12	14
29	4624	4639	4654	4669	4683	4698	4713	4728	4742	4757	1	3	4	6	7	9	10	12	13
30	4771	4786	4800	4814	4829	4843	4857	4871	4886	4900	1	3	4	6	7	9	10	11	13
31	4914	4928	4942	4955	4969	4983	4997	5011	5024	5038	1	3	4	6	7	8	10	11	12
32	5051	5065	5079	5092	5105	5119	5132	5145	5159	5172	1	3	4	5	7	8	9	11	12
33	5158	5198	5211	5224	5237	5250	5263	5276	5289	5302	1	3	4	5	6	8	9	10	12
34	5315	5328	5340	5353	5366	5378	5391	5403	5416	5428	1	3	4	5	6	8	9	10	11
35	5441	5453	5465	5478	5490	5502	5514	5527	5539	5551	1	2	4	5	6	7	9	10	11

36	5563	5575	5587	5599	5611	5623	5635	5647	5658	5670	1	2	4	5	6	7	8	10	11
37	5682	5694	5705	5717	5729	5740	5752	5763	5775	5786	1	2	3	5	6	7	8	9	10
38	5798	5809	5821	5832	5843	5855	5866	5877	5888	5899	1	2	3	5	6	7	8	9	10
39	5911	5922	5933	5944	5955	5968	5977	5988	5999	6010	1	2	3	4	5	7	8	9	10
40	6021	6031	6042	6053	6064	6075	6085	6096	6107	6117	1	2	3	4	5	6	8	9	10
41	6128	6138	6149	6160	6170	6180	6191	6201	6212	6222	1	2	3	4	5	6	7	8	9
42	6232	6243	6253	6263	6274	6284	6294	6304	6314	6325	1	2	3	4	5	6	7	8	9
43	6335	6345	6355	6365	6375	6385	6395	6405	6415	6425	1	2	3	4	5	6	7	8	9
44	6435	6444	6454	6464	6474	6484	6493	6503	6513	6522	1	2	3	4	5	6	7	8	9
45	6532	6542	6551	6561	6571	6580	6590	6599	6609	6618	1	2	3	4	5	6	7	8	9
46	6628	6637	6646	6656	6665	6675	6684	6693	6702	6712	1	2	3	4	5	6	7	7	8
47	6721	6730	6739	6749	6758	6767	6776	6785	6794	6803	1	2	3	4	5	5	6	7	8
48	6812	6821	6830	6839	6848	6857	6866	6875	6884	6893	1	2	3	4	4	5	6	7	8
49	6902	6911	6920	6928	6937	6946	6955	6964	6972	6981	1	2	3	4	4	5	6	7	8
50	6990	6998	7007	7016	7024	7033	7042	7050	7059	7067	1	2	3	3	4	5	6	7	8
51	7076	7084	7093	7101	7110	7118	7126	7135	7143	7152	1	2	3	3	4	5	6	7	8
52	7160	7168	7177	7185	7193	7202	7210	7218	7226	7235	1	2	2	3	4	5	6	7	7
53	7243	7251	7259	7267	7275	7284	7292	7300	7308	7316	1	2	2	3	4	5	6	6	7
54	7324	7332	7340	7348	7356	7364	7372	7380	7388	7396	1	2	2	3	4	5	6	6	7
55	7404	7412	7419	7427	7435	7443	7451	7459	7466	7474	1	2	2	3	4	5	5	6	7
56	7482	7490	7497	7505	7513	7520	7528	7536	7543	7551	1	2	2	3	4	5	5	6	7
57	7559	7566	7574	7582	7589	7597	7604	7612	7619	7627	1	2	2	3	4	5	5	6	7
58	7634	7642	7649	7657	7664	7672	7679	7686	7694	7701	1	1	2	3	4	4	5	6	7
59	7709	7716	7723	7731	7738	7745	7752	7760	7767	7774	1	1	2	3	4	4	5	6	7
60	7782	7789	7796	7803	7810	7818	7825	7832	7839	7846	1	1	2	3	4	4	5	6	6
61	7853	7860	7868	7875	7882	7889	7896	7903	7910	7917	1	1	2	3	4	4	5	6	6
62	7924	7931	7938	7945	7952	7959	7973	7973	7980	7987	1	1	2	3	3	4	5	6	6
63	7993	8000	8007	8014	8021	8028	8041	8041	8048	8055	1	1	2	3	3	4	5	5	6
64	8062	8069	8075	8082	8089	8096	8102	8109	8116	8122	1	1	2	3	3	4	5	5	6
65	8129	8136	8142	8149	8156	8162	8169	8176	8182	8189	1	1	2	3	3	4	5	5	6

66	8195	8202	8209	8215	8222	8228	8235	8241	8248	8254	1	1	2	3	3	4	5	5	6
67	8261	8267	8274	3280	8287	8293	8299	8306	8312	8319	1	1	2	3	3	4	5	5	6
68	8325	8331	8338	8344	8351	8357	8363	8370	8376	8382	1	1	2	3	3	4	4	5	6
69	8388	8395	8401	8407	8414	8420	8426	8432	8439	8445	1	1	2	2	3	4	4	5	6
70	8451	8457	8463	8470	8476	8482	8488	8494	8500	8506	1	1	2	2	3	4	4	5	6
71	8513	8519	8525	8531	8537	8543	8549	8555	8561	8567	1	1	2	2	3	4	4	5	5
72	8573	8579	8585	8591	8597	8603	8609	8615	8621	8627	1	1	2	2	3	4	4	5	5
73	8633	8639	8645	8651	8657	8663	8669	8675	8681	8686	1	1	2	2	3	4	4	5	5
74	8692	8698	8704	8710	8716	8722	8727	8733	8739	8745	1	1	2	2	3	4	4	5	5
75	8751	8756	8762	8768	8774	8779	8785	8791	8797	8802	1	1	2	2	3	3	4	5	5
76	8808	8814	8820	8825	8831	8837	8842	8848	8854	8859	1	1	2	2	3	3	4	5	5
77	8865	8871	8876	8882	8887	8893	8899	8904	8910	8915	1	1	2	2	3	3	4	4	5
78	8921	8927	8932	8938	8943	8949	8954	8960	8965	8971	1	1	2	2	3	3	4	4	5
79	8976	8982	8987	8993	8998	9004	9009	9015	9020	9025	1	1	2	2	3	3	4	4	5
80	9031	9036	9042	9047	9053	9058	9063	9069	9074	9079	1	1	2	2	3	3	4	4	5
81	9085	9090	9096	9101	9106	9112	9117	9122	9128	9133	1	1	2	2	3	3	4	4	5
82	9138	9148	9149	9154	9159	9165	9170	9175	9180	9186	1	1	2	2	3	3	4	4	5
83	9191	9196	9201	9206	9212	9217	9222	9227	9232	9238	1	1	2	2	3	3	4	4	5
84	9243	9248	9253	9258	9263	9269	9274	9279	9284	9289	1	1	2	2	3	3	4	4	5
85	9294	9299	9304	9309	9315	9320	9325	9330	9335	9340	1	1	2	2	3	3	4	4	5
86	9345	9350	9355	9360	9365	9370	9375	9380	9385	9390	1	1	2	2	3	3	4	4	5
87	9395	9400	9405	9410	9415	9420	9425	9430	9435	9440	0	1	1	2	2	3	3	4	4
88	9445	9450	9455	9460	9465	9469	9474	9479	9484	9489	0	1	1	2	2	3	3	4	4
89	9494	9499	9504	9509	9513	9518	9523	9528	9533	9538	0	1	1	2	2	3	3	4	4
90	9542	9547	9552	9557	9562	9566	9571	9576	9581	9586	0	1	1	2	2	3	3	4	4
91	9590	9595	9600	9605	9609	9614	9619	9624	9628	9633	0	1	1	2	2	3	3	4	4
92	9638	9643	9647	9652	9657	9661	9666	9671	9675	9680	0	1	1	2	2	3	3	4	4
93	9685	9689	9694	9699	9703	9708	9713	9717	9722	9727	0	1	1	2	2	3	3	4	4
94	9731	9736	9741	9745	9750	9754	9759	9763	9768	9773	0	1	1	2	2	3	3	4	4
95	9777	9782	9786	9791	9795	9800	9805	9809	9814	9818	0	1	1	2	2	3	3	4	4
96	9823	9827	9832	9836	9841	9845	9850	9854	9859	9863	0	1	1	2	2	3	3	4	4
97	9868	9872	9877	9881	9886	9890	9894	9899	9903	9908	0	1	1	2	2	3	3	4	4
98	9912	9917	9921	9926	9930	9934	9939	9943	9948	9952	0	1	1	2	2	3	3	4	4
99	9956	9961	9965	9969	9974	9978	9983	9987	9991	9996	0	1	1	2	2	3	3	3	4

ANTILOGARITHMS

	0	1	2	3	4	5	6	7	8	9	Mean Differences								
											1	2	3	4	5	6	7	8	9
.00	1000	1002	1005	1007	1009	1012	1014	1016	1019	1021	0	0	1	1	1	1	2	2	2
.01	1023	1026	1028	1030	1033	1035	1038	1040	1042	1045	0	0	1	1	1	1	2	2	2
.02	1047	1050	1052	1054	1057	1059	1062	1064	1067	1069	0	0	1	1	1	1	2	2	2
.03	1072	1074	1076	1079	1081	1084	1086	1089	1091	1094	0	0	1	1	1	1	2	2	2
.04	1096	1099	1102	1104	1107	1109	1112	1114	1117	1119	0	1	1	1	1	2	2	2	2
.05	1122	1125	1127	1130	1132	1135	1138	1140	1143	1146	0	1	1	1	1	2	2	2	2
.06	1148	1151	1153	1156	1159	1161	1164	1167	1169	1172	0	1	1	1	1	2	2	2	2
.07	1175	1178	1180	1183	1186	1189	1191	1194	1197	1199	0	1	1	1	1	2	2	2	2
.08	1202	1205	1208	1211	1213	1216	1219	1222	1225	1227	0	1	1	1	1	2	2	2	3
.09	1230	1233	1236	1239	1242	1245	1247	1250	1253	1256	0	1	1	1	1	2	2	2	3
.10	1259	1262	1265	1268	1271	1274	1276	1279	1282	1285	0	1	1	1	1	2	2	2	3
.11	1288	1291	1294	1297	1300	1303	1306	1309	1312	1315	0	1	1	1	2	2	2	2	3
.12	1318	1321	1324	1327	1330	1334	1337	1340	1343	1346	0	1	1	1	2	2	2	2	3
.13	1349	1352	1355	1358	1361	1365	1368	1371	1374	1377	0	1	1	1	2	2	2	3	3
.14	1380	1384	1387	1390	1393	1396	1400	1403	1406	1409	0	1	1	1	2	2	2	3	3
.15	1413	1416	1419	1422	1426	1429	1432	1435	1439	1442	0	1	1	1	2	2	2	3	3
.16	1445	1449	1452	1455	1459	1462	1466	1469	1472	1476	0	1	1	1	2	2	2	3	3
.17	1479	1483	1486	1489	1493	1496	1500	1503	1507	1510	0	1	1	1	2	2	2	3	3
.18	1514	1517	1521	1524	1528	1531	1535	1538	1542	1545	0	1	1	1	2	2	2	3	3
.19	1549	1552	1556	1560	1563	1567	1570	1574	1578	1581	0	1	1	1	2	2	3	3	3
.20	1585	1589	1592	1596	1600	1603	1607	1611	1614	1618	0	1	1	1	2	2	3	3	3
.21	1622	1626	1629	1633	1637	1641	1644	1648	1652	1656	0	1	1	2	2	2	3	3	3
.22	1660	1663	1660	1671	1675	1679	1683	1687	1690	1694	0	1	1	2	2	2	3	3	3
.23	1698	1702	1706	1710	1714	1718	1722	1726	1730	1734	0	1	1	2	2	2	3	3	4
.24	1738	1742	1746	1750	1754	1758	1762	1766	1770	1774	0	1	1	2	2	2	3	3	4
.25	1778	1782	1786	1791	1795	1799	1803	1807	1811	1816	0	1	1	2	2	2	3	3	4
.26	1820	1824	1828	1832	1837	1841	1845	1849	1854	1858	0	1	1	2	2	3	3	3	4
.27	1862	1866	1871	1875	1879	1884	1888	1892	1897	1901	0	1	1	2	2	3	3	3	4
.28	1905	1910	1914	1919	1923	1928	1932	1936	1941	1945	0	1	1	2	2	3	3	4	4
.29	1950	1954	1959	1963	1968	1972	1977	1982	1986	1991	0	1	1	2	2	3	3	4	4
.30	1995	2000	2004	2009	2014	2018	2023	2028	2032	2037	0	1	1	2	2	3	3	4	4

.31	2042	2046	2051	2056	2061	2065	2070	2075	2080	2084	0	1	1	2	2	3	3	4	4
.32	2089	2094	2099	2104	2109	2113	2118	2123	2128	2133	0	1	1	2	2	3	3	4	4
.33	2138	2143	2148	2153	2158	2163	2168	2173	2178	2183	0	1	1	2	2	3	3	4	4
.34	2188	2193	2198	2203	2208	2213	2218	2223	2228	2234	1	1	2	2	3	3	4	4	5
.35	2239	2244	2249	2254	2259	2265	2270	2275	2280	2286	1	1	2	2	3	3	4	4	5
.36	2291	2296	2301	2307	2312	2317	2323	2328	2333	2339	1	1	2	2	3	3	4	4	5
.37	2344	2350	2355	2360	2366	2371	2377	2382	2388	2393	1	1	2	2	3	3	4	4	5
.38	2399	2404	2410	2415	2421	2427	2432	2438	2443	2449	1	1	2	2	3	3	4	4	5
.39	2455	2460	2466	2472	2477	2483	2489	2495	2500	2506	1	1	2	2	3	3	4	5	5
.40	2512	2518	2523	2529	2535	2541	2547	2553	2559	2564	1	1	2	2	3	4	4	5	5
.41	2570	2576	2582	2588	2594	2600	2606	2612	2618	2624	1	1	2	2	3	4	4	5	5
.42	2630	2636	2642	2649	2655	2661	2667	2673	2679	2685	1	1	2	2	3	4	4	5	6
.43	2692	2698	2704	2710	2716	2723	2729	2735	2742	2748	1	1	2	3	3	4	4	5	6
.44	2754	2761	2767	2773	2780	2786	2793	2799	2805	2812	1	1	2	3	3	4	4	5	6
.45	2818	2825	2831	2838	2844	2851	2858	2864	2871	2877	1	1	2	3	3	4	5	5	6
.46	2884	2891	2897	2904	2911	2917	2924	2931	2938	2944	1	1	2	3	3	4	5	5	6
.47	2951	2958	2965	2972	2979	2985	2992	2999	3006	3013	1	1	2	3	3	4	5	5	6
.48	3020	3027	3034	3041	3048	3055	3062	3069	3076	3083	1	1	2	3	4	4	5	6	6
.49	3090	3097	3105	3112	3119	3126	3133	3141	3148	3155	1	1	2	3	4	4	5	6	6
.50	3162	3170	3177	3184	3192	3199	3206	3214	3221	3228	1	2	3	3	4	4	5	6	7
.51	3236	3243	3251	3258	3266	3273	3281	3289	3296	3304	1	2	2	3	4	5	5	6	7
.52	3311	3319	3327	3334	3342	3350	3357	3365	3373	3381	1	2	2	3	4	5	5	6	7
.53	3388	3396	3404	3412	3420	3428	3436	3443	3451	3459	1	2	2	3	4	5	6	6	7
.54	3467	3475	3483	3491	3499	3508	3516	3524	3532	3540	1	2	2	3	4	5	6	6	7
.55	3548	3556	3565	3573	3581	3589	3597	3606	3614	3622	1	2	2	3	4	5	6	7	7
.56	3631	3639	3648	3656	3664	3673	3681	3690	3698	3707	1	2	3	3	4	5	6	7	8
.57	3715	3724	3733	3741	3750	3758	3767	3776	3784	3793	1	2	3	3	4	5	6	7	8
.58	3802	3811	3819	3828	3837	3846	3855	3864	3873	3882	1	2	3	4	4	5	6	7	8
.59	3890	3899	3908	3917	3926	3936	3945	3954	3963	3972	1	2	3	4	5	5	6	7	8
.60	3981	3990	3999	4009	4018	4027	4036	4046	4055	4064	1	2	3	4	5	6	6	7	8
.61	4074	4083	4093	4102	4111	4121	4130	4140	4150	4159	1	2	3	4	5	6	7	8	9
.62	4169	4178	4188	4198	4207	4217	4227	4236	4246	4256	1	2	3	4	5	6	7	8	9
.63	4266	4276	4285	4295	4305	4315	4325	4335	4345	4355	1	2	3	4	5	6	7	8	9
.64	4365	4375	4385	4395	4406	4416	4426	4436	4446	4457	1	2	3	4	5	6	7	8	9
.65	4467	4477	4487	4498	4508	4519	4529	4539	4550	4560	1	2	3	4	5	6	7	8	9

.66	4571	4581	4592	4603	4613	4624	4634	4645	4656	4667	1	2	3	4	5	6	7	9	10
.67	4677	4688	4699	4710	4711	4732	4742	4753	4764	4775	1	2	3	4	5	7	8	9	10
.68	4786	4707	4808	4819	4831	4842	4853	4864	4875	4887	1	2	3	4	6	7	8	9	10
.69	4898	4909	4920	4932	4943	4955	4966	4977	4989	5000	1	2	3	5	6	7	8	9	10
.70	5012	5023	5035	5047	5058	5070	5082	5093	5105	5117	1	2	4	5	6	7	8	9	11
.71	5129	5140	5152	5164	5176	5188	5200	5212	5224	5236	1	2	4	5	6	7	8	10	11
.72	5248	5260	5272	5284	5297	5309	5321	5333	5346	5358	1	2	4	5	6	7	9	10	11
.73	5370	5383	5395	5408	5420	5433	5445	5458	5470	5483	1	3	4	5	6	8	9	10	11
.74	5495	5508	5521	5534	5546	5559	5572	5585	5598	5610	1	3	4	5	6	8	9	10	12
.75	5623	5636	5649	5662	5675	5689	5702	5715	5728	5741	1	3	4	5	7	8	9	10	12
.76	5754	5768	5781	5794	5808	5821	5834	5848	5861	5875	1	3	4	5	7	8	9	11	12
.77	5888	5902	5916	5929	5943	5957	5970	5984	5998	6012	1	3	4	5	7	8	10	11	12
.78	6026	6039	6053	6069	6081	6095	6109	6124	6138	6152	1	3	4	6	7	8	10	11	13
.79	6166	6180	6194	6209	6223	6237	6252	6266	6281	6295	1	3	4	6	7	9	10	11	13
.80	6310	6324	6339	6353	6368	6383	6397	6412	6427	6442	1	3	4	6	7	9	10	12	13
.81	6457	6471	6486	6501	6516	6531	6546	6561	6577	6592	2	3	5	6	8	9	11	12	14
.82	6607	6622	6637	6653	6668	6683	6699	6714	6730	6745	2	3	5	6	8	9	11	12	14
.83	6761	6776	6792	6808	6823	6839	6855	6871	6887	6902	2	3	5	6	8	9	11	13	14
.84	6918	6943	6950	6966	6982	6998	7015	7031	7047	7063	2	3	5	6	8	10	11	13	15
.85	7079	7096	7112	7129	7145	7161	7178	7194	7211	7228	2	3	5	7	8	10	12	13	15
.86	7244	7261	7278	7295	7311	7328	7345	7362	7379	7396	2	3	5	7	8	10	12	13	15
.87	7413	7430	7447	7464	7482	7499	7516	7534	7551	7568	2	3	5	7	9	10	12	14	16
.88	7586	7603	7621	7638	7656	7674	7691	7709	7727	7745	2	4	5	7	9	11	12	14	16
.89	7762	7780	7798	7816	7834	7852	7870	7889	7907	7925	2	4	5	7	9	11	13	14	16
.90	7943	7962	7980	7998	8017	8035	8054	8072	8091	8110	2	4	6	7	9	11	13	15	17
.91	8128	8147	8166	8185	8204	8222	8241	8260	8279	8299	2	4	6	8	9	11	13	15	17
.92	8318	8337	8356	8375	8395	8414	8433	8453	8472	8492	2	4	6	8	10	12	14	15	17
.93	8511	8531	8551	8570	8590	8610	8630	8650	8670	8690	2	4	6	8	10	12	14	16	18
.94	8710	8730	8750	8770	8790	8810	8831	8851	8872	8892	2	4	6	8	10	12	14	16	18
.95	8913	8933	8954	8974	8995	9016	9036	9057	9078	9099	2	4	6	8	10	12	15	17	19
.96	9120	9141	9162	9183	9204	9226	9247	9268	9290	9311	2	4	6	8	11	13	15	17	19
.97	9333	9354	9376	9397	9419	9441	9462	9484	9506	9528	2	4	7	9	11	13	15	17	20
.98	9550	9572	9594	9616	9638	9661	9683	9705	9727	9750	2	4	7	9	11	13	15	18	20
.99	9772	9795	9817	9840	9863	9880	9908	9931	9954	9977	2	5	7	9	11	14	16	18	20

NATURAL SINES

	0′	6′	12′	18′	24′	30′	36′	42′	48′	54′	Mean Differences 1′	2′	3′	4′	5′
0°	.0000	0017	0035	0052	0070	0087	0105	0122	0140	0157	3	6	9	12	15
1	.0175	0192	0209	0227	0244	0262	0279	0297	0314	0332	3	6	9	12	15
2	.0349	0366	0384	0401	0419	0436	0454	0471	0488	0506	3	6	9	12	15
3	.0523	0541	0558	0576	0593	0610	0628	0645	0663	0680	3	6	9	12	15
4	.0698	0715	0732	0750	0767	0785	0802	0819	0837	0854	3	6	9	12	14
5	.0872	0889	0906	0924	0941	0958	0976	0993	1011	1028	3	6	9	12	14
6	.1045	1063	1080	1097	1115	1132	1149	1167	1184	1201	3	6	9	12	14
7	.1219	1236	1253	1271	1288	1305	1323	1340	1357	1374	3	6	9	12	14
8	.1392	1409	1426	1444	1461	1478	1495	1513	1530	1547	3	6	9	12	14
9	.1564	1582	1599	1616	1633	1650	1668	1685	1702	1719	3	6	9	12	14
10°	.1736	1754	1771	1788	1805	1822	1840	1857	1874	1891	3	6	9	11	14
11	.1908	1925	1942	1959	1977	1994	2011	2028	2045	2062	3	6	9	11	14
12	.2079	2096	2113	2130	2147	2164	2181	2198	2215	2233	3	6	9	11	14
13	.2250	2267	2284	2300	2317	2334	2351	2368	2385	2402	3	6	8	11	14
14	.2419	2436	2453	2470	2487	2504	2521	2538	2554	2571	3	6	8	11	14
15	.2588	2605	2622	2639	2656	2672	2689	2706	2723	2740	3	6	8	11	14
16	.2756	2773	2790	2807	2823	2840	2857	2874	2890	2907	3	6	8	11	14
17	.2924	2940	2957	2974	2990	3007	3024	3040	3057	3074	3	6	8	11	14
18	.3090	3107	3123	3140	3156	3173	3190	3206	3223	3239	3	6	8	11	14
19	.3256	3272	3289	3305	3322	3338	3335	3371	3387	3404	3	5	8	11	14
20°	.3420	3437	3453	3469	3486	3502	3518	3535	3551	3567	3	5	8	11	14
21	.3584	3600	3616	3633	3649	3665	3681	3697	3714	3730	3	5	8	11	14
22	.3746	3762	3778	3795	3811	3827	3843	3859	3875	3891	3	5	8	11	14
23	.3907	3923	3939	3955	3971	3987	4003	4019	4035	4051	3	5	8	11	14
24	.4067	4083	4099	4115	4131	4147	4163	4179	4195	4210	3	5	8	11	13
25	.4226	4242	4258	4274	4289	4305	4321	4337	4352	4368	3	5	8	11	13
26	.4384	4390	4415	4431	4446	4462	4478	4493	4509	4524	3	5	8	10	13
27	.4540	4555	4571	4586	4602	4617	4633	4648	4664	4679	3	5	8	10	13
28	.4695	4710	4726	4741	4756	4772	4787	4802	4818	4833	3	5	8	10	13
29	.4848	4863	4879	4894	4909	4924	4939	4955	4970	4985	3	5	8	10	13
30°	.5000	5015	5030	5045	5060	5075	5090	5105	5120	5135	3	5	8	10	13

31	.5150	5165	5180	5195	5210	5225	5240	5255	5270	5284	2 5 7	10 12
32	.5299	5314	5320	5344	5358	5373	5388	5402	5417	5432	2 5 7	10 12
33	.5446	5461	5476	5490	5505	5519	5534	5548	5563	5577	2 5 7	10 12
34	.5592	5606	5621	5635	5650	5674	5678	5093	5707	5721	2 5 7	10 12
35	.5736	5750	5764	5779	5793	5807	5821	5835	5830	5864	2 5 7	9 12
36	.5878	5892	5906	5920	5934	5948	5952	5976	5990	6004	2 5 7	9 12
37	.6018	6032	6046	6060	6074	6088	6101	6115	6129	6143	2 5 7	9 12
38	.6157	6170	6184	6198	6211	6225	6239	6252	6266	6280	2 5 7	9 11
39	.6293	6307	6320	6334	6347	6361	6374	6388	6401	6414	2 4 7	9 11
40°	.6428	6441	6455	6468	6481	6494	6508	6521	6534	6547	2 4 7	9 11
41	.6561	6574	6587	6600	6613	6626	6639	6652	6665	6678	2 4 7	9 11
42	.6691	6704	6717	6730	6743	6756	6769	6782	6794	6807	2 4 6	9 11
43	.6820	6833	6845	6858	6871	6884	6896	6909	6921	6934	2 4 6	8 11
44	.6947	6959	6972	6984	6997	7009	7022	7034	7046	7059	2 4 6	8 10
45	.7071	7083	7096	7108	7120	7133	7145	7157	7169	7181	2 4 6	8 10
46	.7193	7206	7218	7230	7242	7254	7266	7278	7290	7302	2 4 6	8 10
47	.7314	7325	7337	7349	7361	7373	7385	7396	7408	7420	2 4 6	8 10
48	.7431	7443	7455	7466	7478	7490	7501	7513	7524	7536	2 4 6	8 10
49	.7547	7559	7570	7581	7593	7604	7615	7627	7638	7649	2 4 6	8 9
50°	.7660	7672	7683	7694	7705	7716	7727	7738	7749	7760	2 4 6	7 9
51	.7771	7782	7793	7804	7815	7826	7837	7848	7859	7869	2 4 5	7 9
52	.7880	7891	7902	7912	7923	7934	7944	7955	7965	7976	2 4 5	7 9
53	.7986	7997	8007	8018	8028	8039	8049	8059	8070	8080	2 3 5	7 9
54	.8090	8100	8111	8121	8131	8141	8151	8161	8171	8181	2 3 5	7 8
55	.8192	8202	8211	8221	8231	8241	8251	8261	8271	8281	2 3 5	7 8
56	.8290	8300	8310	8320	8329	8339	8348	8358	8368	8377	2 3 5	6 3
57	.8387	8396	8406	8415	8425	8434	8443	8453	8462	8471	2 3 5	6 8
58	.8480	8490	8499	8508	8517	8526	8536	8545	8554	8563	2 3 5	6 8
59	.8572	8581	8590	8599	8607	8616	8625	8634	8643	8652	1 3 4	6 7
60°	.8660	8669	8678	8686	8695	8704	8712	8721	8729	8738	1 3 4	6 7
61	.8746	8755	8763	8771	8780	8788	8796	8805	8813	8821	1 3 4	6 7
62	.8829	8838	8846	8854	8862	8870	8878	8886	8894	8902	1 3 4	5 7
63	.8910	8918	8926	8934	8942	8949	8957	8965	8973	8980	1 3 4	5 6
64	.8988	8996	9003	9011	9018	9026	9033	9041	9048	9056	1 3 4	5 6
65	.9063	9070	9078	9085	9092	9100	9107	9114	9121	9128	1 2 4	5 6

66	.9135	9143	9150	9157	9164	9171	9178	9184	9191	9198	1	2	3	5	6
67	.9205	9212	9219	9225	9232	9239	9245	9252	9259	9265	1	2	3	4	6
68	.9272	9278	9285	9291	9298	9304	9311	9317	9323	9330	1	2	3	4	5
69	.9336	9342	9348	9354	9361	9367	9373	9379	9385	9391	1	2	3	4	5
70°	.9397	9403	9409	9415	9421	9426	9432	9438	9444	9449	1	2	3	4	5
71	.9455	9461	9466	9472	9478	9483	9489	9494	9500	9505	1	2	3	4	5
72	.9511	9516	9521	9527	9532	9537	9542	9548	9553	9558	1	2	3	3	4
73	.9563	9568	9573	9578	9583	9588	9593	9598	9603	9608	1	2	2	3	4
74	.9613	9617	9622	9627	9632	9636	9641	9646	9650	9655	1	2	2	3	4
75	.9659	9664	9668	9673	9677	9681	9686	9690	9694	9699	1	1	2	3	4
76	.9703	9707	9711	9715	9720	9724	9728	9732	9736	9740	1	1	2	3	3
77	.9744	9748	9751	9755	9759	9763	9767	9770	9774	9778	1	1	2	3	3
78	.9781	9785	9789	9792	9796	9799	9803	9806	9810	9813	1	1	2	2	3
79	.9816	9820	9823	9826	9829	9833	9836	9839	9842	9845	1	1	2	2	3
80°	.9848	9851	9854	9857	9860	9863	9866	9869	9871	9874	0	1	1	2	2
81	.9877	9880	9882	9885	9888	9890	9893	9895	9898	9900	0	1	1	2	2
82	.9903	9905	9907	9910	9912	9914	9917	9919	9921	9923	0	1	1	2	2
83	.9925	9928	9930	9932	9934	9936	9938	9940	9942	9943	0	1	1	1	2
84	.9945	9947	9949	9951	9952	9954	9956	9957	9959	9960	0	1	1	1	2
85	.9962	9963	9965	9966	9968	9969	9971	9972	9973	9974	0	0	1	1	1
86	.9976	9977	9978	9979	9980	9981	9982	9983	9984	9985	0	0	1	1	1
87	.9986	9987	9988	9989	9990	9990	9991	9992	9993	9993	0	0	0	1	1
88	.9994	9995	9995	9996	9996	9997	9997	9997	9998	9998	0	0	0	0	0
89	.9998	9999	9999	9999	9999	1.000	1.000	1.000	1.000	1.000	0	0	0	0	0

NATURAL COSINES

N.B.—*Subtract* Mean Differences

	0′	6′	12′	18′	24′	30′	36′	42′	48′	54′	Mean Differences 1′	2′	3′	4′	5′
0°	1.000	1.000	1.000	1.000	1.000	1.000	.9999	9999	9999	9999	0	0	0	0	0
1	.9998	9998	9998	9997	9997	9997	9996	9996	9995	9995	0	0	0	0	0
2	.9994	9993	9993	9992	9991	9990	9990	9989	9988	9987	0	0	0	0	0
3	.9986	9985	9984	9983	9982	9981	9980	9979	9978	9977	0	0	0	1	1
4	.9976	9974	9973	9972	9971	9969	9968	9966	9965	9963	0	0	1	1	1
5	.9962	9960	9959	9957	9956	9954	9952	9951	9949	9947	0	0	1	1	1
6	.9945	9943	9942	9940	9938	9936	9934	9932	9930	9928	0	1	1	1	2
7	.9925	9923	9921	9919	9917	9914	9912	9910	9907	9905	0	1	1	1	2
8	.9903	9900	9898	9895	9893	9890	9888	9885	9882	9880	0	1	1	2	2
9	.9877	9874	9871	9869	9866	9863	9860	9857	9854	9851	0	1	1	2	2
10°	.9848	9845	9842	9839	9836	9833	9829	9826	9823	9820	1	1	2	2	2
11	.9816	9813	9810	9806	9803	9799	9796	9792	9789	9785	1	1	2	2	3
12	.9781	9778	9774	9770	9767	9763	9759	9755	9751	9748	1	1	2	3	3
13	.9744	9740	9736	9732	9728	9724	9720	9715	9711	9707	1	1	2	3	3
14	.9703	9699	9694	9690	9686	9681	9677	9673	9668	9664	1	1	2	3	3
15	.9659	9655	9650	9646	9641	9636	9632	9627	9622	9617	1	1	2	3	4
16	.9613	9608	9603	9598	9593	9588	9583	9578	9573	9568	1	2	2	3	4
17	.9583	9558	9553	9548	9542	9537	9532	9527	9521	9516	1	2	2	3	4
18	.9511	9505	9500	9494	9489	9483	9478	9472	9466	9461	1	2	3	3	4
19	.9455	9449	9444	9438	9432	9426	9421	9415	9409	9403	1	2	3	4	5
20°	.9397	9391	9385	9379	9373	9367	9361	9354	9348	9342	1	2	3	4	5
21	.9336	9330	9323	9317	9311	9304	9208	9291	9285	9278	1	2	3	4	5
22	.9272	9265	9250	9252	9245	9239	9232	9225	9219	9212	1	2	3	4	5
23	.9205	9198	9191	9484	9178	9171	9164	9157	9150	9143	1	2	3	4	6
24	.9135	9128	9121	9114	9107	9100	9092	9085	9078	9070	1	2	3	5	6
25	.9063	9056	9048	9041	9033	9026	9018	9011	9003	8996	1	2	4	5	6
26	.8988	8980	8973	8965	8957	8949	8942	8934	8926	8916	1	3	4	5	6
27	.8910	8902	8894	8886	8878	8870	8862	8854	8846	8838	1	3	4	5	6
28	.8829	8821	8813	8805	8796	8788	8780	8771	8763	8755	1	3	4	5	7
29	.8746	8738	8729	8721	8712	8704	8695	8686	8678	8669	1	3	4	6	7
30°	.8660	8652	8643	8634	8625	8616	8607	8599	8590	8581	1	3	4	6	7

31	.8572	8563	8554	8545	8536	8526	8517	8508	8499	8490	2	3	5	6	8
32	.8480	8471	8462	8453	8443	8434	8425	8415	8406	8396	2	3	5	6	8
33	.8387	8377	8368	8358	8348	8339	8329	8320	8310	8300	2	3	5	6	8
34	.8290	8281	8271	8261	8251	8241	8231	8221	8211	8202	2	3	5	7	8
35	.8192	8181	8171	8161	8151	8141	8131	8121	8111	8100	2	3	5	7	8
36	.8090	8080	8070	8059	8049	8039	8028	8018	8007	7997	2	3	5	7	9
37	.7986	7976	7965	7955	7944	7934	7923	7912	7902	7891	2	4	5	7	9
38	.7880	7869	7859	7848	7837	7826	7815	7804	7793	7782	2	4	5	7	9
39	.7771	7760	7749	7738	7727	7716	7705	7694	7683	7672	2	4	6	7	9
40°	.7660	7649	7638	7627	7615	.7604	7593	7581	7570	7559	2	4	6	8	9
41	.7547	7536	7524	7513	7501	7490	7478	7466	7455	7443	2	4	6	8	10
42	.7431	7420	7408	7396	7385	7373	7361	7349	7337	7325	2	4	6	8	10
43	.7314	7302	7290	7278	7266	7254	7242	7230	7218	7206	2	4	6	8	10
44	.7193	7181	7169	7157	7145	7133	7120	7108	7096	7083	2	4	6	8	10
45	.7071	7059	7046	7034	7022	7009	6997	6984	6972	6959	2	4	6	8	10
46	.6947	6934	6921	6909	6896	6884	6871	6858	6845	6833	2	4	6	8	11
47	.6820	6807	6794	6782	6769	6756	6743	6730	6717	6704	2	4	6	9	11
48	.6691	6678	6665	6652	6639	6626	6613	6600	6587	6574	2	4	7	9	11
49	.6561	6547	6534	6521	6508	6494	6481	6468	6455	6441	2	4	7	9	11
50°	.6428	6414	6401	6388	6374	6361	6347	6334	6320	6307	2	4	7	9	11
51	.6293	6280	6266	6252	6239	6225	6211	6198	6184	6170	2	5	7	9	11
52	.6157	6143	6129	6115	6101	6088	6074	6060	6046	6032	2	5	7	9	12
53	.6618	6004	5990	5976	5962	5948	5934	5920	5906	5892	2	5	7	9	12
54	.5878	5864	5850	5835	5821	5807	5793	5779	5764	5750	2	5	7	9	12
55	.5736	5721	5707	5693	5678	5664	5650	5635	5621	5606	2	5	7	10	12
56	.5592	5577	5563	5548	5534	5519	5505	5490	5476	5461	2	5	7	10	12
57	.5446	5432	5417	5402	5388	5373	5358	5344	5329	5314	2	5	7	10	12
58	.5299	5284	5270	5255	5240	5225	5210	5195	5180	5165	2	5	7	10	12
59	.5150	5135	5120	5105	5090	5075	5060	5045	5030	5015	3	5	8	10	13
60°	.5000	4985	4970	4955	4939	4924	4909	4894	4879	4803	3	5	8	10	13
61	.4848	4833	4818	4802	4787	4772	4756	4741	4726	4710	3	5	8	10	13
62	.4695	4679	4664	4648	4633	4617	4602	4586	4571	4555	3	5	8	10	13
63	.4540	4524	4509	4493	4478	4462	4446	4431	4415	4399	3	5	8	10	13
64	.4384	4308	4352	4337	4321	4305	4289	4274	4258	4242	3	5	8	11	13
65	.4226	4210	4195	4179	4163	4147	4131	4115	4099	4083	3	5	8	11	13

66	.4067	4051	4035	4019	4003	3987	3971	3955	3939	3923	3	5	8	11	14
67	.3907	3891	3875	3859	3843	3827	3811	3795	3778	3762	3	5	8	11	14
68	.3746	3730	3714	3697	3681	3665	3649	3633	3616	3600	3	5	8	11	14
69	.3584	3567	3551	3535	3518	3502	3486	3469	3453	3437	3	5	8	11	14
70°	.3420	3404	3387	3371	3355	3338	3322	3305	3280	3272	3	5	8	11	14
71	.3256	3239	3223	3206	3190	3173	3156	3140	3123	3107	3	6	8	11	14
72	.3090	3074	3057	3040	3024	3007	2990	2974	2957	2940	3	6	8	11	14
73	.2924	2907	2890	2874	2857	2840	2823	2807	2790	2773	3	6	8	11	14
74	.2756	2740	2723	2706	2689	2672	2656	2639	2622	2605	3	6	8	11	14
75	.2588	2571	2554	2538	2521	2504	2487	2470	2453	2436	3	6	8	11	14
76	.2419	2402	2385	2368	2351	2334	2317	2300	2284	2267	3	6	8	11	14
77	.2250	2233	2215	2198	2181	2164	2147	2130	2113	2096	3	6	9	11	14
78	.2079	2062	2045	2028	2011	1994	1977	1950	1942	1925	3	6	9	11	14
79	.1908	1891	1874	1857	1840	1822	1805	1788	1771	1754	3	6	9	11	14
80°	.1736	1719	1702	1685	1668	1650	1633	1616	1599	1582	3	6	9	12	14
81	.1564	1547	1530	1513	1495	1478	1461	1444	1426	1409	3	6	9	12	14
82	.1392	1374	1357	1340	1323	1305	1288	1271	1253	1236	3	6	9	12	14
83	.1219	1201	1184	1167	1149	1132	1115	1097	1080	1063	3	6	9	12	14
84	.1045	1028	1011	0993	0976	0958	0941	0924	0906	0889	3	6	9	12	14
85	.0872	0854	0837	0819	0802	0785	0767	0750	0732	0715	3	6	9	12	14
86	.0698	0680	0663	0645	0628	0610	0593	0576	0558	0541	3	6	9	12	15
87	.0623	0506	0488	0471	0454	0436	0419	0401	0384	0366	3	6	9	12	15
88	.0349	0382	0314	0297	0279	0262	0244	0227	0209	0192	3	6	9	12	15
89	.0175	0157	0140	0122	0105	0087	0070	0052	0035	0017	3	6	9	12	15

Natural Tangents

	0′	6′	12′	18′	24′	30′	36′	42′	48′	54′	Mean Differences 1′	2′	3′	4′	5′
0°	.0000	0017	0035	0052	0070	0087	0105	0122	0140	0157	3	6	9	12	15
1	.0175	0192	0209	0227	0244	0262	0279	0297	0314	0832	3	6	9	12	15
2	.0349	0367	0384	0402	0419	0437	0454	0472	0489	0607	3	6	9	12	15
3	.0524	0542	0559	0577	0594	0612	0629	0647	0664	0682	3	6	9	12	15
4	.0699	0717	0734	0752	0769	0787	0805	0822	0840	0857	3	6	9	12	15
5	.0875	0892	0910	0928	0945	0963	0981	0998	1016	1033	3	6	9	12	15
6	.1051	1069	1086	1104	1122	1139	1157	1175	1192	1210	3	6	9	12	15
7	.1228	1246	1263	1281	1299	1317	1334	1352	1370	1388	3	6	9	12	15
8	.1405	1423	1441	1459	1477	1495	1512	1530	1548	1566	3	6	9	12	15
9	.1584	1602	1620	1638	1655	1673	1691	1709	1727	1745	3	6	9	12	15
10°	.1763	1781	1799	1817	1835	1853	1871	1890	1908	1926	3	6	9	12	15
11	.1944	1962	1980	1998	2016	2035	2053	2071	2089	2107	3	6	9	12	15
12	.2126	2144	2162	2180	2199	2217	2235	2254	2272	2290	3	6	9	12	15
13	.2309	2327	2345	2364	2382	2401	2419	2438	2456	2475	3	6	9	12	15
14	.2493	2512	2530	2549	2568	2586	2605	2623	2642	2661	3	6	9	12	16
15	.2679	2698	2717	2736	2754	2773	2792	2811	2830	2849	3	6	9	13	16
16	.2867	2886	2905	2924	2943	2962	2981	3000	3019	3038	3	6	9	13	16
17	.3057	3076	3096	3115	3134	3153	3172	3191	3211	3230	3	6	10	13	16
18	.3249	3269	3288	3307	3327	3346	3365	3385	3404	3424	3	6	10	13	16
19	.3443	3463	3482	3502	3522	3541	3561	3581	3600	3620	3	7	10	13	16
20°	.3640	3659	3679	3699	3719	3739	3759	3779	3799	3819	3	7	10	13	17
21	.3839	3859	3879	3899	3919	3939	3959	3979	4000	4020	3	7	10	13	17
22	.4040	4061	4081	4101	4122	4142	4163	4183	4204	4224	3	7	10	14	17
23	.4245	4265	4286	4307	4327	4348	4369	4390	4411	4431	3	7	10	14	17
24	.4452	4473	4494	4515	4536	4557	4578	4599	4621	4642	4	7	11	14	18
25	.4663	4684	4706	4727	4748	4770	4791	4813	4834	4856	4	7	11	14	18
26	.4877	4899	.4921	4942	4964	4986	5008	5029	5051	5073	4	7	11	15	18
27	.5095	5117	5139	5161	5184	5206	5228	5250	5272	5295	4	7	11	15	18
28	.5317	5340	5362	5384	5407	5430	5452	5475	5498	5520	4	8	11	15	19
29	.5543	5566	5589	5612	5635	5658	5681	5704	5727	5750	4	8	12	15	19
30°	.5774	5797	5820	5844	5867	5890	5914	5938	5961	5985	4	8	12	16	20

31	.6009	6032	6056	6080	6104	6128	6152	6176	6200	6224	4	8	12	16	20
32	.6249	6273	6297	6322	6346	6371	6395	6420	6445	6469	4	8	12	16	20
33	.6494	6519	6544	6569	6594	6619	6644	6609	6694	6720	4	8	13	17	21
34	.6745	6771	6796	6822	6847	6873	6899	6924	6950	6976	4	9	13	17	21
35	.7002	7028	7054	7080	7107	7133	7159	7186	7212	7239	4	9	13	18	22
36	.7265	7292	7319	7346	7373	7400	7427	7454	7481	7508	5	9	14	18	23
37	.7536	7563	7590	7618	7646	7673	7701	7729	7757	7785	5	9	14	18	23
38	.7813	7841	7869	7898	7926	7954	7983	8012	8040	8069	5	9	14	19	24
39	.8098	8127	8156	8185	8214	8243	8273	8302	8332	8361	5	10	15	20	24
40°	.8391	8421	8451	8481	8511	8541	8571	8601	8632	8662	5	10	15	20	25
41	.8693	8724	8754	8785	8816	8847	8878	8910	8941	8972	5	10	16	21	26
42	.9004	9036	9067	9099	9131	9163	9195	9228	9260	9293	5	11	16	21	27
43	.9325	9358	9391	9424	9457	9490	9523	9556	9590	9623	6	11	17	22	28
44	.9657	9691	9725	9759	9793	9827	9861	9896	9930	9965	6	11	17	23	29
45°	1.0000	0035	0070	0105	0141	0176	0212	0247	0283	0319	6	12	18	24	30
46	1.0355	0392	0428	0464	0501	0538	0575	0612	0649	0686	6	12	18	25	31
47	1.0724	0761	0799	0837	0875	0913	0951	0990	1028	1067	6	13	19	25	32
48	1.1106	1145	1184	1224	1263	1303	1343	1383	1423	1463	7	13	20	27	33
49	1.1504	1544	1585	1626	1667	1708	1750	1792	1833	1875	7	14	21	28	34
50°	1.1918	1960	2002	2045	2088	2131	2174	2218	2261	2305	7	14	22	29	36
51	1.2349	2393	2437	2482	2527	2572	2617	2662	2708	2753	8	15	23	30	38
52	1.2799	2846	2892	2938	2985	3032	3079	3127	3175	3222	8	16	24	31	39
53	1.3270	3319	3367	3416	3465	3514	3564	3613	3663	3713	8	16	25	33	41
54	1.3764	3814	3865	3916	3968	4019	4071	4124	4176	4229	9	17	26	34	43
55	1.4281	4335	4388	4442	4496	4550	4605	4659	4715	4770	9	18	27	36	45
56	1.4826	4882	4938	4994	5051	5108	5166	5224	5282	5340	10	19	29	38	48
57	1.5399	5458	5517	5577	5637	5697	5757	5818	5880	5941	10	20	30	40	50
58	1.6003	8066	6128	6191	6255	6319	6383	6447	6512	6577	11	21	32	43	53
59	1.6643	6709	6775	6842	6909	6977	7045	7113	7182	7251	11	23	34	45	56
60°	1.7321	7391	7461	7532	7603	7675	7747	7820	7893	7966	12	24	36	48	60
61	1.8040	8115	8190	8265	8341	8418	8495	8572	8650	8728	13	26	38	51	64
62	1.8807	8887	8967	9047	9128	9210	9202	9375	9458	9542	14	27	41	55	68
63	1.9626	9711	9797	9883	9970	0057	0145	0233	0323	0413	15	29	44	58	73
64	2.0503	0594	0686	0778	0872	0965	1060	1155	1251	1348	16	31	47	63	78
65	2.1445	1543	1642	1742	1842	1943	2045	2148	2251	2355	17	34	51	68	85

66	2.2460	2566	2673	2781	2889	2998	3109	3220	3332	3445	18	37	55	73	92
67	2.3559	3673	3789	3906	4023	4142	4262	4383	4504	4627	20	40	60	79	99
68	2.4751	4876	5002	5129	5257	5386	5517	5649	5782	5916	22	43	65	87	108
69	2.6051	6187	6325	6464	6605	6746	6889	7034	7179	7326	24	47	71	95	119
70°	2.7475	7625	7776	7929	8083	8239	8397	8556	8716	8878	26	52	78	104	131
71	2.9042	9208	9375	9544	9714	9887	0061	0237	0415	0595	29	58	87	116	145
72	3.0777	0961	1146	1334	1524	1716	1910	2106	2305	2506	32	64	96	129	161
73	3.2709	2914	3122	3332	3544	3759	3977	4197	4420	4646	36	72	108	144	180
74	3.4874	5105	5339	5576	5816	6059	6305	0554	6806	7062	41	81	122	163	204
75	3.7321	7583	7848	8118	8391	8667	8947	9232	9520	9812	46	93	139	186	232
76	4.0108	0408	0713	1022	1335	1653	1976	2303	2635	2972	Mean Differences no longer sufficiently accurate				
77	4.3315	3662	4015	4374	4737	5107	5483	5864	6252	6646					
78	4.7046	7453	7867	8288	8716	9152	9594	0045	0504	0970					
79	5.1446	1929	2422	2924	3435	3955	4486	5026	5578	6140					
80°	5.6713	7297	7894	8502	9124	9758	0405	1066	1742	2432					
81	6.3138	3859	4596	5350	6122	6912	7720	8548	9395	0264					
82	7.1154	2066	3002	3962	4947	5958	6996	8062	9158	0285					
83	8.1443	2636	3863	5126	6427	7769	9152	0579	2052	3572					
84	9.514	9.677	9.845	10.02	10.20	10.39	10.58	10.78	10.99	11.20					
85	11.43	11.66	11.91	12.16	12.43	12.71	13.00	13.30	13.62	13.95					
86	14.30	14.57	15.06	15.46	15.89	16.35	16.83	17.34	17.89	18.46					
87	19.08	19.74	20.45	21.20	22.02	22.90	23.86	24.90	26.03	27.27					
88	28.64	30.14	31.82	33.69	35.80	38.19	40.92	44.07	47.74	52.08					
89	57.29	63.66	71.62	81.85	95.49	114.6	143.2	191.0	286.0	573.0					

SQUARE ROOTS—FROM 1 TO 100
POWERS, ROOTS & RECIPROCALS

n	n^2	n^2	$\sqrt{n}$	$2\sqrt{n}$	$\frac{1}{n}$	$\sqrt{10n}$	$\frac{1}{\sqrt{n}}$	$\frac{1}{\sqrt{10n}}$
1	1	1	1.0000	1.0000	1.00000	3.1623	1.00000	0.31623
2	4	8	1.4142	1.2599	0.50000	4.4721	0.70711	0.22361
3	9	27	1.7321	1.4422	0.33333	5.4772	0.57735	0.18257
4	16	64	2.0000	1.5874	0.25000	6.3245	0.50000	0.15811
5	25	125	2.2361	1.7100	0.20000	7.0711	0.44721	0.14142
6	36	216	2.4495	1.8171	0.16667	7.7460	0.40825	0.12910
7	49	343	2.6458	1.9129	0.14286	8.3666	0.37796	0.11952
8	64	512	2.8284	2.0000	0.12500	8.9443	0.35355	0.11180
9	81	729	3.0000	2.0801	0.11111	9.4868	0.33333	0.10541
10	100	1000	3.1623	2.1544	0.10000	10.0000	0.31623	0.10000
11	121	1331	3.3166	2.2240	0.09091	10.4881	0.30151	0.09535
12	144	1728	3.4641	2.2894	0.08333	10.9545	0.28868	0.09129
13	169	2197	3.6056	2.3513	0.07692	11.4018	0.27735	0.08771
14	196	2744	3.7417	2.4101	0.07143	11.8322	0.26726	0.08452
15	225	3375	3.8730	2.4662	0.06667	12.2474	0.25820	0.08165
16	256	4096	4.0000	2.5198	0.06250	12.6491	0.25000	0.07906
17	289	4913	4.1231	2.5713	0.05882	13.0384	0.24253	0.07670
18	324	5832	4.2426	2.6207	0.05556	13.4164	0.23570	0.07454
19	361	6859	4.3589	2.6684	0.05263	13.7840	0.22942	0.07255
20	400	8000	4.4721	2.7144	0.05000	14.1421	0.22361	0.07071
21	441	9261	4.5826	2.7589	0.04762	14.4914	0.21822	0.06901
22	484	10648	4.6904	2.8020	0.04545	14.8324	0.21320	0.06742
23	529	12167	4.7958	2.8439	0.04348	15.1658	0.20851	0.06594
24	576	13824	4.8990	2.8845	0.04167	15.4919	0.20412	0.06455
25	625	15625	5.0000	2.9240	0.04000	15.8114	0.20000	0.06325
26	676	17576	5.0990	2.9625	0.03846	16.1245	0.19612	0.06202
27	729	19683	5.1962	3.0000	0.03704	16.4317	0.19245	0.06086
28	784	21952	5.2915	3.0366	0.03571	16.7332	0.18898	0.05976
29	841	24389	5.3852	3.0723	0.03448	17.0294	0.18570	0.05872
30	900	27000	5.4772	3.1072	0.03333	17.3205	0.18257	0.05774